T0310862

Seismic Hazard and Risk Analysis

Seismic hazard and risk analyses underpin the loadings prescribed by engineering design codes, the decisions by asset owners to retrofit structures, the pricing of insurance policies, and many other activities. This is a comprehensive overview of the principles and procedures behind seismic hazard and risk analysis. It enables readers to understand best practises and future research directions. Early chapters cover the essential elements and concepts of seismic hazard and risk analysis, while later chapters shift focus to more advanced topics. Each chapter includes worked examples and problem sets for which full solutions are provided online. Appendices provide relevant background in probability and statistics. Computer codes are also available online to help replicate specific calculations and demonstrate the implementation of various methods. This is a valuable reference for upper level students and practitioners in civil engineering, and earth scientists interested in engineering seismology.

Jack W. Baker is a Professor in the Department of Civil and Environmental Engineering at Stanford University. His research focuses on the modelling of extreme loads on structures and infrastructure systems. He has industry experience in seismic hazard and risk assessment, ground motion selection, and modelling of catastrophe losses for insurance and reinsurance companies. He has recieved several awards for his research and teaching, including the Shah Family Innovation Prize from the Earthquake Engineering Research Institute, the CAREER Award from the National Science Foundation, the Early Achievement Research Award from the International Association for Structural Safety and Reliability, the Walter L. Huber Research Prize from the American Society of Civil Engineers, the Helmut Krawinkler Award from the Structural Engineers Association of Northern California, and the Stanford Eugene L. Grant Award for excellence in teaching

Brendon A. Bradley is a Professor in the Department of Civil and Natural Resources Engineering at the University of Canterbury, New Zealand. His areas of interest include engineering seismology, strong ground-motion prediction, seismic response analysis of structural and geotechnical systems, and seismic performance and loss estimation methods. He has received several awards for his technical contributions, including the Royal Society of New Zealand Rutherford Discovery Fellowship, the Shamsher Prakash Foundation Research Award, the TC203 Young Researcher Award from the International Society of Soil Mechanics and Geotechnical Engineering, the Shah Family Innovation Prize from the Earthquake Engineering Research Institute, the Norman Medal from the American Society of Civil Engineers, the New Zealand Prime Ministers Emerging Scientist Prize, the New Zealand Geotechnical Society Geomechanics Award, and the University of Canterbury Teaching Award.

Peter J. Stafford is a Reader in Engineering Seismology and Earthquake Engineering at the Department of Civil and Environmental Engineering, Imperial College London. His research interests are primarily in the areas of Engineering Seismology and Earthquake Engineering, with a particular focus upon the application of advanced probabilistic and statistical approaches in these fields. He has extensive professional consulting experience in international probabilistic seismic hazard and risk analysis projects for critical nuclear and hydro-electric power generation and gas extraction infrastructure. These projects, typically conducted under the SSHAC Level 3 or 4 frameworks, have led to the development of many state-of-the-art analysis approaches in the field. He also has significant experience with the reinsurance sector and providing inputs to the development of portfolio loss models.

Seismic Hazard and Risk Analysis

JACK W. BAKER

Stanford University, California

BRENDON A. BRADLEY

University of Canterbury, Christchurch, New Zealand

PETER J. STAFFORD

Imperial College London, United Kingdom

CAMBRIDGE
UNIVERSITY PRESS

CAMBRIDGE
UNIVERSITY PRESS

University Printing House, Cambridge CB2 8BS, United Kingdom

One Liberty Plaza, 20th Floor, New York, NY 10006, USA

477 Williamstown Road, Port Melbourne, VIC 3207, Australia

314–321, 3rd Floor, Plot 3, Splendor Forum, Jasola District Centre, New Delhi – 110025, India

103 Penang Road, #05–06/07, Visioncrest Commercial, Singapore 238467

Cambridge University Press is part of the University of Cambridge.

It furthers the University's mission by disseminating knowledge in the pursuit of education, learning, and research at the highest international levels of excellence.

www.cambridge.org
Information on this title: www.cambridge.org/9781108425056
DOI: 10.1017/9781108425056

© Cambridge University Press 2021

This publication is in copyright. Subject to statutory exception and to the provisions of relevant collective licensing agreements, no reproduction of any part may take place without the written permission of Cambridge University Press.

First published 2021

A catalogue record for this publication is available from the British Library.

ISBN 978-1-108-42505-6 Hardback

Additional resources for this publication at www.cambridge.org/9781108425056.

Cambridge University Press has no responsibility for the persistence or accuracy of URLs for external or third-party internet websites referred to in this publication and does not guarantee that any content on such websites is, or will remain, accurate or appropriate.

Contents

Preface

Seismic hazard and risk analysis concepts have existed in a formal sense for only 50 years. In parallel with their continued conceptual development, these analyses have been increasingly adopted as a key element to define the seismic hazard used in the design and assessment of structures in seismic regions of the world to achieve societally acceptable risk.

This book is the result of a knowledge gap experienced by the authors during our years of collective research and consulting in this area. In our work, we have observed that:

- Real-world seismic hazard and risk analysis calculations are so complex that the conceptual framework underlying the calculations is not apparent. This complexity can lead to misunderstanding or mistrust of the results.
- Many analysis contributors are domain experts in branches of earthquake physics or engineering, who would like to better understand how their insights influence the calculations.
- Seismic hazard and risk calculations are often performed according to regulations, so some analysts perform their work because it is how they were instructed to do it, rather than because they understand or agree with the ultimate goals.

When faced with these issues, the authors have often been asked the question, "Where can I learn more about this topic?" The answer to that question has been difficult. There is a wide range of research literature on the topics covered in this book, and many consultants and software providers to perform the work. But no contemporary reference book is suitable for a new entrant to this field. Students hoping to learn more are thus faced with either narrowly focused research papers using varying terminology and mathematical approaches, or more accessible texts which only briefly cover these topics.

With this book, we aim to describe the principles and procedures behind probabilistic seismic hazard and risk analysis, enabling readers to understand best practices and research directions, and enabling earth scientists and civil engineers to see the broader implications of their work. With a focus on concepts and procedures rather than the details of specific scientific models used for inputs, we believe that this book will make the topic more broadly accessible and enable it to become more widely taught at universities and in professional practice.

This book is primarily written for upper-level undergraduate or graduate civil engineers and earth scientists interested in engineering seismology. Because many civil engineers are consumers of seismic hazard and risk reports, we anticipate that it will also be useful outside of universities. The authors have taught professional education courses on this topic to companies and learned societies, and the book serves as a good companion for those courses as well.

Given the widespread existence of seismic hazard and risk assessments, technical specialists from several backgrounds can benefit from training on this topic. We expect that most readers have a basic exposure to geotechnical engineering, structural engineering, or engineering seismology and are interested in learning about probabilistic analysis in these fields. Recognizing that many readers may lack a background in one or more areas, we provide background and references for geophysics and earthquake engineering material as it arises, and provide an appendix on prerequisite probability tools.

We do not aim to comprehensively teach readers the state of the art in constituent models or in refined hazard and risk calculations. Instead, we aim to collect and synthesize work from what are currently disparate fields, showing how they fit together and collectively provide valuable insights. The focus is on the analysis process, and critical concepts to consider, rather than telling analysts what is the "correct" model to use for a particular application. The models discussed here are intended to highlight issues that arise in practical hazard and risk calculations, albeit that they may not generally be at the frontier of research. This choice to focus on synthesis and procedures, rather than on specific models and research frontiers, means that this book should remain relevant for a number of years to come even as our knowledge of seismic hazards grows and our modeling components continue to evolve.

Organization

The book chapters are organized into three parts with two appendices. Chapter 1 introduces the essential concepts associated with seismic hazard and risk analysis. Chapters 2–5 address the hazard inputs that are necessary to perform seismic hazard calculations. Chapters 6–8 are directly concerned with these seismic hazard calculations. Chapters 9–12 extend beyond hazard to address seismic risk. The two appendices provide additional probability and statistics concepts relevant to the content discussed within chapters. Figure P.1 provides an illustration of how the chapters are related.

Chapter 2 is devoted to defining the potential location and size of future earthquake sources. Chapter 3 extends beyond Chapter 2 by estimating the likelihood, or rate of occurrence, of potential earthquake ruptures on sources. Empirical ground-motion modeling approaches to predict

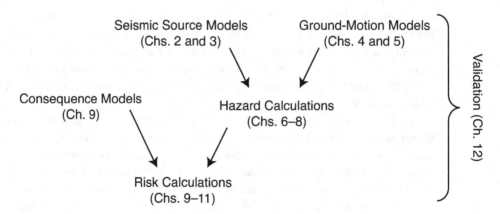

Fig. P.1   Flow chart of seismic hazard and risk analysis calculations. Relevant chapter numbers are noted alongside each step, illustrating the layout of the book content.

ground-motion intensity measures from potential earthquake ruptures at a location of interest are described in Chapter 4. Chapter 5 addresses physics-based ground-motion models that directly simulate earthquake rupture and consequent seismic wave propagation to compute ground-motion time series at a location of interest.

Chapter 6 presents the essential elements of the probabilistic seismic hazard analysis (PSHA) calculation, which integrates the information provided by seismic source and ground-motion models. Additional PSHA calculations and products, such as disaggregation and uniform hazard spectra, are then presented in Chapter 7. Chapter 8 addresses concepts for advanced site-specific seismic hazard analyses that attempt to improve prediction accuracy and precision.

Chapter 9 provides an overview of seismic risk concepts and calculations. The determination of seismic risk for important structures often requires performing response-history analysis, and the selection of ground-motion time series necessary for these analyses is addressed in Chapter 10. Chapter 11 extends concepts from earlier chapters to address spatially distributed systems having multiple structures at different geographical locations or spatially extensive structures. Chapter 12 examines the concept of validation in the context of seismic hazard and risk analysis, in particular, assessing the predictive capability of their constituent models.

Readers may cover the content in this book in different ways based on their background and intent. Chapters 2–4, 6, and 7 provide an essential coverage of seismic hazard analysis inputs and outputs. Portions of Chapter 3, as well as Chapters 5, 8, and 12, address advanced topics in seismic source and ground-motion characterization and may be of interest for the reader looking to understand issues that are driving current research directions. Chapters 9–11 deal with risk considerations and calculations that result from seismic hazards. These chapters, in combination with Chapters 6 and 7, are relevant for engineering-oriented readers looking to apply hazard and risk calculations and understand the link between the two.

Pedagogy

In addition to developing the technical content provided herein, we have aimed to present the material in a manner that supports learning. Several elements are noteworthy. Learning objectives are provided at the start of each chapter. Example calculations are provided throughout the book to assist with comprehension of concepts and the application of equations. A series of examples focused on the San Francisco region, and the nearby 1989 Loma Prieta earthquake and resulting ground motions, are used throughout the book to provide continuity and illustrate how concepts from each chapter are related.

Exercises are provided at the end of each chapter. These exercises focus on basic concepts and are linked to the learning objectives. However, most practical problems in seismic hazard and risk analysis involve calculations requiring external software or bespoke programming that are not well-suited to exercises within a book. To address this, a companion website is provided with example datasets and supplementary materials to help readers perform calculations described here. We have therefore limited the use of specific datasets and specialized software in the book, to keep the material from becoming outdated as tools advance. The website will be a resource for computational tools and other content that will advance over time.

Acknowledgments

This book has benefited directly and indirectly from many people.

A number of technical experts provided insightful feedback, suggestions, and editing on drafts of the manuscript that significantly improved the quality of the completed document. We thank Norm Abrahamson, Trevor Allen, Julian Bommer, Carlo Cauzzi, Yilin Chen, Rodrigo Costa, Helen Crowley, Chris de la Torre, James Dismuke, John Douglas, Jeff Fraser, Matt Gerstenberger, Katsu Goda, Rob Graves, Youssef Hashash, Anne Hulsey, Steve Kramer, Simon Kwong, Robin Lee, Vahid Loghman, Warner Marzocchi, Andy Michael, Mahalia Miller, Gareth Morris, Mark Petersen, Johnny Philpot, Viktor Polak, David Rhoades, Adrian Rodriguez-Marek, Rodrigo Silva Lopez, Jon Stewart, Karim Tarbali, Eric Thompson, Gabriel Toro, Chris Van Houtte, David Wald, and Ádám Zsarnóczay for their assistance.

Several people also provided critical help to deliver this book and improve its readability. Julie King and Hilary Glasman-Deal of Imperial College London provided expert advice regarding writing style and editing. Sarah Lambert, Susan Francis, and Zoe Pruce of Cambridge University Press helped us shepherd the book to completion.

JWB: My contributions to this book would not have been possible without Allin Cornell. His technical expertise, and his emphasis on clarity of communication, influenced me tremendously throughout this project. Norm Abrahamson, Greg Deierlein, Anne Kiremidjian, Helmut Krawinkler, Eduardo Miranda, and Jack Moehle were instrumental at various stages of my career, and first taught me many of the concepts in this book. My parents, Tim and Pat, instilled in me a curiosity and love of learning that was integral to my being prepared to write this book. Finally, my wife Kara provided incredible and reliable support that allowed me to take on this project. Along with my children Leo and Alana, she created a home where I always felt energized and encouraged.

BAB: I greatly appreciate the expertise and mentorship of Misko Cubrinovski, Rajesh Dhakal, Gregory MacRae, and Dominic Lee during my formative years of research in the fields related to this book. The quality and breadth of my research experiences have also been greatly enhanced by many collaborators, and this book has provided an opportunity to synthesize their collective wisdom. Finally, I'd like to thank my wife, Brenda; my parents, Gail and Peter; and my sister, Lauren, for their love and encouragement.

PJS: My understanding of this material reflects the strong influence of two *JB*s on my career. John Berrill first enticed me into the fields of engineering seismology and seismic hazard analysis, and I would not be working in this field without him. Much of my thinking on the topics in this book has benefited from working with Julian Bommer since my PhD, and I will always be grateful for the opportunities he has afforded me. I also owe a great deal to colleagues from major seismic hazard projects. In particular, Adrian Rodriguez-Marek, Ellen Rathje, Frank Scherbaum, and Bob Youngs have all taught me many things, and I've benefited from interactions with many other experts through these projects. I was lucky to be raised by wonderful parents who bestowed upon me attributes necessary for tackling this project. My mother, Noelene, instilled in me a belief that I could achieve anything I set my mind to, and my father, David, gave me the gifts of work ethic, attention to detail, and perseverance. I'm also grateful to my elder sister, Shannon, and younger brother, Matthew, who have always supported me. Finally, my wife Ana, and my children Sophia and Santiago deserve my greatest appreciation. Ana has been unwavering in her support throughout this project, and has sacrificed a lot of her time and energy to enable me to contribute. My children inspire me to learn and grow with them on a daily basis, and always help to keep things in perspective.

Introduction

Earthquakes cause damage to many parts of the natural and built environment, with potentially widespread and devastating impacts. Since 1900, earthquakes have killed approximately 8.5 million people and caused $2 trillion of damage[1] (Daniell et al., 2011). Humanity has learned much from millennia of earthquake observations, and decades of quantitative research, about where and why earthquakes occur, their manifestation at the surface of the earth, and the impacts of induced shaking and other effects. Despite this, much remains uncertain about the timing of earthquake occurrences, the resulting ground shaking, and the performance of structures and infrastructure subjected to shaking. Further, damaging earthquakes at any particular location are infrequent, meaning that experience is a poor guide to understanding and managing earthquake risks. A stakeholder wanting to manage these risks must understand the relevant earthquake processes and their uncertainties, and a framework for making decisions for events with low occurrence probability but high consequences.

This book describes the models needed to assess seismic hazards and risks, and the tools to treat uncertainties in these assessments. The calculations described in this book are performed worldwide in the development of seismic hazard maps, calibration of building codes, pricing of earthquake insurance, catastrophe modeling, and many other applications. The following sections introduce key concepts and provide simple example calculations, which will then be refined in later chapters.

1.1 Hazard and Risk Analysis

Hazard analysis refers to the characterization of natural phenomena resulting from earthquakes. *Risk analysis* refers to the characterization of the consequences of that hazard, such as structural failures, fatalities, or economic costs.

Hazard analysis can consider several physical phenomena produced by earthquakes. Earthquakes create seismic waves, resulting in ground shaking in the surrounding area. Earthquakes also create permanent displacements, in the form of surface fault ruptures, uplift, subsidence, and folding. Ground shaking and permanent displacements are *primary hazards*, and shaking will be a significant focus of this book, as it causes the majority of damage. Earthquake-induced primary hazards can also cause *secondary hazards*, such as tsunamis, landslides, soil liquefaction, and floods. While the primary focus is on the prediction of ground-shaking hazard, this book's approaches can also be applied to model hazard and risk associated with these secondary hazards.

Risk analysis can consider several mechanisms by which adverse consequences from earthquakes occur. Shaking can cause structural damage and collapse, trigger fires, and topple equipment, among other things. Liquefaction can result in significant ground deformation and damage to structures. Landslides and tsunamis can directly damage structures. These physical consequences

---

[1] In 2011 Hybrid Natural Disaster Economic Index-adjusted US dollars.

also affect the safety of building inhabitants. Comprehensive treatment and prediction of these various consequences could fill an entire book, so herein, the focus is upon the general issue of how to specify and utilize models for predicting consequences given ground shaking.

Risk analysis requires a hazard analysis, as well as an understanding of exposure (the presence of assets such as structures and people) and vulnerability (the susceptibility of exposed assets to damage from the hazard). Because of uncertainties in the occurrence of earthquakes and their impacts, the hazard analysis and risk analysis in this book will be performed using probabilistic approaches.

1.2 Uses of Hazard and Risk Information

Before proceeding to evaluate hazard and risk, we should consider the potential uses of the results.

The first category of decisions relates to *resource allocation*. How strong should a structure's earthquake-resisting system be? How much should development in high-hazard locations be restricted? Should an infrastructure system operator retrofit the system to reduce impacts from future earthquakes? How much should an earthquake insurance policy cost? These are the decisions of interest in this book. They involve the possibility of a certain cost now (a more expensive structure, a lost opportunity for development, a costly insurance policy) that may produce benefits in the future, *if there is a damaging earthquake*. An optimal decision must thus balance certain costs against possible benefits. This balancing requires an understanding of how likely it is for the benefits to be seen.

The second category of decisions uses earthquake scenarios for *communication and planning*. Single earthquake scenarios (e.g., a magnitude 7 earthquake on a known fault) are straightforward to define and conceptualize, so scenarios are used in many efforts to plan recovery, build political will for action, and encourage citizen awareness (e.g., Kircher et al., 2006; Porter et al., 2011; Detweiler and Wein, 2017). These scenarios focus attention on specific impacts, and the actions taken based on these studies are not as sensitive to the scenario's likelihood. Omitting probabilistic considerations in these cases may thus be reasonable, to simplify the analysis.

The third category of decisions uses earthquake scenarios for *research studies*. Calculations of scenarios can be used for detailed scientific inquiries into the rupture process and wave propagation. Calculations for historic scenarios can be used to compare predictions to actual observations (Chapter 12). Most research considers individual scenarios before a probabilistic treatment can be considered (and in many cases, a probabilistic treatment may not even be useful).

This book will focus primarily on the first category of decisions, where the likelihoods of adverse outcomes are relevant. The following section will discuss problems with alternative approaches that simplify or omit probabilistic analysis for these asset allocation decisions. The reader should remember that deterministic scenarios may have value in other circumstances, but this book does not address those circumstances.

1.3 Deterministic Analysis

A probabilistic evaluation could perhaps be replaced with a more straightforward calculation if it were possible to identify a "worst-case" earthquake and its resulting consequences. Deterministic evaluations are based on simple rules to define earthquake scenarios and resulting ground motions.

Many deterministic approaches have been proposed, often with the justification that they provide "worst-case" ground motions and are thus more conservative than their probabilistic counterparts. However, we will see that the limitations of this approach arise quickly and that the results are certainly not "worst case."

Consider designing a building, located near two faults, to avoid severe damage from earthquake shaking. What level of shaking should we consider in our design? The following four subsections highlight challenges that arise when attempting to answer this question using deterministic evaluations.

1.3.1 The Considered Earthquake Rupture

A designer looking to choose a worst-case ground motion would need to specify an earthquake scenario, or "rupture." A worst-case analysis should, in principle, consider the rupture producing the strongest ground shaking–presumably the maximum magnitude rupture (i.e., the largest earthquake) that could occur on the closest possible fault. However, several difficulties arise when implementing this in practice. First, the magnitude of an earthquake that a given fault (or a network of faults) can produce is subject to several uncertainties that will be discussed in Chapter 3.

Second, even if the largest possible earthquake on each nearby fault could be determined, there are trade-offs between the magnitudes of various earthquakes and their distances to the site of interest. Consider a hypothetical site located near two earthquake sources. The first source can produce an earthquake (Rupture 1) of magnitude 5 at a distance of 0 km. The second source can produce an earthquake (Rupture 2) of magnitude 7 at a distance of 15 km. We can use a predictive model[2] to estimate the median shaking associated with each rupture. If we are interested in ground shaking in the form of the peak ground acceleration (PGA) from each rupture, Rupture 1 produces a larger value of 0.27 g (versus 0.19 g for Rupture 2). But, if we are interested in the peak ground velocity (PGV), Rupture 2 produces the larger value of 15 cm/s (versus 13 cm/s for Rupture 1).[3] So, the "worst-case" rupture depends upon how ground-motion intensity is measured. One could take the envelope of the PGA and PGV values as a conservative decision, but no single rupture produces these values.

A third challenge arises when the locations of faults near a site are not obvious, and so earthquakes are considered to be possible at any location. In this case, the genuinely worst-case rupture scenario has to be the maximum conceivable magnitude, at a location directly below the site of interest. In many stable continental regions, such as the Eastern United States or Western Europe, one can feasibly hypothesize the occurrence of a large earthquake with magnitude >7 occurring immediately below a site. However, that event may be extremely improbable. Without a formal way to consider the likelihood of extremely rare scenarios, the "worst case" characterization of earthquake sources leads to extreme situations. Chapters 2 and 3 will explore approaches to quantify possible earthquake events for hazard and risk analysis.

1.3.2 Variability of Ground-Motion Intensity

Another challenge with deterministic analysis is the choice of the worst-case ground-motion intensity associated with a specified earthquake rupture. The ground-motion amplitudes discussed

---

[2] Here we use the model of Chiou and Youngs (2014). Many models exist to predict ground motion, based on empirical data and physics concepts. These models are discussed in Chapters 4 and 5.

[3] The physics of how earthquake properties relate to accelerations, velocities, and other characteristics of ground shaking will be considered in Chapters 4 and 5.

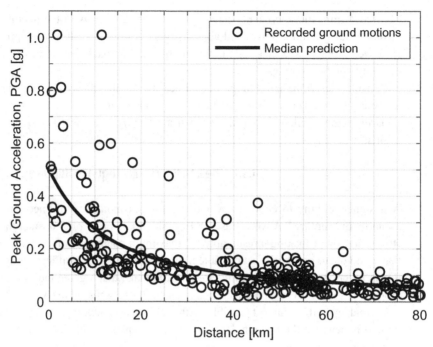

Fig. 1.1 Observed peak ground acceleration values from the 1999 Chi-Chi, Taiwan, earthquake, illustrating variability in ground-motion intensity. Median predicted *PGAs* based on Chiou and Youngs (2014) are also shown.

in Section 1.3.1 were median predictions from an empirical model calibrated to recorded ground motions. However, recorded ground motions show a tremendous amount of variability around those median predictions. To illustrate this point, Figure 1.1 shows *PGA* values observed in the 1999 Chi-Chi, Taiwan, earthquake, plotted versus the closest distance from the earthquake rupture to the recording site. Observations at distances of approximately 10 km, for example, indicate that ground motions varied in amplitude from 0.1 g to more than 1 g.

To account for ground-motion variability, deterministic hazard analyses sometimes specify a mean-plus-one-standard-deviation level of ground motion (e.g., Kramer, 1996), but that amplitude can be exceeded. There is no clear upper bound on the amplitude of ground motion that can be caused by an earthquake with a given magnitude and distance, and so no way to easily define a worst-case amplitude.[4]

1.3.3 Variability of Impacts

Even if it were possible to quantify earthquake sources and ground-motion intensity with certainty, there would still be variability in impacts: the damage to structures and the costs and time associated with recovery. Additionally, considering impacts is a reminder that the calculations in this book

---

[4] There is some true physical upper bound on ground motion amplitude caused by the finite energy release from an earthquake and the inability of the earth to carry more intense seismic waves without being damaged, but this limit is typically far above the intensity of shaking considered in engineering analyses and so is not of practical importance (Andrews et al., 2007; Strasser and Bommer, 2009).

have the ultimate goal of allocating resources. For these decisions, a conservative or deterministic evaluation of impacts is unsatisfactory as it omits critical information about the likelihood of such impacts. When allocating scarce resources for mitigation, or choosing to allocate resources for earthquake risk reduction versus some other beneficial activity, considering only a single impact scenario without understanding its likelihood can lead to inefficient decisions.

1.3.4 Limitations

Given the challenges discussed in this section, it is clear that a deterministic design earthquake rupture and ground-motion intensity is fundamentally limited. It is not factually a "worst-case" event, as a larger earthquake rupture, ground motion, or impact is always possible. Without an actual worst-case event to consider, we are left to identify a "reasonably large" event. That is often done by choosing a scenario of a nearby large-magnitude earthquake rupture scenario and then identifying some level of reasonable ground-shaking intensity associated with this scenario.

Three issues with deterministic analyses follow from the challenges discussed in this section. First, the result is not a "worst-case" ground motion. Further, we will see in Chapter 6 that deterministic rules are not, in general, more conservative than probabilistic rules, so this is not an inherently conservative procedure. Second, the result may be very sensitive to decisions made about the chosen scenario rupture magnitude and ground-motion intensity measure. Since these are decisions of judgment, and the results depend strongly on these judgments, the resulting analysis outputs are not particularly robust. Third, deterministic analyses will produce varying levels of safety across geographic variations and classes of considered structures. The varying safety results because this approach cannot consider factors like earthquake activity rates.

A ground-motion amplitude chosen using a deterministic approach was historically described as a "maximum credible earthquake," or MCE. More recently, however, the acronym has been retained but taken to mean "maximum considered earthquake," recognizing that larger earthquakes and larger ground-motion intensities are also credible. Deterministic thinking will be abandoned for most of this book, although the limitations identified here serve as a useful motivation for thinking about probability-based alternatives.

1.4 Probabilistic Seismic Hazard Analysis

To assess the risk to a structure from ground-motion shaking, we must first determine the annual rate (or probability) of exceeding some level of shaking at a site, for a range of intensity levels. Information of this type can be summarized as shown in Figure 1.2, which indicates that low levels of intensity are exceeded relatively often, while high intensities are rare. This plot is called a *ground-motion hazard curve*, or sometimes merely a hazard curve. The mathematical approach for performing this calculation is known as *probabilistic seismic hazard analysis*, or PSHA.

Because many models and data sources are combined to create results like those shown in Figure 1.2, the PSHA approach can seem opaque. However, when examined more carefully, the approach is rather intuitive. Once understood and implemented, PSHA is flexible enough to accommodate a variety of users' needs, and quantitative so that it can incorporate all knowledge about seismic activity and resulting ground shaking at a site.

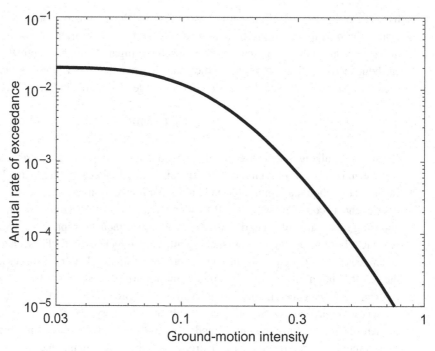

Fig. 1.2   A ground-motion hazard curve, to quantify future ground shaking at a given site.

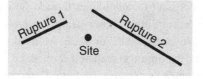

Fig. 1.3   Map of an illustrative site with two earthquake ruptures.

1.4.1 Illustrative Calculation

Consider a site near two earthquake sources, each with an associated type of earthquake rupture (Rupture 1 and Rupture 2), as shown in Figure 1.3. In a given year, the probability of Rupture 1 is 0.05, and the probability of Rupture 2 is 0.01. For simplicity, we assume that occurrences of earthquakes from the two sources are probabilistically independent, and neglect the probability of two earthquakes occurring in one year, so the probability of an earthquake in one year is $0.05 + 0.01 = 0.06$. The information in this paragraph comprises the *source characterization* for this problem.

Now consider the ground motion at the site resulting from an earthquake. Here we will measure ground motion using *PGA*, and for simplicity, we assume that the earthquake produces one of only three possible *PGA* amplitudes. The probabilities of observing a given *PGA*, given an earthquake from either source, are given in Table 1.1. Table 1.1 comprises the *ground-motion characterization* for this problem. These numbers approximate a situation often seen in the real world: Rupture 2 occurs more rarely than Rupture 1 (i.e., a probability of 0.01 per year, versus 0.05 for Rupture 1), but when Rupture 2 occurs it tends to produce larger ground motions.

Table 1.1. Probabilities of observing *PGA* values, given an earthquake from Rupture 1 or Rupture 2		
	$P(PGA = x)$, given an earthquake	
PGA [g]	Rupture 1	Rupture 2
0.1	0.80	0.40
0.2	0.18	0.40
0.3	0.02	0.20

Now we want to compute the probabilities of observing a specified *PGA* value or greater. Let us start with a simple calculation: What is the probability that we observe a *PGA* value of 0.3 g, due to an earthquake from Rupture 1, in a given year? Intuitively, this is the probability of an earthquake from Rupture 1 times the probability that a Rupture 1 earthquake causes *PGA* = 0.3 g ($0.05 \times 0.02 = 0.001$).

Next, we find the probability of observing a *PGA* value of 0.3 g, due to an earthquake from any source, in a given year. Here we add the above probability that this occurs from a Rupture 1 earthquake to the comparable probability from a Rupture 2 earthquake ($0.01 \times 0.2 = 0.002$), for a total annual probability of 0.003 (recall that for now, we are ignoring the probability of two earthquakes in one year—an approximation to simplify this calculation).

$$P(PGA = 0.3 \text{ g}) = 0.05(0.02) + 0.01(0.2) = 0.003. \tag{1.1}$$

Now, let us find the probability that we observe a *PGA* value of 0.2 g *or greater* during a given year.[5] For Rupture 1, there is again a 0.05 probability of observing an earthquake, and a 0.18 + 0.02 = 0.2 probability of observing a *PGA* value of 0.2 g or greater (where the addition considers the probabilities associated with the relevant *PGA* values). Considering both sources, and all relevant *PGA* values, gives us a probability of

$$P(PGA \geq 0.2 \text{ g}) = 0.05(0.18 + 0.02) + 0.01(0.4 + 0.2) = 0.016. \tag{1.2}$$

Finally, repeating the same process for a *PGA* value of 0.1 g or greater gives a probability of 0.06 in a given year (this is the same as the probability of an earthquake, since all earthquakes in this example produce *PGA* $\geq$ 0.1 g). These numerical results are summarized in Figure 1.4. In the more realistic case where the potential *PGA* values for a given source varied continuously, this figure would have probabilities at all *PGA* values, and we would have a continuous curve like that in Figure 1.2. This simple calculation illustrates the primary steps of PSHA: define models for earthquake sources and ground motions, and then combine the two to quantify the likelihood of ground shaking of a given intensity.

Note also that the probabilities associated with each source are combined to get an overall probability, and it is possible to see graphically and numerically how they combine. For this example, Rupture 1 contributes most to the probability of *PGA* $\geq$ 0.1 g, while Rupture 2 contributes more to the probability of *PGA* $\geq$ 0.3 g. We will formalize this type of analysis in Chapter 7.

This example illustrates the general hazard analysis procedure developed in this book. It also points to many issues that require more careful development, such as characterizing earthquake sources,

---

[5] We did not consider the *or greater* condition for the 0.3 g calculations, because there are no possible values of *PGA* greater than 0.3 g in this example.

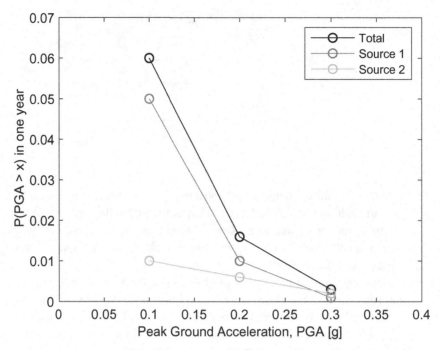

Fig. 1.4 Ground-motion hazard curve for the simple illustrative calculation. Example calculations are shown in black circles and connected by a line. Calculations associated with individual sources are indicated using separate lines. Note that the connecting lines in the figure should not be used for interpolations, as in this example, only discrete *PGA* values are possible. In more realistic calculations, this is not an issue, as real ground motions (and resulting hazard curves) vary continuously in amplitude.

defining metrics for ground shaking intensity, predicting ground shaking intensity at a site, and being precise when calculating event rates and probabilities.

The cover image on this book illustrates these same hazard calculation ingredients, in a more advanced application. The top panel is a map of earthquake sources in the study area, each capable of producing ruptures. The middle panel is a simulation of ground shaking from one rupture, which can be used to predict PGA and other ground-motion amplitude metrics. The bottom panel is a map of seismic hazards for the study area, with coloring indicating the amplitude of ground shaking exceeded with some target probability, obtained when considering all possible ruptures that could occur in the surrounding area. Several models are required to specify these inputs and calculations, but the simple example above contains many of the basic concepts used in more comprehensive calculations.

1.4.2 Hazard Curves from Direct Observation

Conceptually, a result like Figure 1.4 could be obtained from direct observation, rather than with the model-based approach described above. We would observe ground motions at the site of interest for many years, and for each PGA amplitude of interest, compute the fraction of years in which that amplitude was exceeded. As the observation time gets very long, this approach would produce the probabilities shown in Figure 1.4. Another version of this approach would be to directly observe exceedance probabilities of low-amplitude PGA values, and use statistical analysis to infer

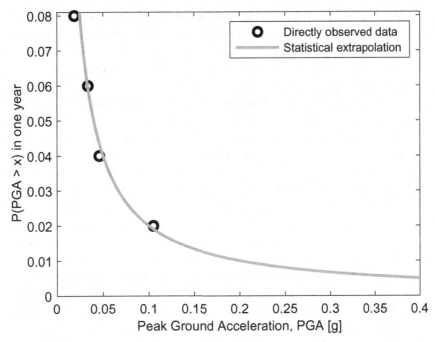

Fig. 1.5 Schematic illustration of a directly observed hazard curve. In this case, we assume a 50-year window of observed ground motions. The largest observed ground motion is assumed to have a 1/50 annual probability of exceedance, the second-largest a 2/50 annual probability, and so on. These probabilities are plotted with circles. The gray line indicates a hypothetical statistical extrapolation.

the implications for exceedance probabilities of higher-amplitude motions. A conceptual illustration of this approach is shown in Figure 1.5. Flood hazard modeling often uses this approach, so there is precedent in related fields.

However, two issues related to available data prevent this approach from being used for seismic hazard analysis. First, this approach requires a very long window of observations. If a PGA amplitude has a probability p of being exceeded in one year, we need at least $1/p$ years of data to have any hope of estimating p from observations. In fact, some quick calculations indicate that $10/p$ years of data will provide an estimate of p with 10% error (see Section 12.4.1). In seismic hazard and risk problems, probabilities of exceedance of approximately $p = 0.001$ per year are often of interest, meaning that this approach would require 10,000 years of direct observation. We could shorten that window somewhat by utilizing the statistical-processing version of the approach, but we cannot shorten it sufficiently to make it work with only a few decades of data. As a qualitative sanity check on this assertion, we note that the majority of locations in the world have not experienced strong ground shaking in the past few decades. It is therefore hopeless to forecast rare ground-motion shaking from a short observational window.

A second, related problem is that the approach requires a ground-motion recording instrument to be present at the site of interest for the duration of the observational window. Unlike flood heights, which can be acquired via simple physical measurements (and can even be measured after the event from high-water marks), the intensity of a strong ground motion requires specialized instruments. Even at locations with recording instruments, the instruments have rarely been present for more than a few decades. Further, the vast majority of locations in the world have no nearby recording instrument.

We might try to circumvent this challenge by using more qualitative ground-motion intensity scales based on human descriptions of felt intensity from earthquakes in the more distant past. However, felt intensity scales are less useful than the quantitative scales used in engineering analysis, and often the observational windows where humans have been present and recorded their impressions of earthquake shaking are still relatively short.

The above discussion of direct observation hazard raises two concepts relevant to model-based hazard calculations. First, while the absence of a recording instrument limits the use of direct observational data to constrain hazard, historical reports and geologic evidence of past earthquakes provide constraints on the earthquake occurrence probabilities used in the model-based calculation of seismic hazard. So the model-based seismic hazard approach can utilize a much broader range of past observational data than the direct-observation approach considered in this section. Second, the discussion of long observation windows with this approach can give a misleading impression that the resulting ground-motion hazard curve is focused on predictions over very long time windows as well. Typical time windows of interest are one year or 50 years: a short period compared with geological time. The long observation windows discussed in this section arise because we are interested in events with a low probability of occurrence within the time window of interest. For example, if we are interested in finding the event with a 1/500 probability of occurring in the next year, a direct empirical estimate of the event will require hundreds or thousands of observations of comparable past years.

In summary, because most locations have no ground-shaking measurements, and the few locations with measurements do not have a suitably long observational window, this direct observation approach cannot be effectively deployed.

1.5 Probabilistic Risk Analysis

Having quantified potential ground shaking from earthquakes using PSHA, we continue to quantify the impacts of earthquakes on the built environment. This requires combining the likelihood of a ground-shaking intensity with the probability of some adverse outcome given that shaking.

1.5.1 Illustrative Calculation

Let us continue the simple illustration from Section 1.4.1 to assess the risk of failure of a structure. Consider a hypothetical structure that may fail due to ground shaking. Given shaking with a PGA of 0.1, 0.2, or 0.3 g, the structure will fail with probabilities of 0.1, 0.5, and 0.9. The information in the previous sentence is often quantified by a *fragility function*.

Our knowledge of this ground-motion hazard and structural fragility is summarized in Table 1.2. The first column lists the three possible PGA values, and the second column contains the hazard results computed above for this illustration. The second column's results are for *exceedances* of PGA values (the standard way of presenting ground-motion hazard), while the risk calculation requires probabilities for *occurrences* of PGA values. We convert to occurrence probabilities in the third column by simple subtraction for this case in which only three discrete values of PGA are considered (e.g., $P(PGA = 0.1 \text{ g}) = P(PGA \geq 0.1 \text{ g}) - P(PGA \geq 0.2 \text{ g})$). The fourth column of the table repeats the failure probabilities assumed here.

Table 1.2.	Hazard and fragility data for the illustrative risk calculation		
PGA [g]	$P(PGA \geq x)$	$P(PGA = x)$	$P(Failure|PGA = x)$
0.1	0.060	0.044	0.1
0.2	0.016	0.013	0.5
0.3	0.003	0.003	0.9

With this data, we multiply and sum to compute the probability of the structure failing in a given year:

$$P(Failure) = \sum_{x} P(Failure|PGA = x)P(PGA = x) \tag{1.3}$$

$$= (0.1)(0.044) + (0.5)(0.013) + (0.9)(0.003) \tag{1.4}$$

$$= 0.0044 + 0.0065 + 0.0027 \tag{1.5}$$

$$= 0.0136. \tag{1.6}$$

We can make a few observations from this calculation. In Equation 1.5, we can see the contributions of each PGA level to failure. In this case, PGA values of 0.2 contribute the most, as they are both relatively frequent and also have a relatively high probability of causing failure when they occur. So the strongest levels of ground shaking may not be of most concern because they occur so infrequently (this result is also informative when recalling the above discussion of designing for the "worst case" ground motion).

For this example, there is slightly more than a 1% chance of the structure failing in a given year. This is the implication of our assumptions about the probabilities of earthquake occurrence, the probabilities of resulting PGA values from each earthquake, and the probabilities that PGAs of different levels would cause failure. We put those three pieces of information together in a systematic way that allows the implications to be clear, and shows the relationships between input assumptions and the resulting risk. The risk calculation is discussed in more detail in Chapter 9.

In Chapters 6 and 9 we will also generalize this calculation to account for the fact that the input models are not perfect. The probabilities considered here are due to random variations in the earthquake and damage processes, and we often call this *aleatory variability*. We often refer to our lack of knowledge about the earthquake and damage processes as *epistemic uncertainty*. These uncertainties are discussed further in Section 1.7.

1.5.2 Risk to Spatially Distributed Systems

When considering spatially distributed systems (e.g., portfolios of buildings, road networks, pipeline networks), the process described above requires modification. The ground motion varies spatially, so shaking intensity needs to be quantified at many locations throughout the system. Probabilistic dependencies in ground-motion predictions will thus need to be quantified. Additionally, when the number of locations of interest is large, it is cumbersome to produce a hazard curve representation quantifying the frequency of occurrence of all permutations of shaking intensity values at the various sites. Instead, it is preferable to represent the ground-motion hazard via a set of Monte Carlo simulations of ground-motion intensity at all locations of interest.

Although the fundamental approach of quantifying earthquake ruptures and predicting ground shaking intensity is the same for spatially distributed systems as for single-site systems, some adjustments in models and calculations are needed. Chapter 11 will discuss these issues.

1.5.3 Hazard versus Risk Calculations

One might naturally ask, Why do PSHA and stop at the ground-motion hazard calculation, before starting again and using that result as input to the risk analysis? A practical reason is that it is typically different people or groups that handle the two stages of the analysis, and modularization allows analysts to focus on their domain of specialty. A deeper reason is that some analysis requires only hazard analysis information. For example, many building codes utilize hazard maps to establish forces to use in designing structures, but do not require explicit risk analysis. Further, a hazard analysis can, to a limited degree, be performed without knowledge of later risk applications (i.e., the structure or system of interest need not be specified). In contrast, a risk analysis requires knowledge of the system of interest and cannot be performed until that system is specified. The hazard analysis still requires an intended eventual risk analysis in order to specify appropriate ground-motion metrics and exceedance rates of interest, among other choices, but in some cases the hazard analysis can be completed without a complete specification of the system. For these reasons, hazard calculations and risk calculations are generally distinct products and are often not tightly coupled.

Despite these separations, there are good reasons to combine the treatment of the two topics in this book. First, when calculations are taken all the way to risk metrics, the necessity of probabilistic treatment is apparent. There is no way to understand the risk if the ground-motion amplitude is specified by an arbitrary rule-based procedure rather than a probabilistic procedure. Second, risk calculations motivate the output metrics produced by hazard calculations. For example, the ground-motion amplitudes of interest in the hazard analysis are the amplitudes relevant for consequence predictions in the risk analysis. The value of the disaggregation and conditional spectrum calculations of Chapter 7 are much more apparent once their utility for risk calculations is clear. Because of this, it is important for readers interested primarily in hazard analysis to have some exposure to risk calculations.

1.6 Benefits of Probabilistic Analysis

Rather than ignoring the uncertainties present in the problem, the probabilistic approach incorporates them into ground-motion intensity calculations. While the incorporation of uncertainties adds complexity to the procedure, the resulting calculations are much more defensible for use in engineering decision-making for reducing risks.

Industrial activities in general have been moving for decades toward probabilistic risk-based regulations (Pate-Cornell, 1994; Bedford and Cooke, 2001; Hartford, 2009). Risk-based regulations facilitate cost–benefit decision-making and provide transparency in targeted design objectives. Risk-based regulations also ensure consistent safety across projects and geographic locations. Almost all societal activities involve an assumption of risk, and probabilistic (as opposed to rule-based) regulatory goals are the only path that allows a broader evaluation and comparison of societal risks from various activities.

Aside from compatibility with other societal activities, there are several reasons to compute risk numbers probabilistically. First, it provides a rational basis for evaluating the performance of the built environment. Second, it provides a performance goal that non-technical stakeholders can understand, compare with other sources of risk, and modify if desired. Third, it provides a transparent basis for evaluating the impacts of model inputs, assumptions, and uncertainties. This can be useful for decision-making, and can also focus future research on topics that will produce the greatest value by reducing key uncertainties. That is, probabilistic risk analysis can be seen as a tool for logical and structured interrogation of analysis results, rather than only a numerical description of the state of nature.

A probabilistic risk analysis is most valuable if it can provide the correct probability of an adverse outcome. However, looking at the advantages noted in the previous paragraph, some of these advantages are still present even when we accept the inevitability that some input models and assumptions will be inaccurate (Melchers, 2007). We should also keep in mind that the limited understanding that may lead to input model errors may also manifest in nonprobabilistic analyses. Explicitly tracking uncertainties is especially valuable for problems where there is a wide range of technical interpretations.

Probability calculations are a critical part of the procedures described here, so a basic knowledge of probability is required to study this topic. Appendix A provides a summary of the concepts and notation used in this document. Readers desiring further education on probability and statistics should refer to one of the many textbooks focusing on these topics (e.g., Ang and Tang, 2007; Benjamin and Cornell, 2014; Ross, 2014).

The value of probabilistic hazard and risk results depends upon combining the input models in an internally consistent manner and performing the hazard and risk calculations following the rules of probability. The value also depends upon the input models being appropriate, in that they reflect the scientific and engineering communities' best understanding of the relevant phenomena.

The choice of appropriate input models evolves as our knowledge and experience grow. Model choice also varies depending upon the circumstances of a particular location's tectonic environment, site conditions, and properties of the considered structures. It is thus not possible to specify appropriate models in a textbook, though we will discuss the pros and cons of model types, and factors affecting the appropriateness of models. It is, however, possible to specify the correct mathematical approaches for calibrating and validating models, and for combining them in hazard and risk calculations. We will focus significant attention on these topics.

1.7 Uncertainties in Probabilistic Analysis

While the preceding discussion has mentioned uncertainty, this section provides some further context on the nature of that uncertainty. Specifically, there are several types of uncertainties present in this analysis, and all types must be considered if the analysis is to represent our knowledge of the system. Here we introduce terminology to precisely refer to types of uncertainties considered in hazard and risk studies and discuss why they are important.

1.7.1 Epistemic Uncertainty and Aleatory Variability

Epistemic uncertainty (from the Greek *episteme*, "relating to knowledge") is used to refer to our lack of knowledge about models for the state of nature. If a parameter has either one value or another, and we do not know which value is correct (e.g., the maximum magnitude of an earthquake that a given seismic source can produce), then the parameter has epistemic uncertainty. These uncertainties often arise in seismic hazard and risk analysis, due to limitations in observed data or limitations in our understanding of the physics underlying the process. Because these uncertainties relate to a lack of knowledge, they are potentially reducible if new knowledge is obtained from additional data or further study. This reducibility feature is important from a hazard and risk analysis perspective, as it will be possible to evaluate whether a particular source of epistemic uncertainty is substantially affecting the analysis results. If it is, the analyst can consider directing further study towards the refinement of those uncertainties.

Aleatory variability (from the Latin *alea*, "game of dice") is used to refer to random outcomes from natural variability in a process, for a given model formulation of the process. If a parameter value varies across multiple observations of the process (e.g., the amplitudes of ground motions observed from earthquakes having comparable rupture and site properties), it has aleatory variability. Because aleatory variability is inherent to the process, it is not reducible through further study.

The terms "uncertainty" and "randomness" are sometimes used to refer to epistemic uncertainty and aleatory variability. But those words are problematic in that they are also often used in a colloquial sense. Thus, it may be ambiguous in some cases whether "uncertainty" is being used to refer specifically to epistemic uncertainty or some more general issue. Because of this potential for misunderstanding, the esoteric epistemic and aleatory terms are often helpful when a clear distinction between the two is needed.

Example

To illustrate the above concepts, consider a component of seismic hazard calculations that will be discussed in some depth in Chapter 3: the occurrence of earthquakes from a given seismic source, within a given period of time.

One model for the earthquake occurrence process is that earthquakes happen with a constant long-term rate and that the occurrence of the next earthquake is independent of the time since the last earthquake (i.e., the occurrences are a Poisson process, Section 3.3.5). In this model, the numerical value of the long-term occurrence rate is a model parameter with epistemic uncertainty due to our imperfect knowledge of the source's behavior. Given knowledge of that rate, the number of earthquakes in any given time period will have aleatory variability, described by the Poisson probability distribution.

A second model for the process is that earthquake occurrence additionally depends upon the time since the last earthquake, motived by the idea that the accumulation and release of stress over earthquake cycles has some memory (see Section 3.8.1). This model requires one or more parameters describing how the rate of earthquakes evolves as a function of time after an earthquake, and knowledge of the time since the previous earthquake. The parameter(s) describing earthquake rate would be an epistemic uncertainty, similar to the Poisson process assumption. In addition, the time of the last earthquake may also be an epistemic uncertainty if we do not have a historical record of the previous event. This second model should have reduced aleatory variability in the prediction of earthquake occurrence relative to the first model, as it puts a stronger constraint on when earthquakes are predicted to occur. However, if the second model's parameters are not well-constrained, its reduced aleatory variability may be offset by increased epistemic uncertainty due to the relatively greater model complexity and need to specify more parameters.

1.7.2 Categorization of Uncertainties

The previous simple example illustrates two points. First, it is most intuitive to consider aleatory variability as some inherent property of nature that we cannot predict with certainty. But more precisely, this aleatory variability is dependent upon our model formulation, as "what we cannot predict with certainty" depends upon the approach we are using to make the prediction. As such, it is more accurate to refer to *apparent aleatory variability* in model-based predictions for hazard and risk analysis, though we often shorten this to "aleatory variability" for brevity.

Second, while epistemic uncertainties are in principle reducible, they may not be practically reducible within the time frame or resources that an analyst has to study the system. In the above example, earthquake occurrence rate parameters may be reducible epistemic uncertainties, but it may take thousands of years to obtain sufficient observational data (as discussed in Section 1.4.2).

The distinction of epistemic versus aleatory is not always clear-cut and proper classification may not be necessary. The necessary issues to address are the following: (1) consider all sources of uncertainty, (2) do not double-count uncertainties, and (3) understand which sources of uncertainty are potentially reducible.

Finally, we note that the careful inclusion of aleatory variability and epistemic uncertainty does not guarantee that the calculation result is error-free (Musson, 2012b). Potential errors in input models can be termed *ontological uncertainties* (Elms, 2004; Marzocchi and Jordan, 2014). These ontological errors can be revealed through a comparison of a hazard or risk forecast with data. Such comparisons are the topic of the next section.

1.8 Validation

The hazard and risk calculations outlined above amount to probabilistic statements about the implications of our knowledge. As such, there are opportunities to validate the results and confirm or reject these implications as consistent with observations from the real world.

Validation of these calculations is far from trivial, however, for a few reasons. First, rare-event calculations are typically of most interest. But our intuitive minds are ill-equipped to evaluate rare-event probabilities, and available observational evidence is rarely sufficient to draw definitive conclusions about the validity of calculations associated with events happening once every hundreds or thousands of years. This limitation can be mitigated by evaluating calculations at many locations, to obtain more data, but this introduces its own set of complications.

Second, there are challenges in comparing past earthquake occurrences, ground motions, and impacts to hazard and risk calculation results. The calculations include both aleatory variability and epistemic uncertainty, and these uncertainties do not precisely align with the variability associated with observations. While aleatory variability is similar to observational variability, the aleatory variability is dependent upon the model formulation. Comparing past events with calculations also has challenges due to uncertainty about historical events.

Third, any criticism of hazard and risk calculations, relative to observations, needs to be categorized as either a concern about model specifications (i.e., "our earthquake source model failed to include rupture scenario X") or a concern about calculation process (i.e., "seismic hazard analysis is a fundamentally flawed procedure"). The former criticism is a possibility with any model specification and can be evaluated. The latter criticism is unreasonable as it is aimed at a mathematical operation that (when correctly implemented) can be proven to achieve the desired calculation.

The above issues are challenging, but they are not insurmountable. Chapter 12 will discuss how to perform validation productively and address some criticisms and misunderstandings related to seismic hazard and risk calculations.

PART I

HAZARD INPUTS

2 Seismic Source Characterization

Chapter 1 demonstrated that in order to determine the effects of earthquakes at a site of interest, earthquake rupture scenarios that can cause shaking at this site must be identified. The most important characteristics of these scenarios are the size of the earthquake and its geographical location. The size and location provide the *what* and *where* attributes, but within PSHA it is also necessary to define the likelihood, or *how often*, of each scenario. This chapter is devoted to understanding how the information about the possible size and location of future earthquakes is obtained. In Chapter 3, focus then moves to the likelihoods, or rates of occurrence, of the hypothesized earthquake scenarios. Other scenario characteristics that influence ground motion are discussed in Chapters 4 and 5.

Learning Objectives

By the end of this chapter, you will be able to do the following:

- Define different types of seismic sources (area, fault, background, and point sources).
- Describe the salient mechanics associated with plate tectonics, from the plate-boundary scale down to earthquake processes on individual faults.
- Obtain a conceptual understanding of large-scale earthquake processes.
- Understand the processes used to identify and characterize seismic sources.
- Compute seismic moments and moment magnitudes for rupture scenarios.
- Estimate the maximum seismogenic potential for an earthquake source.
- Be conversant with geometric and kinematic descriptions of earthquake sources.

2.1 Introduction

A seismic source characterization identifies all potential earthquake rupture scenarios that can generate ground motions of engineering interest. In regions where earthquakes occur frequently, the most important scenarios will often occur within approximately 100 km of the site of interest. In less active regions, important scenarios could occur at distances of 300 km or greater. The reasons why the size of these regions differ will become apparent in Chapters 3–6.

In a geological context, these distances of $\approx$100–300 km are not large. To understand why the potential sources near the site behave as they do, it is usually necessary to consider larger-scale geological processes. The present chapter first discusses some of these processes, fundamental concepts related to what an earthquake is, and the driving mechanisms behind earthquakes on a global scale. The subsequent sections then progressively introduce concepts that enable the characterization of potential rupture scenarios at an arbitrarily local scale. Ultimately, the chapter defines the different types of seismic sources that comprise a seismic source characterization, as well as characteristics of earthquakes in these sources.

The objective of the chapter is to demonstrate how all possible rupture scenarios can be identified. Each rupture scenario is represented generically as rup_i, where i is an index used for bookkeeping purposes. Each of these rupture scenarios must be defined in a manner that is consistent with how ground motions are subsequently computed. The information will, at least, consist of a measure of the size of the earthquake associated with the rupture and a definition of the rupture geometry. More elaborate approaches to estimating ground motions may take into account additional information, as discussed in Chapters 4 and 5. It is therefore necessary to understand why earthquakes are located in particular positions, what controls their geometry in terms of both orientation and physical extent, and how this size is related to the magnitude of the event. Once all of the possible rupture scenarios are identified, see Chapter 3 for calculation of the rates of occurrence of each rupture, $\lambda(rup_i)$.

2.2   Earth Structure and Plate Tectonics

A brief overview of plate tectonics is presented here, to explain why seismic sources exist, and what governs their features. An appreciation of tectonic regimes is useful for understanding why particular earthquake faults have given characteristics, and why they may warrant consideration as unique sources of future earthquakes. The fundamental nature of plate boundaries is replicated to some extent at more regional and local scales of relevance for seismic source characterization. The objective here is to provide some context for subsequent discussions. For readers wishing to gain a more substantial background in earth processes, geodynamics (Jaeger et al., 2007), neotectonics (Yeats et al., 1997), and mechanics of rocks and faulting (Scholz, 2002), a number of good reference texts exist.

Figure 2.1 shows that the Earth is comprised of distinct layers, with the typical divisions being the thin crust, followed by the thick mantle, outer core, and inner core.

The Earth's surface, shielded from space only by our atmosphere, is relatively cold, while the core is very hot. The main differences in temperature take place through the mantle. Although the mantle is solid, over geological time scales it flows as a viscous solid and convective cells exist as a result of the thermal gradient. These cells impose shear drag upon the base of the crust, and this provides the driving mechanism for the motion of tectonic plates (Jaeger et al., 2007). The convergence or divergence of the flow through the mantle causes regions of the crust to be in compression or tension, respectively. However, the Earth's crust is not a continuum, but is rather comprised of a set of discrete tectonic plates (Wegener, 1912).These plates primarily move as rigid bodies but experience significant deformation at their peripheries as they interact with other plates. The effect of the compressive and tensile loading imposed by the viscous drag in the mantle is therefore largely concentrated in relatively narrow bands around the plate margins.

Earthquakes take place within the crust, where temperatures are cool enough to permit the brittle failure of rock. The greatest depth at which earthquakes originate within the crust is referred to as the *seismogenic thickness*. This thickness varies regionally, but is typically on the order of 10–20 km (Scholz, 2002).

The time scale over which the plates move significant distances is very long in comparison with the typical design lives considered within engineering projects. The lateral strain rates imposed through plate tectonics are therefore relatively small in comparison with steady lithostatic stress.

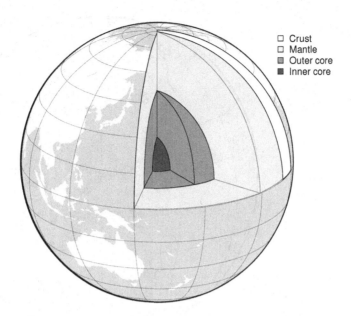

Fig. 2.1   Illustration showing the structure of the Earth using an accurate radial scale. Earth has an average radius of 6371 km. The radius of the inner core is ≈1200–1300 km, and the outer core has a thickness of ≈2200 km. Crustal thickness varies, but 35 km is representative. The mantle therefore has a thickness of ≈2850 km.

It is thus appropriate to view the compressive and tensile stresses as pertaining to deviatoric stresses. In regions where the convective cells of the mantle promote deviatoric tensile stresses laterally within the crust, we should expect that the maximum principal stress direction should be vertical. Conversely, where we have compressive deviatoric stresses, the maximum principal stress direction should be horizontal. This general understanding, coupled with the inferences made using Mohr's circle (Mohr, 1914), allows us to anticipate how faults are likely to be oriented in different regions of compression or tension throughout the crust. However, this theoretical orientation of faults is complicated by existing crustal weaknesses and heterogeneity associated with the evolution of the tectonic plates that may accommodate tectonic deformation.

Our understanding of this driving mechanism for crustal loading allows us to anticipate probable locations of future earthquake events—at least on a global scale. Figure 2.2 shows two views of the Earth with the plate boundaries defined by Bird (2003) overlaid. As well as these plate boundaries, significant global earthquakes are illustrated, color-coded by depth. It is clear from even a cursory inspection of Figure 2.2 that there is a strong correlation between the location of significant earthquakes and the plate boundaries. However, it is also apparent that earthquakes occur within the interiors of tectonic plates, so we cannot focus solely on the plate margins.

2.2.1 Plate Boundaries

The relative motion of the tectonic plates gives rise to distinct types of interaction at their interfaces. The nature of the interaction depends on the relative sense of motion between the plates as well as the type of crust: oceanic crust or continental crust. Oceanic crust is relatively thin in comparison with continental crust, and the temperatures between the two can differ significantly

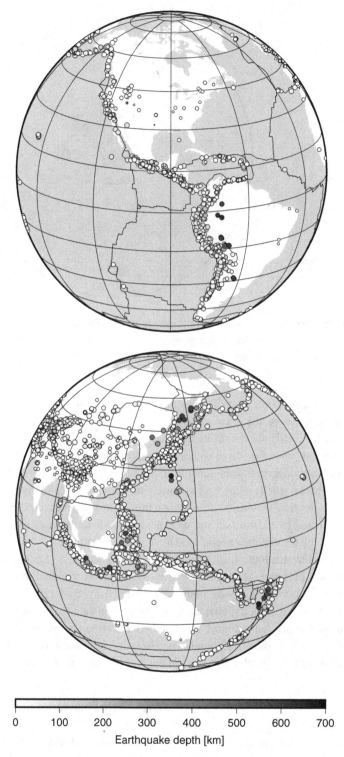

Fig. 2.2   Tectonic plates and their association with major historical earthquakes. The plate margins are from Bird (2003).

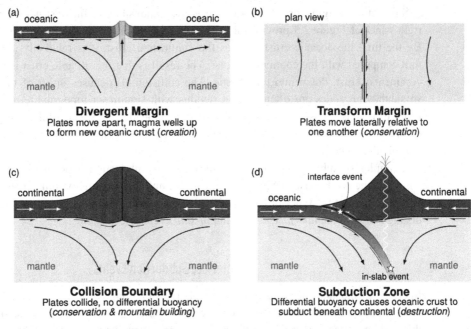

Fig. 2.3 Schematic illustrations of plate margin types. The upper left panel shows a divergent margin associated with oceanic plates moving away from each other, while the upper right panel shows a plan view of a transform margin. The lower panels show two types of convergent margin: collision between two continental plates (*left*), and a subduction zone arising from the collision of oceanic and continental crust (*right*). Example locations for interface and in-slab events are shown in the lower right panel.

as well. These differences in thickness and temperature (or, in effect, buoyancy) govern the types of plate margin interactions. Each of these types of plate margin is discussed in the following subsections and illustrated schematically in Figure 2.3.

Divergent Margins

In regions of the crust that experience tensile deviatoric stresses from the mantle loading, plates move away from each other. A schematic illustration of a divergent plate margin showing the creation of new oceanic crust is provided in the upper left panel of Figure 2.3. The gap left as two plates diverge from each other is filled by upwelling mantle that becomes new crust. The newly formed crust is thin, relatively warm, and buoyant. However, with time as this material travels away from the divergent margin, the material cools and becomes more dense.

Examples of divergent margins are the mid-Atlantic ridge and the East Pacific ridge. Divergent margins can also form within continental crust where tensile stresses accumulate. Examples of such extensional regimes are the Great Rift Valley in Africa and the Taupo Volcanic Zone in New Zealand.

Convergent Margins

Convergent margins occur when two plates move toward each other in a relative sense. When oceanic crust collides with continental crust, a subduction margin occurs. On the other hand, when continental crust collides with other continental crust, a collision margin results.

Subduction margins, or subduction zones, are responsible for the destruction of crust. The lower right panel of Figure 2.3 provides a schematic illustration of what happens in a subduction zone. By the time the oceanic crust reaches the continental crust, it is relatively old, dense, cool, and thin compared with the continental crust. The result is that the oceanic crust is pushed beneath the continental crust, down into the mantle. The collision also causes uplift of the continental crust, so subduction zones are often coastal regions with significant topographic relief. The uplift is a result of isostasy caused by horizontal compression leading to vertical thickening of the overriding continental crust, stacking of the continental and oceanic crust, and volcanism caused by melting of the subducting oceanic crust.

Examples of subduction zones are the Cascadia subduction zone on the west coast of North America, the west coast of South America, and multiple regions within Japan and New Zealand. When discussing the types of earthquakes that can occur within subduction zones it is important to distinguish between *interface* events and *in-slab* events (see Figure 2.3).

Interface Subduction Events

Interface earthquakes occur as a result of relative slip along the low-angle interface as the oceanic crust subducts beneath either continental crust or other oceanic crust. Because of the low angle of this sliding surface interface, earthquakes can have very large rupture widths and are sometimes referred to as low-angle thrust events. Furthermore, many subduction boundaries are long, and the combination of the large available rupture width and length permits very large ruptures to be created. The largest earthquakes ever experienced are associated with interface events at subduction zones.

In-Slab Subduction Events

In-slab events are created by a different mechanism, despite still occurring in subduction zones. Once the subducting oceanic plate is bent and pushed down into the crust, flexural stresses exist, with tensile stresses on the top of the subducting plate and compressive stresses below. Given that the rock is relatively weak in tension, cracks form and grow to create earthquakes in the down-going slab (hence the name "in-slab"). While the interface events are low-angle thrust events associated with compression, the in-slab events tend to be high-angle normal events associated with the tensile components of the bending stresses. The in-slab events also differ from their interface counterparts because they occur in the subducting oceanic seismogenic crust that is surrounded by the ductile upper mantle. As in-slab events occur at greater depths than crustal or subduction interface events, the energy radiated from the source is often strongly attenuated before arriving at the surface. However, because in-slab events involve the rupture of old oceanic crust, they tend to have relatively high stress release at the source (see stress drop in Chapter 5).

Collision Events

Collision margins resulting from convergence of continental crust are less common. The best example of a collision margin is the Himalayan mountain range where the Indian plate is moving northward relative to the Eurasian plate. At these interfaces the lack of a clear difference in buoyancy between the plates dictates that rather than one plate subducting beneath the other the two plates simply meet head-on. At the interface the local thickness of the crust increases through penetration downward, but mainly through increased topographic relief. This process is schematically shown in the lower left panel of Figure 2.3.

Transform Margins

The final type of plate boundary to be considered are transform margins. Transform margins occur when two plates move laterally past each other. This relative lateral movement can result from relative rotation of the two plates or from translation. Classical examples of transform margins are the San Andreas Fault in California and the Alpine Fault in the South Island of New Zealand. A schematic illustration of a transform margin is shown in the upper right panel of Figure 2.3.

In some cases the relative motion between the plates will not be purely lateral and the convergent or divergent component can also be responsible for the generation of topographic relief. This is the case for the Alpine Fault. Transform boundaries at major plate margins can be very long linear features, but they can also arise over shorter length scales in order to satisfy strain incompatibilities arising from differential convergent or divergent motion. A good example of a combination of divergent margins with coupled transform margins is provided by the mid-Atlantic ridge.

2.2.2 Partitioning Tectonic Movement

The relative movement between two tectonic plates is not accommodated by movement over a single interface as suggested by the schematic illustrations in Figure 2.3. Although it is often convenient to consider the tectonic plates as rigid bodies, in reality a significant degree of distributed deformation takes place at their peripheries. A side effect of this deformation is the creation and activation of faults (to be discussed more in the following section), but it is also important to recognize that the large-scale relative movement that we observe is distributed through movements over these smaller-scale internal surfaces to create a zone of distributed deformation in most cases. Therefore, when one asserts that one particular plate is moving at a rate of 30 mm/yr relative to some other plate, it is important to appreciate that this total amount of movement is accommodated through lower amounts of slip across the faults in the vicinity of the plate interface.

2.3 Faults

The nature of the interactions that occur between tectonic plates are replicated at more smaller scales by *faults*. The optimal orientation of a fault plane depends upon the stress field of the region containing the fault. While the general sense of the stress in a region can be known, a number of complicating factors limit our ability to define the state of stress in a region (e.g., Hanks, 1977; Townend and Zoback, 2000; Scholz, 2002). When describing faults we therefore adopt a coordinate system that is independent of the stress field.

Just as plate margins were classified as divergent, convergent, or transform depending upon the relative sense of movement between the two plates, a similar convention is employed to describe individual faults. Figure 2.4 shows the two key parameters that define the orientation of a fault,

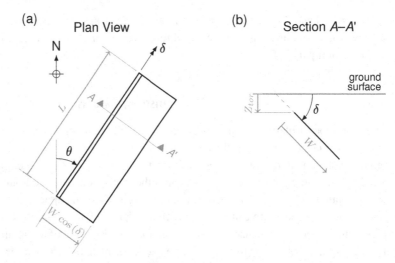

(a) Plan View

(b) Section A–A'

Fig. 2.4 Schematic illustration of a fault with strike angle θ, dip angle δ, rupture length L, rupture width W, and depth to the top of rupture Z_{tor}. A plan view is shown on the left, while a cross-section through A–A' is shown on the right. The double line shown in the left panel indicates the upper edge of the fault. Sometimes the upper edge will be indicated by a series of "teeth" (arrowheads) pointing down-dip from the upper edge.

the *strike* and *dip* angles. The strike angle is defined as the bearing of the fault trace (the surface projection of the upper edge of the fault, as would be seen on a map) measured clockwise from north in degrees. For any given fault trace it can be appreciated that two equivalent angles from north could be defined that are 180° apart. However, when it comes to faults, there is no ambiguity. The strike angle is always defined such that when standing on the fault trace and looking in the direction of strike the fault will always dip down to your right. For a vertical fault the smaller of the two possible strike values is normally taken. In Figure 2.4 the strike angle is $\theta = 35°$.

The dip angle is defined as the angle that the inclined fault plane makes with a horizontal surface. This angle can best be interpreted by viewing a cross-section like Section A–A' in Figure 2.4. A planar fault will appear as a line segment when viewed through this cross section. The range of dip angles is theoretically from $[0, 90]$, although dip angles at the lower end of this range are practically impossible for crustal tectonic events. The dip at the upper end is limited to 90° because the angle of the fault is defined from a position looking in the strike direction. Note that in the plan view of Figure 2.4 the dip angle is indicated using a double arrow symbol to denote a right-hand rotation about the strike direction. This right-hand rule can be useful to ensure no confusion arises with strike and dip angles. If you point the thumb of your right hand in the direction of the strike the dip angle is defined as a rotation in the direction that your fingers are pointing.

A parameter known as the rake is also important for understanding ground motions that arise from a fault rupture. This parameter defines the relative sense of movement between two sides of the fault, as illustrated in Figure 2.5. When looking at the foot wall of the rupture surface, the relative movement of the hanging wall side determines the rake angle. For pure left-lateral (*sinistral*) strike-slip movement, the hanging wall will move horizontally in the direction of the strike. In this case, the rake angle is $\lambda = 0°$. For pure right-lateral (*dextral*) strike-slip movement the hanging wall moves in the opposite direction with $\lambda = 180°$. If the hanging wall moves upward along the *up-dip* direction, then the rake will be $\lambda = 90°$ and we have a reverse mechanism, while we have a normal mechanism for $\lambda = -90°$ and *down-dip* movement of the hanging wall. When the rake angle is

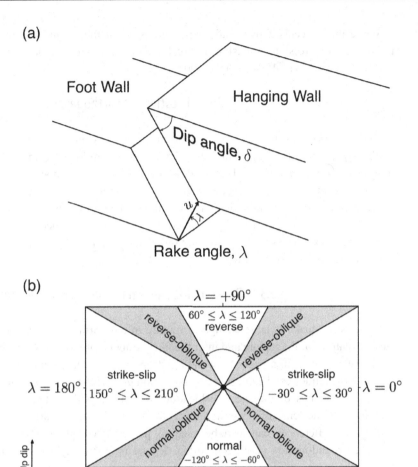

(a)

Foot Wall

Hanging Wall

Dip angle, δ

u

λ

Rake angle, λ

(b)

$\lambda = +90°$

$60° \leq \lambda \leq 120°$
reverse

reverse-oblique

reverse-oblique

$\lambda = 180°$

strike-slip
$150° \leq \lambda \leq 210°$

strike-slip
$-30° \leq \lambda \leq 30°$

$\lambda = 0°$

normal-oblique

normal-oblique

normal
$-120° \leq \lambda \leq -60°$

$\lambda = -90°$

Up dip

Along strike

Fig. 2.5 Demonstration of the rake angle and how different types of earthquake mechanism are defined using the rake. The upper panel shows a perspective view of how the rake angle relates the relative direction of movement between the hanging wall and foot wall. The lower panel shows how ranges of the rake angle are often mapped into generic mechanisms.

consistent with reverse faulting, but the dip angle is shallow, e.g., $\delta \lesssim 45°$, then the dislocation is referred to as *thrust-faulting*. The rake (and sometimes dip) allow a definition of the *style of faulting* that is often used within ground-motion modeling (Chapter 4).

Slip vectors often have both the along-strike and up/down-dip components. The lower panel of Figure 2.5 indicates the ranges of the rake angle where an event would be classified as strike-slip, normal, or reverse. When significant components are combined, it is common to classify the events as reverse-oblique or normal-oblique, corresponding to the shaded regions of Figure 2.5.

Mohr's circle can again be applied to appreciate why certain types of fault tend to have different dip angles. When the principal deviatoric stress is in the horizontal direction, i.e., a compressive regime, the expected dip angle for the formation of a new fault is shallower than 45° by an amount reflecting the frictional strength of the material. On the other hand, in an extensional regime, the deviatoric lateral tension causes the principal stress to be vertical. The dip of a new fault in such a regime would be relatively steep and above 45° by an amount that again reflects the frictional strength of the material.

While the concepts above guide expectations about the orientation of faults (and ruptures) in regions of compression or tension, the real situation is more complex (see, for example, Sibson, 1985; Sibson and Xie, 1998; Scholz, 2002).

2.3.1 Locations of Earthquakes

Earthquakes occur as ruptures over fault surfaces, but they initiate from a particular location known as the *hypocenter*. The hypocenter is a defined as a coordinate in three-dimensional space using longitude, latitude, and depth. This depth is also known as the hypocentral depth.

The hypocenter is sometimes also referred to as the *focus*, or the *focal point*, of the earthquake. Consequently, the hypocentral depth is sometimes referred to as the focal depth.

The surface projection of the hypocenter (the point on the surface of the Earth directly above the hypocenter) is known as the *epicenter* and is defined using longitude and latitude only. Maps showing earthquakes display epicenteral locations, e.g., Figure 2.2.

2.3.2 Determining the Orientation of Earthquakes

When an earthquake is large enough to rupture to the Earth's surface, it is possible to inspect the surface features of the rupture and infer the likely orientation of the responsible fault. However, most earthquakes that occur are too small and too deep for any surface expression to exist. In these cases, rupture orientations can be inferred using seismological techniques (Aki and Richards, 2002).

The orientations of the shear dislocations associated with earthquakes are important because they indicate the sense of stress in the region and where lateral variations in stress exist. Rupture orientation differences can also be used to partition a study region into different subregions because ground-motion predictions often depend upon the style of faulting associated with the earthquake.

Figure 2.6 illustrates how the orientations of ruptures can be defined. On the left side of this figure, a fault plane is shown as a vertical black line. This fault can be regarded as a vertical strike-slip fault (dip angle of 90°) with a strike angle of 0°. The shear stresses acting on this fault surface are shown pointing toward the north on the western side of the fault and pointing to the south on the east side. In the upper portion of this fault, a small square element that straddles the fault is shown. This element has pure shear stresses applied to its left and right faces to represent the stresses acting over the fault. To maintain moment equilibrium, this "couple" of stresses must be balanced by an opposing couple acting on the element's horizontal sides. These horizontal stresses can be thought of as acting on an *auxiliary plane* perpendicular to the actual fault.

In the lower left of Figure 2.6, the same element shown in the upper portion is now rotated by 45° so that the pure shear stresses become an equivalent set of pure normal stresses. One pair of these normal stresses imposes compression on the element, while the other imposes tension. A pressure wave that leaves the source as a result of this dislocation will have a first motion that is positive (compressive) in the radial direction when exiting the second and fourth quadrants, while it will have a negative (dilatational) first motion when leaving the first and third quadrants. The *beach ball* diagram shown on the right of Figure 2.6 summarizes the first motions for all waves leaving the source through the lower half of a sphere centered on the focal point. The compressive first motions are shaded in black, while dilatational first motions are shown in white. Note that the beach ball would look the same for the strike slip rupture shown as well as for a rupture rotated 90° degrees from the real rupture (i.e., the auxillary plane).

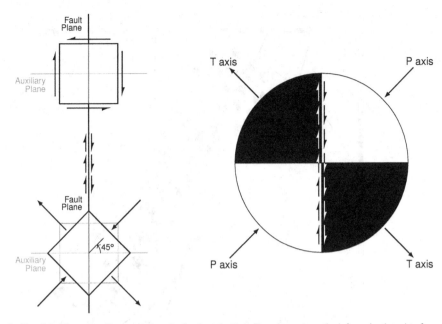

Fig. 2.6 Example of a "beach ball" representing an earthquake focal mechanism. The schematic on the left can be thought of as a plan view of a strike-slip fault. The 'beach ball' shown on the right is consistent with this strike-slip rupture.

The first motions arriving at seismic recording stations are inspected to create a beach-ball diagram. As noted above, the sense of motions leaving the source through different quadrants will either be positive (for compression) or negative (for dilation). The sense of these waves persists as they propagate from the source to the recording stations. Therefore, when multiple recordings are made, using a network of recording instruments with good azimuthal coverage, the positive or negative sense of the first arrivals can be mapped back onto the source to infer the shear dislocation's orientation. The beach-ball diagram then reflects this orientation. The result of inverting observed motions to infer the orientation of the source is known as a *focal mechanism solution*.[1]

Beach-ball diagrams therefore indicate how earthquakes in a given region tend to be oriented. Systematic variations of beach balls across space indicate a change in the stress field and the orientation of faults. If a dislocation is inclined, then the beach ball will have black shading through the center of the ball for a reverse mechanism and white shading for a normal mechanism.

Figure 2.7 shows real-world focal mechanism solutions for events of magnitude > 3 occurring within 200 km of Yerba Buena Island, in San Francisco Bay. The dominant mechanism is strike-slip, but there are also reverse mechanism events (with a black band straddling the center of the ball). The balls themselves suggest two possible orthogonal planes of rupture, but the general sense of faulting in the region helps to distinguish the actual fault plane from the auxiliary plane. In this case, it can be inferred that the strike-slip faulting is generally north-south, along the faults marked with black lines.

The key message here is that seismological observations of small earthquakes can indicate how orientations of shear dislocations vary from point-to-point within a region. Patterns in these beach-ball diagrams are useful for partitioning the study region into areas where particular types of earthquakes (strike-slip, normal, or reverse) preferentially occur.

---

[1] The *moment tensor* obtained as part of this solution can be diagonalized into *double-couple* and *non-douple-couple* components (Shearer, 2009, Chapter 9).

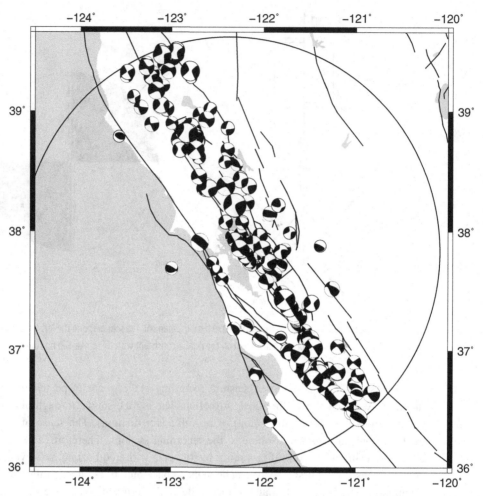

Fig. 2.7 Plot of beach balls for events with magnitudes ≥ 4 within 200 km of Yerba Buena Island in the NCEDC catalogue (NCEDC, 2014). The black circle indicates the 200 km search radius, while the black line segments show active faults.

2.4 Earthquake Processes

Once individual faults are identified, multiple different rupture scenarios *rup* can be associated with a given fault. This can result from changes in the strike and dip angles along the fault, or local zones of reduced frictional strength, among other things. The largest rupture that can occur on a given fault involves displacements over the entire fault. Even larger ruptures can involve the joint rupture of multiple faults within a fault network (e.g. Cesca et al., 2017; Shi et al., 2017). However, faults can also be responsible for the creation of smaller ruptures. The following two subsections introduce conceptual models to explain why earthquakes initiate, how various sections of a fault can rupture independently of others, and why ruptures arrest. Subsequently, Section 2.4.3 discusses mechanisms via which ruptures can interact across different faults. The conceptual models introduced here are not directly used to define different *rup* scenarios, but they help explain why particular scenarios are identified.

2.4.1 Elastic Rebound Theory

Following the 1906 San Francisco earthquake, *elastic rebound theory* was proposed as a simple conceptual model for an *earthquake cycle*. The theory is approximate in several ways, but is useful for explaining the key issues of energy build-up and release during earthquake events.

Elastic rebound theory essentially states that a region of the Earth's crust will be gradually loaded as a result of tectonic processes. The shear stress at a given location will gradually increase as a result of this loading. At some point, this accumulated stress exceeds the frictional resistance of a fault, and an earthquake initiates. Upon initiation of the earthquake a significant amount of the elastic strain energy that was previously stored within the crust[2] is released in the form of elastic seismic waves.

Upon release of this energy, the state of the system in the vicinity of the rupture is effectively reset. The next earthquake in this location will not occur until the stresses again accumulate to their critical value. This cyclic process is illustrated in Figure 2.8.

In Figure 2.8 the general tectonic loading is represented by the deviatoric normal stresses acting over a region containing what will become the fault plane. The relative amplitudes of these deviatoric stresses σ_1 and σ_2 (in addition to the frictional characteristics of the crustal rock) dictate the optimal orientations for a shear dislocation (i.e., a fault plane). As mentioned in the previous section, for any shear couple across a fault plane, one will need an additional couple to satisfy equilibrium; these shear stresses are shown as dashed arrows perpendicular to the fault plane.

Elastic deformation of the crust creates tectonic stresses, causing accumulation of elastic strain energy. At some point, slip occurs over the fault plane, and the stored energy is released through plastic mechanisms at the fault surface and radiation of seismic waves. The elastic medium surrounding the fault will recover much of its prior deformation.

Tectonic strains accumulate very slowly, and very large amounts of energy are released in major earthquake events. This explains the inter-event times between large earthquakes. If stress over the fault plane is uniform and a given event releases all of this stress, then the same type (size and geometry) of event would repeatedly occur on this source. But this is inconsistent with the occurrence of ruptures smaller than the full fault. On the contrary, if the stress over the fault plane varies spatially, then portions of the fault surface may reach a critical level of stress while others have some reserve capacity. This would allow smaller ruptures on the fault, but it is less clear how the larger events can occur.

These issues can be resolved by considering how stresses are transferred spatially due to the dislocation at any given point. The stick-slip models presented in the following section permit consideration of such features.

2.4.2 Stick-Slip Models

A more powerful model for earthquake occurrence and the quasi-cyclic nature of earthquake events was proposed by Burridge and Knopoff (1967). Many model variations have since been proposed,

---

[2] The elastic strain energy associated with a given stress state can be defined as:

$$E_S = \frac{1}{2E}(\sigma_x^2 + \sigma_y^2 + \sigma_z^2) - \frac{\nu}{E}(\sigma_x \sigma_y + \sigma_y \sigma_z + \sigma_x \sigma_z) + \frac{1}{2G}(\tau_{xy}^2 + \tau_{yz}^2 + \tau_{xz}^2)$$

in which the σ_i are normal stresses, τ_{ij} are shear stresses, E is the elastic modulus, G is the shear modulus, and ν is Poisson's ratio. Computation of the total strain energy requires integration of this expression over the stress field around the fault.

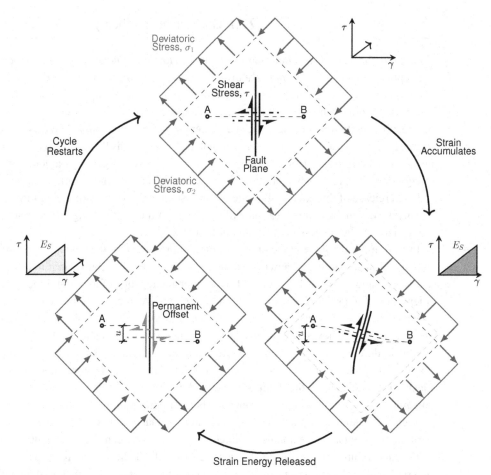

Fig. 2.8 Illustration of the earthquake cycle, as implied by elastic rebound theory. The initial state is shown at the top of the figure, where a cycle of loading has just commenced. Deviatoric stresses in the surrounding region impose shear stress on the fault. A straight line connecting points A and B is perpendicular to the fault. As strain accumulates, the region around, and including, the fault deforms, as shown in the lower-right diagram. The line connecting points A and B is now distorted to reflect this deformation, and the shear stress over the fault is significant. Upon reaching the shear strength, a slip event occurs and releases the strain energy accumulated around the fault. As shown in the lower-left diagram, the slip leads to some permanent offset u, breaking the line previously connecting points A and B.

but they are collectively known as *stick-slip* models. Figure 2.9 illustrates the components of a stick-slip model. The basic idea is that we have a rigid mass that is able to slide along a flat surface under certain conditions. The driving force is provided by a rigid element that moves laterally at a constant (and slow) velocity of v. A spring connecting the mass to the rigid element extends and induces a force $F_k = vtk$ in the spring. The spring stiffness is k, and the extension of the spring is the rigid element's velocity v multiplied by the time over which the spring has been loaded t ($x = vt$). The force in the spring gradually increases until the static frictional force is overcome. This occurs when:

$$xk = F_{\tau,s} = \mu_s mg \qquad (2.1)$$

with μ_s representing the static coefficient of friction between the mass and the sliding surface.

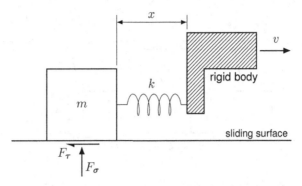

Fig. 2.9 Idealistic representation of a stick-slip model. The rigid body moves to the right with velocity v. The motion of the block of mass m is governed by the relation between the instantaneous spring force $F = kx$ and the frictional resistance F_τ.

When the static frictional resistance is overcome, the rigid mass slides and this sliding is resisted by the dynamic frictional resistance. The equation of motion for the sliding block can then be expressed as:

$$m\ddot{x} + kx = F_{\tau,d} \tag{2.2}$$

where $F_{\tau,d}$ represents the dynamic friction and can be defined in terms of the normal force $F_\sigma = mg$ and the dynamic coefficient of friction μ_d through $F_{\tau,d} = \mu_d mg$. Note that in this simple expression we have neglected any source of damping and also ignore the contribution of the gradual motion imposed through v. Few pedagogical gains are made by relaxing these approximations.

An analogy can be drawn with elastic rebound theory. The initial elongation of the spring, prior to sliding taking place, represents the gradual buildup of elastic strain energy in the medium surrounding a fault. The static friction represents the frictional strength of the fault, and the driving velocity v represents the long-term relative plate motion in the vicinity of the fault.[3]

This simple model suggests why ruptures initiate and why sliding arrests. Equation 2.2 suggests that, once the static frictional resistance is overcome, the sliding mass enters an oscillatory mode of response. However, during the first cycle of this oscillatory behavior, a point is reached where the sliding mass's velocity becomes zero. At this point the friction model being used dictates that an abrupt increase in the frictional resistance takes place where the dynamic friction is replaced by the static friction and the motion terminates.

While elastic rebound theory explains the buildup and release of energy, it does not answer why events of different magnitudes occur on a given fault. Similarly, basic stick-slip models give distributions of event sizes (related to the displacement of the block when it slips) that are inconsistent with empirical observations. However, refined stick-slip models can explain many features of complex seismogenic processes. Figure 2.10 shows one such extension in which the single sliding mass considered in Figure 2.9 is replaced by a series of masses.

In Figure 2.10, each mass is connected to a rigid driving plate moving at a steady velocity of v that represents the tectonic loading. The connection to this plate is provided through flexural springs of a given stiffness, k_F, representing the shear rigidity of the crustal material around the fault surface. Adjacent masses are also connected with axial springs representing the elasticity of the material

---

[3] When the fault is near a plate interface. For intra-plate earthquakes, occurring away from active plate boundaries, differences in regional stress influence the driving motion.

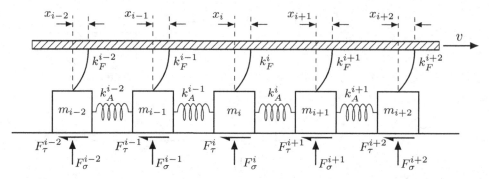

Fig. 2.10 Schematic representation of a multi-degree-of-freedom stick-slip model. The system is loaded by the rigid plate moving to the right with velocity v. Each mass is connected to the rigid plate via a flexural spring with stiffness k_F^i and to the adjacent masses via axial springs with k_A^i and k_A^{i-1}. The motion of individual blocks now depends upon the frictional resistance F_τ^i and the interaction with the local and global stiffnesses.

around the fault. The addition of these axial springs, and the relationship between the flexural and axial stiffnesses, allows individual masses to slip independently of the other masses. Furthermore, the model can generate "ruptures" involving the slip of multiple blocks in a single slip episode. The displacements of each block can differ, replicating differential slip over a rupture surface and heterogeneous stress distributions over the surface following the rupture. The relaxation of local regions of high stress then give rise to "aftershock" type behavior. Models of this type were studied extensively by Carlson and Langer (1989); Carlson et al. (1991) and these authors were able to replicate many features of real earthquake faults.

Further extensions of the stick-slip model include arranging masses in two dimensions (e.g., Olami et al., 1992), introducing nonlinear viscous damping, using more elaborate *rate-and-state* friction laws (e.g., Dieterich, 1979; Ruina, 1983; Bizzarri, 2011), and pushing the number of masses to a continuum limit. For example, the study of Sakaguchi and Okamura (2015) demonstrated that nonlinear viscous damping terms were necessary to replicate typical features of aftershock sequences. Despite the most elaborate models replicating key features of seismic sequences, they still fail to consider fundamental properties such as the fractal nature of faults and the complexity of fault geometry (Kagan, 2006).

To replicate all of the features of observed earthquake processes, the stick-slip models become very complex. However, the basic ideas underlying the models remain quite simple. For source characterization and the identification of ruptures, these stick-slip models provide a conceptual explanation for why particular rupture segments will exist along a fault. For instance, when a fault changes geometry, the effective axial stiffness $k_{A,ij}$ between two blocks (or adjacent regions of the fault) would change. This creates a potential barrier to the propagation of slip from one block to another along the chain.

In cases where a fault is discretized into geometric segments, such as in Figure 2.14 below, a stick-slip representation using blocks to represent each segment can be used to understand the likelihood of single-segment and multi-segment ruptures. Similarly, for a given geometrical description of a fault (or potentially a fault network), stick-slip type models indicate where rupture segments (as opposed to geometric segments) might be located, and the likelihood of ruptures over multiple segments or faults. It should be noted that stick-slip models become conceptually similar to finite-element models as the resolution of the model increases. As such, ruptures arise at different locations as governed by the underlying frictional laws and assumed fault geometries.

2.4.3 Stress Transfer Models

Sections 2.4.1 and 2.4.2 explained earthquake processes by considering an individual fault. However, a region will often have several spatially distributed faults that collectively accommodate tectonic deformation. These faults are collectively referred to as *fault networks* or *fault systems*. Within fault networks, dislocations on one fault will impact the state of stress throughout the wider region.

This concept is illustrated in Figure 2.11 by using the 1989 Loma Prieta earthquake in California as an example. The earthquake caused dynamic and static stress changes throughout the region, and the Coulomb stress change shown in Figure 2.11 is a way of quantifying the static stress change (Okada, 1992; King et al., 1994; Toda et al., 2011). In particular, positive stress change brings regions closer to failure according to the Coulomb failure criterion, while negative stress changes do the opposite. Earthquakes that occurred after the Loma Prieta event are also shown in the figure. The events shown in Figure 2.11b are relatively deep and their spatial extent corresponds closely to the original fault rupture. However, the events shown in Figure 2.11a are shallower, at a depth close to the average depth for the aftershock cluster, and occur over a greater spatial extent. Figure 2.11a provides some evidence for more events occurring in regions of positive stress change. Such observations have been made for many earthquake sequences (e.g., King et al., 1994; Toda and Stein, 2002; Toda et al., 2005).

Elastic rebound theory and stick-slip models capture broad first-order phenomena, but not second-order effects such as the stress perturbations from earthquakes on other nearby faults. In some cases, the stress changes caused by an event on one fault may accelerate or delay the occurrence of an event on another fault.

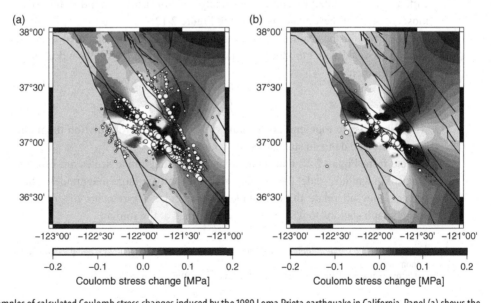

Fig. 2.11 Examples of calculated Coulomb stress changes induced by the 1989 Loma Prieta earthquake in California. Panel (a) shows the stress change at 7.5 km depth, along with aftershocks within ±2.5 km of this depth. Panel (b) shows the stress change at 15 km depth, with aftershocks again within ±2.5 km of this depth. In both panels, black lines represent approximate fault locations. Stress changes are computed for the kinematic source model of Wald et al. (1991) as implemented within the Coulomb software (Toda et al., 2011).

2.5 Earthquake Size

The size of an earthquake event is best characterized by the energy it releases. The *seismic moment*[4] is the standard measure used to quantify the size of an earthquake. The seismic moment, M_0, is formally defined as (Burridge and Knopoff, 1964; Madariaga, 1977):

$$M_0 = \int_A \mu(\boldsymbol{x})u(\boldsymbol{x})dA \approx \mu \int_A u(\boldsymbol{x})dA \tag{2.3}$$

where the *shear rigidity* or *shear modulus*, μ, is assumed here to be constant through the crust, and is hence taken outside of the integral. This expression is more conventionally written as in Equation 2.4 in which the spatial distribution of displacement, $u(\boldsymbol{x})$, has been averaged over the rupture area A to obtain the average displacement $\bar{u}$:

$$M_0 = \mu A\,\bar{u}. \tag{2.4}$$

For large events the rupture area will be very large and the rigidity may vary spatially over the rupture (particularly with depth). However, in practice, the rigidity is usually just assigned a constant representative value such as $\mu = 3.3 \times 10^{10}$ N/m^2. As the rigidity has units of stress and is multiplied by the rupture area we can see that Equation 2.4 can be interpreted as a force (μA) multiplying a displacement $\bar{u}$. Hence, the seismic moment, which will have SI units of Nm, represents the work done by the fault during the rupture.

Most people will be familiar with hearing the size of an earthquake being quantified in terms of its *magnitude*. It is less commonly known that a large number of magnitude scales have been proposed and used over the past several decades. Indeed when the media reports the "Richter scale" size of an earthquake, the quoted magnitude has likely been measured on a different scale. A selection of the most common of these scales is shown in Figure 2.12.

In Figure 2.12 the scaling of each magnitude definition is compared with the *moment magnitude* scale. This scale was introduced by Hanks and Kanamori (1979) and is purely a function of the seismic moment:[5]

$$\boldsymbol{M} = \frac{2}{3}\log_{10} M_0 - 6.0\dot{3} \tag{2.5}$$

where the seismic moment has SI units of Nm. Since its formulation the moment magnitude scale has become the de facto standard scale within PSHA.

Note from Figure 2.12 that while many of the magnitude scales are broadly consistent with the moment magnitude scale, they all depart significantly at large magnitudes. The other scales *saturate* because they all relate the amplitudes of particular components of ground motions observed at seismograph stations to what must have happened at the source.

---

[4] Often called the *scalar seismic moment* in the seismological literature due to it being a constant that scales the terms of the moment tensor.

[5] The original scale defined by Hanks and Kanamori (1979) was defined as:

$$\boldsymbol{M} = \frac{2}{3}\log_{10} M_0 - 10.7$$

with M_0 in units of dyne-cm. Note that 1 dyne-cm is equivalent to 1×10^{-7} Nm and so the $6.0\dot{3}$ in Equation 2.5 arises from $10.7 - \frac{2}{3} \times 7$. The coefficient 10.7 was adopted to ensure that $\boldsymbol{M}$ was similar to surface-wave magnitude M_S for moderate to large events.

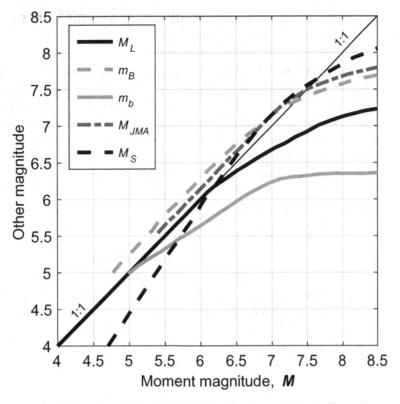

Fig. 2.12 Comparison of common magnitude scales and their relationship to the moment magnitude scale. Here we show moment magnitude, *M*; Richter (or local) magnitude, M_L; broadband body-wave magnitude, m_B; short-period body-wave magnitude, m_b; Japanese Meteorological Agency magnitude, M_{JMA}; and the surface-wave magnitude, M_S.

The main reason why most of these scales saturate relates to the way the source spectrum of earthquakes scales (see Chapter 5); in particular, how the corner frequency of the source spectrum varies with magnitude, as well as upon the finite bandwidth of seismographs (Hanks and Kanamori, 1979). Essentially, while the source spectrum of an earthquake has a corner frequency (see Chapter 5) that is above the frequency used for a given magnitude scale, changes in moment magnitude at the source will be observed at the seismograph stations. However, once the corner frequency becomes close to, or lower than, the frequency used for defining the magnitude scale, then the seismographs do not "see" the effect of changes in source strength. For this reason, low-frequency magnitude scales, such as the surface wave magnitude (M_S), saturate at much higher magnitudes than high-frequency scales such as the body-wave magnitude (m_b). The dependence of the estimated magnitudes upon the source corner frequency, which can vary regionally, also explains why *local* magnitude scales can differ from region to region. Additional discussions of the most common scales encountered in practice are provided in Lay and Wallace (1995) and Kramer (1996).

Unless explicitly stated, hereafter when referring to magnitude we will implicitly mean moment magnitude. The key point here is that this moment magnitude is a function of the seismic moment, which is, in turn, a function of the rupture area and average slip over the fault. Therefore, the sizes of ruptures on faults are connected to their magnitudes. Generally speaking, the larger the rupture, the larger the magnitude of the causative event.

2.6 Definitions of Seismic Sources

A seismic source model (SSM) will typically combine different types of sources representing the probable locations of future earthquakes. A schematic illustration of a source model is shown in Figure 2.13. The overall study region is shown, and an area source and a fault source ar identified.

To justify the presence of these sources, the characteristics of the seismicity (earthquake activity) within the sources must be different than that of the general region. The regional seismicity is represented by the background source in the figure. It represents all activity that is not associated with other source types.[6] In most cases, the identified sources will represent regions of higher activity than the background source, but this is not required. Methods to characterize activity rates for individual sources are the subject of Chapter 3.

In the sections that follow, various types of sources are described.

2.6.1 Fault Sources

The most obvious representation of a seismic source is a fault. A fault can be regarded as a three-dimensional surface representing a zone of weakness[7] in the rocks deep beneath the Earth's surface,

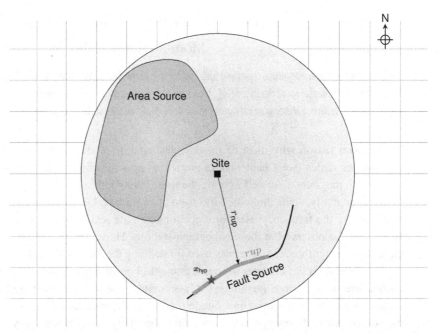

Fig. 2.13 Schematic illustration of a seismic source model comprised of an area source, a fault source, and a background source. The large circle represents the study region and the background source extends beyond this. For the fault source, a possible rupture scenario is shown along with a representation of the distance from the site to the rupture.

---

[6] It is important that events are not represented more than once.
[7] The "strength" of the fault is governed by the frictional resistance to relative motion over the fault surface. This frictional strength is lower than the strength of competent rock surrounding the fault.

over which earthquakes occur. However, as mentioned earlier, it will rarely be possible to constrain the rates of occurrence of events of all sizes that would occur on a given fault. Even if events have previously taken place on a given fault, it is not always clear how the fault might rupture in the future. For example, a long fault may consist of a number of fault segments, typically distinguished by changes in geometry along its length. The changes in geometry that exist may only be surficial (and could relate to topographic loading[8]), but they may also reflect extended changes that persist to greater depths. In the latter case, the segment ends may provide barriers to the growth and propagation of ruptures along the fault. An individual fault source may therefore generate "characteristic" rupture scenarios associated with the activation of individual segments, or combinations of segments. A seismic source model must be able to represent the likelihoods of all of these rupture scenarios.

A fault source is represented within a seismic hazard analysis by a simple polygon in three-dimensional space (usually a quadrilateral or rectangle oriented according to the strike and dip of the fault, as defined in Section 2.3), or a number of such polygons that represent the geometric segments of the fault. Note that the geometric segments need not be the same as the rupture segments. In cases where the variations in geometry along the fault are subtle, it is common to reflect this in the fault's geometry but not suppose that such variations are barriers to the propagation of a rupture of multiple geometric segments.

Figure 2.13 shows how an individual rupture might be generated within a fault source. In some calculations, a rupture of a given size might be generated and then allowed to "float" uniformly along the fault, assuming that all such positions are valid. In other cases, the distribution of hypocentral locations along the fault might be assumed to follow a uniform distribution, and geometrically admissible ruptures are then created. In more involved cases the likelihoods for an event of this magnitude may need to respect the fault segmentation model assumed appropriate for the fault.

An example of a real fault is shown in Figure 2.14. This is the Hayward Fault, which runs along the eastern side of San Francisco Bay and is traditionally characterized by three main segments: the northern, southern, and southern extension. While these three broad segments exist, individual ruptures on this fault may span all three segments, combinations of adjacent segments, or portions of individual segments. These scenarios can be established by discretizing the fault into nominal segments and then looking at possible ruptures as combinations of adjacent segments. An example of such a segmentation-based approach is shown in the lower panel of Figure 2.14.

Geological field evidence of historical ruptures is often used to characterize these large fault segments. For example, evidence may exist that the northern segment preferentially ruptures as a contiguous block rather than in conjunction with portions of the southern segment. In practice, and given that a detailed rupture history is often not available, fault geometry may be used to set fault segment boundaries. These reasons for partitioning a fault into large segments arise from what is commonly referred to as *barriers* to rupture propagation. However, as will be discussed in more detail later, once the fault segments are broken down into the individual segments (as in the central

---

[8] As an earthquake rupture propagates toward the Earth's surface the rupture will tend to grow along the path of least resistance. Berrill (1988) has shown that the presence of building structures, hillsides, mountains, and other types of topographic loading can cause ruptures to divert from their nominal geometries at depth.

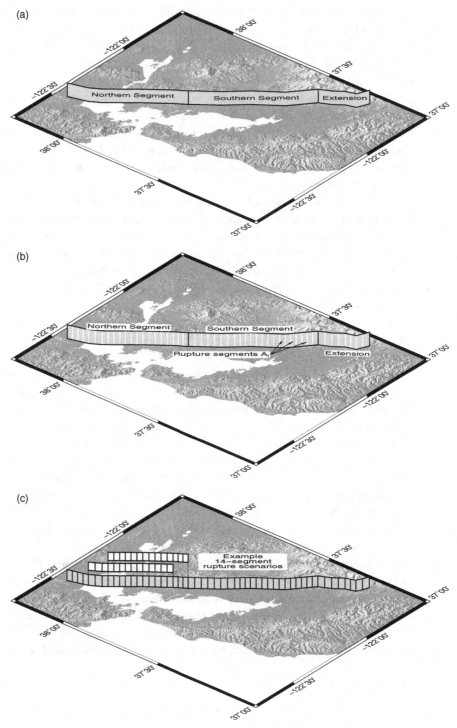

Fig. 2.14 Example decomposition of a fault source (the Hayward Fault, along the east of San Francisco Bay) into segments. Panel (a) shows the northern, southern, and southern extension segments of this fault (Field et al., 2013). Panel (b) shows how each segment may be decomposed into several smaller segments with a nominal area of A_i. Combinations of contiguous segments then relate to distinct rupture scenarios, as shown in panel (c). The magnitude of each rupture is related to the total area of the segments in each scenario.

and bottom panels of Figure 2.14), rules that dictate how these segments interact to form ruptures govern which scenarios may be generated by the fault. These rules may account for changes in geometry along the fault, or they may imply that adjacent segments within a given parent fault section are more likely to rupture together than adjacent segments in different fault sections. The rules may even be extended to govern how individual segments on one fault interact with segments on other nearby faults.

2.6.2 Area Sources

In many regions, it is known that active faults exist, but it is not possible to characterize individual faults as sources. If a region has had only relatively small events that have not breached the surface, the causative fault locations are likely not known with precision.

In these cases, it is common to define an area source that aggregates the seismic activity over some spatial region. In the hazard calculations, individual earthquake ruptures may be modeled, but they have a uniform distribution of spatial locations throughout the source. The general orientations of the ruptures within the source may be known from focal mechanism solutions or in-situ measurements. In this case, ruptures can be generated that are consistent with these orientations. However, such generated ruptures are only occurring on hypothetical faults consistent with the seismic activity over the area source in the aggregate.

Area sources are typically a polygon or a polyhedron. The characteristics of seismic activity within an area source should be substantially uniform within the source, but different from regions outside of it. Differences could be in the level of activity, the distribution of magnitudes of different sizes, the nature of geological structures within the region, or the inferred geometry of the ruptures and style of faulting.

A schematic example of how these fictitious ruptures might be generated for a particular area source is shown in Figure 2.15. In this figure, the area source is a polygon that is discretized and represented by a grid of points. From each point a series of possible ruptures are generated that are consistent with any available constraints. As was discussed in Section 2.5, the sizes of the ruptures tend to increase with the earthquake magnitude.

The total activity rate for the source can be assigned to the grid points or spatial cells, on a *pro rata* basis. Each spatial cell within Figure 2.15 will have some area, A_i (or volume) that is a fraction of the total for the source A_T. The rate of activity attributed to each cell is then A_i/A_T.

One issue that must be considered for area sources is how these fictitious ruptures are generated. With a fault source, the fault and individual segments' spatial extents are defined as part of the model. However, the constraints upon the geometries of individual ruptures in an area source are not as clear. For example, if a focal position is close to a boundary of the area source, is it possible for the rupture to propagate beyond the source zone? If such ruptures are permitted, the boundary that allows propagation is referred to as a *leaky boundary*. As will be discussed in Chapter 4, because of the way in which ground-motion models represent source-to-site distances, the decision to permit or exclude rupture propagation through a source boundary can be important to the resulting hazard calculations. The decision should consider whether there is any physical impediment to prevent a rupture from propagating. In some cases there will be a clear difference in geological units that may work for or against the decision to model a boundary as leaky. Uncertainty about boundaries can be reflected during the hazard calculations via logic tree branches (see Section 6.7).

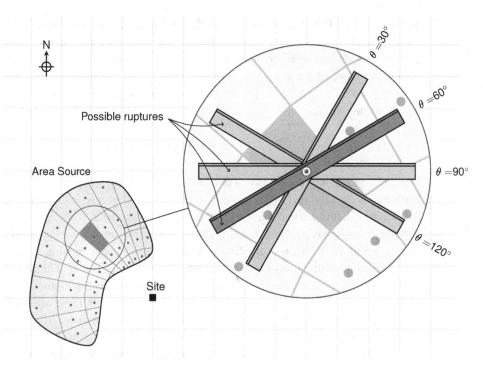

Fig. 2.15 Schematic illustration of how fictitious ruptures might be generated within an area source. Double lines in the expanded view indicate the tops of the ruptures.

2.6.3 Point Sources

Point source models represent seismic activity as a point source where earthquakes are assumed to occur at a single location, and rupture dimensions are not considered. Chapter 4 will highlight the fact that the overwhelming majority of twenty-first-century ground-motion models measure distances to ruptures with finite dimensions. However, some models work with point-to-point measures of distance.

While such sources are strictly only applicable for ground-motion models that adopt point-source distance metrics, there are cases where an analyst may choose to represent a source as a point. For example, imagine a fault that always ruptured with a characteristic magnitude and that this rupture covered the entire fault. In this case, the closest point on the rupture to the site will always be the same point. If ground motions were computed in terms of this closest distance between the rupture and the site, then the analyst could save computational time by simply representing all ruptures as coming from this closest point. In this case, the ground-motion model would still use the rupture distance to account for the finite rupture effects. Note, however, that if multiple sites were considered, this representation would need to be adapted for each new site location.

Another use for point sources is when the analyst wishes to represent spatially distributed seismicity over some area but does not wish to model finite rupture effects. This may occur when the dimensions of the largest possible ruptures are small. In these cases replacing the area source by distributed point sources can be more computationally efficient. This approach will introduce a degree of error when ground-motion models taking finite-distance metrics are used, but the error will often be small, particularly in the case that the area being modeled is some distance from the site (Bommer and Akkar, 2012).

Therefore, while explicit consideration of finite ruptures, either on fault sources or on fictitious faults within area sources, is preferred, there are practical and computational reasons why point sources might be used within hazard calculations.

2.6.4 Background Sources

At any location in the world there will be some background level of seismicity. This level may be very low, but it is extremely difficult to argue that no such level exists. While source model development often focuses on the identification and characterization of fault and area sources, these sources will usually cover only a subset of the study domain. The remaining area must be represented by a background source.

A background source is effectively an area source, but it will often be relatively complex because the geometry is the Boolean difference between the overall study area and any other sources in the model. This means that background sources will often contain holes and may often also need to be represented using multiple polygons. The issues discussed in Section 2.6.2 regarding leaky boundaries are also relatively complex when considering background sources. For example, the analyst must decide whether fictitious ruptures generated within the background source are able to propagate into, or even through, other source zones.

In areas of relatively low seismicity it is common for the source that contains the site to provide a significant contribution to the hazard. It is also common for the site in question to be located within the background source. Thus, background sources can be important to the hazard, though their activity rates tend to be lower than in other area and fault sources. Therefore, it is important to ensure that background sources are modeled with care when developing the source model.

2.6.5 Balancing Specificity and Uncertainty

When developing a seismic source model, one must balance individual sources' specific features and the ability to constrain the model parameters. Some degree of compromise must always be made.

The example calculations in Chapter 1 showed the need to identify individual earthquake scenarios *and* their rate of occurrence. This requires a certain level of information for every identified rupture *rup*. Unfortunately, there is never sufficient information available to constrain the rates of ruptures for large numbers of individual fault sources. So even if all active faults in a region can be identified, the activity rates for all ruptures on all these faults is unlikely to be constrained.

Consider two end-member cases: a very complex source model and a very simple model. A complex source model may represent every fault in the region. For each fault we need to specify the geometry, the rate at which it generates ruptures of a particular style and magnitude, how the various segments interact within the fault, and potentially how ruptures on particular segments of this fault influence ruptures on other faults, among other things. In the case that all of these features can be constrained, we should expect that the future earthquake scenarios predicted by our source model would accurately reflect both the magnitude and spatial distributions of activity in the region. However, our model requires a large number of parameters.

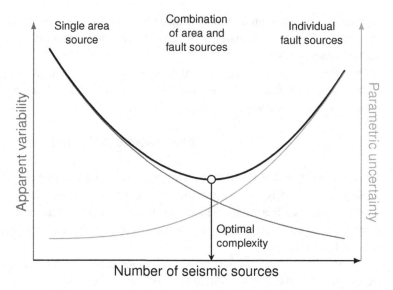

Fig. 2.16 Schematic illustration of the balance between different types of uncertainty as the complexity of the source model varies. Apparent aleatory variability is plotted with the dark gray curve and reduces with the number of seismic sources. Parametric uncertainty is plotted in light gray and increases with increasing numbers of seismic sources. The heavy black line represents the total uncertainty arising from a combination of apparent variability and parametric uncertainty.

In contrast, if the source model was extremely simple and consisted of a single area source, only a small number of parameters would need to be constrained. However, our model would predict an "average" behavior throughout the region. That is, the earthquake likelihoods, distribution of earthquake magnitudes, and associated rupture geometries would be the same at every point within the source. When comparing the predictions of this model with future seismicity we would expect a considerable degree of variation away from this average model, with certain regions having higher or lower activity rates. Similarly, in some locations the distribution of magnitudes would depart from our average predictions.

In both cases, given our model, the departures of the observed (past or future) seismicity from our predictions can be viewed as *apparent variability*. That is, under the assumption that our model is correct, we would observe what appear to be random variations. In reality, many of these apparent variations will actually be biases in the model that result from not adequately representing the physical reality through a simplified model. It is important to recognize that neither of these end-member models is "correct." Rather, they are both imperfect but are also potentially useful (Box, 1980; Field, 2015).

Figure 2.16 illustrates these concepts schematically. As we increase our source model's complexity, we should better represent reality and the apparent variability should decrease. However, we have limited data with which to constrain the parameters of the source model. The uncertainties in each parameter of our source model can be represented collectively by the degree of *parametric uncertainty*.

Any project will have a finite amount of existing data and a finite budget and time frame for collecting additional data. The project source model must balance complexity by reducing apparent variability while being aware of the limitations of the data available to constrain the models.

As noted in Figure 2.16, source characterization models often combine individual fault sources with area sources. These area sources aggregate the features of unknown, or poorly constrained, faults over some spatial region. In these regions it is possible to estimate the characteristics and frequency of earthquakes without explicitly representing faults.

Figure 2.16 suggests that some optimal level of complexity can be found. In practice it is very difficult (if not impossible) to truly quantify these components of uncertainty, and so we consider this concept, but do not formalize the implied optimization of this whole procedure. Because there is no optimal solution to the source characterization problem, a different analyst may develop an alternative source characterization. The existence of these alternative views of the seismic source characterization does not cause problems because it arises as an inevitable consequence of a complex problem constrained by limited data. Provided that each alternative source representation is consistent with the available empirical constraint and is also supported by some underlying physical rationale,[9] then the representation cannot be rejected. To deal with this situation, seismic hazard analysts attempt to ensure that their calculations reflect the epistemic uncertainty implied by these available interpretations of the available data. Historically, the most common way to represent this epistemic uncertainty is through the use of logic trees, which will be presented and discussed later in Section 6.7.

2.7 Source Characteristics

For the purposes of seismic hazard analysis, earthquake ruptures are described in very simple terms. The physics of earthquakes and the generation of ground motions is more advanced than that implied by these simple source representations. Studies following major events reveal a great amount of information about the physics. However, PSHA makes assertions about the future, not what has already happened. It is in this context where our capabilities are limited. While many physical phenomena influence ruptures and resulting ground motions from a given rupture, for predictive purposes we need to be able to robustly specify these conditions a priori. This degree of specification is currently beyond our means and so we revert to the principles portrayed through Figure 2.16 in which we seek a balance between apparent variability and parametric uncertainty.

For now then we make a conscious modeling decision to attempt to develop robust simple models rather than elaborate models that are very difficult to constrain. By way of an example, a rupture scenario that could be prescribed as an input to kinematic ground-motion simulations (see Chapter 5) might consist of the full spatiotemporal distribution of slip over every point of a rupture surface. However, even for this complex source description we would still determine the magnitude of the event using Equation 2.5.

If we then considered another realistic and plausible rupture scenario with the same moment magnitude, we would find another possible spatiotemporal distribution of slip over the same rupture surface, yet the precise rupture history might be quite different. The ground motions coming from this alternative rupture scenario may be quite different even though the scenario has the same magnitude and rupture geometry. The point is that we can use precise models of many detailed rupture scenarios

---

[9] Which is particularly important when using the source characterization model to make inferences about rupture scenarios for which no empirical evidence exists.

(which collectively produce a range of outcomes) or we can use a single simple rupture model (which may in the end have comparable uncertainty). In practice this latter approach is adopted and so the description of the rupture scenarios does not need to be very complex.

The sections that follow discuss some of the key ingredients used to characterize seismic sources and their associated ruptures.

2.7.1 Source-Scaling Relations

While earthquakes initiate at singular points, the rupture area expands to have significant dimensions. We have seen that the size of an earthquake measured using the seismic moment scales linearly with the rupture area, while the moment magnitude scales linearly with the logarithmic seismic moment. Hence, moment magnitude will scale linearly with the logarithmic rupture area.[10]

When estimating the seismogenic potential of a given fault, the above considerations are very useful. If we are able to identify an active fault, we can infer the magnitude of an earthquake rupture spanning the entire fault. There are also cases where we may estimate the length of a fault, but be less certain about the depth of the seismogenic layer or the fault dip. In these cases, it will be straightforward to estimate the rupture length, but it will be more challenging to estimate the area accurately.

Similarly, it is possible to estimate the magnitude of a rupture of any given size. To be linked most directly with the definition of moment magnitude this size will preferably be the rupture area. However, because of the self-similar manner[11] in which many attributes of earthquake scale, the magnitude can also be estimated from the rupture length, rupture width, and average or maximum slip.

A number of source-scaling relations have been proposed in the literature for linking observed or inferred rupture dimensions to moment magnitude. Some relationships relate the displacements over a fault to magnitude as well. The following subsections provide an overview of these relations.

Rupture Area Scaling

Immediately prior to a rupture the level of stress acting on the fault surface can be represented as σ_0. Following the rupture, this stress will have reduced to a lower level of σ_1. The difference in these levels is the amount of stress "dropped" over the course of the earthquake, and is referred to as the *static stress drop*, $\Delta\sigma$. During a rupture the relationship between static and dynamic friction may evolve with both slip and slip rate. These variations cause the levels of stress acting over the fault source to change dynamically *during* the rupture itself (Ruff, 1999). Hence, it is possible (and generally expected) for the maximum stress during rupture to exceed the value of σ_0 (Madariaga, 1977). This *dynamic stress drop* is a very important parameter for earthquake ground-motion prediction, as we will see later in Chapter 5. However, in what follows we are only referring to static stress drop.

We can relate the change in stress (the static stress drop, $\Delta\sigma$) over the fault to the strain associated with the typical deformation in the surrounding medium using Equation 2.6:

$$\Delta\sigma \propto \mu \frac{\bar{u}}{W}. \tag{2.6}$$

---

[10] Empirical data show that area increases by a factor of ≈ 10 for each unit change in magnitude.
[11] The relationships among physical properties remain stable regardless of scale.

Here, μ is the shear rigidity (or shear modulus), $\bar{u}$ is the average displacement over the rupture, and $\mathcal{W}$ is a characteristic length scale for the rupture. The ratio $\bar{u}/\mathcal{W}$ can be interpreted as a measure of shear strain because the elastic deformations in the vicinity of the rupture extend in three spatial dimensions to a distance comparable to this length scale. The exact region that is influenced by the rupture will depend upon the specific geometry and sense of slip for the rupture. Therefore, a more precise expression can be written as:

$$\Delta\sigma = C\mu\frac{\bar{u}}{\mathcal{W}}, \tag{2.7}$$

in which C is a coefficient that reflects the specific geometry and sense of slip of a given rupture. For example, the value of C would differ between a circular rupture and a rectangular rupture. It would also differ between a rectangular rupture that is entirely embedded in the crust versus one that breaches the surface. This is because C reflects the relationship of the average displacement to the distribution of displacement over the rupture, and the rupture's geometry and boundary conditions influence this relationship.

The expression for the seismic moment (Equation 2.4) can then be rearranged so that it defines the average displacement as:

$$\bar{u} = \frac{M_0}{\mu A}. \tag{2.8}$$

Combining Equations 2.7 and 2.8 allows us to write an expression for the static stress drop that depends only upon the seismic moment and other parameters related to the geometry of the dislocation:

$$\Delta\sigma = \frac{CM_0}{A\mathcal{W}}. \tag{2.9}$$

If we recognize that the area of the dislocation will typically scale as $A \propto \mathcal{W}^2$, then we can express Equation 2.9 in terms of the rupture area:

$$\Delta\sigma \propto \frac{CM_0}{A^{3/2}} \quad \rightarrow \quad A \propto \left(C\frac{M_0}{\Delta\sigma}\right)^{2/3}. \tag{2.10}$$

Taking logarithms of both sides of this expression leads to a general form for a scaling relationship between the area of a rupture and the seismic moment (and consequently moment magnitude):

$$\log A = \frac{2}{3}\log M_0 - \frac{2}{3}\log\Delta\sigma + \beta_C. \tag{2.11}$$

However, recalling Equation 2.5 allows us to write:

$$\boldsymbol{M} = \log A + \frac{2}{3}(\log\Delta\sigma + \beta'_C). \tag{2.12}$$

In Equation 2.12, the term β'_C is a constant that aggregates the dependence upon the geometric coefficient via $\log C$ as well as the constant term from the M_0–$\boldsymbol{M}$ relation (Equation 2.5). Therefore, the logarithm of the rupture area should scale directly with the moment magnitude. Furthermore, given a rupture area, a larger static stress drop, $\Delta\sigma$, will produce a larger magnitude value.

A number of empirical source-scaling relationships have shown strong agreement with this theoretical scaling between rupture area and magnitude (Wells and Coppersmith, 1994; Hanks and Bakun, 2002, 2008; Shaw, 2009; Stafford, 2014b). For example, Wells and Coppersmith (1994) provide pairs of equations, varying with style of faulting, predicting magnitude as a function of rupture area, or rupture area as a function of magnitude. For strike-slip earthquakes in active crustal regions, their expressions are:

$$M = 3.98(\pm 0.07) + 1.02(\pm 0.03)\log A \qquad \sigma_M = 0.23 \tag{2.13}$$

$$\log A = -3.42(\pm 0.18) + 0.90(\pm 0.03)M \qquad \sigma_{\log A} = 0.22 \tag{2.14}$$

where the ± values are the standard errors in the model parameters obtained from the regression analysis. Note that Equation 2.14 is not simply the inverse of Equation 2.13. Generally, one should not invert empirical source-scaling relations because the independent variable in the regression analysis is different in each case.[12] Equation 2.13 provides the expected magnitude *given* a known area, while Equation 2.14 provides the expected value of log *A given* a known magnitude. Equation 2.13 defines a straight line between magnitude and logarithmic area, and the empirical estimate of the slope of this line, 1.02 ± 0.03, is consistent with the theoretical slope of 1.0 from Equation 2.12.

Interestingly, the study of Anderson et al. (1996), which deals with the closely related problem of linking moment magnitude and rupture length, also includes a term to account for the fault slip rate. They find that for a given observed rupture length the moment magnitude will be greater for faults with the lowest slip rates. This result is consistent with the dependence upon stress drop shown in Equation 2.12, as faults with low slip rates (and hence long recurrence intervals) undergo greater healing than more active faults and this results in greater static and dynamic stress drops (Kanamori et al., 1993; Anderson et al., 1996; Marone, 1998).

The general scaling relation of Equation 2.12 arose from arguments that do not necessarily hold for very large events. Once ruptures become very large, the boundary conditions for the rupture change, the aspect ratio[13] of the rupture increases, and the scaling of slip consequently changes. These factors combine to cause breaks in the scaling of the magnitude-area expressions. The main reasons for the break in scaling are explained in the following sections.

Rupture Width Scaling

Although the rupture area continues to grow as the magnitude increases, the down-dip rupture width is limited by the seismogenic thickness of the crust. For relatively small magnitude events the logarithmic rupture width scales linearly with magnitude, but for larger events the widths saturate. A general scaling model linking rupture width (W) to magnitude would then have the form (Stafford, 2014b):

$$\log W = \min\left[\beta_0 + \beta_1 M, \log\left(\frac{Z_{seis}}{\sin\delta}\right)\right] \tag{2.15}$$

where Z_{seis} is the seismogenic thickness and δ is the dip of the rupture. The seismogenic thickness and the dip combine to define an upper limit upon the rupture width. Again, using Wells and Coppersmith (1994) as an example, the logarithmic width for strike-slip earthquakes can be estimated from:

$$\log W = -0.76(\pm 0.12) + 0.27(\pm 0.02)M \qquad \sigma_{\log W} = 0.14. \tag{2.16}$$

However, the corresponding expressions for dip-slip events predict larger widths for a given magnitude.

---

[12] An exception can be made when orthogonal regression techniques are used to develop the equations.
[13] The ratio of the rupture length to the rupture width.

The rupture widths of large reverse or low-angle thrust faulting events tend to be large because of both the shallow dip angles and the fact that compressive regimes often have greater seismogenic thicknesses. In contrast, normal faulting events in extensional regimes have relatively low seismogenic depths and high dip angles, so their rupture widths saturate earlier.

For inferring seismogenic potential, the fact that widths saturate means that they are not very useful for discriminating between magnitudes of large events. For faults with the same dip angle, all shallow crustal events that are large enough to rupture through the full seismogenic depth (requiring magnitudes of around 7 and above; Stafford, 2014b) will have the same widths. However, recognizing that widths saturate is important for understanding why breaks in scaling are observed for both rupture area and length.

Rupture Length Scaling

For a rectangular rupture, the relationship between the area (A), length (L), and width (W) of the rupture is obvious. However, the scaling of these dimensions with magnitude is a little more complicated. For small magnitudes, the problem is relatively simple as all rupture dimensions appear to scale in a log-linear manner (Hanks, 1977). That is, the magnitude scales linearly with the logarithm of the dimension (Wells and Coppersmith, 1994). However, the previous section described a break from linearity in the scaling of the rupture area as magnitudes become large.

Equation 2.12 suggested linear scaling of the logarithmic area with magnitude, while Equation 2.15 indicated that the rupture width saturates. Empirically, a break in scaling of both the rupture area and rupture length relations with magnitude is observed at dimensions that correspond to the rupture width saturating (see Figure 2.17 for rupture area).

For large-magnitude earthquakes it is important to consider these breaks in scaling. Consider the difference in implied magnitudes in Figure 2.17, showing the models of Wells and Coppersmith (1994), which has no scaling break, and Hanks and Bakun (2008), which includes a break. The general scaling is very similar for small to moderate rupture areas, but significant differences arise for the largest ruptures. Although the differences in magnitude may not appear to be large, the

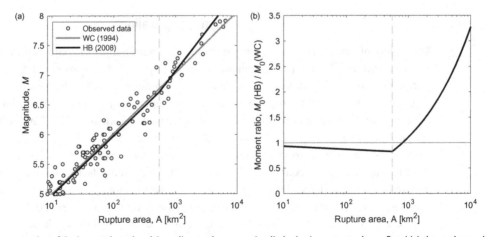

Fig. 2.17 Demonstration of the impact that a break in scaling can have upon implied seismic moment release. Panel (a) shows observed data, and a comparison of the log-linear relation of Wells and Coppersmith (1994) and the piecewise log-linear relation of Hanks and Bakun (2008). Panel (b) shows the ratio of the seismic moment implied by Hanks and Bakun (2008) over that implied by Wells and Coppersmith (1994).

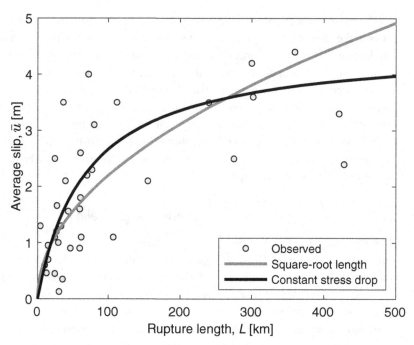

Fig. 2.18   Nonlinear scaling of fault slip as a function of rupture length. Two models from Shaw (2013) are shown along with data from the same study.

implication of these differences can be seen in Figure 2.17b where the ratios of the seismic moments (energy release) associated with these magnitudes are shown. For the largest events, the log-linear model of Wells and Coppersmith (1994) suggests moment release that is less than half of that implied by Hanks and Bakun (2008). In Chapter 3, particularly Section 3.5.2, the importance of such differences will become clear in the context of seismic moment release rates.

Rupture Slip Scaling

The scaling of fault displacement with magnitude has been a topic of controversy over the years. For small magnitudes, the moment magnitude scales linearly with the logarithm of displacement, like the area, width, and length (Hanks, 1977; Wells and Coppersmith, 1994). The controversy arises for very large earthquakes where the rupture width saturates and the aspect ratio L/W increases significantly. For these cases, and when focusing upon shallow crustal earthquakes, Shaw and Scholz (2001) infer that displacements saturate as the aspect ratio (and hence magnitude) continues to increase. More recent work tends to support this finding (Leonard, 2010; Shaw, 2013). Figure 2.18 shows data for a number of events considered by Shaw (2013) that indicate that saturation occurs for large rupture lengths (and hence large aspect ratios). However, there is some evidence that such scaling may not apply for large subduction zone earthquakes. Allen and Hayes (2017) found that average displacements increased significantly for large subduction interface earthquakes with magnitudes above 8.5.

This fault slip behavior also has important implications for the scaling of rupture area when aspect ratios become very large. Recall that the expression for the seismic moment (Equation 2.4)

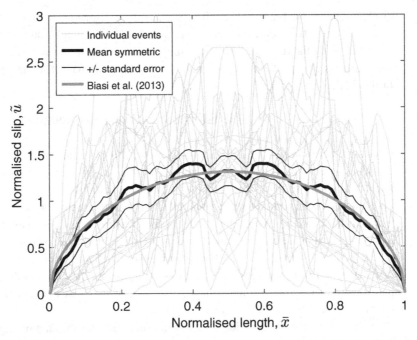

Fig. 2.19   Average distribution of slip along the strike of a rupture. Figure based upon the work of Biasi et al. (2013) and using data from Biasi and Weldon (2006).

involved the product of the rupture area and the average displacement. As the seismic moment increases for very large-magnitude events, the rupture area would have to increase even faster once the displacement saturates. The magnitude-area scaling relation of Shaw (2009) is trilinear because of this fact. At small magnitudes the model of Shaw (2009) possesses linear scaling consistent with the considerations of Section 2.7.1. Once the rupture width saturates as per Section 2.7.1, a change in the area scaling arises in a manner consistent with Hanks and Bakun (2008). A second break then enters once the aspect ratio L/W becomes very large, corresponding to the displacement saturation.

Figure 2.18 shows the mean slip over the length of the fault rupture based on field measurements over many events. The distribution of the slip varies over the rupture, with slip tending to be greatest near the center and tapering to zero at the ends. This slip distribution arises from the kinematic constraints upon the fault surface.

An example of the shape of this slip distribution is provided in Figure 2.19. The model shown of Biasi et al. (2013) has the form:

$$\tilde{u}(\bar{x}) = 1.3 \sin^{1/2}(\pi\bar{x}) \tag{2.17}$$

where $\tilde{u} = u/\bar{u}$ is the slip normalized by the average slip, and $\bar{x}$ is the normalized position along the length of the rupture. While the square-root sine function replicates the observed data well, the specific details of the slope of the function as the slip approaches zero have important physical implications for the fracture energy (Scholz, 2002).

When a point-estimate of the displacement is made at some location along a fault, often from offset of surface features, or through paleoseismic investigations, it is simply one point on this distribution. Empirical scaling relations are generally provided in terms of the average or maximum displacement for the rupture. There is, consequently, a degree of uncertainty associated with making inferences about the average or maximum displacement from individual, or small numbers of, point estimates along the rupture. This uncertainty can map into uncertainties in the sizes of past events and estimates of slip rates for faults, both of which have impacts for hazard assessments (discussed further in Section 3.4.1).

2.7.2 Estimation of Maximum Magnitude

Once a seismic source has been classified as active, the potential of the source for generating large earthquakes must be determined. In particular, the largest potential earthquake magnitude is of interest. For fault sources, this potential is estimated using the known characteristics of that fault. For area sources, the estimated potential reflects assumptions about the type of ruptures that can arise within the source. This hypothetical event is known as the maximum magnitude m_{max}, and from the preceding sections we can estimate this parameter through the use of source-scaling relationships.

In a practical setting, if a fault was mapped in the field, or characterized via remote sensing techniques,[14] then it becomes possible to estimate the maximum possible length of the rupture. If an estimate of the seismogenic depth and the dip of the fault is available, then estimates of the maximum possible rupture area can also be made.

Expressions such as Equation 2.12 provide the basis for prediction equations. Source-scaling relations provide equations that relate moment magnitude to source dimensions, such as area, using the form:

$$M = \alpha_0 + \alpha_1 \log A. \tag{2.18}$$

The specific values of the α_i coefficients vary slightly from study to study, faulting mechanism, and tectonic environment. However, values of $\alpha_0 = 4.0$ and $\alpha_1 = 1.0$ are representative for crustal earthquakes (e.g., Wells and Coppersmith, 1994; Hanks and Bakun, 2002; Leonard, 2014).

To obtain Equation 2.12, we made some assumptions about the shape of the rupture and how stress was relieved through the process of rupturing. However, the spatial distribution of stress change over the rupture surface will vary over the rupture. Further, the actual shapes of ruptures depend upon the geometry of the fault, the nature of the stress field, the medium undergoing deformation, and the specific dynamics associated with the rupture propagation (Heaton, 1990). Therefore, a more realistic representation of Equation 2.18 is given by:

$$M \sim \mathcal{N}(\alpha_0 + \alpha_1 \log A, \sigma_M^2) \equiv \mathcal{N}(\mu_M(A), \sigma_M^2). \tag{2.19}$$

In this equation, we use the expression of Equation 2.18 to define the mean magnitude of a normal distribution, $\mathcal{N}$, that has a variance of σ_M^2. Wells and Coppersmith (1994) found that values of σ_M were on the order of 0.2–0.3 (depending upon the style of faulting, and rupture dimension used to infer the magntiude). That is, even when the geometry of a mapped fault is extremely well-known, there will be uncertainty in the maximum magnitude of potential ruptures.

The same process applies to the assignment of magnitude values to hypothetical ruptures defined on a geometric basis. When rupture scenarios are defined by first identifying potential rupture geometries, then it is necessary to use relations like Equation 2.19 to define their associated magnitude. When rupture scenarios are defined by first hypothesizing the existence of an event with a given magnitude, different source-scaling relations are used that predict geometrical properties as a function of the assumed magnitude.

---

[14] In particular, LiDAR can reveal the location and extent of faults via the geomorphological effects they induce.

Similar expressions can be written to define the distribution of magnitude given the rupture length, rupture width, and different types of slip. Each such expression defines a different plausible distribution for the magnitude value. Maximum magnitude estimates are addressed again in Section 3.5.4 from a statistical perspective where the implications for estimates of rates of rupture scenarios are also exposed.

2.7.3 Seismogenic Potential

In many locations, geological maps will indicate the presence of faults that have been identified from past field investigations and desk studies. However, when developing a SSM for PSHA the focus is upon identifying all potential sources of *future* earthquake events. The emphasis has been placed on the word "future" here because a geological map may contain identified faults that would not be considered fault sources within a seismic hazard study.

The deformation of the crust resulting from tectonic processes has been taking place for a very long time, and during certain periods of time, particular faults may have been active. However, as plates translate and rotate, and as the related stress fields change orientation, these faults may cease to become active. It is therefore important to assess whether or not a fault is *seismogenic*. That is, is the fault *currently* able to generate earthquakes? This question is usually answered by looking at the time since the fault last ruptured. If evidence suggests that a fault has not ruptured within the past $\approx 10,000$ years (i.e., within the Holocene epoch), this is ordinarily sufficient to classify the fault as not being seismogenic. Note that saying that a fault is not seismogenic does not strictly mean that it *cannot* generate earthquakes. In the context of PSHA, it simply means that even if the source does create ruptures, they occur so infrequently that they will have a negligible impact upon probabilistic hazard calculations at return periods of typical interest.

2.8 Conceptual Development of SSMs

We have seen that a source characterization involves the specification of spatial regions with a similar ability to generate certain types of rupture. The approach taken to identify and constrain these spatial regions must depend upon the tectonic setting, the available data, and the ability to acquire new data during the project. As a result it is not possible to simply provide a set of unique instructions that should be followed to develop a SSM for any given analysis. However, some general concepts and principles can be applied to guide the process. This section outlines these concepts.

As will be seen later in Chapter 6, the various parameters that ultimately enter a seismic hazard calculation have different impacts upon the results of the calculations (Molkenthin et al., 2015). A general understanding of these sensitivities permits an analyst to appreciate the attributes of a SSM likely to have the greatest impact upon hazard and risk calculations. This intuition should be tested through later sensitivity calculations, and these calculations also act to strengthen one's intuition.

This understanding of hazard sensitivity is very important for the construction of a SSM. The overarching consideration that should be made at each step when developing the source model is "Does this matter for the hazard and risk?" Two important considerations are:

- Is the feature an input to the models used for predicting ground motions?
- If the feature is used, are the hazard calculations sensitive to how it is modeled?

To provide a concrete example of these two points, consider the characterization of an area source. Two features of an area source that frequently need to be modeled are the depth distribution of hypocenters within the source and the source-scaling relationship to represent finite ruptures. Now imagine the case in which the adopted ground-motion model is parameterized purely in terms of the distance to the epicenter. In such a case, the depths and sizes of the events do not influence the ground-motion predictions. As such, depth distributions and source-scaling relations are simply not relevant to the hazard calculation. This example is somewhat contrived given the present state of ground-motion modeling, which generally accounts for the finite size of ruptures. Nonetheless, there exist many examples of interesting scientific aspects of seismic sources that do not influence hazard results, and this should be considered when devoting effort to model development.

2.8.1 General Processes for SSM Development

In most cases the source model developed for a given study is not the first to have been developed in the region of investigation. The conceptual approach of trying to identify the optimal balance shown schematically in Figure 2.16 therefore begins by evaluating previous efforts that have been made. In particular, it is important to assess whether the passage of time enables additional specificity to be included, or whether past source model attributes are still consistent with available data.

These initial investigations will broadly result in one of three outcomes:

1. Few previous hazard studies exist in the region. Furthermore, the relevant literature is limited for the region.
2. Some hazard studies have been carried out, but aspects of the studies warrant revision in light of modern standards of acceptability. Relevant information regarding source models exists in the literature, but has not yet been assimilated within hazard studies.
3. Extensive precedent exists and several high-quality studies have been conducted. The relevant literature and data resources for the region are extensive and are continually improving.

The approach that an analyst adopts will differ depending upon which of the above cases is encountered.

Seismic Source Model Development in Data-Poor Regions

Cases where limited precedent exists are likely because the region has relatively low seismicity or limited historical occupation. This simply reflects the fact that the chances of previous hazard studies having been carried out increase with increasing population and seismic activity. When the region for the study is characterized by low rates of seismic activity, programs focusing upon compiling existing data or collecting new data are unlikely to yield significant benefits. In light of this reality, a common approach is to make use of global analogs and to tune these to the specific region of interest. When adopting global analogs, the primary assumption is that at least one region on the planet is analogous from a neotectonic perspective. This similarity will initially be based upon generic tectonic classifications. For instance, if the region is a stable tectonic region, then data from other stable regions can be combined under the assumption that this combined data would look similar to the data that would have been gathered in the study region.

Information that can be "imported" through the use of global analogs includes background rates of earthquake activity, distributions of maximum regional magnitudes, source-scaling relations, depth distributions, and distributions of rupture mechanism. In cases where no local data are available to assess these imported model components' appropriateness, then (large) uncertainties must be defined that reflect this lack of knowledge.

When some limited amount of local data does exist, it is important to use. The data might be used to identify whether a particular global analog is appropriate or indicate which of some possible alternatives might be most relevant. However, the limited data are often insufficient to make conclusive distinctions, and uncertainties must be defined that reflect this status. The existence of some data should reduce the uncertainty, even if by only a small amount.

In these data-poor regions, the source model's complexity is likely to be limited and would be located to the left-hand side of Figure 2.16. The model will likely include primarily areal sources and few, if any, fault sources. A good general approach in these data-poor regions is to begin with a homogenous background source and then identify regions with the potential for departures from the types or frequencies of ruptures in this background region. This approach may lead to the subdivision of the overall study region into multiple area sources, or to the introduction of particular fault sources.

Seismic Source Model Development in Data-Rich Regions

In data-rich regions the initial challenge is to identify and critique all existing studies. Prior studies may have been conducted for different purposes, at different points in time, and with varying levels of resources. Therefore, some of their modeling decisions may reflect these constraints rather than the state of the art at the time. Furthermore, in regions of high activity (which are often data-rich regions) more earthquake events may have occurred, and more research studies may have been carried out since the publication of the previous hazard analyses. The review of any existing work should also consider these new data.

The literature may provide a good starting source model with at least one areal source (even if only the background source) and several fault sources. The subsequent analyses can focus upon the refinement of this initial source model and upon the representation of earthquake ruptures arising from these sources.

In data-rich regions it is always useful to keep the end objective of the study in mind, i.e., the definition of hazard through seismic hazard curves. Sensitivity studies using source models from previous studies can enable critical sources to be identified. Studies in data-rich regions will generally be located toward the right-hand side of Figure 2.16, with a number of fault sources, but large parametric uncertainties. The primary objective in data-rich regions should be to identify which uncertainties have the greatest influence on the hazard and focus on these critical parameters or model features. Uncertainty reduction can be achieved through additional data acquisition or a more focused critique of the literature and existing data.

Seismic Source Model Development in Data-Neutral Regions

The final scenario falls between the data-poor and data-rich regions. In this intermediate case it is likely that existing studies will be identified, but that these studies often have limitations or problems. It is common in data-neutral regions for previous studies to have developed bespoke models using insufficient data. While local data are desirable, in these situations they often suffer from sampling biases and are not robust with respect to future events.

In data-rich regions the approach of starting with global analogs and updating these using local data is not common because the local data are already sufficient. However, in data-neutral regions this approach can be advantageous. The local data may not be sufficient to calibrate a model entirely, but sufficient to adjust a global analog starting point. Additionally, the available data will often be sufficient to distinguish between alternative models imported from other regions.

Global analogs are typically useful for area sources or general regional attributes (e.g., maximum regional magnitude, or the relative activity rates of small and large events). Data-neutral regions will often also include at least one fault source within the model, however, and these are less amenable to being treated with analogs. While individual fault sources can often be identified in data-neutral regions, fault segmentation details may not be available. As such, sources will often use ruptures whose locations "float" over the identified faults, rather than distinct rupture scenarios that reflect some segmentation model.

Exercises

2.1 An active fault has a mapped length of 80 km and a strike of 35°. The dip of the fault plane is 40°. If an earthquake nucleates on this fault at a depth of 10 km and the base of the seismogenic crust is located at $Z_{seis} = 15$ km depth, and assuming a surface-rupturing event, what will be the minimum and maximum possible distances between the surface trace of the fault and the epicentral location of the earthquake?

2.2 If an earthquake occurring on the fault described in Exercise 2.1 results in rupture of the full width and 70% of the mapped fault length, produces a 1.5-m-high fault scarp, and features exposed at the fault scarp indicate that the last rupture occurred with a rake of 115°, estimate the moment magnitude for this event. The rigidity of the crust may be taken as 3.3×10^{10} N/m².

2.3 What type of fault is described in the previous question? Explain in what type of tectonic environment you would expect to find this type of faulting.

2.4 Using sketches, explain why the dips of reverse faults are generally lower than those for normal faults.

2.5 What difference in moment magnitude is associated with a factor of 2 in seismic moment?

2.6 When inferring the seismic potential of a fault, source-scaling relations are used to relate geometric properties of the fault to the distribution of possible of magnitudes. Making reference to the result of the previous question, comment upon the implications for seismic moment release for such relations.

2.7 Classify the style of faulting of the following ruptures:

(a) strike 45°, dip 30°, rake 90°
(b) strike 90°, dip 90°, rake 180°
(c) strike 320°, dip 60°, rake −70°
(d) strike 0°, dip 50°, rake 45°

Comment upon the role played by each parameter in defining the style of faulting.

2.8 Using the relationship between seismic moment and moment magnitude, compute the value of the magnitude that would be required to rupture the circumference of the Earth if we assume that the seismogenic thickness of the crust is 15 km, that the average displacements scale according to $\bar{u} = \alpha L$ where $\alpha = 10^{-5}$ and L is the rupture length in km, and a crustal rigidity of $\mu = 3.3 \times 10^{10}$ N/m². Take the average Earth radius as $R_{\oplus} = 6371$ km.

2.9 Normalized displacements, $\bar{u} = u/u_{max}$, along the normalized position, $\bar{x} = x/L$, of a rupture are thought to scale according to $\bar{u}(\bar{x}) = \sin^{1/2}(\pi \bar{x})$ as shown in Figure 2.19. However, for the purposes of this exercise, assume that the relation can be defined as $\bar{u}(\bar{x}) = \sin(\pi\bar{x})$, and that the slip distribution is independent of depth.

A paleoseismic investigation on a vertical strike-slip fault is undertaken 20 km from the end of an 80-km rupture and reveals a displacement of 3.54 m. Assuming that the seismogenic thickness is 15 km and that the full width is ruptured, compute the moment magnitude for this event. Assume a crustal rigidity of 3.3×10^{10} N/m^2.

2.10 The distribution of displacements for a rectangular earthquake rupture of length L and width W can be defined by:

$$u(x, z) = \sin\left(\frac{\pi x}{L}\right) \tag{2.20}$$

where x is the position along strike, and z is the position down the dip of the fault, with both coordinates being local to a given rupture.

Consider two ruptures, both of length $L = 50$ km, that rupture the same vertical strike-slip fault. The ruptures are nonoverlapping and collectively rupture the entire seismogenic crust. The seismogenic depth in the region is $Z_{seis} = 20$ km, and the shear modulus of the crust varies linearly with depth and is equal to $\mu = 1.7 \times 10^{10}$ N/m^2 at the Earth's surface and $\mu = 3.3 \times 10^{10}$ N/m^2 at Z_{seis}. Assume that one rupture starts from the top of the fault and propagates downward, and the other starts at the bottom of the fault and propagates upward. Find the ratio of the areas of the two ruptures required to ensure that they have the same moment magnitude, and report the value of this magnitude.

2.11 Seismic source models will use some combination of individual fault sources and areal sources. Explain the conceptual process that one would go through to determine the appropriate level of complexity for a source model. Make explicit reference to apparent aleatory variability and parametric uncertainty.

2.12 When estimating average rates of occurrence of individual rupture scenarios, a geology-based approach or a seismicity-based approach can be adopted. Explain the main features of these two approaches and describe situations in which each method might be preferred over the other.

2.13 One conceptual model for the earthquake cycle is based upon stick-slip dynamics. Describe the key features of stick-slip models, and particularly emphasize what mechanisms lead to initiation and termination of an individual rupture.

2.14 An active normal fault striking NE-SW has a mapped surface trace of 23 km. High-resolution location of microseismic events indicates a planar fault plane dipping 60° to the NW. Earthquake hypocenters are observed at a maximum depth of 10 km. The fault is marked by a series of NW-facing fault scarps developed on a 7400-year-old alluvial fan surface. These scarps are up to 1.8 m high. A historical event in 1843 displaced a stone wall vertically by 0.4 m.

(a) What is the magnitude of the maximum earthquake expected on this fault? Assume that the rigidity modulus is 3.3×10^{10} N/m^2.
(b) What is the fault slip rate?
(c) What is the recurrence interval for surface-rupturing events?

2.15 Consideration of finite ruptures allows certain approximations to be made when generating ruptures for a given seismic source. In this context, answer the following questions:

(a) Under what circumstances is it appropriate to consider a single hypocentral depth for a seismic source?
(b) Under what circumstances could a fault source be replaced by a single point source?

(c) How would your responses to parts (a) and (b) change if uncertainty in static stress drop were to be considered for these sources?

2.16 The source-scaling relationships of Wells and Coppersmith (1994) for moment magnitude, M, rupture area, A, and strike-slip earthquakes are:

$$\mu_M = 3.98 + 1.02 \log_{10}(A), \quad \sigma_M = 0.23$$

and

$$\mu_{\log_{10}(A)} = -3.42 + 0.9M, \quad \sigma_{\log_{10}(A)} = 0.22.$$

These parameters define normal distributions with mean μ and standard deviation σ.

(a) A rupture is thought to have an area of 10 km^2. Compute the expected value of the magnitude for this rupture, and determine the probability that the magnitude for the rupture exceeds $M5.5$.

(b) Using the expected scaling relationship of Wells and Coppersmith (1994), determine how many $M5$ earthquakes could fit within the rupture area of an $M7$ event.

(c) Using the expected scaling relationship of Wells and Coppersmith (1994), estimate the magnitude for a rupture that has the size of a typical door.

(d) What is the probability that the rupture area for an $M5$ earthquake is greater than that for an $M6$ earthquake?

3 Characterization of Earthquake Rates and Rupture Scenarios

Chapter 2 discussed the conditions under which earthquakes occur and the characteristics that these events have at their source. We will refer to the collective representation of seismic sources in a particular region as a *seismic source model* (SSM). The present chapter moves to consider specific earthquake events as distinct *rupture scenarios* and to explain how to estimate their average rate of occurrence and their probability of occurring within a particular time interval. How each rupture scenario is characterized depends upon what information is required to make predictions of ground motions for the scenario (discussed in Chapters 4 and 5). However, the minimum information typically required is the magnitude of the event and a geometric description of the rupture. The collection of all possible future rupture scenarios in a region of interest and their likelihood (or probability of occurrence in a specific time interval) will be referred to as an earthquake rupture forecast (ERF).[1] These earthquake rupture forecasts are a vital component of any seismic hazard assessment (Chapter 6).

Learning Objectives

By the end of this chapter, you will be able to do the following:

- Understand the general approaches that are used to constrain rupture rates.
- Analyze seismicity datasets, including harmonizing magnitudes, declustering, associating events to sources, and fitting magnitude-frequency distributions.
- Understand the fundamental concepts of paleoseismology and how field evidence can constrain large-earthquake recurrence rates.
- Combine seismicity and geological information, in a unified manner, to constrain rupture rates.
- Calculate rates of occurrence and exceedance for levels of earthquake magnitude in a seismic source using magnitude-frequency distributions.
- Generate rupture scenarios that are compatible with a given seismic source.
- Compute time-dependent probabilities of occurrence for rupture scenarios.

3.1 Introduction

In the initial development of PSHA (Cornell, 1968), the characterization of a seismic source comprised: (1) a geometric description of its location and (2) a magnitude-frequency distribution defining how many times per year, on average, earthquakes of a given magnitude occurred. With time we have observed earthquakes that involve complex interactions among different faults and have

---

[1] An analogy for the relationship between SSM and ERF in weather prediction is that a *global circulation model* is used to obtain a *weather forecast* for a specific region and time period.

developed a better appreciation of which source properties influence ground motions. Therefore, modern source characterization reflects these developments so that advances in source modeling can be linked appropriately to ground-motion modeling, and ultimately reflected in hazard and risk calculations. To that end, this chapter begins by providing an overview of approaches to determine rates of occurrence for rupture scenarios. With this high-level picture established, the chapter moves on to describe the various components required as inputs to these methods. A unified approach that combines the best of traditional methods is then presented. The chapter then explains how rupture scenarios are generated once a seismic source characterization has been completed. Finally, the characterization of time-dependence for rupture scenarios is presented.

3.2 Approaches to Determining Rupture Rates

Several approaches have been proposed for estimating the rates of occurrence of rupture scenarios. The three most commonly considered approaches are:

- Seismicity-based approaches
- Geological approaches
- Earthquake simulators.

In the present section, these three approaches are briefly outlined.

3.2.1 Seismicity-Based Approaches

The first approach for defining rupture rates uses primarily *seismicity data* (the historical record of earthquake occurrence) to develop a magnitude-frequency distribution that prescribes the average rates of occurrence of events as a function of magnitude. Here, and frequently hereafter, the term "events" is used as shorthand for "earthquake events" The actual rupture scenarios are then created by selecting a magnitude represented by this distribution and generating a geometric rupture compatible with the source geometry. The rate of occurrence of the scenario is obtained by partitioning the total rate of occurrence of the given magnitude over all plausible rupture geometries that can exist for the respective source. This approach is schematically illustrated in Figure 3.1.

3.2.2 Geological Approaches

The second approach for defining rupture rates uses primarily *geological data* to identify geometric ruptures that are consistent with the parent source's geometry. For each geometric rupture, a magnitude can be obtained from a source-scaling relation (see Section 2.7.1) or from the definition of the seismic moment (if an associated estimate of the average slip for the rupture is available). Each scenario's rate of occurrence is then constrained by the overall slip rate for the fault. This approach is illustrated in Figure 3.2.

3.2.3 Earthquake Simulators

The final approach discussed here for defining rupture rates uses earthquake simulators (Tullis et al., 2012b). Section 2.4 gave an overview of conceptual models that are often used to explain key aspects of earthquake initiation and recurrence. In particular, Section 2.4.1 introduced elastic

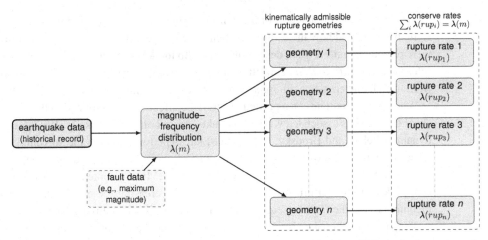

Fig. 3.1 Process via which rates of occurrence of rupture scenarios are estimated using seismicity data. While the dominant source of data is the earthquake data, information related to the fault is often still used. The rate of occurrence of a given magnitude $\lambda(m)$ is partitioned over the ruptures consistent with this magnitude. When multiple magnitudes are considered, the steps on the right are replicated for each magnitude.

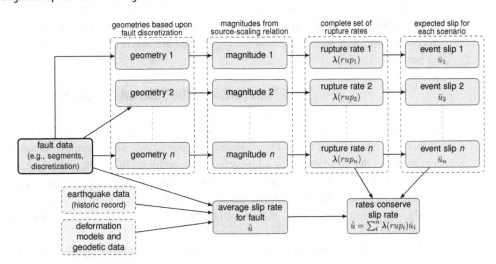

Fig. 3.2 Process via which rates of occurrence of rupture scenarios are estimated using geological data. While the dominant data source is geological data, the historic earthquake record and deformation models consistent with geodetic data can also be used to support the process. The average slip rate needs to be consistent with the rates of occurrence of the individual slip events.

rebound theory and the basic principles of loading and relaxation of stress over fault surfaces, and Sections 2.4.2 and 2.4.3 described stick-slip models and how earthquakes on one fault can influence the occurrence of events on other faults. Earthquake simulators can be thought of as combining the general attributes of these conceptual models and applying them over a fault network.

Studies of long-term relative plate motion (e.g., DeMets et al., 1990) permit constraints to be imposed upon large-scale regional tectonic loading. This loading, driven by the relative motion of plates, leads to crustal strain and stresses on surfaces throughout the crust, e.g., Figure 2.8. In an earthquake simulator, the known faults within a region are identified and represented as finite surfaces. Given a regional tectonic loading, stresses on these faults are determined from the strain field near the fault and the relative orientations of the faults to the regional loading (Sibson, 1985).

The frictional characteristics of the faults determine at what point earthquakes initiate (governed by static friction), and how these ruptures then develop (governed by dynamic friction). These concepts were previously raised when describing stick-slip models in Section 2.4.2. A family of friction laws, referred to as *rate and state* friction models (Ruina, 1983), are usually used to represent these frictional characteristics, with many variants existing within the literature (e.g., Bizzarri, 2011).

When rupture initiates on a particular fault, the effects are not isolated to that fault only. Rather, the shear dislocations associated with earthquakes cause both static (Okada, 1992; King et al., 1994) and dynamic (Felzer and Brodsky, 2005, 2006) stress perturbations in the surrounding medium. When these stress perturbations interact with faults with some level of preexisting stress, the perturbation can move the fault "toward" or "away" from failure. In the context of elastic rebound theory, one can think of these perturbations as increasing or decreasing the elastic strain energy stored in the medium around the fault (or, equivalently, the springs in stick-slip models).

Therefore, earthquake simulators generate earthquake ruptures on faults through the combined mechanisms of long-term regional strain and short-term stress interactions. It is then possible to simulate sequences of events throughout a region that can replicate many of the features of observed seismicity (e.g., Toda et al., 2005; Robinson et al., 2011). In some cases, these earthquake simulator approaches have been shown to produce comparable, or superior, performance relative to models based upon statistical calibration to seismicity data (e.g., Tullis et al., 2012a; Mancini et al., 2020). That said, the development of earthquake simulators is an active area of research and the formerly discussed seismicity- and geology-based approaches are most common in hazard studies at present.

3.2.4 Comparison of the Different Approaches

In Figures 3.1 and 3.2 both earthquake (seismicity) and fault (geological) data appear. Therefore, rather than representing distinct ways to obtain rupture rates, the different approaches reflect different perspectives that can be used to address the same problem. In practice, the approach that dominates tends to depend upon the available data. Earthquake simulators have not received widespread usage in practical applications thus far, but require well-defined fault sources and preferably geodetic data. In the remaining comparison, the focus is entirely upon the traditional seismicity and geological approaches.

The geological approach works well in active areas with a significant history of earthquake occurrence and geological investigations. The seismicity-based approach is often the only option in regions where fault sources are hard to characterize due to a limited seismic record and limited geological investigations. Furthermore, the different approaches tend to constrain rupture scenarios over different magnitude ranges. Generally speaking:

- The seismicity-based approach is particularly useful for constraining rates for small to moderate magnitude events that do not provide surface evidence of rupturing, for both fault and area sources.
- The geological approach is particularly useful for constraining the rates of the largest events on fault sources. The events need to be sufficiently large to have left evidence of their occurrence in the geological landscape (surface expression of faulting, offset of geological features, paleoseismic evidence, etc.).

As they each offer constraints upon different types of events, it is preferable to use both approaches when determining rupture rates. Recently, approaches have been developed that merge both the geological and seismicity-based approaches into a consistent framework. This relatively new development is discussed later in this chapter. Regardless of the approach that is adopted, the end goal remains the same. For every rupture scenario, *rup*, that arises from the SSM, we seek to define the rate of occurrence $\lambda(rup)$ that will form an essential part of the hazard calculations.

Before progressing on to discuss more specific details of these approaches in the sections that follow, the conceptual approaches depicted in Figures 3.1 and 3.2 are implemented within a simple example in order to clarify the key practical steps that need to be followed.

Single Scenario Example

Consider a vertical strike-slip fault of with a length of 40 km in a region where the seismogenic depth is $Z_{seis} = 11$ km. From a seismicity analysis, the rate of occurrence of events of magnitude equal to 6.7 is $\lambda(m = 6.7) = 0.02$. From geological analyses, it is found that the slip rate for the fault is 10 mm/year. It is assumed that this fault only ruptures in this single scenario. We wish to determine the rupture scenario's attributes corresponding to these conditions using each of the two approaches described previously.

Seismicity-Based Approach

The rate of occurrence of *all* ruptures that have a magnitude of 6.7 is defined as $\lambda(m = 6.7) = 0.02$ events/year. To investigate what types of rupture are potentially consistent with this magnitude, we can use a source-scaling relation to infer the area, A, of the ruptures. For this purpose, Equation 2.14 gives:

$$A = 10^{-3.42+0.90\times6.7} = 407 \text{ km}^2 \tag{3.1}$$

while the median rupture width, W, is found from Equation 2.16:

$$W = 10^{-0.76+0.27\times6.7} = 11.2 \text{ km.} \tag{3.2}$$

As this rupture width exceeds the seismogenic thickness, and we have a vertical fault (the dip, $\delta = 90°$), the implication is that an event of this size will rupture through the seismogenic thickness. The length, L, of the rupture can then be found from the ratio of the rupture area to the available rupture width:

$$L = \frac{A}{\min[Z_{seis}/\sin(\delta), W]} \quad \rightarrow \quad L = \frac{407 \text{ km}^2}{\min[11 \text{ km}, 11.2 \text{ km}]} = 37.0 \text{ km.} \tag{3.3}$$

Therefore, an event of this size effectively ruptures the entire fault. Technically, one could say that an event of this size could "float" laterally along the fault[2] given that its rupture length is less than the available fault length of $L_{max} = 40$ km. Varying the rupture's position would mean considering multiple rupture scenarios and is at odds with our assumption that the fault ruptures in a single scenario. Therefore, for this example, as well as practical purposes, it is reasonable to consider a single scenario rupturing the entire fault.

In light of these source dimensions, we can then say that the rate of occurrence of a rupture covering the entire fault is equal to the rate of occurrence of magnitude 6.7 events. Therefore, using

---

[2] If we did this, we would place the 37×11 km^2 rupture at various positions along the fault and would partition the overall rate of $\lambda(m = 6.7)$ among the various alternatives (see Section 3.6.1).

the seismicity-based approach in which we begin with the rate of occurrence of events with a given magnitude, we can arrive at the result that $\lambda(rup) = \lambda(m = 6.7) = 0.02$ times/year. This rate of occurrence corresponds to a recurrence interval of $1/\lambda(rup) = 50$ years for this rupture, and the geometry of the rupture is a vertical surface with length 37 km and width 11 km (or effectively the rupture of the entire fault).

Geological Approach

Addressing the same problem using the geological information regarding the slip rate, we usually need to make some assumptions about the types of geometric ruptures that are possible for this fault. There are many possibilities for a generic fault with the dimensions provided.[3] Nevertheless, assuming that the fault generates only a single scenario, we can assert that the entire fault will rupture.

Under this assumption, our *rup* is characterized by a rupture length of 40 km and a rupture width of 11 km, giving a total rupture area of 440 km². We can now compute an estimate of the associated magnitude using a source-scaling relation. From Equation 2.13, the median magnitude corresponding to a rupture of this size is:

$$m = 3.98 + 1.02 \times \log\left(440 \text{ km}^2\right) = 6.68. \tag{3.4}$$

Therefore, we have assumed that the rupture scenario corresponds to a magnitude 6.68, vertical planar event with a length of 40 km, and a down-dip dimension of 11 km. To obtain the rate of occurrence, we need to compute the typical amount of slip that occurs in this type of event, and there are two main options available to us. The first is to recall the definition of the seismic moment as $M_0 = \mu A \bar{u}$, from which an estimate of the average slip can be obtained if one assumes a value of the crustal rigidity μ. Recalling that $M_0 = 10^{1.5(m+6.03)}$ in SI units of Nm (Equation 2.5) and adopting a value for the rigidity of $\mu = 3.3 \times 10^{10}$ N/m² allows an estimate of the average displacement to be found as:

$$\bar{u} = \frac{M_0}{\mu A} \qquad \rightarrow \qquad \bar{u} = \frac{10^{1.5\left(6.68+6.03\right)} \text{ Nm}}{\left(3.3 \times 10^{10} \text{ N/m}^2\right)\left(440 \times 10^6 \text{ m}^2\right)} = 0.80 \text{ m}. \tag{3.5}$$

The second approach is to use a source-scaling relationship that provides an estimate of the average displacement for a given magnitude. Making use of Wells and Coppersmith (1994), we can obtain:

$$\bar{u} = 10^{-6.32+0.90\times m} \qquad \rightarrow \qquad \bar{u} = 10^{-6.32+0.90\times6.68} = 0.49 \text{ m}. \tag{3.6}$$

Although these two estimates appear to be quite different, the estimate from the seismic moment relationship is well within one standard deviation from the median estimate of Wells and Coppersmith (1994).[4] Differences of this order are not unexpected in this field. They should be kept in mind when considering the validity of the assumption made earlier that a rupture length of 37 km was practically the same as 40 km.

---

[3] For example, small portions of the fault may rupture in relatively small events. At the same time, larger areas may also rupture, causing alternative scenarios.

[4] This corresponds to the 78th percentile of their distribution, which has a standard deviation of 0.28 for $\log \bar{u}$.

Table 3.1. Rupture attributes obtained using alternative methods discussed in the text. For the two columns corresponding to $\dot{\bar{u}}$ as the starting point, the first estimates the average slip using the seismic moment relationship while the second uses the source-scaling relation of Wells and Coppersmith (1994)

Attribute	Given $\lambda(m)$	Given $\dot{\bar{u}}$ (M_0)	Given $\dot{\bar{u}}$ (WC94)
Magnitude, m	6.7	6.68	6.68
Rupture area, A	407	440	440
Rupture length, L	37.0	40.0	40.0
Rupture width, W	11.0	11.0	11.0
Average slip, $\bar{u}$	–	0.80	0.49
Rupture rate, $\lambda(rup)$	0.020	0.0125	0.0205
Recurrence interval, $\lambda(rup)^{-1}$	50.0	79.9	48.8

With these estimates of average slip per event now available, along with the provided slip rate estimate of $\dot{\bar{u}} = 10$ mm/yr (0.01 m/yr), the rate of occurrence of these full-fault rupturing events can be computed. Using the single-event slip estimate from the seismic moment relation gives:

$$\lambda(rup) = \frac{\dot{\bar{u}}}{\bar{u}} = \frac{0.01 \text{ m/yr}}{0.80 \text{ m}} = 0.0125 \text{ yr}^{-1}. \tag{3.7}$$

Similarly, using the slip estimate from Wells and Coppersmith (1994) gives:

$$\lambda(rup) = \frac{\dot{\bar{u}}}{\bar{u}} = \frac{0.01 \text{ m/yr}}{0.49 \text{ m}} = 0.0205 \text{ yr}^{-1}. \tag{3.8}$$

The average time between repeat occurrences of an event, known as the *recurrence interval*, can be found from the reciprocal of the rate of occurrence. The associated recurrence intervals implied by the rates in Equations 3.7 and 3.8 are approximately 80 and 50 years, respectively. A summary of the results obtained from each of the above approaches is presented in Table 3.1. Note that the fact the recurrence interval of 48.8 years from the $\bar{u}(WC94)$ case is closer to the 50-year value from the $\lambda(m)$ case than for the $\bar{u}(M_0)$ approach is coincidental and does not imply some advantage of using the source-scaling relations.

3.3 Constraints from Seismicity Data

Seismicity data is an input to both the seismicity- and geology-based approaches for estimating rupture rates. The present section describes the nature of seismicity data and discusses the routine processing steps required when working with the data. These data are ultimately used in the calibration of magnitude-frequency distributions subsequently discussed in Section 3.5.

To calibrate statistical models representing rates of future rupture scenarios, databases of past earthquake activity first need to be compiled. This information is broadly referred to as seismicity data. The data may come in the form of historical reports of past events, consist of geological evidence for the occurrence of events, and generally include instrumentally recorded events. The amount of information within a given seismicity database can vary considerably and will depend upon whether the source of the data is historical, geological, or instrumental. However, certain

critical pieces of information are always required in order for the data to be useful. These fundamental attributes are:

- The origin time of the event
- The hypocentral location of the event
- A measure of the size of the event.

Ideally, the size of the event will be provided using a magnitude scale (preferably the moment
magnitude scale). However, it is also common for sizes of historical events to be defined in terms of
epicentral intensity.

Each of the above-listed attributes has some associated uncertainty. The level of this uncertainty
depends upon the source of the data. Ideally, these uncertainty estimates should also be provided
within the database, but this is not always the case (particularly in a consistent manner). In the
following subsections, the various sources of information that may contribute to a seismicity database
are briefly described.

3.3.1 Historical Data

When an earthquake occurs anywhere around the world, the seismic waves generated will be
recorded on high-sensitivity seismograph networks. However, in most parts of the world, this
situation has existed only for several decades, at best.[5] For the majority of human history, there
were no seismographs and no instrumental observations of earthquakes.[6]

Due to the relatively long interval between significant earthquakes in any given region, the
likelihood of observing a large event during the period of instrumental observation is low. If only
the instrumental catalog were used, the largest, most infrequent, rupture scenarios would be poorly
constrained. For this reason, information from recent instrumentally recorded events is combined
with historical information about events occurring before the instrumental period.

A historical event is one documented in the literature, artwork, or any other non-instrumental way.
The concept of a magnitude scale did not exist when these events occurred, and, as a result, we
generally infer the size and the location of the event based on reports of the earthquake effects.
The relative strength of the waves experienced depends upon both the source strength and the
distance between the source and observation points. When multiple observations have been made,
one can make an informed guess as to the likely origin and size of the event. However, there will
be significant uncertainties associated with both of these pieces of information. As this historical
information depends upon the historical record for a given location, the time span over which useful
information can be obtained varies considerably from region to region. For example, in seismically
active parts of Asia, the historical record can span centuries. In other seismically active regions, such
as the United States or New Zealand, where written records are limited to those following Spanish
or British occupation, the relevant time span is much shorter.

---

[5] Significant expansion of the World-Wide Standardized Seismic Network (WWSSN) took place in the early 1960s
in response to nuclear test ban discussions in the late 1950s. The development of the network aimed to enable
discrimination between natural events and underground explosions but has played a significant role in understanding global
seismology (Peterson and Hutt, 2014).

[6] While other devices technically existed, the information provided cannot be compared to modern instruments.

3.3.2 Geological Evidence for Rates of Occurrence

Once we move beyond the time span covered by the historical record, we can only gain information to constrain rupture rates from geological sources. Approaches for constraining rupture rates using geological sources fall within the domain of paleoseismology. Paleoseismology is introduced in Section 3.4.1, and is covered extensively by McCalpin (2009). For now, it suffices to say that paleoseismic investigations lead to estimates of magnitudes of past significant events, long-term slip rates, individual event displacements, and rupture segmentation. This information can then be used to develop recurrence interval estimates for rupture scenarios that activate particular combinations of fault segments. Where this information is available it can help constrain the values of $\lambda(rup_i)$ that are depicted in Figure 3.2.

The two most important differences between geological information and historical information are:

- The geological (paleoseismic) information can be obtained directly from the source location of an event,[7] while the historical information is based upon observations at some distance from the actual source.
- Paleoseismic investigations can be performed in any region where the natural environmental conditions permit, regardless of human occupation history.

The nature of the geological constraints is discussed in more detail in Section 3.4.

3.3.3 Instrumental Seismicity

Both of the previously discussed sources of information on earthquake occurrence are most relevant for large earthquake events. However, we can also learn about rates of large events by studying the statistics of small events. When focusing upon small events, instrumental recordings of earthquakes are most relevant. In addition to the event time, position, and magnitude, high-quality networks also produce error estimates for each parameter.

Instrumental seismicity datasets will often be compiled from multiple data sources and networks, often using different recording instruments, location algorithms, and magnitude scales. Such differences impact inferences about the occurrence rates of earthquakes. Therefore, to understand more about the nature of seismicity data, we need to understand how we locate earthquakes and estimate their size.

3.3.4 Earthquake Location

An earthquake rupture generates different types of seismic waves, which travel at different velocities, and follow different travel paths to reach a given location. Chapter 5 explains this in more detail. The speed and direction of travel depend upon the elastic properties of the propagation medium. Using a model for the crust's velocity and density, it is possible to compute the travel paths that will be taken by the various types of waves, and the time for each wave type to travel from the source to some observation point.

---

[7] This primarily refers to techniques such as trenching or dating of offset terraces; see Section 3.4.1. Other techniques, such as paleoliquefaction analysis, dating of tsunami deposits, or dating of precariously balanced geological features, rely upon making inferences at spatial positions somewhat removed from the earthquake source location.

When this exercise is performed for a given crustal model, it becomes possible to construct *travel time curves* (e.g., Aki and Richards, 2002; Shearer, 2009). These curves define the time taken for a particular type of wave to travel to a specific location at some distance from the rupture. If the propagation medium were homogeneous and isotropic, then the "curves" would be straight lines with slopes equal to the reciprocal of the velocity of the wave.[8] However, as the Earth consists of layers of different impedance and is approximately spherical, travel paths of waves are curved, and hence the travel time curves are nonlinear.

At a particular recording station, distinct wave arrivals (known as *phases*) can be identified, and the relative arrival times of these phases will indicate the distance the waves must have traveled. With multiple estimates obtained from multiple stations, the event origin location and time can be determined.

As a model for the Earth is used within the location algorithm, the estimates of distance traveled to each station will have some uncertainty associated with them, and this uncertainty maps over to the spatial location and origin time of the event. These uncertainties are influenced by several factors, including the number of recording stations, their azimuthal coverage, and the quality of the velocity model used. Even when an event occurs within a dense regional network of instruments where the azimuthal coverage is good, one should still expect location uncertainties to be approximately 1 km laterally, and 2–3 km vertically (Husen and Hardebeck, 2010). It is not uncommon for uncertainties to be significantly ($\approx$2–10 times) greater in less-constrained situations.

We also need an estimate of the event's size. This comes from the amplitudes of the seismic waves recorded at a given location. When combined with the information about the distance between the source and the observation point, one can estimate how strong the waves were when they left the source. There are many different ways to estimate the size, as previously discussed in Section 2.5. Section 3.3.5 looks briefly at the implications of using these different scales for seismicity analyses.

3.3.5 Data Preparation

Raw seismicity data obtained from a seismograph network must be processed for use in seismicity analysis. The processing of seismicity databases is time-consuming, especially when information from multiple data providers needs to be combined. This section outlines the main steps to prepare raw datasets for use in a seismicity analysis for PSHA.

Throughout the section, a running example using a real seismicity dataset demonstrates the steps typically taken during a seismicity analysis. The seismicity data have been obtained from the Northern Californian Earthquake Data Center (NCEDC).[9] The search criteria used were to identify any earthquake with a magnitude of at least 2.0 within 200 km of Yerba Buena Island (122.3660°W 37.8099°N), in the San Francisco Bay area. This site is shown in Figure 3.3. The NCEDC provides instrumental data as part of the Northern Californian Seismic Network (NCSN) from 1967; additional seismicity data for earlier events are also taken from the Uniform Californian Earthquake Rupture Forecast 3 (UCERF3) seismicity database (Felzer, 2013a).

The spatial distribution of the raw data is shown in Figure 3.4. In this figure, the depth of the events is denoted using different shades, while the size of the marker represents the magnitude.

---

[8] The travel time curves are plots of time taken against distance traveled. The slope represents a change in travel time over a change in distance traveled and is known as the slowness.

[9] http://quake.geo.berkeley.edu/ncedc/catalog-search.html.

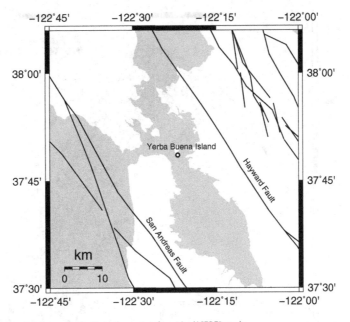

Fig. 3.3   Example site location showing the major faults in the region from the UCERF3 study.

The seismicity is not uniformly distributed over the region. It is also clear that there is a strong correlation with the topography (see the inset in Figure 3.4).

Our ultimate goal is to associate each historical event with a rupture scenario and consequently estimate the average rates of occurrence of each of these scenarios. In order to estimate these rates of occurrence, several preparatory steps need to be taken, including:

- **Harmonization:** ensure that all events in the catalog have their size quantified according to a common magnitude scale.
- **Duplicate removal:** ensure that each event is represented only once (this is particularly an issue when merging multiple raw catalogs).
- **Depth redistribution:** many catalogs make crude assumptions regarding the depths of events (e.g., fixing them at common reference levels), and it can be important to consider the implications of these assumptions.
- **Declustering:** for compatibility with the assumption that seismicity follows a Poisson process, dependent events, namely, foreshocks and aftershocks, are usually removed.
- **Association to sources:** assign each event to a seismic source, or, more precisely, to a particular rupture scenario.

Each of the steps is briefly described with reference to the example in the following sections. Note that the particular approach taken to model the rates of occurrence determines which of the above steps is considered for any given analysis.

Harmonization

Moment magnitude is now the preferred measure of earthquake size for PSHA and related disciplines. However, this is a recent development, and for other areas of seismology, other magnitude

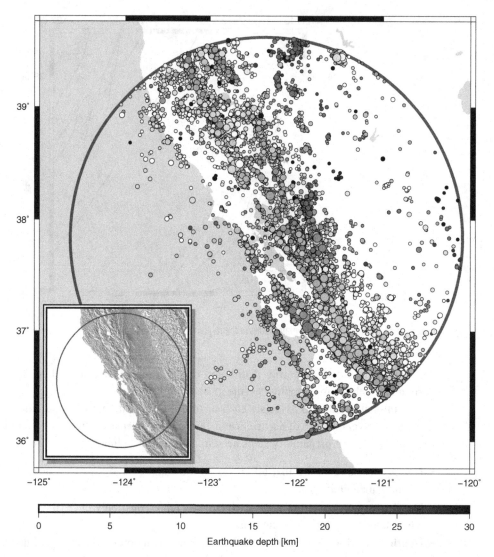

Fig. 3.4 Events of magnitude at least 2.0 within 200 km of Yerba Buena Island. The inset in the lower left shows the regional topography.

scales are still preferred. For example, for routine determination of earthquake size and location, it is far more common for seismological agencies to adopt a *local* magnitude scale. Therefore, seismicity datasets commonly need to be converted to moment magnitude. It is preferable to work with moment magnitude as the ground-motion models discussed in Chapters 4 and 5 work with this measure of earthquake size, and it is essential to have consistency when specifying rates and ground motions for a rupture.

The characteristics of the raw catalog are summarized in Table 3.2, where six different magnitude scales are utilized. Fewer than 1% of the events are reported on the moment magnitude scale. Roughly 94% of the events have their magnitudes reported on a coda magnitude scale (Eaton, 1992), because we have considered very small events (Klein, 2006). Larger events, often of greatest interest, are more likely to have a moment magnitude assessment. This is reflected in Table 3.2 where the largest mean and median magnitudes correspond to the entries for the moment magnitude scale.

Table 3.2. Statistics of the example NCEDC seismicity catalogue in terms of magnitude type. Magnitude types are those reported in the raw NCEDC catalogue (Klein, 2006)

Type	Portion (%)	Minimum	Maximum	Mean	Median
M_L	4.77	2.06	7.00	3.33	3.20
M_d	93.62	2.00	4.73	2.43	2.33
M_{dl}	0.03	2.00	2.41	2.12	2.07
M_h	0.05	2.10	2.90	2.62	2.70
M	0.90	2.87	6.02	3.73	3.70
M_x	0.63	2.00	4.32	2.40	2.32

The various scales that are adopted will typically provide magnitude estimates comparable to moment magnitude over a given range (see Figure 2.12). However, for very small and very large magnitudes, this approximate equality will break down. It is essential to account for differences between the scales, as even small differences in magnitude have nontrivial differences in terms of implied energy release. Recall that seismic moment scales with moment magnitude according to $M_0 \propto 10^{\frac{3}{2}M}$. Hence, a 0.2 unit difference in magnitude corresponds to a factor of $10^{0.3} \approx 2.0$ on the seismic moment (Equation 2.5).

A large number of *magnitude conversion equations* exist in the literature. The functional forms of the equations are usually simple first- or second-order polynomials such as

$$\hat{M} = g(M_i) = \alpha_0 + \alpha_1 M_i \tag{3.9}$$

that take a magnitude estimate in a given scale M_i and return an estimate of the corresponding moment magnitude $\hat{M}$. However, none of these conversion equations defines a deterministic mapping from one scale to another. Every time a conversion is made, the variability associated with the conversion itself, which we denote as $\sigma_{g(M_i)}$, must be taken into consideration. When the uncertainty in the estimate of the magnitude is already significant for these older instrumental events,[10] adding more uncertainty can act to dilute the utility of the observations. Nevertheless, this error propagation should be undertaken to reflect the legitimate uncertainty associated with the moment magnitude for each event. Using the first-order second-moment approach (Equation A.49), the uncertainty in the converted magnitude is:

$$\sigma_M = \sqrt{\sigma_{g(M_i)}^2 + \left(\frac{\partial g(M_i)}{\partial M_i}\sigma_{M_i}\right)^2} \tag{3.10}$$

where the magnitude uncertainty in the original scale is σ_{M_i}. Given the basic forms of the conversion equation, the partial derivative in Equation 3.10 is readily evaluated. For the example form in Equation 3.9, $\partial g(M_i)/\partial M_i = \alpha_1$.

For our example, the local magnitude events are converted to moment magnitude using the relationship presented in Wang et al. (2009). The other scales have primarily been used for very small events. Many of these are filtered out when we consider catalog completeness levels, and so we retain their reported magnitude at face value.

---

[10] Errors of 0.2–0.3 magnitude units are common for older instrumental seismicity, but this is still much smaller than the 0.5–1.0 unit uncertainty associated with historical events.

Duplicate Removal

Many countries operate their own local instrumental networks and record the events detected by these networks. However, the period of observation of these local networks will often differ from regional or global databases. Therefore, a common activity when compiling a seismicity database is to compare potential sources of seismicity data to ensure that no earthquakes are missed. Additionally, when multiple databases have a temporal overlap, the analyst will find events that are included in each catalog and others that appear to be unique. There are two reasons why these *unique* entries might exist:

1. The entries are unique, because one or more recording networks were unable to identify the event.
2. Multiple entries across different catalogs correspond to the same earthquake. However, the different location algorithms used by the different agencies have resulted in differences in the magnitude (potentially using different scales), time, or position.

Naturally, if we look to merge databases, we do not want to include multiple variations of the same event, so we must distinguish between the above two cases.

To identify duplicates, the potential culprits must first be identified and then considered on a case-by-case basis. Usually, one will have a hierarchy of databases, and when there is doubt, the information in the primary database will be used. Otherwise, one needs to rule out basic errors in data entry, and then consider the differences in the various attributes of the event. As a general rule, the timing information should be the most reliable attribute, so differences in this field are often the most informative. Differences in magnitudes, position, and, particularly, depth are less useful because the estimates have significant uncertainties. Thus, it is difficult to distinguish real differences from estimation uncertainty when deciding whether events are duplicates.

For our example, we do not use overlapping catalogs, so duplication removal is not relevant.

Depth Redistribution

It is more challenging to constrain an event's depth than its lateral position. Simultaneously inverting the observed waveforms at the seismograph stations in order to determine lateral coordinates and depth generally makes it difficult to constrain the lateral position. Thus, many agencies historically assigned standard depth values for events that are assumed to be of a given nature, such as shallow crustal earthquakes. This practice is less common now due to increased network densities in most parts of the world. However, the former practice can still cause problems because older catalogs will be used if possible. For example:

1. When assigning events to individual seismic sources, dipping faults may not be assigned events they generated, and
2. Some declustering algorithms will overestimate the size of clusters because interevent distances will be underestimated when all events are assigned the same depth.

To quantify the sensitivity of seismicity results to assumed depths, the depth distribution of the events not assigned fixed values can be used as a sampling distribution from which random depths are assigned to previously fixed-depth events (McGinty, 2001). Multiple datasets can be sampled in this way; the variation in implied rupture rates will indicate the importance of the depth assignments (Stafford et al., 2008).

For our running example, there are no fixed depths, and so redistribution is not required.

Start year	End year	Completeness M	Start year	End year	Completeness M
1850	1855	6.0	1932	1942	4.5
1855	1860	5.8	1942	1967	4.1
1860	1870	5.7	1967	1997	4.0
1870	1885	5.6	1997	2000	2.6
1885	1895	5.5	2000	2007	2.4
1895	1932	5.3			

Table 3.3. Completeness levels for the San Francisco region (Felzer, 2013b)

Completeness Levels

Reliable estimates of the rates of occurrence of rupture scenarios can be obtained from seismicity data only if the number of events in the seismicity catalog is representative of the actual rate. There are two main reasons why the number of events in a catalog may differ from this actual rate:

1. Events may have gone undetected by the instrumental network (usually an issue for small events), and

2. The recurrence interval for an event may be long in comparison with the duration of the catalog (mainly an issue for larger events).

To detect an earthquake, a person or instrument, must be in the vicinity of the event, as seismic waves attenuate with distance. Before the existence of highly sensitive seismographs, an event would be detected only if humans felt the waves it generated. Historical databases, in particular, consequently have a higher rate of events near populated areas. This same phenomenon is relevant for instrumental databases where an event is more likely to be detected if it occurs near a recording station.

For a catalog to be regarded as complete, all events above a certain magnitude level must be observed. As will be discussed further in Chapter 4, ground motions generally increase with magnitude and decay with distance. Therefore, it is more likely for a large event to be identified than a small event taking place at the same position.

For historical seismicity, completeness levels change as a result of new regions becoming populated. As the average spacing between population centers decreased, the likelihood of an earthquake being felt increased. Similarly, as instruments become more dense and more sensitive, smaller events are increasingly likely to be recorded.

Although changes in completeness are linked to changes in the regional seismograph network, it is common to define completeness levels using statistical techniques. The primary approach in this vein is to assume that the regional seismicity is well-described by a Poisson process with exponential interevent arrival times. Completeness levels are then determined by identifying the magnitude level for which a significant departure from the expected rates of occurrence is observed. The best approach is to combine such statistical arguments with the information about network coverage and station spacing.

For our running example, we make use of the completeness levels suggested within the UCERF3 study by Felzer (2013b), which are shown in Table 3.3. The limiting levels listed in Table 3.3 are also displayed in Figure 3.5, where seismicity data from the NCEDC catalog are combined with the data used in the UCERF3 seismicity analyses. The UCERF3 catalogue (Felzer, 2013a) is only used

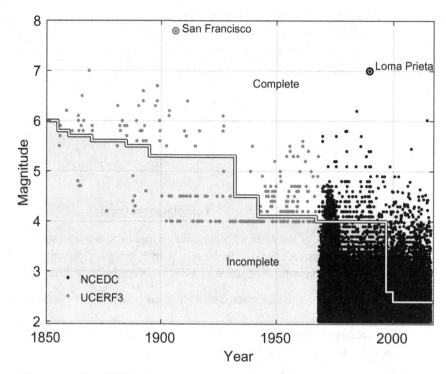

Fig. 3.5 Completeness levels adopted for the UCERF3 seismicity analyses (Felzer, 2013b). Events in the shaded area marked 'incomplete' are generally excluded from seismicity analyses. The 1906 San Francisco and 1989 Loma Prieta events are also annotated.

to supplement the NCEDC data before the NCEDC network becoming operational. For the period of overlap, the NCEDC data are assumed to take precedent over the UCERF3 data.

In Figure 3.5, a number of observed events are located in the gray shaded region, and these events would be excluded from any seismicity analysis. In many locations with more limited data than this case, it is frustrating for the analyst to remove this incomplete data. However, while many events have been recorded over the period 1900–1950, their attributes are not well-known. This is why most events in this period are assigned magnitude values in 0.5 unit increments.

Declustering

A common assumption within seismic hazard analysis is that seismicity can be well-described by a Poisson process (Cornell and Winterstein, 1988). A fundamental property of Poisson processes is that the instantaneous rate of events is constant and does not depend upon the occurrence of other events located close in either space or time. However, earthquake sequences feature significant numbers of aftershocks and these events are dependent upon the mainshock. The purpose of declustering seismicity data is to remove these dependent events so that the underlying long-term average rate of occurrence can be estimated.

Declustering a catalog, and removing foreshocks and aftershocks, is done purely to enable an estimate of the homogeneous Poisson rate of occurrence. The act of declustering does not imply that these dependent events are not important for hazard and risk (Boyd, 2012). Later, in Section 7.7.1, methods for accounting for the effects of these events within hazard analysis are presented.

Figure 3.6a shows the cumulative number of events of all magnitudes for the NCEDC dataset in the running example before imposing the completeness levels. Figure 3.6b then splits the data

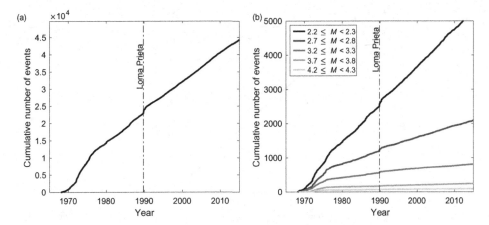

Fig. 3.6 Cumulative number of events with respect to time for the NCEDC seismicity data within 200 km of Yerba Buena island. Panel (a) shows numbers of events of any magnitude. Gray lines display numbers of events for the individual magnitude bins in panel (b).

to show data for individual magnitude bins. If the seismicity data in this catalog arose from a pure Poisson process, then we would expect to see lines of constant slope in both figure panels and for all magnitudes. However, consider the deviation coinciding with the 1989 Loma Prieta earthquake in Figure 3.6. The sudden increase in event rates occurring immediately after the Loma Prieta event reflects its aftershocks. Changes in the observed rates of events can also arise from changes in the region's completeness level. While aftershock sequences influence rates for all events with magnitudes lower than the mainshock, changes in completeness only impact the numbers of observed small magnitudes.

Declustering a seismicity catalog requires identifying clusters of dependent events in some systematic manner. Declustering algorithms perform this identification, and they make use of common attributes of aftershock sequences:

1. The modified Omori relation describes the rates of occurrence.
2. Magnitudes are distributed according to a bounded exponential distribution.
3. The largest aftershock has a magnitude roughly one unit lower than the mainshock.
4. The spatial density of events is inversely proportional to the distance from the mainshock.

These relations are discussed in the following sections.

Omori's Law and Aftershock Rates

One of the oldest relations in statistical seismology dates back to the end of the nineteenth century when Omori (1894) published an expression defining the temporal change in the rate of occurrence of aftershocks. Omori's expression was later modified by Utsu (1961) to the form shown in Equation 3.11 and is commonly called the "Omori–Utsu law" or the "modified Omori law".[11]

$$\lambda(t) = \frac{K}{(t + c)^p}.$$
(3.11)

---

[11] Strictly, Equation 3.11 does not represent a *law*, but this expression, along with Båth's *law*—introduced in the following section—is commonly refered to as such.

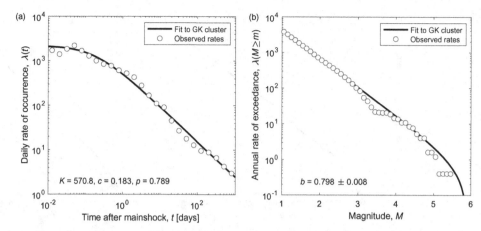

Fig. 3.7 Panel (a) shows the modified Omori law fit to the Loma Prieta aftershock sequence. Observed rates are obtained from the cluster of events identified using a minimum magnitude of 1.0 and the Gardner and Knopoff (1974) (GK) algorithm. Panel (b) shows the magnitude-frequency distribution for the cluster.

The expression states that the rate of occurrence at time t, $\lambda(t)$, of earthquakes of any magnitude within an aftershock sequence begins (for $t \ll c$) at a level equal to Kc^{-p} and remains roughly constant until time c. Beyond this time, the rate asymptotically scales as Kt^{-p}. An example of this model is provided in Figure 3.7 in which the aftershock sequence from the 1989 Loma Prieta earthquake is plotted.

The modified Omori law has been shown to behave very well for replicating the rates of aftershocks following significant events. In these cases, the parameters K, c, and p are calibrated to the particular event in question. Although the values of K and c vary from event to event quite considerably, the parameter p is usually around a value of 1. While Equation 3.11 is usually effective in describing the rate of occurrence of aftershocks in a cluster, there are some cases where relatively large aftershocks trigger additional aftershock clusters that may deviate from the basic scaling shown (Gospodinov, 2017; Spassiani and Marzocchi, 2018).

The parameter K primarily controls the maximum rate of events that occur immediately after the mainshock as $t \to 0$, i.e., $\lambda(t = 0) \propto K$. However, this maximum rate depends upon the magnitude range considered. The lower the magnitude, the greater the observed number of events. In practice, a given region will have a level of magnitude completeness (see Section 3.3.5) that will exist at the time of the mainshock. However, when a major event occurs, it is common for a temporary seismic network to be rapidly deployed in the vicinity of the mainshock. So the actual completeness level for a given seismic sequence may be lower.

Magnitude-Frequency Distribution of Aftershock Sequences

It has been consistently observed that the distribution of event magnitudes within the sequence is well-represented by a doubly bounded exponential distribution (to be discussed in detail in Section 3.5.1). An example fit of this distribution to the Loma Prieta aftershock sequence is shown in Figure 3.7. The depicted magnitude range for the exponential fit spans from 1.0 to 5.9. This upper magnitude bound is one magnitude unit lower than the Loma Prieta mainshock (an $\boldsymbol{M}6.93$ event). The use of this one magnitude unit difference is consistent with the assumption that Båth's *law* is appropriate (Båth, 1965).[12] In Figure 3.7, the agreement between the observed magnitude distribution and the fitted model is excellent.

Combined Model for Aftershock Magnitudes and Rates

Based on the consistent observation that aftershock sequences have temporal rates that agree with the modified Omori equation, and magnitude distributions consistent with both a doubly bounded exponential distribution and Båth's law, Shcherbakov et al. (2004) proposed a unified model that brings these three attributes together.

To achieve this unification, Shcherbakov et al. (2004) allowed K from Equation 3.11 to become magnitude-dependent in a manner that satisfied the first and third criteria above. The resulting expression is shown as follows in which m_{ms} is the mainshock magnitude, and Δm represents the difference between the mainshock and the largest aftershock:

$$\lambda(t; M > m) = \frac{K(m_{ms}, \Delta m)}{(t + c)^p} = \frac{10^{a+b(m_{ms} - \Delta m - m)}}{(t + c)^p}.$$ (3.12)

Note that this expression provides the total rate of events having a magnitude of at least m at time t. This formulation is also consistent with the earlier work of Reasenberg and Jones (1989, 1990), with the exception that this earlier work did not incorporate Båth's law.

Spatial Distribution of Aftershock Sequences

In addition to their temporal and size distribution, the spatial distribution of these events is important. In the declustering algorithm of Gardner and Knopoff (1974) the temporal and spatial windows defined to identify aftershocks are increasing functions of magnitude. These windows define ranges in time and space, and any event occurring within these ranges is associated with the mainshock. Felzer and Brodsky (2006) considered the decay of aftershock rates with distance, focusing upon the decay from small to moderate events. They found that the decay of aftershock rate density could be described using a function of the form:

$$\rho(r) = cr^{-n}$$ (3.13)

with c being a constant that varies with the total number of aftershocks (and hence with the mainshock magnitude and considered magnitude range for the aftershocks). The parameter n controls the rate of decay with distance.

Figure 3.8 shows how the aftershocks of the 1989 Loma Prieta earthquake decay with distance from the finite rupture model of Wald et al. (1991). Also shown in this figure is a least squares fit to this data using a function of the form:

$$\rho(r) = c \max(1, r)^{-n}$$ (3.14)

that truncates the power law scaling for distances less than 1 km.

Equations 3.13 and 3.14 work well in general, but imply that aftershock densities decay in an isotropic manner, radially away from the mainshock. However, the spatial distribution of fault sources that may host aftershocks will generally cause nonisotropic seismicity. Finally, it should also

---

[12] Båth actually found that the average difference between the mainshock magnitude and the magnitude of the largest aftershock was $\Delta M \approx 1.2$.

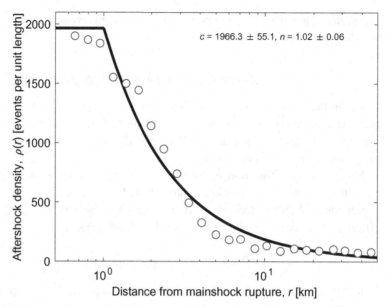

$c = 1966.3 \pm 55.1$, $n = 1.02 \pm 0.06$

Fig. 3.8 Distance decay of the 1989 Loma Prieta aftershock sequence. The annotated parameters c and n relate to Equation 3.14.

be mentioned that these equations represent the event densities over the full time period spanning the aftershock sequence. In some cases, event densities at given distances will evolve in time due to effects such as transient migration of pore fluid (Sibson, 1973).

Effects of Declustering

Several declustering algorithms have been proposed, and the key features of the most common approaches (e.g., Gardner and Knopoff, 1974; Reasenberg, 1985) are outlined below.

The basic concept is that a dependent event can be recognized because it occurs shortly after some prior event at a position that is relatively close to that prior event. Therefore, the various algorithms define spatiotemporal windows to link events to each other. In both space and time, larger magnitude events can have larger windows. The classification of events as dependent or otherwise can be done in a deterministic (e.g., Gardner and Knopoff, 1974) or probabilistic manner (e.g., Reasenberg, 1985).

A comparison of the effect of declustering using the common algorithms of Gardner and Knopoff (1974) and Reasenberg (1985) is shown in Figure 3.9. The region shown here is to the southeast of San Francisco Bay, approximately where the Calaveras Fault system merges into the San Andreas Fault. In this figure, it is evident that the Gardner and Knopoff (1974) algorithm is far more aggressive in removing events deemed to be dependent.

Another way to look at the impact of declustering is to count the numbers of events falling in magnitude bins for the raw and declustered catalogs. Figure 3.10 shows such a comparison for a declustered catalogue obtained using the Gardner and Knopoff (1974) algorithm. Average rates of occurrence, $\lambda(M = m_i)$, are shown by making direct counts of the numbers of events in bins of 0.1 magnitude units width, and dividing these by the period of observation corresponding to the catalog completeness of Table 3.3.

The logarithmic scale of the plot means that the differences between the raw and declustered catalogs are large, even if they do not look striking. For the smallest events, the declustered catalog

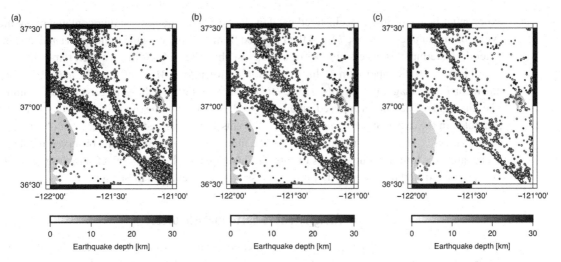

Fig. 3.9 Comparison between raw and declustered catalogues over a small spatial region. Panel (a) shows the raw catalogue, while panels (b) and (c) correspond to the results obtained using Reasenberg (1985) and Gardner and Knopoff (1974), respectively.

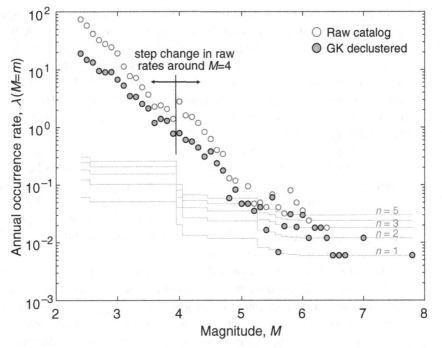

Fig. 3.10 Annual rates of occurrence for the Yerba Buena Island case study obtained for complete catalogs before and after declustering using the Gardner and Knopoff (1974) (GK) method. The thin gray lines represent the rates obtained from observing the numbers of events, n (noted on the figure), in light of the corresponding period of completeness for each magnitude. The robustness of the occurrence rate estimates increases with the distance between the observed data and these lines.

contains about 25% of the events in the raw catalog. In Section 3.5.1, we will explain that it is typical to observe that earthquake magnitudes are exponentially distributed. An exponential distribution would be represented by a straight line when plotting $\ln \lambda(M = m)$ against M. Figure 3.10 shows that the rates for magnitudes in the approximate range $M \in [2.4, 4.5]$ are closer to this expected exponential behavior in the declustered catalog. In particular, the raw catalog contains many more events of magnitude just above $M = 4$ than implied from extrapolating the trend from smaller events. After declustering, this excess number of events just above $M = 4$ has mostly been removed.

In Figure 3.10, the approximate linear relation between $\ln \lambda (M = m)$ and M appears to break down for larger magnitude events. However, the sample size for these larger events is minimal. The thin gray lines superimposed upon Figure 3.10 indicate the annual rates that would be computed in the case that particular numbers of events were observed in the period of observational completeness.[13] These lines illustrate that for the moderate to large events, the observed numbers of events are very low, and fluctuations of rates in this range are likely to reflect sampling bias rather than physical differences in activity rates. This issue will be revisited in Section 3.5.

Assignment to Earthquake Sources

Thus far, we have considered procedures for preparing and inspecting a dataset associated with a general study region. However, we need to estimate $\lambda (rup)$ and so must move from a regional scale down to a source, and ultimately segment or rupture geometry, scale. To this end, we need to associate the events in the seismicity catalog with the seismic sources identified in the source model.

In principle, this is a straightforward exercise of looking at the earthquake locations and the geometry of the sources. In practice, because each event in the catalog will have a spatial uncertainty associated with it and the source geometry will also have some uncertainty, it may not be obvious which source was responsible for each event. For the largest events, this is generally not an issue as faults large enough to generate such events will leave surface expressions over many tens of kilometers that can be correlated to various felt or observed effects. However, for smaller events, the errors in the position are often more substantial than the dimensions of the rupture itself, and they leave no permanent surface evidence.

Several approaches have been adopted to assign events to sources. The optimal approach was presented by Wesson et al. (2003) and allows individual events to be associated with identified seismic sources, while formally accounting for the uncertainties in the geometry of the individual sources and locations of observed events, among other things. The general structure of this framework is shown in Figure 3.11. Through the application of this framework, each event has a set of probabilities computed, $P(S_i|O)$, representing the probability of source S_i being responsible for the event given the observations O.

While the approach of Wesson et al. (2003) is optimal, it is also relatively involved, and in many cases, more straightforward, approximate approaches are adopted. For example, Powers and Field (2013) addressed the problem by creating *fault-zone polygons*. These polygons are defined on the Earth's surface. They are formed from the union of polygons that represent uncertainty in the location of the fault trace, the dip of the fault, and geological features in the vicinity of the fault that are attributed to activity on it. Each of these polygons is defined in a deterministic way, but collectively reflect uncertainties in both fault location and event position in an approximate manner. Any events that fall within the fault-zone polygon are then assigned to the fault.

The approach taken to assign events to sources will influence resulting rupture rates, though this sensitivity is not straightforward to determine a priori. One should thus try to account for the relevant uncertainties, even if only approximately.

---

[13] For example, if n_i events of a given magnitude are observed, then the implied rate of occurrence is $\lambda_i = n_i/t_{obs,i}$ where $t_{obs,i}$ is the period of complete observation for this size event.

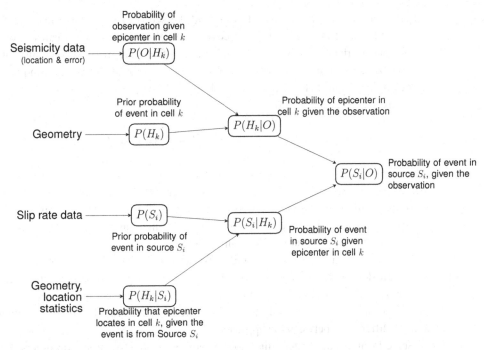

Fig. 3.11   General framework for associating observed earthquake events to distinct seismic sources. Adapted from Wesson et al. (2003)

3.4 Geological Constraints on Activity

With the passage of time and an increase of available information, seismic source characterization in tectonically active regions has shifted from being dominated by relatively coarsely defined area and fault sources toward source models dominated by fault sources and individual fault segments. While small magnitude events have a relatively high likelihood of occurring throughout a region, the largest events are far more likely to occur on existing faults due to their reduced frictional strength. A consequence of trying to represent specific faults or fault segments, however, is that very little seismicity data will be available to constrain the rates of occurrence of the associated rupture scenarios. To constrain the rates of occurrence, it is, therefore, necessary to use information beyond that provided by seismicity datasets. In the following subsections, sources of additional constraint upon rates of occurrence are briefly introduced. Several texts (e.g. Yeats et al., 1997; McCalpin, 2009; Burbank and Anderson, 2011) provide significantly more comprehensive treatments of this material and should be consulted for further information.

The first step in any geological investigation is to "map" the faults within the study region—identifying the geographical location of the faults and then determining their level of activity. Additionally, faults arise on a local scale due to larger-scale driving mechanisms. As a result, the identified faults often portray a partial picture of a broader network of inter-related faults that collectively accommodate tectonic strain. Therefore, it is often necessary to move from factual (but incomplete) information about the locations and characteristics of mapped faults (or fault segments) to an interpretation of the broader fault network. This interpretation step is vital as it moves the focus away from locations where earthquakes are known to have occurred in the past and to locations where future events may occur. It is beyond the scope of the present book to discuss how faults are identified and mapped, but good coverage of the relevant issues can be found in Yeats et al. (1997) and Burbank and Anderson (2011). Once faults are mapped, several geological techniques can impose constraints on the timing and size of past events. Generally, the approaches are classified as belonging to the field of paleoseismology, and comprehensive resources exist that detail these methods (e.g., McCalpin, 2009).

3.4.1 Paleoseismology

The field of paleoseismology concerns the study of earthquakes, and their effects, that took place well before the existence of any instrumental or historical records. Paleoseismic investigations provide information of direct relevance to seismic hazard analysis, such as constraints on fault segmentation, per-event slip, time since the last event, recurrence intervals for particular ruptures, and magnitudes of past events (see Chapter 9 of McCalpin, 2009).

Paleoseismological methods identify distinct geological offsets that can be attributed to abrupt coseismic deformation.[14] The type of information used depends upon the stress-regime of the region. McCalpin (2009) details the methods adopted in different circumstances and supports this with specific examples of faults and regions from around the world. The present section highlights some critical differences between the approaches.

Some locations are better suited to a paleoseismic investigation than others. For example, many investigations cannot be performed along offshore subduction trenches. In other areas, such as deserts, other geomorphological processes destroy evidence of past earthquake activity and make paleoseismic investigations extremely challenging, if not impossible.

When conditions are suitable for paleoseismic investigations, various techniques can be applied. For example, paleoseismic trench investigations involve excavating a narrow linear trench across an active fault and interpreting offsets in exposed strata on either side of the fault. In cases where carbon-rich material has been deposited on either side of a soil horizon reflecting the timing of an earthquake, this material can be dated using laboratory techniques, and bounds can be imposed upon the timing of the offset. By undertaking multiple such trench investigations along a mapped fault, information about the segmentation of faults can be deduced, and a picture of the rupture history for the fault can be composed.

Paleoseismic Constraint in Dip-Slip Environments

Dip-slip faulting arises in both extensional and compressional regimes. While particular details influence paleoseismic interpretation in these different regimes, there are also common features unique to dip-slip environments. Primarily, it is often possible to identify past events from vertical offsets either at the surface (such as offset river terraces) or in trenches dug across a fault.

For both normal and reverse faults, coseismic displacement will cause sedimentary deposits in the near-surface to be offset vertically (while the actual slip occurs in the up- or down-dip direction). Sediments that are subsequently deposited will form new, unbroken, layers that will then

---

[14] Not all crustal strain is released by earthquakes. In some cases, slow creep along faults, or folding and general deformation of the crust, will also contribute to balancing the overall moment budget (Section 3.5.2). In subduction zones, *episodic tremor and slip* can occur (e.g., Rogers and Dragert, 2003). These latter phenomena are typically referred to as *aseismic*, while moment expended during earthquake ruptures are *coseismic*.

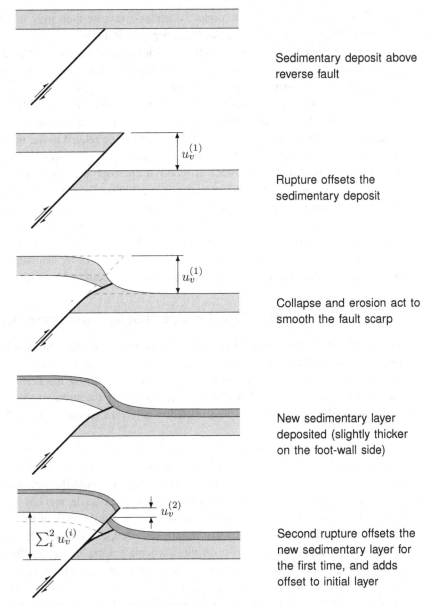

Sedimentary deposit above
reverse fault

Rupture offsets the
sedimentary deposit

Collapse and erosion act to
smooth the fault scarp

New sedimentary layer
deposited (slightly thicker
on the foot-wall side)

Second rupture offsets the
new sedimentary layer for
the first time, and adds
offset to initial layer

Fig. 3.12   Schematic illustration showing the cross-section of a reverse fault interacting with sedimentary deposits. The passage of time
flows from top to bottom.

be offset in future surface-rupturing events. In Figure 3.12, a schematic illustration shows the general evolution of a reverse fault.[15] This illustration is highly idealized but serves to explain how scarps of reverse faults are eroded and flattened between coseismic events. Any sedimentary deposits placed above the eroded fault scarp are offset in future events, but will importantly be offset by less than the total cumulative offset of the older deposits that have already experienced coseismic displacement.

---

[15] Normal faults have many similarities, but McCalpin (2009) also highlights a number of important differences.

Therefore, trench investigations conducted across a reverse fault may reveal a rupture history by observing differential offsets in sedimentary deposits of different ages.

The information described thus far identifies the number of events that have displaced sedimentary layers to a particular depth. However, we also need size and timing information to reconstruct the rupture history. The stereotypical way to obtain this full rupture history in dip-slip environments is to undertake active fault investigations through trenching. By excavating a trench that crosses a fault, it is possible to observe the offsets in sedimentary strata located on either side of the fault plane mentioned previously. The size of the offset can be related to the magnitude of the causative event using source-scaling relations, and constraints upon the timing of past events are obtained by dating carbon-rich material (such as wood or charcoal) on either side of the offset (e.g., Biasi et al., 2002; Atwater et al., 2003; Klinger et al., 2003). The size and timing estimates will have significant uncertainties associated with them, particularly when only point-estimates from a single trench location are available. However, trenching at multiple locations along a mapped fault makes it possible to construct a complete representation of the distribution of per-event displacement along a fault (e.g., Biasi and Weldon, 2009). This, in turn, allows better estimates of the magnitude to be obtained, and potentially provides information about the segmentation of ruptures on a fault.

Paleoseismic Constraint in Strike-Slip Environments

For dip-slip faults, the offsets will be in the up- and down-dip directions, while for strike-slip faults one may find evidence for lateral displacement along the fault strike. In particular, linear features that cross the fault (natural or human-made), such as railway lines, roads, riverbanks, terraces, and streams, may reveal evidence of displacement along the strike of the fault. Figure 3.13 shows a schematic illustration of a strike-slip fault where past ruptures have offset stream channels at particular points.

The first challenge to reconstruct the rupture history is to determine how many distinct events have taken place and the spatial extent of these ruptures. How easy this is depends upon several factors such as how well preserved the offset features are, how large the initial offsets were, the spatial resolution at which point-estimates of displacement can be made, and how active the fault is. Taking point 2 in Figure 3.13 as an example, the stream crossing the fault at this point initially continued directly across the fault. When the first rupture occurred, the downstream channel was shifted along the fault by some amount (1.46 m in this particular example). Using the source-scaling relation of Wells and Coppersmith (1994) and assuming that this was the maximum surface displacement along the rupture, an estimate of the magnitude for the causative event is $M6.94$.[16] Now, this offset of approximately 1.5 m could easily be smaller than the width of the stream itself, which would lead to a partially offset channel. In this case, the downstream channel would continue to carry the stream's flow, and the offset would be scoured and eroded, making its later detection challenging. For events of magnitude even smaller than this, the chances of observing surface evidence of faulting decline quite quickly, but so do the surface offsets. For example, the second rupture event shown in Figure 3.13 has a displacement of 1.04 m at this same location and would correspond to an $M6.82$ event using the same assumptions adopted previously.

---

[16] The expression is: $M = 6.81 + 0.78 \log_{10}(MD)$, with MD being the maximum observed surface displacement.

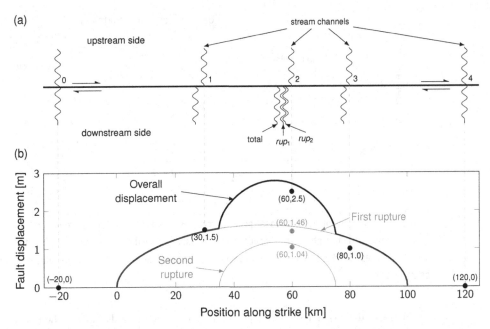

Fig. 3.13 Illustration of information available from offset stream channels crossing a strike-slip fault. The schematic in panel (a) shows a plan view of streams that have not been offset at points 0 and 4, streams with a single offset at points 1 and 3, and stream channels associated with two distinct offsets at point 2. The markers in panel (b) show measured offsets at particular spatial positions along the strike. The solid lines in panel (b) show the actual distribution of displacement caused by two distinct events (in gray) as well as the total cumulative displacement (in black).

For the particular ruptures discussed above and shown in Figure 3.13, it is likely that flow would be slightly diverted at the location of the fault, but that the flow would stay within the same channel.[17] Once the magnitude becomes larger, or the effect of multiple events accumulates, the channel could be offset entirely. In this case, the stream may travel along strike slightly (depending upon the terrain) and then continue down the original channel, or it may create a new channel that is a direct downstream extension of the upstream portion. In the latter case, one may find evidence of old river channels that are no longer used at some distance along the fault strike. Any organic material that is fluvially deposited (or entrained with fluvial deposits) in the near-surface sediments of such channels has presumably been placed before the rupture and can be used to estimate the timing of the event.

Challenges Constraining Earthquake Size and Timing

Thus far, the discussion implies that reconstructing a paleoseismic history is relatively straightforward. However, consider again the scenario shown in Figure 3.13. In this case, several pieces of information for constraining rupture rates can be extracted. The first issue influencing our inferences is how clear the rupture history is around the offsets of the stream channel at point 2. It is plausible that the first rupture offset this point by 1.46 m, diverting the channel, and that the second smaller rupture took place "shortly" after and continued to divert the channel to the final offset of 2.5 m.

---

[17] This depends upon both the width of the original stream channel as well as the integrity of the material the stream is flowing through, i.e., how susceptible it is to scour.

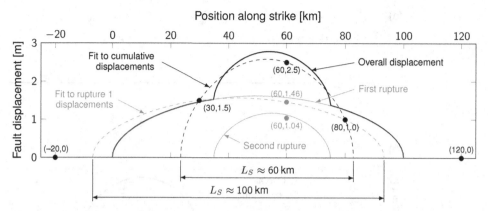

Fig. 3.14   Illustration of displacement profiles fit to the point-estimates using either the black points corresponding to the cumulative displacements from both rupture events, or just the point-estimates from the first rupture event. The fitted profiles are shown with dashed lines while the true displacements are shown with solid lines. The implied surface rupture lengths are shown below the plot.

In this case, the smaller second rupture may not be identified, causing the entire 2.5 m offset to be attributed to a single event.

The evidence for this single-event interpretation is that the surface rupture is at least as long as the distance between points 1 and 3 (that were both offset), but shorter than the distance between points 0 and 4 with no observed offset. Using the source-scaling relation of Leonard (2014), the limits upon the magnitude of this single-event scenario are $M \in [7.18, 7.72]$.[18] If the maximum single-event displacement is assumed to be the 2.5 m cumulative offset at point 2, the relation of Wells and Coppersmith (1994) gives $M 7.12 \pm 0.29$. We could also obtain an estimate from the average surface displacement, which, if using the arithmetic average of the three point-estimates, gives 1.67 m and corresponds to a magnitude estimate of $M 7.24 \pm 0.28$. However, we have no way of knowing what the distribution of displacement is, and Figure 2.19 showed that significant variability exists.

However, if we assumed that the displacement profile conformed to the square-root sine function of Equation 2.17, then the three point-estimates of the cumulative displacement can be used to identify the expected maximum displacement and surface rupture length. Figure 3.14 shows this profile using the dashed black line, and the magnitude then implied from the surface rupture length of $L_S \approx 60$ km gives $M 7.29$, while the maximum displacement of 2.58 m implies $M 7.13$. Therefore, assuming the entire offset at point 2 is attributed to a single event, a range of different magnitude estimates will be obtained. The key point is that we cannot somehow reconcile all of these different magnitude estimates in order to obtain the "true" magnitude of the event.

If we were able to identify the two distinct ruptures (with displacements of 1.46 m and 1.04 m), the picture of the rupture history changes significantly. For rupture 1, the magnitude bounds from the limiting lengths remain $M^{(1)} \in [7.18, 7.72]$, while the fit shown in Figure 3.14 implies a length of 100 km that corresponds to $M^{(1)} 7.54$, using Leonard (2014). However, the displacement-based constraint now implies a magnitude of $M^{(1)} 6.94$–6.96, depending upon whether the fitted maximum displacement or the point-estimate of 1.46 m is used. For rupture 2, we can only say that we have

---

[18]   The expression being used here is: $\log(M_0) = 9.15 + 2.27 \log(L_S)$, for $L_S \leq 65 \times 10^3$, with L_S being the surface rupture length in meters, and $\log(M_0) = 12.86 + 1.5 \log(L_S)$, for $L_S > 65 \times 10^3$.

no evidence for the rupture extending to points 1 and 3, implying a maximum length of 50 km and a magnitude of $M^{(2)}7.18$. Taking the only displacement estimate as the maximum implies a magnitude of $M^{(2)}6.82$.

The discussion above illustrates the challenges paleoseismologists face when interpreting their field evidence. In cases where the information does not paint a clear picture, it may be possible to look for further evidence to reject certain hypotheses. For example, one could move along the fault, between points 0 and 4, and reduce the upper limit for the length of the first rupture. However, significant uncertainties will still remain, and these must be recognized when making use of paleoseismic constraints.

Constraints upon the timing of these events suffer from similar challenges. In particular, it is possible to date only certain types of material, and datable material will not always be found. Additionally, assumptions must be made about the depositional history that impacts how tightly one can constrain the event's date. For example, if organic material in an intact sedimentary layer is shown to be 3500 ± 60 years old,[19] then we can say only that the previous rupture must have taken place at least this long ago. To impose a tighter constraint, we would need to know how much time elapsed between the previous event and the deposition of the organic material that has been dated. Therefore, while the point-estimate of the date for the material can be constrained with reasonable accuracy, there will still be an interval (potentially an open interval if only one event is identified) assigned to the time of occurrence of the events. When multiple events are identified, a reasonable estimate of the fault slip rate can be obtained. However, it remains to partition this overall rate over the particular ruptures that the fault tends to generate.

3.4.2 Fault Slip Rates

Slip rates for faults carry units of length per unit time (often mm/yr) and imply some continuous rate of deformation. In reality, the vast majority of displacement over a fault occurs during coseismic rupture, although faults that *creep* will also have some contribution from aseismic deformation. The slip rate for a given fault can be determined by dividing the total cumulative displacement by the time period over which the displacement accumulated. This approach assumes that the displacements accumulated over the time period being considered provide an unbiased estimate of long-term displacements, i.e., the time period is long enough to average out temporal fluctuations. These cumulative offsets can be estimated in several ways, but generally relate to the identification of geological features that have been displaced over some large scale. While the paleoseismic investigations in the previous section focused on identifying per-event offsets of at most a few meters, the estimation of long-term slip rates usually focuses on the cumulative effect of many such events.

The average slip rate can then be determined as:

$$\dot{u} = \frac{u_T}{T} = \frac{\sum_i u_i}{T} \tag{3.15}$$

where $\dot{u}$ represents the average slip rate, u_i is the slip associated with a particular earthquake rupture, and T is the total time period over which the displacements are aggregated. Note that the estimates of both the total displacement and the time period will have uncertainties associated with them, and

---

[19] The dating process will have measurement uncertainty associated with it, which in this case is ± 60 years.

the computed slip rate should reflect these. When the displacement is characterized as $u_T \pm \sigma_{u_T}$ and the time as $T \pm \sigma_T$, the slip rate estimate becomes $\dot{\bar{u}} \pm \sigma_{\dot{\bar{u}}}$ where:

$$\sigma_{\dot{\bar{u}}} = \dot{\bar{u}} \sqrt{\left(\frac{\sigma_{u_T}}{u_T}\right)^2 + \left(\frac{\sigma_T}{T}\right)^2}. \qquad (3.16)$$

For example, Lienkaemper and Borchardt (1996) obtained estimates of the slip rate of the Hayward Fault in the San Francisco Bay area using offsets in an alluvial fan arising from a well-defined stream crossing the fault. They were able to obtain estimates of times and offsets from two distinct trench locations of 42 ± 6 m over a time period of 4.58 ± 0.05 ka (ka = thousand years) and 66 ± 6 m over 8.27 ± 0.05 ka. From Equations 3.15 and 3.16 these estimates correspond to slip rates of 9.17 ± 1.31 mm/year and 7.98 ± 0.73 mm/year, respectively.

A slip rate for a given fault does not constrain activity rates for individual ruptures, unless some reasonably strong assumptions are made. Specifically, if the fault is assumed only to generate particular characteristic ruptures, the slip rate combined with an estimate of the average slip per characteristic rupture will enable the recurrence interval for these ruptures to be computed.

In the more general case where multiple ruptures of different sizes can occur on the fault, and the relative likelihood of these ruptures is unknown, the slip rate alone provides a weak constraint. To explain why, recall that the seismic moment for a single event is defined as $M_0 = \mu A \bar{u}$. If we assume that the rupture area is fixed and is equal to the total area of the fault, and that the shear modulus, μ, is constant, then the seismic moment release *rate* can be related to the long-term average slip rate as:

$$\dot{M}_0 = \frac{d}{dt} M_0 = \mu A \frac{d\bar{u}}{dt} = \mu A \dot{\bar{u}}. \qquad (3.17)$$

Therefore, knowledge of the overall slip rate for the fault is similar to knowing the seismic-moment release rate. However, these values relate to the cumulative slip arising from all types of ruptures that have occurred, and there are infinitely many combinations of rupture scenarios and rates that would lead to the same slip rate or moment release rate. Section 3.6.2 will show that our final predictive models describing the rates of seismic activity on each source must be consistent with these long-term slip rates on the faults.

3.4.3 Geodetic and GNSS Strain Rates

Geodetic surveys and analysis of global navigation satellite system (GNSS) data can indicate relative displacements among fixed points on the Earth's surface with time. From this information, it is possible to determine regional strain rates that can, in turn, be related to regional seismic-moment release rates (e.g. Holt and Haines, 1995; Tu et al., 1998, 1999). These results provide important source model constraints, as the seismic-moment release rates implied by the individual sources within the region should be consistent with the regional strain rates.

Despite this potential, geodetic data also have some shortcomings. Geodetic information typically covers a few decades at most, which is relatively short compared with recurrence intervals for major events. Therefore, the possibility of temporal fluctuations in these regional strain rates in time must be considered. In practice, this means that it is not necessary for the cumulative moment release rates for all individual sources (obtained from long-term slip-rate estimates) to precisely match the present-date regional moment release rate implied by the geodetic observations. However, any significant deviation between the two implied rates warrants serious consideration. Differences may arise as a result of ruptures predicted by an ERF not having occurred during the relatively short period of observation covered by the geodetic survey. In this case, the geodetic moment release rate will be lower than that implied by the ERF. On the other hand, if a major event has occurred during the geodetic survey period, significantly higher levels of moment release would be implied by the geodetic survey. In the case of a moment deficit (predicted events not having occurred), it is possible to infer particular spatial regions where future events may occur in order to equilibrate this deficit (e.g., González Ortega et al., 2018).

Geodetic information is also useful for understanding what degree of deformation takes place aseismically. Accepting that the chance of observing a significant rupture over the short period of geodetic observation is low, the geodetic surveys effectively provide continuous strain measurements over a region between major seismic events. This continuous strain may have contributions from both elastic strain accumulation (see Figure 2.8), which eventually leads to coseismic strain release, and aseismic creep. The overall seismic moment release rate for a region must be partitioned between both coseismic and aseismic release. While the coseismic component dominates, a 5–10% contribution from aseismic release is not uncommon.

As noted above, geodetic data describe relative movements of particular points on the Earth's surface. Therefore, a modeling step is required to translate the geodetic data to regional deformation and consequently fault slip rates. The spacing between locations where geodetic measurements are made can exceed the distance between fault sources within a fault network. In such situations, the modeling step needs to distribute observed deformation across multiple faults onto the individual faults. While geodetic data are precise in defining the movement of individual points on the Earth, the data also have relatively low resolution in comparison with geological approaches that focus upon the specific location of faults. Therefore, it is useful to combine geological and geodetic data to conduct a joint inversion.

Several approaches have been proposed for translating the observed regional deformation into concentrated deformation on particular fault structures and distributed deformation away from faults. The US National Seismic Hazard Mapping Project documentation (Petersen et al., 2014) discusses the combined use of geological and geodetic data to constrain activity rates, and makes comparisons among different methods. Another summary of the most common approaches is provided in Parsons et al. (2013), along with a discussion of their relative performance when applied to Californian data.

3.5 Magnitude-Frequency Distributions

The combination of seismicity data and geological constraints discussed in Sections 3.3 and 3.4 constrains rates of occurrence of ruptures over a broad magnitude range. However, an ERF provides rates of occurrence for all possible events that could occur, and many such events will not have empirical constraints from seismicity or geological data. A *magnitude-frequency distribution* is a mathematical model that describes the relative likelihoods of all events that can be generated by a particular source or region. Both approaches to determining rupture rates outlined in Section 3.2 make use of these magnitude-frequency distributions.

The present section presents the two most common magnitude-frequency distributions: the Gutenberg–Richter (GR) distribution; and the characteristic earthquake distribution. Brief sentiments on other distributions are also provided.

3.5.1 Gutenberg–Richter Distribution

Following an analysis of global and regional earthquake statistics, Gutenberg and Richter (1944) proposed a linear relationship between magnitude and the logarithm of the number of earthquakes *of that given magnitude*. In subsequent publications, the same log-linear relationship has been adopted to represent the total number of events per year with a magnitude *greater than or equal to a particular value*:

$$\log_{10} N(M \geq m) = a - bm. \tag{3.18}$$

In this formulation, the total number of earthquakes per year having $M \geq 0$ is $N(M \geq 0) = 10^a$, and is referred to as the *activity rate*. The parameter b is known as the Gutenberg–Richter b-value, and Gutenberg and Richter (1944) found this b-value to be close to unity in their original study.[20] Over the decades since their original contribution, many more analyses have confirmed that it is rare for this b-value to differ much from unity.

Equation 3.18 is not directly used in practice for two main reasons. First, the expression is unbounded and implies a finite rate of occurrence of earthquakes of any magnitude, including physically impossible events. Second, it is possible to define the total activity rate only for magnitudes that are at least as large as the level of completeness for a given seismicity catalog (see Section 3.3.5). The modern equivalent of the Gutenberg–Richter distribution is the doubly bounded exponential distribution (Cornell and Vanmarcke, 1969) whose PDF is presented as

$$f_M(m) = \frac{\beta \exp\left[-\beta\left(m - m_{\min}\right)\right]}{1 - \exp\left[-\beta\left(m_{\max} - m_{\min}\right)\right]}, \qquad m_{\min} \leq m \leq m_{\max} \tag{3.19}$$

where β, $m_{\min}$, and $m_{\max}$ are parameters defined below. This expression is valid for $m_{\min} \leq m \leq m_{\max}$, with $f_M(m) = 0$ for magnitudes outside this range. Equation 3.19 starts with an unbounded exponential distribution $f_M(m) = \beta \exp\left(-\beta m\right)$ and then truncates the distribution for $m < m_{\min}$ and $m > m_{\max}$.

For seismicity analyses, the value of $m_{\min}$ is linked to the magnitude of completeness of the seismicity catalog. For forward application within PSHA, $m_{\min}$ reflects the minimum magnitude deemed to be of relevance for the particular engineering application (see Section 6.5.2). The upper bounding value $m_{\max}$ represents the largest magnitude deemed physically plausible for the source or region under consideration. Approaches for estimating $m_{\max}$ from a statistical perspective are introduced in Section 3.5.4. For fixed values of $m_{\min}$ and $m_{\max}$, an estimate of the parameter β as well as the activity rate $\lambda(M \geq m_{\min})$ can be obtained following the procedure outlined in Section 3.5.9.

While Equation 3.19 no longer contains the parameter b, it is related directly to the parameter β via $\beta = \ln(10) \times b$. Given that usually $b \approx 1.0$, one should expect that $\beta \approx 2.3$. It is common to refer to the b-value and talk of the Gutenberg–Richter distribution when working with the doubly bounded exponential.

---

[20] When combining relatively well-constrained counts of large global events with smaller Californian events they obtained $\hat{b} = 0.97 \pm 0.15$.

The CDF (see Equation A.22) of the doubly bounded exponential distribution is obtained from integrating the expression in Equation 3.19:

$$F_M(m) = \frac{1 - \exp\left[-\beta\,(m - m_{\min})\right]}{1 - \exp\left[-\beta\,(m_{\max} - m_{\min})\right]}, \qquad m_{\min} \le m \le m_{\max}. \tag{3.20}$$

Note that $F_M(m) = 0$ for $m < m_{\min}$ and $F_M(m) = 1$ for $m > m_{\max}$.

3.5.2 Seismic-Moment Release Rates

A magnitude-frequency distribution, such as the doubly bounded exponential distribution, defines the rate of occurrence of events of various magnitudes. However, as moment magnitudes also represent values of seismic moment, knowing the rate of occurrence of events of a given magnitude also enables the determination of rates of occurrence of coseismic moment release. To do this, we can express seismic moment as a function of the moment magnitude (Equation 2.5) as:

$$M_0(m) = 10^{cm+d} \equiv e^{\gamma m + \delta} \tag{3.21}$$

where $c = 1.5$ and $d = 9.0$ when seismic moment is computed in units of Nm, and the parameters in the equivalent exponential version are related to c and d via $\gamma = c\ln(10)$ and $\delta = d\ln(10)$. As a result of this one-to-one mapping, rates of occurrence of magnitude are directly equivalent to rates of occurrence of the corresponding seismic moment values. Noting that $\lambda(M = m) = \lambda(M \ge m_{\min})f_M(m)$, an expression for the total seismic moment release rate (or "moment rate," for short) from a source can then be written as

$$\begin{aligned}
\dot{M}_0 &= \int_{m_{\min}}^{m_{\max}} M_0(m)\lambda(M = m)\,dm \\
&= \lambda\,(M \ge m_{\min}) \int_{m_{\min}}^{m_{\max}} M_0(m)f_M(m)\,dm.
\end{aligned} \tag{3.22}$$

The exponential nature of Equation 3.21 dictates that this integral, and hence $\dot{M}_0$, is heavily influenced by the occurrence rates of the largest events.

For the case of the doubly bounded exponential (Gutenberg–Richter) distribution, Equation 3.19 is substituted into Equation 3.22 and a closed form solution is obtained as:

$$\dot{M}_0 = \lambda\,(M \ge m_{\min}) \left(\frac{\beta}{\gamma - \beta}\right) \frac{M_0\,(m_{\max})\,e^{-\beta(m_{\max}-m_{\min})} - M_0\,(m_{\min})}{1 - e^{-\beta(m_{\max}-m_{\min})}}. \tag{3.23}$$

It is wise to compute the seismic moment release rate implied by any magnitude-frequency distribution. In many cases, an individual source's distribution will involve extrapolation (if no empirical constraint upon the largest events exists) or interpolation (if the empirical data are sparse between the seismicity and geological data). Therefore, it is important to ensure that the combined moment release rate from the individual sources is consistent with expectations of release rates from the overall study region. As noted in Section 3.4.3 the overall moment release rates for a spatial region can be estimated independently by looking at geodetic deformations and information from long-term plate motion models (DeMets et al., 1990; Holt and Haines, 1995; Bird, 2003).

The moment release from the overall SSM is defined using the following equation, in which the subscript i reflects each source, and the subscript T signifies the total release rate over all $n_{sources}$ seismic sources:

$$\dot{M}_{0,T} = \sum_{i=1}^{n_{sources}} \dot{M}_{0,i}. \tag{3.24}$$

$\dot{M}_{0,T}$ can alternatively be obtained from the overall magnitude-frequency distribution for the study region. The total regional rate of exceedance of the minimum magnitude, $\lambda(M \geq m_{min})_T$, is the sum of the rates for the individual sources (fault sources, area sources, and the background source):

$$\lambda(M \geq m_{min})_T = \sum_{i=1}^{n_{sources}} \lambda(M \geq m_{min})_i. \tag{3.25}$$

The PDF for the region is the weighted sum of the PDFs for the individual sources:

$$f_M(m)_T = \sum_{i=1}^{n_{sources}} \frac{\lambda(M \geq m_{min})_i}{\lambda(M \geq m_{min})_T} f_M(m)_i. \tag{3.26}$$

With these components defined, the total moment release rate is evaluated using the following equation:

$$\dot{M}_{0,T} = \lambda(M \geq m_{min})_T \int_{m_{min}}^{m_{max}} M_0(m) f_M(m)_T \, dm. \tag{3.27}$$

These rates will match externally constrained moment rate estimates only when all sources of moment release are appropriately considered. Specifically, aseismic moment release contributions and background sources must be included.

When the individual $f_M(m)_i$ are well-described by parametric distributions, such as the GR distribution, this does not mean that $f_M(m)_T$ will similarly be described by the same, or any other, parametric distribution. When the magnitude-frequency distributions for the individual sources are defined with confidence, the magnitude-frequency distribution for the background source may need to be back-calculated to preserve the observed magnitude-frequency distribution for the overall region. This back-calculation can be achieved using the following equation, in which the subscript B denotes contributions from the background source:

$$\lambda(M \geq m_{min})_B f_M(m)_B = \lambda(M \geq m_{min})_T f_M(m)_T - \sum_{\forall i \neq B}^{n_{sources}} \lambda(M \geq m_{min})_i f_M(m)_i. \tag{3.28}$$

For most locations, the uncertainty associated with which, if any, parametric distribution is most appropriate for each source will be quite large. Rather than a formal requirement, Equation 3.28 is a *conceptual* reminder that the $f_M(m)_i$ of the individual sources should amalgamate to form a $f_M(m)_T$ that is consistent with the regional empirical data.

3.5.3 Implied Fault Slip Rates

A similar approach to Section 3.5.2 can be applied to determine the slip rate, $\dot{u}$, implied by a magnitude-frequency distribution for a fault source, given a link between average per-event slip and moment magnitude, $\bar{u}(m)$. Several authors present expressions to estimate $\bar{u}(m)$ (e.g., Wells and Coppersmith, 1994; Leonard, 2014). These expressions can be written as:

$$\bar{u}(m) = 10^{a_1 m + a_2} \equiv e^{\alpha_1 m + \alpha_2} \tag{3.29}$$

where the values of the parameters $\alpha_1 = a_1 \ln(10)$ and $\alpha_2 = a_2 \ln(10)$ vary with the style-of-faulting. As the form of Equation 3.29 is identical to Equation 3.21, it is tempting to make a similar substitution as in Section 3.5.2 and obtain a closed-form solution for the average slip rate for the special case of a GR distribution. However, two issues complicate this matter:

1. Equation 3.29 reflects only the median slip for a given event. Models usually provide a positively skewed distribution of slip values for a given magnitude, and this should be accounted for to avoid underestimating the slip rate.
2. For small magnitude events that occur on a given fault source, only a small portion of the fault source undergoes slip, and most events will not produce any surface evidence.

The second point is particularly important when comparing with estimates of fault slip-rate from paleoseismic investigations. Paleoseismic estimates are usually for a point on the fault but are assumed to reflect the slip rate over the entire fault. Estimates of the slip rate implied by a magnitude-frequency distribution should also reflect slip over the entire fault. One direct way of computing the slip rate is to recognize the relation between seismic moment release rate and slip rate (Youngs and Coppersmith, 1985):

$$\dot{M}_0 = \mu A_f \dot{\bar{u}} \quad \rightarrow \quad \dot{\bar{u}} = \frac{\dot{M}_0}{\mu A_f} \tag{3.30}$$

where here A_f represents the area of the entire fault. The other way that is more analogous with the approach of Section 3.5.2 is to weight slip contributions by the relative area of the fault associated with each slip event. The slip rate implied by a magnitude-frequency distribution that accounts for both this relative rupture area and the variability in the displacements for a given magnitude event is given by

$$\dot{\bar{u}} = \lambda \, (M \geq m_{\min}) \int_\varepsilon \int_{m_{\min}}^{m_{\max}} \frac{A(m)}{A(m_{\max})} \hat{\bar{u}}(m) e^{\varepsilon \sigma_{\ln \bar{u}}} f_M(m) f_E(\varepsilon) dm d\varepsilon. \tag{3.31}$$

Here, the average displacement is assumed to follow a lognormal distribution and ε is a standard normal variate such that the density $f_E(\varepsilon)$ is equivalent to a standard normal density function $\phi(\varepsilon)$. The term $A(m)$ is the estimate of the rupture area from a source scaling relation, while the median average displacement is denoted by $\hat{\bar{u}}(m)$ for a given magnitude. The distribution of average displacement is therefore $\ln \bar{u} \sim \mathcal{N}(\ln \hat{\bar{u}}, \sigma_{\ln \bar{u}}^2)$, or $\ln \bar{u} \sim \mathcal{N}(\alpha_1 m + \alpha_2, \sigma_{\ln \bar{u}}^2)$.

These approaches, where slip rates are determined to be consistent with the magnitude–frequency distribution for a source, are sometimes referred to as *moment balancing* approaches. However, this expression usually relates to mapping external estimates of slip rate to seismic moment release using Equation 3.30. That approach constrains the overall energy released from a fault, but does not constrain how the deformation is accommodated over all ruptures arising over the fault.

To demonstrate the impact of the variability in the displacement values, consider a strike-slip fault that is assumed to have a Gutenberg–Richter magnitude-frequency distribution with a b-value of 1.0, bounding magnitude values of $m_{\min} = 5.0$ and $m_{\max} = 7.5$, and a rate of exceeding $m_{\min}$ of $\lambda(M \geq m_{\min}) = 1.0$. The corresponding parameters for the Wells and Coppersmith (1994) model are $\alpha_1 = -14.55$, $\alpha_2 = 2.07$, and $\sigma_{\ln \bar{u}} = 0.64$.[21] Using these parameters in Equation 3.31 gives a

---

[21] The original Wells and Coppersmith (1994) article uses base-10 logarithms and so presents $a_1 = -6.32$, $a_2 = 0.90$, and $\sigma_{\log_{10} \bar{u}} = 0.28$ for strike-slip events. The corresponding expression used for the rupture area is $\log_{10} A(m) = -3.42 + 0.9m$.

slip rate of $\dot{\bar{u}} = 0.013$ m/year, while ignoring the variability in the same expression gives $\dot{\bar{u}} = 0.011$ m/year. The corresponding slip rate estimated from Equation 3.30 with $\mu = 3.3 \times 10^{10}$ N/m^2 is $\dot{\bar{u}} = 0.015$ m/year.

3.5.4 Estimation of Maximum Magnitude

Section 2.7.2 discussed how source-scaling relations could constrain the maximum magnitude produce by a source. Now, we will consider statistical techniques to provide additional constraints.

Several methods have been developed to estimate the maximum magnitude for a region (or area sources). The two most common are a statistical approach driven by local data only (Kijko and Sellevoll, 1989; Kijko, 2004) and a Bayesian approach where a regional analog is updated with local data (Cornell, 1994; USNRC, 2012). These approaches are described briefly here, and additional details are provided in Appendix B.3.

Note that both the statistical and Bayesian approaches are most well-suited for assessing the maximum magnitude for regional databases representative of areal sources. The methods also apply in theory for a single fault source, but the available datasets for these cases tend to be prohibitively small. As a result, the maximum magnitudes for individual fault sources tend to be estimated using available paleoseismic information and the mapped geometry (see Section 2.7.2). In reality, the approach taken to estimate maximum magnitudes for individual fault sources is essentially an informal application of the Bayesian approach. The analyst would use source-scaling relations along with the mapped geometry to create an informative prior distribution and then adjust this based upon any available information about the history of ruptures on the fault.

Statistical Estimation of m_{max}

The procedure of Kijko (2004), to estimate the maximum magnitude from a statistical point of view, starts by taking a declustered seismicity catalog that is complete above some level m_{min} and identifying how many events, n, exist. The estimate of m_{max} is then obtained by taking the largest observed event and adding an increment related to the nature of the magnitude-frequency distribution for the region. The increment itself depends upon the value of maximum magnitude as well, and so an iterative procedure is required to define m_{max}. Kijko and Graham (1998) and Kijko (2004) provide expressions for the increment (Δm_{max}) that should be added to the maximum observed event from a source or region. The general expression is

$$m_{max} = m_{max}^{obs} + \int_{m_{min}}^{m_{max}} F_{M_n}(m)dm \equiv m_{max}^{obs} + \Delta m_{max} \qquad (3.32)$$

where m_{max}^{obs} is the largest magnitude in the catalog and $F_{M_n}(m)$ is the probability that n events would all have magnitudes below m, given an assumed magnitude-frequency distribution (see Appendix B.3).

In addition, these authors introduce approximate expressions for the uncertainty in the estimate of m_{max}. Kijko (2004) suggests that the variance of m_{max} can be estimated as

$$\text{var}(\hat{m}_{max}) = \sigma_{m_{max}^{obs}}^2 + \Delta m_{max}^2 \qquad (3.33)$$

where $\sigma_{m_{max}^{obs}}$ is the standard error in the estimate of the magnitude for the maximum observed earthquake.

As Equation 3.32 shows m_{max} appearing on the left-hand side and in the upper limit of the integral on the right-hand side, an iterative procedure is required to determine the value of m_{max}. This approach does not always converge, or converge to sensible values, as briefly explained in Appendix B.3.

Bayesian Estimation of m_{max}

In regions of low seismicity, databases will be too sparse to reliably estimate m_{max} using the statistical approach of the previous section. This is because the maximum observed magnitude is often much lower than the true maximum ($m_{max}^{obs} \ll m_{max}$), and the number of events n tends to be low. Furthermore, if there is one large observed event, the additional increment based upon the small sample will be too large and will produce an unrealistically large estimate of m_{max}.

The alternative Bayesian approach to estimating m_{max} was developed to address these problems (Cornell, 1994; USNRC, 2012). The approach relies on the ergodic assumption and regional analogs. Initially, the analyst must classify the tectonic setting for the study region and gather statistics of the largest events from similar regions throughout the world. The distribution of these m_{max} values from analogous regions worldwide is used as the prior distribution. Region-specific data are n observed magnitudes greater than some m_{min}, and are denoted $\boldsymbol{m}$ (where bold indicates a vector).

A Bayesian framework is then used to update the prior distribution using the region-specific data

$$f'_{M_{max}}(m_{max}|\boldsymbol{m}) = \frac{\mathcal{L}(\boldsymbol{m}|m_{max})\,f_{M_{max}}(m_{max})}{\int \mathcal{L}(\boldsymbol{m}|m_{max})\,f_{M_{max}}(m_{max})\,dm_{max}} \tag{3.34}$$

where $f_{M_{max}}(m_{max})$ is the prior distribution and $\mathcal{L}(\boldsymbol{m}|m_{max})$ is the likelihood of observing $\boldsymbol{m}$, given some m_{max} value. The denominator of Equation 3.34 is a normalizing constant that ensures the posterior distribution, denoted by $f'_{M_{max}}(m_{max}|\boldsymbol{m})$, is a legitimate PDF.

The form of the likelihood function depends upon the underlying magnitude-frequency distribution. However, the Bayesian approach is most commonly applied using a doubly bounded exponential distribution (i.e., Equation 3.19). In this case, the likelihood function is

$$\mathcal{L}\left(m_{max}|\boldsymbol{m},\beta\right) = \begin{cases} 0 & \text{for } m_{max} < m_{max}^{obs} \\ \left(\frac{1}{1-\exp\left[-\beta(m_{max}-m_{min})\right]}\right)^n & \text{for } m_{max} \geq m_{max}^{obs} \end{cases} \tag{3.35}$$

where β is the parameter of the GR distribution and n is the number of observed earthquakes in $\boldsymbol{m}$. Note that m_{min} in Equation 3.35 need not be the same value as used for the rest of the seismicity analysis, but the combination of m_{min} and n should be defined consistently.

Figure 3.15a shows an example of the likelihood function for the Yerba Buena Island site. The function is similar to a Heaviside step function located at m_{max}^{obs}, but has a peak just above this value. The sharpness of this peak depends upon n. As n increases, the peak narrows to reflect that we are more likely to have observed an event close to the true maximum. For the example shown in Figure 3.15, $m_{min} = 6$, and there are $n = 6$ events in the Gardner–Knopoff declustered catalog that have at least this magnitude.

Figure 3.15b shows both the prior and posterior distributions for the Yerba Buena Island site. The prior distribution is chosen to be a normal distribution with a mean of 8 and a standard deviation of 0.4 magnitude units. In practice, we should use as much information as possible to develop the

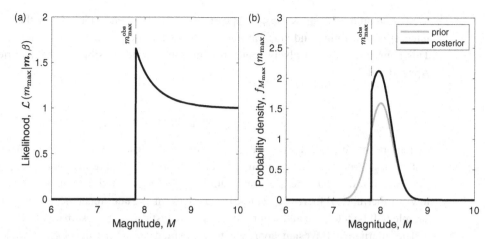

Fig. 3.15 Example application of Bayesian estimation of m_{max} for the Yerba Buena Island site. Panel (a) shows the likelihood function computed from the empirical data using Equation 3.35. Panel (b) shows a normally distributed prior with a mean of 8 and a standard deviation of 0.4, and the posterior distribution from Equation 3.34.

most suitable prior distribution. The likelihood function truncates the prior, so that the posterior distribution always exceeds the maximum observed event for the region.

In cases where we have very few events, the likelihood function effectively preserves the prior distribution for magnitudes above the maximum observed event. Thus, it is vital to use as much information as possible to construct an informative prior distribution in cases where the empirical data are limited.

Equation 3.34 produces a distribution of possible m_{max} values. This distribution reflects our inability to truly know the m_{max} for a given source. In most cases, this uncertainty will be reflected by discretizing the posterior distribution and using a logic tree (to be discussed in Section 6.7). If a single value of m_{max} is desired, the most common options are to adopt either the posterior mean or the posterior mode. The posterior distribution shown in Figure 3.15 has a mean of 8.07 and a mode of 7.95.

3.5.5 Example Gutenberg–Richter Calculations

Several example calculations and sensitivity analyses are presented in this section to illustrate the behavior of the Gutenberg–Richter (GR, or doubly bounded exponential) distribution. While the section focuses upon the Gutenberg–Richter distribution, many of the insights gained from these example calculations are relevant for other magnitude-frequency distributions.

To begin, consider a seismic source with a GR distribution describing the activity rates. To fully specify the distribution, the minimum, m_{min}, and maximum, m_{max}, magnitudes must be defined along with the b-value (β value) and the activity rate, $\lambda(M \geq m_{min})$. The corresponding values for the first example are $m_{min} = 5.0$, $m_{max} = 8.0$, $b = 1.0$ (hence $\beta = \ln(10)$), and $\lambda(M \geq m_{min}) = 0.05$. That is, on average, an event with a magnitude of at least $M = 5.0$ occurs every 20 years.

Table 3.4 provides the probabilities and rates of occurrence, and the rates of exceedance, for this particular source, using a discretization of the magnitude range with $\Delta m = 0.2$. While values are provided for $P(M = m_i)$ in this discretized example, the GR distribution is continuous and the

Table 3.4. Example calculations for a Gutenberg–Richter distribution with $\lambda(M > m_{min}) = 0.05$, $m_{min} = 5.0$, $m_{max} = 8.0$, and $b = 1.0$. The distribution is discretized using $\Delta m = 0.2$ such that $P(M = m_i)$ should be interpreted as $P(m_i - \Delta m/2 \leq M < m_i + \Delta m/2)$, and likewise for $\lambda(M = m_i)$.

i	m_i	$P(M = m_i)$	$\lambda(M = m_i)$	$\lambda(M \geq m_i - \Delta m/2)$
1	5.1	0.3694	0.01847	0.05
2	5.3	0.2331	0.01165	0.03153
3	5.5	0.1471	0.007353	0.01988
4	5.7	0.09279	0.004640	0.01252
5	5.9	0.05855	0.002927	0.007882
6	6.1	0.03694	0.001847	0.004955
7	6.3	0.02331	0.001165	0.003108
8	6.5	0.01471	0.0007353	0.001942
9	6.7	0.009279	0.0004640	0.001207
10	6.9	0.005855	0.0002927	0.0007432
11	7.1	0.003694	0.0001847	0.0004505
12	7.3	0.002331	0.0001165	0.0002657
13	7.5	0.001471	7.353e-05	0.0001492
14	7.7	0.0009279	4.640e-05	7.567e-05
15	7.9	0.0005855	2.927e-05	2.927e-05
		Sum = 1.0	Sum = 0.05	

probability of the random variable M *equaling* any particular value of $M = m_i$ is zero. The tabulated values therefore represent:

$$P(M = m_i) = P(m_i - \Delta m/2 \leq M < m_i + \Delta m/2). \tag{3.36}$$

For the rate of exceedance we make use of the expression:

$$\lambda(M \geq m) = \lambda(M \geq m_{min})G_M(m) \tag{3.37}$$

where $G_M(m) = 1 - F_M(m)$ is the CCDF obtained from Equation 3.20.

Table 3.4 shows that the probabilities of the magnitude falling within a particular bin must always sum to unity, while the rates of occurrence must always sum to the total activity rate $\lambda(M \geq m_{min})$.

While m_{min} is a modeling choice (to be discussed further in Chapter 6), the remaining parameters estimated using statistical techniques will thus be uncertain. The following examples perturb some of these parameters to illustrate the impact of their uncertainties upon subsequent seismic hazard calculations.

Variation in *b*-Value

Figure 3.16 shows the effect of varying the b-value by 10% about a nominal value of $b = 1.0$. All other parameters remain fixed. Table 3.5 tabulates the rates of exceedance for each case and presents ratios of these rates with respect to the central value.

The ratios in Table 3.5 demonstrate the extent to which the rates of exceedance can differ as the magnitude increases. Although the individual curves shown in Figure 3.16 do not appear drastically different, the tabulated values show differences in rates that approach a factor of two for the

Table 3.5. Impact of variations in the b-value on rates of exceedance using a GR distribution with $\lambda(M > m_{min}) = 0.05$, $m_{min} = 5.0$, and $m_{max} = 8.0$. The nominal b-value is $b = \mu_b = 1.0$ and the variations represent a 10% increase ($b^+ = 1.1$) and 10% decrease ($b^- = 0.91$) of this value. The distribution is discretized using $\Delta m = 0.2$ such that $\lambda(M \geq m_i)$ should be interpreted as $\lambda(M \geq m_i - \Delta m/2)$

i	m_i	$\lambda(M \geq m_i \vert \mu_b)$	$\lambda(M \geq m_i \vert b^+)$	$\lambda(M \geq m_i \vert b^-)$	$\frac{\lambda(M \geq m_i \vert b^+)}{\lambda(M \geq m_i \vert \mu_b)}$	$\frac{\lambda(M \geq m_i \vert b^-)}{\lambda(M \geq m_i \vert \mu_b)}$
1	5.1	0.05	0.05	0.05	1	1
2	5.3	0.03153	0.03012	0.03286	0.9552	1.042
3	5.5	0.01988	0.01814	0.02159	0.9126	1.086
4	5.7	0.01252	0.01092	0.01417	0.8720	1.132
5	5.9	0.007882	0.00657	0.009293	0.8334	1.179
6	6.1	0.004955	0.003949	0.006082	0.7969	1.227
7	6.3	0.003108	0.002369	0.003969	0.7623	1.277
8	6.5	0.001942	0.001418	0.002579	0.7298	1.328
9	6.7	0.001207	0.0008443	0.001665	0.6994	1.379
10	6.9	0.0007432	0.0004988	0.001063	0.6711	1.431
11	7.1	0.0004505	0.0002906	0.0006675	0.6451	1.482
12	7.3	0.0002657	0.0001651	0.0004071	0.6213	1.532
13	7.5	0.0001492	8.953e-05	0.0002357	0.6000	1.580
14	7.7	7.567e-05	4.398e-05	0.0001230	0.5812	1.625
15	7.9	2.927e-05	1.654e-05	4.88e-05	0.5649	1.667

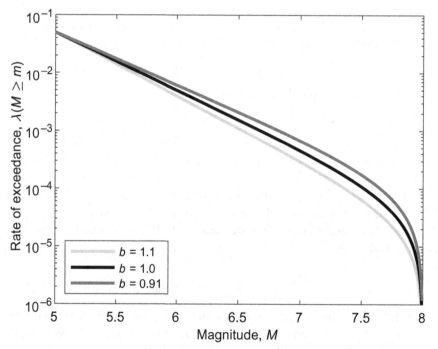

Fig. 3.16 Demonstration of the impact of 10% changes in the b-value on the rates of exceedance of magnitude. The nominal b-value is $b = 1.0$, and the other GR parameters are $m_{min} = 5.0$, $m_{max} = 8.0$, and $\lambda(M \geq m_{min}) = 0.05$.

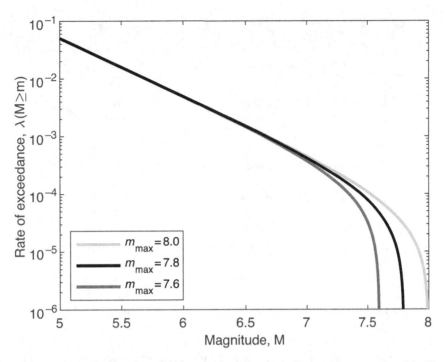

Fig. 3.17 Rates of exceedance of various magnitude values for three different values of m_{max}. The remaining parameters of the GR distribution are fixed at $b = 1.0$, $m_{min} = 5.0$, and $\lambda(M \geq m_{min}) = 0.05$.

largest events. An important implication of these differences is that the energy release rates implied by the three distributions are very different (see Section 3.5.2). Specifically, the distribution with the b-value 10% lower than the nominal value has a seismic moment release rate 46% higher. Similarly, the distribution with a b-value 10% higher releases only two-thirds of the energy of the nominal case. Therefore, relatively small numerical differences in the b-value can cause significant differences in the rates of occurrence of large earthquakes and, hence, substantial differences in the implied energy release rates for a given source.

Sensitivity to m_{max}

Figure 3.17 shows the effect of changing the value of m_{max} over three values spanning a total of 0.4 magnitude units. Table 3.6 demonstrates that changes to just m_{max} have negligible effects on the rates of exceedance of small to moderate events. However, there are significant differences in the upper tail as M approaches m_{max}.

As with the b-value, the changes in rates of exceedance due to m_{max} are most significant for large magnitudes, and these events contribute significantly to energy release. In the particular case shown here, moving m_{max} from the central value of 7.8 up to $m_{max} = 8.0$ increases the overall seismic moment release rate by 27%. Similarly, dropping to $m_{max} = 7.6$ reduces the seismic moment release 21%. Therefore, changes in m_{max} of just 0.2 magnitude units imply >20% differences in the energy released by the source, and the impact is not symmetric.

Changing m_{max} in isolation simply permits the source to generate larger events. It does not, however, change the rates of occurrence of smaller, more frequent earthquakes. The reason why this is important will become more apparent in Chapters 6 and 7. However, there are many cases

Table 3.6. Impact of variations in m_{max} on rates of exceedance using a GR distribution with $\lambda (M > m_{min}) = 0.05$, $m_{min} = 5.0$, and $b = 1.0$. The distribution is discretized using $\Delta m = 0.2$ such that $\lambda (M \geq m_i) \equiv \lambda (M \geq m_i - \Delta m/2)$

i	m_i	$\lambda (M \geq m_i \vert m_{max} = 7.6)$	$\lambda (M \geq m_i \vert m_{max} = 7.8)$	$\lambda (M \geq m_i \vert m_{max} = 8.0)$
1	5.1	0.05	0.05	0.05
2	5.3	0.0315	0.03152	0.03153
3	5.5	0.01983	0.01986	0.01988
4	5.7	0.01247	0.0125	0.01252
5	5.9	0.007819	0.007858	0.007882
6	6.1	0.004887	0.004929	0.004955
7	6.3	0.003037	0.003080	0.003108
8	6.5	0.00187	0.001914	0.001942
9	6.7	0.001133	0.001179	0.001207
10	6.9	0.0006685	0.0007143	0.0007432
11	7.1	0.0003753	0.0004214	0.0004505
12	7.3	0.0001904	0.0002366	0.0002657
13	7.5	7.364e-05	0.0001200	0.0001492
14	7.7	0	4.642e-05	7.567e-05
15	7.9	0	0	2.927e-05

when the small to moderate events dominate hazard estimates due to their higher rates of occurrence. Particularly in regions of low seismicity, it is not uncommon for hazard results to be relatively insensitive to m_{max} for relatively short return periods.

Constrained Seismic Moment Release Rate

Figure 3.18 shows what happens when the same three values of m_{max} from the previous example are considered, but when the activity rates $\lambda (M \geq m_{min})$ are adjusted to ensure that all three distributions have the same seismic moment release rate, $\dot{M}_0$ (Equation 3.22). To conserve the moment release rate, the rates of occurrence of the small to moderate earthquakes must be increased when the m_{max} value is decreased, and vice versa. The numerical differences in these rates can be seen for a range of magnitude values in Table 3.7.

The implications of changing m_{max} may be counterintuitive in this case. In the previous section, changing m_{max} in isolation had little or no effect upon the rates of occurrence of small to moderate magnitude earthquakes. However, when the moment release rate for the source is conserved, increasing the value of m_{max} requires a reduction in the rates of occurrence for these small to moderate events. Figure 3.18 shows this, and illustrates why it is a mistake to suppose that adopting a large value of m_{max} is conservative. In regions of relatively high seismicity, and for analyses where relatively short return periods are of interest, it is often the case that attempting to be conservative by using a large m_{max} value has the opposite effect and leads to lower hazard estimates. The lower hazard arises from the reduction in the rates of occurrence of the smaller events that dominate the hazard. This issue is revisited in Chapters 6 and 7.

3.5.6 Characteristic Earthquake Distribution

When performing seismicity analyses, there are many situations where one observes departures from the scaling implied by the GR distribution, particularly at large magnitudes. In some cases, this

Table 3.7. Impact of variations in m_{max} on rates of exceedance using a Gutenberg–Richter distribution with $m_{min} = 5.0$, $b = 1.0$, and $\lambda(M > m_{min})$ constrained to ensure equal seismic moment release rates. The distribution is discretized using $\Delta m = 0.2$ such that $\lambda(M \geq m_i) \equiv \lambda(M \geq m_i - \Delta m/2)$

i	m_i	$\lambda(M \geq m_i \mid m_{max} = 7.6)$	$\lambda(M \geq m_i \mid m_{max} = 7.8)$	$\lambda(M \geq m_i \mid m_{max} = 8.0)$
1	5.1	0.06357	0.05	0.0394
2	5.3	0.04005	0.03152	0.02485
3	5.5	0.02521	0.01986	0.01566
4	5.7	0.01585	0.01250	0.009868
5	5.9	0.00994	0.007858	0.006212
6	6.1	0.006213	0.004929	0.003905
7	6.3	0.003861	0.003080	0.002449
8	6.5	0.002377	0.001914	0.001531
9	6.7	0.001441	0.001179	0.0009513
10	6.9	0.0008500	0.0007143	0.0005857
11	7.1	0.0004772	0.0004214	0.0003550
12	7.3	0.0002420	0.0002366	0.0002094
13	7.5	9.363e-05	0.0001200	0.0001176
14	7.7	0	4.642e-05	5.963e-05
15	7.9	0	0	2.307e-05

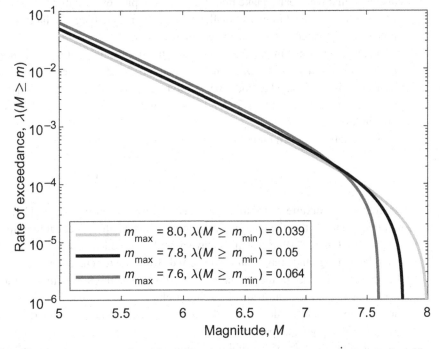

Fig. 3.18   Influence of changes in m_{max} upon rates of exceedance when the seismic moment release rate, $\dot{M}_0$, is constrained. All distributions share the same $m_{min} = 5.0$ and $b = 1.0$ values, but have $\lambda(M \geq m_{min})$ values set to ensure equal $\dot{M}_0$.

departure can reflect poorly calibrated historical data or misinterpreted paleoseismic information. For example, imagine that an event has a true recurrence interval of 10,000 years, but the period of observation for a catalog is shorter, at just 100 years. If this event happened to occur during our period of observation, then a naïve empirical estimate of 1/100 for the rate of occurrence is far too high.

But if no evidence can be found for other prior events, it is challenging to obtain a more realistic rate estimate. The resulting overestimated rate will cause an apparent departure from the doubly bounded exponential distribution for larger-magnitude events.

However, there are also legitimate physical reasons why specific segments of a fault may rupture preferentially in certain sized events. So, there is also empirical evidence for these so-called characteristic events. Schwartz and Coppersmith (1984) recognized consistent observational departures from the recurrence rates implied by the GR distribution at large magnitudes. They also hypothesized that one event's occurrence could result in the heterogeneous stress distribution over the ruptured area becoming more uniform compared with other portions of a fault. Therefore, when the next event is triggered, it is more likely that the same region will rupture again, compounding the difference in stress and frictional characteristics between the ruptured and non-ruptured portions of the fault. This process, of gradual smoothing results in regions of a fault that tend to rupture in a repeated manner.

The concept of *characteristic earthquakes* that was proposed by Schwartz and Coppersmith (1984) triggered a significant amount of research, with evidence for and against characteristic fault behavior presented (e.g., Wesnousky, 1994; Stirling et al., 1996; Parsons and Geist, 2009; Parsons et al., 2012; Page and Felzer, 2015; Field et al., 2017a; Stirling and Gerstenberger, 2018; Wyss, 2020). However, testing the hypothesis of characteristic events is complicated by the fact that there are relatively few faults where sufficient events have occurred to obtain robust statistical results. Wesnousky (1994) concluded that when earthquake faults appear to display exponential behavior, this may be the result of the composite effect of many smaller rupture segments along this fault. Individual fault segments in the study appeared to have characteristic behavior.

Youngs and Coppersmith (1985) discussed the implications of using a *characteristic earthquake distribution* within seismic hazard analyses and presented a mathematical model for the magnitude-frequency distribution. The model has a mixture of a doubly bounded exponential distribution for the small to moderate earthquakes, and a uniform distribution for the "characteristic events" at large magnitudes.

We can define a general mixture distribution in terms of its PDF as

$$f_X^{\text{mix}}(x) = w f_X^{(1)}(x) + (1 - w) f_X^{(2)}(x) \tag{3.38}$$

where the two mixture distributions are denoted by $f_X^{(i)}(x)$ where $i \in \{1, 2\}$, and the weight, w, is required to ensure that the integral of the mixture distribution is equal to unity.

Consider Figure 3.19 in which we have a doubly bounded exponential distribution over the small to moderate magnitude range, and a uniform distribution for characteristic events. Suppose we define the upper magnitude for the exponential distribution as the lower limit of the characteristic events. In that case, the probability of observing a characteristic event is $p_{\text{char}} = \Delta m_{\text{char}} f_c$, where f_c is the uniform probability density for the characteristic scenarios. As there is no overlap between the uniform and exponential parts we have $p_{\text{char}} = w$.

Combining these two distributions using Equation 3.38 results in the following PDF for the mixture distribution (also shown in Figure 3.19):

$$f_M(m) = \begin{cases} (1 - \Delta m_{\text{char}} f_c) \dfrac{\beta \exp[-\beta(m - m_{\min})]}{1 - \exp[-\beta(m_{\text{char}}^{LO} - m_{\min})]} & \text{for } m_{\min} \leq M < m_{\text{char}}^{LO} \\ f_c & \text{for } m_{\text{char}}^{LO} \leq M \leq m_{\text{char}}^{HI} \end{cases} \tag{3.39}$$

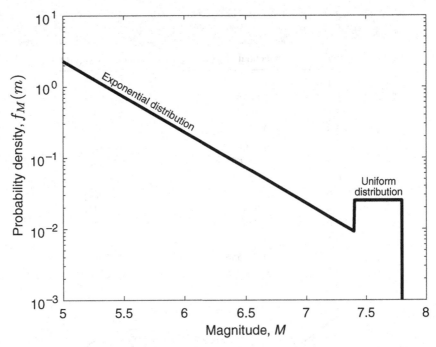

Fig. 3.19 PDF of the characteristic earthquake model with $m_{\min} = 5.0$, $b = 1.0$, $m_{\mathrm{char}} = 7.6$, $\Delta m_{\mathrm{char}} = 0.4$, and $f_c = 0.025$.

where $f_M(m)$ is defined for $m_{\min} \leq m \leq m_{\mathrm{char}}^{HI}$, and is zero otherwise. Here, we have introduced $m_{\mathrm{char}}^{LO} = m_{\mathrm{char}} - \frac{\Delta m_{\mathrm{char}}}{2}$ and $m_{\mathrm{char}}^{HI} = m_{\mathrm{char}} + \frac{\Delta m_{\mathrm{char}}}{2}$ to simplify the expressions.

The associated CDF is:

$$
F_M(m) = \begin{cases}
(1 - \Delta m_{\mathrm{char}} f_c) \dfrac{1 - \exp\left[-\beta(m - m_{\min})\right]}{1 - \exp\left[-\beta\left(m_{\mathrm{char}}^{LO} - m_{\min}\right)\right]} & \text{for } m_{\min} \leq M < m_{\mathrm{char}}^{LO} \\[4mm]
(1 - \Delta m_{\mathrm{char}} f_c) + f_c\left(m - m_{\mathrm{char}}^{LO}\right) & \text{for } m_{\mathrm{char}}^{LO} \leq M \leq m_{\mathrm{char}}^{HI}
\end{cases}
\tag{3.40}
$$

with $F_M(m) = 0$ for $M < m_{\min}$, and $F_M(m) = 1$ for $M > m_{\mathrm{char}}^{HI}$.

Figure 3.20 shows the probabilities of exceedance implied by this characteristic magnitude–frequency distribution. For comparison, a GR distribution with the same b-value and overall magnitude range is also shown. Even though the probability of observing a characteristic event, in this case, is just $p_{\mathrm{char}} = 0.01$, the departure from the GR distribution is significant for the largest events. To quantify this departure, Table 3.8 presents the probabilities of exceedance for each distribution and ratio of their values. The characteristic distribution specifies probabilities (and hence rates) of exceedance of the largest events that are more than five times higher than the GR distribution. These differences at large magnitudes have significant implications for the seismic moment release rate, as shown in Figure 3.20b.

The way the characteristic earthquakes are represented varies. In some cases, a source can be regarded as being purely characteristic, and only events of a certain magnitude (or magnitude range) are generated (discussed next in Section 3.5.7). In other cases, the mixture distribution of Equation 3.39 is used. While the expressions in this section have used a uniform distribution

Table 3.8. Difference in exceedance probabilities between the Gutenberg–Richter and characteristic distributions with $m_{min} = 5.0$, $b = 1.0$, $m_{char} = 7.6$, $\Delta m_{char} = 0.4$, and $f_c = 0.025$

i	m_i	Exponential $G_M^E(m_i)$	Characteristic $G_M^C(m_i)$	Ratio $G_M^C(m_i)/G_M^E(m_i)$
1	5.1	1	1	1
2	5.3	0.6066	0.6096	1.005
3	5.5	0.3677	0.3725	1.013
4	5.8	0.2226	0.2286	1.027
5	6.0	0.1346	0.1412	1.049
6	6.2	0.08108	0.08808	1.086
7	6.4	0.04861	0.05586	1.149
8	6.6	0.02889	0.03629	1.256
9	6.8	0.01692	0.02441	1.443
10	7.0	0.009651	0.01720	1.782
11	7.3	0.005236	0.01281	2.447
12	7.5	0.002556	0.01015	3.973
13	7.7	0.0009285	0.00500	5.385

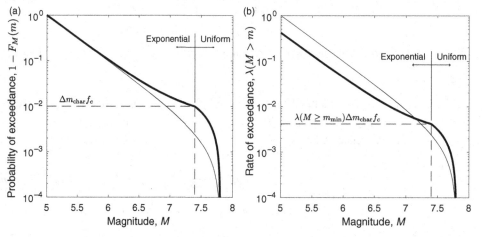

Fig. 3.20 Panel (a) shows the CCDF of the characteristic distribution with $m_{min} = 5.0$, $b = 1.0$, $m_{char} = 7.6$, $\Delta m_{char} = 0.4$, and $f_c = 0.025$. For comparison, the thin line shows a GR distribution with the same b and m_{min} values, and an $m_{max} = m_{char} + \Delta m_{char}/2$. Panel (b) shows the rate of exceedance for these distributions with $\lambda(M \geq m_{min})$ set to ensure the same moment release rate for both cases.

for the characteristic events, other distributions can be used. A truncated normal distribution for characteristic events is also common (e.g., Bommer and Stafford, 2016).

3.5.7 Maximum Magnitude Model

Section 3.5.6 presented a magnitude-frequency distribution with a mixture of exponential and characteristic components. In some cases, analysts prefer not to include the exponential component and opt for only a characteristic component. Such representations are referred to as either pure characteristic distributions or a *maximum magnitude model*.

The maximum magnitude model arises when analysts can constrain recurrence intervals for large characteristic events on a fault, but cannot adequately constrain rates of occurrence for smaller events (e.g., Stirling et al., 2012). The constraint upon these large events typically comes from geological investigations (which, as noted previously, cannot also resolve smaller events). Smaller events seen in seismicity analyses are then attributed to an area source or the background seismicity.

The parameterization of the maximum magnitude model is very flexible. In the most extreme case, a delta function could represent the occurrence of an event with a single specific magnitude. More commonly, some uncertainty in the event size will be reflected through the use of a distribution, such as the uniform distribution presented in the previous section, or a truncated normal distribution. Regardless of the parameterization, the maximum magnitude model will have nonzero contributions around the characteristic magnitude, and zero probability density for smaller magnitudes.

3.5.8 Alternative Magnitude-Frequency Distributions

While the GR (doubly bounded exponential) and characteristic magnitude-frequency distributions are by far the most common, many other alternatives have been proposed. A summary of alternative distributions has been provided by Utsu (1999, 2002). The main differences among the proposals relate to the behavior of the distributions at large magnitudes. Given that this is the most incomplete portion of the seismicity databases, it is often difficult to distinguish between physically meaningful departures and statistically implied departures arising from an incomplete empirical record.

The alternative models discussed by Utsu (1999) are primarily parametric magnitude-frequency distributions. Parametric models are convenient, but there is no requirement to adopt a parametric form. Ultimately, we require a specification of the rate of occurrence for each rupture scenario in the source characterization model, and the magnitude-frequency distribution can adopt complicated non-parametric forms (e.g. Field et al., 2017a).

3.5.9 Fitting Magnitude-Frequency Distributions

Once a form for the magnitude-frequency distribution has been identified, it is necessary to determine the parameters of this distribution using available data. As magnitude-frequency distributions are commonly defined in terms of rates of exceedance, exceedances rates for multiple magnitude values will share some observed data and thus are not independent. This dependency means that parameters of magnitude-frequency distributions cannot be appropriately estimated using least-squares fitting or optimization methods using an L2 norm. For this, and other reasons not mentioned here, maximum likelihood estimation (MLE) methods are employed (Myung, 2003).

Several approaches for determining magnitude-frequency distributions parameters have been proposed, accounting for effects such as

- Magnitude uncertainties (Tinti and Mulargia, 1985; Rhoades, 1996)
- Discretization of the magnitude range (Weichert, 1980; Bender, 1983)
- Uneven observation intervals for different completeness levels (Weichert, 1980)
- Prior information regarding likely b-values (Veneziano and Van Dyck, 1985)

The most commonly adopted methods are those of Weichert (1980), Veneziano and Van Dyck (1985), and Johnston et al. (1994). The latter is effectively a Bayesian extension of Weichert (1980).

Table 3.9. Event counts and rates for the combined NCEDC/UCERF3 seismicity data using magnitude bins of 0.1 units. Every second entry is shown to conserve space

i	m_i	$n(m_i)$	$n_{GK}(m_i)$	$t_{obs,i}$	$\lambda(M = m_i)$	$\lambda_{GK}(M = m_i)$	$\lambda(M \geq m_i)$	$\lambda_{GK}(M \geq m_i)$
1	2.4	1211	311	16.38	73.92	18.98	330.1	106.9
3	2.6	806	259	19.38	41.58	13.36	198.2	73.13
5	2.8	538	175	19.38	27.76	9.029	124.2	50.28
7	3.0	370	130	19.38	19.09	6.707	72.13	32.17
9	3.2	149	67	19.38	7.687	3.457	41.75	20.20
11	3.4	95	49	19.38	4.901	2.528	26.89	13.39
13	3.6	44	23	19.38	2.270	1.187	18.32	8.743
15	3.8	40	25	19.38	2.064	1.290	13.68	6.163
17	4.0	137	39	49.38	2.774	0.7898	10.22	4.099
19	4.2	112	42	74.38	1.506	0.5647	5.848	2.704
21	4.4	62	23	74.38	0.8335	0.3092	3.159	1.696
23	4.6	34	20	84.38	0.4029	0.2370	1.698	1.008
25	4.8	11	5	84.38	0.1304	0.05925	0.9509	0.5929
27	5.0	4	4	84.38	0.0474	0.04740	0.7021	0.4507
29	5.2	4	3	84.38	0.0474	0.03555	0.5599	0.3559
31	5.4	5	2	121.38	0.04119	0.01648	0.4630	0.2792
33	5.6	6	1	146.38	0.04099	0.006831	0.3533	0.2018
35	5.8	13	5	161.38	0.08055	0.03098	0.2804	0.1758
37	6.0	6	5	166.38	0.03606	0.03005	0.1503	0.1262
39	6.2	3	3	166.38	0.01803	0.01803	0.09015	0.08414
41	6.4	3	2	166.38	0.01803	0.01202	0.05409	0.04808
43	6.6	1	1	166.38	0.00601	0.00601	0.03005	0.03005
45	6.8	0	0	166.38	0	0	0.01803	0.01803
47	7.0	2	2	166.38	0.01202	0.01202	0.01803	0.01803
49	7.2	0	0	166.38	0	0	0.00601	0.00601
51	7.4	0	0	166.38	0	0	0.00601	0.00601
53	7.6	0	0	166.38	0	0	0.00601	0.00601
55	7.8	1	1	166.38	0.00601	0.00601	0.00601	0.00601

Figure 3.21 shows the empirical rates of occurrence for both the raw and declustered seismicity catalogs discussed in Section 3.3.5 (specifically Figure 3.10). The fitted models shown in Figure 3.21 use the approach of Johnston et al. (1994) in Appendix B. Table 3.9 provides numerical values for the data and fitted models.

In Figure 3.22, the corresponding rates of exceedance are shown. This figure suggests some deviations from the expected log-linearity for magnitudes around 5.0. These deviations may be due to departures of the data from the long-term seismicity rates, incorrect estimates of completeness levels, inappropriate declustering, or the Gutenberg–Richter distribution being inappropriate for this case. However, it is always challenging to reconcile the relative contribution of each possible source.

It is also worth noting, from both Figures 3.21 and 3.22, that the b-value for the raw catalog (including aftershocks) is very close to unity, and that the value drops slightly after declustering. These numerical results are consistent with those obtained from many studies all over the world. Apparent significant departures from $b = 1$ should be treated with great caution.

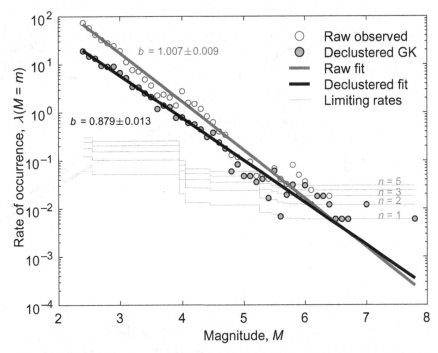

Fig. 3.21 Rates of occurrence computed from the combined NCEDC/UCERF3 data within 200 km of Yerba Buena island. The declustered catalogue is obtained using the Gardner and Knopoff (1974) (GK) algorithm, and the fitted models using the approach of Johnston et al. (1994) (Appendix B).

Table 3.9 highlights how sparse the data are for the moderate to large events. These event counts were obtained for a relatively large region in an area of high seismicity by global standards. For many practical cases, the event counts will be far lower than those presented here. The statistical errors in the parameters will be consequently much larger than the values presented in Figure 3.22. An approach for obtaining the uncertainties in the seismicity parameters is provided in Appendix B.

3.5.10 Spatial Variation of Seismicity Parameters

The discussion thus far has assumed that we can identify seismic source zones a priori and then derive the parameters of source-specific magnitude-frequency distributions using the seismicity data assigned to each source. However, it is often challenging to identify where the boundary between two adjacent source zones should lie.

Consider the circular region shown in Figure 3.23. The overall region's magnitude-frequency characteristics can be described using $m_{min} = 4.0$, $m_{max} = 7.0$, and $b = 1.0$. However, the activity rate, $\lambda(M \geq m_{min})$, varies systematically with position throughout the circular region. In particular, the sector at high values of the Y coordinate, identified by the straight lines in Figure 3.23, has a lower rate of exceedance $\lambda(M \geq m_{min}) = 1.0$ than the rest of the region, which has $\lambda(M \geq m_{min}) = 2.0$. This area of lower activity is named "Area 1" and accounts for one-third of the region's total area, while the remainder of the region is named "Area 2." The seismicity shown in Figure 3.23 has been simulated for 60 years with the above rates of exceedance. The magnitude of each event has been drawn from the corresponding GR distributions.

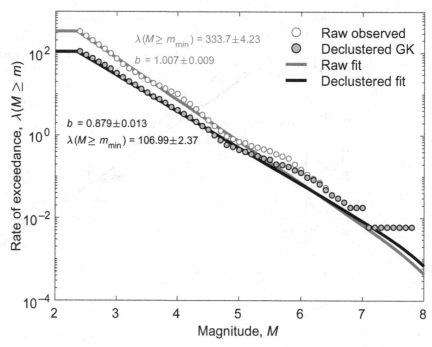

$\lambda(M \geq m_{\min}) = 333.7 \pm 4.23$
$b = 1.007 \pm 0.009$

○ Raw observed
◐ Declustered GK
— Raw fit
— Declustered fit

$b = 0.879 \pm 0.013$
$\lambda(M \geq m_{\min}) = 106.99 \pm 2.37$

Fig. 3.22 Rates of exceedance computed from the combined NCEDC/UCERF3 data within 200 km of Yerba Buena island. The declustered catalogue is obtained using the Gardner and Knopoff (1974) (GK) algorithm, and the fitted models using the approach of Johnston et al. (1994) (Appendix B).

If the lines showing the boundary between Areas 1 and 2 in Figure 3.23 were not displayed, it would be difficult to identify these two distinct areas, despite the factor of 2 difference in the underlying activity rates. For less-active regions, this problem is even more challenging as stable spatial patterns take longer to emerge.

Instead of specifying spatial boundaries between area sources a priori, an alternative approach that is often adopted is to work with a boundary-less source model. Under this approach, the historical seismicity is spatially smoothed, and activity rates are computed at a grid of locations throughout the analysis domain. An example of this smoothing is shown in Figure 3.24 using the raw seismicity shown in Figure 3.23.

At each spatial location, $\mathbf{x}_i$, the smoothed rate, $\bar{\lambda}(M \geq m_{\min}|\mathbf{x}_i)$, that is plotted in Figure 3.24 is computed as a weighted mean of the rates at surrounding locations $\mathbf{x}_j$:[22]

$$\bar{\lambda}(M \geq m_{\min}|\mathbf{x}_i) = \sum_j w_j(\mathbf{x}_j|\mathbf{x}_i)\lambda(M \geq m_{\min}|\mathbf{x}_j). \tag{3.41}$$

[22] Equation 3.41 is equivalent to smoothing the event occurrence and then computing spatially varying rates at each location. For example, a single earthquake that occurs in spatial cell i is not treated as one event in cell i and zero events in all other cells. Instead, each cell is assigned a fractional value w_j, the sum of which equals 1, and the total number of events is thus conserved

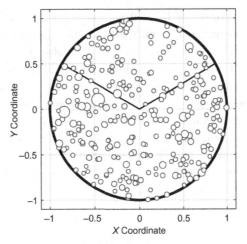

Fig. 3.23 Seismicity catalog for the activity rate example. The marker size scales with the magnitude of the event. "Area 1" is the sector identified toward the top of the figure and represents one-third of the area of the region with half of the seismicity rate.

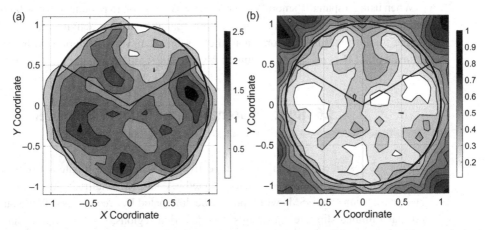

Fig. 3.24 Smoothed interpretation of the seismicity presented in Figure 3.23. Panel (a) shows spatially smoothed activity rates obtained using a Gaussian kernel. Panel (b) shows the coefficients of variation (standard errors in weighted mean rates, divided by these weighted mean rates) for the smoothed activity rates.

In Equation 3.41, the weights are defined according to an appropriate smoothing kernel. To generate Figure 3.24, the Gaussian kernel used has the form

$$w_j(\boldsymbol{x}_j|\boldsymbol{x}_i) = \frac{1}{K} \exp\left(-\frac{\|\boldsymbol{x}_j - \boldsymbol{x}_i\|^2}{2c^2}\right). \tag{3.42}$$

Here, K is a normalizing factor that ensures that the overall rate of occurrence is preserved before and after smoothing, and c is the spatial length scale for the smoothing. Other kernels, aside from the Gaussian kernel shown in Equation 3.42, have also been explored (e.g., Stock and Smith, 2002).

Naturally, the precision with which one can constrain these spatially varying activity rates will depend upon the amount of available data in the vicinity of each location. To quantify this, Figure 3.24a shows how the coefficient of variation for the weighted mean rate varies with position. This coefficient of variation is defined as the ratio of the standard error in the mean estimate and the mean estimate itself. There is no standardized definition for the standard error for a weighted mean, but here the standard error is computed using the expression from Cochran (1977), which has been shown by Gatz and Smith (1995) to perform very well. This expression is given in the following, in which $w_j \equiv w_j(x_j|x_i)$, $\bar{w} \equiv \bar{w}(x_i)$, $\lambda(x_j) \equiv \lambda(M \geq m_{\min}|x_j)$, $\bar{\lambda} \equiv \bar{\lambda}(M \geq m_{\min}|x_i)$, and n is the total number of observations:

$$\text{s.e.}\left[\bar{\lambda}(x_i)\right] = \frac{n}{(n-1)\left(\sum_j w_j\right)^2}\left\{\sum_j \left[w_j\lambda(x_j) - \bar{w}\bar{\lambda}\right]^2\right.$$

$$\left. -2\bar{\lambda}\sum_j \left(w_j - \bar{w}\right)\left[w_j\lambda(x_j) - \bar{w}\bar{\lambda}\right] + \bar{\lambda}^2\sum_j \left(w_j - \bar{w}\right)^2\right\}. \tag{3.43}$$

When using a spatially smoothed approach, it is common to pool data over a large spatial region to estimate b, $m_{\min}$, and $m_{\max}$, and then obtain spatially varying activity rates. Each effective point source within the analysis domain is then assigned a GR distribution with the location-specific estimate of $\bar{\lambda}(M \geq m_{\min}|x_i)$, a common b-value, and common values of $m_{\min}$ and $m_{\max}$.

3.6 Rupture Scenarios and Computation of Rates

Section 3.2 described two common perspectives for obtaining estimates of rupture rates. Chapter 2 also discussed cases where PSHA is conducted in data-poor, data-rich, and data-neutral regions. For data-poor regions, the SSMs will typically be dominated by area sources; fault sources, if they exist, will have limited slip rate constraints. For data-rich regions, it is common to have relatively well-defined slip rate estimates for several active faults and additional information about the timing and spatial extent of major events on these faults. Naturally, for the data-neutral cases, the analyst will have some intermediate level of information.

The perspectives in Section 3.2 relate to methods that can be adopted depending upon the type of source model and level of data-richness. The following sections explain how these perspectives might be implemented in practice.

3.6.1 Rupture Rates from Seismicity Models

This chapter's ultimate objective is to define rates of occurrence of earthquake *ruptures*. Section 3.5 focused upon defining how frequently events of different *magnitudes* occur. While this information

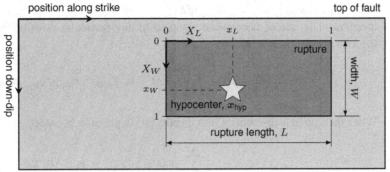

Fig. 3.25   Schematic illustration of a rupture on a simple fault. Rupture dimensions are conditioned on the magnitude assigned to the hypocentral position shown by the star.

is vital for the definition of a rupture scenario, it is only one attribute. A rupture must be defined in a manner that enables ground motions to be predicted. In Chapter 4, the most commonly used GMMs within seismic hazard studies are discussed. These represent the source of an earthquake in a relatively straightforward manner (in contrast to the more detailed descriptions in Chapter 5). For the models of Chapter 4, the definition of a rupture typically consists of the magnitude of the earthquake, the rake angle or style of faulting, and the geometric dimensions of the rupture. To generate a rupture consistent with this requirement, the typical process is as follows:

1. Consider a particular location, x, within a source and regard this a hypocenter x_{hyp}.
2. For this position select a value of earthquake magnitude, m.[23]
3. Given this magnitude, and information about the types of earthquakes the source can produce, compute rupture dimensions.
4. Position this rupture in a manner that satisfies the geometric constraints of the source.

This process is illustrated schematically in Figure 3.25 for a simple fault source. The dark gray rectangle shows the portion of the fault that ruptures in this particular event. The relative position of the hypocenter within the rupture is defined using the variables x_L and x_W which define the hypocentral coordinates in terms of relative distances in the along-strike and down-dip directions.

To determine the rate of occurrence for the rupture shown in Figure 3.25, it is first necessary to define how likely an event is to occur with the hypocentral position shown. The probability of a hypocenter at this position is defined as $P\left(X_{\text{hyp}} = x_{\text{hyp}}\right)$. In reality, this probability reflects the chance that a hypocenter lies in some discrete volume or area (depending upon the representation of the source: volume for area sources, area for fault sources). We can represent this asymptotically

---

[23] In most studies, the position will not influence the magnitude-frequency distribution. However, there are cases where the magnitude-frequency distribution can vary with depth within a source. This arises as very large events are less likely to nucleate at shallow depths than smaller events (Mai et al., 2005)

as $P\left(X_{\mathrm{hyp}} = x_{\mathrm{hyp}}\right) \approx f_{X_{\mathrm{hyp}}}(x_{\mathrm{hyp}})dX_{\mathrm{hyp}}$, where $f_{X_{\mathrm{hyp}}}(x_{\mathrm{hyp}})$ represents the PDF for hypocentral locations. The rate of occurrence assigned to any particular position in the source must preserve the total rate for the source:

$$\lambda\left(M = m\right) = \int_{X_{\mathrm{hyp}}} \lambda(M = m | X_{\mathrm{hyp}} = x_{\mathrm{hyp}}) f_{X_{\mathrm{hyp}}}(x_{\mathrm{hyp}})dX_{\mathrm{hyp}}. \tag{3.44}$$

When the source is discretized into n cells, this expression becomes equivalent to

$$\lambda\left(M = m\right) = \sum_{i=1}^{n} \lambda(M = m | X_{\mathrm{hyp}} = x_{\mathrm{hyp},i}) P(X_{\mathrm{hyp}} = x_{\mathrm{hyp},i}). \tag{3.45}$$

When hypocentral locations are assumed equally likely throughout the source (Cornell, 1968), $P\left(X_{\mathrm{hyp}} = x_{\mathrm{hyp}}\right)$ equals $\Delta A_i/A$ for fault sources with total area A that are discretized into patches of area ΔA_i, and $\Delta V_i/V$ for "area" sources of total volume V that are discretized into discrete volumes of ΔV_i. The rate of events of a given magnitude occurring at a particular location is then given by

$$\lambda(M = m | X_{\mathrm{hyp}} = x_{\mathrm{hyp},i}) = \lambda\left(M = m\right) \frac{\Delta A_i}{A}. \tag{3.46}$$

Other assumptions that reflect departures from this uniform likelihood can also be considered, if information is available to constrain them (e.g., Mai et al., 2005).

Source dimensions must be obtained using source-scaling relations to define the geometric rupture (Section 2.7.1). As these source dimensions are a function of the magnitude of the event, we can assert that

$$\lambda(M = m | X_{\mathrm{hyp}} = x_{\mathrm{hyp}}) = \int_{\Theta} \lambda(M = m | \theta, x_{\mathrm{hyp}}) f_{\Theta}(\theta | m, x_{\mathrm{hyp}})d\theta \tag{3.47}$$

where the vector θ includes all parameters that define the geometry of the rupture. Equation 3.47 states that to obtain the overall rate of occurrence of events of a particular magnitude with some hypocentral location, we must consider all possible ways a rupture could be represented, for that m and x_{hyp}, and partition the total rate among them. Each different rupture is characterized by an instance of $\Theta = \theta$.

For example, to define the rupture shown in Figure 3.25, the vector θ would include the rupture's strike, dip, length, width, and the relative position of the hypocenter. For a rupture within a fault source, the strike and dip angle will typically be fixed. For a rupture within an areal source, the strike and dip angles may be allowed to vary to reflect a range of possible source mechanisms. The joint distribution $f_{\Theta}\left(\theta | M = m, X_{\mathrm{hyp}} = x_{\mathrm{hyp}}\right)$ is the product of marginal distributions for individual rupture dimensions and hypocenter locations (although multivariate models exist, e.g., Stafford, 2014b).

To provide an example of the joint distribution $f_{\Theta}\left(\theta | M = m, X_{\mathrm{hyp}} = x_{\mathrm{hyp}}\right)$, consider a case in which the source-scaling relationship of Wells and Coppersmith (1994) is adopted to represent the source dimensions and we assume that the positions of the hypocenter within the rupture are independent and uniformly distributed. To obtain a rupture length, L, and width, W, it is common to start with predictions of the rupture area A. This is because scaling relations for rupture areas are relatively stable and relate closely to moment magnitude. One can then either assume that $L = W = \sqrt{A}$ or obtain $L = A/W$ with W coming from a source-scaling relation. Wells and Coppersmith (1994) represent the distribution of A and W for a given magnitude value as independent lognormal distributions $f_A^{LN}(a|M = m)$ and $f_W^{LN}(w|M = m)$, respectively, and combining these with our assumption of uniformly distributed hypocenters yields[24]

$$f_\Theta\left(\theta|M = m, X_{\text{hyp}} = x_{\text{hyp}}\right) = f_A^{LN}(a|m)f_W^{LN}(w|m)f_{X_L}^U(x_L)f_{X_W}^U(x_W) \qquad (3.48)$$

where $\theta = \{a, w, x_L, x_W\}$, and the superscripts LN and U are used to denote lognormal and uniform distributions, respectively. This formulation is complicated in practice as not all combinations of $\{a, w, x_L, x_W\}$ are kinematically admissible for a given hypocentral location. However, the concepts remain valid.

In many real-world applications, the hazard results are not particularly sensitive to the assumptions embedded within $f_\Theta\left(\theta|M = m, X_{\text{hyp}} = x_{\text{hyp}}\right)$, so the representation of these ruptures can be simplified. For example, finite-fault effects are most important when modeling ground motions from large crustal and subduction events (e.g., Goda and Atkinson, 2014). However, very large events can rupture the entire fault. In such cases, as most GMMs do not consider the relative location of the hypocenter, the location of the hypocenter becomes irrelevant as all origin positions result in the entire fault rupturing. As far as these GMMs are concerned, all such ruptures are equivalent. Similar arguments can be made for small events that are distant from the site of interest. In this case, although a given hypocenter can have multiple rupture geometries, the geometries would all produce comparable ground motions.

The important point is that the integrand of Equation 3.47 is really equivalent to $\lambda(rup)$. To obtain the rates of occurrence of particular ruptures, we can begin with a magnitude-frequency distribution that defines how often all events of magnitude m occur within a source. A given source could generate many different ruptures with that magnitude. Expressions like Equations 3.44 and 3.47 demonstrate how this overall rate for a given magnitude is decomposed into scenarios with this magnitude occurring at particular hypocentral locations and causing ruptures of a given size and orientation. The rate of occurrence of a given rupture can then be defined as

$$\lambda(rup) = P\left(\Theta = \theta|M = m, X_{\text{hyp}} = x_{\text{hyp}}\right)\lambda\left(M = m|X_{\text{hyp}} = x_{\text{hyp}}\right)P\left(X_{\text{hyp}} = x_{\text{hyp}}\right). \qquad (3.49)$$

When more elaborate descriptions of the rupture are required as input to a GMM, the process outlined above continues along similar lines. For example, it may be necessary to prescribe slip vectors over the rupture surface for kinematic ground-motion modeling to be possible (e.g., Section 5.3.2). In this case, information about the rupture discretization and the slip vectors is included within the parameter vector θ. We must also define the probability of that particular combination of parameters in $f_\Theta(\theta)$. Naturally, for complex rupture descriptions, specifying the probability distribution is more challenging.

3.6.2 Rupture Rates from Geological Constraints

The previous section represented ruptures by beginning with a magnitude-frequency distribution and then identifying kinematically admissible rupture scenarios consistent with a given magnitude event. This approach is relatively traditional and is most commonly used in practical seismic hazard studies. However, as more information is gathered about individual fault sources, and particularly

---

[24] Large events will rupture the entire seismogenic depth and rupture width will saturate (Stafford, 2014b). It is therefore necessary for the the rupture length to compensate for the width saturation, or for a break in the displacement scaling to arise. Stafford (2014b) demonstrated that nontrivial correlations exist between these rupture parameters, and a more elaborate joint distribution of area and width may be warranted.

about how these sources may be segmented, it becomes possible to work in the opposite direction. We begin by identifying a set of plausible rupture scenarios, constrained by geometric bounds on rupture segments, historically observed slip-rates, and observed per-event slip amplitudes. Using these plausible scenarios, and the overall rate of coseismic energy release, or crustal deformation, it is possible to perform an inversion to determine how frequently each of these scenarios occurs. This approach is advantageous when the source model is dominated by a well-studied, but potentially complex, fault network, and we wish to partition energy release rates over the possible rupture scenarios that are compatible with this network. These scenarios can include ruptures of individual segments on a fault, multiple segments on a fault, or even multiple faults.

Field and Page (2011), building upon earlier work by Andrews and Schwerer (2000), developed a methodology to estimate rates of these different types of earthquake rupture. The key feature of the approach is that all sources of fault activity information can be included, and an inversion produces consistent rates of rupture scenarios. This approach differs importantly from the traditional method presented in the previous section. There, seismicity analyses (that potentially include several external constraints) are used to characterize rates of events on each seismic source. The analyst hopes that when these individual sources are combined into an overall source model, the system of sources results in regional rates of occurrence that are sensible. The results can be checked for consistency with any external constraints, but this is done in a post-hoc manner. The advantage of the inversion approach is that the overall process can be viewed as a constrained optimization problem. Any available external constraints may be imposed upon the solution from the outset. However, the disadvantages of the approach relate to the requirement to sufficiently constrain the inversion so that unique (or a small number of nonunique) solutions can be found. In most regions of the world, there is currently insufficient information to produce unique solutions.

The approach of Field and Page (2011) was adopted with minor modification for the UCERF3 study (Field et al., 2013). The motivation for this change in approach to ERF development was the desire to reconcile inconsistencies associated with previous rupture forecasts made for California (Working Group on California Earthquake Probabilities, 2003, 2008). In Section 3.4, we discussed how several sources could provide constraints upon activity rates outside the seismicity catalog for a region. For fault-based source models, or fault-based rupture forecast models, information such as fault slip rates, fault segmentation, slip partitioning, and paleoseismic constraints can play a stronger role in defining earthquake activity than the limited seismicity catalog. Field et al. (2013) used all of these sources of information regarding the activity rates of fault segments throughout California. Also, they imposed constraints on regional-level activity rates. For example, the regional seismicity of California arising from all sources can be well described using a GR distribution. On the other hand, the magnitude-frequency distributions for the individual sources can depart significantly from GR, the characteristic distribution, or any other distribution. The inversion framework enables the net result of the inverted rates to be consistent with GR at a regional level while leaving freedom for the data to govern the nature of magnitude-frequency distributions on the individual sources.

The specific constraints considered by Field and Page (2011) are detailed in what follows. In this specification, it is assumed that the regional fault network is represented by discretizing all fault segments into patches with some nominal area. In the original Field and Page (2011) representation, the patches had a down-dip width matching the fault width, such that each fault is effectively discretized into patches along strike in a one-dimensional manner. While the original presentation and the implementation in Field et al. (2013) both followed this one-dimensional approach, the method itself is easily extensible to higher dimensions. The effect of discretizing the faults using

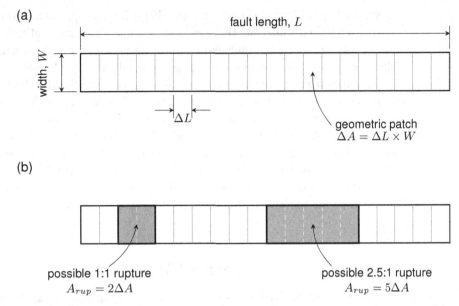

Fig. 3.26 Schematic illustration of how a fault might be discretized for inversion-based rupture rate estimates. The upper panel shows a view looking parallel to a normal vector to the fault. The overall length L is discretized into several elements of length ΔL where $\Delta L = W/2$. The lower panel shows two possible ruptures that might occur on this fault with different aspect ratios. Each possible rupture involves the complete rupture of a number of discrete geometric patches.

a relatively long down-dip patch length is that the ruptures considered within the inversion are effectively assumed to be of at least moderate size. This can be appreciated from Figure 3.26, which schematically illustrates how faults might be discretized.

In Figure 3.26a, the fault is discretized into geometric patches of length ΔL, equal to half of the fault width; i.e., $\Delta L = W/2$. This discretization defines the locations of possible rupture scenarios, as shown in Figure 3.26b. In Figure 3.26b, two possible ruptures are shown, and their areas are integer multiples of the geometric patch areas. The spatial resolution of rupture positions along strike is therefore governed by ΔL in this example.

To specify possible ruptures given this discretization, it is necessary to account for plausible aspect ratios for these ruptures. Given that small to moderate ruptures that do not rupture the full seismogenic thickness tend to have aspect ratios of around 1.0, on average, ruptures of individual geometric patches are not considered.[25] The implication is that the minimum size of a possible rupture that would be considered under this discretization is $2\Delta A$, corresponding to an aspect ratio of 1.0. Any other aspect ratio higher than 1.0 arising from a contiguous rupture of geometric patches is also considered. The magnitude of the smallest event that is considered within this inversion procedure is the magnitude of a rupture with area = W^2. Seismogenic thicknesses encountered worldwide are usually at least 10 km, which means that this one-dimensional approach to discretization can be used to consider ruptures with magnitudes of approximately 6 and above. If one wished to resolve rates of ruptures from smaller events, then a two-dimensional discretization (involving discretization down-dip) would be required.

---

[25] Aspect ratios of 1:2 are assumed, under this condition, to be too unlikely.

Despite the implied lowest magnitude resolvable using this approach, smaller ruptures are not ignored in the hazard calculations. They can be represented within area and background sources, and can also be used to constrain magnitude-frequency distributions for the known faults.

Following the discretization of the entire fault network, all possible ruptures represented by the model are known. This discretization is achieved using a scheme like that illustrated in Figure 3.26. The ruptures mainly consist of contiguous ruptures of geometric patches within a single fault. However, they may also consist of complex ruptures involving geometric patches on multiple (proximally located) faults. The analyst must specify the relative likelihood of such complex multifault ruptures based on constraints associated with fault geometry (Field et al., 2013).

With all possible ruptures identified, the governing equations for the inversion process are set up. Recall that the objective is to eventually obtain an estimate of $\lambda(rup)$ for every possible rup encapsulated by the model. The following sections describe each inversion constraint in further detail.

Constraint 1: Average Slip Rate

Constraints associated with estimates of average slip-rates, $\dot{\bar{u}}_s$, on each subsection of every fault are imposed by taking the average slip over a particular subsection, s, of a rupture, r, termed $\bar{u}_{sr}$, and multiplying this by the average occurrence rate for that rupture $\lambda(rup_r)$. The total slip rate for each subsection is found by summing these contributions over all possible ruptures R that involved that subsection:

$$\sum_{r=1}^{R} \bar{u}_{sr} \lambda(rup_r) = \dot{\bar{u}}_s. \tag{3.50}$$

The *subsections* referred to here correspond to the geometric patches previously defined (see Figure 3.26).

The estimates of slip rates on each subsection of the fault can come from several sources, including plate-motion studies, geodetic monitoring, coseismic slip estimates from paleoseismic investigations, and finite-element modeling of the fault network, among others. The spatial resolution used to discretize the fault is made at a high resolution to entertain the possibility of ruptures occurring at many different positions along the strike (replicating the effect of *floating* ruptures discussed with reference to Figure 3.25). Usually, the information available regarding slip rates will not be available at the same spatial resolution. Therefore, the same slip rate will often be specified for many adjacent subsections of a given fault, with a possible departure to smooth slip rates down to zero at fault ends.

Application of Equation 3.50 provides S equations (one for each subsection slip rate estimate) for the R unknown rupture rates. When all ruptures are assumed to consist of the rupture of at least two subsections, as considered by Field and Page (2011), this first equation set represents an underdetermined system of equations where the number of subsections, S, is less than the number of rupture scenarios, R. For a simple case where no interaction is permitted between individual fault sources, the number of ruptures R that will exist within the model depends upon the number of segments for each fault S_i according to:

$$R = \sum_{i}^{n_{\text{faults}}} \frac{S_i (S_i - 1)}{2}. \tag{3.51}$$

Additional constraints are therefore needed.

Constraint 2: Paleoseismic Information

Event frequencies estimated from paleoseismic investigations produce the next set of constraints. From a rupture identified through paleoseismic investigations (Section 3.4.1), it is possible to identify (or assume) all subsections involved in this rupture. However, paleoseismic investigations do not always identify ruptures that have taken place in the past (e.g., small-magnitude events, non-surface-rupturing events, or surface-rupturing events that do not present evidence in a particular trench location). For this reason, it is necessary to account for the likelihood that a particular rupture would have been detected from an investigation given that the rupture did involve the location of the paleoseismic investigation using the term P_r^{paleo}:

$$\sum_{r=1}^{R} G_{sr} P_r^{\text{paleo}} \lambda \left(rup_r \right) = \lambda_s^{\text{paleo}} \tag{3.52}$$

where the constraint on the right-hand-side, λ_s^{paleo}, is the rate of occurrence of events (from any rupture) that involve this particular subsection of the fault, inferred from the paleoseismic studies. The term G_{sr} is an $S \times R$ matrix that contains zeros and ones, with the ones being used to indicate that subsection s was involved in rupture r, and zeroes indicating no involvement. Therefore, for a case where a paleoseismic rupture was certain of being detected ($P_r^{\text{paleo}} = 1$), and only one type of rupture occurred on this fault, then G_{sr} would have row s full of ones and Equation 3.52 would state that $\lambda(rup_r) \equiv \lambda_s^{\text{paleo}}$. However, when multiple types of rupture are possible, the sum of the rates of these ruptures needs to equal the observed rate from the paleoseismic investigations λ_s^{paleo} (assuming that all relevant ruptures were detected, i.e., $P_r^{\text{paleo}} = 1$ for all r).

Constraint 3: Prior Estimates of Rupture Rates

Prior assumptions on rupture rates produce additional constraints. This set of equations is similar to Equation 3.52:

$$\sum_{r=1}^{R} G_{sr} \lambda \left(rup_r \right) = \lambda_s^{\text{prior}} \tag{3.53}$$

where λ_s^{prior} is an a priori estimate of the rate of ruptures that involve subsection s. The matrix G_{sr} is the same as for Equation 3.52 and simply indicates membership of segment s to rupture r. In order for this third set of equations to provide any unique constraint, the a priori estimates of the segment rupture rates should not replicate the paleoseismic rates imposed in the previous set of equations.

Equation 3.52 should be applied to segments for which paleoseismic investigations have been conducted. However, Equation 3.53 may reflect assumptions based upon these specific investigations and can, therefore, be applied to segments for which no paleoseismic results have been directly obtained. As will be seen shortly, not all constraints are given equal power within the inversion. It makes sense that if paleoseismic results are used to infer prior rates within this third equation set, they should be down-weighted in comparison with the actual measurements.

Constraint 4: Rate Smoothness

The next constraint encourages similar-sized ruptures at various positions along a fault to have the same rates. Referring to Figure 3.26, this condition says that a rupture of the third and fourth geometric patches should have the same rate of occurrence as a rupture of the fourth and fifth patches. Mathematically, this constraint is represented as

$$\lambda(rup_r) - \lambda(rup_{r+1}) = 0. \tag{3.54}$$

This *smoothness* constraint is not strongly implied by physical conditions with current databases. That is, databases involving multiple ruptures on individual faults are rare and so variations in rupture rates along a given fault are not well understood. However, while the constraint is not strong in a physical sense, it is essential for the inversion procedure because it imposes many constraints. If single-fault ruptures are considered, then this smoothness constraint introduces $\sum_i^{n_{\text{faults}}} (S_i - 2)(S_i - 1)/2$ equations. Therefore, this smoothness constraint is often critical for establishing an overdetermined system of equations.

Constraint 5: Regional Seismicity

The final constraints ensure that the total rate of ruptures of a given magnitude match the regionally observed rates from seismicity analysis. This equation set is represented as in Equation 3.55, where M_r indicates whether rupture r has a magnitude falling in the range $m \pm \Delta m$:

$$\sum_{r=1}^{R} M_r \lambda(rup_r) = \lambda_{m \pm \Delta m}^{\text{prior}} \tag{3.55}$$

where $\lambda_{m \pm \Delta m}^{\text{prior}}$ is the prior estimate of the rates of occurrence of events with magnitudes in the discrete range $m \pm \Delta m$.

Solution to Constrained Inversion

The combined sets of constraints can be represented in matrix form:

$$X_{rup}\lambda(rup) = d \tag{3.56}$$

where $\lambda(rup)$ represents an $R \times 1$ vector of all of the rupture rates to be inverted for, X_{rup} is a matrix reflecting all of the equation sets previously introduced, and d are all the data, or a priori constraints, imposed on the system. That is, d includes slip-rate estimates, paleoseismic event rates, prior estimates of segment rupture rates, magnitude-frequency constraints, and smoothness constraints.

Some imposed constraints may conflict with others. For example, paleoseismic evidence may suggest that certain size ruptures are more likely to occur at one end of a fault than others (either for physical reasons or due to sampling bias), which is inconsistent with the smoothness constraints. The overdetermined system of equations, and the possible conflict between certain constraints dictates that Equation 3.56 is an incomplete description of reality. We should not expect that one unique vector of $\lambda(rup)$ will exist for a given problem. For this reason Field and Page (2011) formulated the system of equations as

$$d_{\text{obs}} = X_{rup}\lambda(rup) + \varepsilon \tag{3.57}$$

in which now d_{obs} denotes the observed data constraints and ε is a vector of error terms, reflecting that no solution will perfectly satisfy all constraints. Under this framework, the inversion for the set of rupture rates can be found from a constrained optimization, or non-negative least squares problem.[26] The optimization problem to obtain the rupture rate estimates $\hat{\lambda}(rup)$ is formulated as

$$\min_{\lambda} \quad \varepsilon^T W \varepsilon \tag{3.58}$$
$$\text{subject to} \quad \lambda(rup) \geq 0 \quad \forall rup$$

where W is a diagonal[27] matrix of weights, with elements of $1/\sigma_{d_i}^2$ reflecting the combined effect of possible data errors as well as subjective degrees of belief regarding the strength of the constraints being considered. The solution of this set of equations can be obtained relatively easily using commonly available numerical libraries. However, the results are quite sensitive to the various modeling decisions regarding the weighting of data, discretization of the faults, and how slip is distributed spatially during a rupture, among other things. Therefore, care should be taken to ensure that these modeling decisions are fully documented and that the results' sensitivity to these decisions is explored and understood.

The general framework of Field and Page (2011), and its refinements in Field et al. (2013), is likely to garner more widespread adoption in the future. However, several challenges remain. As mentioned previously, the approach requires a certain level of information to identify the plausible rupture segments and estimate slip rates and other properties for these segments.

From a practical perspective, there is also subjectivity in assigning weights to multiple, potentially contradictory, constraints. For example, the magnitude required to rupture a given rupture segment can be estimated using source-scaling relations. However, there are several such models and some account for parameters not considered by others. For example, Anderson et al. (1996) consider the fault slip rate directly within their scaling relation. If the scaling relations of Wells and Coppersmith (1994) and Anderson et al. (1996) are both used in combination, it is necessary to assign weights to their predictions, but also to judge whether the latter model's slip-rate scaling is appropriate for the region of application. These problems do not have simple answers and add to the inversion's computational demands.

Finally, the inversion often requires constraints to be imposed that reflect a fault system, or fault network. This network will often span a geographic region far more extensive than normally considered within a traditional site-specific hazard analysis. For example, an inversion to determine rupture rates for a site within the San Francisco Bay area would require modeling the interactions among fault segments that occur over the surrounding fault network. This issue arises because the inversion framework imposes constraints at both local (fault segment level) and regional (fault network level) scales to represent interactions among segments and estimate rates of multisegment ruptures. These issues will all receive more attention from the research community as this inversion-based approach becomes more widespread.

---

[26] Rupture rates $\lambda(rup)$ cannot be negative, and this constraint needs to be imposed. However, this complicates the actual estimation procedure.

[27] In applications thus far, this matrix has been assumed to be diagonal, but this implies that all data constraints are independent of one another. There will be cases in which the constraints will have some correlation. For example, multiple paleoseismic studies undertaken by one group of people may differ systematically from analyses of another group. In such cases, the weights corresponding to constraints from a given group could be correlated and reflected with off-diagonal terms in this weight matrix.

3.7 Generation of Rupture Scenarios

While Section 3.6.2 started by specifying the geometry of all major ruptures from the outset, the more common approach, discussed in Section 3.6.1, explained that ruptures and their rates relate to the output of a magnitude-frequency distribution. For this latter case, various practical considerations associated with generating the actual rupture scenarios have not been discussed thus far. The generation of rupture scenarios depends strongly upon the GMMs' requirements regarding the rupture's specification. The present section provides a discussion and some examples of generating rupture scenarios in a manner consistent with the source characterization and seismicity analyses that would have been undertaken.

3.7.1 Finite Ruptures on Fault Sources

We have seen from the source-scaling relationships that rupture dimensions can become very large for large-magnitude events. While an event hypocenter may be a significant distance from the site, the rupture may pass very near to, or even through, the site. Therefore, as will be discussed in Chapter 4, many GMMs will require that the rupture geometry is characterized so that the closest distance to the rupture can be used. This closest distance is commonly measured to the rupture itself, or the surface projection of the rupture. These distance measures are finite-source equivalents of hypocentral and epicentral distances for point sources.

When the horizontal distance to the surface projection of the rupture is considered, then three particular cases arise, as shown in Figure 3.27. The upper panel of Figure 3.27 shows a case where a bilateral rupture takes place away from the end of the fault, so it is easily contained within the fault. The closest point to this bilateral rupture is associated with the end point of the rupture. The central panel shows a case where the epicenter is within a distance of $l_m/2$ of the fault's end. In this case, a pure bilateral rupture cannot occur unless the fault is permitted to grow. Here, the closest point to the site is still measured from the rupture's end point, but now this point is not merely a distance of $l_m/2$ along the strike from the epicenter. In the last case shown in the bottom panel, the epicenter is located at a point such that $|X| \le l_m/2$, which guarantees that the closest distance to the rupture will always be $R_c = r_y$. For each of these three cases, the relationship defining the closest distance R_c in terms of the epicentral position (via coordinate X) will be different. This contrasts with the case where epicentral distance is employed, and a simple geometric equation exists to connect R_e to X.

As shown in Figure 3.27, a given rupture will now have some finite length that we can call l_m. The distance from the site to the closest point of the surface projection of the rupture will be

$$r_c = \sqrt{x'^2 + r_y^2}. \tag{3.59}$$

Here, the variable x' defines the along-strike distance associated with the closest point to the rupture, as shown in Figure 3.27. Because of the three different cases that can arise, this variable must be defined in a rather inelegant manner according to

$$x' = \begin{cases} \min(0, -\alpha l + l_m) & \text{for } x \le -\alpha l + l_m/2 \\ \min(0, x + l_m/2) & \text{for } -\alpha l + l_m/2 < x < -l_m/2 \\ 0 & \text{for } -l_m/2 \le x \le l_m/2 \\ \max(0, x - l_m/2) & \text{for } l_m/2 < x < (1-\alpha)l - l_m/2 \\ \max(0, (1-\alpha)l - l_m) & \text{for } x \ge (1-\alpha)l - l_m/2. \end{cases} \tag{3.60}$$

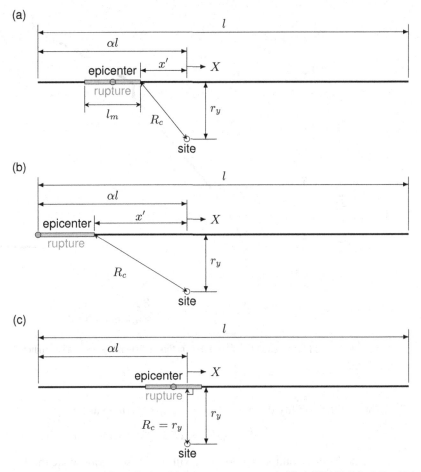

Fig. 3.27 Plan view showing three distinct cases that can arise with finite ruptures on a vertical fault (thick black line). Panel (a) shows an unconstrained rupture (very thick gray line), where the closest distance is measured to one end of the rupture. Panel (b) shows a case where a bilateral rupture cannot exist due to the epicenter being located at the end of the fault. Panel (c) shows a case where the closest distance is measured to some point within the rupture.

Now, because the length of the rupture l_m will depend upon the magnitude of the earthquake scenario, the probability distribution of the distance measure is complicated and is coupled with the magnitude:

$$P[R_c \leq r_c \mid M = m] = P\left[|X'| \leq |x'(r_c)| \mid M = m\right] = P\left[|X'| \leq \sqrt{r_c^2 - r_y^2} \mid M = m\right]. \quad (3.61)$$

Casting the conditional distances into the form of absolute deviations from the projection of the site, and considering only one side of the projection, gives

$$|x'| = \begin{cases} 0 & \text{for } |x| \leq l_m/2 \\ \min\left(0, |x| - l_m/2\right) & \text{for } l_m/2 < |x| < \beta l - l_m/2 \\ \min(0, \beta l - l_m) & \text{for } |x| \geq \beta l - l_m/2. \end{cases} \quad (3.62)$$

Note that $\beta \in \{\alpha, (1 - \alpha)\}$ depending upon whether the $|x|$ parameter corresponds to negative or positive values of x, respectively.

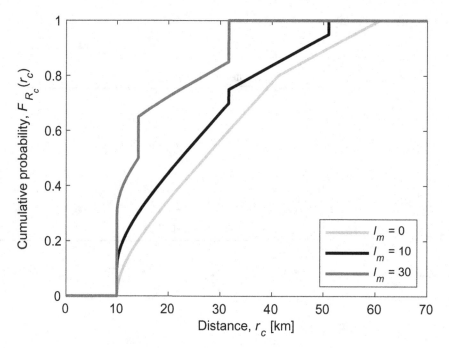

Fig. 3.28 Comparison of the CDF for distances to finite ruptures of three different lengths on a line source ($\alpha = 0.4$, $l = 100$ km, $r_y = 10$ km).

The total probability distribution can finally be written as

$$F_{|X'|}(|x'|) = \alpha F^{-}_{|X'|}(|x'|) + (1 - \alpha)F^{+}_{|X'|}(|x'|) \tag{3.63}$$

where the F^{-} contribution is associated with $x' < 0$ scenarios while the F^{+} contribution is for $x' \geq 0$.

Unlike the distribution of epicentral distances,[28] the distribution for a finite-distance metric will depend directly upon the rupture geometry, which in this example is defined using l_m and the assumption that bilateral ruptures are preferred. As such, the actual distance distribution will change as shown in Figure 3.28 because the link between x' and x varies with l_m; see Equation 3.62.

When $l_m = 0$, the distribution in Figure 3.28 is consistent with that for epicentral distance. When l_m becomes large, we can appreciate that more epicentral locations will result in ruptures that include the closest point between the source and the site. An increasing concentration of probability density occurs at the closest distance as a result.

When the rupture length l_m is defined as a function of the earthquake magnitude, using a source-scaling relation, the relative likelihood of magnitude-distance combinations can be computed. If each magnitude level is considered equally likely, then the likelihoods of different distances would be represented as shown in Figure 3.29.

However, once we recognize that the magnitudes have different rates of occurrence, then the complete joint distribution of magnitude and distance can be computed as shown in Figure 3.30. When the rupture scenarios are defined simply in terms of earthquake magnitude and the closest distance between the rupture and the site, this figure represents the probabilities of all rupture scenarios that are relevant for this particular source.

---

[28] This can be derived following the approach of Cornell (1968), and is left to the reader as an exercise.

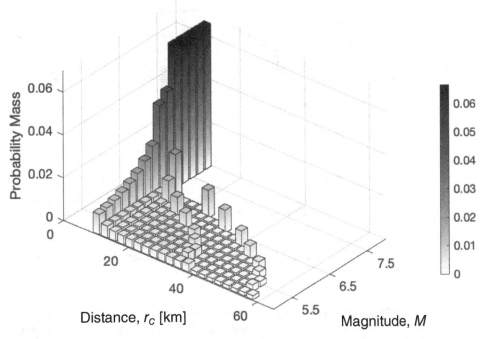

Fig. 3.29 Probability distribution of magnitudes and distances accounting for finite rupture dimensions. Earthquake rates are not considered.

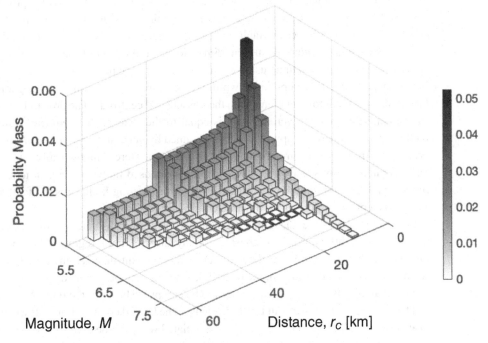

Fig. 3.30 Joint distribution of magnitude and distance accounting for finite rupture dimensions and Gutenberg–Richter event probabilities.

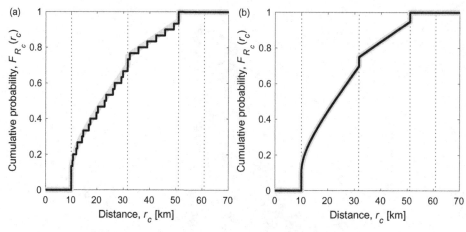

Fig. 3.31 Comparison of analytical and discrete numerical evaluation of the cumulative distribution function for a line source with finite ruptures of length $l_m = 10$ km. Panel (a) shows discrete results for 30 uniformly spaced hypocentral locations, while panel (b) shows 300 such hypocenters.

In practice, the ability to develop analytical expressions for distance distributions when finite rupture effects are considered is complicated by the rupture dimensions being a function of the earthquake size and by generic faults existing in 3D rather than the 2D cases arising from the use of a vertical fault in these examples. Rather than developing analytical expressions for the rupture distributions, it is more common to compute these using discrete numerical approximation.

Consider the vertical strike-slip fault again and retain the assumption that hypocentral locations are equally likely to occur over the fault surface. If ruptures are characterized using the distance to the closest point on the rupture, the up-dip dimensions and location of the rupture must be considered. However, suppose we continue using the horizontal distance to the surface projection of the rupture. In that case, the depth and up- and down-dip dimensions of the rupture can be ignored. The fault can then be discretized into n segments with their associated centroids. The probability that an earthquake has an epicenter in any given segment is equal to the ratio of that segment's length and the total fault length. The source's ruptures are then obtained by generating *kinematically admissible* ruptures originating from each centroid (hypocentral location). Here, kinematically admissible means that the generated ruptures must obey the constraints imposed upon the fault in terms of its allowable ruptures; i.e., must a rupture be contained within the existing fault, or is the fault allowed to grow? These concepts were previously illustrated in Figure 3.27.

Figure 3.31 shows the CDFs that are obtained using this discrete approach. For these cases, the process is straightforward. One considers many discrete fault segments and uses their centroids to generate the ruptures. From these ruptures, the relevant distance metric is computed. Once all segments have been considered, the overall distribution can be obtained.

Note that this distribution does not need to be computed because the rupture scenarios themselves, and their associated rates, are all that is needed for the hazard computations. However, it can be useful to amalgamate all of the rupture scenarios associated with a respective source to understand the most common types of scenarios. Once the fault has been discretized, then, in many ways, each segment can be regarded as being a new source. For uniformly spaced segments and the assumption of a uniform likelihood of an event occurring over the fault, each of the fault segments has an activity rate apportioned from the source's total activity rate on a pro rata by area basis. That is, if ten segments are considered, then each has an activity rate that is one-tenth of the total rate for the fault.

Table 3.10. Example calculations for rupture scenarios generated for a simple fault ($\alpha = 0.4$, $l = 100$ km, $r_y = 10$ km) discretized into five equal-length segments. m_i is a magnitude representing a bin of width $\Delta m = 0.2$ centered on the listed value, $l_m(m_i)$ is the median rupture length for this magnitude predicted by Wells and Coppersmith (1994), x_j is the epicentral location, measured along strike from the closest point to the site, x'_{ij} defines the closest point on the rupture to the site, and $r_{c,ij}$ is the corresponding closest distance. The final three columns represent the probability of a magnitude of this size, the probability of the distance value (equal to $P[X = x_j]$), and the overall probability of the rupture scenario

m_i	$l_m(m_i)$	x_j	x'_{ij}	$r_{c,ij}$	$P(M = m_i)$	$P(R_c = r_{c,ij}\|m_i)$	$P(rup)$
5.5	6.92	−30.0	−26.54	28.36	0.1471	0.2	0.02941
5.5	6.92	−10.0	−6.54	11.95	0.1471	0.2	0.02941
5.5	6.92	10.0	6.54	11.95	0.1471	0.2	0.02941
5.5	6.92	30.0	26.54	28.36	0.1471	0.2	0.02941
5.5	6.92	50.0	46.54	47.60	0.1471	0.2	0.02941
6.5	28.84	−30.0	−11.16	14.98	0.0147	0.2	0.00294
6.5	28.84	−10.0	0.00	10.00	0.0147	0.2	0.00294
6.5	28.84	10.0	0.00	10.00	0.0147	0.2	0.00294
6.5	28.84	30.0	15.58	18.51	0.0147	0.2	0.00294
6.5	28.84	50.0	31.16	32.73	0.0147	0.2	0.00294
7.5	120.23	−30.0	0.00	10.00	0.0015	0.2	0.00029
7.5	120.23	−10.0	0.00	10.00	0.0015	0.2	0.00029
7.5	120.23	10.0	0.00	10.00	0.0015	0.2	0.00029
7.5	120.23	30.0	0.00	10.00	0.0015	0.2	0.00029
7.5	120.23	50.0	0.00	10.00	0.0015	0.2	0.00029

To demonstrate the underlying computations that give rise to the distributions shown in Figure 3.31, Table 3.10 tabulates the intermediate calculations associated with discretizing the fault into a number of segments. While Figure 3.31 shows the results for a large number of possible epicentral locations, Table 3.10 shows results for a very coarse discretization whereby the 100-km-long fault is divided into five equal-length segments. Also, while the magnitude probabilities are computed for the GR distribution previously shown in Table 3.4, which spans magnitudes from 5.0 to 8.0, in Table 3.10 example calculations are shown for just three values corresponding to small, moderate, and large events. This coarse discretization still shows how the relationship between x_j and x'_{ij} varies as a function of epicentral position and rupture length. Most importantly, discretization enables straightforward computation of the probabilities associated with a given distance. In this example, a single finite rupture scenario is used for each distinct epicentral location. The probability associated with this rupture scenario (and hence with the corresponding distance) is, therefore, equal to the probability of the epicentral location.

Thus far, the examples considered in this section have dealt only with simple geometric faults consisting of a single fault segment. In more realistic applications, the analyst must also generate rupture scenarios that include adjacent fault segments, or even nearby faults. In these cases, the rupture generation model must include rules for when a rupture initiating within a particular fault segment will propagate to adjacent segments.

Figure 3.32 demonstrates the issue associated with the propagation of ruptures along a segmented fault. If fault segments correspond only to geometric boundaries, potentially arising from a simplified representation of a fault, there may be no reason to prohibit propagation. However, if large geometric changes occur, then these may represent physical impediments to propagation. An example of how these issues might be dealt with within a hazard analysis is shown in Figure 3.32c and d. Here, the

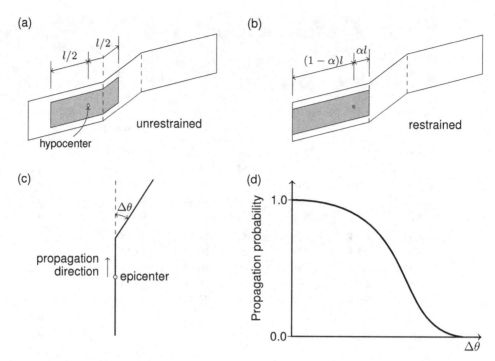

Fig. 3.32 Issues associated with rupture propagation along segmented faults. Panel (a) shows a rupture initiating on one geometric segment being permitted to propagate unrestrained to adjacent segments. In panel (b) the rupture is restrained within its own segment. The lower panels illustrate how propagation might be parameterized. Panel (c) shows the geometry, and panel (d) shows the propagation probability.

probability that a rupture will propagate is defined as a function of the change in strike across segments. When this change is relatively small, then the probability of propagation is very close to unity, while for values beyond a certain angular change, the probabilities drop considerably.

A similar issue arises with triggering of ruptures over multiple faults. Figure 3.33 illustrates this situation and a particular approach to model the likelihood that a rupture initiating on Fault A will trigger dislocation on an adjacent fault (Fault B). In this particular case, the probability of triggering is described in terms of the parallel and transverse offsets of the fault to be triggered. In more elaborate cases, additional factors such as the stress history on the fault to be triggered could be considered. For example, for the same geometric offset, the probability of triggering will be increased if the fault to be triggered is at a critical state itself. In contrast, if the fault to be triggered has recently been relaxed, then the probability should reflect this. These concepts can be incorporated through Coulomb stress transfer models (e.g., King et al., 1994; Toda and Stein, 2002).

3.7.2 Finite Ruptures in Area Sources

The above process for fault sources highlights issues also of relevance for area sources. The complexity of a general area source geometry dictates that analytical distance distributions are generally not available (even when point source distance metrics are used). The area source is therefore discretized into smaller spatial regions. Like the case of fault sources, once a discrete cell

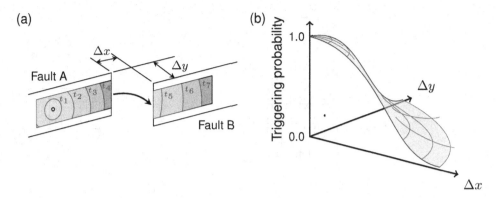

Fig. 3.33 Schematic illustration of issues associated with triggering of ruptures on nearby faults. In panel (a) an event originates on Fault A at a location marked by a small circle. The rupture then grows about this point as shown by the isochrone contours marked by time stamps t_i. Panel (b) provides the likelihood that a rupture on Fault B will be triggered by the initial event, as a function of the spatial coordinates Δx and Δy.

has been defined, then in many ways it acts as a new source with an activity rate reflecting its relative size compared with the original source.[29]

Figure 3.34 shows some examples of how an area source might be discretized. In Figure 3.34a, a simple rectangular source is shown, and the discretization is performed using a Cartesian grid. The major axes of the rectangular source are aligned with the coordinate directions. It is thus possible to subdivide the source into equal-sized geographic cells. Depending upon the requirements for the specification of the rupture scenarios, an additional discretization with depth may also be required.

The area source's geometry will usually prevent a simple gridded discretization. Figure 3.34b shows the common situation where the source is a polygon with no obvious, or simple, local coordinate system. In these cases, a decision must be made as to how the source is subdivided. Figure 3.34c shows the discretization of the source using a polar coordinate system rather than a Cartesian system. The polar system is useful in that the overall study region is typically defined in terms of a radial distance from the site of interest.

So far, the decisions influencing the type of discretization have related to geometry only. In all cases, one should expect the results to converge as the spatial discretization becomes sufficiently fine. This was demonstrated, to some extent, for a fault source in Figure 3.31. However, if computational efficiency is important, additional factors should be considered, as some discretization approaches are more efficient than others.

The most appropriate discretization scheme for a given application will also depend upon the type of analysis undertaken. The advantage of using a regular discretization, as shown in the upper row of Figure 3.34, is that the spatial cells would not need to change if the location of the site was altered. Therefore, when one is interested in computing hazard for several sites influenced by a given area source, the geometric computations need to be performed only once in this case. In contrast, the approaches based on polar coordinate systems shown in the bottom row of Figure 3.34 are naturally anchored to the site in question. Therefore, while the scheme shown in Figure 3.34d

---

[29] This pro rata by area partitioning of the overall activity rate for the source is consistent with the assumption that event hypocenters are equally likely to occur throughout the volume defined by the source boundaries.

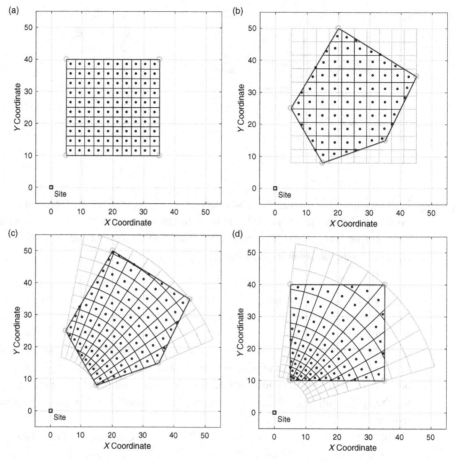

Fig. 3.34 Examples of options for discretizing area sources. Panel (a) shows Cartesian discretization of a rectangular source. Panel (b) shows Cartesian discretization of a generic polygon. Panel (c) shows a polar discretization of the same generic polygon. Panel (d) shows a polar discretization of a rectangular source, but using logarithmically spaced radii.

may be computationally efficient for this particular site, the computation of cell areas and centroids needs to be repeated if one wishes to maintain the same efficiency for any other site.

Once a cell is identified, the nature of the rupture scenarios associated with the cell depends upon the requirements of the GMMs used for the hazard calculations. If a point-source distance metric is used, only the cells' centroids need to be defined. For each cell, the same distance distribution can then be used for events of all magnitudes. In the more common case, where the dimensions of the rupture are considered, the centroids of the discretized cells are regarded as hypocentral locations from which ruptures are generated.

Figure 3.35 illustrates this process for a particular centroid location. For each event originating in this cell, rupture dimensions are generated using source-scaling relations. A strike and dip angle for the rupture must be specified to define the rupture geometry. For any given area source, the source characterization process will result in distributions of strikes, dips, and depths being defined. These distributions can then be used to allocate the geometric properties that define the rupture orientation.

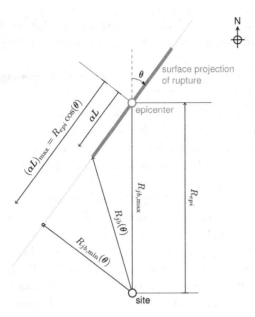

Fig. 3.35 Plan view of the geometry associated with a vertical strike-slip rupture (thick gray line) arising from a given epicentral location as a function of the strike of the rupture θ.

Figure 3.35 shows a case where a vertical strike-slip rupture is generated as a function of the strike angle θ. For the same epicentral location (cell centroid), this strike angle's specification will dictate the closest distance between the rupture and the site. In the case shown in the figure, the maximum distance between the source and site will take place when $\theta = \pi/2$ or $\theta = 3\pi/2$, while the minimum possible distance occurs when $\theta = 0$ or $\theta = \pi$. The particular distance for all strike angles, aside from $\theta = \pi/2$ and $\theta = 3\pi/2$, depends upon the total length of the rupture and what portion (represented by α) of the rupture propagates in the forward or backward strike direction. The more general case of computing distances to dipping ruptures is discussed in Section 3.7.3.

The final consideration that must be made for area sources, which is often more challenging than the equivalent consideration for fault sources, is whether or not ruptures can permeate the source boundaries. For known fault sources, the geometry of the source governs potential rupture locations. However, for many area sources, while the general sense of faulting within the spatial region may be known to some degree, the specific locations of the faults are not. Therefore, if the source's boundaries are based upon seismicity analyses (which are dominated by small events with small source dimensions), the size and physical extent of the parent faults will not be known. In these cases, it may be appropriate to allow generated ruptures to breach the source's geometric boundaries. In other cases, where the boundaries of the source zone are associated with significant geology changes, it may be appropriate to confine all ruptures inside the area source.

These assumptions, or decisions, can have a significant impact upon the hazard calculations. As discussed in Section 2.7.1, the relationship between a rupture's area and magnitude has implications related to the change in stress that occurs during a rupture. Because of this, it can be appropriate to prohibit large magnitude events with relatively small ruptures as this would imply unrealistically large levels of stress change. This may restrict large magnitude events to only locations where they can fit inside the source. The resulting spatial distribution of seismicity may be quite different from initial assumptions made before the rupture generation process.

3.7.3 Distances to Finite Ruptures

Key concepts related to defining distances to finite ruptures were discussed in Sections 3.7.1 and 3.7.2 by making use of vertical faults. However, finite ruptures within both fault and area sources will often have nonvertical dip angles. Several different distance metrics are used by the GMMs discussed in Chapters 4 and 5, and the present section defines the most common of those associated with finite ruptures.

The two most common distance metrics are the *rupture distance*, R_{rup}, and the *Joyner–Boore distance*, R_{jb}. The rupture distance is the shortest distance between a site and the finite rupture in three-dimensional (3D) space. The Joyner–Boore distance is the shortest distance between a site and the three-dimensional rupture's surface projection. That is, to compute R_{jb}, one first projects the rupture up to the surface of the Earth and then finds the shortest distance between the site and this projection in two-dimensional (2D) space. Another distance metric that is often used to determine a site's location relative to the hanging-wall/foot-wall boundary is the *strike distance*, R_x. Figure 3.36 illustrates these distance metrics for a rupture that has a strike angle of $\theta = 0$ and a dip angle of δ.

Figure 3.36 shows lines connecting the sites with the closest point on the rupture (either in 3D for R_{rup} or 2D for R_{jb}). For some sites, such as A, B, C, and H, these lines appear coincident for both R_{rup} and R_{jb}. However, in the elevation views, it is clear that these lines have an additional depth component relevant to the R_{rup} calculation. Note that the sites A, B, C, and H fall either to the west or east of the vertical black dashed lines. For all sites to the west of the leftmost black dashed line, the shortest distance to the rupture will be between the site and a point on the rupture's top edge. For all sites to the east of the rightmost black dashed line, the shortest distance is measured to the rupture's bottom edge.

For sites D and F, the plan view representation of these lines connecting the sites with the closest points on the rupture (or projected rupture) differs significantly. These differences arise due to the dipping rupture. In the elevation view looking north, one can appreciate that the lines for R_{rup} are perpendicular to the rupture surface. In contrast, the elevation view looking west reveals that there is an additional slant component as well.

In the plan view of Figure 3.36, the strike distance corresponds to the shortest distance between a site and a line containing the surface projection of the rupture's top edge. As $\theta = 0$ in this figure, R_x is the difference in east coordinates of a site and a point on the rupture's top edge, with sites on the hanging wall having $R_x > 0$.

To compute the distance metrics R_{rup}, R_{jb}, and R_x for a general rectangular rupture oriented with strike θ and dip δ, one can make use of the coordinate transformations illustrated in Figure 3.37. The starting point is to position a Cartesian coordinate system at the hypocentral location and use a right-hand triad of north, east, and depth, $\{n, e, z\}$. Two coordinate transformations are then applied. The first transformation is a rotation of θ about the z-axis and creates axes u and v' that lie in the horizontal plane, but point along and perpendicular to the strike direction, as shown in Figure 3.37a. This first step uses the transformation matrix $\boldsymbol{T}_\theta$. A second transformation then involves a rotation of δ about the u-axis, as shown in Figure 3.37b. This second step uses the transformation matrix $\boldsymbol{T}_\delta$. The result of these transformations is a coordinate system $\{u, v, w\}$, where the u and v axes lie in the rupture plane, and w is an out-of-plane coordinate. Once in the $\{u, v, w\}$ space, finding the point on the rupture closest to the site is straightforward.

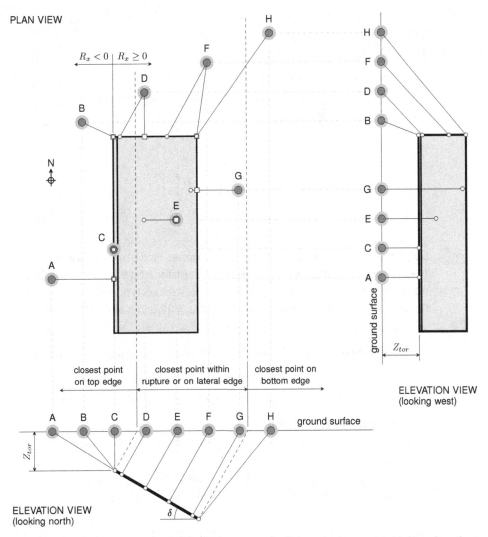

Fig. 3.36 Illustration of the R_{rup}, R_{jb}, and R_x distance metrics for dipping ruptures. In all views, the sites annotated A–H are shown by gray circles. Thin black lines from the sites to the small open circles on the rupture show the shortest distance between the site and the rupture, R_{rup}. In the plan view, the open square markers show the closest point between the surface projection of the rupture and the site, and thin lines connecting these points represent R_{jb}.

The transformations can be chained together to apply an overall transformation of $\boldsymbol{T}_\delta \boldsymbol{T}_\theta$:

$$
\begin{Bmatrix} u \\ v \\ w \end{Bmatrix} = \boldsymbol{T}_\delta \boldsymbol{T}_\theta \begin{Bmatrix} n \\ e \\ z \end{Bmatrix} = \underbrace{\begin{bmatrix} 1 & 0 & 0 \\ 0 & \cos(\delta) & \sin(\delta) \\ 0 & -\sin(\delta) & \cos(\delta) \end{bmatrix}}_{\boldsymbol{T}_\delta} \underbrace{\begin{bmatrix} \cos(\theta) & \sin(\theta) & 0 \\ -\sin(\theta) & \cos(\theta) & 0 \\ 0 & 0 & 1 \end{bmatrix}}_{\boldsymbol{T}_\theta} \begin{Bmatrix} n \\ e \\ z \end{Bmatrix}
$$

$$
= \underbrace{\begin{bmatrix} \cos(\theta) & \sin(\theta) & 0 \\ -\sin(\theta)\cos(\delta) & \cos(\theta)\cos(\delta) & \sin(\delta) \\ \sin(\theta)\sin(\delta) & -\cos(\theta)\sin(\delta) & \cos(\delta) \end{bmatrix}}_{\boldsymbol{T}_\delta \boldsymbol{T}_\theta} \begin{Bmatrix} n \\ e \\ z \end{Bmatrix}.
$$

(3.64)

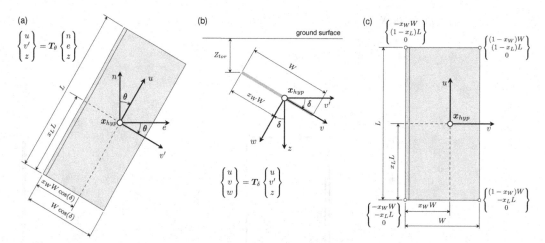

Fig. 3.37 Coordinate systems used for computing distances to finite ruptures. Panel (a) shows the rupture geometry in the original $\{n, e, z\}$ system, looking along the z-axis, and also shows the rotation about the positive z-axis to $\{u, v', z\}$. Panel (b) views the rupture looking along the positive u-axis, and the rotation about this axis to $\{u, v, w\}$. Panel (c) views the rupture looking along the positive w-axis. The coordinates of the hypocenter are $\boldsymbol{x}_{hyp} = \{0, 0, 0\}$ in all systems.

The closest point on the rupture, $\{u_c, v_c, 0\}$, to the site is found from the pair of expressions

$$u_c = \max\left[\min\left(u, (1 - x_L)L\right), -x_L L\right]$$
$$v_c = \max\left[\min\left(v, (1 - x_W)W\right), -x_W W\right]. \tag{3.65}$$

The rupture distance, R_{rup}, is then

$$R_{rup} = \sqrt{(u - u_c)^2 + (v - v_c)^2 + w^2}. \tag{3.66}$$

For the Joyner–Boore distance, R_{jb}, and strike distance, R_x, only the first coordinate transformation using $\boldsymbol{T}_\theta$ is required. First, one rotates to obtain $\{u, v', z\}$:

$$\begin{Bmatrix} u \\ v' \\ z \end{Bmatrix} = \boldsymbol{T}_\theta \begin{Bmatrix} n \\ e \\ z \end{Bmatrix}. \tag{3.67}$$

The first expression for u_c in Equation 3.65 remains valid, but the second expression for v_c is replaced by Equation 3.68, in which $W' = W\cos(\delta)$:

$$v'_c = \max\left[\min\left(v', (1 - x_W)W'\right), -x_W W'\right]. \tag{3.68}$$

The Joyner–Boore distance is

$$R_{jb} = \sqrt{(u - u_c)^2 + (v' - v'_c)^2} \tag{3.69}$$

and the strike distance is

$$R_x = v' + x_W W'. \tag{3.70}$$

For complex rupture geometries, an efficient approach to computing the finite distance metrics is first to triangulate the rupture surface, using as few triangles as possible. Once triangulated, one can make use of the efficient algorithm of Eberly (2020) for finding the shortest distance between a point and a triangle in 3D space. The shortest distances among those computed for all of the triangles then gives the solution for the overall rupture surface.

3.8 Time-Dependent Rupture Rates

Section 2.4.1 introduced elastic rebound theory and explained that as the occurrence of a particular earthquake event relaxes the stress acting over the rupture surface, a certain amount of time is typically required for these stresses to build up again before another similar event can take place. This working hypothesis of earthquake "cycles" has been in place for more than a century. Naturally, the real physical system is more complicated, and it is common to observe significant variation in the time between similar events on a given source. The fundamental reason for this is that the external tectonic loading is not the only cause of stresses over the rupture surface. The occurrence of a rupture changes the physical properties of the rupture surface, can potentially change the surface's geometry, influence pore pressures, and, most importantly, redistributes stresses throughout the crust in the vicinity of the rupture. Similarly, other ruptures that occur may also influence the state of stress on the fault in question (from both static and dynamic stress changes).

Therefore, a model for the time-dependence of ruptures of a single type needs to provide the distribution of recurrence times for this event. This distribution will be characterized by some mean recurrence interval and a measure of the variation about this mean interval. A common use case for such models is when considering the characteristic earthquake distribution previously introduced in Section 3.5.6. For such magnitude-frequency distributions, the characteristic scenario has a heightened average rate of occurrence, and possibly also a time-dependent rate. In contrast, the exponential portion of the characteristic distribution used to represent the rates of smaller events is assumed to follow a Poisson process.

The following section introduces the two most common models to represent time-dependent behavior for specified (often characteristic) rupture scenarios and compares these to the time-independent Poisson model. For this Poisson case, the average rate of occurrence is constant and equal to the reciprocal of the mean recurrence interval. For the time-dependent models representing elastic rebound behavior, the expected rate of occurrence drops shortly after an event has taken place, and then increases as time elapses. Exactly how these models increase, and how they behave for long interevent times, depends upon the model and is discussed in the following section.

3.8.1 Time-Dependent Models for Individual Scenarios

The default assumption that is usually made in seismic hazard studies is that a Poisson process describes earthquake occurrence. This assumption implies that interevent times are exponentially distributed. A time-dependent extension of such a model is an inhomogeneous Poisson process where rates of occurrence of different events can vary $\lambda(M = m; t)$.[30] However, a more common approach is to directly model the time-dependence of particular ruptures (often characteristic ruptures). If elastic rebound is a valid theory, the likelihood of an event occurring within a given time window should depend strongly upon the elapsed time since the previous major event on this fault. In contrast, the representation of seismicity as a Poisson process dictates that the likelihood of observing an event

---

[30] Aftershock PSHA, presented in Section 7.7.1, is a typical example of such an approach.

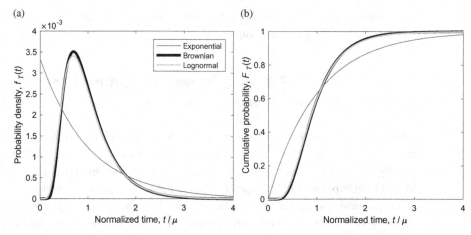

Fig. 3.38   Properties of temporal distributions used to represent earthquake occurrence. Panel (a) shows probability density functions for the next rupture time, while panel (b) shows the corresponding cumulative distribution functions. All time scales are normalized by the mean recurrence interval for the source.

in a given time window is independent of the time of the last event. Time-dependent models specify that occurrence times near the average recurrence interval for the rupture are more likely, while short times are less likely.

Two common models for arrival times for these characteristic events are the lognormal distribution and the Brownian Passage Time (BPT) distribution (Ellsworth et al., 1999; Matthews et al., 2002).[31] While the form of the lognormal distribution is well known, the BPT distribution is less so, and its PDF and CDF are defined as

$$f_T(t) = \sqrt{\frac{\mu}{2\pi\alpha^2 t^3}}\ \exp\left[-\frac{\left(t-\mu\right)^2}{2\mu t \alpha^2}\right]\tag{3.71}$$

$$F_T(t) = \Phi\left(\frac{t-\mu}{\alpha\sqrt{\mu t}}\right) + \exp\left(\frac{2}{\alpha^2}\right)\Phi\left(-\frac{t+\mu}{\alpha\sqrt{\mu t}}\right).\tag{3.72}$$

In these equations, μ is the mean recurrence interval for the rupture scenario, and α is known as the aperiodicity parameter, which plays a similar role to the logarithmic standard deviation of a lognormal distribution. Ellsworth et al. (1999) suggested that a value of $\alpha = 0.5$ is consistent with their analysis of 37 recurrent earthquake sequences.

As shown in Figure 3.38a, the PDFs for these distributions appear very similar. This figure demonstrates the scaling of the exponential distribution for interevent times (corresponding to a Poisson process) and compares this with the lognormal and Brownian passage time distributions. The PDFs for the common time-dependent models are far more intuitive for events that repeat themselves with some degree of regularity.

Matthews et al. (2002) present the conceptual benefits of the BPT model in terms of it having a more realistic instantaneous failure rate for long elapsed times since the past rupture. Therefore, while Figure 3.38 shows that the BPT and lognormal models are similar, the BPT model is generally favored in practice.

---

[31] Other distributions have been proposed (e.g., Cornell and Winterstein, 1988; Rundle et al., 2006), but are less commonly adopted. Additional time-dependent models for general earthquake sequences are briefly discussed in Section 3.8.4.

Although time-dependent distributions like the BPT model appear to reflect our theoretical and conceptual understanding of earthquake processes, they are always challenging to calibrate. The finding of Ellsworth et al. (1999) that $\alpha \approx 0.5$ for several sequences in locations around the world is promising in that if this degree of variability is relatively stable, then we may be able to make a reasonable assumption about the aperiodicity and focus upon the estimation of the mean recurrence interval.[32] However, this still presents a significant challenge as few locations worldwide have sufficiently long records to constrain average recurrence interval estimates. Indeed, Sykes and Menke (2006) considered many crustal and subduction regions and observed variations in the coefficient of variation of recurrence intervals.

However, it is not always necessary to have information about contiguous sequences of repeat ruptures. Section 3.4.2 showed that average long-term slip rates could be determined by averaging the total accumulated slip over some time period. Now, *if* some typical rupture size can be assumed appropriate for a given fault, source-scaling relations can be used to estimate the expected extent of slip from an individual characteristic event. Combining the long-term slip rate with these per-event slip estimates provides a means to estimate the hypothesized rupture scenario's average recurrence interval.

3.8.2 Time-Dependent Hazard for a Single Source

The models discussed in the previous section provide the probability distributions for recurrence times. These distributions describe the statistics for sequences of characteristic events on a given source. However, when performing time-dependent hazard calculations for a given site, it is essential to know *where* we currently are in a particular sequence. Three key pieces of information are required when looking to model time-dependent hazard from a given fault (or rupture scenario):

• The mean recurrence interval for the scenario;
• The variation in recurrence intervals for the scenario(s)
• When the last rupture consistent with this scenario took place.

In some situations, the previous rupture associated with this scenario will have taken place during the observation period, so that a precise estimate will exist. Paleoseismic evidence may also provide an estimate of the time of the past rupture. However, such estimates will also have associated uncertainty. In many other cases, the only available constraint is the knowledge that the scenario has *not* occurred for at least some period (often the historical period of observation). These different types of constraints have distinct influences on the probability that the next scenario will occur in some future period of interest (Field and Jordan, 2015).

As discussed by Field and Jordan (2015), the probability that a characteristic rupture takes place in a time interval Δt, given that an amount of time t_p has elapsed from the time of the previous rupture to the beginning of the interval Δt, is defined by

$$P\left(\Delta t \mid t_p\right) = \frac{\int_{t_p}^{t_p+\Delta t} f_T(t)dt}{\int_{t_p}^{\infty} f_T(t)dt} = \frac{F_T(t_p + \Delta t) - F_T(t_p)}{1 - F_T(t_p)}. \tag{3.73}$$

---

[32] This is an example of the *ergodic assumption* to be discussed in Chapter 8.

When the elapsed time since the previous event is not precisely known, it must be represented by a distribution $f_{T_p}(t_p)$. In this case, the expected value of the probability that the event occurs in the interval Δt is defined by[33]

$$E\left[P\left(\Delta t \mid t_p\right)\right] = \int_0^\infty P\left(\Delta t \mid t_p\right) f_{T_p}(t_p) dt_p \tag{3.74}$$

where $P\left(\Delta t \mid t_p\right)$ corresponds to the expression previously presented in Equation 3.73.

For the case where the time of the previous event is unknown, it is necessary to consider all possible times for the previous event and to assume that each of these possibilities is equally likely. One might suspect that the probability of some event occurring then becomes similar to the equivalent Poisson probability, but this is not the case. The average rate of occurrence, λ, is defined in terms of the mean recurrence interval as $\lambda = 1/\mu$. According to the Poisson model, the probability that at least one event occurs in some time period Δt is defined by

$$P(\Delta t) = 1 - e^{-\Delta t/\mu}. \tag{3.75}$$

However, Field and Jordan (2015) demonstrate that the appropriate probability for the case where the time of the previous event is unknown is given by

$$P\left(\Delta t \mid t_{p,u}\right) = \frac{\Delta t - \int_0^{\Delta t} F_T(\tau) d\tau}{\mu}. \tag{3.76}$$

Equation 3.76 is valid, irrespective of the particular form of the temporal distribution $F_T(t)$, provided that it is a time-dependent distribution; i.e., $F_T(t + \Delta t) \neq F_T(t)$. Here, $t_{p,u}$ is used to represent the case that the time since the previous event is unknown.

Figure 3.39 compares these probability estimates using both the Poisson and BPT distributions. For the BPT distribution, two different values of the aperiodicity are assumed. The value of $\alpha = 0.2$ corresponds to a model where events occur relatively periodically, while the $\alpha = 0.7$ case represents a model with more variation between successive events. The figure demonstrates that the extent to which the time-dependent BPT models provide estimates that differ significantly from a Poisson assumption varies considerably with both the forecast interval and recurrence interval. When the forecast interval is very short compared with the recurrence interval, the Poisson and BPT estimates do not vary significantly from one another. Similarly, as the forecast interval becomes large compared with the recurrence interval (of less practical relevance), then the differences between the time-dependent and time-independent predictions diminish.

The scenario outlined above, where the time since the previous event is entirely unknown, does not often occur in practice. More common is that the time since the previous event is unknown, but a constraint exists because no characteristic events have occurred on this fault throughout the historical period of observation. In this case, rather than treating t_p as being entirely unknown, we can impose the constraint that $t_p \geq t_H$, where t_H is the period of historic observation. In this case, the probability that an event occurs in some future interval Δt is (Field and Jordan, 2015)

$$P\left(\Delta t \mid t_{p,u} > t_H\right) = \frac{\Delta t - \int_{t_H}^{t_H + \Delta t} F_T(\tau) d\tau}{\int_{t_H}^\infty \left[1 - F_T(t)\right] dt}. \tag{3.77}$$

---

[33] The expectation is evaluated over the positive real domain from 0 to ∞ as t_p is strictly positive. The distribution $f_{T_p}(t_p)$ must provide support only over this same domain.

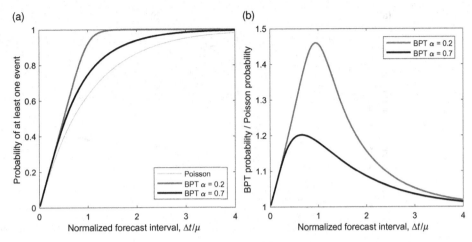

Fig. 3.39 Probability of at least one event occurring in some forecast interval Δt as a function of the recurrence interval, μ. The horizontal axes show $\Delta t/\mu$. Panel (a) shows the actual probabilities obtained using two BPT distributions along with the time-independent estimates from a Poisson model. Panel (b) shows the extent to which the time-dependent probabilities exceed the Poisson estimates.

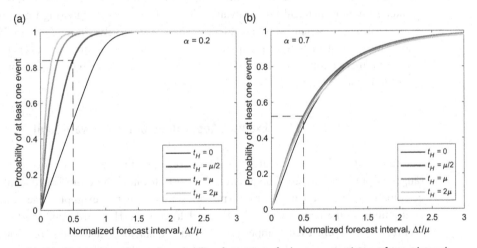

Fig. 3.40 Influence of the historic open interval upon the probability of occurrence of at least one event in some forecast interval. Panels (a) and (b) show $\alpha = 0.2$ and $\alpha = 0.7$, respectively.

Figure 3.40 shows the influence of the length of the "open historical interval" t_H, for two cases of aperiodicity. For example, when both the length of the historical interval, t_H, and the forecast interval, Δt, are equal to $t_H = \Delta t = \mu/2$, then the probability of observing at least one event is 84% for $\alpha = 0.2$, while it is only 52% for $\alpha = 0.7$. Figure 3.40a, corresponding to relatively periodic rupture cycles with $\alpha = 0.2$, shows a strong dependence upon the length of the open interval. In particular, as ruptures typically have successive intervals close to the mean interval, the probability of at least one event increases drastically as the length of the open interval first approaches and then exceeds the mean recurrence interval. In contrast, Figure 3.40b, corresponding to a far less periodic cycle with $\alpha = 0.7$, is much less sensitive to the value of t_H. In this case, having a long open interval is not particularly unusual for a sequence where there is a significant degree of variation in successive intervals.

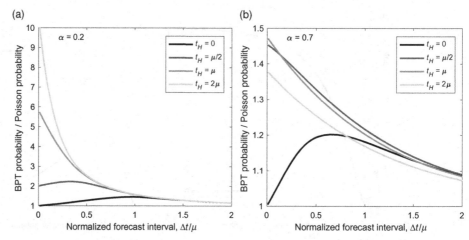

Fig. 3.41 Ratios of BPT:Poisson probabilities as a function of the normalized forecast interval $\Delta t/\mu$. Panels (a) and (b) show $\alpha = 0.2$ and $\alpha = 0.7$, respectively, using very different scales.

Figure 3.41 demonstrates the extent of this dependence upon the interval in another way by plotting the time-dependent probabilities shown in Figure 3.40 as ratios with respect to the corresponding time-independent Poisson values. The consequence of ignoring time-dependent effects is particularly severe when rupture sequences show a relatively strong degree of periodicity, and a significant amount of time has elapsed since the previous rupture. However, these consequences diminish rather rapidly as the aperiodicity increases, as can be seen through a comparison of the two panels in Figure 3.41.

3.8.3 Example Application for the Hayward Fault

The sensitivities shown in Figures 3.39–3.41 correspond to generic situations, but cover the typical ranges of practical interest. This section demonstrates how these time-dependent probabilities may be computed for a realistic scenario to provide a more concrete example. One of the primary sources contributing to the hazard for Yerba Buena Island is the Hayward Fault, particularly the southern section of this fault. Lienkaemper (2002) and Lienkaemper and Williams (2007) have presented an extensive history of past earthquake ruptures on this fault spanning more than 1800 years. Estimates of the ruptures' magnitudes vary, but $M = 6.9$ is a reasonable average. Furthermore, both Lienkaemper and Williams (2007) and Parsons (2008) have estimated the BPT parameters of the mean recurrence interval and aperiodicity for this fault.

Written evidence confirms that the most recent major event on this fault took place in 1868. The penultimate event is believed to have occurred early in the eighteenth century, but estimates of this event's timing are available only through paleoseismic investigations. Lienkaemper and Williams (2007) estimate that the penultimate event most likely occurred in 1725. They also provide a 68% confidence interval of [1695,1777] and a 95% confidence interval of [1658,1786]. They find that the mean recurrence interval for the fault is $\mu = 170 \pm 82$ years and that the fault has an aperiodicity of $\alpha = 0.48$. On the other hand, Parsons (2008), using a very different technique, find a mean recurrence interval of $\mu = 210$ years and an aperiodicity of $\alpha = 0.6$.

For these example calculations, two hypothetical scenarios are considered. In the first case, it is assumed that the most recent event to have occurred is the circa 1725 event, for which paleoseismic

Table 3.11. Time-dependent calculations for the Southern Hayward Fault. LW07 = Lienkaemper and Williams (2007) and P08 = Parsons (2008). Probabilities are for a forecast interval of $\Delta t = 50$ years

Model	Start of interval	μ	α	$\Delta t/\mu$	μ_{t_p}	σ_{t_p}	$P(\Delta t \mid t_p)$	$E[P(\Delta t \mid t_p)]$
LW07	1850	170	0.48	0.294	125	41	0.422	0.409
P08	1850	210	0.60	0.238	125	41	0.301	0.293
LW07	2020	170	0.48	0.294	152	–	0.455	–
P08	2020	210	0.60	0.238	152	–	0.321	–

analyses have provided only an approximate estimate of the timing. With this information, the challenge is to compute the probability that at least one event will occur in a 50-year forecast interval that begins in 1850, given the knowledge that no rupture has occurred since the 1725 event.[34] In the second case, a more contemporary example is considered where we know that the most recent event occurred in 1868 and we are interested in the probability that at least one event on this fault will occur within a 50-year forecast interval commencing in 2020 (and assuming that the current historical open interval extends to this time). In both scenarios, we assume that the "future" event will have similar properties to the previous characteristic events on the Hayward Fault. We will make use of the BPT parameters of both Lienkaemper and Williams (2007) and Parsons (2008).

Table 3.11 presents a summary of the overall calculations that are made in these two cases. When the time since the past event is uncertain (for the interval starting in 1850), a distribution for the time of the past event is constructed from the information provided by Lienkaemper and Williams (2007). Specifically, it is assumed that the standard deviation of the elapsed time σ_{t_p} is equal to half of the range from their 68% confidence interval, equal to $\sigma_{t_p} = 41$ years. A normal distribution for the time of the past event is then assumed, with a mean equal to $\mu_{t_p} = 1850 - 1725 = 125$ years and a standard deviation of σ_{t_p}. Also, the 95% confidence interval is used to truncate this normal distribution so that we use an asymmetric truncated normal distribution. This distribution is selected to provide a degree of realism to the computations required when the actual time of the most recent event is uncertain. As discussed in Biasi et al. (2002) and Biasi (2013), real PDFs for past event times are often relatively complicated and cannot always be described in terms of a common probability distribution.

Table 3.11 demonstrates how sensitive the probability estimates can be to the parameters of the time-dependent distribution. Recall that the estimate of the mean recurrence interval from Lienkaemper and Williams (2007) was 170 ± 82 years and so the alternative interval considered in Table 3.11 from Parsons (2008) is well within this range. However, this difference in mean recurrence interval, coupled with the 0.12 difference in aperiodicity, leads to a roughly 40% difference in the computed probability that at least one event will occur in the forecast interval of $\Delta t = 50$ years. To appreciate why such a large difference can arise, it is worth visualizing the distributions that underlie these calculations.

Figure 3.42 shows the PDFs for the two considered BPT distributions. The lower aperiodicity of $\alpha = 0.48$ proposed by Lienkaemper and Williams (2007) gives a distribution that is far more peaked around the mode than the case for $\alpha = 0.60$ from Parsons (2008). In Figure 3.42a, the full marginal PDFs are shown, while in Figure 3.42b, due account is taken of the fact that no subsequent event has occurred in the period between 1725 and the start of the forecast interval in 1850. This knowledge allows us to truncate and renormalize the density functions. Truncation is applied at the "present"

---

[34] Retrospectively, we know this interval contains the time of the next event that occurred in 1868.

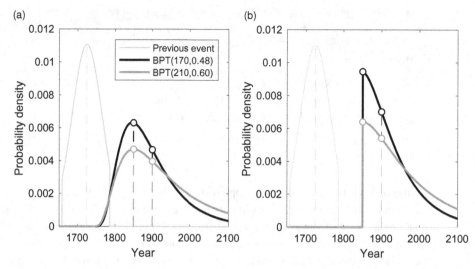

Fig. 3.42 Panel (a) shows the marginal PDFs for two BPT distributions relevant as soon as an event occurs in 1725. Panel (b) shows the corresponding conditional PDFs given that the previous event occurred in 1725 and that no subsequent event occurred prior to 1850. Previous event times are shown as a truncated normal distribution.

time (the start of the forecast interval), and the marginal density values are divided by the probability content remaining in the upper tail of the marginal distribution ($1 - F_T (t = 1850)$).

The lower aperiodicity causes higher probability density values in the region of interest, and hence a steeper gradient in the cumulative distribution function. This steeper gradient then means that the probability of at least one event in the forecast interval is significantly higher for the $\alpha = 0.48$ case. This relationship is visualized in Figure 3.43, where the probability of at least one event in the forecast interval of $\Delta t = 50$ years is plotted against time. In this figure, the probability values are plotted against year, where year is interpreted as being the year at the start of the forecast interval. Therefore, even for 1725, there is a non-zero probability because we are reflecting the chance (albeit still small) that at least one event will occur over the next 50-year period. The particular values highlighted by the markers correspond to the values presented in Table 3.11 and are associated with this roughly 40% difference in probability. Note that this difference increases further as the year corresponding to the forecast interval's start increases.

The example calculations presented correspond to the simple case where a well-defined characteristic event occurs with some degree of periodicity on a given fault. In many practical situations, the reality will be more complicated than this due to fault segmentation. The issue that arises, in this case, is that the same portion of a given fault may rupture in multiple different characteristic scenarios. For example, the Southern Hayward Fault may rupture by itself, but it may also rupture in conjunction with the Northern Hayward Fault, or the southern extension of this fault. Each of these rupture scenarios can be considered a characteristic scenario with its own mean recurrence interval and degree of aperiodicity. However, the details of the segmentation are not always known as clearly as for the Hayward Fault, and paleoseismic information will not always be able to resolve a clean history of ruptures and accurate segmentation information. In models like UCERF3 (Field et al., 2015), where all ruptures are represented by combinations of segments that rupture together, this issue of how to define appropriate times since the last event, mean recurrence intervals, and aperiodicity is far more complicated. Treatment of such cases is not discussed here, but Field et al. (2015) provide a discussion of these issues and a suggested modeling approach for a highly segmented fault system.

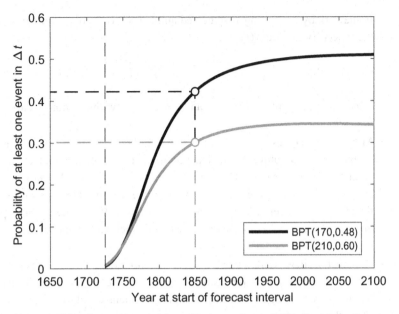

Fig. 3.43 Probability of at least one event in a forecast interval of $\Delta t = 50$ years as a function of the elapsed time since the previous event. Here, the previous event is assumed to have taken place in 1725 and the probabilities are shown in each year assuming that this year is the start of a $\Delta t = 50$ year forecast interval.

3.8.4 Time-Dependence within Earthquake Sequences

Sections 3.8.1–3.8.3 focused on modeling the time-dependence of particular rupture scenarios. These approaches are conceptually consistent with elastic rebound theory introduced in Section 2.4.1 in that the likelihood of an event depends upon the time since the previous event. However, it was also noted in Section 2.4.3 that events occurring within fault networks cause interactions that complicate the simple cycle implied by elastic rebound. The aperiodicity α within the BPT model of Section 3.8.1 allows for variations away from cyclic repetitions of characteristic ruptures. However, the model itself is still applied to that specific rupture scenario.

Many models have been proposed to accommodate more general time-dependence over a range of different rupture scenarios within an earthquake sequence. Models that explicitly include time-varying rates of earthquake occurrence are referred to as *nonhomogeneous* models and can be used to compute nonhomogeneous hazard estimates. A number of nonhomogeneous models for earthquake occurrence have been proposed (e.g., Anagnos and Kiremidjian, 1984, 1988; Cornell and Winterstein, 1988), and these are briefly summarized in the following section.

General Classes of Time-Dependent Models

The main groups of time-dependent models are the following (Anagnos and Kiremidjian, 1988; Cornell and Winterstein, 1988):

- **Nonhomogeneous Poisson processes**: These models receive further attention in the context of aftershock hazard analysis (Section 7.7.1). The basic framework is that a constant Poisson rate λ is replaced with a time-dependent rate $\lambda(t)$. They may be embedded within a more traditional framework by computing an expected number of events over some time window ΔT by integrating the time-varying rate $\lambda(t)$.

$$\Lambda(\Delta T) = \int_{t}^{t+\Delta T} \lambda(\tau)d\tau. \tag{3.78}$$

This expected number, $\Lambda(\Delta T)$, can then be used within a homogeneous Poisson process for this time window.

- **Markov processes**: Markov and semi-Markov processes encompass models for which the probability of a future event depends upon information about the most recent event, but not additional earlier events. The core difference between the Markov and semi-Markov variants is related to the assumptions of how this future event probability varies over the elapsed time since the most recent event (Anagnos and Kiremidjian, 1988). The so-called *slip-predictable* and *time-predictable* classes of models discussed by Cornell and Winterstein (1988) can be viewed as types of semi-Markov models.

- **Renewal models**: Within these models, the earthquake process "resets" after each event, and there are various alternatives for the distribution of interevent times. The renewal models' common feature is that the interevent time is an independently and identically distributed random variable. Thus, the time of the most recent event is essential, but other information, such as the magnitude of that event, is not.

- **Triggering models**: Triggering models encapsulate the concepts previously discussed within Section 2.4.3 where the occurrence of events influences the likelihoods of future events due to static and dynamic stress transfer effects. These models can operate as deterministic predictors, but may also include stochastic components.

While many proposals have been put forward within the above classes, these models are not often utilized in seismic hazard analysis practice for two reasons. First, they require the specification of more parameters than their Poisson counterparts and are generally more challenging to calibrate as a consequence. This is compounded by the relatively short observational periods for instrumental seismicity, and the significant uncertainties associated with timing estimates of large events from paleoseismological investigations (e.g., Mosca et al., 2012). Second, and as explained further in Chapter 6, in many hazard analyses, the objective is to compute the time-independent average rate of occurrence, so that hazard estimates relate to arbitrary time windows rather than specific time windows. That does not imply that time-dependence can be ignored. However, it does mean that analysts often assume that their observational periods are sufficiently long that the average through the time-dependent fluctuations of event rates does not deviate significantly from the underlying average rate.

Despite the above complications, two time-dependent models that have received a reasonable level of attention in practice are discussed in the following section.

ETAS and EEPAS Models

The Epidemic Type Aftershock Sequence (ETAS) (Hawkes and Adamopoulos, 1973; Ogata, 1988) and Every Earthquake a Precursor According to Size (EEPAS) (Evison and Rhoades, 2004; Rhoades and Evison, 2004) models share the common attribute that they enable a rate of activity to be defined that is spatiotemporally varying. ETAS models have been shown to outperform Poisson models (Console and Murru, 2001), and have been successfully implemented within significant hazard and risk projects (e.g., Field et al., 2017b; Bourne et al., 2018). The ETAS model can be thought of as representing short-term clustering processes (Gerstenberger et al., 2016). On the other hand, EEPAS models are more relevant for medium- to long-term clustering processes. They have been shown to perform well in multiple active regions around the world (e.g., Rhoades and Evison, 2006; Rhoades, 2007; Gerstenberger et al., 2016; Rhoades and Christophersen, 2019).

The basic assumption of ETAS is that the current rate of earthquake activity at any particular point in space and time is the cumulative result of a background activity rate and the influence of previous events that have occurred in the vicinity of the point in question. In this framework, it is not essential to distinguish between foreshocks, mainshocks, and aftershocks, because every event is effectively treated as though it is a mainshock. Each event will have an effect on future rates that will scale with the size of the event and will decay in time. Therefore, larger events will have effects that persist for much longer than smaller events.

Mathematically, ETAS models can be represented as

$$\lambda(t, \boldsymbol{x}) = \lambda_0 \mu(\boldsymbol{x}) + \sum_{i|t_i < t} \lambda_T(t, \boldsymbol{x} | m_i, t_i, \boldsymbol{x}_i) \tag{3.79}$$

where $\lambda(t, \boldsymbol{x})$ is the earthquake rate density (the number of events per unit time per unit spatial volume) of events having magnitudes greater than $m_{\min}$, t represents time, and $\boldsymbol{x}$ is a location in space. The term λ_0 is the total background rate of events with a magnitude above $m_{\min}$, and $\mu(\boldsymbol{x})$ represents the spatial density of the background earthquakes.

The summation term represents the modification to the rate associated with events that have occurred before time t. The influence of these events is defined by

$$\lambda_T(t, \boldsymbol{x} | m_i, t_i, \boldsymbol{x}_i) = \frac{k \, 10^{a(m_i - m_{\min})}}{(t - t_i + c)^p} D(\boldsymbol{x}, \boldsymbol{x}_i) \tag{3.80}$$

with $D(\boldsymbol{x}, \boldsymbol{x}_i)$ defining the spatial distribution of triggered events. Note that these contributions effectively combine a term that is consistent with the modified Omori rate with a distance taper.

The EEPAS model was developed after recognizing the *precursory scale increase* phenomenon (Evison and Rhoades, 2004), in which major shallow earthquakes are preceded by smaller events over a medium to long period (Rhoades and Evison, 2004). Just as the ETAS model assumes that every earthquake is effectively a mainshock event that spawns aftershocks, the EEPAS model effectively assumes that every earthquake is a precursor to subsequent seismicity.

The mathematical framework for the EEPAS model is similar to the ETAS model of Equation 3.79 in that they both define an event rate density as a function of time, t, and position, $\boldsymbol{x}$:

$$\lambda(t, m, \boldsymbol{x}) = \mu \lambda_0(t, m, \boldsymbol{x}) + \sum_{t_i \geq t_0} \eta(m_i) \lambda_i(t, m, \boldsymbol{x}; m_i). \tag{3.81}$$

However, in Equation 3.81, the term μ is a model constant that can be interpreted as the portion of events that occur without a precursor, while $\lambda_0(t, m, \boldsymbol{x})$ is the baseline rate density. The function $\eta(m)$ is defined so that the long-term rate density is consistent with the overall magnitude-frequency distribution for the region (which will often be assumed as a Gutenberg–Richter distribution). The final term $\lambda_i(t, m, \boldsymbol{x})$ is a contribution to the future rate density near $\boldsymbol{x}$ due to the ith earthquake.

This future rate density contribution is defined using

$$\lambda_i(t, m, \boldsymbol{x}; m_i) = w_i f_{1i}(t; m_i) g_{1i}(m; m_i) h_{1i}(\boldsymbol{x}; m_i) \tag{3.82}$$

where the functions $f_{1i}(t; m_i)$, $g_{1i}(m; m_i)$, and $h_{1i}(\boldsymbol{x}; m_i)$ can be thought of as kernels that distribute the precursory influence of each event in a manner depending upon the size m_i of the event. Larger earthquakes have a more substantial impact on the transient rate. That is, the functions f, g, and h increase in time, magnitude, and space with increasing m_i (Rhoades and Evison, 2004). Finally, the term w_i is a weight factor that could be employed to downgrade the effects of known aftershocks.

Exercises

3.1 A fault has a seismic moment release rate of $\dot{M}_0 = 6 \times 10^{17}$ Nm/yr. If it creates only earthquakes with average displacements of 4.0 m, a rupture length of $L = 50$ km, and a rupture width of $W = 15$ km, what is the average recurrence interval for the event?

3.2 A fault generates only magnitude 6.0 events when it ruptures. If the seismic moment release rate for this fault is 8.6×10^{16} Nm/yr, what is the average slip rate for the fault?

3.3 Paleoseismic investigations of active faults can constrain recurrence intervals for large earthquakes. Such studies can suggest that faults exhibit characteristic behavior. Discuss reasons why a fault may have characteristic behavior and why paleoseismic investigations may incorrectly infer such behavior.

3.4 Seismicity is generally assumed to be described by a Poisson process in seismic hazard studies. The plots of earthquake counts against time in panels (a) and (b) of the figure below show departures from what would be expected for a Poisson process. Explain what these departures are and provide rational explanations for why such departures may be observed in reality.

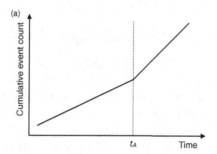 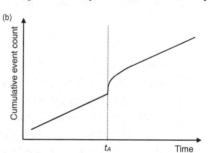

3.5 A geologist advises you to be conservative and to adopt a large value to represent the maximum magnitude a source can produce. Explain circumstances under which this advice makes sense as well as situations where the advice is poor.

3.6 A seismic source has seismicity characterized by a doubly bounded exponential distribution with a minimum magnitude of $m_{min} = 5$ and a maximum magnitude of $m_{max} = 7.5$. The total number of events per year with magnitude greater than or equal to m_{min} is 10.

(a) Explain why the exponential distribution is usually truncated at m_{min} and m_{max}.

(b) A seismicity analysis finds an estimate of the b-value for the source of 0.95, but with a standard deviation of 0.1. What are the rates of exceedance of $m = 7$ events corresponding to plus and minus one standard deviation on the b-value? Comment upon the symmetry of these rates with respect to the expected rate.

3.7 In PSHA, seismicity is usually assumed to be described by a Poisson process. Discuss the primary characteristics of a Poisson process and compare these with the assumptions that form the basis of elastic rebound theory.

3.8 Explain the fundamental difference between the moment magnitude scale and all other magnitude scales.

3.9 You must process a seismicity catalogue that contains a mixture of magnitude estimates in both the surface-wave and moment magnitude scales. An equation for converting from surface-wave magnitude to moment magnitude is given as

$$M = \begin{cases} 0.67M_S + 2.07 & \text{for } 3.0 \leq M_S \leq 6.1 \text{ and } \sigma_M = 0.17 \\ 0.99M_S + 0.08 & \text{for } 6.1 < M_S \leq 8.2 \text{ and } \sigma_M = 0.20. \end{cases}$$

If an event in your catalogue has a surface-wave magnitude estimate of $M_S = 5.7$ and this estimate is associated with a standard deviation of $\sigma_{M_S} = 0.15$, what is the mean converted moment magnitude and its associated standard deviation?

3.10 Consider the vertical strike-slip fault of length l shown in the figure below. Assuming that earthquake epicenters are uniformly distributed along the fault, derive an expression for the PDF of epicentral distance, r_e, with respect to the site shown in the figure.

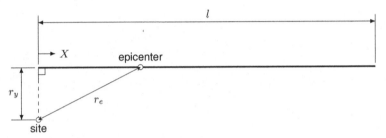

3.11 An earthquake in a region where $V_p = 6$ km/s and $V_s = 3.5$ km/s is recorded at two ground-motion instruments, A and B, the first being located exactly 100 km north of the second. If the P-S wave arrival difference is measured as 7.5 seconds at A and 5.0 seconds at B, and it is known that the epicenter of the earthquake lies on the north-south line joining the two instruments, compute the epicentral position and focal depth of the earthquake.

3.12 A seismic source has its magnitude-frequency distribution characterized by a doubly bounded exponential distribution. The bounding magnitude values are $m_{min} = 5.0$ and $m_{max} = 7.5$. The b-value for the source is 1.0 and the rate of events with magnitude greater than or equal to m_{min} is $\lambda(M \geq m_{min}) = 0.3$ events per year. Answer the following:

(a) How many events per year have magnitudes greater than or equal to $M6$?
(b) How many events per year have magnitudes less than 5.5?
(c) How many events per year have magnitudes between 5.5 and 6?
(d) If one observed this seismic source for an infinitely long time, what portion of the events generated by this source would have magnitudes between 6.2 and 6.7?
(e) What is the recurrence interval for earthquakes with magnitude greater than 7.25?

3.13 In Section 3.5.5 the parameters of the Gutenberg–Richter distribution were varied and activity rates and seismic moment release rates were computed. Repeat this exercise using a characteristic earthquake distribution consisting of an exponential distribution for small to moderate events and a uniform distribution for the characteristic events. Adopt the parameters from Figure 3.19, along with $\lambda(M \geq m_{min}) = 0.01$, as the base case.

3.14 Repeat the exercise in the previous question, but substitute a truncated normal distribution in place of the uniform distribution to represent characteristic earthquakes. Center the truncated normal distribution on m_{char}, and truncate over the same Δm_{char} magnitude range. Ensure that the probability of observing a characteristic event is preserved.

3.15 Compute and plot the rates of occurrence and exceedance for magnitude-frequency distributions with the following properties:

(a) Doubly bounded Gutenberg–Richter distribution with parameters: $m_{min} = 4$, $m_{max} = 8$, $b = 1.0$, $\dot{M}_0 = 1 \times 10^{19}$ Nm/yr.

(b) Characteristic distribution with parameters the same as (a), and additionally: $m_{char} = 7.75$, $\Delta m_{char} = 0.5$, and $f_c = 0.1$.

Compare and contrast the relative rates as a function of magnitude. Given the trends, and considering the moment rate release as a function of magnitude, discuss the size of earthquakes that release the majority of elastic energy stored across fault surfaces.

4   Empirical Ground-Motion Characterization

As discussed in Chapter 1, probabilistic seismic hazard analysis requires models for predicting ground-motion intensity that can be coupled with the seismic source models addressed in Chapters 2 and 3. In this chapter, we focus on a particular class of models for ground-motion characterization referred to as *empirical ground-motion models* (GMMs) . Empirical GMMs take a rupture scenario (Chapter 3) and compute the probability distribution of an intensity measure that describes the ground motion. This probability distribution is then used in the calculation of seismic hazard in Chapter 6.

Learning Objectives

By the end of this chapter, you will be able to do the following:

- Evaluate a range of intensity measures used in the engineering characterization of ground motion.
- Describe available ground-motion observations used in the development of empirical ground-motion models.
- Describe the mathematical representation of empirical ground-motion models and their scaling with key predictor variables.
- Compute the distribution of ground motion for a given rupture scenario and determine probabilities of exceeding various levels of ground-motion intensity.
- Describe modeling uncertainty in empirical ground-motion models and apply multiple models and parametric uncertainties to quantify these epistemic uncertainties.

4.1   Introduction

Empirical GMMs predict the distribution of ground-motion intensities and are principally calibrated using observational data. Theoretical considerations also play a role in their development, but ultimately their consistency with observational data is a primary consideration. Figure 4.1 illustrates several of the key concepts in this chapter. The figure plots measures of ground-motion intensity from a particular database against the distance between the earthquake rupture and the recording station. Superimposed on these empirical data are predictions from an empirical GMM, represented by median values and 16th and 84th percentiles of the ground-motion distribution.

Several features of empirical GMMs are notable with reference to Figure 4.1.

- An empirical GMM predicts a particular ground-motion intensity measure, such as $SA(1 \text{ s})$, rather than the ground-motion time series (i.e., the variation of acceleration with time).
- Because the focus is on intensity measures, little attention is given to the underlying nature of the seismic waves (e.g., body or surface waves; see Section 5.4) that comprise the acceleration time series.

- The ground-motion intensity values generally increase as the earthquake magnitude increases and as the source-to-site distance decreases.
- Despite a large amount of data in total, there is a relative paucity of data for large magnitudes and short source-to-site distances.
- There is large variability in the ground-motion intensity values for a given magnitude and source-to-site distance, as illustrated by the significant difference between the 84th and 16th percentiles of the empirical GMM distribution.

4.1.1 Historical Terminology

Parametric equations to predict ground-motion intensity measures were first developed in the 1960s (Douglas, 2019). For several decades these equations were referred to as *attenuation relations*, the name being derived from the attenuation of ground-motion intensity with source-to-site distance. The growing complexity of model functional forms subsequently led to a shift in naming to *ground-motion prediction equations* (GMPEs), to acknowledge that such equations account for features beyond simply attenuation (e.g., magnitude, local soil conditions; see Section 4.5). More recently, the phrase *empirical ground-motion models* (empirical GMMs) is increasingly used. The GMM convention, used exclusively in this book, is a general phrase that applies equally to both empirically based models, which are the focus of this chapter, and physics-based models, which are the focus of Chapter 5.

This section is also a useful juncture to clarify the phrases *empirical* and *physics-based* GMMs to avoid perceptions of any implied hierarchy among these two approaches. Both approaches to

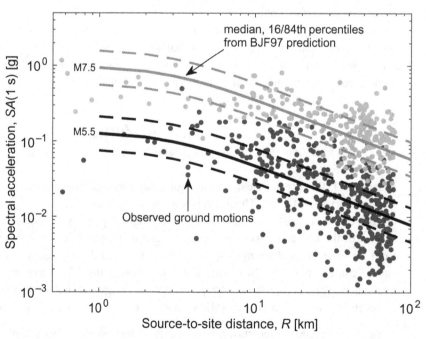

Fig. 4.1 Observed (pseudo) spectral acceleration, *SA*(1 s), data from the NGA-West2 ground-motion database and the empirical prediction of the Boore et al. (1997) (BJF97) model. The observed data is colored based on binned magnitude ranges centered at $M = 5.5$ (black) and $M = 7.5$ (gray), and with ± 0.2 magnitude unit widths. The solid and dashed lines represent the median and 16th/84th percentiles of the BJF97 prediction, respectively.

ground-motion modeling draw heavily upon physical theory and empirical calibration. As discussed further in Chapter 5, so-called physics-based methods seek to explicitly address many aspects of the underlying physics that give rise to ground motions, and predict the ground-motion time series directly. Despite this, physics-based methods require constitutive relationships that have been calibrated using empirical observations. In the same vein, as discussed later in this chapter, empirical models adopt mathematical forms guided by considerations of earthquake source, wave propagation, and site response physics. General advances in GMM prediction will, therefore, occur as a result of symbiotic progress in *both* empirical and physics-based methods.

4.2 Engineering Characterization of Ground Motion

Figure 4.2 illustrates several ground-motion acceleration time series that have been recorded from historical earthquakes. Because earthquake-induced ground motions are a complex seismic signal, it is often desirable to describe them through parameters, or summary statistics, referred to as ground-motion *intensity measures* (IMs). In this section, a variety of commonly used IMs are presented.

In general, ground-motion severity can be considered as a function of amplitude,[1] frequency content, and duration. All other things being equal, the larger the ground-motion amplitude, the more damage it will cause. Frequency content is important because the dynamic response of a system is a function of the similarity of the ground-motion excitation frequencies with the natural frequencies of the system of interest. Finally, ground-motion duration influences demands for both linear and nonlinear systems. For example, if the ground-motion amplitude and frequency content cause nonlinear response of the system, a longer duration will generally result in a greater accumulation of damage.

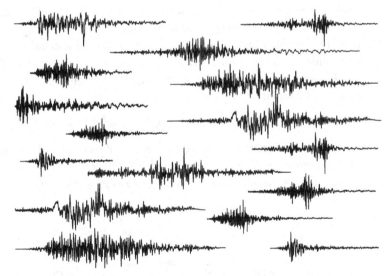

Fig. 4.2   Examples of representative ground-motion acceleration time series recorded in past earthquakes, illustrating their complexity.

[1] *Intensity* is sometimes used as a synonym for *amplitude* in this context, but we use the latter to avoid confusion with the broad definition of *intensity measure* (IM).

Ground-motion IMs typically represent one or more of the above three general aspects of ground-motion severity, and the following subsections outline commonly used IMs. From the outset, it is important to appreciate that various IMs focus on different aspects of the ground-motion time series. Therefore, multiple IMs are generally considered necessary to provide a comprehensive description of ground-motion severity.

4.2.1 Peak Amplitudes

The simplest quantification of a ground-motion time series is the maximum absolute amplitude. Because the ground motion is usually represented by acceleration, velocity, or displacement, there are three such quantities. We refer to these maximum amplitudes as the peak ground acceleration (PGA), peak ground velocity (PGV), and peak ground displacement (PGD), respectively.

Figure 4.3 illustrates the north-south (NS) time series of the ground motions recorded during the 1989 Loma Prieta earthquake at Yerba Buena Island and Treasure Island, which are two nearby stations located on rock and soil, respectively[2] (Finn et al., 1993). The acceleration time series of the ground motion was measured directly, and the velocity and displacement time series are obtained through single and double integration, respectively. For each of the figure panels, the corresponding PGA, PGV, and PGD values are also noted. Different vibration frequencies dominate the acceleration, velocity, and displacement time series due to the predominant frequency reducing with integration. Therefore PGA, PGV, and PGD indicate the ground-motion amplitude for high-, moderate-, and low-frequency bandwidths. The different nature of the two depicted ground motions also means that the relative PGA, PGV, and PGD amplitudes vary. For example, the PGD for Treasure Island is 3.1 times that at Yerba Buena Island, while the PGV value is 3.8 times larger.

4.2.2 Response Spectra

The peak response of a simplified structure—the single-degree-of-freedom (SDOF) system—is an important metric to understand ground-motion severity for structural systems. Figure 4.4 illustrates the basic process by which pseudo-acceleration response spectral ordinates are computed.[3] The ground-motion acceleration time series, $a(t)$, is imposed at the base of the SDOF system, and the relative displacement response, $x(t)$, of the single degree-of-freedom is computed as the solution to the equation of motion:

$$\ddot{x}(t) + 2\xi\omega_n\dot{x}(t) + \omega_n^2 x(t) = -a(t) \tag{4.1}$$

where ω_n and ξ are the angular natural frequency and damping ratio of the SDOF,[4] and the overdots denote time derivatives. The oscillator period, T, is given by $T = 2\pi/\omega_n$. The spectral displacement, $SD(T)$, is defined as the maximum absolute value of the displacement of the SDOF:

$$SD(T) = \max_t |x(t)| . \tag{4.2}$$

---

[2] These ground motions are repeatedly used as examples in this section.
[3] Additional background information on SDOF dynamics is given in Chopra (2016).
[4] The angular natural frequency is equal to $\omega_n = \sqrt{k/m}$, where k is the lateral stiffness of the oscillator and m is the mass. The damping ratio is the ratio of the damping coefficient, c, to the critical level of damping for the system. The critical damping is equal to $2m\omega_n$. Therefore, $\xi = c/(2m\omega_n)$.

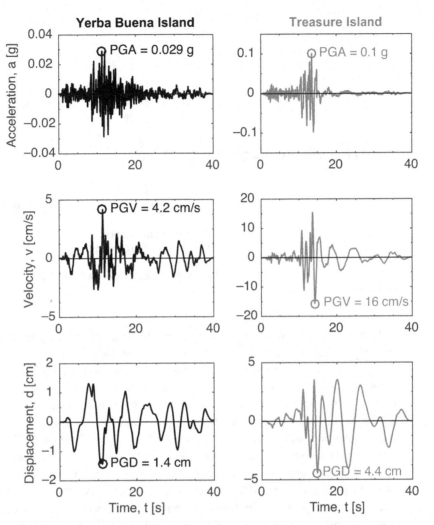

Fig. 4.3 Illustration of peak ground acceleration (PGA), velocity (PGV), and displacement (PGD) intensity measures computed for the north-south (NS) component ground motions recorded at Yerba Buena Island (rock) and Treasure Island (soil) during the 1989 Loma Prieta earthquake. The two sites are 75–77 km from the earthquake source and separated by a distance of approximately 2 km (Finn et al., 1993).

The pseudo-spectral acceleration, $SA(T)$, is then computed from the spectral displacement as

$$
\begin{aligned}
SA(T) &= \omega_n^2\, SD(T) \\
      &= \omega_n^2\, \max_t |x(t)|\,.
\end{aligned}
\tag{4.3}
$$

Spectral displacement and pseudo-spectral acceleration are both a function of the angular natural frequency and damping ratio. Therefore, strictly speaking, we should define them as $SD(T, \xi)$ and $SA(T, \xi)$, respectively. Because most spectral acceleration ordinates are computed for $\xi = 0.05$ (i.e., 5% of critical damping), we generally suppress the explicit notation of dependence on critical damping ratio, and sometimes even the dependence on T for brevity.

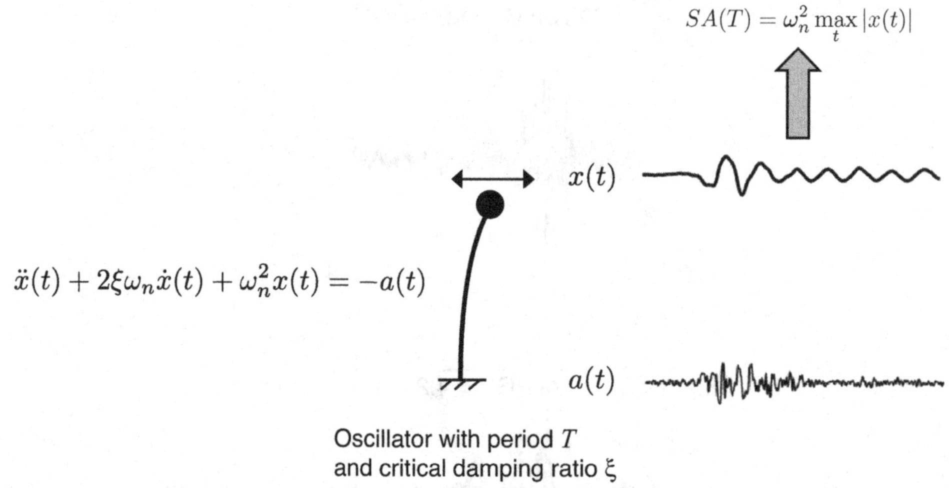

$$SA(T) = \omega_n^2 \max_t |x(t)|$$

$x(t)$

$$\ddot{x}(t) + 2\xi\omega_n\dot{x}(t) + \omega_n^2 x(t) = -a(t)$$

$a(t)$

Oscillator with period T
and critical damping ratio ξ

Fig. 4.4   Illustration of the response spectrum concept based on the peak relative displacement response, $x(t)$, of a SDOF system to the ground-motion excitation, $a(t)$.

The pseudo-spectral acceleration SA is useful in a structural engineering context because the maximum absolute displacement multiplied by the stiffness k yields the maximum base shear force (i.e., $V_b = k \times SD$) that the SDOF system must resist, and this is equal to the $m \times SA$ from Equation 4.3. SA is not strictly the total acceleration of the SDOF mass, which is the combination of the acceleration relative to the base and the acceleration of the base itself, hence the use of the term "pseudo-acceleration." For small levels of damping that are commonly considered, pseudo-acceleration and total acceleration are approximately equivalent (Clough and Penzien, 1975; Chopra, 2016); thus, for brevity, it is common to refer to the pseudo-spectral acceleration simply as the spectral acceleration.

The benefit of response spectral ordinates (e.g., SA and SD) is that they provide a direct estimation of the peak response of a SDOF system which approximates the specific (real) system of interest. For most natural periods of engineering interest, the response of a SDOF depends largely upon how much energy the earthquake excitation has near the natural frequency of the oscillator. Thus, spectral ordinates account for both the amplitude and frequency content of a ground motion.

Figure 4.5 illustrates the spectral acceleration ordinates of the Yerba Buena Island and Treasure Island ground motions as a function of oscillator period. For all oscillator periods, the SA values are larger for the Treasure Island ground motion, consistent with the ground-motion time series amplitudes shown in Figure 4.3. However, the two motions' relative spectral amplitudes vary as a function of the oscillator period. The maximum SA values for the Yerba Buena motion occur over a period range of approximately $T = 0.15$–0.8 s. In contrast, the corresponding maximum values for the Treasure Island motion occur over a period range of $T = 0.25$–1.5 s. This observation of longer "predominant" periods for a ground motion on soil, compared with that on rock, is common. Also, the SA values for very short oscillator periods (i.e., $SA(0.01$ s) values of 0.029 and 0.1 g) are equal to the PGA values depicted in Figure 4.3.[5]

While SA ordinates based on SDOF response are appealing in their simplicity, structural and geotechnical systems are typically sensitive to excitation at multiple vibration periods. If the system

---

[5] As $T \to 0$, the system is very stiff, $k \to \infty$, and the mass effectively moves as though it were rigidly connected to the ground. Therefore, the maximum absolute displacement of the system mass and the ground become equivalent.

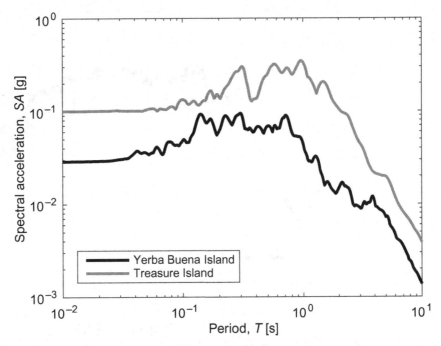

Fig. 4.5 Spectral acceleration ($\xi = 0.05$) intensity measures computed for the north-south component ground motions recorded at Yerba Buena Island and Treasure Island during the 1989 Loma Prieta earthquake.

response is dominated by the first-mode vibration period, T_1, then $SA(T_1)$ is often a valuable IM. However, shorter vibration periods are of interest for considering higher modes of vibration, and nonlinear seismic response leads to an elongation of the effective period associated with the first mode of vibration, so vibration periods greater than T_1 are also of interest.

4.2.3 Fourier Amplitude

The application of the Fourier transform to a ground-motion acceleration time series, $a(t)$, enables the ground motion to be expressed as a complex-valued function of frequency:

$$a(f) = \int_{-\infty}^{\infty} a(t)e^{-i(2\pi f)t}\,dt \tag{4.4}$$

where $a(f)$ is the Fourier transform of $a(t)$, often referred to as the "frequency-domain" representation of the acceleration. In practice, this equation is solved using the discrete Fourier transform:

$$a(f_k) = \sqrt{\frac{\Delta t}{N_T}} \sum_{n=0}^{N_T-1} a(n\Delta t)e^{-i(2\pi f_k)n\Delta t} \tag{4.5}$$

where f_k are the discrete frequencies at which the Fourier transform is computed, and Δt and N_T are the time increment and number of time steps, respectively. Rather than viewing the ground motion as an ordered set of acceleration values at particular times, in the frequency domain, we view the ground motion as the superposition of sinusoidal components with particular amplitudes and phase angles at each frequency.[6] Just as considering the entire set of acceleration–time pairs

[6] Noting Euler's identity, $e^{i\pi} + 1 = 0$, implies that $a(t) = \sum_k |a(f_k)|\sin[2\pi f_k t + \phi(f_k)]$.

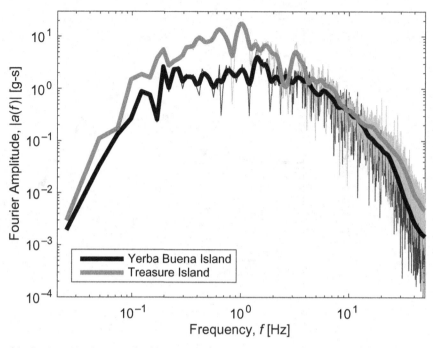

Fig. 4.6 Illustration of the Fourier amplitude spectra ($|a(f)|$) computed for the north-south component ground motions recorded at Yerba Buena Island and Treasure Island during the 1989 Loma Prieta earthquake. Thin lines result from the direct application of Equation 4.5, with frequency-smoothed values in bold lines.

provides a complete description of the ground motion, so too does the consideration of the entire set of amplitude–phase–frequency triplets. Because $a(f)$ is complex-valued, the absolute value (i.e., complex modulus, $|a(f)|$) corresponds to the amplitude of the sinusoid for a given frequency, while its complex argument, $\arg(a(f))$, is related to the phase angle associated with that frequency. Figure 4.6 illustrates the Fourier amplitude spectra of the Yerba Buena Island and Treasure Island ground motions. The "raw" amplitude spectra vary significantly with small changes in frequency, and therefore it is conventional to compute frequency-smoothed spectra,[7] which are shown in boldface lines in the figure.

Fourier amplitude spectra and response spectra are similar in that they both indicate the ground-motion intensity as a function of frequency or oscillator period. However, there are some important differences between the two.

A first difference is that the Fourier spectrum is a property of the ground-motion time series, as compared with the response spectrum that represents the peak response of a SDOF system to the ground-motion excitation. Because SDOF response is a simple proxy for the response of a more complex multi-degree-of-freedom (MDOF) structure, the use of response spectra by structural engineers is common. However, because response spectra are not a direct reflection of ground-motion acceleration records, comparison of response spectra can pose difficulties. In particular, when comparing empirical GMMs from different regions, the direct link between seismological

---

[7] Using the smoothing function of Konno and Ohmachi (1998) with $b = 40$.

theory and Fourier spectra make it more informative than response spectra (Bora et al., 2016). This is why Fourier spectra are more commonly adopted in physics-based simulation approaches (see Section 5.5.3).

A second difference is that Fourier spectral amplitudes vary significantly over small frequency ranges—reflecting physical phenomena and/or the finite frequency resolution for a given Δt. Therefore, smoothing is implemented for some applications. Smoothed spectra are illustrated in Figure 4.6. On the other hand, with response spectra, damped oscillators respond to excitation over a range of frequencies, producing an inherent smoothing of amplitudes with frequency.

4.2.4 Duration

Strong-motion duration is an important factor in ground-motion severity. All other things equal, a longer duration of shaking will produce greater resonance and result in greater cumulative damage during nonlinear dynamic response. Despite this, strong-motion duration is a secondary intensity measure, because it is only important if the amplitude and frequency content of the ground motion produce a large (i.e., nonlinear) response.

There are many definitions of duration, and a detailed discussion of their merits can be found in Bommer and Martinez-Pereira (1999). Here, three common duration measures are discussed.

Bracketed and Uniform Durations

Both the bracketed and uniform duration definitions require the specification of a threshold acceleration, A_T, which indicates ground-motion acceleration levels of practical concern. The bracketed duration, D_B, is defined as the time interval between the first and last exceedances of the threshold acceleration, which can be mathematically expressed as:

$$D_B = \max_t(t_{|a(t)|>A_T}) - \min_t(t_{|a(t)|>A_T}) \tag{4.6}$$

where $t_{|a(t)|>A_T}$ signifies time instants where the condition $|a(t)| > A_T$ is satisfied.

The uniform duration, D_U, is defined as the cumulative time over which the acceleration amplitude exceeds the threshold acceleration, which can be mathematically expressed as

$$D_U = \int_t I[|a(t)| > A_T]dt \tag{4.7}$$

where $I[.]$ is an indicator function, which equals 1 if the condition is true, and 0 otherwise.

Figure 4.7a illustrates uniform and bracketed durations for the Treasure Island ground motion using a threshold of $A_T = 0.05$ g. In practice, A_T should be specified based on the system being analyzed (e.g., the yield acceleration in a slope stability context; Bray and Travasarou, 2007). In this example, the bracketed duration is $D_B = 3.99$ s, and the uniform duration is $D_U = 1.09$ s. Because the Yerba Buena Island ground motion amplitude is less than 0.05 g (see Figure 4.3), it has zero uniform and bracketed durations for this threshold acceleration.

Significant Duration

An alternative definition of duration that does not require the specification of a threshold acceleration is to consider the time over which a specific percentage of the total energy imparted by the ground-motion time series occurs. It can be shown that the energy contained in the ground-motion time series

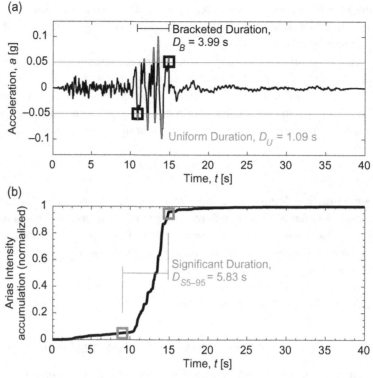

Fig. 4.7 Illustration of (a) uniform and bracketed, and (b) significant duration for the Treasure Island ground motion during the 1989 Loma Prieta earthquake. The uniform and bracketed durations use a threshold of $A_T = 0.05$ g, while significant duration is based on 5% and 95% thresholds.

is proportional to the cumulative squared acceleration. If we define t_X as the time at which $X\%$ of the total squared acceleration has accumulated, that is,

$$\frac{\int_0^{t_X} [a(t)]^2 dt}{\int_0^{\infty} [a(t)]^2 dt} = \frac{X}{100}, \tag{4.8}$$

then the significant duration, D_S, is the time difference between the accumulation of two different percentages of the cumulative squared acceleration, $X\%$ and $Y\%$ (where $Y > X$), mathematically:

$$D_{SX-Y} = t_Y - t_X. \tag{4.9}$$

The value of X is almost universally considered to be 5%, and the two most common values of Y are 75% and 95%. These two metrics are subsequently referred to as the 5–75% significant duration, D_{S5-75}, and the 5–95% significant duration, D_{S5-95}, respectively. Although the specific percentages (X and Y) considered are empirical, there is a loose physical basis for their selection. Generally, the P-wave arrivals for ground motions of engineering interest comprise 5% or less of the total accumulated squared acceleration, so t_5 approximately coincides with the S-wave arrival time. The time at which 75% of the total squared acceleration is accumulated, t_{75}, generally includes the arrival of the majority of S-wave phases, and t_{95} includes the majority of the S-wave and surface wave phases (Bommer et al., 2009). Hence D_{S5-75} and D_{S5-95} loosely relate to the duration of significant body and body+surface wave arrivals, respectively.

Figure 4.7b illustrates the 5–95% significant duration for the Treasure Island ground motion. In this instance, the value of $D_{S5-95} = 5.83$ s is similar to the bracketed duration in Figure 4.7a, except the time interval starts approximately 2 s earlier. Looking at the acceleration time series in Figure 4.7a, D_{S5-95} approximately reflects the duration of significant body and surface waves (see Section 5.4), with the surface waves after $t = 15$ s small in amplitude.

4.2.5 Cumulative Intensity Measures

Several ground-motion IMs consider all three general aspects of ground-motion severity (i.e., amplitude, frequency content, and duration). Two in particular that are frequently utilized are Arias intensity (Arias, 1970) and cumulative absolute velocity (EPRI, 1991), which are discussed together below to compare and contrast their similarities and differences.

Arias intensity, AI, is related to the integral of the squared acceleration:[8]

$$AI = \frac{\pi}{2g} \int_0^{t_{max}} a(t)^2 dt. \tag{4.10}$$

Cumulative absolute velocity, CAV, is defined as the integral of the absolute value of acceleration.

$$CAV = \int_0^{t_{max}} |a(t)| dt. \tag{4.11}$$

Other than their difference in the use of squared acceleration (AI) versus the absolute value of acceleration (CAV), these IMs vary in two other ways. First, AI has a scalar coefficient in front of the integral because it physically represents the amount of energy, per unit weight, of an infinite number of undamped SDOF oscillators with frequencies uniformly distributed in $(0, \infty)$; whereas CAV has no direct theoretical representation. Second, AI has historically been computed using Equation 4.10 with acceleration in m/s^2, giving AI units of m/s; whereas, when originally proposed, CAV utilized acceleration in g, giving CAV units of g-s. Since the metrics could potentially have the same units despite representing different aspects of ground-motion severity, this arbitrary difference in unit ensures that the resulting numbers have different orders of magnitude.

Both AI and CAV consider duration because of the use of time integration. They both also consider frequency content because, for a given amplitude, a lower frequency will lead to a larger value of the integral over a single cycle. Figure 4.8 illustrates the temporal evolution of AI and CAV for the two ground motions from the Loma Prieta earthquake previously considered. The Treasure Island record has larger AI and CAV because of its larger acceleration amplitudes over a greater time interval and broad range of vibration frequencies. Because AI is based on squared acceleration, and acceleration peaks tend to be of high frequency, then AI is more strongly correlated to high-frequency ground-motion IMs, whereas CAV is more correlated to moderate-frequency ground-motion IMs (Bradley, 2012a, 2015). Figure 4.8 illustrates this dependence via the large accumulation of AI during the strongest accelerations, while the CAV accumulation occurs over a longer time period. Therefore, AI is a more important IM for quantifying the seismic response of structures with higher natural vibration frequencies, while CAV has greater utility for structures with lower natural frequencies.

---

[8] From Parseval's theorem this integral is also equivalent to that over the squared Fourier amplitude in the frequency domain.

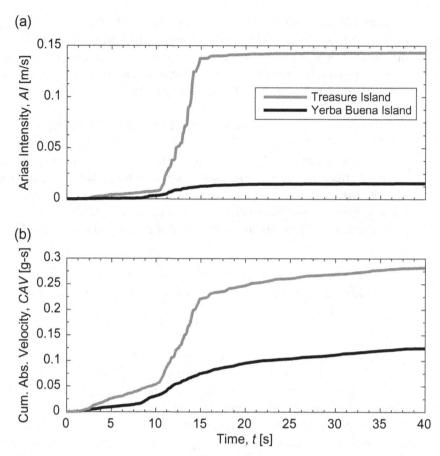

Fig. 4.8 Illustration of the temporal accumulation of (a) Arias intensity, and (b) cumulative absolute velocity for the north-south components of the Yerba Buena Island and Treasure Island ground motions during the 1989 Loma Prieta earthquake.

Finally, while AI and CAV consider all three aspects of ground-motion severity in a qualitative sense, this does not mean that they are superior to the previous IMs that consider only amplitude or frequency content. For example, neither AI nor CAV provides a precise IM for systems that respond to a narrow band of excitation frequencies.

4.2.6 Intensity Measures for Multi-Component Ground Motion

The discussion of IMs to this point has focused on a single component of ground motion; however, ground motions involve three translational and three rotational components. Typically in engineering analysis, the rotational components are neglected, and the vertical component receives less attention than its horizontal counterparts. We must still consider, however, horizontal shaking in all orientation directions. As a result, there is merit in obtaining a single IM metric that quantifies the intensity of shaking in a range of horizontal orientations.

The approach for representing two horizontal ground-motion components as a single IM has evolved over the past decade, but retracing these developments is instructive to understand the rationale for the current consensus. Traditionally, when a ground motion was recorded in two orthogonal horizontal directions, the IM in each recorded direction was computed and the geometric mean of the two values was taken as the IM for the ground motion:

$$IM_{\mathrm{GM}} = \left(IM_x \times IM_y\right)^{0.5} \tag{4.12}$$

where IM_x and IM_y are the IM values for the two orthogonal components (x and y) of the recorded ground motion, and IM_{GM} is the geometric mean IM value. This approach is simple, and the averaging of the two IM values reduces the ground motion database size by half, which is advantageous for empirical GMM calibration (Boore et al., 1997).

A limitation of the geometric mean calculation is that the IM_{GM} value depends on the in-situ orientation of the ground-motion recordings.[9] If the recording instrument were rotated by several degrees, the IM_x and IM_y values would change and consequently IM_{GM} also. The preferred refinement to address this is to compute the IM value for all horizontal orientations of the ground motion. Two common definitions are used when considering orientation-dependent ground-motion IMs. The first is the median IM value obtained across all orientations, which yields the so-called *RotD50* metric:

$$IM_{\mathrm{RotD50}} = \underset{\theta}{\mathrm{median}}\left[IM(\theta)\right] \tag{4.13}$$

where $IM(\theta)$ is the IM value of the horizontal ground motion along orientation θ. The "RotD" subscript denotes that the computation is being performed over a range of rotation angles,[10] and the "50" denotes the 50th percentile (i.e., median) value over these orientations is being computed (Boore, 2010). Alternatively, the maximum value over all orientations is:

$$IM_{\mathrm{RotD100}} = \underset{\theta}{\max}\left[IM(\theta)\right]. \tag{4.14}$$

To illustrate, Figure 4.9 shows the SDOF oscillator response in the horizontal plane, for two different oscillator periods, when subjected to the Yerba Buena ground motion from the 1989 Loma Prieta earthquake. The spectral acceleration based on the time series for orientation θ, SA_θ, divided by SA_{RotD100}, is also shown with a dashed line. The orientation-dependence of the ground-motion intensity is referred to as *directionality*. Comparing the two figure panels, the response is very polarized for $SA(T = 2.0\text{ s})$ in Figure 4.9b, and relatively less so for $SA(T = 0.3\text{ s})$ in Figure 4.9a. One numerical measure of polarity is the ratio $IM_{\mathrm{RotD100}}/IM_{\mathrm{RotD50}}$, which has values of 1.17 and 1.41 for the two examples in Figure 4.9. An unpolarized ground motion would have the same intensity in all orientations, and therefore a ratio of 1.0. A perfectly polarized ground motion would have spectral response along a single orientation, and this would yield a ratio of $\sqrt{2} = 1.41$ (Boore, 2010; Shahi and Baker, 2014).

While the rotation-independence of both IM_{RotD50} and IM_{RotD100} avoid limitations of past formulations, the decision of which definition to use will be a function of the properties of the particular structure being considered (e.g., Stewart et al., 2011). Importantly, the specific definition used in seismic hazard and subsequent seismic response analysis calculations must be consistent (Baker and Cornell, 2006c).

---

[9] The arithmetic mean for AI (i.e., $(AI_x + AI_y)/2$), and quadratic mean for Fourier amplitude (i.e., $([a_x(f)]^2 + [a_y(f)]^2)^{0.5}$), are both orientation-independent and therefore commonly adopted. Otherwise, a RotD-type definition (e.g., Equation 4.13) is most common for other IMs.

[10] The "D" term indicates that the orientation varies for the different IMs calculated, specifically in the context of SA ordinates for different oscillator periods (Boore et al., 2006; Boore, 2010).

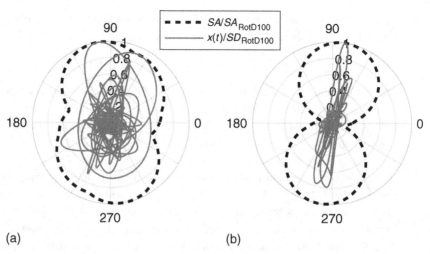

Fig. 4.9   Directionality of the Yerba Buena Island ground-motion spectral response in the horizontal plane as a function of time, and orientation-dependent spectral acceleration: (a) $T = 0.3$ s ($IM_{RotD100}/IM_{RotD50} = 1.17$) and (b) $T = 2.0$ s ($IM_{RotD100}/IM_{RotD50} = 1.41$). Orientations are in degrees, and the spectral accelerations (SA) are normalized by the maximum value over all orientations $SA_{RotD100}$, while the oscillator displacement response $x(t)$ is normalized by the maximum spectral displacement $SD_{RotD100}$.

4.2.7 Intensity Measures for Specific Problem Types

Inevitably, the selection of IMs is problem-specific. As will become evident in subsequent chapters on seismic risk (Chapter 9) and ground-motion selection (Chapter 10), IM choice influences the variability and uncertainty in the prediction of: (1) this IM via GMMs and (2) the risk metric of interest conditioned on the ground-motion IM. Thus, it is preferable to identify and adopt IMs that are precisely predicted using GMMs and strongly correlated with the risk metric of interest.

Example applications where the IMs introduced are frequently used include the following:

- PGA: Simplified liquefaction and other geotechnical assessments (e.g., Idriss and Boulanger, 2008)
- PGV: Deformation of underground structures and infrastructure subject to transient ground displacements (e.g., Jeon and O'Rourke, 2005)
- $SA(T)$: The most common intensity measure, ubiquitous in structural applications, with the period T based on the first-mode of vibration (e.g., Shome et al., 1998)
- D_S: Quantification of the effects of long-duration ground motions on structural response (e.g., Hancock and Bommer, 2006)
- AI: Seismic slope displacements (e.g., Bray and Travasarou, 2007)
- CAV: Triggering of liquefaction (specifically, a modified version, CAV_5) (e.g., Kramer and Mitchell, 2006).

The above are common choices for their respective problems, but ultimately the analyst has the freedom to adopt alternatives. An additional consideration is that IMs are simplified representations of ground-motion severity. It is possible to develop more specific IMs that contain additional information about the problem of interest, and therefore predict a specific risk metric with lower uncertainty (e.g., Cordova et al., 2001; Luco and Cornell, 2007; Marafi et al., 2016). However, as these IMs become more problem-specific, they also become less generally applicable. As a result, problem-specific IMs may not have empirical GMMs necessary to characterize the hazard in this form. Furthermore, GMMs often require extrapolation beyond the constraints of empirical data (e.g., to large magnitudes and small source-to-site distances), and problem-specific IMs are less likely to have a seismological basis for their extrapolation (see Section 4.7.2). Thus, there is a tension between using general IMs (which are widely used, and for which empirical GMMs exist and are constrained by theoretical considerations) and more specific IMs (for which empirical GMMs may need to be specifically developed, and may have extrapolation complications).[11] In the discussion and examples that follow, emphasis is given to general IMs for these reasons.

Given the limitation of simplified IMs for describing a complex ground-motion time series, another option is to utilize a vector of IMs to describe ground-motion severity. Ideally, each component of the vector would enhance the prediction of response and describe a unique aspect of ground-motion severity. Seismic hazard analysis can be performed directly for a vector of IMs (Chapter 7). Then fragility or vulnerability functions that utilize this vector of IMs can be used to compute seismic risk metrics (Chapter 9). Alternatively, seismic hazard analysis can be performed for a single (scalar) IM (Chapter 6). Then, the distribution of a vector of IMs conditioned on the scalar IM can be utilized to obtain seismic fragility or vulnerability functions (most commonly via ground-motion selection and response-history analyses, discussed in Chapter 10).

4.3 Ground-Motion Databases

Instruments capable of recording strong motions for quantitative analysis, first developed in the early 1900s, have been critical to characterizing ground motions. Recordings from historical earthquakes are archived in several databases, such as the following:

1. Pacific Earthquake Engineering Research Center (PEER): Several databases of worldwide earthquakes for shallow crustal, stable continental, and subduction zone earthquakes (`http://peer.berkeley.edu/peer_ground_motion_database`)
2. National Research Institute for Earth Science and Disaster Resilience (NIED): Surface (K-Net) and subsurface (Kik-Net) databases of earthquakes occurring in Japan (`http://www.kyoshin.bosai.go.jp/`)
3. Pan-European Engineering Strong Motion (ESM): Earthquakes from the European-Mediterranean and Middle-East regions (`https://esm-db.eu`)
4. New Zealand GeoNet: Earthquakes occurring in New Zealand (`http://www.geonet.org.nz/`)
5. Center for Engineering Strong Motion Data (CESMD): Ground motions from several different national networks (`https://strongmotioncenter.org/`).

The availability in the past two decades of public strong-motion databases has enabled a significant increase in research on empirical ground-motion characterization, through ease of access and greater consistency in the underlying metadata. The process by which "raw" measurements from instruments are signal-processed to optimize the *signal* (from the earthquake-induced ground-motion excitation)

---

[11] One avenue is the development of *conditional* GMMs, which construct a model for a specific IM based on an existing model of a more general IM (e.g., Tothong and Cornell, 2006; Macedo et al., 2019).

from the *noise* (all other forms of vibration) is a specialized topic (e.g., Boore and Bommer, 2005) that is particularly important for the use of older recordings from analog instruments, but less of a concern for modern digital instruments. Many (but not all) databases now provide ground-motion time series and IMs that have already been signal-processed following best-practice procedures (e.g., Ancheta et al., 2014). As a result, we do not discuss ground-motion processing and filtering explicitly in this chapter.

4.3.1 Worldwide Distribution of Earthquakes and Recorded Ground Motions

As discussed in subsequent sections, ground motions observed in different locations worldwide are often grouped to develop empirical GMMs. Because there is often a paucity of data in a single region, grouping provides more data to constrain GMMs. However, this does come at the cost of the developed model being generic for a type of region, and not necessarily representing the specific conditions near the site of interest (see Chapter 8).

Figure 4.10 shows the global distribution of earthquakes from four of the databases listed in the previous subsection. In viewing Figure 4.10, it is noted that ground-motion databases for engineering applications are a subset of all earthquakes that have been recorded in that region. Specifically, observations from small-magnitude earthquakes may be excluded if such earthquake events are deemed unimportant for engineering applications. Similarly, small-amplitude ground motions may also be excluded (for having low signal-to-noise ratios). The minimum magnitude and ground-motion amplitude limits of interest have, however, been decreasing over time as instruments become increasingly sensitive and small-amplitude ground motions are used to understand regional variations in ground-motion prediction.

4.3.2 Increase in Ground-Motion Database Size over Time

The size of ground-motion databases is increasing exponentially with time. This exponential trend is a result of increases in (1) the number of instruments that are deployed; (2) the sensitivity of instruments, allowing ground motions of lower amplitudes to be routinely recorded; and (3) regional and international collaborations resulting in databases that are no longer specific to single countries, with an increased emphasis on smaller-magnitude events to understand regional variations. Figure 4.11 illustrates the increase in the number of recorded ground motions as a function of year, as found in the NGA-West2 database (Ancheta et al., 2014). The linear trend of logarithmic numbers of ground motions against time indicates an approximately 4.5-fold increase in ground motions every 10 years.

To understand the relative contributions of the three cited causes of the exponential increase in empirical ground-motion database size, Figure 4.12 illustrates the number of ground motions recorded with time, separated by their *PGA* value. A significant portion of the ground motions since approximately the year 2000 are associated with small-amplitude ground motions (that have $PGA < 0.01$ g).[12] In contrast, before approximately the year 2000, there are very few ground motions with small amplitudes. In part, this reflects the transition from analog to digital ground-motion measurement. Older small-amplitude ground motions from analog instruments were either not recorded or are considered poor-quality and not included in databases.

---

[12] The large increase in 1999 results from the dense recording of the 1999 Chi-Chi, Taiwan, earthquake sequence.

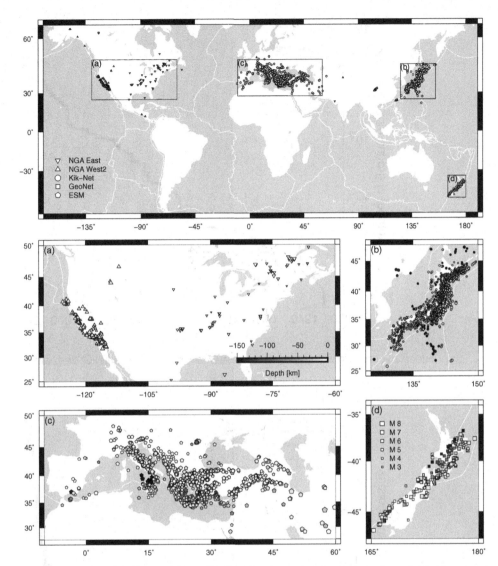

Fig. 4.10 Earthquake locations of several publicly available earthquake-induced ground-motion databases. In panels (a)–(d) earthquake depth is indicated by the amount of shading, and magnitude by the marker size.

4.3.3 Magnitude and Distance Distribution

The predominance of small-amplitude ground motions seen in Figure 4.12 is the result of the distribution of ground-motion records as a function of their magnitude and distance from the earthquake source (referred to as *source-to-site distance*; Section 3.7.3). Figure 4.13 illustrates the magnitude and source-to-site distance distribution of the ground motions in the NGA-West2 database. For context, the magnitude and distance values that result in *PGA* values of approximately 0.05 g and 0.01 g are also shown. The majority of the ground motions result from smaller magnitude earthquakes and are recorded at larger source-to-site distances. These two trends should be intuitive

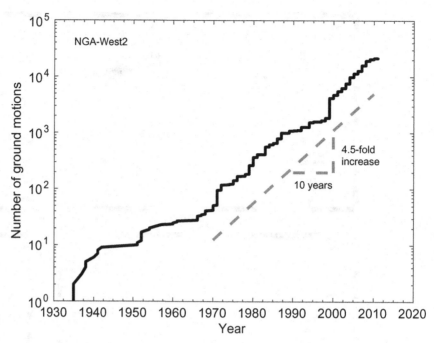

Fig. 4.11   Increase in the size of empirical ground-motion databases over time illustrated for the NGA-West2 database.

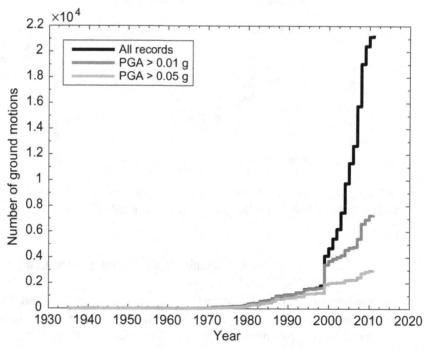

Fig. 4.12   Number of ground motions in the NGA-West2 database as a function of the year recorded and their PGA value. The majority of ground motions have small amplitudes, with only 2993 of 21,298 ground motions in the database having $PGA > 0.05$ g.

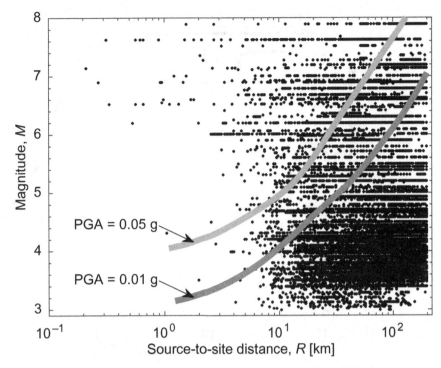

Fig. 4.13 Magnitude and source-to-site distance distribution of recorded ground motions in the NGA-West2 database. Magnitude-distance thresholds of $PGA = 0.01$ and 0.05 g are based on the median predictions from Chiou and Youngs (2014)

for the following reasons. First, we know that large-magnitude events are less likely to occur,[13] so the likelihood of recording a ground motion from a large magnitude earthquake is relatively low. Second, as the source-to-site distance doubles, the area on the ground surface increases by a factor of four. As a result, it is much more likely to observe a ground motion at a large distance from an earthquake than at a small distance. These trends imply that we will always be data-poor in the top-left-hand corner of a magnitude-distance distribution such as Figure 4.13, and our GMMs will have more uncertainty for these conditions. These are also often the scenarios which dominate the seismic hazard, as we will see in Chapter 6.

4.3.4 Maximum Usable Response Spectral Period

The quality (e.g., signal-to-noise ratio) of a ground-motion time series is a function of the specific IMs of interest (i.e., both the signal and noise amplitudes vary with frequency). As a result, it may not be appropriate to compute a specific IM type for a given time series, while other IM types may be computed without complication. This notion can be most easily illustrated by examining the appropriateness of computing different SA ordinates as a function of oscillator period, T.

A decreasing signal-to-noise ratio at long vibration periods is problematic for ground motions that were recorded on older analog instruments and do not have reliable signals at longer vibration periods (Chiou et al., 2008; Ancheta et al., 2014). As a result, the notion of *maximum usable period* refers to the maximum T for which SA ordinates can be reliably computed due to signal-to-noise

---

[13] As discussed in Chapter 3, generally, there is a 10-fold decrease in the rate of occurrence of events for every unit increase in magnitude.

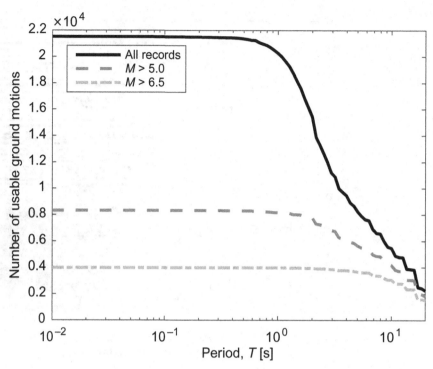

Fig. 4.14 Available number of ground-motion recordings as a function of oscillator period, T, from the NGA-West2 database. Ground motions from larger magnitude earthquakes are more likely to be usable at longer oscillator periods.

considerations. The usable period range is related to properties of the bandpass filter applied to isolate signal from noise (e.g., Akkar and Bommer, 2006; Douglas and Boore, 2011). For example, Figure 4.14 illustrates the number of usable ground motions as a function of the SA period of interest for the NGA-West2 database. For $T < 1$ s, generally, all ground motions provide usable SA values, but for $T > 1$ s this number decreases rapidly, with only $\approx 20\%$ of ground motions providing usable SA values for $T = 10$ s. However, the maximum usable period is strongly linked to the earthquake magnitude.[14] The trend for $M > 5.0$ and $M > 6.5$ events in Figure 4.14 illustrates the reduction in usable ground motions is less significant for moderate- and large-magnitude earthquakes.

4.4 Mathematical Representation

Having discussed ground-motion metrics of interest, and available empirical data, we now discuss the calibration and use of empirical GMMs.

4.4.1 Probabilistic Form

Empirical GMMs take the following basic probabilistic form:

$$IM \sim \mathcal{LN}(\mu_{\ln IM}, \sigma^2_{\ln IM}), \tag{4.15}$$

[14] Specifically, the ground-motion corner frequency, which is correlated with earthquake magnitude; see Section 5.5.3.

which states that IM is a lognormal random variable with mean and standard deviation of $\mu_{\ln IM}$ and $\sigma_{\ln IM}$, respectively. The lognormal distribution is almost universally adopted for empirical GMMs because empirical studies have shown that this is an appropriate assumption (Section 4.4.3).

Because of the relationship between lognormal and normal distributions (Section A.5.2), $\ln IM$ has a normal distribution. The normal random variable $\ln IM$ is commonly expressed in the following form:

$$\ln IM = \mu_{\ln IM}(rup, site) + \sigma_{\ln IM}(rup, site) \cdot \varepsilon \qquad (4.16)$$

where $\mu_{\ln IM}$ and $\sigma_{\ln IM}$ are explicitly noted as functions that depend on the earthquake rupture, rup, and the site considered, $site$; and, ε is a standard normal random variable that represents the variability in $\ln IM$.

The functional dependence of Equation 4.16 on rup and $site$ is specific to individual empirical models (see Section 4.5), but as concrete examples, rup can be represented with variables such as magnitude (M), $site$ by variables such as the 30-m time-averaged shear-wave velocity ($V_{S,30}$), and the geographical relationship between rup and $site$ by the source-to-site distance (R). Positive and negative values of ε result in larger- and smaller-than-average values of $\ln IM$, respectively.

A property of the lognormal distribution is that $\exp(\mu_{\ln IM})$ is the median of the distribution of IM (Section A.5.2). As a result, reference is often made to GMMs providing models for "the median and the standard deviation." This notion may appear initially confusing because it is the median of IM and the standard deviation of $\ln IM$. However, this property of lognormal distributions provides equivalence between the mean of $\ln IM$ and the median of IM. Because of the assumption of lognormality, these two parameters are sufficient to provide a complete description of the distribution of IM for any given rupture scenario and site characterization. Further discussion on this apparent variability is provided in Section 4.4.3.

4.4.2 Example GMMs: BJF97 and CY14

To illustrate the development of empirical GMMs, as well as their use in prediction, we will make use of two example GMMs. Over decades of development and refinement, the prediction models for $\mu_{\ln IM}$ and $\sigma_{\ln IM}$ in Equation 4.16 have become complex, consisting of many functional terms and tables containing dozens of coefficients. These modern models are not easily implemented via hand calculations, so here we will use an older and simpler (but obsolete) model to provide insight via example calculations. We will also consider one modern and more complex model. Its application is identical in concept, but we will avoid discussing details associated with its complete implementation.

The Boore, Joyner, and Fumal (1997) Model (BJF97)

The mean and standard deviation of the BJF97 model (Boore et al., 1997) takes the form[15]:

$$\mu_{\ln SA} = a_0 + a_1 (M - 6) + a_2 (M - 6)^2 + a_3 \ln\left(\sqrt{R^2 + a_4^2}\right) + a_5 \ln(V_{S,30}) \qquad (4.17)$$

$$\sigma_{\ln SA} = a_6 \qquad (4.18)$$

---

[15] Different variable names for the coefficients than in the original BJF97 have been used to maintain simplicity. In this instance, R is the Joyner–Boore definition on source-to-site distance (R_{jb}; see Section 3.7.3).

where $a_0 - a_6$ are SA period-dependent empirical coefficients which are obtained from regression analysis on the ground-motion database that was used by BJF97. Thus, there are three predictor variables (M, R, and $V_{S,30}$) and seven empirical coefficients at each period.

To simplify the use of this model even further for the prediction of $SA(1\text{ s})$ from strike-slip earthquakes,[16] substituting in the values of the coefficients from Boore et al. (1997) gives:

$$\mu_{\ln SA(1\text{ s})} = -3.4415 + 1.42M - 0.032M^2 - 0.798\ln\left(\sqrt{R^2 + 8.41}\right) - 0.698\ln(V_{S,30}) \tag{4.19}$$

$$\sigma_{\ln SA(1\text{ s})} = 0.52 \tag{4.20}$$

Chiou and Youngs (2014) Model (CY14)

The CY14 model is a modern equivalent of the BJF97 model developed using the NGA-West2 empirical database. It uses 12,244 records from 300 different earthquakes. There are a total of 11 predictor variables (i.e., M, R, ...), and 45 coefficients (either empirically determined or theoretically constrained) in the model. For comparison, the BJF97 model used 112 records, 14 earthquakes, three predictor variables, and seven coefficients. Furthermore, as well as the manyfold increase in data, state-of-the-art empirical GMMs (such as CY14) increasingly use insights from ground-motion simulation (see Chapter 5) to account for phenomena that are not adequately constrained through empirical data alone (see Section 4.7.2). Although we do not provide explicit expressions for this model here, we will make use of the model throughout this book within numerical examples.

4.4.3 Apparent Variability in Empirical Models

Figure 4.1, and other comparisons in this chapter, shows that there is significant scatter in observed ground-motion intensities, even after accounting for the effect of predictor variables such as magnitude, source-to-site distance, and site conditions, among others. In the context of Equation 4.16, the difference between a ground-motion observation ($\ln im$) and the mean prediction ($\mu_{\ln IM}$) is referred to as the prediction residual, $\Delta = \sigma_{\ln IM} \cdot \varepsilon$. Visually this can be thought of as the vertical difference between the data and the median model prediction in Figure 4.1.

An important characteristic of prediction residuals is the distribution that they conform to. Figure 4.15 illustrates a typical quantile-quantile plot of normalized prediction residuals (ε) in comparison to the quantiles from a standard normal distribution for two different IMs. The data plot approximately on one-to-one lines beyond two standard deviations from the mean, and within the statistical rejection bounds to nearly four standard deviations. Such empirical results indicate that the assumption of a normal distribution for $\ln IM$ is justified up to 3.5–4 standard deviations (Bommer et al., 2004; Jayaram and Baker, 2008; Strasser et al., 2008).

Because the lognormal distribution is theoretically unbounded, truncation of the prediction distribution is sometimes employed for practical convenience. Figure 4.15, however, illustrates that there is no empirical basis for truncation below approximately four standard deviations. The details of the distribution so far into its tails are practically immaterial for most applications that are not concerned with seismic hazard at very low exceedance rates.

---

[16] Coefficient a_0 is a function of the earthquake style of faulting.

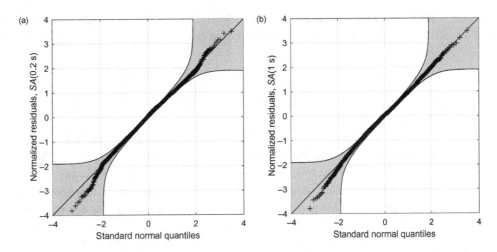

Fig. 4.15 Quantile-quantile plots comparing normalized residuals (ε) against standard normal quantiles, illustrating the appropriateness of the normality assumption of $\ln SA(T)$ for (a) $T = 0.2$ s and (b) $T = 1$ s. The data lying within the gray "acceptance" region indicate the normality assumption cannot be rejected at the 95% confidence level, using the Kolmogorov–Smirnov criterion (Aldor-Noiman et al., 2013). Data are from $M > 5$ earthquakes in the NGA-West2 database.

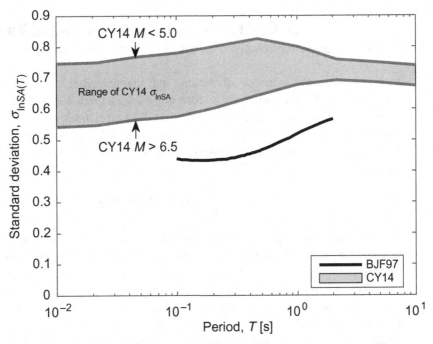

Fig. 4.16 Apparent aleatory standard deviation of the BJF97 and CY14 GMMs as a function of oscillator period. The BJF97 model $\sigma_{\ln SA}$ is only a function of oscillator period, T, whereas the CY14 model is a function of several variables, principally T and M.

In addition to the distribution type, the size (i.e., standard deviation) of the variability is also of primary interest. Figure 4.16 illustrates the standard deviation of the BJF97 and CY14 GMMs as a function of oscillator period. The (lognormal) standard deviation of the BJF97 model varies from approximately 0.44 to 0.57. To get a sense for the significance of this level of variability, if the median ground-motion intensity was $IM_{50} = \exp(\mu_{\ln IM}) = 0.5$ g and the standard deviation was $\sigma_{\ln IM} = 0.6$, then 68% (i.e., $\mu \pm 1\sigma$) of intensity measure values would be expected to fall in the interval:

$$\exp(\mu_{\ln IM} \pm \sigma_{\ln IM}) = [0.275 \text{ g}, 0.911 \text{ g}] \tag{4.21}$$

which represents approximately a factor of 3.3 in the ground-motion intensity. Extending the example to cover 95% of IM values (i.e., $\mu \pm 1.96\sigma$) leads to a factor of 10.5. Given the size of this variability, it is critical to consider it in the predictive model. Hence, these models must provide a probability distribution, rather than just a single value.[17] Later PSHA calculations in Chapter 6 will account for the possibility of unlikely outcomes, such as extreme intensities much larger than the predicted mean.

The BJF97 model considers standard deviation solely as a function of the oscillator period. Contemporary GMMs predict standard deviations that also vary as a function of magnitude, source-to-site distance, and site parameters. For example, the CY14 standard deviation is a function of earthquake magnitude, as shown in Figure 4.16, with larger variability for smaller magnitude events. This larger variability for smaller magnitudes can be attributed to several causes: (1) the larger uncertainty in the estimate of earthquake location and size for smaller events, (2) more "averaging" of randomness in the source rupture properties for larger-sized events, and (3) the increased influence of near-surface attenuation at smaller magnitudes.

4.5 General Trends in Empirical Data and Models

In this section, the influence of rupture and site properties on IMs is considered. We will consider these effects qualitatively by comparing several ground-motion time series, and then more quantitatively by comparing median empirical GMM predictions with observed data. The intent of this section is to develop an empirical understanding of the effects of these parameters on ground-motion intensity. Additional physics-based insights into the reasons for these trends are presented in Chapter 5.

4.5.1 Magnitude

All other factors equal, ground-motion amplitude increases with increasing magnitude, since a larger amount of energy is released from the earthquake source. Ground-motion duration also increases as a result of a longer duration of rupture on the causative fault plane. Figure 4.17 illustrates several ground-motion time series corresponding to earthquake events with different magnitudes, in which the acceleration amplitude and duration increase as magnitude increases. Although not apparent from the acceleration time series, the longer rupture duration of larger magnitude earthquakes also decreases the so-called corner frequency of the radiated ground-motion spectrum, resulting in an increase in the low-frequency ground-motion amplitudes (see Section 5.5.3).

Building on the intuition gained from Figure 4.17, Figure 4.18 illustrates the variation in $SA(1 \text{ s})$ amplitudes of a subset of ground motions in the NGA-West2 database as a function of M, for two different source-to-site distance ranges. In addition to the $SA(1 \text{ s})$ values from observations, the median

---

[17] This is one of the fundamental problems with so-called 'deterministic' seismic hazard analysis, see Sections 1.3 and 6.10.3

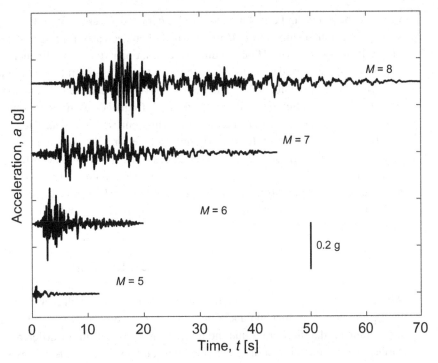

Fig. 4.17 Effect of magnitude on ground-motion time series characteristics illustrated through example recorded ground motions. Other than the variation in magnitude, the recordings have source-to-site distances of approximately $R = 20$ km and soil conditions associated with $V_{S,30} = 500$ m/s.

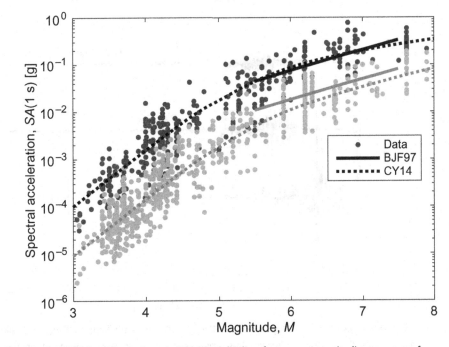

Fig. 4.18 Effect of magnitude on $SA(1\,\text{s})$ amplitudes from the NGA-West2 database for two source-to-site distance ranges of $R = 10$–20 km (black) and 80–100 km (gray), and $V_{S,30} = 400$–600 m/s. The median empirical predictions from the BJF97 and CY14 models are shown as lines for the mid-point distances of the data ($R = 15$ and 90 km), $V_{S,30} = 500$ m/s, and strike-slip faulting conditions. The BJF97 model is shown only over its stated $M5.5$–7.5 range of applicability.

values of the BJF97 and CY14 models are also depicted. Over the $M = 3$–5 range, there is a relatively steep gradient in the $\ln SA(1\ \text{s})$–M trend, which then decreases for larger M values. This decrease in the gradient of $\ln IM$ with M occurs due to the variation in the corner frequency of the ground-motion spectrum with magnitude (Fukushima, 1996), and further saturation for large-magnitude near-source scenarios (Bommer et al., 2007). The extent of this slope change is a function of the IM considered, with the slope reduction most significant for high-frequency ground-motion IMs.

Figure 4.18 shows that both models are consistent with the data. However, the median of the BJF97 model is shown only for the magnitude range for which this model was developed (5.5–7.5). The approximately linear nature of the model in $\ln SA(1\ \text{s})$–M space (the quadratic M scaling in Equation 4.19 is very weak at $T = 1$ s) means that it would lead to overprediction of the $SA(1\ \text{s})$ amplitudes if extended to lower and higher M values. In contrast, the CY14 model exhibits the "concave from below" trend that is seen in the empirical data and is also applicable over the entire $M3$–8 range of the data.

4.5.2 Source-to-Site Distance

All other factors equal, ground-motion amplitude decreases with increasing distance as a result of both geometric spreading and anelastic attenuation (see Section 5.4.4). While amplitude decreases, the significant duration of ground motion increases, as a result of different waves (i.e., P-, S-, and surface-waves) traveling at different velocities, and also the effects of wave scattering. Figure 4.19

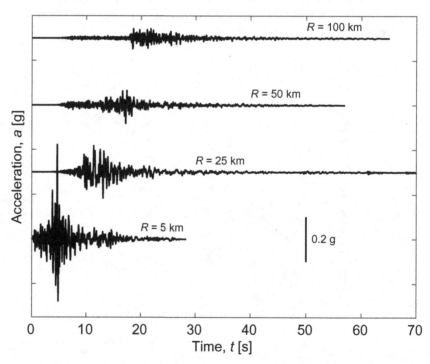

Fig. 4.19 Effect of source-to-site distance on ground-motion time series characteristics illustrated through example recorded ground motions. Other than the variation in source-to-site distance, the recordings have magnitudes of approximately $M = 6.2$ and soil conditions associated with $V_{S,30} = 500$ m/s.

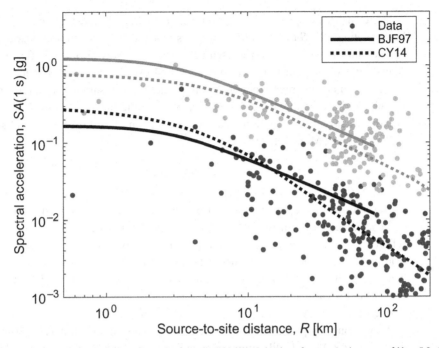

Fig. 4.20 Effect of source-to-site distance on $SA(1\text{ s})$ amplitudes from the NGA-West2 database for magnitude ranges of $M = 5.3–5.7$ (black) and 7.3–7.7 (gray), and $V_{S,30} = 400–600$ m/s. The median empirical predictions from the BJF97 and CY14 models are shown as lines for $M = 5.5$ and 7.5, $V_{S,30} = 500$ m/s, and strike-slip faulting. The BJF97 model is shown only over its stated range of applicability of $R < 80$ km.

illustrates several ground-motion time series with different source-to-site distances, for which the acceleration amplitude decreases and duration increases with increasing source-to-site distance. Although not apparent from the acceleration time series, a larger source-to-site distance also changes the ground motion's relative frequency content. High-frequency components of motion experience a greater number of oscillations for a given distance and are consequently attenuated more significantly due to energy loss per cycle (referred to as *anelastic attenuation*; see Section 5.4.4).

Figure 4.20 illustrates the variation in $SA(1\text{ s})$ amplitudes of a subset of ground motions in the NGA-West2 database as a function of R for two different magnitude ranges. The median values of the BJF97 and CY14 models are also depicted. As will be discussed in Chapter 5, from a fundamental physics perspective, there are three principal effects of source-to-site distance reflected in the slope of the empirical model predictions in Figure 4.20. First, for the approximate range of $R = 10$–100 km, *geometric spreading* (see Section 5.4.4) is dominant and $\ln SA$ decays linearly with $\ln R$. Second, at longer source-to-site distances, anelastic attenuation becomes increasingly important, and the rate of amplitude decay with distance increases. Because anelastic attenuation occurs due to material adsorption, it is a function of the number of vibration cycles and therefore increases with distance and frequency. As a result, the effects of anelastic attenuation are not pronounced in this figure for $SA(1\text{ s})$ but are increasingly significant for SA at shorter oscillator periods. Finally, at short source-to-site distances, the finite rupture geometry is important, and there is a "saturation" of the ground-motion amplitude as the source-to-site distance approaches zero (e.g., Abrahamson et al., 2008).

In viewing the empirical functional form of the BJF97 model given in Equation 4.19 the simple linear scaling of $\ln SA$ with $\ln R$[18] illustrates that geometric spreading is accounted for, but there is no anelastic attenuation term represented.[19] The a_4 term is responsible for the near-source distance saturation. Contemporary models such as CY14 explicitly account for anelastic attenuation. They also have a more complex functional form for near-source saturation as a function of the earthquake magnitude (because this indicates the size of the finite fault), and lower geometric spreading rates at larger distances due to the increasing contribution of surface waves and significant body wave reflections (e.g., Moho reflection).

4.5.3 Site Conditions

As seismic waves propagate upward from the earthquake source to the surface, a reduction in the material impedance[20] in the near-surface causes an increase in ground-motion amplitude. This is referred to as site amplification (e.g., Equation 5.7). Softer near-surface soil deposits can, therefore, lead to ground-motion amplification relative to that observed on rock sites. One exception to that general statement is when intense ground motion causes nonlinear deformation in surficial soils. Nonlinear response reduces the surface ground-motion amplitude and significantly changes its frequency content and duration.

A commonly adopted metric that correlates with site amplification is the 30-m time-averaged shear wave velocity, $V_{S,30}$. Unlike magnitude and source-to-site distance, which have direct theoretical links to ground-motion intensity (see Chapter 5), the $V_{S,30}$ metric is a pragmatic choice. The choice is popular because surficial velocity estimates are often available to 30 m depth, and this metric correlates reasonably well with site amplification effects (e.g., Choi et al., 2005). It does not imply an underlying causality between this particular metric and site response effects other than representing an average change in impedance from the crustal properties at the source to those in the near-surface.

Figure 4.21 illustrates several ground-motion time series with different soil conditions, as represented through $V_{S,30}$. Loosely speaking, $V_{S,30}$ values in the order of $V_{S,30} < 250$ m/s are regarded as soft soils, $250 < V_{S,30} < 500$ m/s as stiff soils, and $V_{S,30} > 500$ m/s as highly weathered through to intact rock. Comparing the rock ($V_{S,30} = 1000$ m/s) and stiff soil ($V_{S,30} = 400$ m/s) ground motions in Figure 4.21 illustrates that the latter does exhibit a modest increase in ground-motion amplitude, and also a longer duration of strong shaking.

The amplitudes at the two $V_{S,30} = 150–250$ m/s sites in Figure 4.21 are appreciably larger. The frequency content of these ground motions is also dominated by lower frequencies due to the resonance of the near-surface soils at their natural vibration modes. The amplitude at the $V_{S,30} = 150$ m/s site is lower than at the 250 m/s site due to significant nonlinear soil response (liquefaction) leading to reduced amplitudes. The 150 m/s site also has a short significant duration because the soil is no longer capable of propagating strong ground motions to the surface (Ishihara and Cubrinovski, 2005). This very soft soil ground motion illustrates the challenges of modeling site effects using simple predictor variables such as $V_{S,30}$.

Figure 4.22 illustrates the variation in $SA(1\text{ s})$ amplitudes of a subset of ground motions in the NGA-West2 database as a function of $V_{S,30}$ for both large and small amplitude ground motions

---

[18] Note in Equation 4.19 the R^2 term is within a square root.

[19] In which $\ln SA$ would scale linearly with R (Chiou and Youngs, 2014).

[20] Seismic impedance is the product of material density, ρ, and wave velocity, V; see Equation 5.9.

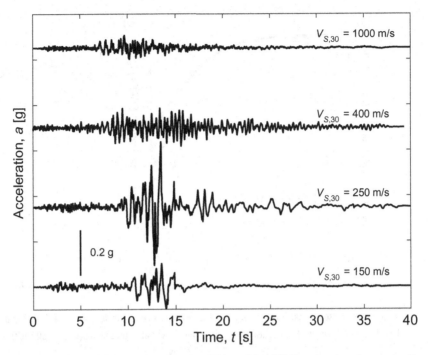

Fig. 4.21 Effect of site conditions on recorded ground motions from the 1989 Loma Prieta (M6.9) earthquake at four nearly adjacent locations. Other than the variation in 30-m time-averaged shear wave velocity, $V_{S,30}$, the recordings all have source-to-site distances of approximately $R = 75$ km.

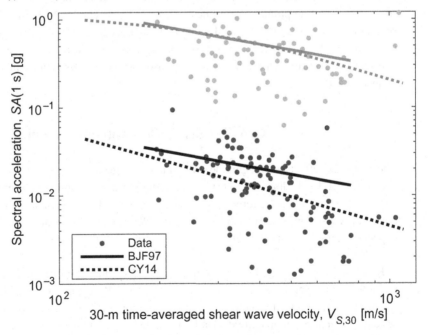

Fig. 4.22 Effect of near-surface site conditions (i.e., $V_{S,30}$) on SA(1 s) amplitudes from the NGA-West2 database for larger and smaller amplitude ground motions. The larger motions are from events with $M = 6.8$–7.2 and distances $R = 0$–15 km, and are depicted in gray. The smaller motions are from events with $M = 5.3$–5.7 and distances $R = 40$–60 km, and are depicted in black. The median empirical predictions from the BJF97 and CY14 models are shown as lines for $M = 7.0/R = 5$ km and $M = 5.5/R = 50$ km, respectively, and a vertical strike-slip fault.

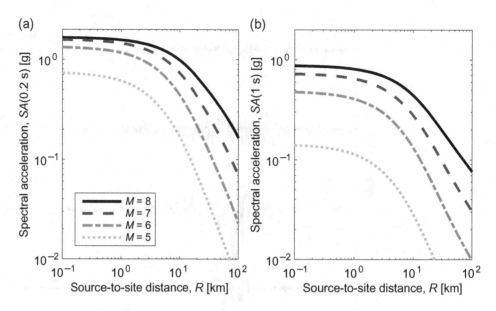

Fig. 4.23 Scaling of response spectral ordinates with magnitude and source-to-site distance for (a) $SA(0.2\,s)$, and (b) $SA(1\,s)$. Results are based on the CY14 model for a vertical strike-slip fault observed at a stiff soil site ($V_{S,30} = 500$ m/s).

(associated with different magnitude and source-to-site distance ranges). In addition to the $SA(1\,s)$ values from observations, the median values of the BJF97 and CY14 models are also depicted. In general, $\ln SA$ amplitudes scale linearly with $\ln V_{S,30}$. The CY14 model departs from this linear scaling for larger amplitude ground motions, which occur at low $V_{S,30}$ values during strong shaking (i.e., for the $M = 7.0/R = 5$ km scenario). This departure occurs because the CY14 model considers nonlinear near-surface site response, which is not represented in the BJF97 model (Equation 4.19).

4.5.4 Variation in Effects with Spectral Period

The empirical effects of magnitude and source-to-site distance in the previous subsections were shown for $SA(1\,s)$ ordinates. However, such effects are also a function of ground-motion frequency and therefore manifest differently for each IM depending on what ground-motion frequencies contribute most significantly to that IM. Because SA is a common ground-motion IM and is explicitly dependent on period, it is insightful to examine the variation in these effects at different oscillator periods.

Figure 4.23 illustrates the variation of $SA(0.2\,s)$ and $SA(1\,s)$ amplitudes as a function of magnitude and source-to-site distance. These two IMs can be considered as representative of high- and moderate-frequency ground motion, respectively. For a given source-to-site distance, increases in magnitude lead to greater relative increases in $SA(1\,s)$ amplitudes than in $SA(0.2\,s)$ amplitudes. Both IMs exhibit "magnitude saturation" in that each unit increases in magnitude causes decreasing increases in amplitude (as also observed in Figure 4.18). This

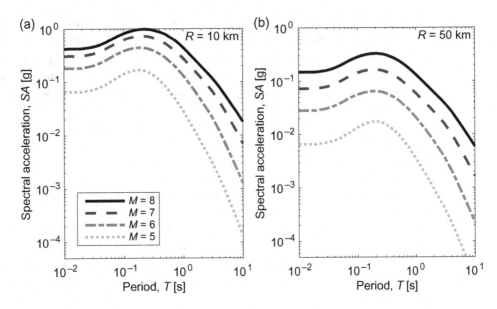

Fig. 4.24 Scaling of response spectral ordinates as a function of magnitude for source-to-site distances of (a) $R = 10$ km and (b) $R = 50$ km. Results are based on the CY14 model for a vertical strike-slip fault observed at a stiff soil site ($V_{S,30} = 500$ m/s).

effect is most pronounced for small source-to-site distances, and is stronger for $SA(0.2\ \text{s})$. For example, at $R < 1$ km the median $SA(0.2\ \text{s})$ amplitudes are practically insensitive to magnitude in the range of $M = 6$–8.

Figure 4.24 provides similar results to Figure 4.23, but instead via plotting response spectral ordinates for oscillator periods from $T = 0.01$–10 s, for two different source-to-site distances. Figure 4.24a further illustrates that the effect of magnitude on SA amplitudes is most pronounced at long periods, with the $SA(T = 10\ \text{s})$ amplitudes spanning a factor of nearly 200 from $M = 5$–8, while the $SA(T = 0.01\ \text{s})$ amplitudes span a factor of less than 10. This trend is also seen in Figure 4.24b for $R = 50$ km, but the difference is less pronounced, illustrating the link between magnitude and distance saturation.

4.5.5 Additional Factors

Besides the general trends with magnitude, source-to-site distance, and site conditions, additional factors have received explicit attention with increases in observation data and theoretical understanding. Modern GMMs generally consider the majority of the following factors, several of which are illustrated empirically using the CY14 GMM in Figure 4.25.

Tectonic type: Active shallow crustal, stable crustal, subduction slab, subduction interface, volcanic forearc, and volcanic backarc earthquakes have differing rock mineralogy, stress conditions, seismogenic depths, and wave propagation paths, among other factors. Contemporary empirical GMMs are generally developed separately for these different tectonic region types, but sometimes treat tectonic type as a predictor variable within a single GMM (e.g., Zhao et al., 2006).

Style of faulting: High-frequency ground-motion amplitudes from relatively large-magnitude events exhibit a correlation with style of faulting, with reverse ruptures on the order of 10–20% larger, and normal ruptures 10–20% smaller, than strike-slip ruptures (e.g., Bommer et al., 2003). Much of this effect appears to be a result of differences in fault dip and rupture depth among these ruptures. For example, Abrahamson et al. (2014) note that when the tendency of reverse ruptures to occur at greater depths than strike-slip ruptures (for a given M) is accounted for, there is no statistically significant effect directly due to style of faulting. Similarly, Chiou and Youngs (2014) discuss the effect of fault dip on ground-motion amplitudes, also reducing the direct dependence on style of faulting.

Rupture depth: Because confining stress increases with depth, deeper ruptures tend to release a greater amount of radiated energy for a given moment magnitude (Chiou et al., 2008). Figure 4.25a illustrates the variation in $SA(T)$ amplitudes for three different values of the depth to the top-of-rupture, Z_{tor} (see Figure 3.36). The dependence on rupture depth increases for shorter oscillator periods, with a difference of 30% for $Z_{tor} = 6$ km compared with $Z_{tor} = 0$ km for $SA(0.01 \text{ s})$. This resulting increase in amplitude with increasing Z_{tor} is, however, partially offset by an increase in source-to-site distance for a given source and site location, which geometrically is most significant in the near-source region.

Rupture hanging-wall effects: For a given R_{rup}, more of the rupturing fault is near a site on the hanging wall, relative to a site on the footwall (Figure 2.5). Figure 4.25b illustrates the variation in $SA(1 \text{ s})$ amplitudes as a function of the distance metric R_x (positive on the hanging wall; negative on the foot wall; see Figure 3.36), and its dependence on fault dip angle, δ. The hanging wall amplitudes increase with decreasing fault dip, relative to that of a vertically dipping fault ($\delta = 90°$), for which hanging wall effects are absent. At $R_x = 10$ km in Figure 4.25b, the $SA(1 \text{ s})$ amplitudes for $\delta = 45°$ are 60% greater than for $\delta = 90°$. Although not shown in Figure 4.25b, hanging-wall effects are most significant at short oscillator periods.

Rupture directivity: Because seismic rupture speeds are similar to the shear-wave velocity of crustal rocks (Section 5.3), the superposition of seismic waves leads to the phenomenon of source directivity. This effect is strongest when the earthquake rupture direction, slip direction, and direction from the fault to the site are colinear and result in the majority of the seismic energy radiated from the fault arriving at the site in a condensed time period. The resulting ground motions have an increased amplitude at low frequencies and shorter duration at locations of positive (or "forward") directivity, and lower amplitudes and longer durations at locations of negative (or "backward") directivity (Somerville et al., 1997; Spudich et al., 2014). Forward-directivity ground motions can be particularly damaging to some structures (Alavi and Krawinkler, 2004). Figure 4.25c illustrates the variation in $SA(T)$ amplitudes for conditions of neutral, positive, and negative directivity (as quantified by the parameter ΔDPP; Chiou and Youngs, 2014), with the effects most pronounced at long oscillator periods.

Long-period site response effects: Deep geological structures, such as sedimentary basins, can lead to large ground-motion amplification at long periods due to wave reverberations (Chapter 5; Day et al., 2008). Metrics describing deeper geologic structure are commonly used in empirical GMMs to implicitly address this phenomenon (e.g., Abrahamson et al., 2008). Figure 4.25d illustrates the variation in $SA(T)$ amplitudes as a function of the depth to the 1.0 km/s shear-wave velocity horizon, $Z_{1.0}$. The impact of increasing $Z_{1.0}$ (which is often indicative of deeper sedimentary soils) is greatest at long oscillator periods. For example, $SA(5 \text{ s})$ amplitudes are 20% larger for a deeper basin site ($Z_{1.0} = 800$ m) than at an average $Z_{1.0} = 300$ m depth, which is typical for stiff soil conditions (Chiou and Youngs, 2014). Similarly, $SA(5 \text{ s})$ amplitudes are nearly 30% smaller for a shallow sediment site ($Z_{1.0} = 50$ m) relative to this typical $Z_{1.0} = 300$ m site.

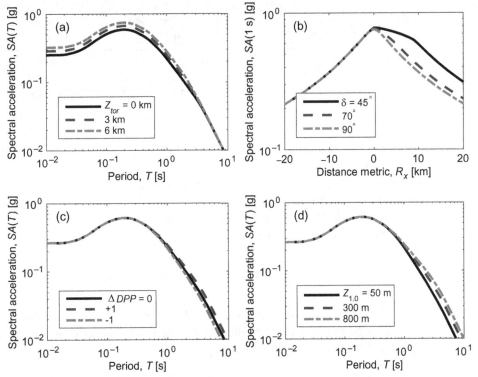

Fig. 4.25 Illustration of the effects of additional factors on predicted ground-motion amplitudes in the CY14 GMM. (a) Depth to top-of-rupture (Z_{tor}) effect as a function of vibration period. (b) Hanging-wall effect on $SA(1\ s)$ amplitudes as a function of source-to-site distance, R_x, for three different fault dip angles δ. (c) Rupture directivity effect on SA amplitudes for neutral, forward, and backward directivity conditions ($\Delta DPP = 0, +1, -1$, respectively). (d) Effect of depth to 1 km/s shear wave velocity horizon, $Z_{1.0}$, on SA amplitudes. Unless noted, the primary predictor variable values used are $M = 7.5$, $R_{rup} = 20$ km, and $V_{S,30} = 500$ m/s, with other distance metrics (e.g., R_{jb}, R_x) calculated geometrically from these values.

4.6 Prediction Using Empirical GMMs

With an understanding from previous sections on the mathematics of empirical GMMs, their general features relative to observed ground-motion data, and the variability in the observations that is considered as apparent variability in the models, we are now in a position to use them for ground-motion prediction.

4.6.1 Evaluating the Ground-Motion Distribution

Because ground-motion IMs can be considered as lognormally distributed (Section 4.4.3), we can compute the probability of exceeding any IM level using the mean and standard deviation obtained from an empirical GMM via

$$P(IM > x \mid rup, site) = 1 - \Phi\left(\frac{\ln x - \mu_{\ln IM}}{\sigma_{\ln IM}}\right) \tag{4.22}$$

$$= 1 - \Phi(\varepsilon) \tag{4.23}$$

Table 4.1. Probability of $SA(1\ s) > 0.4$ g for three source-to-site distances

R [km]	$\mu_{\ln IM}$	im_{50} [g]	$\sigma_{\ln IM}$	$P[SA(1\ s) > 0.4\ g]$
3	−1.041	0.353	0.52	0.4050
10	−1.771	0.170	0.52	0.05014
30	−2.619	0.073	0.52	5.294×10^{-4}

where $\Phi\,()$ is the standard normal cumulative distribution function, and $\mu_{\ln IM}$, $\sigma_{\ln IM}$, and ε are as defined in Equation 4.16.

Equation 4.23 uses the cumulative distribution function to compute $P(IM > x \,|\, rup, site)$, but sometimes it may be useful to use an alternate formulation incorporating the probability density function (PDF) for IM. Noting that the cumulative distribution function is equivalent to an integral of the PDF (Equation A.22), we can also write:

$$P(IM > x \,|\, rup, site) = \int_{x}^{\infty} f_{IM}(u)\, du \qquad (4.24)$$

where $f_{IM}(u)$ is the PDF of IM, given rup and $site$. Unlike the cumulative distribution function $\Phi\,()$, $f_{IM}(u)$ can be written out analytically. Substituting in this PDF gives an integral form, equivalent to Equation 4.23, that can be evaluated using contemporary programming languages:

$$P(IM > x \,|\, rup, site) = \int_{x}^{\infty} \frac{1}{\sigma_{\ln IM} \sqrt{2\pi}} \exp\left(-\frac{1}{2}\left(\frac{\ln u - \mu_{\ln IM}}{\sigma_{\ln IM}}\right)^2\right) du. \qquad (4.25)$$

To connect these equations to a visual display of ground-motion predictions, consider Figure 4.26, which shows $SA(1\ s)$ predictions on stiff soil ($V_{S,30} = 500$ m/s) for an $M6.5$ strike-slip earthquake, as a function of source-to-site distance, R. Suppose we were interested in the probability of $SA(1\ s) > 0.4$ g. The median and $\pm$ one standard deviation of the BJF97 prediction are plotted for distances between 1 and 100 km. At distances of $R = 3, 10$, and 30 km, the PDF of the predicted distribution of $SA(1\ s)$ is also superimposed.[21] At those three distances, Equation 4.19 gives predicted means of $\ln SA(1\ s) = -1.041, -1.771$, and -2.619, respectively.[22] At all three distances, the standard deviation of $\ln SA(1\ s)$ is 0.52.

With knowledge of the mean and standard deviation for each scenario, we can use Equation 4.23 to compute the probabilities of exceedance. For example, at a distance of $R = 3$ km,

$$P[SA(1\ s) > 0.4\ g \,|\, M = 6.5, R = 3\ km] = 1 - \Phi\left(\frac{\ln(0.4) - (-1.041)}{0.52}\right) = 0.4050. \qquad (4.26)$$

Table 4.1 provides details of the resulting probabilities for all three source-to-site distances. These probabilities correspond to the fraction of the corresponding PDFs in Figure 4.26 that are shaded. To aid visual comparison with Figure 4.26, the BJF97 median prediction, $im_{50} = \exp(\mu_{\ln IM})$, is also provided in the table.

---

[21] Because the y-axis is in log-scale, the lognormal distribution appears as a symmetric normal distribution.

[22] All of the example calculations will provide answers with more significant figures than should reasonably be used or reported. This is done to allow reproduction of the example calculations exactly, and because many answers are intermediate results for later calculations.

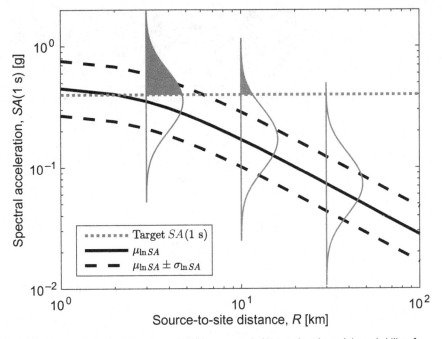

Fig. 4.26 Graphical depiction of the example ground-motion model for a magnitude *M*6.5 earthquake, and the probability of *SA*(1 s) > 0.4 g at several source-to-site distances *R* annotated via the PDF.

Consider a second related example, which will provide intermediate results for calculations that follow in Section 6.3.1. We want to predict the ground-motion probability distribution for a site with $V_{S,30} = 500$ m/s that is a distance of $R = 10$ km from an *M*6.5 earthquake. The problem is similar to the previous example, but we now consider only a single source-to-site distance and examine how the probability of exceedance varies due to changing the threshold value x.

Using the mean and standard deviation values of the BJF97 model in the previous example, the first and second columns of Table 4.2 provide a range of threshold values, $x = 0$–0.6 g, in increments of 0.05 g, and the corresponding exceedance probabilities. As would be expected, the values of $P(IM > x_i)$ decrease as x increases, and vary in the range of $[0, 1]$.

Further intuition can be gained by considering the difference between exceedance probabilities for sequential threshold values, which yields a result proportional to the distribution PDF:

$$P(x_i < IM \leq x_{i+1}) = P(IM > x_i) - P(IM > x_{i+1}). \tag{4.27}$$

$P(x_i < IM \leq x_{i+1})$ values from this equation are shown in the third column of Table 4.2. The largest $P(x_i < IM \leq x_{i+1})$ values occur near the median value of the distribution, $im_{50} = 0.170$ g, as would be anticipated.

4.6.2 Comparison of Model Prediction with Observed Ground Motions

Comparing observed ground motions with model predictions provides a means to examine model performance and also interpret the observations. Such comparisons are also performed iteratively during GMM development.

Table 4.2. Exceedance probabilities, $P[IM > x_i]$, for $M = 6.5$, $R = 10$ km, $V_{S,30} = 500$ m/s, and $IM = SA(1\,\text{s})$. $P(x_i < IM \leq x_{i+1})$ values in the third column are based on Equation 4.27

$x_i[g]$	$P(IM > x_i)$	$P(x_i < IM \leq x_{i+1})$
0.00	1.000	0.009
0.05	0.991	0.144
0.10	0.847	0.251
0.15	0.596	0.218
0.20	0.378	0.148
0.25	0.230	0.092
0.30	0.138	0.055
0.35	0.083	0.033
0.40	0.050	0.019
0.45	0.031	0.012
0.50	0.019	0.007
0.55	0.012	0.004
0.60	0.008	0.008

Residual Partitioning

For comparison of predictions against observations it is conventional to recast the predictive functional form of Equation 4.16 as

$$\ln im_{i,j} = \mu_{\ln IM}(rup_i, site_j) + \Delta_{i,j}$$
$$\Delta_{i,j} = \delta B_i + \delta W_{i,j}$$
(4.28)

where $\ln im_{i,j}$ is the natural logarithm of the observed ground-motion IM from the ith earthquake (rup_i) at the jth site ($site_j$); $\mu_{\ln IM}(rup_i, site_j)$ is the prediction of the mean $\ln IM$ value; and $\Delta_{i,j}$ is the *total* residual for the jth site from the ith earthquake. By comparing Equations 4.16 and 4.28 we can see that $\sigma_{\ln IM} \cdot \varepsilon = \Delta_{i,j}$.

The total residual is then further partitioned into δB_i, the between-event residual for the ith earthquake; and $\delta W_{i,j}$, the within-event residual of the observation at the jth site from the ith earthquake. Denoting the standard deviations of δB_i and $\delta W_{i,j}$ as τ and ϕ, respectively, they are related to the standard deviation of the total residual, $\sigma_{\ln IM}$ (as defined in Equation 4.16) by

$$\sigma_{\ln IM}^2 = \tau^2 + \phi^2.$$
(4.29)

The reason for the "partitioning" of the total residual into between- and within-event terms is that the observed ground motions from a single earthquake i are not independent observations (they are all causally related to the same earthquake).[23] In order to achieve statistically independent data from the direct observations, the between-event residual for the specific IM and event i is determined, and the within-event residuals are then obtained by rearranging Equation 4.28 as

$$\delta W_{i,j} = \Delta_{i,j} - \delta B_i.$$
(4.30)

---

[23] Chapter 8 provides further extensions of the form given by Equation 4.28 in which other effects are considered.

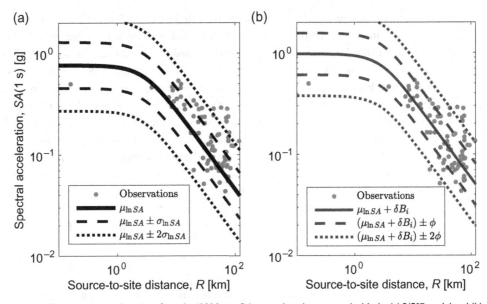

Fig. 4.27 *SA*(1 s) values of observed ground motions from the 1989 Loma Prieta earthquake compared with the (a) BJF97 model and (b) including adjustment to the mean and standard deviation resulting from the between-event residual.

The between-event residual, δB_i, is estimated via mixed-effects regression, as discussed in Section B.6. This process ensures that individual earthquakes do not unreasonably dominate the model development. For example, if an individual earthquake provides a very large number of recordings,[24] and all of these are influenced by a stronger than average source strength, this systematic effect will be reflected in a positive δB_i term for that event. But that positive δB_i should be treated as only a single observation (from a single earthquake) rather than many observations (as would be the case if it was left in the residual terms for each recording).

Example: Loma Prieta Ground-Motion Observations

Figure 4.27 shows the 82 observed ground motions from the 1989 Loma Prieta earthquake within the NGA-West2 database in comparison with the BJF97 model. The observations and mean prediction exhibit a similar attenuation with source-to-site distance. The variability in the observations is consistent with the BJF97 model, as indicated by the $\pm 1\sigma$ and $\pm 2\sigma$ lines. However, 57 of the 82 observed SA(1 s) amplitudes are greater than the mean prediction ($\mu_{\ln SA}$). The residuals are computed from Equation 4.28, and have a mean value of $\bar{\Delta}_{i,j} = 0.26$, indicating that the observations are, on average, $\exp(0.26) = 1.3$ times larger than the mean prediction.

The comparison of observations to various numbers of standard deviations on either side of the mean prediction is a visual way to interpret the ε value (Equation 4.16). For example, all the observations that plot between the μ and $\mu + \sigma$ lines have ε values in the interval $[0, 1]$. From Equation 4.23, we also know that this is directly related to the predicted exceedance probability for this amplitude.

---

[24] Such as the 1999 Chi-Chi earthquake (Figure 1.1), whose contribution to ground-motion databases is visible in Figures 4.11 and 4.13.

To compare observations and prediction in Figure 4.27 more robustly, we partition the total residuals into their corresponding between- and within-event terms (via Equation 4.28). To do so we use the between- and within-event standard deviations for the BJF97 model, which are $\tau = 0.214$ and $\phi = 0.474$, respectively (Boore et al., 1997). Because both τ and ϕ are constants, Equation B.42 (with $n_i = 82$) can be used to determine the between-event residual as $\delta B_i = 0.245$. Equation B.42 shows that for an increasing number of observations for a given earthquake, the between-event residual approaches the arithmetic mean of the total residuals (with a 5% difference between 0.245 and the arithmetic mean of 0.26 in this example).

Note that these $n_i = 82$ observed ground motions yield only a single value for the between-event residual, $\delta B_i = 0.245$. Thus, ground motions from many earthquakes are required to estimate δB_i values and ascertain whether a predictive model is consistent with observed event-to-event variability, as well as examine biases with parameters such as M and Z_{TOR} (see Section 12.5.2).

With the between-event residual now computed for the specific earthquake, Equation 4.28 can be reordered as:

$$\ln im_{i,j} = \left(\mu_{\ln IM}(rup_{i,j}, site_j) + \delta B_i\right) + \delta W_{i,j}. \tag{4.31}$$

$\mu_{\ln IM}$ and δB_i are now grouped together as known terms in what is commonly referred to as an "event-adjusted" mean, and $\delta W_{i,j}$ represents the residual between the observations and the event-adjusted mean.

Figure 4.27b illustrates the comparison between the observed $SA(1\ s)$ amplitudes and the event-adjusted prediction using the BJF97 model. Relative to the (unadjusted) BJF97 model (Figure 4.27a), the adjusted model has a median which is larger by the factor $\exp(\delta B_i) = 1.278$, and because the event-term is a constant (rather than a random variable with standard deviation τ) the standard deviation of the event-adjusted prediction is ϕ, the standard deviation of $\delta W_{i,j}$ (which is 9% smaller than $\sigma_{\ln SA(1\ s)} = 0.52$). The event-adjusted model shown in Figure 4.27b provides a visibly good representation of the observed data. The mean of the within-event residuals is 0.0147, and the standard deviation is 0.477. Both of these values are very similar to the assumption that the within-event residual is a random variable with zero mean and standard deviation of $\phi = 0.47$ and can be formally rejected as different using conventional hypothesis testing (Ang and Tang, 2007).

Despite the distribution of the observed within-event residuals being consistent with the model, we can also explore the dependence of these observed residuals with other variables. Such explorations are instructive because the BJF97 model assumes that all the parametric dependencies are accounted for in the mean model and that the within-event residual is a random variable without any parameter dependence. Figure 4.28 illustrates the dependence of the within-event residuals as a function of source-to-site distance, R, and 30-m time-averaged shear-wave velocity, $V_{S,30}$. In both figures, the residuals cluster relatively evenly about the zero residual lines, but display dependence on the x-axis variable, as indicated by the least-squares regression fit. The slope of the lines[25] suggests that the functional dependence with R and $V_{S,30}$ leads to a relative overestimation of observations at smaller parameter values and an underestimation at larger parameter values.

---

[25] P-values of 0.0066 and 0.0049, respectively, imply that the null hypothesis of zero slope can be rejected at the 99% confidence level.

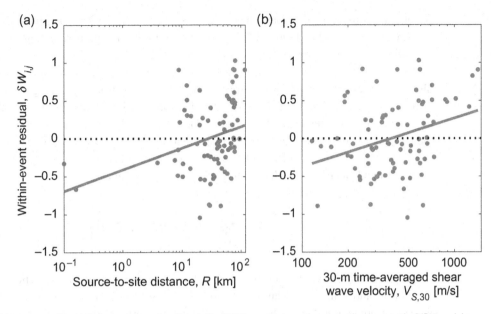

Fig. 4.28 Within-event residuals of observed ground motions in the 1989 Loma Prieta earthquake with respect to the BJF97 model as a function of (a) R and (b) $V_{S,30}$.

Primary conclusions from this comparison of the BJF97 prediction with the Loma Prieta $SA(1\ \text{s})$ amplitudes are as follows:

1. The observations are, on average, larger than the mean prediction, with a between-event residual of $\delta B_i = 0.26$. Given that $\tau = 0.214$, δB_i is 1.2 standard deviations from the zero mean of the between-event random variable.
2. The within-event residuals, $\delta W_{i,j}$, are consistent with the theoretical distribution of mean zero and standard deviation ϕ.
3. There are statistically significant dependencies of the within-event residuals with R and $V_{S,30}$.
4. Whether these dependencies in $\delta W_{i,j}$ reflect a flaw in the model for use in general conditions, or are specific to this earthquake event and regional setting, would require additional observations to confirm.

It is important to point out that the biases and imprecisions evident through the above comparison are the result of the simplistic BJF97 model adopted, and are often less pronounced in contemporary models (such as the CY14 model presented previously). Larger datasets are also generally required to identify systematic biases in contemporary models, so the above example and discussion should be viewed as simply illustrating relevant considerations (discussed further in Section 12.5.2).

4.6.3 Identifying "Rare" Observations

The previous discussion focused on the consistency of an empirical model compared with observational data. However, the use of empirical models to identify "rare" ground motions is also particularly fruitful. For example, there are eight observed ground motions in Figure 4.27a that lie above the $+2\sigma$ prediction—something that should occur only 2.3% of the time if the model is unbiased. However, because the observations are correlated due to the positive event residual, it is not correct to simply compare the 8 out of 82 ratio with the 2.3% prediction. Once the event-adjustment is made, then Figure 4.27b illustrates that only five ground motions are above the $+2\sigma$ prediction.

While the model treats all variation about the mean prediction as purely random, there are ultimately physical reasons for such observations departing from the average by such an amount. Careful examination can help identify why such observations occur, leading to the discovery of important physical phenomena (see Chapter 5). Alternatively, there may be effects associated with the particular earthquake source, path, or site that deviate from the "global" mean and that can be incorporated into non-ergodic models (see Chapter 8). In either case, the study of observational data is critical to advancing our understanding of strong ground motions.

4.7 Epistemic Uncertainty

Empirical GMMs, described in the form of Equation 4.16, provide a prediction as a function of their mean and standard deviation. The standard deviation represents the apparent variability in the model prediction due to the complex phenomenon and inherent simplifications in any model. In addition to the representation of apparent variability, we have another form of uncertainty as a result of knowledge limitations, *epistemic uncertainty* (Section 1.7), which is uncertainty that the adopted empirical GMM adequately represents reality. In this section, we consider the manifestations of such uncertainty, its root sources, and methods to account for it. Section 6.8 provides additional discussion on epistemic uncertainties in the context of PSHA and necessitates the issues raised in this section.

4.7.1 Example: Multiple Model Predictions

Consider $SA(1\text{ s})$ amplitudes predicted by four plausible models. Figure 4.29 illustrates the medians and standard deviations predicted for $M = 6$ and 7.5 vertically dipping strike-slip earthquakes, as a function of source-to-site distance, R, using four NGA-West2 GMMs (Abrahamson et al., 2014; Boore et al., 2014; Campbell and Bozorgnia, 2014; Chiou and Youngs, 2014) (denoted as ASK14, BSSA14, CB14, and CY14, respectively). For a given model, the general trends of increasing amplitude with increasing magnitude and decreasing source-to-site distance are evident, as previously discussed. However, there are appreciable variations among the predictions provided by the different models. For example, at a distance of $R = 0.1$ km for the $M = 7.5$ case, the median prediction values range from 0.52 g (ASK14) to 0.80 g (CY14)—a factor of 1.5. The standard deviation values range from 0.67 to 0.72. These four models were developed using the same underlying ground-motion database. Therefore the differences in their predictions result from alternative decisions regarding the selected subset of ground motions from the database and model functional form.

In the context of PSHA (Chapter 6), differences in the exceedance probabilities from the different models are of ultimate interest. Figure 4.30 illustrates the PDFs (Section 4.6.1) from the four GMMs for an $M = 7.5$ event at respective source-to-site distances of $R = 0.1$, 20, and 100 km. There is an appreciable variation in the PDFs for $R = 0.1$ km. In contrast, there is relatively little difference between the four models for $R = 20$ km. For $R = 100$ km, three of the four models are similar, with one (ASK14) notably different. The similarities and differences in the PDFs are reflected in

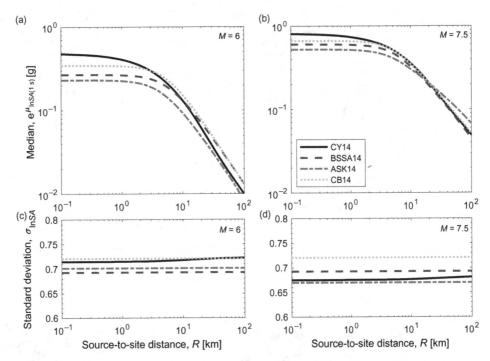

Fig. 4.29 Illustration of epistemic uncertainty in ground-motion prediction via four plausible models for active shallow crustal earthquakes from the NGA-West2 project. Panels (a) and (b) are median $SA(1\ \text{s})$ predictions, while (c) and (d) standard deviations for two different magnitude values.

calculated exceedance probabilities. For example, the probabilities of $SA(1\ \text{s}) > 1.0$ g for the $R = 0.1$ km scenario range from 0.17 to 0.37, whereas the probabilities of $SA(1\ \text{s}) > 0.3$ g for the $R = 20$ km scenario range from 0.26 to 0.30.

This example illustrates that model uncertainty often varies with the scenario being considered. Observational data are more abundant for some scenarios than others. Hence, we should expect the range of model predictions to vary depending on the amount of data constraining the models. Figure 4.13 illustrated the sparsity of ground-motion data in the magnitude-distance parameter space around $M = 7.5$, $R = 0.1$ km, which could explain the large variation in model predictions for this scenario (Figure 4.30a). However, this same logic does not explain why there is little variation in the four model predictions for the $R = 20$ km scenario relative to the $R = 100$ km scenario, since Figure 4.13 illustrates that there is appreciably more data at larger source-to-site distances. This same issue is present in the context of the larger difference between median predictions for the $M6$ event at $R = 20$ km versus $M7.5$ at $R = 20$ km, with more observational data for the former than the latter scenarios. The consideration of alternative models alone is not sufficient to capture epistemic uncertainty.

4.7.2 Sources of Model Uncertainty

Three predominant sources of epistemic uncertainty in empirical GMMs are the adopted ground-motion dataset, the model formulation, and parametric uncertainty from the regression analysis. These sources are depicted schematically in Figure 4.31 and discussed in the subsections below.

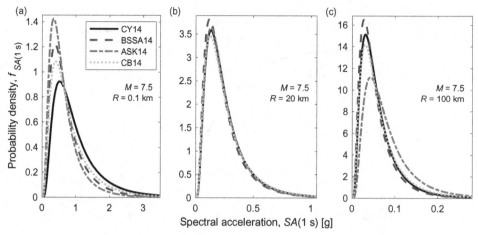

Fig. 4.30 Probability density functions (PDFs) from model predictions in Figure 4.29 at three source-to-site distances.

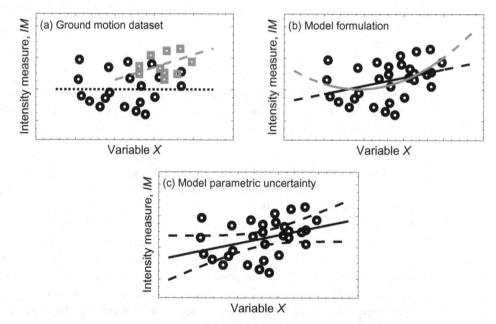

Fig. 4.31 Conceptual illustration of sources of epistemic uncertainty in empirical ground-motion modeling through the dependence of *IM* prediction with respect to a generic variable *X*: (a) ground-motion dataset selection, (b) alternative model parametric forms, and (c) parametric uncertainty from regression analysis. In (a) and (b), the black and gray lines illustrate two alternative models. In (c), the dashed lines indicate a range of possible models based on parametric uncertainty.

Ground-Motion Dataset

Ground-motion time series and their metadata are of variable quality and relevance to a given application area, which leads to subjective decisions regarding the specific data that should be used for empirical GMM development. Figure 4.31a illustrates the effect of different dataset inclusion criteria. The larger data subset indicates no apparent trend between *IM* and *X*, whereas a smaller subset indicates that IM increases with X. Thus, the choice of the subset will ultimately influence the model regression results.

Factors that may determine whether a ground motion is retained for model development include the quality of the ground-motion time series, the "free-field" nature of the instrument location, availability and quality of metadata (e.g., M, R, $V_{S,30}$), maximum allowable source-to-site distance (generally a function of magnitude), classification of an event as a "mainshock" or "aftershock," the minimum allowable number of records from a specific event, and exclusion of events that are "overrepresented" or considered as outliers. The four models depicted in Figures 4.29 and 4.30 illustrate the differences that result from alternative views on the above considerations. All four models started with the same database (Ancheta et al., 2014) of 21,366 potential ground motions from 600 earthquakes. Through exercising different decisions with respect to the above factors, they ended with datasets having 12,244–15,750 ground motions from 300–326 earthquakes (Abrahamson et al., 2014; Boore et al., 2014; Campbell and Bozorgnia, 2014; Chiou and Youngs, 2014; Gregor et al., 2014). Decisions on retaining data are particularly challenging where data of marginal quality exist for conditions that are poorly covered in the observational dataset, as these strongly influence decisions on model functional form and consequent regression parameters, as discussed next.

Model Formulation

Figure 4.31b illustrates the effect of different model formulations (i.e., functional forms) on the prediction of a ground-motion IM for a fixed dataset. While physics-based insights are considered in the interpretation of ground-motion observations (such as described in Section 4.5), there is a range of functional forms that can fit the data to a similar degree. Figure 4.29 provides an example of the effect of different functional forms for four alternative models. The effect of the functional forms is particularly evident in how the medians and standard deviations vary with source-to-site distance and magnitude.

The differences between alternative formulations become particularly significant as a model is used for predictions poorly constrained by data or extrapolating beyond the data.[26] Examples of poorly constrained phenomena include large-magnitude scaling, hanging-wall effects, directivity, sedimentary basin effects, and nonlinear site response. Because sufficient observational data cannot be obtained quickly, insights from theoretical considerations and numerical simulations can guide the nature of functional forms. For example, the following theory or simulations have been used by NGA-West2 models: Chiou et al. (2010) (magnitude scaling), Donahue and Abrahamson (2014) (hanging-wall), Spudich and Chiou (2008) (directivity), Day et al. (2008) (sedimentary basin), and Seyhan et al. (2014) and Kamai et al. (2014) (nonlinear site response).

The adopted functional form and dataset also influence the extent to which an empirical GMM is generic for use globally or tailored for region- or site-specific application. As discussed in Chapter 8, contemporary empirical GMMs seek to incorporate additional data and model components that reflect the specific conditions at the site of interest. Epistemic uncertainty will exist in the application of a model to a specific site (Campbell, 2003).

---

[26] The schematic example in Figure 4.31b is oversimplified in being one-dimensional. In practice, models have many input parameters and a high-dimension parameter space, meaning that the extrapolation problem is pervasive.

Model Parametric Uncertainty

When empirical models are fit to a dataset using regression, or similar techniques, there is uncertainty in the regression parameters resulting from the finite dataset considered. This leads to uncertainty in model parameters and resulting predictions (Arroyo and Ordaz, 2011). Figure 4.31c illustrates that a linear model for the variation of IM with X has uncertainty in the intercept and slope parameters, and these uncertainties lead to a confidence interval for median predictions. Parametric uncertainty is most important in areas where there is little observational data.

Al Atik and Youngs (2014) calculate model parametric uncertainty for the four NGA-West2 models presented in Figure 4.29, highlighting that parametric uncertainty leads to median predictions with standard deviations of 0.05–0.30 natural logarithmic units. In this case, parametric uncertainty is smaller than the uncertainty between the median predictions, but the opposite can be true for predictions with poor data constraint.

Determination of model parameter uncertainty is also complicated by the increasing use of theoretical and simulation-based constraints used in model functional forms. These constraints have associated modeling uncertainties, but they are less amenable to conventional statistical techniques for determining model parametric uncertainty.

4.7.3 Treatment of Modeling Uncertainty

Several approaches to account for modeling uncertainty in empirical GMM prediction can be adopted with varying levels of complexity. This section outlines three approaches that range from state of practice to state of the art. The first two approaches are based on the use of existing published empirical GMMs and the determination of appropriate "weights," while the third approach centers around the use of meta-modeling.

Multiple Models with Constant Weights

When collecting alternative models for ground-motion prediction, one must ask the question, "Which models are most applicable for the target site?" Lists of minimim requirements have been proposed that determine whether a model can be considered a plausible candidate for application (e.g. Cotton et al., 2006; Bommer et al., 2010). For models that pass such criteria, comparisons can then be made with locally observed ground-motion data, and various tests used to evaluate "model applicability" (Scherbaum et al., 2004).

The main challenge in comparing plausible models with locally observed ground-motion data is the misalignment between the data used for comparison and that of interest for prediction. Specifically, the observed ground motions are likely to be from relatively small-magnitude earthquakes at large source-to-site distances, in contrast to the magnitudes and distances of most interest in PSHA (Section 4.3.3 and Beauval et al., 2012).

The result of this process is that each model is assigned a "weight" (see Section 6.6). For example, consider each of the four models for active shallow crustal earthquakes in Figure 4.29. If the result of the above examination was that each model was equally plausible for all rupture scenarios, then weights of 0.25 would be assigned to each of the four models, and used in a logic tree methodology for hazard analysis, as discussed in Section 6.7.

Multiple Models with Rupture-Specific Weights

The use of models with a single (constant) weight is appealingly simple and common in practice. The primary limitation with that approach is the implication that alternative models are equally valid for all possible rupture scenarios that are considered in the hazard analysis calculation (e.g., all ranges of possible magnitude, source-to-site distance). It also implies that the range of predictions from the models is consistent with observational data (illustrated in Section 4.7.1 to be incorrect). Since the models are generally developed using different datasets and model formulations, the model weights should be a function of rupture-specific parameters. For GMMs that predict *SA* for multiple oscillator periods, this notion also implies that weights should be a function of the oscillator period.

Using weights that vary with predictor variables is laborious in terms of the book-keeping, and in how the weights are derived in a consistent manner. However, the conflict noted in the previous section between model performance against locally observed ground motions and those ground motions that dominate the seismic hazard analysis can be overcome with the ability to consider weights as a function of predictor variables. Furthermore, an approach to simplify defining model weights for the full range of all predictor variables is to first perform the hazard analysis based on nominal weights for the alternative models and determine the seismic sources that dominate the hazard calculation from disaggregation (Section 7.2). Due attention can then be devoted to assigning model weights for the scenarios which dominate the hazard calculation.

While rupture-specific weights avoid several limitations with the constant-weight approach, they too have problems. First, with regard to the three sources of model uncertainty in Figure 4.31, the weights (constant or varying with scenario) do not address the model parameter uncertainty. Second, the use of existing available models can lead to the situation in which many models are available (and considered) in well-studied regions, but few models are available (and considered) in poorly studied regions. This is likely to violate a basic premise that more knowledge should result in less uncertainty and vice versa. That is, more models often lead to a greater diversity in predictions, while a few models based on very limited data will often provide similar results, and provide an inappropriate sense of small model uncertainty. Third, the use of constant or variable weights on existing models never directly deals with the problem that we actually wish to characterize: the epistemic uncertainty on the ground-motion exceedance probability (Equation 4.23), which is used in the seismic hazard calculation (Equation 6.1) – rather than the GMM mean or standard deviation.

Meta-Models

The above two approaches consider that a collection of existing models represents model uncertainty, but does not directly focus on representing the uncertainty in ground-motion exceedance probability. Instead of weighting models, suppose we focus on this overall objective of characterizing the exceedance probability uncertainty. We can represent the uncertainty in both the model mean and aleatory standard deviation through a meta-model, also referred to as a "backbone" model (Atkinson et al., 2014). A meta-model, or backbone model, can be thought of as a new model that can be configured to directly represent the uncertainty in the probability of exceedance. Its functional form can be adopted or adapted from an existing model (Bommer and Stafford, 2020), or can be defined to represent the scaling implied by an ensemble of models (Abrahamson et al., 2019). The benefit of the meta-model approach is that it allows for consideration of all three primary sources of modeling uncertainty, not simply the variation among existing models. A practical example of using meta-models for handling GMM uncertainty can be found in Abrahamson et al. (2019).

Meta-modeling approaches are particularly useful in cases where we are trying to tailor predictions to a target site. As empirical GMMs are often generic (i.e., they contain few site-specific features), adjustments are needed to make them more region- and site-specific in order to obtain predictions with greater accuracy and precision (Bommer and Stafford, 2020). This is more easily achieved with a single meta-model, rather than a collection of alternative existing models. Further details on site-specific modeling are provided in Chapter 8.

4.8 Limitations of Empirical GMMs

Empirical GMMs have been developed for over 50 years to predict earthquake-induced ground motions. Despite the increasing refinement of empirical GMMs over time, they are still a highly simplified representation of reality. Relative to a more comprehensive attempt to model the physics of ground motions (Chapter 5), notable simplifications include:

1. The earthquake rupture is represented by simply the earthquake magnitude, and possibly its style of faulting and depth. Explicit consideration is typically not given to the location of the hypocenter (except when directivity models are included), or the spatial and temporal evolution of slip on the fault.
2. The effects of wave propagation are represented simply via one or more source-to-site distance metrics. No explicit consideration is given to the stratigraphy of the Earth's crust in the region considered. The geometry of sedimentary basins can be particularly important, but basin effects are considered only by a simplified basin depth parameter (if at all).
3. The effect of surficial (local) site response is often considered simply in terms of the 30-m time-averaged shear-wave velocity of the site, $V_{S,30}$. No consideration is given to the particular soil stratigraphy in this top 30 m, or depths greater than this. The particular stress-strain properties of the soils, which influence nonlinear response, are also not considered.

The consequence of the above simplifications (among others) is both a lack of precision in ground-motion prediction for a specific location and large apparent aleatory variability. Strasser et al. (2009) illustrate that over the approximately 40-year period until 2009, there has been almost no reduction in the apparent aleatory variability in empirical GMMs. This trend can also be reflected in the outcomes from the suite of models over the past decade, including the NGA-West2 models (Figure 4.29). This lack of apparent aleatory variability reduction has occurred despite the massive growth in recorded strong ground motions (Figure 4.11) and advancements in the scientific understanding of earthquake-induced ground-motion phenomena.

However, stable apparent aleatory variability does not imply a lack of progress in empirical GMM development. As noted in Section 4.7.2, significant progress has been made using theoretical or numerical simulations to constrain empirical GMMs for scenarios that are poorly represented by empirical data. However, precisely because of a lack of empirical data, such advancements do not have any practical impact on the resulting apparent aleatory variability of resulting models. In emphasizing the importance of this progression, Abrahamson and Silva (2008) refer to the fact that contemporary empirical GMM development is not merely "curve fitting" with observational data.

Instead, it is a "model-building" process that synthesizes such data with seismological and geotechnical information, part of the transition toward the use of physics-based simulations of ground motion (Chapter 5).

The simplifications of an inherently complex problem and the use of global (rather than site-specific) datasets are the principal reasons for the large apparent aleatory variability in the empirical GMMs. Such models also have a lack of precision in the predicted median ground-motion intensity at a given location as a result of being derived by historical data recorded from many earthquake events at an even larger number of locations. Each earthquake, and each recording location, has its own specific (source, path, and site) "features" that deviate from the group "average." Those features are repeatable effects that are not explicitly modeled with this approach. The explicit consideration of such repeatable effects and the development of "site-specific" empirical GMMs is discussed further in Chapter 8.

The use of physics-based ground-motion simulation methods (Chapter 5) is another alternative with the potential to overcome many of the simplifications of empirical GMMs presented in this chapter and enable predictions with greater accuracy and precision. Achieving this, however, requires that the simulation input models and parameters be probabilistically characterized. Otherwise, the physics-based models will simply transfer the empirical GMMs' apparent aleatory variability into equivalent parametric uncertainty in the simulation model parameters (equivalently described in Figure 2.16 for seismic source modeling). There would be no benefits to such an exercise. With careful characterization, physics-based models do hold great promise, however, both for informing empirical GMMs and as predictive models on their own.

Exercises

4.1   Fourier spectra (FAS) and response spectra (SA) are intensity measures that both reflect ground-motion amplitude as a function of frequency. Compare and contrast these two measures, in particular, the situations in which one is perceived as advantageous over the other from theoretical and/or practical perspectives. Select a ground motion from a public database and compute FAS and SA values to support your answer.

4.2   Define the bracketed, D_B, uniform, D_U, and significant, D_S, durations of a ground-motion time series. As the absolute acceleration threshold for D_B increases from zero, how would you expect the values of bracketed and uniform duration to vary? For an earthquake of a particular magnitude, explain how you would expect D_B and D_S to vary with distance?

4.3   Not all observed ground motions are suitable for the reliable determination of some intensity measures (IMs). Section 4.3.4 discussed this idea in the context of the maximum period for response spectra that are suitable. How does this concept apply for the determination of all the common IMs discussed in Section 4.2?

4.4   Explain why empirical ground-motion databases will always have a relative paucity of observational data from large-magnitude earthquakes at small source-to-site distances.

4.5   Name at least three factors that limit the number of ground motions in an empirical database.

4.6   An empirical ground-motion model for spectral acceleration, PGA [g], has a prediction defined by the function:

$$\mu_{\ln PGA} = c_0 + c_1 M + c_2 \ln(R_{rup} + c_3)$$

Table 4.3. Coefficients of GMM in Exercise 4.6		
Coefficient	Description	Value
c_0	Source/units term	−0.868
c_1	Magnitude scaling	0.50
c_2	Geometric spreading	−1.10
c_3	Near-source saturation	10.0
τ	Between-event standard deviation	0.3823
ϕ	Within-event standard deviation	0.4085

for which the coefficients are provided in Table 4.3.

(a) Compute the median PGA for the following earthquake scenario: $M = 6$ and $R_{rup} = 20$ km.

(b) For this same scenario, what is the probability that shaking of 0.35 g is exceeded?

(c) If an earthquake of $M = 6$ occurs and produces shaking that is observed to be 63% higher than average, across many stations, what is the probability of exceeding 0.4 g at the specific location at $R_{rup} = 20$ km?

(d) The location of an earthquake source is not known with absolute certainty. The $M = 6$ scenario is thought most likely to occur at a distance of 20 km, but it is also considered possible that it could be closer at 15 km (with a 40% chance), or more distant at 25 km (with a 15% chance). Using a logic tree approach, determine the expected value of the median PGA.

4.7 Explain why ground-motion models predict spectral amplitudes that tend to saturate at large magnitudes, and why the degree of this saturation varies with spectral period.

4.8 When reviewing a seismic hazard report, you note that the authors have used a ground-motion model for predicting the spectral acceleration, SA, that is defined as:

$$\mu_{\ln SA} = 1.0 + 0.9M - 1.1 \ln\left(R_{jb}\right) + 0.4 \ln\left(V_{S,30}\right).$$

The equation for the total variance (reflecting the aleatory variability) is given as $\sigma^2 = \tau^2 + \phi^2$, with the between-event standard deviation being $\tau = 0.5$ and the within-event standard deviation being $\phi = 0.3$.

Note that M is the moment magnitude, R_{jb} is the Joyner–Boore distance, and $V_{S,30}$ is the average shear-wave velocity measured over the uppermost 30 m of the site.

There are at least four problems with this equation. Identify these problems and explain what is wrong in each case.

4.9 The equation used for predicting PGA [g] in a particular region can be defined as:

$$\mu_{\log_{10}(PGA)} = -1.34 + 0.77M - 0.074M^2 + (-3.16 + 0.32M) \log_{10} \sqrt{R_{jb}^2 + 7.7^2}$$

with

$$\sigma_{\log_{10}(PGA)}(M) = 0.75 - 0.08M$$

where M is the moment magnitude, R_{jb} is the closest distance to the surface projection of the rupture.

A site is influenced by two equally likely earthquake scenarios. Scenario 1 consists of an $M5.0$ event, while Scenario 2 consists of an $M7.5$ event. Both of these scenarios have R_{jb} distances of 10 km.

(a) Calculate the median prediction (i.e., $10^{\mu_{\log_{10}(PGA)}}$) of PGA corresponding to each of these scenarios.

(b) Under what circumstances does Scenario 1 result in higher PGA values than Scenario 2?

4.10 Consider a contemporary GMM of your choice for predicting (1) response spectra (at multiple vibration periods) and (2) Arias intensity. For each model, undertake the following tasks.

(a) Compute the median IM values for an $M = 7$ earthquake at $R = 50$ km and an $M = 6$ earthquake at $R = 20$ km, observed at a site with $V_{S,30} = 200$ m/s (make reasonable assumptions for all other "secondary" input parameters). Discuss the relative values of the two different scenarios, including differences in spectral amplitudes as a function of vibration period (*spectral shape*).

(b) What are the effects, in terms of predicted IM values, of minor variations in the secondary parameters considered?

(c) Consider a site located only 10 km from the $M = 7$ event. Compute the IM values for this scenario.

(d) Compute the ratio of IM values, for the $M = 7$ event at the two sites ($R = 10$ km or 50 km), for $V_{S,30} = 200$ m/s versus $V_{S,30} = 500$ m/s. Discuss the extent to which nonlinear site response is modeled in the GMM that you have selected, and the significance of this effect.

Physics-based ground-motion models (GMMs) take a rupture scenario (Chapter 3) and utilize ground-motion simulation methods to derive a distribution of ground-motion intensity that can be used in the calculation of seismic hazard (Chapter 6). These physics-based models are, therefore, an alternative to the empirical GMMs for ground-motion characterization presented in Chapter 4. While empirical models predict ground-motion intensity measures (IMs), physics-based GMMs explicitly predict earthquake-induced ground-motion time series. The predicted ground-motion time series can then be subsequently summarized via IMs for use in seismic hazard analysis.

As with the comparison of empirical- and physics-based approaches for every modeling problem in science, each modeling paradigm has strengths and weaknesses. This chapter aims to present physics-based ground-motion characterization from the perspective of its utilization in seismic hazard analysis. A particular aspect of this is the consideration of parametric and modeling variability and uncertainties to provide a robust description of the ground-motion distribution obtained using physics-based GMMs.

Learning Objectives

By the end of this chapter, you will be able to do the following:

- Explain the latent need and opportunities associated with physics-based ground-motion characterization.
- Describe the process of kinematic rupture generation to define the slip, rise time, rake, and rupture velocity over a fault source and understand their numerical representation in simulations as a collection of point sources on subfaults.
- Describe the salient phenomena of seismic wave propagation used in physics-based ground-motion models, describe how they are represented in three-dimensional crustal models, and compute solutions to idealized problems associated with these phenomena.
- Critique the relative merits and computational constraints of comprehensive, simplified, and hybrid physics-based ground-motion simulation methods.
- Classify uncertainties in ground-motion simulation methods and compute the distribution of ground-motion intensity that is utilized in seismic hazard analysis.

5.1 Introduction

Physics-based ground-motion simulation methods explicitly model a seismic rupture as a dislocation on a causative fault and the consequent seismic waves that result from momentum conservation. Figure 5.1 illustrates the intensity of ground motion at several time instances from a physics-based

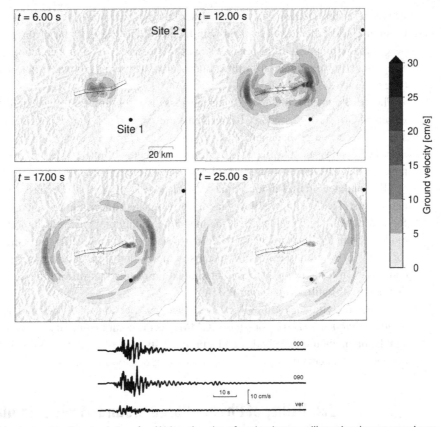

Fig. 5.1 Physics-based ground-motion simulation of an $M6.8$ earthquake at four time instances illustrating the vector maximum of the horizontal ground velocity at the Earth's surface. The causative fault is indicated by the three contiguous planes with the thick edge denoting that which is closest to the ground surface and the epicenter denoted with a star. The three-component ground-motion velocity time series simulated at Site 1 is also depicted (000, 090, ver are the north-south, east-west, and vertical components, respectively).

simulation of an $M6.8$ strike-slip rupture. The temporal evolution of rupture on the fault results in explicit modeling of the effect of rupture direction on seismic wave propagation, and the effect of the specific geophysical and geotechnical properties of the crust on seismic wave propagation through reflection, refraction, amplification, and attenuation. At a single location, the result from a single physics-based ground-motion simulation is a three-component ground-motion time series, from which the desired ground-motion IMs (Section 4.2) can be computed.

As indicated via Figure 5.1, the two key ingredients for physics-based ground-motion simulation methods are (1) a description of the earthquake rupture and (2) a 3D crustal model in which the consequent seismic waves propagate. Mathematically, this can be expressed as:[1]

$$u(x, t; \xi, \tau) = M(t; \xi, \tau) * \nabla G(x, t; \xi, \tau) \tag{5.1}$$

[1] This expression is strictly applicable when the rupture can be represented as a point source located at ξ. As discussed in Section 5.3, rupture over a finite-fault geometry can be expressed as a collection of point sources, and hence this convolution also occurs spatially for all locations ξ on the rupture surface Σ (Aki and Richards, 2002, Section 3.3).

where M is the *moment tensor* that describes the rupture's shear dislocation, which initiates at location ξ and time τ, and evolves with time t; ∇G is the gradient of the *Green's function*, which describes the propagation of seismic waves to location x, and as a function of time t; the operator $*$ represents the convolution in time; and u is the resulting ground displacement at location x and time t.

The development, and use, of physics-based ground-motion simulation methods (relative to empirical GMMs discussed in Chapter 4) naturally requires a greater understanding of source rupture and seismic wave propagation physics, which are addressed in Sections 5.3 and 5.4, respectively. Section 5.5 subsequently outlines the principal features of physics-based ground-motion simulation methods that utilize Equation 5.1 to obtain a simulated ground motion.

The simulated ground-motion time series depicted in Figure 5.1 required specific values of model parameters. These values will always have associated uncertainties when performing simulations for a particular target region. Also, the modeling methodology itself is imperfect. Utilizing different model parameters and model methodologies will give rise to different simulated ground motions for a given location. Section 5.6 discusses these modeling uncertainties and how their explicit consideration leads to an ensemble of simulation results that inform the definition of a probability distribution for utilization within PSHA.

Consistent with the terminology introduced in Section 4.1.1, we will refer to *physics-based ground-motion models* (physics-based GMMs) as the combination of a physics-based ground-motion simulation method (or methods), and a probabilistic description of the joint distribution of all model parameters, that enables a probabilistic prediction of ground motion.

5.2 Utility of Physics-Based Ground-Motion Simulation

The practical utilization of simulated ground motions from physics-based GMMs takes two principal forms, as illustrated in Figure 5.2. Physics-based GMMs can be used to undertake seismic hazard analysis to obtain design ground-motion IMs (i.e., left-hand side of Figure 5.2). Also, the simulated ground-motion time series that result from a physics-based GMM can be used in response-history analysis of structures (i.e., right-hand side of Figure 5.2).

The latent need for physics-based GMMs is motivated by three intertwined factors: (1) lack of observed large-amplitude ground motions that are of principal interest in seismic hazard analysis; (2) increasing utilization of ground-motion time series, which are not directly predicted via the use of empirical GMMs; and (3) intrinsic advantages of a physics-based approach over empirical alternatives. The three subsections below elaborate on each of these needs.

5.2.1 Lack of Large-Amplitude As-Recorded Ground Motions

As will be appreciated from Chapters 6 and 7, design ground motions are often associated with large magnitude earthquakes at relatively close source-to-site distances—and ground motions from such earthquake scenarios are poorly represented in databases of ground-motion recordings (Section 4.3). The global databases of historical ground-motion recordings, therefore, do not provide a sufficient representation of events of this nature in general, especially for specific locations of interest.[2]

---

[2] Where region- and site-specific effects may lead to a departure from the average ergodic observations (Chapter 8).

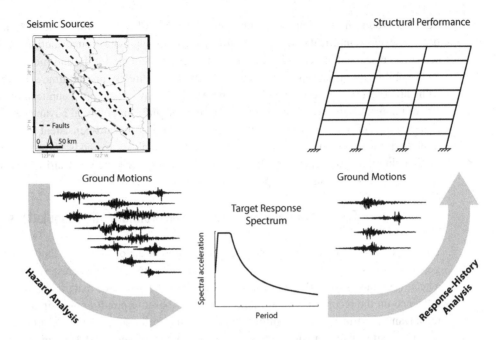

Fig. 5.2   The two principal means for utilization of simulated ground motions: In the ground-motion prediction portion of seismic hazard analysis to determine design ground-motion parameters (left), and as input ground-motion time series in dynamic response-history analyses (right). After Bradley et al. (2017)

The relative paucity of recorded ground motions results in large epistemic uncertainty in empirical GMMs for these (important) scenarios because the mathematical regression equations used in empirical GMM development are not well constrained by data and their extrapolation may not even be physically realistic (Section 4.7.2). Physics-based GMMs still require calibration with the same limited data. However, the assumption is that their physical bases allow for better extrapolation and site-specific calibration.

5.2.2 Increasing Utilization of Ground-Motion Time Series

Empirical GMMs were initially developed in the 1960s (Cornell, 1968; Douglas, 2003) when seismic design used only force-based methods to determine seismic loading (via the response spectrum concept discussed in Section 4.2.2). While the functional complexity of empirical GMMs has increased over the past 50 years, they still ultimately provide a prediction in the form of a ground-motion IM distribution, not the ground-motion time series itself.

Presently, there is a rapid increase in the proportion of seismic design and assessment that utilizes response-history analysis of structures, in which ground-motion time series are applied as a boundary condition and the dynamic response of the structure is evaluated as a function of time using numerical tools, such as the finite element method (Clough and Penzien, 1975; Chopra, 2016). Since empirical GMMs provide only ground-motion IMs, there is a disconnect with the input requirements of response-history analyses. As a result, many procedures have been developed for *ground-motion selection* (Chapter 10) in which representative ground-motion time series are selected that are consistent with the seismic hazard, as defined through one or more ground motion IMs.

Selected ground motions are conventionally those recorded from the same historical earthquakes used to develop empirical GMMs. Therefore, the lack of available ground-motions for certain rupture scenarios, discussed in Section 5.2.1, also becomes problematic for ground-motion selection.

The ability for ground-motion simulation methods to directly predict time series, as compared to using historical records from a global database (which may have limited applicability for the specific site of interest, as discussed in Chapter 8), offers a significant potential advantage. Finally, ground-motion selection is a nontrivial exercise to undertake in a hazard-consistent manner, which likely inhibits the widespread adoption of response-history analysis methods. Advancements in ground-motion simulation are, therefore, likely to have a spill-over benefit toward increasing adoption of response-history analysis for the prediction of structural response.

5.2.3 Advantages of Physics-Based Ground-Motion Simulations

Earthquake rupture occurs as a complex spatio-temporal dislocation on a fault source with variable geometry and stress conditions. Seismic waves arrive at the site of interest via propagation through a heterogeneous crust, and local geology (particularly, near-surface materials) can have an important influence on the recorded ground motions. Physics-based ground-motion simulation methods have the advantage of incorporating information about the earthquake source, seismic wave propagation, and local site characteristics that are specific to the region, and site, in question. Several of these salient phenomena are discussed in Sections 5.3 and 5.4. In contrast, empirical GMMs often simplistically represent these effects, for example, representing the earthquake source via magnitude, wave propagation path via source-to-site distance, and site effects via the 30-m time-averaged shear-wave velocity (Section 4.5). While recent empirical GMMs utilize more parameters, there is a limit on a statistical regression equation's ability to represent a complex process.

The explicit consideration of salient physics implies that physics-based simulations have the *potential* to generate more accurate and precise predictions of ground motion at the site than those available using empirical models based on globally recorded historical ground motions. This is particularly relevant for large magnitudes, small source-to-site distances, and for other phenomena (sedimentary basin response, hanging-wall effects, etc.) that are poorly constrained in empirical GMMs (see Section 4.7.2). In such cases, physics-based GMM extrapolations beyond empirical observations should outperform empirical models. However, this is contingent upon the specific methods having the capability to accurately reproduce the characteristics of recorded ground motions, using model representations of the earthquake rupture and crustal structure, and the availability of reliable information for such models to be robustly developed in the region of interest. Thus, model validation (Chapter 12) is central to understanding the predictive capability of simulation-based methods, and hence realizing the conceptual advantages of physics-based GMMs.

5.3 Earthquake Source Representation

Earthquakes occur as a result of slip across a fault surface (Chapter 2). In this section, we expand on this general concept by quantifying earthquake-induced slip on faults and how it varies in space and time over the fault surface. We begin with a kinematic description of slip at a point (on the fault), generalize to the variation over the fault surface, and, finally, describe it through equivalent

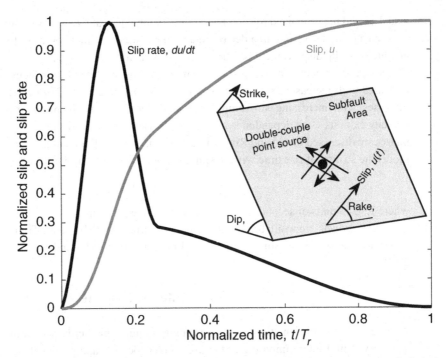

Fig. 5.3 Slip-time function describing the temporal variation of rupture slip, and its derivative, slip rate, at a point on the fault plane (Liu et al., 2006). Over this small subfault area, the rupture can be kinematically represented as a double-couple moment tensor at a point for wave propagation simulations.

double-couple moment tensors. Insights from dynamic rupture modeling for use in kinematic representations are also discussed.

5.3.1 Point Source Rupture

A *point source* approximation is commonly considered when the length scale of the rupture is small relative to the distance at which ground motions are recorded.[3] When such an approximation is not valid for an entire rupture dimension, then the rupture area can be subdivided into smaller subfaults, over which the rupture occurs, for the approximation to hold.

Consider the "point source" depicted in the inset of Figure 5.3. Over the (small) subfault area (dA), the planar geometry of the fault surface (θ, δ) as well as the slip direction (λ), are considered as constants. The seismic moment resulting from rupture of the subfault area is equal to $M_0 = \mu(dA)u$ (Equation 2.4). Mathematically, the slip on the fault can be comprehensively described through a moment tensor, M, referenced in Equation 5.1. This moment tensor is discussed subsequently in Section 5.3.3, and also reflected schematically as a double-couple in Figure 5.3.

In addition to the amplitude of slip, its temporal variation is also necessary to describe the rupture. Figure 5.3 illustrates an analytical model (Liu et al., 2006) for the variation of slip with time. For generality, the model is described in both normalized time and slip amplitude, where the total duration over which slip occurs is referred to as the rise time, T_r. The time derivative of slip, the slip rate, is

---

[3] Loosely speaking, the rupture "appears" as a point when the distance of the observer is large.

also depicted in normalized form. Once rupture commences at $t/T_r = 0$, the slip rate increases rapidly to reach a peak value, and then decays in a broadly two-stage manner until the slip stops at $t/T_r = 1$, at which the slip displacement reaches its maximum value, $u = u_{max}$.

Because of the link between seismic moment and rupture area (Section 2.5), slip amplitude increases with increasing seismic moment (e.g., Equation 2.8 and Figure 2.18). The rise time also increases with increasing slip amplitude (and therefore seismic moment), even though rupture velocity exhibits some dependence on slip amplitude (further details on rupture growth and stopping phases are discussed in Aki and Richards, 2002, Chapter 11). For example, Graves and Pitarka (2015) model the scaling of rise time with seismic moment for strike-slip faults as[4]

$$T_r = 3.12 \times 10^{-7} M_0^{1/3} \tag{5.2}$$

where T_r is in units of seconds, and M_0 in Newton-meters (Nm).

Several models are available for the temporal variation of slip (e.g., Tinti et al., 2005; Liu et al., 2006) and the scaling of slip amplitude and rise time (e.g., Mai and Beroza, 2002; Aagaard et al., 2008).

5.3.2 Finite-Fault Rupture

A more general representation of source rupture requires explicitly accounting for the fault's finite geometry, with rupture initiating at the hypocenter location and propagating over the fault surface. A *kinematic* description of (finite) fault rupture requires the a priori specification of the fault slip at each location, via (1) the slip initiation time and (2) its evolution with time.

Examination of past earthquake ruptures, using *source inversion* solutions from observed ground motions, illustrates that they exhibit complexity over all spatial scales (Mai and Beroza, 2002). By way of example, Figure 5.4 illustrates the kinematic description of the fault rupture that was previously highlighted in Figure 5.1. Each of the figure panels respectively illustrates the spatial variation of slip amplitude, rise time, and rake direction over the fault area obtained from one realization of a stochastic rupture generation algorithm (Graves and Pitarka, 2010, 2015). At each point on the rupture surface, the slip amplitude and rise time scale the slip-time function previously presented in Figure 5.3, while the rake direction orients the double-couple point source. The triplet of three numbers in the top left of each figure panel indicates the minimum/mean/maximum values of that quantity. The following depicted features provide useful insights into the salient aspects of fault rupture complexity.

Slip amplitude: The mean slip displacement is 100.1 cm, which is a direct consequence of the prescribed moment magnitude, fault area, and shear rigidity (i.e., Equation 2.8). The slip is spatially correlated, with a large variance in the slip amplitude indicated by values ranging from 0 to 334.9 cm.

Rise time: The rise time also exhibits large variability ($T_r = [0, 10.4]$ s), and is correlated with slip amplitude; i.e., higher rise times occur generally in areas of higher slip amplitude (Liu et al., 2006; Graves and Pitarka, 2016). Rise times generally increase near the ground surface, representing fault weakening, and at depths near the brittle-ductile transition zone (Graves and Pitarka, 2015).

Rake: The variation in rake is spatially correlated about the mean specified value ($\lambda = 0°$ in this example), with a standard deviation of 15°. For the particular realization in Figure 5.4, a range of 112° occurs.

---

[4] The direct relation is $T_r = 1.45 \times 10^{-9} M_0^{1/3}$ for M_0 in dyne-cm (see Section 2.5).

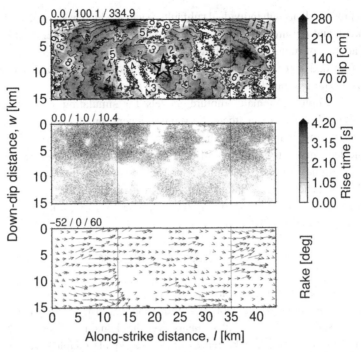

Fig. 5.4   Kinematic rupture model of the *M*6.8 earthquake depicted in Figure 5.1 illustrating slip amplitude (top), rise time (center), and rake direction via the slip vector (bottom). Hypocenter location and rupture front time contours are also depicted (top panel). The triplet of numbers at the top left of each panel indicates the minimum, mean, and maximum values of the respective quantity depicted. The vertical lines at *l* = 12.8 and *l* = 35 km delineate the three contiguous planar segments of the rupture with different strike direction in Figure 5.1. Rake vectors are down-sampled over a coarse spatial grid for clarity.

Rupture velocity: The top panel of Figure 5.4 also illustrates contours of rupture initiation time over the fault area. Considering that concentric circles would be associated with a constant rupture velocity, the complexity of these contours indicates a spatial variation in the velocity at which the rupture front propagates over the fault area. Generally, locations of higher-than-average slip amplitude have higher rupture velocities.

The nature of the rupture over the fault directly influences the ground motions observed at a site. One principal effect is rupture directivity (see Section 4.5.5 in the context of empirical GMMs). The bilateral rupture and strike-slip mechanism depicted in Figure 5.4 represent near-ideal conditions for source directivity, which is illustrated in Figure 5.1 in the form of significant ground velocities at *t* = 12 s at locations in the vicinity of the ends of the causative fault.

The complexity of fault rupture also directly influences the coherence of propagated seismic waves. Early parametric models for kinematic rupture (so-called *kinematic rupture generators*) were generally too simplistic, leading to coherent ground motions that were excessively strong at low frequencies (as a result of rupture directivity) and relatively weak at higher frequencies. In this vein, further generalizations beyond that depicted in Figure 5.4 include the representation of geometrical complexity (fault roughness) in the planar fault geometry—allowing local variations in strike and dip angles of the fault in the same manner as the rake angle, and reducing ground-motion coherence to realistic levels (Shi and Day, 2013; Trugman and Dunham, 2014; Graves and Pitarka, 2016).

The kinematic representation of rupture over a fault surface (coupled with the slip-time function) provides the necessary description for ground-motion simulation. Over the fault surface, rupture is represented through individual subfaults, each of which has a specified geometry (strike, dip, subfault area), and rupture kinematics (rake, slip amplitude, rise time, rupture initiation time, and temporal variation of rupture). This allows each subfault on the fault surface to be represented with a moment tensor, and hence can be applied within the elastodynamic equation of motion to numerically compute the resulting ground motion via simulation (Section 5.5).

Developing kinematic rupture models for simulating ground motions from potential future earthquakes requires a stochastic rupture generation algorithm and its necessary input parameters. By way of example, the method of Graves and Pitarka (2010, 2015) requires the specification of the fault geometry (fault length and width), moment magnitude, and hypocenter location.[5] Two-dimensional correlated random fields for the slip, rise time, and rake (as shown in Figure 5.4) can then be generated, providing alternative realizations of the potential rupture. Also, these specified parameters will have uncertainty (e.g., hypocenter location), which can be varied when generating multiple rupture realizations.

5.3.3 Moment Tensor Representation

The physical nature of an earthquake as a shear dislocation across a fault surface cannot directly be used in modeling ground motion via Equation 5.1. Instead, we can use a set of *equivalent body forces* that result in the same displacements over the domain of interest. The inset of Figure 5.3 indicated that slip at a point can be represented by a double-couple (see Section 2.3.2 and Shearer, 2009, Chapter 9). A force couple is a pair of opposing forces f separated by a distance d. To generalize, when the forces point in the i direction and are separated in the j direction, the moment caused by this force couple is $M_{ij} = f_i d_j$. For the three dimensions ($i, j = 1, 2, 3$) there are nine such M_{ij} components, and these comprise the moment tensor, M, in Equation 5.1 that completely describes the possible set of equivalent body forces.

In prescribing the moment tensor for each subfault it is most convenient to use a local coordinate system (u_1, u_2, u_3), where u_1 is the slip direction (indicated in Figure 5.3), u_2 is 90° anticlockwise of u_1 in the plane of the subfault (i.e., both u_1 and u_2 are tangent to the subfault surface), and u_3 is the outward normal to the subfault surface. For the common case in which slip occurs over the subfault with constant rake, then only u_1 has nonzero slip, and it can be shown that the moment tensor is given by (Aki and Richards, 2002, Equation 3.25):

$$M = \begin{bmatrix} 0 & 0 & M_0 \\ 0 & 0 & 0 \\ M_0 & 0 & 0 \end{bmatrix}. \tag{5.3}$$

These two nonzero terms, $M_{13} = M_{31} = M_0$, represent the two couples depicted in the inset of Figure 5.3, hence the name "double couple."

The moment tensor in the local coordinate system for each subfault can then be transformed into the global (x, y, z) coordinate system based on the subfault planar geometry (i.e., θ, δ). As a result, the nine components of the moment tensor in global coordinates are generally nonzero. The specification of the subfault moment tensor in numerical solutions of the wave equation is discussed in Section 5.5.1.

---

[5] Additional optional parameters can be specified, but default to calibrated values otherwise.

5.3.4 Kinematic versus Dynamic Rupture Modeling

The previous subsections have addressed the kinematic description of rupture over a fault surface for use in ground-motion simulation. This kinematic rupture description is predominantly used in physics-based ground-motion simulation methods for engineering applications. Therefore, it is important to understand the physics that is, and is not, considered in their development and utilization.

A *dynamic* representation of fault rupture requires specification of the in-situ stress conditions in the fault vicinity and a constitutive relation for the effective stress-strain behavior of the fault surface (Aki and Richards, 2002, Chapter 11). Once nucleation of rupture at a point occurs, a dynamic rupture model then explicitly solves the elastodynamic equation of motion to determine how rupture propagates in a physically consistent manner over a fault surface in space and time. Thus, the output of a dynamic rupture model is the spatial and temporal variation of rupture on the fault, which is also the same output produced from a (statistical) kinematic rupture generator.

Compared with dynamic rupture models, there are several features of kinematic rupture generators that can be nonphysical. For example, kinematic slip distributions can be inconsistent with fracture energy requirements at rupture edges because of their parametric definition. Kinematic rupture generators can also be hindered by the spatial resolution limitations of any source inversion data of past events that they utilize.

Despite their potential, dynamic rupture models are rarely used for physics-based ground-motion simulation at present, due to challenges in quantifying in-situ stresses and constitutive behavior, as well as high computational costs. Nonetheless, insights from dynamic rupture modeling are used to refine kinematic rupture modeling. Kinematic rupture generators that incorporate such insights are typically referred to as *pseudo-dynamic* rupture generators (e.g., Guatteri et al., 2004; Liu et al., 2006; Graves and Pitarka, 2016). Examples include the use of slip-time functions (e.g., Figure 5.3) that produce physically consistent rupture starting and stopping phases and energy spectra; the correlation of slip amplitude, rise time, and rupture velocity; and the fractal features of geometrical fault roughness (Guatteri et al., 2004; Schmedes et al., 2013; Shi and Day, 2013; Trugman and Dunham, 2014; Mai et al., 2017).

5.4 Seismic Wave Propagation

Earthquakes cause seismic waves to propagate through the Earth's crust (Chapter 2). This section examines the salient physics that govern the propagation of seismic waves through the Earth that are relevant to seismic hazard. We will discuss different seismic waves of importance (P-, S-, Rayleigh, and Love waves); how these waves are transmitted, reflected, and refracted at material boundaries; and attenuation in amplitude during propagation. The ability to represent these phenomena in simulations via crustal models is also discussed.

The material presented in this section is intentionally brief. More extensive introductory coverage is provided from a seismological perspective in Shearer (2009) and a geotechnical earthquake perspective in Kramer (1996, Chapter 5), among others. Aki and Richards (2002) provide a comprehensive treatment of the subject matter.

5.4.1 The Wave Equation

The propagation of seismic stress waves is governed by the so-called wave equation. In one spatial dimension (x), the wave equation takes the form

$$\frac{\partial^2 u}{\partial t^2} = V^2 \frac{\partial^2 u}{\partial x^2} \tag{5.4}$$

where u is the particle displacement (i.e., the movement of a given point in the Earth's crust), and V is the propagation velocity of the wave.

Two important variables define the spatial properties of a propagating wave. The first is the wavelength, λ, which is the spatial length over which one full oscillation of a wave occurs. The second is the wavenumber, k, which is related to the wavelength of the motion by

$$\lambda = VT = \frac{V}{f} = \frac{2\pi}{\omega}V = \frac{2\pi}{k} \tag{5.5}$$

where T, f, and ω are the period, frequency, and angular frequency of the wave, respectively. From this expression it can be understood that wavelength decreases with decreasing velocity.[6] It is useful to consider that wavelength λ is to wavenumber k as circular frequency ω is to vibration period T. Equation 5.5, therefore, provides an important link between temporal and spatial properties of seismic waves.

5.4.2 Body and Surface Waves

Earthquake-induced ground motions measured at the Earth's surface are comprised of body and surface waves, and understanding their different properties is important for both interpreting recorded ground motions and predicting them through ground-motion simulation.

Particle Motions

Figure 5.5 illustrates the particle motions of the four seismic waves that are of principal importance for seismic hazard considerations. Primary (P) and secondary (S) waves are *body waves* that propagate directly from the seismic source. These are also commonly referred to as pressure and shear waves, which emphasizes their particle motions. P-waves produce particle motion that is parallel to the direction of wave propagation, whereas S-wave particle motion is transverse to the direction of propagation. Consequently, P-waves result in volumetric compression and dilation, whereas S-waves result in shearing distortions, but not volumetric dilation. The transverse particle motion of S-waves is commonly decomposed into its components of motion, in the horizontal $(x-y)$ and vertical $(x-z)$ planes, perpendicular to the direction of propagation, and these are referred to as SH and SV waves, respectively.

Rayleigh and Love waves are *surface waves* that are produced from the interaction of the body waves with the Earth's surface (Shearer, 2009, Section 8). Rayleigh waves result from the constructive interference of P- and SV-waves at the Earth's surface. The particle motion is retrograde elliptical near the surface[7] (i.e., anticlockwise in the $x-z$ plane shown in Figure 5.5), with an amplitude that decreases exponentially with depth. Love waves occur from the constructive

---

[6] Thus, seismic wavelengths generally reduce as they approach the Earth's surface where lower velocity materials exist.
[7] With increasing depth the extent of retrograde motion decreases until the motion becomes vertical, and eventually prograde.

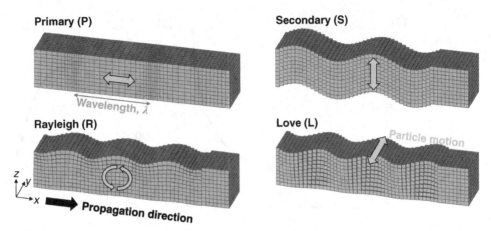

Fig. 5.5 Particle motions of body (P, S) and surface (Rayleigh, Love) waves.

interference of refracted SH waves at the Earth's surface as a result of a depth-dependent velocity gradient.[8] Because of their origin from SH-wave interference, Love wave particle motion occurs in the horizontal plane transverse to the direction of propagation, with an amplitude that decreases exponentially with depth.

Propagation Velocities

Figure 5.6 provides a comparison of the velocities of the body and surface waves examined in Figure 5.5. The ratio of the P- and S-wave velocities is a function of Poisson's ratio, v:

$$\frac{V_P}{V_S} = \sqrt{\frac{2 - 2v}{1 - 2v}} \tag{5.6}$$

where V_P and V_S are the respective P- and S-wave velocities. Since $0 < v < 0.5$ for typical geomaterials, then $V_P > V_S$. The larger velocity of P-waves implies that they are the first to arrive in recorded ground motions, and hence the name "primary" waves, with the relatively slower S-waves arriving "second." For a typical value of $v = 0.25$, $V_P = \sqrt{3}V_S$.

The velocity of Rayleigh waves, V_R, relative to P- and S-waves in a homogenous half-space is also a function of v, as depicted in Figure 5.6. Rayleigh wave velocity is obtained by numerically solving a characteristic equation with no explicit analytical solution (Shearer, 2009, Section 8.2). For $v = 0.25$, $V_R = 0.92V_S$, such that the rule-of-thumb, $V_R \approx 0.9V_S$ is often used.

Unlike P-, S-, and Rayleigh waves in a homogeneous medium, Love wave velocity in a velocity gradient is a function of circular frequency (ω). Figure 5.6 illustrates how the variation in Love wave velocity is a function of frequency for a simple two-layer model with the surficial layer having an S-wave velocity of V_{S1} and thickness H, and a half-space S-wave velocity of V_{S2}. The frequency-dependence of the propagation velocity is a property referred to as *dispersion*. To appreciate the limiting velocities as a function of frequency, consider that at high frequencies the wavelength is very small (relative to the low-velocity layer thickness, H) and therefore the Love wave is significant only in the surficial layer (and hence travels at approximately the S-wave velocity of this layer).

---

[8] Seismic velocities generally decrease near the Earth's surface, so such conditions are typical; however, the extent of Love wave generation is a function of the velocity gradient.

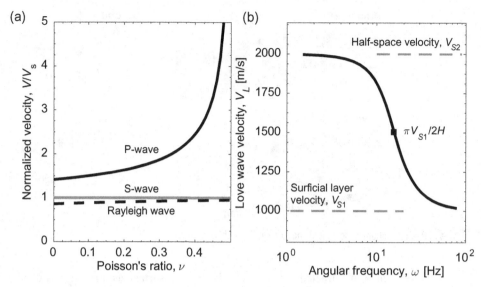

Fig. 5.6 (a) Relative variation in P-, S- and Rayleigh-wave velocities as a function of Poisson's ratio and (b) variation in Love wave velocity as a function of frequency for a surficial layer of thickness $H = 100$ m.

In contrast, for low frequencies, the wavelength is very large (relative to H) and therefore the Love wave essentially only "sees" the half-space, and hence propagates at this velocity. The transition between these two asymptotic cases occurs near the angular natural frequency of the surficial soil layer, $\omega = 2\pi f = 2\pi V_{S1}/4H = \pi V_{S1}/2H$, which is the frequency at which the Love wave velocity is the average of the half-space and surficial layer V_S values.

While Rayleigh waves are nondispersive in a homogenous half-space, the Earth's surface is generally comprised of materials that increase in propagation velocity with depth. Similar to the description above for Love waves, as the depth over which Rayleigh waves are significant is a function of their wavelength (in turn a function of frequency from Equation 5.5), Rayleigh waves with longer wavelengths (low frequency) can propagate faster[9] than those with shorter wavelengths (high frequency). Therefore, in a heterogeneous medium, Rayleigh waves are also dispersive, and the dispersive nature of Rayleigh and Love waves can be used to evaluate subsurface velocity structure through nondestructive site investigation (e.g., Park et al., 1999). Another implication of the dispersive property of surface waves is that low-frequency surface waves propagate faster and arrive at a particular site before their high-frequency counterparts.[10]

5.4.3 Wave Behavior at Material Interfaces

To this point, we have generally considered a homogenous medium in examining the propagation of waves. However, realistic conditions in the Earth are much more complicated, with materials of different elastic properties. To understand the behavior in heterogeneous media it is necessary to understand the behavior of waves at material interfaces.

---

[9] For the common case of a "normally dispersive" deposit with velocity increasing with depth.
[10] Body waves are not dispersive in elastic media, but slightly dispersive in anelastic media, though significantly less than surface waves.

Seismic Impedance

Consider a seismic P-wave propagating in one dimension (x) that encounters, at $x = 0$, a boundary between two different materials. The harmonic wave traveling in the $+x$ direction and approaching the material boundary is referred to as the incident wave. When the incident wave hits the boundary, part of it will be transmitted through the boundary (the transmitted wave), and part will be reflected (the reflected wave). The incident and reflected waves both travel in Material 1 and so will have wavelengths $\lambda_1 = 2\pi/k_1$ (see Equation 5.5). The transmitted wave travels in Material 2, so it will have a wavelength of $\lambda_2 = 2\pi/k_2$.

Through compatibility of stresses and displacements at the material interface, the properties of the reflected and transmitted waves can be determined as (Kramer, 1996, Section 5.4)

$$A_R = \frac{1-\alpha}{1+\alpha}A_I; \qquad A_T = \frac{2}{1+\alpha}A_I \tag{5.7}$$

and

$$\sigma_R = \frac{\alpha-1}{1+\alpha}\sigma_I; \qquad \sigma_T = \frac{2\alpha}{1+\alpha}\sigma_I \tag{5.8}$$

where A and σ are the wave displacement and stress amplitudes, respectively; the subscripts I, R, and T represent the incident, reflected, and transmitted waves, respectively; and α is the *impedance ratio*, defined as

$$\alpha = \frac{\rho_2 V_{p_2}}{\rho_1 V_{p_1}}. \tag{5.9}$$

The impedance ratio is a fundamental quantity that determines the displacement and stress amplitudes associated with transmitted and reflected waves at a material boundary.[11]

Figure 5.7 provides an illustration of the variation in displacement and stress amplitudes as a function of the impedance ratio, as given by Equations 5.7 and 5.8. For example, when $\alpha < 1$, the incident wave can be thought of as approaching a "softer" material (this is typically the case as seismic waves travel from the deep crust upward to the surface). In this case, the transmitted wave's displacement amplitude becomes larger than that of the incident wave. However, the maximum stress of the transmitted wave will be less than that of the incident wave because of the longer wavelength. The reflected wave will have a smaller amplitude than the incident wave, with the negative value of σ_R indicating a change in polarity.

The two extreme impedance cases of $\alpha = 0$ and $\alpha = \infty$ are also of interest. The first represents a free surface (i.e., $\rho_2 V_2 = 0$ at the surface), while the second represents a rigid boundary (i.e., $\rho_2 V_2 = \infty$). In the free-surface case, the incident wave is reflected with the same displacement and stress amplitude, but opposite polarity ($A_R = A_I, \sigma_R = -\sigma_I$). The displacement amplitude of the wave at the boundary (the transmitted displacement) is also equal to two times the incident wave ($A_T = 2A_I$). In the rigid-boundary case, the incident wave is reflected with the same displacement and stress amplitude and polarity ($A_R = -A_I, \sigma_R = \sigma_I$), and the stress amplitude "transmitted" at the boundary is twice that of the incident wave ($A_T = 0, \sigma_T = 2\sigma_I$).

As previously noted, since the impedance of materials typically decreases as one approaches the surface, the displacement amplitude of the motion tends to increase. Therefore, based on elastic wave propagation, soft soils will give rise to larger-amplitude motions than stiffer soils or rock.

---

[11] While the 1D case considered here is in terms of P-waves, the specific impedance for S-waves is the same in concept and given by $\alpha = \rho_2 V_{s_2}/\rho_1 V_{s_1}$.

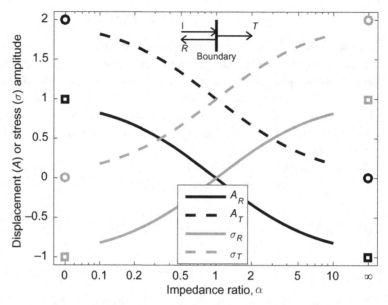

Fig. 5.7 Displacement (A) and stress (σ) amplitudes of transmitted (T) and reflected (R) waves as a function of impedance ratio (due to an incident (I) wave with $A_I = \sigma_I = 1$). The impedance ratio is plotted on log scale, with the limiting cases of $\alpha = 0$ and ∞ also added for values outside the range $0.1 < \alpha < 10$.

Refraction of Inclined Waves

Waves propagate in a 3D medium and generally do not approach interfaces at 90-degree angles, as was assumed in the previous subsection.

Let θ define the angle of an inclined wave, as measured normal to a material interface. Figure 5.8 illustrates an inclined incident wave propagating in Material 1 with velocity V_1, approaching an interface at an angle θ_I to Material 2 with velocity V_2. Snell's law states that $\sin\theta/V$ is a constant, where V can be either the P- or S-wave velocity (for these respective waves), and thus the following relationship exists between the incident, reflected, and transmitted waves (Shearer, 2009, Section 4.1):

$$\frac{\sin\theta_I}{V_1} = \frac{\sin\theta_R}{V_1} = \frac{\sin\theta_T}{V_2}. \tag{5.10}$$

Because the incident and reflected waves travel in the same medium (i.e., with the same velocity), the incident and reflected wave angles are equal (i.e., $\theta_I = \theta_R$). The angle of the transmitted wave can be obtained from rearrangement as

$$\theta_T = \sin^{-1}\left(\frac{V_2}{V_1}\sin\theta_I\right). \tag{5.11}$$

The fact that $\theta_T \neq \theta_I$, when $V_1 \neq V_2$, means that the direction of wave propagation changes; i.e., the wave ray path is "bent." This is referred to as *refraction*. When $V_1 > V_2$, for which the wave is approaching softer material, $V_2/V_1 < 1$, therefore $\theta_T < \theta_I$ (and the opposite for $V_1 < V_2$).

Figure 5.8b illustrates the multiple reflections and refractions that occur in a three-layer velocity profile based on SH-wave propagation. In this common case of materials with reducing stiffness toward the surface, Snell's law indicates that an upwardly propagating inclined wave will be continually refracted toward a vertical path. Simultaneously, the effects of impedance also result

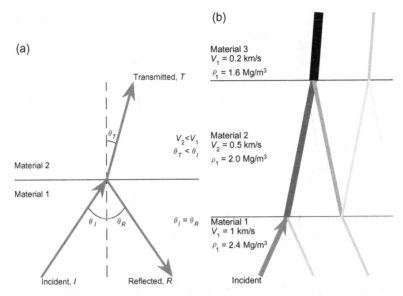

Fig. 5.8 Inclined wave behavior at a material interface: (a) single interface with incident, reflected, and (refracted) transmitted waves, and (b) multiple reflected and transmitted waves in a three-layer profile, with increasing line width and darkness indicating increasing wave displacement amplitude.

in the displacement amplitude of the refracted wave increasing. Furthermore, the effect of multiple reflections within Material 2 leads to subsequent "scattered" waves propagating in Material 3, which both have a smaller amplitude than the 'direct' wave and also will be delayed in time (due to the additional propagation time within Material 2).

Figure 5.8 implies that a nearly vertically propagating wave that reaches the ground surface will have horizontal particle-motion predominantly due to S-waves, and vertical motion predominantly due to P-waves.[12] This assumption is relied upon in many site response analysis methods in geotechnical earthquake engineering, which typically consider only the response of a 1D soil deposit to vertically propagating S-waves (e.g., Kramer, 1996, Chapter 7).

Wave Conversion at Material Interfaces

The previous discussion of wave refraction focused on the effects of material velocities on propagation direction and ignored displacement and stress continuity at the material interface. Because of the different amplitudes and angles of the incident, reflected, and transmitted waves, the summation of displacements and stresses due to a single wave type (e.g., P-wave) will generally not be continuous. Consequently, incident waves with particle displacements and stresses in a direction that is not parallel or perpendicular with the material interface will result in "converted" seismic waves to maintain continuity (Shearer, 2009, Section 6.3). Figure 5.9 provides a schematic illustration of this "mode" conversion that occurs due to the inclined incidence of P, SV- and SH-waves. Figure 5.9a illustrates that for an incident P-wave, the angles of reflection and refraction of the P-wave can be computed as previously discussed (e.g., Equation 5.11). Reflected and refracted SV-waves (i.e., S waves in the "vertical" plane of the page) are required to provide continuity. Because of the lower

---

[12] The contribution of surface waves is ignored in this discussion.

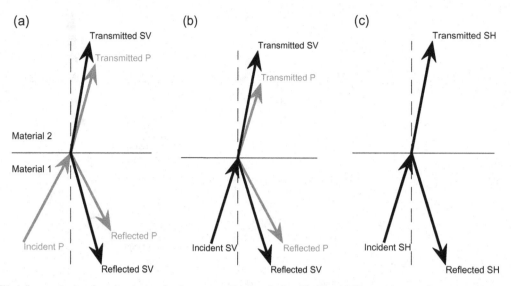

Fig. 5.9 "Mode" conversion in reflected and transmitted waves resulting from inclined P-, SV-, and SH-waves intersecting a material boundary.

S-wave velocity, relative to the P-wave, the reflected and transmitted SV waves have a smaller angle normal to the material interface than the corresponding P-waves (via Snell's law).

The mode conversion in Figure 5.9b for an incident SV-wave follows the same principles as for the incident P-wave discussed above. The incident SH-wave in Figure 5.9c represents a case where particle displacement and stresses (which occur "into" the page of the 2D image of Figure 5.9c) are parallel to the material boundary. Therefore, displacement and stress compatibility due to the incident, reflected, and transmitted SH-waves is maintained without wave conversions. In a similar vein, incident P- and SV-waves with an incidence angle of $\theta_I = 0°$ also produce no wave conversion.

Critical Incidence Angle and Total Internal Reflection

The critical incidence angle, θ_{crit}, is the angle for which the refracted wave travels parallel to the material boundary. Setting $\theta_T = 90° = \pi/2$ in Equation 5.11 and rearranging gives

$$V_2 \frac{\sin \theta_I}{V_1} = \sin \theta_T = \sin \left(\frac{\pi}{2} \right)$$
$$\rightarrow \theta_{crit} = \theta_I = \sin^{-1} \left(\frac{V_1}{V_2} \right).$$

(5.12)

Furthermore, if the incident angle is greater than the critical angle, $\theta_I > \theta_{crit}$, then Equation 5.12 requires $\sin \theta_T > 1$, and hence θ_T is undefined—that is, there is no transmitted wave. As a result, all of the incident wave energy is contained in the reflected wave, and this is referred to as *total internal reflection*. Figure 5.10a illustrates these three different cases for incident angles that are precritical, critical, and postcritical.

Figure 5.10b illustrates the variation in the critical angle as a function of the velocity ratio, V_1/V_2.[13] For $V_1/V_2 = 1$, there is no velocity change across the material boundary, and $\theta_{crit} = 90°$, implying

---

[13] This relationship applies for P- and S-waves, as well as mode conversions, such as illustrated in Figure 5.9.

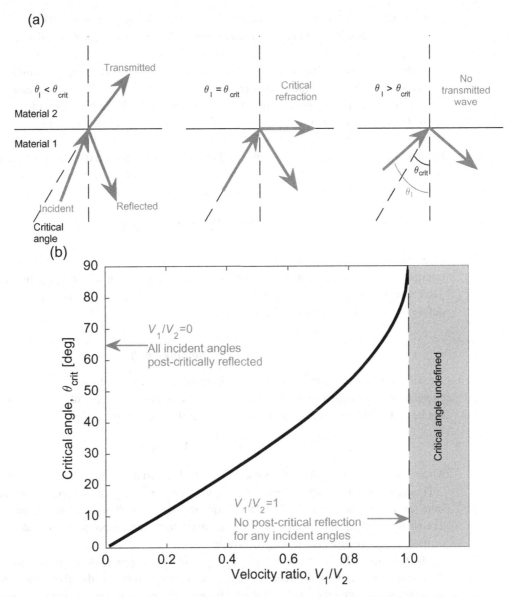

Fig. 5.10 (a) Effect of incidence angle on whether subcritical, critical, and postcritical refraction occurs ($V_1/V_2 = 0.67$); (b) effect of interface velocity ratio on critical angle. In critical refraction the transmitted wave travels along the material interface (at velocity V_2); postcritical refraction results in no transmitted wave and is thus also referred to as total internal reflection.

that all incidence angles are subcritical (i.e., total internal reflection cannot occur). Similarly, for $V_1/V_2 > 1$ the critical angle is undefined, implying that a wave propagating from a "stiff" to "soft" material will always result in a transmitted wave. In contrast, during propagation from a soft to stiff material (i.e., $V_1/V_2 < 1$) the critical angle exists, and decreases as the velocity ratio decreases. For example, given a velocity ratio of 0.5 (e.g., $V_1 = 300$ m/s and $V_2 = 600$ m/s) the critical angle is 30°, meaning that total internal reflection will occur for incidence angles in the range $30° < \theta_I < 90°$.

The above discussion is particularly relevant in the case of surficial sedimentary soil deposits overlying stiff rock, which can have a large velocity ratio. Based on impedance considerations, waves propagating into a soft soil layer will increase in amplitude (e.g., Figure 5.8). After reflection from the ground surface, these down-going waves may subsequently interact with the soil–rock interface at postcritical angles, leading to total internal reflection. These waves then become "trapped" in the surficial soil layer, which leads to a "guiding" effect of the wave energy, and thus such effects are often referred to as *waveguide* effects (Shearer, 2009, Section 4.4). Because waveguide effects cause seismic waves to propagate in a specific confined medium, guided waves suffer significantly less geometric spreading attenuation (see Section 5.4.4) than unguided waves. One pervasive example is basin-edge effects (e.g., Graves, 1993; Kawase, 1996), in which body waves enter a sedimentary basin through its thickening edge, reflect off the Earth's surface, and then are post-critically reflected at the basin–rock interface.[14] Waveguide effects consequently lead to ground motions with larger amplitudes and longer duration (e.g., Choi et al., 2005).

5.4.4 Wave Attenuation

In Section 5.4.3 wave amplitudes varied as a result of reflection and transmission at material interfaces. Wave amplitudes also reduce during propagation in a homogeneous medium due to geometric spreading of the wavefront and energy loss due to anelastic processes and plastic deformation, which are discussed in this section.

Geometric Spreading

As seismic energy radiates from the earthquake source, wave amplitudes decrease due to geometric attenuation, or so-called geometric spreading. The energy associated with seismic wave propagation can be represented via the energy flux (Shearer, 2009, Section 6):

$$E^{flux} = \frac{1}{2}V\rho A^2 \omega^2 S \tag{5.13}$$

where S is the surface area of the wavefront, and other variables are previously defined.

During body-wave propagation in three dimensions from a point source, the surface area of the spherical wavefront is given by $S = 4\pi r^2$. Equating the energy flux over a wavefront of radius r with a reference condition of radius $r_0 = 1$ implies that body-wave amplitudes attenuate as $A \propto 1/r$, that is, inversely proportional to propagation distance.

The wavefront during surface-wave propagation from a point source occurs over a cylinder at the Earth's surface, with a circumference of $S = 2\pi r$. Similar to the above logic for body wave propagation, equating the energy flux at a distance of r to a reference distance implies that surface-wave amplitudes attenuate as $A \propto 1/\sqrt{r}$.

Figure 5.11 shows the normalized amplitude reductions from geometric spreading of body and surface waves. While Figure 5.11 implies that surface-wave amplitudes could be large at short distances, note that all earthquake rupture-induced wave energy is initially transmitted as body waves, and surface waves occur from body-wave converted energy. As a result, body waves dominate most ground-motion amplitudes at short distances. However, at larger distances, the generation of surface waves combined with their more gradual geometric attenuation leads to an increase in their amplitudes relative to those of body waves. Ultimately, surface waves generally dominate ground-motion amplitudes at very large distances (i.e., >1000 km).

---

[14] The inclined geometry of the basin edge facilitates refraction at a low angle relative to the horizontal, and thus the subsequent potential for postcritical reflection at the thickening basin interface with underlying rock.

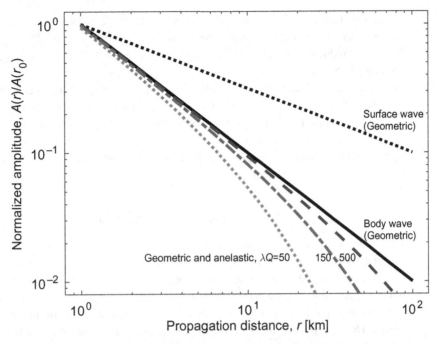

Fig. 5.11 Effect of geometric spreading and anelastic attenuation on ground-motion amplitudes as a function of ray path propagation distance, r, and the product of wavelength and quality factor, λQ. Amplitudes considering anelastic attenuation are shown for body waves only.

Note that both body- and surface-wave attenuation rates discussed above are for propagation in a homogenous medium. In a heterogeneous medium, seismic refraction and reflection (Section 5.4.3) make the ray path complex. As a result, the convenient substitution of the ray path distance, r, for a simple source-to-site distance metric, R (e.g., Figure 4.25) can lead to empirical observations as a function of R deviating from these theoretical attenuation rates. In addition to empirical observations, such effects can also be studied via numerical solutions, since no analytical solutions exist for geometric attenuation in complex heterogeneous media.

Anelastic Attenuation

Even under small-strain deformations, seismic wave amplitudes decay as a result of material adsorption due to anelastic processes and internal friction (Shearer, 2009, Section 6.6). It is typical to describe the degree of anelastic attenuation by the dimensionless *quality factor*, Q:

$$\frac{1}{Q(f)} = -\frac{\Delta E}{2\pi E} \tag{5.14}$$

where E is the peak strain energy, and ΔE is the energy loss per cycle, for the wavefield (in an Eulerian frame of reference) at a given frequency.[15] This attenuation can be approximated as exponential decay, which at a propagation distance x can be expressed as:

$$A(x) = A_0 \exp\left(-\frac{\pi f}{VQ}x\right) = A_0 \exp\left(-\frac{\pi}{Q}\frac{x}{\lambda}\right) \tag{5.15}$$

where A_0 is the initial amplitude.

---

[15] The quality factor is related to the material damping ratio (e.g., Kramer, 1996) by the expression $\xi = 1/2Q$.

The implications of Equation 5.15 are that amplitude reduction due to anelastic attenuation increases exponentially with propagation distance, and the rate of decay grows with increasing frequency, reducing seismic velocity, and reducing quality factor. The rightmost expression in Equation 5.15, based on the expression $\lambda = V/f$ (Equation 5.5), illustrates that attenuation is a function of the number of wave cycles (x/λ), and the energy loss per cycle ($1/Q$). Figure 5.11 illustrates the effect of anelastic attenuation (in addition to geometric spreading) for three different values of the product λQ. Note that quality factor is approximately linearly related to seismic velocity (i.e., $Q \propto V$), so anelastic attenuation is particularly significant for low-velocity materials near the Earth's surface. However, the propagation distance through near-surface materials is generally small because of the curved ray path from the earthquake source.[16]

Other Attenuation Mechanisms

Two other mechanisms produce attenuation of ground-motion amplitudes—wave scattering and material plasticity—and are briefly discussed here.

Seismic wave scattering refers to the modification of seismic waves resulting from propagation through material heterogeneities (Wu and Aki, 1988). As with material interfaces (e.g., Figures 5.8 and 5.9), material heterogeneities result in reflection, refraction, and mode conversion of seismic waves. They are most evident in observed ground-motion time series in the form of later-arriving seismic waves, or *coda*. However, the specific behavior is more nuanced as a function of the length scale of the heterogeneities, a, and the seismic wavenumber k (Equation 5.5) of consideration. When the wavelengths are several orders of magnitude larger than the scale of heterogeneities ($ka \ll 1$), the seismic waves effectively do not see them and propagate as if in a homogeneous medium. The most significant scattering occurs when the seismic wavelength and heterogeneity length scale are similar ($ka \approx 1$), and can produce large refraction angles relative to the incident angle. Finally, when the wavelengths are small relative to the heterogeneity length scale ($ka \gg 1$), scattering occurs, but primarily in the "forward" direction; that is, the scattering angle is small. For a given character of heterogeneities, the above dependence on ka implies that scattering is most significant for large k, that is, for higher frequencies. It follows that this scattering attenuation can be explicitly modeled if the heterogeneities in the crust are directly represented in 3D wave propagation (e.g., Hartzell et al., 2010; Imperatori and Mai, 2013), indirectly modeled using scattering theory (Zeng, 1993), or implicitly accounted for as apparent anelastic attenuation via the specification of the quality factor, Q (Aki, 1980).

Nonlinear plasticity refers to the ability of a geomaterial to undergo permanent deformation, resulting in a changing material stiffness and hysteretic energy dissipation. The effects of nonlinear soil behavior on ground-motion characteristics have been well recognized by geotechnical engineers for over 50 years (Matasovic and Hashash, 2012). Methods for treating this behavior range from simple equivalent linear approaches (Idriss and Seed, 1968) to advanced plasticity theory-based approaches (Prevost, 1978). However, such effects have been routinely observed in strong-motion records only for the past 30 years (Aki, 2003). Nonlinear soil response is generally most important

---

[16] The net attenuation through heterogeneous material is given by $A = A_0 e^{-\pi f t^*}$, where $t^* = \int dr/Q(r)V(r)$ along the ray path, r (Hough and Anderson, 1988).

at higher frequencies ($f > 0.5$ Hz). Therefore, it has only recently attracted significant attention in the seismological community, as interest has grown in the prediction of high-frequency ground motions. Explicitly representing nonlinear near-surface material behavior in 3D wave propagation is complicated by the regional scale of the problem (typically tens to hundreds of kilometers) compared with the small scale (meters to tens of meters) below the ground surface for which these effects are significant. This range of scales poses challenges for both computation and the specification of material plasticity parameters throughout the modeled domain. Several efforts have been made to consider near-surface plasticity explicitly in 3D wave-propagation problems (e.g., Taborda et al., 2012). However, more commonly, 3D wave-propagation (in anelastic media) would be conducted, and subsequent nonlinear near-surface site response performed separately in an uncoupled fashion (e.g., Roten et al., 2012; de la Torre et al., 2020), or coupled through the domain reduction method (Bielak et al., 2003).

In addition to material plasticity associated with surficial deposits, plasticity also occurs in the immediate fault zone during rupture. The effects of such "on-fault" plasticity on ground motions is currently an active research topic (e.g., Roten et al., 2014). An important component of this research is to ensure consistency between the rupture's kinematic description (Section 5.3), usually based on inferences of past events that use inversion methods based on elasticity, and how forward wave propagation occurs.

5.4.5 Interpretation of Ground-Motion Time Series

The seismic wave propagation concepts discussed above enable interpretation of ground-motion time series (either simulated or observed) to understand the salient causative phenomena. By way of example, Figure 5.12 illustrates the three-component time series simulated at Site 1 and Site 2 previously shown in Figure 5.1. Site 1 is located closer to the fault on relatively soft soils and is in a sedimentary basin. Site 2 is located farther from the fault than Site 1 and is on rock conditions. Because of the smaller source-to-site distance, the first arrival of P- and S-waves, annotated on each time series, occurs at an earlier time at Site 1 than Site 2. The time difference between the P- and S-first arrivals is also smaller at Site 1.

As discussed in Section 5.3.2, Figure 5.1 illustrates that significant rupture directivity occurs in the vicinity of the ends of the rupturing fault (most evident for $t = 12$ s), and this manifests in the Site 1 and Site 2 ground motions as a long-period signal immediately following the S-wave arrival, which can be particularly damaging (e.g., Wald and Heaton, 1994; Alavi and Krawinkler, 2004).

The largest velocity amplitudes usually result from the S-wave arrivals, as S-waves typically represent 90–95% of the total body wave energy, and P-waves the remaining 5–10%.[17] Because the rupture occurs over a finite duration, and because of the multiple ray paths from source to site

---

[17] The P- and S-wave components of the far-field displacement from a double-couple dislocation are proportional to V_P^{-3} and V_S^{-3}, respectively (Aki and Richards, 2002, Equation 4.32). From Equation 5.6, for $v = 0.25$, $V_P = \sqrt{3}V_S$, which results in S-wave amplitudes that are $3^{3/2} = 5.2$ times those of P-wave amplitudes. Further, at the Earth's surface, Poisson's ratio increases for soils (e.g., a saturated stiff soil might have $V_P = 1.5$ km/s, $V_S = 0.35$ km/s); and impedance effects, therefore, amplify the up-going S-wave approximately 2.5 times more than those of P-waves. The combination of these two factors is $5.2 \times 2.5 = 12.9$, i.e., an S-wave amplitude around 13 times larger than that of the P-wave.

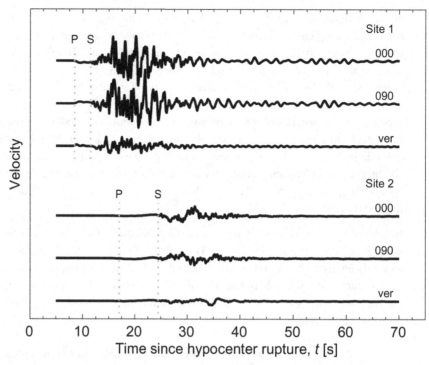

Fig. 5.12 Simulated three-component velocity time series for the rupture shown in Figure 5.1 at two locations. Site 1 is located closer to the source and in a sedimentary basin, while Site 2 is located further from the source. Both time series are shown with the same velocity amplitude scale. First direct arrivals of P- and S-waves are annotated for each time series.

(due to refraction and reflection at significant changes in the media, as well as small-scale scattering), the body waves arrive over an extended period of time. Geometric spreading causes lower ground-motion amplitudes at Site 2, due to its larger distance from the rupture.

Site 1's location in a sedimentary basin also results in greater impedance-related amplification of ground motion and the development of postcritical reflections of body waves at the basin edge, resulting in significant "basin-induced" surface waves. These basin-induced surface waves[18] are evident in Figure 5.1 at $t = 17$ and 25 s, and also in the long duration of reverberations from $t = 30$–70 s in Figure 5.12. The vertical ("ver") component of the ground motion during this time window indicates the relative size of the Rayleigh-wave portion of the basin waves, since Love waves do not produce vertical ground motions, whereas the horizontal components contain the combined effect of Rayleigh and Love waves. In contrast to Site 1, at Site 2, no significant surface-wave amplitudes are observed, and the variation in the S-wave-dominated ground-motion amplitudes with time is mostly the result of the evolution of the rupture on the fault, and the reflected and refracted wave arrivals. Because of the relatively small source-to-site distances, the surface wave amplitudes (other than the basin-induced surface waves at Site 1) are small relative to body waves, and anelastic attenuation has little effect.

---

[18] Such effects also occur in a different basin at the rightmost extent of the rupture in Figure 5.1.

This example illustrated how wave propagation physics underpins the interpretation of complex ground motions. Of course, a thorough analysis requires more than a cursory glance of the ground-motion time series. For example, from the time series alone, it is difficult to isolate the first arrivals of the basin-induced surface waves at Site 1 because they overlap with the later-arriving S-waves having longer ray paths and/or resulting from later rupture of the fault. Signal processing and other advanced seismological techniques are useful in such instances but are beyond the scope of the present discussion.

5.4.6 Crustal Models for Ground-Motion Simulation

Crustal models provide the computational domain within which the seismic wave equations are solved in ground-motion simulation methods. Therefore, the 3D crustal properties strongly influence seismic wave propagation, behavior at material interfaces, and attenuation. The theory examined in the previous sections of this chapter predicts these phenomena in idealized circumstances. However, numerical methods allow consideration of arbitrarily complex 3D crustal models.

Three-dimensional crustal models require the specification, at every location in the geographic region of interest, of the properties required for solving the seismic wave equations. For elastic wave propagation, this requires specification of the P- and S-wave velocities and density. Density is typically inferred from P-wave velocity (e.g., Brocher, 2005) because it provides a reasonable approximation with a precision that has little impact on seismic wave propagation. S-wave velocity can also be inferred from P-wave velocity (e.g., Brocher, 2005), but these relationships have high uncertainty for sedimentary deposits and may produce imprecise simulation results. When anelastic attenuation is also considered (e.g., Equation 5.15), then parameters to define the quality factor, Q, are also needed for P- and S-waves. Q_S is typically specified as a parametric function of V_S, and Q_P as a function of Q_S (possibly also dependent on the V_P/V_S ratio) (Taborda and Roten, 2015, Table 2). There are appreciable uncertainties in these Q_S–V_S relationships, as well as their frequency dependence (Withers et al., 2015). Because the most well-constrained parameters in 3D crustal models are usually the P- and S-wave velocities, with density and anelastic attenuation parameters obtained through correlations, such 3D crustal models are also frequently referred to colloquially as *velocity models*. Figure 5.13 provides an example of the S-wave velocity in the Canterbury, New Zealand, region for the model of Lee et al. (2017).

The development of crustal models, as with any model, should focus on the aspects of the crustal structure that most influence simulation results. Typically, imaging of sedimentary basin geometry and velocity structure are the most challenging tasks. These tasks are also likely the most important, as basin-induced surface waves and basin-directivity coupling are critical phenomena for ground motion at the frequencies of engineering interest (Olsen, 2000; Bielak et al., 2010). Also, the required level of spatial resolution varies with depth. At significant depths, say, $z = 10$ km, a spatial resolution of 1 km may be sufficient because the crustal properties at this depth principally affect the low-frequency ground motion, which has long wavelengths. On the other hand, near the ground surface (say, $z = 200$ m), crustal structure at shorter wavelengths is needed to account for its significant influence on high-frequency ground motion. Consequently, 3D crustal models are typically developed using a multidisciplinary set of data, including crustal tomography, seismic reflection and refraction; and geophysical and hydrological borehole logs, among others (Magistrale and Day, 1999; Süss and Shaw, 2003; Ramírez-Guzmán et al., 2012; Taborda and Roten, 2015; Lee et al., 2017).

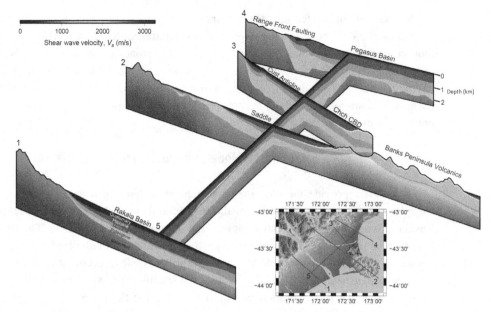

Fig. 5.13 Fence diagram, illustrating multiple vertical transects describing the 3D variation in shear wave velocities in the Canterbury, New Zealand, region. (After Lee et al., 2017)

5.5 Methods for Physics-Based Ground-Motion Simulation

5.5.1 Comprehensive Physics-Based Simulation

Seismic wave propagation in three dimensions can be directly solved via the following two coupled equations (Shearer, 2009, Section 3):

$$\rho \frac{\partial^2 u_i}{\partial t^2} = \partial_j \sigma_{ji} + f_i \tag{5.16}$$

$$\sigma_{ij} = \lambda \delta_{ij} \partial_k u_k + \mu(\partial_i u_j + \partial_j u_i) - m_{ij} - \xi_{ij} \tag{5.17}$$

where u is displacement, t is time, ρ is material density, σ is the stress tensor, f is a body force, λ and μ are the elastic Lamé parameters, m_{ij} is the moment-density tensor (moment tensor per unit volume) describing the rupture, ξ_{ij} represents energy losses due to anelastic or plastic attenuation, ∂ represents the partial derivative, and the subscripts refer to *Einstein summation notation*.[19] Representations of both the moment tensor and anelastic attenuation were discussed previously in Section 5.3.3.

Equation 5.16 is the elastodynamic equation of motion (or momentum equation). Equation 5.17 reflects the stress-strain constitutive equation, with the first two terms on the right-hand side representing a linear isotropic solid, ξ_{ij} reflects anelastic and plastic behavior, and m_{ij} is nonzero only on the fault surface. To provide a connection to prior discussions, the solution of Equations 5.16 and 5.17 is equivalent to the solution of Equation 5.1, and represents the wave equation in 3D as a generalization of the linear-elastic 1D version presented in Equation 5.4.

[19] Indices i, j take values $1, 2, 3$; $\delta_{ij} = 1$ for $i = j$ and zero otherwise, and repeated indices in a product are summed over, e.g., $\partial_k u_k \equiv \frac{\partial u_1}{\partial x_1} + \frac{\partial u_2}{\partial x_2} + \frac{\partial u_3}{\partial x_3}$.

Numerical Solution of the Wave Equation

Equations 5.16 and 5.17 can be solved analytically under specific conditions to understand the fundamental wave propagation properties examined in Section 5.4 (Shearer, 2009, Section 3). However, in reality, the Earth's crust is composed of geologic and geotechnical materials with variable properties in three dimensions. It is necessary to utilize numerical simulations to account for arbitrarily complex crustal characteristics.

The wave equation is typically solved numerically using finite difference (FD), finite element (FE), spectral element (SE), or discontinuous Galerkin (DG) methods. SE and DG methods (e.g., Komatitsch and Tromp, 1999; de la Puente et al., 2008; Mazzieri et al., 2013) share basic concepts with FE, therefore only the FD and FE methods are discussed here briefly. Further details can be found in Taborda and Roten (2015).

Solutions of the seismic wave equation using the FD method have been common in physics-based ground-motion simulation for three decades (Frankel and Vidale, 1992; Olsen et al., 1995; Graves, 1996), because of its simplicity in implementation and computational efficiency in parallelization. These methods use a FD approximation of the partial derivatives in the governing wave equation at grid points (or "nodes") throughout the computational domain, resulting in a sparse system of equations that are typically solved using an explicit time-stepping solution. For the wave equation, in particular, "staggered grid" FD methods (Virieux, 1984; Graves, 1996) are frequently used where stress and velocity are computed on a staggered grid (the two different grids are offset by half a grid width in space and time), which yields greater accuracy for the same computational demand. The practical limitations of FD are that the method becomes complex to implement and computationally less efficient for the consideration of free-surface topography and nonuniform spatial discretization.

Solutions of the wave equation using FE methods have also been widely employed for the past 20 years (e.g., Bao et al., 1998; Aagaard et al., 2001). FE methods (and hence the SE and DG methods) offer the benefits of unparalleled flexibility in handling complex geometry between material layers, surface topography, and spatially variable discretization, which are more challenging with the FD method. While the FD method approximates the partial derivatives directly, FE methods approximate the partial differential equation's solution. However, for an equivalent spatial and temporal discretization, FE methods are generally less computationally efficient than FD methods for the same spatial accuracy.

The representation of the source rupture via subfault moment tensors (Section 5.3.3) depends upon the numerical scheme adopted, but most commonly is considered directly via the moment-density tensor, m_{ij}, in Equation 5.17. As previously noted, the moment-density tensor is the moment tensor (M in Equation 5.1) per unit volume.[20] Alternatively, the moment tensor can be specified via body forces, f, in Equation 5.16 (e.g., Graves, 1996), but this entails numerical discretization approximations.[21]

Anelastic attenuation in the numerical solution of the seismic wave equation is conventionally represented through memory-variable approaches via ξ_{ij} in Equation 5.17 (Day and Bradley, 2001; Bielak et al., 2011). These methods require specification of the anelastic quality factors for P- and S-wave propagation, Q_p, and Q_s (Taborda and Roten, 2015), as well as their frequency dependence (Withers et al., 2015). The consideration of plasticity can also be easily handled by FD

---

[20] In numerical implementations the relevant volume will be a function of the adopted spatial discretization (e.g., Olsen and Archuleta, 1996).

[21] The body-force terms are typically used to compute Green strain tensors for reciprocal simulations (e.g., Graves et al., 2011a) or to insert adjoint sources for waveform inversion (e.g., Chen et al., 2007).

and FE methods (e.g., Taborda et al., 2012). However, it has significant computational implications because of the need to undertake local iteration to ensure compatibility among stress and strain at the material level.

Despite the differences in theory and implementation resulting from alternative numerical approaches, verification exercises illustrate that results for different numerical methods are consistent (Bielak et al., 2010; Chaljub et al., 2015). The differences between the alternative implementations are practically negligible relative to the differences between such simulations and observations.

Effect of Temporal and Spatial Discretization on Maximum Simulation Frequency

Numerical simulations must ensure that the discretization of the continuum does not practically affect the resulting solution. For ground-motion simulations, the principal impact of spatial and temporal discretization is on the maximum wave frequency that can be simulated without producing aliasing (Proakis and Manolakis, 1996, Chapter 4) and excessive numerical dissipation (Thomas, 1995, Chapter 5). Depending on the particular simulation method used, and the truncation order of the method, a certain number of grid points are required per wavelength. Typically, a minimum of $N = 5$ nodes or grid points are required per wavelength in fourth-order FD methods (i.e., consider three at the zero nodes and two at the maxima and minima of a sine wave over one vibration cycle), and $N = 10$ for second-order FD or FE methods (Graves, 1996; Bielak et al., 2010). Given that wavelength, $\lambda = V/f$, where V is velocity and f is frequency, it follows that

$$\lambda = N\Delta x = \frac{V_{min}}{f_{max}}. \tag{5.18}$$

Thus

$$f_{max} = \frac{V_{min}}{N\Delta x} \tag{5.19}$$

where f_{max} is the maximum frequency without significant spatial aliasing within the material with the minimum velocity, V_{min}. For example, a minimum (S-wave) velocity of $V_{min} = 1$ km/sec, a grid spacing of $\Delta x = \Delta y = \Delta z = 0.2$ km, and $N = 5$ yields $f_{max} = 1$ Hz.

Explicit numerical methods are typically conditionally stable. For example, the second-order temporal discretization in staggered-grid FD methods (Graves, 1996) leads to the stability criteria

$$\Delta t < 0.495 \frac{\Delta x}{V_{max}} \tag{5.20}$$

where Δt is the time step, Δx ($= \Delta y = \Delta z$) is the spatial step, and V_{max} is the maximum (P-wave) velocity. Similar criteria are required for FE methods (Watanabe et al., 2017). Taking the prior example ($V_{min} = 1$ km/sec, $\Delta x = 0.2$ km, $f_{max} = 1$ Hz), with a typical value of $V_{max} = 8$ km/sec, yields $\Delta t < 0.012$ s. In order to simulate ground motions up to $f_{max} = 1$ Hz, the Nyquist frequency ($f_N = 1/(2\Delta t)$) to prevent temporal aliasing requires that $\Delta t < 0.5$ s. Hence, the stability criterion is significantly more stringent than the aliasing criterion.

Equations 5.19 and 5.20 show the important link between the maximum frequency f_{max} to consider in the simulation and the corresponding spatial and temporal discretization. From Equation 5.19, f_{max} scales with Δx, which, for a uniform spatial discretization in 3D, means that the total number of points in the computational domain scales with $(\Delta x)^3$. Furthermore, from Equation 5.20, Δt scales linearly with Δx in order to maintain stability. Hence, the computational demand due to increasing f_{max} scales with $(\Delta x)^4$. In the above example, doubling the desired maximum frequency to $f_{max} = 2$ Hz would require a reduction to $\Delta x = 0.1$ km and $\Delta t < 0.006$ s, corresponding to a $(2)^4 = 16$-fold increase in computational demand.

5.5.2 Hybrid-Broadband Ground-Motion Simulation

Most regional-scale ground-motion simulations that explicitly solve the wave equation have been limited to modeling the propagation of low-frequency waves (e.g., $f_{max} \leq 1$ Hz) and relatively high values of minimum shear-wave velocity (e.g., $V_{S,min} = 500$ m/s). This has been due, in part, to limitations of simulation algorithms and computer codes, and high-performance computing capacity. However, it is also due to limitations of theories and models describing source rupture (Section 5.3) and wave propagation with high-resolution regional crustal models (Section 5.4) for high-frequency simulation.

The challenges with physics-based ground-motion simulations for $f > 1$ Hz are problematic, as high-frequency motions drive seismic risk for most of the built environment. From an engineering perspective, ground-motion simulation needs to provide time series that have a broad range of vibration frequencies. Typically, a frequency range of $f = 0.1 - 10$ Hz is sufficient for engineering utilization.[22]

To produce simulations with the frequency bandwidth required for realistic IM computation, engineering applications, and structural design, *hybrid* methods are primarily utilized (Hartzell et al., 1999). Hybrid methods combine the rigor of the comprehensive physics-based solution (Equations 5.16 and 5.17) of the wave equation for low frequencies (LF) with simplified physics-based methods for high frequencies (HF), to produce a *broadband* (BB) ground motion.[23] The frequency that separates the two LF and HF approaches is typically referred to as the transition frequency. Figure 5.14 illustrates the ground motion that was depicted in Figure 5.1 in both the time and frequency domains. The simulation was produced by first using separate LF and HF simulation methods. The BB ground motion was then obtained by transforming the results from the LF and HF methods into the frequency domain and combining them using a matched filter (e.g., Mai and Beroza, 2003). In this example, the matched filter has a transition frequency of $f = 1$ Hz. The resulting BB ground motion in the frequency domain is then transformed back to the time domain. For the time series shown in Figure 5.14, it is evident that the BB ground motion contains both S-waves from the HF simulation (see Section 5.5.3), as well as coherent body waves and long-duration surface-wave reverberations from the LF simulation.

A common transition frequency is $f = 1$ Hz (Graves and Pitarka, 2010, 2015; Mai et al., 2010; Olsen and Takedatsu, 2015; Frankel et al., 2018). This is a common choice because of evidence that ground motions for $f > 1$ Hz exhibit sufficient complexity that they can be represented by a simulated time series with apparently random phase, and observational evidence that seismic waveforms computed using the comprehensive LF approaches often, at present, do not realistically predict observations beyond such frequencies.

---

[22] Although for near-rigid safety-critical structures, and sensitive structural components, frequencies up to 100 Hz are often desired.

[23] It is common in the literature for the LF comprehensive and HF simplified physics-based methods to be referred to as "deterministic" and "stochastic" methods, respectively. These terms are a misnomer since the HF simplified physics-based methods have deterministic attributes in their parametric description of the Fourier spectra and ground-motion duration, and the LF comprehensive physics-based methods have stochastic elements in their description of the source (e.g., Figure 5.4) and crustal models.

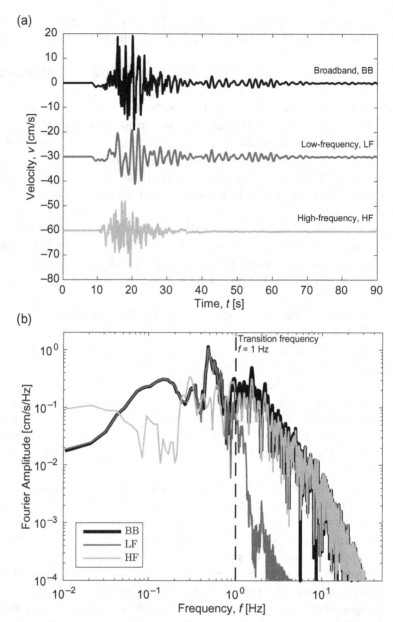

Fig. 5.14 Illustration in the (a) time domain and (b) frequency domain of how a hybrid ground-motion simulation method combines the simulations from low-frequency and high-frequency methods to produce a broadband ground motion. The ground motion is the north-south ("000") component of the time series from Figure 5.1.

The maximum frequency in LF approaches has been slowly increasing over time due to increasing computational resources, improved representation of small-scale source and crustal structure, and new models of physics relevant at higher frequencies. For example, simulation validation against historical events using the comprehensive approach for frequencies up to $f = 4$ Hz were undertaken by Graves and Pitarka (2016) and Taborda and Bielak (2013) for the Imperial Valley and Chino Hills earthquakes, respectively. Similarly, simulations of potential future events on the Hayward and San Andreas Faults by Rodgers et al. (2019) and Roten et al. (2016) considered similar maximum frequencies (f = 5 Hz and f = 4 Hz, respectively). Continued progress in this area should continue to increase high-frequency predictive capability of LF methods, and thus increase the transition frequency in BB methods. This increase in the transition frequency over time is gradual, as the physics of ground motions at high frequencies is significantly more complex than that at low frequencies, with phenomena such as fault roughness (Shi and Day, 2013; Graves and Pitarka, 2016), wave scattering due to small-scale heterogeneities (Hartzell et al., 2010; Imperatori and Mai, 2013), frequency-dependent attenuation (Withers et al., 2015), on-fault and off-fault material nonlinearity (Taborda et al., 2012; Roten et al., 2014), and topographic free surface effects (Imperatori and Mai, 2015; Restrepo et al., 2016) all becoming increasingly important.

5.5.3 Simplified Physics-Based Simulation of High-Frequency Ground Motion

Because high-frequency ground motion is important to many (short period) structural and geotechnical systems, an alternative to the comprehensive physics approach (Section 5.5.1) is presently needed to adequately model ground motion at high frequencies.

Many simulation methods do not directly solve the 3D equations governing seismic wave propagation (Equations 5.16 and 5.17), and thus are referred to here as *simplified physics-based simulation* methods.[24] A summary of such methods is provided by Douglas and Aochi (2008). Here only the classical "stochastic simulation" approach (Hanks and McGuire, 1981; Boore, 2003) is discussed because of its widespread application.

The basis of the method is to predict the Fourier amplitude spectrum of the ground motion produced by a point source. Ground-motion time series are then obtained by coupling the Fourier amplitude spectrum with a random-phase spectrum. The remainder of this section discusses the details of this approach and its extension to finite faults.

The Fourier amplitude spectrum, $A(f)$, for the ground motion at a site resulting from a point-source rupture can be described in the frequency domain as a direct product of terms that reflect the source $E(f, M; \theta_E)$, path $P(f, R; \theta_P)$, and site, $S(f, \theta_S)$:

$$A(f; \theta) = E(f, M; \theta_E) \times P(f, R; \theta_P) \times S(f; \theta_S) \tag{5.21}$$

where M is magnitude, R is hypocentral distance, and the vector $\theta = (\theta_E, \theta_P, \theta_S)$ represents additional source, path, and site parameters beyond the primary dependence on M and R. These additional inputs are discussed in the context of each term below.

It is insightful to compare Equation 5.21 with the general representation given by Equation 5.1. In this regard, the source spectrum, E, reflects the approximate effect of the moment tensor, whereas the combined path and site effects ($P \times S$) approximate the Green's function, G. Finally, the convolution in time in Equation 5.1 is reflected by the multiplicative product in the frequency domain of Equation 5.21.

---

[24] While these approaches can also be applied at low frequencies, comprehensive physics-based approaches are often preferred for their potential ability to characterize site-specific effects.

Source, E

The source (acceleration) spectrum is commonly assumed to be a so-called ω^2 spectrum[25] with the form:

$$E(f, M; \boldsymbol{\theta}_E) = \frac{\mathbb{C} M_0 (2\pi f)^2}{1 + (f/f_c)^2} \qquad (5.22)$$

where f_c is the corner frequency, and the term $\mathbb{C}$ contains physical and geometric parameters related to the source:

$$\mathbb{C} = \frac{R_{\Theta\Phi} V F}{4\pi \rho_{src} V_{S,src}^3 R_0} \qquad (5.23)$$

where $R_{\Theta\Phi}$ is the radiation pattern (Shearer, 2009, Section 9.3), describing the strength of waves radiated from a shear dislocation with azimuth (Θ) and take-off angle (Φ); V represents the partitioning of shear-wave energy into two horizontal components ($= 1/\sqrt{2}$); F is the free-surface effect ($= 2$ for vertically propagating SH waves); ρ_{src} and $V_{S,src}$ are the density and S-wave velocity at the source depth from the crustal model; and R_0 is the reference distance (conventionally $R_0 = 1$ km). It is common to use a generic value of $R_{\Theta\Phi} = 0.55$ based on averaging over a suitable range of azimuths and take-off angles (Boore and Boatwright, 1984); however, this is not a strict requirement (e.g., the comprehensive approach in Section 5.5.1 considers $R_{\Theta\Phi}$ explicitly through the moment tensor). From Equation 5.21, the source spectrum specifies a reference amplitude to which the path and site effects are subsequently applied.

The relevance of the corner frequency f_c can be understood by noting that for high frequencies $E(f \rightarrow \infty, M; \boldsymbol{\theta}_E) = \mathbb{C} M_0 (2\pi)^2 f_c^2$. Therefore, the greater the corner frequency, the greater the amplitudes of high-frequency waves. Following Brune (1970, 1971), it is common to relate the corner frequency to properties of the source geometry and stress release through the expression[26]

$$f_c = 4.9 \times 10^{-1} V_{S,src} \left(\frac{\Delta\sigma}{M_0} \right)^{1/3} \qquad (5.24)$$

where $\Delta\sigma$ is the Brune stress parameter. This term is frequently referred to as "stress drop", however, this is a model-specific parameter, and not a theoretical property of the rupture (Atkinson and Beresnev, 1997).

Figure 5.15 illustrates the source spectra for moment magnitudes $M6.0$ and $M7.5$ and Brune stress parameter values of $\Delta\sigma = 5$ and 10 MPa. Magnitude, and hence seismic moment, M_0 (Equation 2.5), controls the low-frequency source spectral amplitudes, while $\Delta\sigma$ controls the corner frequency, f_c, and thus the high-frequency amplitudes. The corner frequency values for the $M6.0$ spectra with $\Delta\sigma = 5$ and 10 MPa are $f_c = 0.28$ and 0.36 Hz, respectively. Similarly, the values for the $M7.5$ spectra are $f_c = 0.050$ and 0.063 Hz.

The larger magnitude $M7.5$ event has corner frequencies that are approximately 5.6 times lower than the $M6.0$ event, illustrating the appreciably greater low-frequency energy that larger magnitude earthquakes possess. Because of the link between low-frequency Fourier spectra and long-period

---

[25] Other functional forms of the spectrum have been proposed (e.g., Boore, 2003), with forms possessing multiple corner frequencies necessary to model larger magnitude ruptures as point sources, rather than treat such ruptures as the summation of multiple subfaults (see Section 5.5.3).

[26] For V_S, $\Delta\sigma$, and M_0 in units of km/s, Pa, and Nm, respectively. In the original expression of Brune, the coefficient is 4.9×10^6 based on respective units of km/s, bar, and dyne-cm.

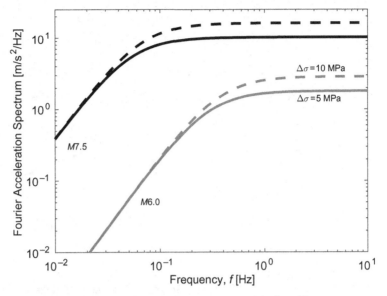

Fig. 5.15 Source spectrum of Equation 5.22 as a function of frequency for two moment magnitude and Brune stress parameter values.

response spectra (Bora et al., 2016), the greater low-frequency energy of large magnitude earthquakes is seen in empirical models for long-period response spectra (Section 4.5.4).

The exponent in Equation 5.24 also results in the factor of two difference in $\Delta\sigma$ shown in Figure 5.15, leading to only a factor of $2^{1/3} = 1.26$ difference in f_c. Conversely, small changes in corner frequency have a large effect on the stress parameter, which is important when inferring $\Delta\sigma$ from observations, as errors in f_c lead to large errors in $\Delta\sigma$. The significant variability in unsmoothed Fourier spectra (e.g., Figure 4.6) therefore makes determining f_c directly from observations difficult. It is more appropriate to estimate f_c (and thus $\Delta\sigma$) from a broadband fitting approach (e.g., Edwards and Fäh, 2013) so that it reflects the general shape of the spectrum. This is particularly true for small M events, where it is difficult to discern f_c due to path and site attenuation of high frequencies. All methods, however, have trade-offs between parameters that influence spectral shape (Boore et al., 1992).

Path, P

The path contribution, $P(f, R; \boldsymbol{\theta}_P)$, is composed of frequency-independent geometric spreading and frequency-dependent anelastic attenuation:

$$P(f, R; \boldsymbol{\theta}_P) = G(R; \boldsymbol{\theta}_P) \times D(f, R; \boldsymbol{\theta}_P). \tag{5.25}$$

The geometric spreading term, $G(R; \boldsymbol{\theta}_P)$, is typically a piecewise function. Figure 5.16 provides an example model for California based on empirical analysis of observations (Raoof et al., 1999). The rate of attenuation is $1/R$ for $R < 40$ km and $1/\sqrt{R}$ for $R > 40$ km.[27] In this example, the geometric attenuation rates follow the theoretical body and surface wave rates. However, as noted in Section 5.4.4, the use of source-to-site distance, R, to describe complex ray paths involving refraction

---

[27] Some regions will exhibit no apparent geometric spreading at moderate distances (roughly twice the crustal thickness, typically $R \approx 60$–140 km) as a result of constructive interference of Moho-reflected waves (Boore, 2003).

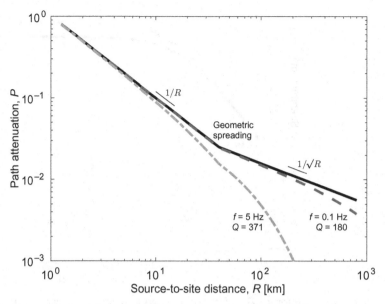

Fig. 5.16 Path attenuation resulting from geometric spreading and anelastic attenuation illustrated for two different frequencies and respective quality factors.

and reflection in heterogeneous crust generally means that attenuation rates $R^{-\gamma}$ will have $\gamma \neq 1.0$ and 0.5 for the body- and surface-wave-dominated ranges of geometric spreading, respectively.

The anelastic attenuation term, $D(f, R; \boldsymbol{\theta}_P)$, is defined as

$$D(f, R; \boldsymbol{\theta}_P) = \exp\left[\frac{-\pi f R}{Q(f)V_Q}\right] \qquad (5.26)$$

where $Q(f)$ is the frequency-dependent quality factor (Section 5.4.4), and V_Q is the shear wave velocity used in the determination of $Q(f)$. Note the similarity between Equation 5.26 and Equation 5.15, and that from observational data amplitudes we can determine the combined effect of QV, rather than Q or V independently. The frequency-dependence of Q typically takes the form $Q(f) = Q_0 f^{\eta}$, such that Q_0 represents the reference "damping" for $f = 1$ Hz, and often $0.3 \leq \eta \leq 0.6$ (Boore, 2003), but the frequency-independent case of $\eta = 0$ is also frequently assumed. The overall effect of $D(f, R; \boldsymbol{\theta}_P)$ is to reduce the high-frequency ground-motion amplitudes relative to those at low frequencies.

Figure 5.16 illustrates the effect of the anelastic attenuation term on the path contribution (Equation 5.25). The material velocity is $V_Q = 3.5$ km/s, and the quality factor model is $Q(f) = 180 f^{0.45}$ for $f \geq 1$ Hz (Raoof et al., 1999) and $Q(f) = 180$ for $f < 1$ Hz (Graves and Pitarka, 2010). For a frequency of $f = 0.1$ Hz, the effect of anelastic attenuation is negligible until distances beyond approximately $R = 200$ km. For $f = 5$ Hz, anelastic attenuation leads to amplitudes reducing by a factor of two at a distance of approximately $R = 60$ km, and by an order of magnitude at $R = 200$ km.

The examples in Figure 5.16 illustrate that the effects of anelastic attenuation on amplitudes increase exponentially with distance. Typically, the effects are not noticeable within 50 km from an earthquake source, but become increasingly significant, particularly at distances greater than 100 km. The impact of attenuation is a function of quality factor, with stiffer materials usually having higher quality factors and lower damping. The amount of attenuation is also a function of the wave frequency. Higher frequency waves attenuate faster, because they have a smaller wavelength, λ, and thus undergo a greater number of cycles for a given propagation distance.

An alternative treatment of anelastic attenuation via Equation 5.26 is to explicitly model the ray path and consequent anelastic attenuation through a layered, typically 1D, crustal model. In this instance $D(f, R; \boldsymbol{\theta}_P)$ takes the form (Ou and Herrmann, 1990; Graves and Pitarka, 2010)

$$D(f, R; \boldsymbol{\theta}_P) = \exp\left[-\pi f \sum_{i=1}^{L} \frac{r_i}{Q_i V_{Q,i}}\right] \tag{5.27}$$

where L is the number of layers, and r_i, Q_i, and $V_{Q,i}$ are the ray length, quality factor, and shear wave velocity in layer i, respectively. In these 1D layered models, it is common to use a frequency-independent Q_i, because the layer quality factors, thicknesses, and depths will still produce frequency-dependent attenuation of the surface ground motion.

Site, S

The final term in Equation 5.21 reflects *linear* site response, which is comprised of two parts:

$$S(f; \boldsymbol{\theta}_S) = I(f; \boldsymbol{\theta}_S) \times K(f; \boldsymbol{\theta}_S). \tag{5.28}$$

$I(f; \boldsymbol{\theta}_S)$ represents the impedance effects associated with the change in density and velocity of crustal layers at the site, and $K(f; \boldsymbol{\theta}_S)$ represents the diminution (or "site attenuation") that takes place during wave propagation beneath the site.

The effect of site impedance, $I(f; \boldsymbol{\theta}_S)$, is most commonly approximated by the square-root impedance between the source and the site based on a 1D crustal model[28] (Boore and Joyner, 1997). Mathematically, the site impedance is obtained in this manner via

$$I(f; \boldsymbol{\theta}_S) = \sqrt{\frac{\rho_{src} V_{src}}{\overline{\rho}(f) \overline{V}(f)}} \tag{5.29}$$

where $\overline{\rho}$ and $\overline{V}$ are the travel-time weighted average density and velocity, respectively, from the surface to a depth, z, equivalent to a quarter-wavelength for the frequency in question. The condition of a quarter-wavelength, $z(f) = \overline{V}/(4f)$, makes the determination of $\overline{\rho}$ and $\overline{V}$ dependent on an implicit function between f and $z(f)$. Given a 1D crustal model it is practically expedient to specify z and then compute $\overline{\rho}$ and $\overline{V}$ from the following equations:

$$\overline{\rho}(f) = \frac{1}{z(f)} \int_0^{z(f)} \rho(z) dz \tag{5.30}$$

$$\frac{1}{\overline{V}(f)} = \frac{1}{z(f)} \int_0^{z(f)} \frac{1}{V(z)} dz. \tag{5.31}$$

The site attenuation term typically takes the form

$$K(f; \boldsymbol{\theta}_S) = \exp(-\pi f \kappa_0) \tag{5.32}$$

where κ_0 is the site attenuation parameter.

---

[28] Conceptually, it is possible to use the 1D model to obtain a transfer function describing these impedance effects. However, the simplicity of the 1D model gives rise to large amplifications over narrow frequency bands, which are not seen in empirical observations (Boore, 2013). On the other hand, the square-root impedance approach ignores resonance, and therefore will not account for such effects that can exist at soft soil sites.

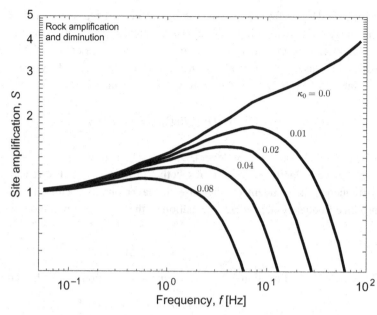

Fig. 5.17 Site amplification for a generic 1D crustal model (with $V_{S,30}$) for various values of κ_0. The 1D crustal model is from Boore (2016) and the source density and velocity are $\rho_{src} = 2800$ kg/m^3 and $V_{src} = 3.5$ km/s, respectively.

Figure 5.17 illustrates the combined effects of site impedance and attenuation on the site amplification factor $S(f; \boldsymbol{\theta}_S)$ for a generic 1D crustal model (Boore, 2016) and various levels of values of κ_0. The effect of the site impedance can be isolated for the depicted case of $\kappa_0 = 0.0$, where the site amplification continues to increase with increasing frequency. This continued increase in amplification reflects the fact that higher frequencies have shorter wavelengths, and hence shallower depths are considered in Equations 5.30 and 5.31. As shallower depths are associated with lower shear-wave velocities, the ratio of the impedance at the source depth to that of Equation 5.29 generally increases with frequency. The effects of κ_0 grow exponentially and dominate over the impedance effects at high frequencies.

Considering the path and site attenuation models, represented in Equations 5.27 and 5.32, the total attenuation along the path and at the site can be expressed as

$$D(f, R; \boldsymbol{\theta}_P) \times K(f; \boldsymbol{\theta}_S) = \exp\left[-\pi f \kappa\right] \tag{5.33}$$

where

$$\kappa = \Delta\kappa + \kappa_0 = \sum_{i=1}^{L} \frac{r_i}{Q_i(f)V_{Q,i}} + \kappa_0. \tag{5.34}$$

This representation is instructive because it illustrates the apportionment of the attenuation into path-dependent and path-independent components. Given the previous comments concerning the various approaches by which the path attenuation can be quantified (e.g., Equation 5.26 versus Equation 5.27), it is important to ensure that the site attenuation is represented such that the total path + site attenuation is self-consistent (Ktenidou et al., 2014). In a physical sense, this framework assumes that the uppermost layer in the crustal model (i.e., $i = 1$ in Equations 5.27 and 5.34) lies beneath the soil or weathered rock profile associated with κ_0.

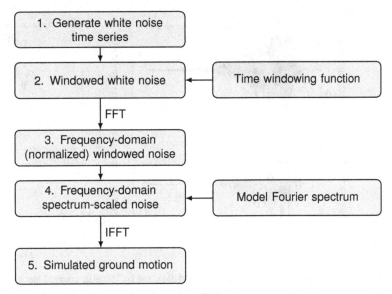

Fig. 5.18   Flowchart for obtaining ground-motion time series from a Fourier amplitude spectrum model.

It was previously noted that the site amplification model given by Equation 5.29 only explicitly models the average effects of linear site response and ignores resonance. As previously seen from empirical observations in Section 4.5.3, large-amplitude ground motions are also influenced by the nonlinear response of surficial soils. It is possible to empirically include the effects of nonlinear response into the site amplification factor, $S(f; \theta_S)$. However, nonlinearity changes not only the frequency content of surface ground motion, but also its duration and phasing in time (which is inconsistent with the stationarity assumption between the Fourier and temporal domains in this HF method). Thus, it is preferable to use this method to obtain high-frequency simulated ground motions for a reference condition at which nonlinearity and resonance in site response are not significant, and then subsequently incorporate these effects through other methods (e.g., Section 8.6).

Obtaining Ground-Motion Time Series

Through the combination of source, path, and site models, Equation 5.21 provides the predicted Fourier amplitude spectrum of a ground motion. Obtaining ground-motion time series requires coupling the Fourier amplitude spectrum with a random phase spectrum. Hanks and McGuire (1981) demonstrated that, at relatively high-frequencies (e.g., $f > 1\,\mathrm{Hz}$), phase angles are effectively random and uncorrelated between frequencies. Figure 5.18 outlines the process of obtaining time series, and Figure 5.19 illustrates several steps for an $M6$ earthquake at $R = 20$ km hypocentral distance. The process is as follows. Normally distributed white noise (i.e., a time series with random phase angles and constant power over all frequencies) is first generated for the desired duration (Step 1). The white noise is then windowed (Step 2) through multiplication by a time windowing function (discussed further below). The windowed noise is transformed into the frequency domain with the fast fourier transform (FFT)[29] and normalized by its root-mean-squared (RMS)

---

[29] Which is the numerical method to implement the discrete Fourier transform given by Equation 4.5.

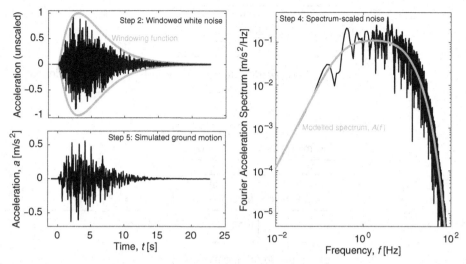

Fig. 5.19 Example of windowed white noise, spectrum-scaled noise, and final simulated ground-motion time series corresponding to an $M6$ event at a distance of $R = 20$ km. Other model properties include $\Delta\sigma = 5$ MPa, $Q(f) = 180f^{0.45}$, $V_Q = V_{src} = 3.5$ km/s, $\rho_{src} = 2800$ kg/m³, and $\kappa_0 = 0.045$.

value[30] (Step 3). This is then multiplied by the modeled Fourier amplitude spectrum predicted via Equation 5.21 (Step 4). Finally, the inverse FFT (IFFT) is applied to obtain the predicted ground-motion time series.

The time windowing function describes the temporal variation in acceleration amplitudes as a function of time. The exponential windowing function of Saragoni and Hart (1973) is commonly adopted, and is depicted in Figure 5.19. Other windowing models are available that have a stronger theoretical or empirical basis (e.g., Stafford et al., 2009), but Boore (2003) argues that such differences have only a minor average effect on ground-motion intensity measures such as spectral accelerations. However, the impact on other intensity measures, e.g., AI, D_{S5-75}, and D_{S5-95}, can be more pronounced.

One principal parameter in the windowing function is the duration of ground motion. It is typical for this duration to be expressed as $D_T = D_E + D_P$, where D_E is the source duration (equivalent to *rise time* discussed in Section 5.3) and D_P is the path duration. The source duration is related to the inverse of the source corner frequency. Depending on the specific source spectrum (and thus corner frequency definition) used, it is typical to consider $D_E = 1/f_c$ (e.g., Frankel, 1995) or $D_E = 0.5/f_c$ (e.g., Atkinson and Silva, 2000). The path duration increases with distance due to the increasing time delays in the arrival of reflected and refracted body waves with increasing distance. It is typically calibrated based on observed ground motions, and varies appreciably by region (Boore and Thompson, 2014, 2015).

The simplicity and sufficient accuracy of this method are part of its appeal and popularity (Boore, 2003, Table 5). Despite this, the method does have several limitations, some of which can be partially overcome (e.g., using interfrequency correlations of Fourier spectra; Stafford, 2017), but many of

---

[30] From Parseval's theorem, the total energy (average power times duration) is the same in the time and frequency domains, so normalizing by the RMS value in the time domain is equivalent to normalizing by the square root of the mean power in the frequency domain. Therefore, the normalization in the frequency domain is done so that the signal before applying the FAS filter will have a unit power.

which are a result of the underpinning assumptions. Limitations such as ignoring directivity, long-period basin and surface waves, and spatial correlation of ground motion can be avoided by applying the method only at high-frequencies and using the hybrid approach to couple it with a low-frequency simulation (Section 5.5.2).

Consideration of Finite-Fault Effects

The Fourier acceleration spectrum defined by Equation 5.21 is conventionally used for a point source rupture at a given hypocentral distance. With increasing magnitude, the effects of finite-fault geometry become important and require modification of this basic approach. This modification can take one of three different approaches, which depends on the desired level of sophistication.

The first, and most straightforward, approach to account for finite-fault effects is to consider alternative source spectrum shapes and use an effective distance metric, R_{eff}, rather than the hypocentral distance (Boore, 2009). A limitation of the use of the finite-fault equivalent R_{eff} is that it is still a point-source approximation, and therefore does not account for finite-fault rupture effects. On this basis, extensions of the basic point-source stochastic methods have been developed to handle finite-fault geometry (e.g., Motazedian and Atkinson, 2005; Boore, 2009). In this second approach, finite-fault effects are handled through a planar rupture geometry, hypocenter location, and assumed constant rupture velocity and stress parameter at each subfault to specify the time lag for rupture on each fault and the subfault acceleration spectrum, respectively. The explicit consideration of rupture over the finite fault allows for the temporal evolution of subfault ruptures, and thus consideration of directivity and directionality. The final approach, most commonly adopted in hybrid-broadband simulation approaches (Frankel, 1995; Graves and Pitarka, 2010), uses further information about the rupture description that is already adopted in the LF comprehensive solution (see Section 5.3). Specifically, the variation in rupture velocity and slip amplitude is used to scale the stress parameter at each subfault, and define the rupture initiation time of each subfault reflecting the non uniform rupture velocity over the fault surface.

5.6 Prediction Using Physics-Based GMMs

PSHA (Chapter 6) requires that GMMs provide the distribution of one or more ground-motion IMs (Section 4.2) for a given rupture, $f_{IM|rup}$. For this reason, significant emphasis in Chapter 4 was given to the probabilistic description of ground motion from an empirical GMM resulting from apparent aleatory variability (Section 4.6) and epistemic uncertainty (Section 4.7).

To achieve compatibility between SSMs (Chapters 2 and 3) and the prediction of $f_{IM|rup}$, the rupture description variable, rup, will encapsulate only a few attributes, such as rupture geometry and magnitude.[31] The variability in the rupture geometry (Section 3.6) and magnitude (Section 3.5) is treated within the seismic source modeling. Given this conditioning, empirical GMMs directly provide the mean and apparent aleatory variability necessary to describe $f_{IM|rup}$, under the assumption of lognormality. In contrast, a physics-based ground-motion simulation method for a

---

[31] The empirical GMM may require other source-related parameters, such as the average rake angle, but this is not usually considered a random variable.

fixed set of input parameters produces essentially[32] a deterministic ground-motion time series (and consequent IMs of interest). However, there are many plausible kinematic rupture descriptions (i.e., Section 5.3) that can exist given the specific rupture geometry and magnitude described by *rup*. By generating an ensemble of such kinematic rupture descriptions, and consequent ground-motion simulations, a distribution of the ground-motion IM can then be obtained.

In addition to the kinematic rupture specification, uncertainties also exist in the "path" and "site" components of wave propagation modeling. Consequently, a probabilistic prediction using a physics-based simulation method requires a joint distribution of all model parameters. Additional modeling uncertainty to describe the misfit with observations, if not adequately addressed through parameter uncertainty, is also necessary. As mentioned in the introduction of this chapter, a probabilistic *physics-based GMM* prediction requires both a physics-based ground-motion simulation method and a description of its parametric and modeling uncertainties.

The following subsections outline the sources and essential features of uncertainties for physics-based GMMs and how they are integrated into ground-motion prediction. Comprehensive treatment of uncertainty in physics-based GMMs is an active research area, so present challenges in adopting physics-based GMMs for PSHA are also articulated.

5.6.1 Types of Uncertainties

Table 5.1 shows two dimensions of uncertainties in physics-based ground-motion prediction (Toro et al., 1997). Concepts and examples of these uncertainties are discussed below. Section 5.6.2 subsequently discusses how the uncertainty terms (defined by variables in Table 5.1) can be combined.

From a computational perspective, uncertainty in the estimated ground motion will arise from either (1) variation in the specific values of simulation method parameters or (2) the consideration of different simulation methods.[33] These are referred to as *parametric* and *modeling* uncertainties, respectively. From an uncertainty classification perspective, as discussed in Section 1.7, parametric and modeling uncertainties can be further segregated as to whether they represent (apparent) aleatory variability or epistemic uncertainties.

The following subsections discuss the essence of parametric and modeling uncertainties and the distinction between their aleatory and epistemic components. Notably, the distinction between parametric and modeling uncertainty is model-dependent (Toro et al., 1997), as is also the case for aleatory and epistemic uncertainties (Section 1.7). For example, a simulation method parameter, such as the Brune stress parameter $\Delta\sigma$, may be considered a (known) deterministic constant in a specific tectonic region, or an uncertain model parameter. In the former case, there is no parametric uncertainty, by definition. In the latter case, there is aleatory variability in the value of $\Delta\sigma$ for future earthquakes in the region, as well as epistemic uncertainty in the estimation of $\Delta\sigma$ based on observational data (as discussed further in this section). While this latter case results in parametric variability and uncertainty, the potential benefit is an improved prediction of observational data, and

---

[32] Strictly speaking, there are stochastic elements in both the simplified and comprehensive-physics-based methods. In the simplified-physics method, the windowed white noise has random phase angles (e.g., Figure 5.19). In the comprehensive-physics method, the kinematic "slip" distribution over the fault surface is typically randomly generated (e.g., Figure 5.4). These components can be fixed, i.e., the sampled phase angles or slip fields are reproducible, by specifying an (arbitrary) "seed" value in the pseudo-random number generators. However, within the simplified-physics method, the use of random phase angles leads to variability in response spectral ordinates that is generally small, on the order of 0.1–0.2 logarithmic standard deviation units for $f > 1$ Hz (Stafford, 2017).

[33] Either entirely different methods, or different component models within a broader simulation method.

Table 5.1. Partitioning of uncertainty in ground-motion prediction using physics-based GMMs, and variable definitions related to the resulting ground-motion prediction

	Aleatory variability	Epistemic uncertainty
Parametric	Event-to-event variation in source, path, and site-specific parameters of the model for future events (σ^2_{param})	Uncertainty in the probabilistic description of model parameters (e.g., mean, variance, and distribution shape for each parameter, and correlation among parameters) ($Var[\mu_{param}]$, $Var[\sigma^2_{param}]$)
Modeling	Unexplained variability between observations and simulations due to physical processes imperfectly represented (or omitted entirely) from the model; and chaotic processes that are inherently random (σ^2_{model})	Uncertainty in the probabilistic description (e.g., mean, variance, distribution shape) of the model due to the finite number of observations ($Var[\mu_{model}]$, $Var[\sigma^2_{model}]$)
Total	Inherent variability in the prediction of future events associated with variability in model parameters and limitations in the modeling approach itself ($\sigma^2_{\ln IM}$)	Uncertainty in the probabilistic description of both model parameters and model distribution due to finite observation and calibration data ($Var[\mu_{\ln IM}]$, $Var[\sigma^2_{\ln IM}]$)

consequently less modeling uncertainty. This trade-off between model complexity and parametric uncertainty shares the same parallels with empirical GMMs, and also seismic source characterization as described in Figure 2.16.

Parametric Uncertainty

Figure 5.20 illustrates the determination of parametric uncertainty for physics-based GMMs. Typically, parameters will be either theoretically constrained (and equal to a constant) or calibrated empirically against observational data, for example, through the examination of Fourier spectra from ground-motion recordings to determine parameters used in the simplified-physics method (e.g., Anderson and Hough, 1984; Edwards and Fäh, 2013). There will be uncertainty in the specific parameter estimates (e.g., the Brune stress parameter, $\Delta\sigma$, or site attenuation, κ_0), which is epistemic uncertainty, and depicted as the error bars on each point in Figure 5.20. This parameter uncertainty arises from (1) different methods for parameter determination from data, (2) the covariance between different model parameters that are being inverted for, and (3) the finite data sample that is used to estimate the parameters.

To model future events, the collection of all available parameter estimates can be used to construct a probabilistic model(s) for each parameter. Figure 5.20 illustrates two parameters, X and Y (e.g., $\Delta\sigma$ and κ_0), for which PDFs are estimated ($\hat{f}_X$ and $\hat{f}_Y$, respectively) based on the parameter estimates from past events. These distributions represent aleatory variability in the value of the associated parameters for future events.[34] Also, the limited data used to estimate these distributions results in the distribution parameters (e.g., mean, variance) having epistemic uncertainty.

---

[34] When determining these distributions, the uncertainty in the parameter estimates of past events should be explicitly accounted for to avoid over-estimation of the aleatory variability.

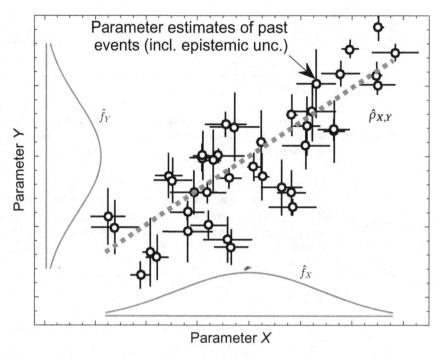

Fig. 5.20 Conceptual illustration of uncertainty in parameters that consequently results in parametric uncertainty in physics-based GMMs. The marginal distributions of parameters, and their correlation, represent aleatory variability. Epistemic uncertainty in the distributions of these parameters exists due to the finite data used in their calibration, and the error bars illustrating uncertainty on parameter estimates from past events also reflect another source of epistemic uncertainty.

Figure 5.20 also illustrates that the empirical observations of parameters X and Y are correlated, and characterized by the estimated correlation coefficient, $\hat{\rho}_{X,Y}$. It is essential to appropriately consider such correlations[35], to ensure that physically plausible realizations are generated when sampling the parameter distributions (e.g., Crempien and Archuleta, 2015; Graves and Pitarka, 2016).

Once the probabilistic description of all parameters is quantified, it is possible to determine the consequent parametric variability and uncertainty in the resulting ground-motion IMs of interest. Because of the nonlinear coupling of parameter effects on simulated ground motions (due to the physics-based nature of the methods), Monte Carlo simulation is commonly used to sample realizations of the parameters and perform the ground-motion simulation for each realization. The resulting ground-motion realizations can then describe the parametric uncertainty distribution at each location of interest. While it is possible to describe this distribution in a parametric or nonparametric fashion (see Section 5.6.2), at present, for convenience, we will refer to the distribution of the predicted ground-motion IM via its mean, μ_{param}, and variance, σ^2_{param}, respectively, and the corresponding epistemic uncertainty (variance) in the ground-motion IM mean and variance, resulting from the propagation of parametric epistemic uncertainty, as $Var[\mu_{param}]$ and $Var[\sigma^2_{param}]$, respectively.

---

[35] For simplicity, this example implicitly assumes that parameters X and Y have a multivariate normal distribution with a corresponding linear correlation coefficient. More generally, the determination of the appropriate multivariate distribution is necessary.

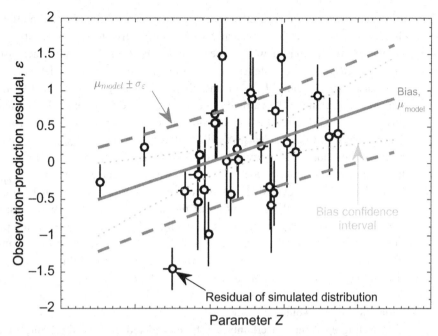

Fig. 5.21 Conceptual illustration of modeling uncertainty in physics-based GMMs. The points are residuals describing the difference between an IM observation and the simulation distribution, with error bars depicting the distribution of each residual resulting from parametric uncertainty. The bias is the average value of the residuals, and in this case μ_{model} has a dependence upon Z. The depicted residual standard deviation (σ_ε) is related (as explained in the text) to the aleatory modeling variability (σ_{model}). The epistemic uncertainty in the estimate of the model bias is illustrated via a 68% confidence interval. The epistemic uncertainty that exists in the residual standard deviation is not shown. Residuals may exhibit trends with one or more variables, as illustrated via Parameter Z.

Modeling Uncertainty

With the consideration of parametric uncertainty, physics-based GMMs can provide a probabilistic ground-motion prediction. However, this probabilistic prediction assumes that the model is a perfect representation of reality. The modeling component of physics-based GMM uncertainty (Table 5.1) reflects the misfit to, and extrapolation beyond, observational data.

Assessing modeling uncertainty requires the direct comparison between simulations and observational data through validation (Chapter 12), as depicted in Figure 5.21. The misfit will conventionally be described with respect to one or more ground-motion IMs, specifically using the "residual," ε, between the (logarithmic) observation and prediction of the IM.[36] Figure 5.21 depicts features associated with the modeling bias (and its parameter dependence), the modeling aleatory variability, and the epistemic uncertainty as a result of the limited observational data (which is assumed to be a representative sample for the future events to be predicted).

First, the mean trend in the residuals is an indicator of the bias of the physics-based GMM with respect to the observations. This bias, μ_{model}, may also be a function of one or more parameters

---

[36] Here, the variable ε is used to denote the "residual," as different from the total residual, Δ, in Chapter 4. Because of the inherent correlation of total residuals across multiple events and sites, a rigorous determination of modeling uncertainty would be based upon partitioned residuals (Section 4.6.2), and ε is therefore intended to represent any such residual term.

(e.g., "Parameter Z" in Figure 5.21) associated with source, path, or site effects. In some instances, the bias term may not be statistically significant relative to the variability in the residuals and the number of observations, in which case it can be ignored. However, in other cases the model bias may indicate significant inconsistencies that warrant a revision to the model itself.

Second, the variability between the observations and physics-based GMM is reflective of the modeling uncertainty. The vertical error bars in Figure 5.21 reflect the epistemic uncertainty in the physics-based GMM resulting from parametric uncertainty, as discussed in the previous section. The variability in the residuals, σ_ε^2, illustrated via the term $\mu_{model} \pm \sigma_\varepsilon$ in Figure 5.21, therefore represents the total variability as a result of both the modeling aleatory variability, σ_{model}^2, and the propagated uncertainty due to parameter epistemic uncertainty, $\sigma_{param,epistemic}^2$ (which is also applicable to empirical GMMs, as discussed by Stafford, 2013). In Figure 5.21, the size of the error bars, indicating the distribution of the simulated residuals (with variance $\sigma_{param,epistemic}^2$), is relatively small compared with the variability of the mean values of the prediction residuals. In this example, the modeling aleatory variability σ_{model}^2 will therefore be similar to σ_ε^2. However, it is possible to have a situation in which the total variability in the prediction residuals is similar to the ground-motion uncertainty resulting from parameter epistemic uncertainty (i.e., $\sigma_\varepsilon^2 \approx \sigma_{param,epistemic}^2$). In this case, the relative size of the required "additional" modeling aleatory variability σ_{model}^2 would be small.

Modeling epistemic uncertainty can be considered most easily through the use of two or more different approaches for the component models within a ground-motion simulation method, or two or more methods themselves. This notion is parallel to the idea of considering modeling uncertainty in empirical GMMs through the use of multiple models (Section 4.7.1). As discussed in Section 4.7.3, the consistent treatment of modeling uncertainty using multiple models requires considering that (1) the observational data used to assess model performance, and weight alternative models, is likely to be from relatively small-magnitude earthquakes at large source-to-site distances, in contrast to the magnitudes and distances of most interest; and (2) model uncertainty can be systematically underestimated if there is a lack of alternative models.

An alternative treatment of modeling epistemic uncertainty, also discussed in Section 4.7.3 in the context of empirical GMMs, is the use of meta-models. In the context of physics-based GMMs, a meta-model would be developed by increasing the epistemic uncertainty in the ground-motion simulation parameters as a surrogate for the uncertainty that results from the model idealizations itself. This consideration will increase the parametric uncertainty in the physics-based GMM,[37] increasing the uncertainty in the distribution of prediction residuals. Ideally, the increase would match that of the residual variance itself (i.e., $\sigma_{param,epistemic}^2 \approx \sigma_\varepsilon^2$), and therefore remove the need to consider additional modeling variability. While this approach will broaden the prediction distribution, it assumes that the single underlying model is correct, which is problematic if the model is missing key physics, particularly for extrapolation beyond observational data.

Finally, similar to that for parameter uncertainty, there also exists epistemic uncertainty in the estimates of the modeling bias and variability as a result of the limited observational data used. This is partially illustrated in Figure 5.21 in the form of the 68% confidence interval in the estimated bias, which is related to the variance in the bias estimate, $Var[\mu_{model}]$. This confidence interval also broadens near the peripheries of the observational data. Uncertainty in the estimate of the model variability (i.e., $Var[\sigma_{model}^2]$) also exists, but for brevity this is not shown in Figure 5.21.

---

[37] Specifically, $Var[\mu_{param}]$ and $Var[\sigma_{param}^2]$, defined in Equations 5.39 and 5.40.

Aleatory Variability versus Epistemic Uncertainty

A discussion of aleatory variability and epistemic uncertainty in seismic hazard analysis was initially presented in Section 1.7. Table 5.1 subsequently distinguished between aleatory and epistemic components of parametric and modeling uncertainties in the context of physics-based GMMs. In essence, aleatory variability represents parametric and model details of the source, path, and site effects that cannot be quantified before the event, whereas epistemic uncertainty results from the uncertainty in the quantification of the model and its parameters, including their aleatory variability. Section 4.7.2 summarized the three principal sources of epistemic uncertainty in model-based predictions of ground motion via empirical GMMs: ground-motion dataset, model formulation, and model parametric uncertainty. These same three sources are also applicable to model uncertainty in physics-based GMMs, as well as for the probabilistic models that describe their parameters (e.g., replacing the y-axes of Figure 4.31 with "Parameter X").

The determination of whether uncertainty is aleatory or epistemic is also context-dependent (Der Kiureghian and Ditlevsen, 2008). This is illustrated in the distribution of physics-based GMM parameters, X and Y, shown in Figure 5.20. The estimated PDFs of these parameters, $\hat{f}_X$ and $\hat{f}_Y$, represent the aleatory variability in the parameter values for future events (i.e., the distribution from which a random value for the next event is realized). When such a future event occurs, it has a specific parameter value, which is also the case for past events. However, the specific values for these past events are uncertain (they have epistemic uncertainty), because we have finite data from which to infer them, and our inference methods are imperfect (i.e., we measure ground motions as $a(t)$ or $v(t)$, from which model-based parameters such as $\Delta\sigma$ and κ_0 are then inferred). This epistemic uncertainty in past event values is also illustrated using the vertical error bars in Figure 5.21. This uncertainty partially explains the misfit between observations and the mean of the simulation, but generally not without the addition of further modeling variability, σ_{model}. For example, the hypocenter locations for past events are often estimated to within 1 km accuracy (epistemic). However, the random (aleatory) variability in the location of the hypocenter for a future rupture[38] will have a variability much larger than 1 km.

Another important consideration in the distinction between aleatory and epistemic uncertainties is the degree to which the *ergodic* assumption has been invoked to combine data from different spatial regions, as discussed in Section 8.5.1.

5.6.2 Prediction Distribution

The prediction distribution obtained from a physics-based GMM can be constructed based on the concepts introduced in the previous subsection, including the variables defined in Table 5.1. As in Section 4.6 for empirical GMMs, it is initially assumed in the discussion below that the distribution of IM is lognormal (so that $\ln IM$ is normal), and then generalizations are subsequently presented.

Consider that the ground-motion IM distribution predicted as a result of parametric aleatory variability is defined as

$$\ln IM_{param} \sim \mathcal{N}(\mu_{param}, \sigma^2_{param}). \tag{5.35}$$

---

[38] Described by a probability distribution that is a function of the location along-strike and down-dip on the fault surface (e.g., Mai et al., 2005).

Through validation against observations (e.g., Figure 5.21), the modeling bias μ_{model} and apparent variability σ^2_{model} have also been assessed.[39] In particular, the appropriate model aleatory variability is obtained from the residual variance by subtracting the propagated parameter epistemic uncertainty via[40]

$$\sigma^2_{model} = \sigma^2_\varepsilon - \overline{\sigma}^2_{param,epistemic} \qquad (5.36)$$

where $\overline{\sigma}^2_{param,epistemic}$ is the mean value of the propagated epistemic parameter uncertainty, which will have a different variance for every observation (as indicated through the variable error bar sizes in Figure 5.21). Equation 5.36 indicates that σ^2_{model} is the variability we expect if the model parameters are known exactly.

The mean and variance of the ground-motion IM obtained from the physics-based GMM, accounting for both parametric and modeling uncertainty, can then be obtained from:

$$\mu_{lnIM} = \mu_{param} + \mu_{model} \qquad (5.37)$$

$$\sigma^2_{lnIM} = \sigma^2_{param} + \sigma^2_{model}. \qquad (5.38)$$

These mean and variance terms provide the necessary parameters to evaluate the ground-motion prediction distribution in the same manner as for empirical GMMs, as discussed in Section 4.6.1.

The epistemic uncertainty, present in the mean and variance estimates above, can also be quantified by

$$Var[\mu_{lnIM}] = Var[\mu_{param}] + Var[\mu_{model}] \qquad (5.39)$$

$$Var[\sigma^2_{lnIM}] = Var[\sigma^2_{param}] + Var[\sigma^2_{model}]. \qquad (5.40)$$

Equations 5.36–5.40 make the reasonable assumption of no correlation in the parametric and modeling components of uncertainty.

Nonparametric Distributions

It was previously mentioned that parametric uncertainty in physics-based GMMs is most commonly quantified using Monte Carlo simulations for random realizations of model parameters. The resulting ground-motion simulation realizations produce a nonparametric description of the ground-motion distribution without the need to assume a distribution type.

Figure 5.22a illustrates this idea via a histogram of IM values at a given location obtained from $N = 100$ simulations, each of which is considered equally plausible (i.e., each has a probability of $1/N$). These N simulations yield a distribution of the predicted IM with a median of approximately 20 and a lognormal standard deviation of approximately 0.3, without the need to assume the distribution shape. Figure 5.22b presents an empirical cumulative distribution function, $\hat{F}_{IM}(im)$, of the N simulation results shown in Figure 5.22a. The probability of exceeding a specific level of IM can be estimated as the fraction of simulation realizations that exceed this value. For example, the probability of exceeding 30 is $P(IM > 30) = 1 - \hat{F}_{IM}(30) = 9/100$, or 0.09.

The attractiveness of avoiding the lognormal distribution assumption has led to its adoption in proof-of-concept regional PSHA using physics-based GMMs (e.g., Graves et al., 2011b;

---

[39] Discussion in this section refers to variance, rather than standard deviation, because of the ability to directly add variances (see Appendix A).

[40] This "removal" of propagated parameter uncertainty is also commonly overlooked in the equally applicable context of empirical GMMs (Rhoades, 1997; Moss, 2011; Kuehn and Abrahamson, 2018).

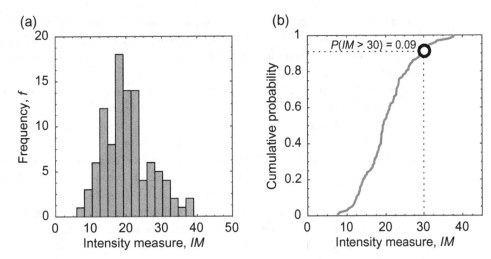

Fig. 5.22 (a) Histogram of simulated ground-motion intensity at a specific location from an ensemble of $N=100$ simulations, and (b) empirical cumulative distribution function, $\hat{F}_{IM}$, of the same simulation ensemble illustrating how exceedance probability is determined.

Tarbali et al., 2019). There are, however, several challenges with adopting the nonparametric approach. The first challenge is the coarse nature of the tails of the distribution, resulting from the finite number of simulations performed. Thus, probabilistic seismic hazard results for low exceedance rates require a large number of simulations, which can be computationally prohibitive. The second challenge is to incorporate modeling bias and uncertainty into the non-parametric distribution resulting from parameter uncertainty alone. Proof-of-concept examples to date (Graves et al., 2011b; Tarbali et al., 2019) have not explicitly included modeling uncertainty, instead considering only limited treatment of parametric uncertainties (e.g., hypocenter location and random realizations of the kinematic rupture over the fault surface, with no uncertainty in the crustal model). One approach to incorporate "additional" modeling uncertainty (described by μ_{model} and σ^2_{model}), while retaining the nonparametric distribution due to parameter uncertainty, is to use kernel density estimation techniques (e.g., Wasserman, 2006, Chapter 6), with the kernel variance equal to σ^2_{model}.

The relative size of explicitly represented uncertainties in the physics-based GMM (due to parameter uncertainty, and also the consideration of two or more simulation approaches resulting in modeling uncertainty), compared with the "additional" modeling uncertainty that is required to reflect the misfit to observations, also has implications for ground-motion selection (Chapter 10). Specifically, the explicit representation of uncertainties is reflected in an ensemble of simulated ground-motion time series, whereas the "additional" modeling uncertainty is simply in the form of numerical adjustments (μ_{model} and σ^2_{model}) to the IM distribution of the physics-based GMM. Therefore, it will be challenging to select ground-motion time series representative of the IM distribution if the latter term is large relative to the former.

5.6.3 Current Challenges in Uncertainty Quantification

There remain significant unresolved and actively researched aspects of uncertainty quantification in physics-based GMM development. This subsection provides an overview of such issues, principally from the perspective of hybrid-broadband ground-motion simulation methods.

Computational capacity constraints, principally concerning comprehensive physics-based simulation methods (Section 5.5.1), hinder research examining parameter and modeling uncertainties at frequencies of interest. Such capacity constraints, therefore, have inhibited the breadth and depth of uncertainty examination.

A large body of research has examined the variability in simulated ground-motion IMs resulting from changes to parameter values, as well as the variability when simulation results are regressed against causal parameters, such as magnitude, source-to-site distance, and azimuth, among others (e.g., Hartzell et al., 2011; Imtiaz et al., 2015; Douglas and Aochi, 2016; Vyas et al., 2016; Sun et al., 2018; Withers et al., 2018). In most instances, the considered variation in the simulation inputs reflects an attempt to undertake sensitivity studies, because the probabilistic distribution of the input parameters is often not adequately constrained. This challenge is exacerbated by the fact that the necessary parameter distributions are model-dependent. For example, the uncertainty in the modeling of anelastic attenuation to implicitly account for wave scattering in a 1D crustal model will be different than in a 3D crustal model with large- and small-scale material heterogeneities. The regression and interrogation of simulated results as a function of magnitude, source-to-site distance, etc., also contributes to understanding of the aleatory variability in empirical GMMs that use such predictor variables (Section 4.5) but does not directly describe the uncertainty in simulated ground motions at a site-specific location, which is needed when using physics-based GMMs in PSHA. The averaging-based factorization concept of Wang and Jordan (2014) provides a useful partitioning of this apparent variability between events, sites, and ray paths, from that which exists at a single site, and is one means to understand uncertainties in simulations further.

Fewer studies have extensively examined the predictive capability of physics-based GMMs against observational data for validation purposes, and to assess modeling uncertainty. Such studies (e.g., Graves and Pitarka, 2010; Mai et al., 2010; Dreger and Jordan, 2015; Lee et al., 2020) generally have also not considered parameter uncertainty in the prediction of observational data. Beauval et al. (2009) and Hartzell et al. (2011), among others, discuss the validation against observations as well as parameter uncertainty, but do not subtract the effects of parameter epistemic uncertainty in assessing modeling uncertainty (Equation 5.36).

Several studies have also examined the effect of different ground-motion simulation methods (Ameri et al., 2009; Hartzell et al., 2011), although the explicit combination of multiple methods has generally not been addressed.

The above sentiments reflect progression in the science of physics-based GMMs, from initially obtaining reliable simulation results that reflect the mean ground-motion properties for IMs of interest (e.g., Dreger and Jordan, 2015) toward an eventual comprehensive treatment of the uncertainty on a site-specific basis partitioned into the various components described in Table 5.1. Through this progression, various applications of the insights from simulations have arisen. For example, physics-based simulations have been used to inform the functional scaling of empirical GMMs for rupture scenarios that are poorly constrained by empirical observations (Abrahamson et al., 2008). The increasing consideration of, and constraints on, uncertainties in physics-based GMMs can then allow their direct use in PSHA (e.g., Graves et al., 2011b; Tarbali et al., 2019), albeit with the need to account for modeling uncertainty that is not present in the ensemble of simulations (Section 5.6.2).

Finally, physics-based GMMs utilize the basic ingredients of a rupture and crustal model and produce ground-motion time series from which multiple ground-motion IMs can be computed. The varying realism of these models means that the fidelity of physics-based GMMs will vary geographically and even spatially within a given geographic region. The fidelity will also be a function of the specific IM(s) of interest. Consequently, the rate of adoption of physics-based GMMs will vary on a region- and problem-dependent basis (Bradley et al., 2017).

Exercises

5.1   What are the conceptual benefits of using simulated ground motions in seismic hazard analysis and/or seismic response history analysis?

5.2   Empirical and physics-based GMMs provide two conceptual approaches for ground-motion prediction.

    (a) Discuss the physical considerations that are reflected in empirical GMMs, and the empirical calibration used in physics-based GMMs.

    (b) Given the relative merits of both approaches, how might they be used together to attain their respective benefits for ground-motion prediction?

5.3   Describe the role that simulated ground motions, from methods validated for generic conditions, can play in the further development of empirical ground-motion models, and utilization in ground-motion selection.

5.4   A seismic wave with frequency $f = 0.5$ Hz is propagating in a medium with velocity $V_S = 3$ km/s and density $\rho = 2.5$ Mg/m^3.

    (a) What is the wavelength, λ, and wavenumber, k, of this wave?

    (b) What would be the minimum spatial discretization Δx of the medium, if this wave were to be accurately simulated using a finite difference method that is fourth-order in space (requiring $N = 5$ points per wavelength)?

    (c) What is the maximum frequency that would be accurately simulated in this medium if the spatial discretization was $\Delta x = 0.1$ km?

    (d) How does the computational requirements to solve the 3D equation of motion scale with the selected spatial discretization?

5.5   Consider an earthquake of along-strike length, $L = 100$ km, and down-dip width, $W = 15$ km.

    (a) What is the median moment magnitude that would be estimated from the magnitude-area scaling relation of Equation 2.18 using $\alpha_0 = 4.0$ and $\alpha_1 = 1.0$?

    (b) Assuming a shear rigidity of $\mu = 3.3 \times 10^{10}$ N/m^2, what is the average slip displacement u over the fault surface?

    (c) Using Equation 5.2 estimate the average rise time.

    (d) Using the model of Liu et al. (2006) (Figure 5.3), what is the maximum slip rate (velocity) for the average slip amplitude computed in (c)?

    (e) Considering that the distribution of slip at a point on the fault surface has a lognormal standard deviation of $\sigma_{\ln u} = 0.85$, with the mean value μ_u computed in step (b), compute the 10th and 90th percentile values of the slip amplitude distribution and the corresponding maximum slip rates.

5.6   Explain the mechanisms of propagation of P-waves and S-waves, noting which of the two waves would be expected to appear first on the seismograms for a given earthquake.

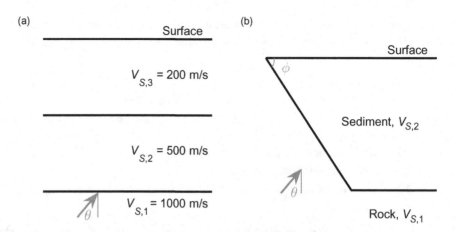

Fig. 5.23   (a) Three-layer deposit and (b) sedimentary basin with inclined basin edge for Exercises 5.8–5.10. The incident wave has an inclination of θ_I relative to vertical, and the basin edge an angle ϕ relative to the ground surface.

5.7   Describe what Love wave dispersion means. Given this dispersive characteristic, would the frequency content of Love waves at an instrument location increase or decrease with time? Describe the conditions required for Rayleigh waves to be dispersive.

5.8   Consider a deposit with two sedimentary layers overlying a half-space (Figure 5.23a). The surficial layer is 10 m thick with a shear-wave velocity of $V_{S,3} = 200$ m/s. The lower sedimentary layer is 40 m thick with a shear-wave velocity of $V_{S,2} = 500$ m/s. The half-space has a shear-wave velocity of $V_{S,1} = 1000$ m/s. For simplicity, assume that the densities are approximately constant for all three layers.

   (a) Compute the impedance ratio α at the two layer boundaries for an upward-propagating wave.
   (b) If a vertically incident wave ($\theta = 0$) within the half-space, with unit displacement and stress amplitude, intersects the horizontally layered material boundaries, what will be the amplitudes of the reflected and transmitted waves at the two boundaries and the ground surface?
   (c) If the time of the wave in the half-space intersecting the lower sedimentary layer is $t = 0$ s, what is the arrival time of the first three waves (one direct, the other two reflected) that reach the ground surface?

5.9   Consider the same three-layer problem in the above question, but now the incident wave is inclined at an angle θ from the vertical.

   (a) If $\theta = 45°$, compute the angle of the refracted waves that are transmitted into the two sedimentary layers above, and then also those associated with the down-going wave that is reflected off the ground surface.
   (b) What is the critical angle θ_{crit} of the incident half-space wave that leads to total internal reflection in any layer, and at what layer boundary does it occur?

5.10   Consider the sedimentary basin with an inclined basin edge depicted in Figure 5.23b.

(a) For a vertically incident wave ($\theta = 0$), with velocities $V_{S,1} = 1000$ m/s and $V_{S,2} = 400$ m/s, and basin-edge inclination $\phi = 30°$, determine the angle of the transmitted wave that reaches the ground surface through (i) the basin edge, and (ii) the horizontal base of the basin.

(b) What is the critical angle beyond which a wave within the sedimentary basin will undergo total internal reflection at the boundary with the underlying rock?

(c) Derive the expression for the minimum inclination angle θ of the incident wave in the rock which will result in propagation through the basin edge, reflection off the ground surface, and total internal reflection at the base of the sedimentary basin. Compute this minimum angle for the parameter values stated in (a).

(d) Using the expression from (c), how does this minimum inclination angle change for (i) a two-fold reduction in basin sediment velocity (i.e., $V_{S,2} = 200$) and (ii) a two-fold increase in the basin-edge inclination (i.e., $\phi = 60°$)?

(e) Based on the results in this question, comment on the ability of basin edge geometry to lead to total internal reflection.

5.11 Consider a body wave in a homogeneous half-space with velocity $V_S = 2000$ m/s and frequency $f = 0.2$ Hz.

(a) At a propagation distance $r = 50$ km, what will be the relative amplitude reduction as a result of geometric spreading?

(b) Using the empirical expression for the quality factor, $Q_S = 50V_S$ (where V_S is in km/s), what is the relative amplitude reduction resulting from anelastic attenuation?

(c) How do the results for (a) and (b) change for a body wave of $f = 5$ Hz propagating in the same half-space?

(d) Using the same expression as (b), for both $f = 0.2$ Hz and $f = 5$ Hz, determine the amplitude reduction due to anelastic attenuation for a material with velocity $V_S = 500$ m/s.

(e) How much larger will the geometric spreading and anelastic attenuation be at a propagation distance of $r = 200$ km than $r = 50$ km? (Consider both $f = 0.2$ and 5 Hz, and $V_S = 2000$ and 500 m/s cases.)

5.12 Hybrid broadband ground-motion simulation makes use of comprehensive- and simplified-physics methods at low and high frequencies.

(a) Discuss and relative merits and drawbacks of the comprehensive and simplified solutions.

(b) How would one determine the appropriate transition frequency at which the "strengths" of each respective method should be combined to produce the broadband waveform?

5.13 For the simplified-physics method of ground-motion simulation (e.g., Equation 5.21):

(a) What is the dependence of the high-frequency ground-motion amplitudes (FAS) on the stress parameter $\Delta\sigma$? How much do these amplitudes change if $\Delta\sigma$ is doubled?

(b) For a source with $V_{S,src} = 3.6$ km/s and $\Delta\sigma = 6$ MPa, compute the FAS (over the range $f = 0.01–10$ Hz) for two earthquakes of M5 and M6.5. How does the difference in magnitude affect the (i) corner frequency and (ii) FAS amplitudes above and below the corner frequency? Use reasonable values for other parameters required.

(c) Discuss the combined effect of the anelastic attenuation along the ray path (Q) and the near-surface attenuation (κ_0), and the need to ensure self-consistency.

5.14　Compare and contrast the manner in which modeling uncertainty in physics-based GMMs is determined relative to that for empirical GMMs in the following.

(a) What is the equivalent of the mean (bias) term associated with modeling uncertainty for empirical GMMs?

(b) How does the standard deviation of modeling uncertainty relate to apparent aleatory variability in empirical GMMs?

(c) If a physics-based GMM contains no parametric uncertainty, how is its modeling uncertainty related to the apparent aleatory uncertainty in an empirical GMM?

PART II

HAZARD CALCULATIONS

6

PSHA Calculation

This chapter introduces tools for integrating information presented earlier in the book into a forecast of the likelihood of observing ground motion using probabilistic seismic hazard analysis (PSHA). Specifically, the rupture scenarios defined in Chapter 3 will be combined with ground-motion models from Chapters 4 and 5. The basic tools for incorporating input information, and considering epistemic uncertainty in that information, are developed here. Several example calculations are provided to illustrate the mechanics of these calculations and to demonstrate how input parameter values affect the output hazard results. Chapter 7 then presents other calculations related to PSHA, and Chapter 8 relaxes some common assumptions in PSHA to allow for more refined predictions.

Learning Objectives

By the end of this chapter, you will be able to do the following:

- Compute a ground-motion hazard curve.
- Evaluate the sensitivity of a hazard calculation to changes in input models or input parameters.
- Relate an event's annual rate to the probability of that event during a specified period of time.
- Determine the ground-motion intensity with a specified exceedance rate or return period, and the slope of a ground-motion hazard curve.
- Specify and utilize a logic tree to incorporate epistemic uncertainty into a hazard calculation.

6.1 Introduction

This chapter utilizes models from previous chapters to compute the likelihood of observing ground motion of a given intensity at a particular site. At a basic level, PSHA is composed of five steps:

1. Specify the ground-motion intensity measure (IM) of interest (Section 4.2).
2. Specify the site properties that help predict ground-motion intensity (Section 4.5.3).
3. Compute the locations, characteristics, and occurrence rates of all rupture scenarios capable of producing damaging ground motions (Chapter 3).
4. Predict the resulting distribution of ground-motion intensity as a function of the site characteristics and each rupture scenario's properties (Chapters 4 and 5).
5. Consider all possible ruptures, and uncertainty in resulting ground-motion intensity (Figure 6.1 and Equation 6.1).

The result of these calculations is a *ground-motion hazard curve*, which quantifies the full distribution of levels of IM, and their associated rates of exceedance. While this conceptual procedure is rather simple, numerical implementation can make the fundamentals of the process more challenging to

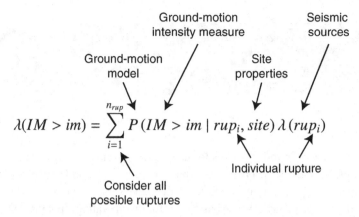

$$\lambda(IM > im) = \sum_{i=1}^{n_{rup}} P(IM > im \mid rup_i, site)\, \lambda(rup_i)$$

Fig. 6.1   Inputs to a probabilistic seismic hazard analysis calculation.

discern when using complex modern input models. Therefore, this chapter begins with very simple formulations and examples to introduce the principles and procedures of PSHA. Later in the chapter, more realistic examples will address issues arising from greater complexity.

6.2 The PSHA Calculation

PSHA is a mathematical calculation that performs the steps outlined in Section 6.1. Figure 6.1 indicates how these steps relate to the following equation:

$$\lambda(IM > im) = \sum_{i=1}^{n_{rup}} P(IM > im \mid rup_i, site)\, \lambda(rup_i) \tag{6.1}$$

where $\lambda(IM > im)$ is the occurrence rate of ground motions with IM greater than im, $P(IM > im \mid rup_i, site)$ comes from the ground-motion model (GMM), $\lambda(rup_i)$ comes from the seismic source model (SSM), and we sum over all considered ruptures. The set of considered ruptures should include all ruptures of potential interest for the hazard and risk problems of interest. The summation operation adds up the rates of occurrence of $IM > im$ from all possible earthquake ruptures.

Rather than have a single list of rupture scenarios as in Equation 6.1, it can sometimes be useful to index ruptures according to their seismic source:

$$\lambda(IM > im) = \sum_{j=1}^{n_{sources}} \sum_{i=1}^{n_{rup_j}} P\left(IM > im \mid rup_{i,j}, site\right) \lambda\left(rup_{i,j}\right) \tag{6.2}$$

where $n_{sources}$ is the number of sources considered, n_{rup_j} is the number of ruptures from source j, and $rup_{i,j}$ denotes rupture i on source j. Explicit indexing of sources was traditionally preferred because, in most SSMs, the ruptures were associated with distinct sources.[1]

---

[1] Though some newer SSMs have less distinctly defined sources, and include ruptures involving multiple fault sources (e.g., Field et al., 2014).

A second modification to Equation 6.1 is to use an explicit parameter to capture the variability associated with a GMM. Recall from Section 4.4 that the variate ε is the normalized deviation of $\ln IM$ from a mean prediction:

$$\varepsilon = \frac{\ln IM - \mu_{\ln IM}(rup, site)}{\sigma_{\ln IM}}. \tag{6.3}$$

This variate is independent of rup, so its probability can be represented by a multiplicative term in the hazard equation. Additionally, conditional upon $rup, site, \varepsilon$, and a GMM, exceedance of a given im is deterministic. So the $P(IM > im \mid rup, site)$ term in Equation 6.2 can be replaced with an indicator function, $I(IM > im \mid rup, site, \varepsilon)$, equal to 1 if the argument is true and 0 otherwise. The hazard can then be calculated as

$$\lambda(IM > im) = \sum_{j=1}^{n_{sources}} \sum_{i=1}^{n_{rup_j}} \sum_{k=1}^{n_\varepsilon} I(IM > im \mid e_k, rup_{i,j}, site) f_\varepsilon(e_k) \Delta e \, \lambda\left(rup_{i,j}\right) \tag{6.4}$$

where e_k are a discrete set of ε values with intervals of width Δe, and $f_\varepsilon()$ is the probability density function for ε (generally, a standard normal PDF). Note that Δe should be quite small for this equation to be numerically equivalent to the above formulation. In the limit, as Δe tends to zero, it is exact.

Equations 6.1, 6.2, and 6.4 are mathematically equivalent and produce the same numerical result for a given site. All three equations serve to integrate information about the locations, characteristics, and rates of occurrence of possible earthquake ruptures, and the distribution of ground-motion intensity at the location of interest due to those earthquakes. The distinction is merely that Equations 6.2 and 6.4 add explicit indexing related to ruptures and ε values that will facilitate later analysis to understand the ruptures that contribute most to hazard (see Section 7.2). The simple Equation 6.1 more directly indicates the inputs to the calculation, however, so we use that formulation for convenience as well.

6.2.1 Alternate Formulations

The notation of Equations 6.1 and 6.2 differs from presentations of the PSHA integral presented in some other documents, so here we comment on the relationship among these formulations.

In the original PSHA formulation by Cornell (1968), GMMs were dependent only upon earthquake magnitude and distance, thus earthquake rupture rup_i required specification of only m_i and r_i. For a single-source model, the rate of a given rupture can be computed as the rate of earthquakes on the source, $\lambda(EQ)$, multiplied by the probability of m_i and r_i given an earthquake:

$$\lambda(rup_i) = \lambda(EQ) \, f_{M,R}(m_i, r_i) \, dm \, dr \tag{6.5}$$

where dm and dr are increments of magnitude and distance, respectively, and $f_{M,R}(m_i, r_i)$ is a joint probability density function for M and R.

Cornell (1968) further assumed earthquakes to be point sources, so the distance distribution was independent of an earthquake's magnitude. Therefore the joint probability density can be computed as a product of marginal probabilities:

$$f_{M,R}(m_i, r_i) = f_M(m_i) \, f_R(r_i). \tag{6.6}$$

Finally, because Cornell (1968) solved the PSHA calculation analytically rather than numerically, an integral over continuous magnitude and distance values is used instead of a summation over a discrete set of ruptures. This leads us to a re-expression of the PSHA calculation for a single source case as

$$\lambda(IM > im) = \lambda(EQ) \int_R \int_M P(IM > im \mid m, r)\, f_M(m_i)\, f_R(r_i)\, dm\, dr. \qquad (6.7)$$

For the multiple-source case, exceedance rates for each source are summed as in Equation 6.2:

$$\lambda(IM > im) = \sum_{j=1}^{n_{sources}} \lambda(EQ_j) \left(\int_R \int_M P(IM > im \mid m, r)\, f_M(m)\, f_R(r)\, dm\, dr \right)_j \qquad (6.8)$$

where $\lambda(EQ_j)$ denotes the rate of earthquakes on source j, and the integrals are in parentheses subscripted by j, indicating that the M and R distributions are for source j. This equation may be familiar to readers of other documents describing PSHA (e.g., Reiter, 1990; Kramer, 1996; Bazzurro and Cornell, 1999; McGuire, 2004).

The classic formulation of Equation 6.8 needs several modifications to be compatible with more modern calculation procedures. First, modern GMMs use distance metrics that account for fault finiteness, such as the closest distance to the rupture plane or the surface projection of the rupture plane (Figure 3.35). The earthquake magnitude and rupture distance are thus no longer independent (Section 3.6), so a joint probability distribution for M and R is needed instead of the product of their individual probabilities:

$$\lambda(IM > im) = \sum_{j=1}^{n_{sources}} \lambda(EQ_j) \left(\int_R \int_M P(IM > im \mid m, r)\, f_{M,R}(m, r)\, dm\, dr \right)_j. \qquad (6.9)$$

Another permutation of Equation 6.9 uses a conditional probability distribution for R given M and a marginal probability distribution for M, using the property that $f_{M,R}(m, r) = f_M(m)\, f_{R|M}(r \mid m)$.

Second, as discussed in Chapter 4, modern GMMs characterize fault ruptures by many more parameters than simply a magnitude and distance measure (e.g., rupture mechanism, rupture depth, site conditions). Therefore the $P(IM > im \mid rup)$ term requires many parameters to specify rup, and the probability distributions associated with rupture-related terms also need to be included. This is often handled by denoting the additional parameters collectively as $\boldsymbol{\theta}$, and writing the PSHA equation as

$$\lambda(IM > im) = \sum_{j=1}^{n_{sources}} \lambda(EQ_j) \left(\int_{\theta} \int_R \int_M P(IM > im \mid m, r, \boldsymbol{\theta})\, f_{M,R,\Theta}(m, r, \boldsymbol{\theta})\, dm\, dr\, d\boldsymbol{\theta} \right)_j. \qquad (6.10)$$

As with Equation 6.4, some formulations make ground-motion variability given M, R, and $\boldsymbol{\theta}$ as an explicit variable in the calculation. Using the variate ε from Equation 6.3, Equation 6.10 becomes

$$\lambda(IM > im)$$

$$= \sum_{j=1}^{n_{sources}} \lambda(EQ_j) \left(\int_{\varepsilon} \int_{\theta} \int_R \int_M I(IM > im \mid m, r, \boldsymbol{\theta}, \varepsilon)\, f_{M,R,\Theta}(m, r, \boldsymbol{\theta})\, f_{\varepsilon}(\varepsilon)\, dm\, dr\, d\boldsymbol{\theta}\, d\varepsilon \right)_j. \qquad (6.11)$$

Finally, modern PSHA calculations are performed using computer software, with numerical summations over rupture parameter values, rather than with analytical integrals. For example, Equation 6.9 would be implemented numerically as

$$\lambda(IM > im) = \sum_{j=1}^{n_{sources}} \lambda(EQ_j) \left(\sum_{k=1}^{n_r} \sum_{l=1}^{n_m} P(IM > im \mid r_k, m_l) \, f_{M,R}(m_l, r_k) \, \Delta m \, \Delta r \right)_j \qquad (6.12)$$

where k and l are indices to sum over distance and magnitude values (with increments of Δr and Δm, respectively), and n_r and n_m are the number of such values. We can convert Equations 6.10 and 6.11 to a similar summation form.

The above context shows that these various formulations are mathematically equivalent. A primary limitation of these alternate formulations, in terms of modern PSHA, is that rupture and site parameterization require many variables (Section 4.5.5), which are cumbersome to list explicitly. Accommodating the many parameters motivated the introduction of θ. However, the differing treatment of M and R from other factors is due to historical precedent rather than a fundamental difference in the parameters. The $M/R/\theta$ formulation is particularly incompatible with simulation-based GMMs and hazard analysis, where R is not considered explicitly, and the rupture realizations require more elaborate parameterization. A second, related limitation is that the emphasis on integration variables distracts from insight into the rupture and site parameterization for a given problem. The formulation of Equation 6.1 is thus appealing because of its generality and its greater emphasis on the underlying SSM and GMMs, rather than intermediate integration variables. We note that the Equation 6.1 formulation is adopted from Field et al. (2003).

6.2.2 Return Periods and Probabilities

Equations 6.1, 6.2, and 6.4 (and the equivalent formulations discussed in Section 6.2.1) compute the mean rate of IM exceedance. Sometimes it is preferable to know the return period or the probability of an IM exceedance, in a specific window of time. The probability of an IM exceedance in particular, requires more information about the time-varying nature of exceedances. We will discuss these conversions and related assumptions in this section.

The *return period* of an IM exceedance is defined as the reciprocal of the rate of exceedance:

$$RP = \frac{1}{\lambda} \qquad (6.13)$$

where λ is the exceedance rate and RP is the return period. Typically, λ has units of events per year, so RP has units of years per event. For example, if a given IM has a 0.01 annual rate of exceedance, then the return period is equal to $1/(0.01 \text{ years}^{-1}) = 100$ years. This does not imply that the ground motion will be exceeded exactly once every 100 years, but rather that the average (or mean) time between exceedances is 100 years. Thus, the reciprocal of the exceedance rate is more precisely termed the *mean* return period. While the *mean return period*, or simply *return period*, is commonly used to refer to the reciprocal of the rate of exceedance, it is often clearer to report rates rather than return periods.

For a given rate of exceedance, one can also compute a *probability of exceeding* a given IM within a given window of time. This calculation requires further information regarding the probability distribution of time between occurrences of earthquakes. This distribution is nearly always assumed to be *Poissonian*, for three reasons: it results in simple mathematical equations, it appears to match

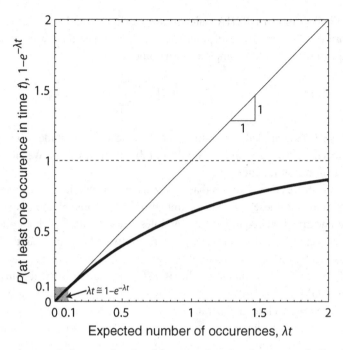

Fig. 6.2 Probability of occurrence of an event in time t, as a function of the expected number of occurrences, λt (from Equation 6.16).

observations in many cases, and more complicated models often do not impact the final results significantly.[2] The Poisson model assumes that the rate of earthquakes is constant in time, that occurrences of earthquakes are independent in time (that is, independent of anything such as the time since the most recent occurrence), and that the probability of more than one occurrence in a very short interval is negligible.

Under the assumption of Poissonian occurrences, the probability of observing at least one event in a period of time t is equal to

$$P(N \geq 1) = 1 - e^{-\lambda t} \tag{6.14}$$

where N is the number of events in time t, and λ is the rate of occurrence of events of interest, e.g., $\lambda(IM > im)$ or $\lambda(rup_i)$. Figure 6.2 shows a plot of this relationship. It can also be of interest to determine the rate of occurrence based on specified probability and time period. From Equation 6.14 it follows that

$$\lambda = -\frac{1}{t} \ln[1 - P(N \geq 1)]. \tag{6.15}$$

For example, using a commonly adopted time period of $t = 50$ years, a 10% probability of observing at least one exceedance is equivalent to $\lambda = (-1/50) \ln(1 - 0.1) = 2.1 \times 10^{-3}$ (or $RP = 475$ years). Similarly, for $t = 50$ and $P = 0.02$, $\lambda = 4.04 \times 10^{-4}$ or $RP = 2475$ years.

If λt is small (less than approximately 0.1), then the probability can also be approximated by

$$P(N \geq 1) = 1 - e^{-\lambda t} \cong \lambda t. \tag{6.16}$$

---

[2] Further details and alternative models are discussed in Sections 3.8.4 and 7.7.

This approximation comes from taking a first-order Taylor series expansion of $1 - e^{-\lambda t}$. Figure 6.2 indicates the accuracy of the approximation, where the plot follows a straight line with a slope of 1 for λt values less than 0.1. For the examples considered in the previous paragraph of $t = 50$ years and $P = 0.1$ or 0.02, the approximation of Equation 6.16 yields $\lambda = 2 \times 10^{-3}$ ($RP = 500$ years) and $\lambda = 4 \times 10^{-4}$ ($RP = 2500$ years), respectively, which have approximation errors of 5% and 1%.

A typical assumption is that earthquake occurrences are Poissonian, and IM exceedances conditional on rup are independent of rup. If the rup occurrences are Poissonian with rate λ, and if each $IM > im$ given rup is independent, with probability p, then $IM > im$ events associated with rup are a *Poisson process with random selection*, having rate $p\lambda$. With this assumption, we can interchangeably talk about earthquakes or IM exceedances as being Poissonian.

The above equations can convert PSHA results between exceedance rates, probabilities of exceedance, and return periods. The following are several important considerations related to these conversions:

1. Conversion between exceedance rates and probabilities of exceedance is almost always made by assuming a Poissonian occurrence of earthquakes.
2. Probabilities of exceedance and exceedance rates are equivalent only if the probability value of interest is small (i.e., less than 0.1).
3. Rates of exceedance, or the expected numbers of exceedances, can exceed unity. However, probabilities of exceedance are always ≤ 1.

There are, of course, alternate calculations available if we predict the occurrence of earthquakes using a model other than the Poisson model. This will be discussed further in Section 7.7.

6.3 Example Calculations

To illustrate the PSHA procedure, several numerical examples are provided below, starting from simple calculations and building to more complex cases. These examples will compute rates of exceeding varying values of $SA(1 \text{ s})$, using the procedures described above.

6.3.1 Example: One Rupture Scenario

We first start with a simple site shown in Figure 6.3. There is a single fault that produces strike-slip earthquakes at a distance of $R = 10$ km, and the site has a $V_{S,30} = 500$ m/s. Consider a simple case where the fault produces only magnitude $M = 6.5$ earthquakes at a rate of 0.01 per year. There is only a single rupture scenario to consider in this case, denoted here as rup_1. Although the rupture

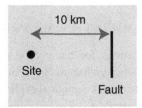

Fig. 6.3 Map view of the example site, with one fault capable of producing earthquakes at a distance of 10 km.

Table 6.1. Intermediate calculations to compute $\lambda(SA(1\,s) > 0.2\,g)$ for the example of Section 6.3.1

i	m_i	$\lambda(rup_i)$	$P(SA(1\,s) > 0.2\,g \mid rup_i)$	$P(SA(1\,s) > 0.2\,g \mid rup_i)\lambda(rup_i)$
1	6.5	0.01	0.378	0.00378
				Sum = 0.00378

description of future earthquakes is assumed fixed in this idealized example, we still expect variations in observed $SA(1\,s)$ at the site due to differences from event to event that are not captured by the simple rupture parameters.

Using the Boore et al. (1997) model discussed earlier in Section 4.4.2 (with $M = 6.5$, strike slip fault, $R = 10$ km, and $V_{S,30} = 500$ m/s) we predict a median $SA(1\,s)$ of 0.170 g, and logarithmic standard deviation of 0.520. We can use these results to find the probability that a rup_1 earthquake will cause an $SA(1\,s)$ greater than 0.2 g, as discussed in Section 4.6.1:

$$P(SA(1\,s) > 0.2\,g \mid rup_1) = 1 - \Phi\left(\frac{\ln(0.2) - \ln(0.170)}{0.520}\right)$$

$$= 0.378. \tag{6.17}$$

The annual rate of exceeding 0.2 g is thus $0.378 \times 0.01 = 0.00378$ per year. The return period of $1/0.00378 = 264$ years should be intuitive: the earthquake rupture occurs on average once every 100 years, and there is approximately a 1/3 chance that it causes an $SA(1\,s) > 0.2$ g, so $SA(1\,s) > 0.2$ g is observed approximately once every 300 years.

This quick example is provided to develop intuition regarding the core PSHA calculations and is consistent with the more formal equations presented above. Adopting Equation 6.1 to perform the calculation, the summation is replaced by a single multiplication of the rupture rate and the GMM probability (for the single potential rupture):

$$\lambda(SA(1\,s) > 0.2\,g) = P(SA(1\,s) > 0.2\,g \mid rup_1)\,\lambda(rup_1)$$

$$= 0.378 \times 0.01$$

$$= 0.00378. \tag{6.18}$$

Table 6.1 puts these numbers into a tabular form that will be useful for later, more complex calculations.

Repeating this calculation for another $SA(1\,s)$ value, we find the following for $SA(1\,s) > 0.5$ g:

$$P(SA(1\,s) > 0.5\,g \mid rup_1) = 1 - \Phi\left(\frac{\ln(0.5) - \ln(0.170)}{0.520}\right)$$

$$= 0.0191 \tag{6.19}$$

Substituting this into Equation 6.1, the annual rate of exceeding 0.5 g is

$$\lambda(SA(1\,s) > 0.5\,g) = 0.0191 \times 0.01 = 0.000191. \tag{6.20}$$

Repeating these calculations for many $SA(1\,s)$ values, one can construct the curve shown in Figure 6.4. This ground-motion hazard curve for $SA(1\,s)$ summarizes the rates of exceeding a variety of $SA(1\,s)$ values. The two rates computed explicitly above are labeled on this figure as well. Because both axes often cover several orders of magnitude, we often plot one or both axes of ground-motion hazard curves using a log scale.

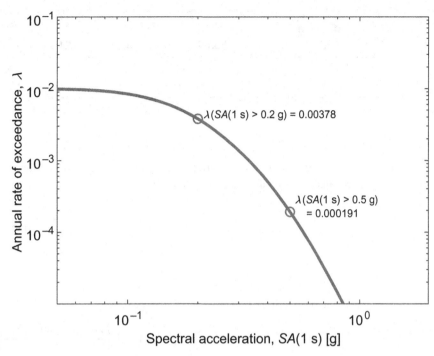

$\lambda(SA(1 \text{ s}) > 0.2 \text{ g}) = 0.00378$

$\lambda(SA(1 \text{ s}) > 0.5 \text{ g})$
$= 0.000191$

Fig. 6.4 $SA(1 \text{ s})$ hazard curve for the example of Section 6.3.1.

In this case, the hazard curve is merely the Gaussian complementary cumulative distribution function for $\ln SA(1 \text{ s})$ multiplied by the rate of earthquakes. Because of the logarithmic scaling of the plot, the normal distribution shape may not be recognizable, but this result should be clear from looking at the form of Equation 6.19.

6.3.2 Example: Two Rupture Scenarios

Let us slightly generalize the previous example by considering a fault capable of producing two rupture scenarios. The first, rup_1, is identical to the scenario in the immediately preceding example (i.e., $M = 6.5$, $R = 10$ km, $\lambda(rup_1) = 0.01$). The second, rup_2, is a magnitude 7.5 earthquake, also at a distance of 10 km, which occurs at a rate of $\lambda(rup_2) = 0.002$ times per year. We will continue using the Boore et al. model to predict $SA(1 \text{ s})$ for all ruptures.

The previous example quantified the hazard from rup_1, so let us focus on calculating the hazard from rup_2. We predict a median $SA(1 \text{ s})$ of 0.450 g if the rup_2 rupture occurs, and a logarithmic standard deviation of 0.520. Now, consider the two $SA(1 \text{ s})$ values considered in the previous example. The probability of $SA(1 \text{ s}) > 0.2$ g, given earthquake scenario rup_2, is

$$P(SA(1 \text{ s}) > 0.2 \text{ g} \mid rup_2) = 1 - \Phi\left(\frac{\ln(0.2) - \ln(0.450)}{0.520}\right)$$

$$= 0.940. \tag{6.21}$$

We can then multiply this probability by the rate of occurrence of rup_2, to get the rate of $SA(1 \text{ s}) > 0.2$ g due to rup_2 earthquakes. The PSHA formula of Equation 6.1 includes a summation over all

Table 6.2. Intermediate calculations to compute $\lambda(SA(1\,s) > 0.2\,g)$ for the example of Section 6.3.2

i	m_i	$\lambda(rup_i)$	$P(SA(1\,s) > 0.2\,g \mid rup_i)$	$P(SA(1\,s) > 0.2\,g \mid rup_i)\lambda(rup_i)$
1	6.5	0.01	0.378	0.00378
2	7.5	0.002	0.940	0.00188
				Sum = 0.00566

sources, so we add this rate to the corresponding rate for rup_1 to find the overall rate of $SA(1\,s) > 0.2\,g$ (the results are also tabulated in Table 6.2):

$$\lambda(SA(1\,s) > 0.2\,g) = \overbrace{0.378 \times 0.01}^{rup_1} + \overbrace{0.940 \times 0.002}^{rup_2}$$

$$= 0.00378 \quad + 0.00188$$

$$= 0.00566. \tag{6.22}$$

Similarly, the probability of $SA(1\,s) > 0.5\,g$, given rup_2 is

$$P(SA(1\,s) > 0.5\,g \mid rup_2) = 1 - \Phi\left(\frac{\ln(0.5) - \ln(0.450)}{0.520}\right)$$

$$= 0.419. \tag{6.23}$$

We then combine this result with our other information to compute the overall rate of $SA(1\,s) > 0.5\,g$:

$$\lambda(SA(1\,s) > 0.5\,g) = \overbrace{0.0191 \times 0.01}^{rup_1} + \overbrace{0.419 \times 0.002}^{rup_2}$$

$$= 0.000191 + 0.000838$$

$$= 0.00103. \tag{6.24}$$

The two rates computed in Equations 6.22 and 6.24 are plotted in Figure 6.5, along with rates for a range of other $SA(1\,s)$ values. Also shown in Figure 6.5 are hazard curves for the two individual rupture scenarios, and the intermediate rate calculations from Equations 6.22 and 6.24. These results have a few interesting features. First, the hazard curve for rup_1 in the figure is identical to the curve in Figure 6.4. We have simply added the additional rates of exceedance due to rup_2 to get the total exceedance rates from all ruptures. Second, the relative contributions of the two ruptures to the ground-motion hazard vary depending upon the $SA(1\,s)$ value of interest. At relatively lower $SA(1\,s)$ values, such as in the calculation of Equation 6.22, rup_1 contributes much more to the overall rate of exceedance. At larger $SA(1\,s)$ values, such as in the calculation of Equation 6.24, the rup_2 scenario becomes more likely to produce the $SA(1\,s)$ exceedance. This is because rup_1 has a low probability of causing large $SA(1\,s)$ values, even though rup_1 occurrences are much more frequent than the larger-magnitude earthquakes from rup_2. Specifically, for $SA(1\,s)$ values greater than approximately 0.3 g, rup_2 makes a greater contribution to the hazard than rup_1, even though $\lambda(rup_2) = 1/5\,\lambda(rup_1)$. This is a common situation in real-world PSHA calculations: frequent small-magnitude earthquakes generally dominate low-intensity ground motion. In contrast, high-intensity ground motion is relatively more likely to be caused primarily by rare, large-magnitude earthquakes.

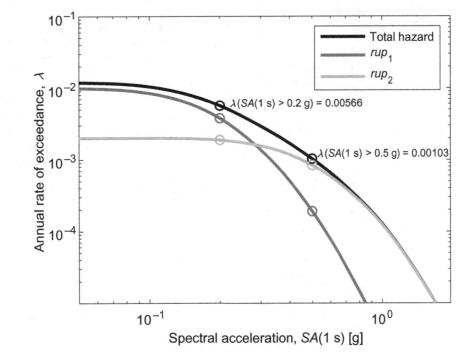

Fig. 6.5 *SA*(1 s) hazard curve for the example two-rupture case.

6.3.3 Example: Source with Gutenberg–Richter Magnitudes

In this example, we consider a source capable of producing earthquakes with a variety of magnitudes. The earthquake magnitudes follow the bounded Gutenberg–Richter model with $m_{min} = 5$, $m_{max} = 8$, $b = 1$, and an earthquake rate of $\lambda(M \geq 5) = 0.05$ events per year. Exceedance and occurrence rates for ruptures of various magnitudes, given this model, were computed in Section 3.5.5 and reported in Table 3.4. We again assume that all earthquakes occur at a distance of 10 km, and use the Boore et al. GMM from the previous examples to compute exceedance rate of an SA(1 s) value.

To compute the rate of exceeding some SA(1 s) value, we calculate the rates of observing various earthquake rupture scenarios, compute the probabilities of exceeding the SA(1 s) value given each rupture scenario, and then sum the products of the two terms over the range of ruptures. Table 6.3 shows those probabilities for calculations of $\lambda(SA(1 \text{ s}) > 0.2 \text{ g})$. For illustration, the potential magnitudes have been discretized into increments of 0.2, leading to 15 discrete earthquake magnitudes, each defining a unique rupture scenario. The first column enumerates the values of the index i. The second column lists the magnitude of each rupture (specifically, the midpoint of the magnitude range). The third column lists the rates of observing each rupture, taken from Table 3.4. The fourth column lists the probability of SA(1 s) > 0.2 g, given an occurrence of the rupture scenario with the given magnitude. The fifth column lists the products of the third and fourth columns. Equation 6.1 is a summation of the terms in the fifth column, so the rate of SA(1 s) > 0.2 g is $\lambda(SA(1 \text{ s}) > 0.2 \text{ g}) = 0.0027$.

We continue the hazard analysis by repeating the calculations of Table 6.3 for SA(1 s) > 0.5 g, and reporting results in Table 6.4. By repeating this calculation for many more SA(1 s) values, we can create the ground-motion hazard curve shown in Figure 6.6. The two individual rates of exceedance calculated above are labeled on this figure.

Table 6.3. Intermediate calculations to compute $\lambda(SA(1\,s) > 0.2\,g)$ for the example of Section 6.3.3

i	m_i	$\lambda(rup_i)$	$P(SA(1\,s) > 0.2\,g \mid rup_i)$	$P(SA(1\,s) > 0.2\,g \mid rup_i)\lambda(rup_i)$
1	5.1	0.0185	0.001	1.59×10^{-5}
2	5.3	0.0117	0.003	3.85×10^{-5}
3	5.5	0.0074	0.011	7.84×10^{-5}
4	5.7	0.0046	0.029	1.35×10^{-4}
5	5.9	0.0029	0.068	1.99×10^{-4}
6	6.1	0.0018	0.137	2.54×10^{-4}
7	6.3	0.0012	0.242	2.82×10^{-4}
8	6.5	7.35×10^{-4}	0.378	2.78×10^{-4}
9	6.7	4.64×10^{-4}	0.529	2.45×10^{-4}
10	6.9	2.93×10^{-4}	0.674	1.97×10^{-4}
11	7.1	1.85×10^{-4}	0.795	1.47×10^{-4}
12	7.3	1.17×10^{-4}	0.884	1.03×10^{-4}
13	7.5	7.35×10^{-5}	0.940	6.91×10^{-5}
14	7.7	4.64×10^{-5}	0.972	4.51×10^{-5}
15	7.9	2.93×10^{-5}	0.988	2.90×10^{-5}
				Sum $= 0.00212$

Table 6.4. Intermediate calculations to compute $\lambda(SA(1\,s) > 0.5\,g)$ for the example of Section 6.3.3

i	m_i	$\lambda(rup_i)$	$P(SA(1\,s) > 0.5\,g \mid rup_i)$	$P(SA(1\,s) > 0.5\,g \mid rup_i)\lambda(rup_i)$
1	5.1	0.0185	0.000	9.04×10^{-9}
2	5.3	0.0117	0.000	4.39×10^{-8}
3	5.5	0.0074	0.000	1.77×10^{-7}
4	5.7	0.0046	0.000	5.93×10^{-7}
5	5.9	0.0029	0.001	1.67×10^{-6}
6	6.1	0.0018	0.002	3.98×10^{-6}
7	6.3	0.0012	0.007	8.07×10^{-6}
8	6.5	7.35×10^{-4}	0.019	1.40×10^{-5}
9	6.7	4.64×10^{-4}	0.046	2.12×10^{-5}
10	6.9	2.93×10^{-4}	0.095	2.78×10^{-5}
11	7.1	1.85×10^{-4}	0.175	3.23×10^{-5}
12	7.3	1.17×10^{-4}	0.285	3.32×10^{-5}
13	7.5	7.35×10^{-5}	0.419	3.08×10^{-5}
14	7.7	4.64×10^{-5}	0.562	2.61×10^{-5}
15	7.9	2.93×10^{-5}	0.695	2.04×10^{-5}
				Sum $= 0.00022$

Comparing Table 6.3 with Table 6.4, we can make several observations. The first three columns of both tables are identical, as they are describing only the rupture properties and so are not affected by changes in the $SA(1\,s)$ value of interest. All probabilities in the fourth column are much larger in Table 6.3 than in Table 6.4: the $SA(1\,s)$ threshold was lower in Table 6.3, so the probability of exceeding the threshold is therefore higher.

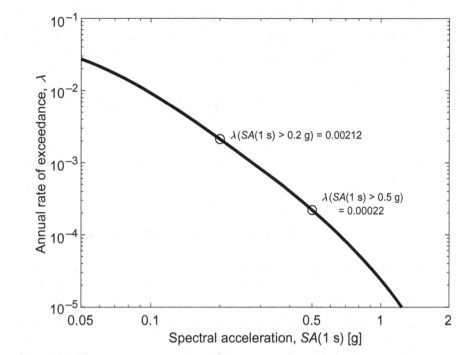

$\lambda(SA(1\ s) > 0.2\ g) = 0.00212$

$\lambda(SA(1\ s) > 0.5\ g)$
$= 0.00022$

Fig. 6.6   *SA*(1 s) hazard curve for the example site with one source and a Gutenberg-Richter magnitude distribution.

In Table 6.4, the probability of *SA*(1 s) > 0.5 g is effectively zero for the smallest magnitude considered. Therefore, considering even smaller-magnitudes would have no impact on the final answer because smaller magnitude earthquakes have effectively zero probability of causing an *SA*(1 s) > 0.5 g. In Table 6.3, however, even magnitude 5 earthquakes have a nonzero probability of causing *SA*(1 s) > 0.2 g; this is somewhat concerning because lower-magnitude earthquakes could also cause *SA*(1 s) > 0.2 g, so including them would have changed the final answer. This suggests that the choice of the minimum considered earthquake magnitude can be important in some cases. We will return to this issue in Section 6.5.

By looking at the right-hand column of these tables, we can also identify the magnitudes that make the most significant contribution to the probability of exceeding the *SA*(1 s) value of interest. Each number in this column is the rate of the given rupture causing an exceedance of the *SA*(1 s) value of interest. In Table 6.3, the rates in the right-hand column are largest for magnitudes in the middle of the range: these ruptures are more likely to occur than the largest-magnitude earthquakes (as seen in column 3), *and* because they have a reasonable probability of causing *SA*(1 s) > 0.2 g. In Table 6.4, on the other hand, the largest rates are associated with the largest-magnitude ruptures. Even though these ruptures are relatively rare, they are the only ones with significant probabilities of causing *SA*(1 s) > 0.5 g. We will revisit this information more quantitatively in Section 7.2.

For more complex sites than the simple cases shown in the above examples, the PSHA summations can quickly get lengthy. For this reason, PSHA is performed using computer software in all practical analysis cases. The software's purpose is to perform the calculations shown here, but for more complicated cases involving many earthquake sources, while also using modern complex GMMs. Keep in mind, however, that there is no conceptual change in the calculation. The bookkeeping is simply a bit more involved. Additionally, in the example above, we used a relatively wide magnitude spacing of 0.2 units in order to shorten Tables 6.3 and 6.4. However, when performing these calculations in a computer program, it is both appropriate and easy to use a finer discretization of the rupture scenarios of interest.

6.4 Hazard Curve Metrics

This section discusses several metrics that are useful for the evaluation of ground-motion hazard data. To illustrate, we use the numerical data from the example hazard curve of Figure 6.6, as reported in Table 6.5.

6.4.1 *IM* Value with a Given Exceedance Rate

In building codes and other standards, an engineer must often evaluate a structure's performance given an IM exceeded with some rate (e.g., the IM with a 1000-year return period). Finding this value involves finding the inverse of the function in Equation 6.1, which cannot be computed directly. Typically, we solve this problem by computing the exceedance rates for a variety of IM values and then interpolating the results.

For example, using Table 6.5, let us find the $SA(1\text{ s})$ value with a 10^{-3} annual exceedance rate. $SA(1\text{ s}) = 0.2$ g has too large of an exceedance rate, and $SA(1\text{ s}) = 0.3$ g has too small of an exceedance rate, so we interpolate between the two to find that $SA(1\text{ s}) = 0.268$ g has the desired rate. This result is annotated on Figure 6.7.

An Aside: Logarithmic Interpolation

Logarithmic interpolation is common for seismic hazard and risk data, so we provide an additional explanation here. To interpolate between two data points, $\{x_1, y_1\}$ and $\{x_2, y_2\}$, and find the value y

Table 6.5. Numerical values of the hazard curve shown in Figure 6.6	
$SA(1\text{ s})$ [g]	$\lambda(SA(1\text{ s}) > x)$
0.0	5.00×10^{-2}
0.1	9.20×10^{-3}
0.2	2.12×10^{-3}
0.3	8.19×10^{-4}
0.4	4.00×10^{-4}
0.5	2.20×10^{-4}
0.6	1.31×10^{-4}
0.7	8.15×10^{-5}
0.8	5.27×10^{-5}
0.9	3.50×10^{-5}
1.0	2.38×10^{-5}

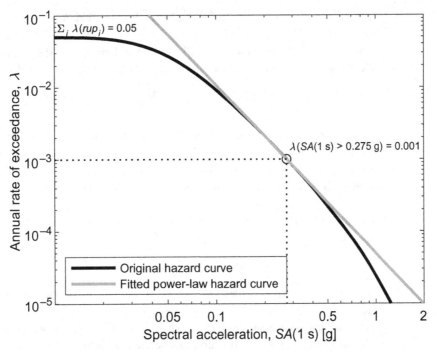

Fig. 6.7   Hazard curve for the example of Section 6.3.3, annotated with the metrics computed in Section 6.4. The plot has been extended to lower $SA(1\text{ s})$ values than other figures in this chapter, to illustrate the $\lambda(M \geq m_{\min})$ calculation.

corresponding to a given x (with $x_1 \leq x \leq x_2$), we can use the following formula, which is a simple linear interpolation applied to $\{\ln x, \ln y\}$ values:

$$\ln y = \ln y_1 + (\ln x - \ln x_1)\frac{\ln y_2 - \ln y_1}{\ln x_2 - \ln x_1}. \tag{6.25}$$

Using the property that $\ln a - \ln b = \ln(a/b)$, and exponentiating both sides, Equation 6.25 can be reformatted as follows for convenience:

$$y = y_1 \exp\left(\ln(x/x_1)\frac{\ln(y_2/y_1)}{\ln(x_2/x_1)}\right). \tag{6.26}$$

Taking the example of the $SA(1\text{ s})$ value with a 10^{-3} annual exceedance rate from the previous section, with the points $\{8.19 \times 10^{-4}, 0.3\}$ and $\{2.12 \times 10^{-3}, 0.2\}$, and evaluating for $x = 10^{-3}$, we compute

$$y = 0.3 \times \exp\left(\ln(10^{-3}/8.19 \times 10^{-4})\frac{\ln(0.2/0.3)}{\ln(2.12 \times 10^{-3}/8.19 \times 10^{-4})}\right) = 0.275. \tag{6.27}$$

Note that $\{x, y\}$ values interpolated in this way will plot as a straight line on a logarithmically scaled plot such as Figure 6.7.

6.4.2 Hazard Curve Derivative

Later calculations in Chapters 8 and 9 will require the derivative of the ground-motion hazard curve. Specifically, the absolute value of the derivative with respect to im is needed:

$$\left| \frac{d\lambda(IM > x)}{dx} \right| = \left| \frac{\lambda(IM > x + dx) - \lambda(IM > x)}{dx} \right| \tag{6.28}$$

where dx is an infinitesimal increment of x. The term $|d\lambda(IM > x)/dx|$ can be described as a rate density (Bazzurro and Cornell, 2002). The operation is analogous to taking the derivative of a cumulative distribution function to obtain a probability density function, though here it is for rates rather than probabilities, and the absolute value is required because the hazard curve rates are for rates of $IM > x$ rather than rates of $IM < x$.

Noting that $|\lambda(IM > x + dx) - \lambda(IM > x)| = \lambda(x < IM \leq x + dx)$, and cancelling the two instances of dx, Equation 6.28 is often written simply as

$$|d\lambda(IM > x)| = \lambda(x < IM \leq x + dx). \tag{6.29}$$

In words, $|d\lambda(IM > x)|$ provides the occurrence rate of ground motions with IM between x and $x + dx$.

Often, a discrete approximation is made when calculations are performed numerically. This is done by taking differences of the hazard curve at discrete IM values:

$$\Delta\lambda_i = \lambda(IM > x_i) - \lambda(IM > x_{i+1}) \tag{6.30}$$

where $x_1, x_2, \ldots, x_n$ are the discrete IM values of interest, ordered from smallest to largest, and we define $\lambda(IM > x_{n+1}) = 0$ so that the difference can be evaluated for x_n.

For the hazard curve data of Table 6.5, $\Delta\lambda_1 = 0.05 - 0.0092 = 0.0408$, $\Delta\lambda_2 = 0.0092 - 0.00212 = 0.00708$, etc. These values are plotted in Figure 6.8a. The same hazard curve, differentiated using finer discrete intervals, is plotted in Figure 6.8b.

6.4.3 Hazard Curve Slope Parameter

Because power-law functions can locally approximate typical hazard curves, we estimate a logarithmic slope by computing the value of k in the following equation:

$$\lambda(IM > im) = k_0 im^{-k} \tag{6.31}$$

where k_0 and k are constants to be computed. The slope parameter k indicates the relative frequencies of smaller- and larger-amplitude ground motions and is a useful metric in some risk calculations (Chapter 9).

A convenient way to solve for k is to take the logarithm of both sides of Equation 6.31:

$$\ln\lambda(IM > im) = \ln k_0 - k(\ln im) \tag{6.32}$$

and then select two values—$\{im_1, \lambda_1\}$ and $\{im_2, \lambda_2\}$—from the hazard curve to perform a fit. Substituting each pair of values into Equation 6.32, subtracting the two equations, and solving for k gives

$$k = -\frac{\ln(\lambda_1/\lambda_2)}{\ln(im_1/im_2)}. \tag{6.33}$$

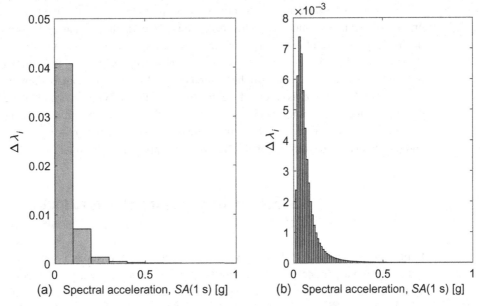

Fig. 6.8 Discretized hazard curve differences using Equation 6.30 and the data of Table 6.5. (a) Results when taking $SA(1\,s)$ increments of 0.1 g. (b) Results when taking $SA(1\,s)$ increments of 0.01 g, using an interpolated hazard curve. Each plotted bar spans the associated interval of $SA(1\,s)$ values.

Using the results for $SA(1\,s) = 0.2$ g and $SA(1\,s) = 0.3$ g in Table 6.5, Equation 6.33 gives $k = 2.35$. Substituting either pair of the $\{im, \lambda\}$ values back into Equation 6.32 gives $k_0 = 4.86 \times 10^{-5}$, though that numerical value is generally of less interest.

Figure 6.7 includes a plot of this estimated power-law hazard curve. Note that it passes through the $SA(1\,s) = 0.2$ g and $SA(1\,s) = 0.3$ g points used for fitting. Note also that the slope of the real hazard curve changes with $SA(1\,s)$ values (and displays the typically seen result that the hazard curve is steeper for larger IM).

There is a potential here for confusion relative to Section 6.4.2, which defined a derivative. While a derivative of a function is often noted as the slope of that function, the slope parameter k is related to a specific function approximation. It is not equal to the derivative of the function. This k parameter is often referred to as a slope parameter in practical seismic hazard studies, so we retain that terminology in this section. However, be careful to distinguish between the distinct concepts of a hazard curve derivative and a hazard curve "slope" parameter k.

6.4.4 Rate of Earthquakes

One simple diagnostic from a hazard curve plot is that the intercept of the hazard curve is equal to the rate of all earthquake ruptures considered in the SSM. This result can be shown using Equation 6.1: $P(IM \geq 0 \mid rup_i, site) = 1$ for all rup_i, so in this special case Equation 6.1 simplifies to

$$\lambda(IM \geq 0) = \sum_{i=1}^{n_{rup}} \lambda\,(rup_i). \tag{6.34}$$

We can see this in Figure 6.7, where the y intercept of the hazard curve is equal to 0.05, the rate of $M \geq m_{\min}$ earthquakes assumed in the example calculation. In this simple example, we knew

$\lambda(M \geq m_{\min})$ because it was specified in the problem we just solved. However, for practical problems with many sources, or when reviewing a hazard curve produced by someone else, viewing the intercept is a useful quick check on the reasonableness of this aspect of the SSM.

When the the hazard curve is plotted in log scale, the $im = 0$ point will not be plotted, but the earthquake rate can still be estimated by looking at a small enough im value so that $P(IM > im \mid rup_i, site)$ is essentially equal to 1 for all rup_i. We can verify that we have a sufficiently small im value to perform this check by checking that the hazard curve is flat near the y-intercept. This indicates that further reductions in im are not substantively increasing $P(IM > im \mid rup_i, site)$, and thus indicates that those probabilities must be approximately equal to 1.

6.5 Sensitivity of Hazard Results to Inputs

Repeatedly solving the PSHA equation with varying input values is one way to understand the impact of input parameters on the resulting ground-motion hazard. This section provides several such examples, with comments indicating often-observed features from these calculations.

6.5.1 Seismicity Rate

Figure 6.9 shows calculations for the site discussed in Section 6.3.3, but with the occurrence rates for all ruptures multiplied by a constant. From Equation 6.1, we see that a multiplicative factor on $\lambda(rup_i)$ will translate directly into the same multiplicative factor on $\lambda(IM > im)$. Figure 6.9 illustrates this relationship: the hazard curves scale up and down by a fixed factor in proportion to the scaling on $\lambda(M \geq m_{\min})$.

6.5.2 Minimum Magnitude

Figure 6.10 shows calculations for the site discussed in Section 6.3.3, but with minimum considered magnitude varied. In this case, $\lambda(rup_i)$ is set to 0 for all ruptures with $M < m_{\min}$, and is unchanged for all ruptures with $M \geq m_{\min}$. Increasing the $m_{\min}$ threshold both decreases the exceedance rate of the smallest $SA(1\ s)$ values and increases the range of $SA(1\ s)$ values whose rates are affected. Keep in mind that minimum magnitudes are not a physical phenomenon, as smaller-magnitude events will undoubtedly happen. Instead, they are a choice of numerical convenience, under the assumption that the omitted small-magnitude events will not affect the resulting hazard and risk calculations (Bommer and Crowley, 2017). For all input parameters, including magnitudes and distances, it is numerically helpful to omit consideration of cases with an inconsequential contribution to seismic hazard and, ultimately, risk. Results such as Figure 6.10 can be used to evaluate the suitability of chosen limits on these parameters.

6.5.3 Maximum Magnitude

Consider the example of Section 6.3.3, which had a Gutenberg–Richter distribution of earthquake magnitudes, and now consider alternate values for the maximum magnitude from this source. Table 3.6 provided $\lambda(rup_i)$ results for the cases $m_{\max} = 7.6, 7.8$, and 8.0.

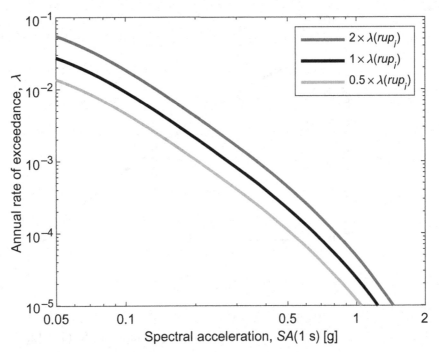

Fig. 6.9   Hazard curves for the example of Section 6.3.3, when $\lambda(rup_i)$ is multiplied by a constant for all ruptures.

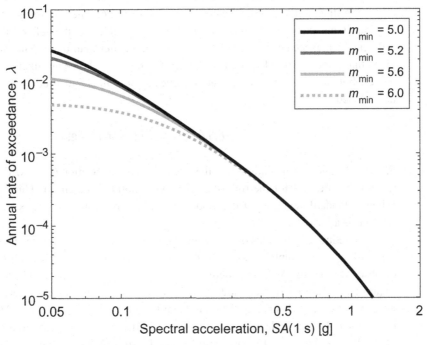

Fig. 6.10   Hazard curves for the example of Section 6.3.3, for varying choices of m_{min}.

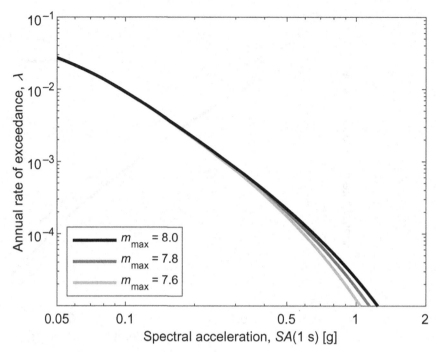

Fig. 6.11   Hazard curves for the example of Section 6.3.3, for varying choices of m_{max}.

Figure 6.11 shows the hazard curves from using the GR distribution with each m_{max} value. The $m_{max} = 8$ line in the figure is equivalent to Figure 6.6. These choices of m_{max} have effectively no impact on exceedance rates of small $SA(1\ s)$ values at the left of the figure but do cause differences in exceedance rates of $SA(1\ s)$ values greater than 0.3 g. This is intuitively reasonable, given that large magnitude ruptures affected by the m_{max} choice are likely to produce large $SA(1\ s)$ values, but the frequent smaller rupture events most likely to produce small $SA(1\ s)$ values are not affected.

6.5.4 Ground-Motion Model σ

Figure 6.12 shows calculations for the site discussed in Section 6.3.3, with σ in the GMM varied (refer to Section 4.4 for the role of σ in ground-motion prediction). This is done by replacing the original standard deviation in the model, σ, with a new value $\sigma' = x \times \sigma$, where x is a factor by which σ is adjusted.

As we reduce σ', the rates of exceeding large $SA(1\ s)$ values decrease. This is because the smaller standard deviation implies less variability in $SA(1\ s)$ for a given rup, and thus a lower probability of observing a large $SA(1\ s)$. There is less discrepancy for small $SA(1\ s)$ values, and in fact, the hazard curves are nearly equivalent for $SA(1\ s) > 0.05$ g in this case; recall that for $SA(1\ s) > 0$ we expect the hazard curves to be equivalent for all cases and equal to $\lambda(M > m_{min})$, as discussed in Section 6.4.4. But at large $SA(1\ s)$ values the difference is substantial: a factor of five in the rate of $SA(1\ s) > 0.6$ g for the base case versus the (unrealistically extreme) case $\sigma' = 0.1 \times \sigma$. At small $SA(1\ s)$ values, the trend will reverse, and small σ values can produce higher exceedance rates

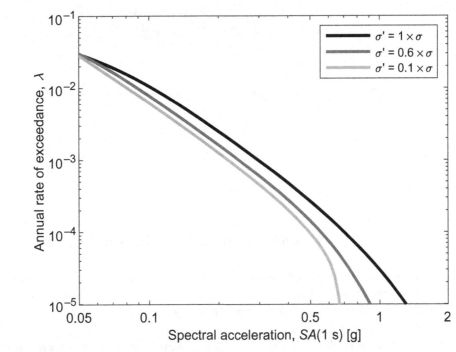

Fig. 6.12 Hazard curve for the example of Section 6.3.3, for varying values of σ in the GMM.

(there is some hint of this cross-over near 0.1 g in Figure 6.12), but this is rarely of any engineering relevance and the trend for large $SA(1\ s)$ values is much more critical. This sensitivity to σ is typical and motivates studies to reduce σ, as will be discussed further in Chapter 8.

6.6 Model Uncertainty

The PSHA integral relies on models for earthquake rupture occurrence and resulting ground-motion intensities, both of which are subject to uncertainty due to our lack of knowledge about the exact processes that govern each. Given this, it should be clear that the above rate-of-exceedance results are *conditional upon the adopted input models*. Recognizing that those inputs are models and not reality, the previous calculations cannot be considered complete in the sense of reflecting all uncertainties about future ground motions.

We now consider methods for treatment of that uncertainty. The goal here is to consider both the earthquake process and our predictive models to be represented by probabilities rather than deterministic statements. Ideally, we enumerate all possible models and assign them degrees of belief, under the assumption that the models are mutually exclusive and collectively exhaustive. Then we can consider those alternative models within the general probabilistic framework we have established. It can be quite challenging to enumerate possible models, and even more challenging to assign them probabilities, but the difficulty of this task does not prevent the effort from being beneficial.

Table 6.6. Probabilities associated with die roles, for two models		
	Probability	
Value	Model 1	Model 2
1	1/6	0
2	1/6	1/4
3	1/6	1/2
4	1/6	1/4
5	1/6	0
6	1/6	0

6.6.1 Motivation: Die with Unknown Properties

Let us consider an example to build intuition regarding model uncertainty (this example is adapted from Norman Abrahamson). We are interested in building a model for the outcome of rolling a die. We have not observed the die, but we know that the outcomes of the previous four rolls of the die were 2, 3, 3, and 4.

One model for this die is to assume that it is like other dice that we have previously observed. Most dice are six-sided, with probability 1/6 of observing the numbers 1 through 6. The data observed in this case are consistent with such a die, so this is a plausible model.

A second model is to assume that this is a four-sided die, or that it is a loaded die such that 2, 3, and 4 are the only possible outcomes of a roll. For argument's sake, let us assume it is a loaded die and that numbers come up exactly in proportion to what we have observed in the previous four rolls. Table 6.6 shows an example set of probabilities for Model 1 (it is a typical six-sided die) and Model 2 (only 2, 3, and 4 can be rolled, with probabilities consistent with past observations). Although Model 2 is more consistent with the observed data, Model 1 is more broadly relevant, and judgment suggests that it may be applicable here—similar situations often arise in seismic hazard studies where limited data are available.

Now, we are offered an opportunity to gamble on the outcome of the next roll of the die, so we need to estimate probabilities of various outcomes to decide whether the gambling odds are appealing. Let us assume for simplicity that the two models in Table 6.6 are the only plausible models for this die and that each is equally likely to be correct, so each is assigned a weight of 0.5 (we require these weights to sum to 1). Then we could estimate the probability of a given outcome, $Event$, occurring to be the probability of $Event$ given a model, multiplied by the weight for that model, and summed over all possible models:

$$P(Event) = \sum_{k=1}^{n_{models}} P(Event \mid \text{model } k)w_k \tag{6.35}$$

where n_{models} is the number of models and w_k is the weight associated with model k.

For the die example of Table 6.6, we can compute the probability of rolling a 3 as

$$P(3) = (1/6)(0.5) + (1/2)(0.5) = 1/3. \tag{6.36}$$

This probability is higher than the 1/6 probability associated with a fair six-sided die, and lower than the 1/2 probability implied by the four prior observations. In this calculation, we have accounted for the random outcomes of rolling a die with known properties (i.e., the probabilities in a given column of Table 6.6), as well as uncertainty about the properties of the die (i.e., the 0.5 weight we gave to each model). While the two sources of uncertainty differ in nature, they both reflect uncertainty about future outcomes.

The analogy between this example and modeling future earthquake consequences using recent observations is clear, it is hoped. The PSHA calculations earlier in this chapter produced probabilities reflecting random outcomes of a process, but they are model-dependent. When formulating models for earthquake occurrence and ground-motion intensity, we can rely on a general understanding of earthquakes (similar to Model 1 in this example that considered "typical" dice) and on site-specific observational data (similar to Model 2 that matched prior observations). In reality, there are many more than two plausible models, which in turn reflect a theoretically infinite number of model permutations.

Finally, we note that additional data would reduce the uncertainty regarding the appropriate model. With a few more observed outcomes, it would likely be possible to determine whether Model 1 or Model 2 is correct (e.g., observation of a 6 would confirm that Model 1 is correct, if these are the only two possible models). Moreover, if the die is in fact fair, then with more data, an empirical-based model like Model 2 will converge toward Model 1. However, no amount of additional data would reduce uncertainty about the outcome of an individual die roll given a model. For example, even if we knew with certainty that Model 1 was correct, additional data would not influence the estimated probability of 1/6 associated with any specific value on the die. This distinction between reducible versus irreducible uncertainty can be important in hazard and risk calculations, as discussed below.

6.6.2 Epistemic Uncertainty and Aleatory Variability

Recall from Section 1.7 that we use the terms *epistemic* and *aleatory* to distinguish the two types of uncertainty. Epistemic uncertainty refers to (potentially reducible) lack of knowledge about models for the state of nature. Aleatory variability refers to random outcomes from natural variability in a process (for a given model formulation of the process).

If a parameter sometimes has one value and sometimes has another (e.g., the outcome from a roll of a die, or the IM value observed from earthquakes described by a set of rupture parameters), it has aleatory variability. If a parameter either has one value or another, and we do not know which value is correct (e.g., the probability that a die roll will be a 3 in the previous example, or the m_{max} for a seismic source), then the parameter has epistemic uncertainty.

While the distinction between these two is straightforward in many cases, in hazard calculations the distinction can be less clear (or is dependent upon model formulation). For example, much of GMM uncertainty, which we typically treat as aleatory variability, can be viewed as epistemic under some models. More discussion of this is presented in Chapter 8. Further, some conceptually epistemic uncertainties (e.g., the influence of an underlying sedimentary basin on ground motions at a given location) may not be reducible within a reasonable amount of time and are hence treated as aleatory. Finally, when we compute mean hazard curves, the classification of a given uncertainty often has no impact on numerical results or decisions (as discussed further below).

The distinction of epistemic versus aleatory is not always clear-cut and proper classification may not be necessary. The necessary issues to address are (1) to consider all uncertainties, (2) to not double-count uncertainties, and (3) to understand what sources of uncertainty are potentially reducible.

6.7 Logic Trees

With the Section 6.6 motivation in mind, we now revisit the basic PSHA equation to incorporate model uncertainty. Rather than compute a single ground-motion hazard curve, we will compute a family of hazard curves, each with its own set of underlying models. A ubiquitous tool to achieve this is the *logic tree* (Kulkarni et al., 1984), the structure of which is shown in Figure 6.13.

6.7.1 Logic-Tree Structure

The tree is organized into columns, each of which addresses one uncertain aspect of the overall hazard model. Typically a label above the column defines the model aspect associated with the branches. In Figure 6.13 these are labeled "Model Component A," "Model Component B" and "Model Component C." In each column, a discrete set of branches emerges from that column's node, indicating possible models (or possible parameters of a model). In Figure 6.13 these are labeled "Model A1," "Model A2," etc. Most PSHA calculations[3] use a discrete set of branches, which may be a discretization of a continuous distribution (e.g., for m_{max}).

If model aspects in two columns are dependent, the right-hand column has branches conditional on the left-hand branches to which it connects. For example, if column A has branches for a fault-based source and an areal source, the column B branches for the fault-based source could be parameters for

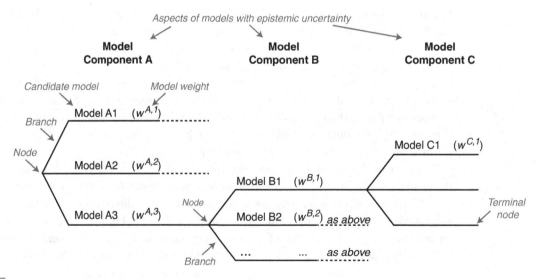

Fig. 6.13   Structure of a logic tree, with key components labeled.

---

[3] If Monte Carlo techniques are used to perform the PSHA calculation, then the parameter can be continuous, and samples drawn from the conditional distribution during the calculation stage. This approach is not common and so is not discussed further here; Section 6.9 provides further discussion.

that model (e.g., fault dips and segmentation). In contrast, the column B branches for the areal source would differ (e.g., the spatial smoothing kernel). For this reason, the logic tree typically has SSM aspects first (i.e., to the left of the tree) so that GMM aspects are conditional on the SSM aspects. If model aspects in two columns are independent, the branches are typically listed for only one case, with other branches labeled "as above" or "as below."

Traversing the logic tree from the left through a set of branches to the right provides a complete set of models and model parameters required to perform a PSHA calculation. One complete set of models associated with a traversal of the tree is denoted a *branch*, and the node at the right-hand-side termination is denoted a terminal node.

6.7.2 Logic-Tree Weights

Each branch of the logic tree has an associated weight, as shown in Figure 6.13. In each column, weights are assigned to each branch, and the weights of branches connecting to a common parent node sum to 1. The weights assigned to branches with a common parent node are conditional upon the models to the left of that parent node. For example, the weights $w^{B,1}$ and $w^{B,2}$ in Figure 6.13 would be weights conditional on Model A3. If a given model parameter is independent of the parent branch, no distinct conditional branches and weights need to be specified.

The weight associated with a complete set of branches (i.e., a complete path transversing from left to right through the tree) is the product of weights associated with each branch in the set. For example, the weight associated with the combination of models A1, B1, and C1 in Figure 6.13 is $w^{A,1} \times w^{B,1} \times w^{C,1}$.

Because the sum of weights for each set of branches sharing a parent node sums to 1, the weights associated with all terminal nodes will also sum to 1. Collectively, the terminal branches and weights specify a discrete probability distribution of epistemic uncertainties on the model set (Scherbaum and Kuehn, 2011). The actual distribution of epistemic uncertainties is likely continuous, but this discrete approximation is more convenient, in terms of both specifying models and numerically evaluating hazard. When we view these model weights as probabilities representing epistemic uncertainties, these requirements and results should make sense: (1) the model weights associated with a specific set of branches sharing a common parent node represent a discrete probability distribution for the associated uncertain parameter (conditional on the parameters specified by the parent node), and so should have probabilities that sum to 1, and (2) the weights associated with terminal nodes represent the joint probability distribution for the full family of uncertain model aspects. Again, the probabilities associated with outcomes of this distribution sum to 1.

6.7.3 Specifying Logic Trees

A PSHA calculation requires specifying the epistemic uncertainty in models and model parameters, via a set of logic-tree branches and associated weights. For a given analysis, satisfying that goal requires problem-specific knowledge and knowledge that is evolving over time. Thus, it is not possible to "solve" this issue in a textbook format, but general principles and approaches to this problem can be discussed. As an example of a real-world calculation, Figure 6.14 shows the US Geological Survey's logic tree for fault-based seismicity in California. For SSMs, epistemic uncertainty is typically quantified using a set of alternative models (e.g., alternate geometries for areal sources or alternate segmentation rules for faults), as well as distributions for each model's parameters (e.g., m_{max} or the Gutenberg–Richter b value). In Figure 6.14, the left portion of the figure lists a range of such alternative models (subgrouped as "Fault models," "Deformation models," and "Earthquake-rate models").

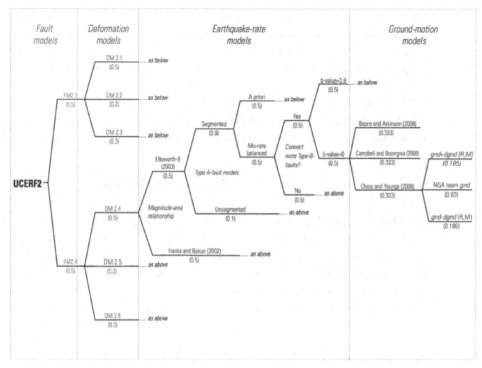

Fig. 6.14   US Geological Survey logic tree for fault-based seismicity in California (Petersen et al., 2008).

For ground-motion prediction, epistemic uncertainty is typically quantified using a set of alternative GMMs. The alternatives can be a set of distinct models and/or modifications to a baseline GMM that reflect model uncertainty (e.g., Atkinson et al., 2014). The use of distinct models allows the analyst to capture alternate views of how IM scales with rupture parameters. However, especially when the distinct models were calibrated from similar datasets or with similar model formulations, a set of published models is unlikely to represent the complete epistemic uncertainty associated with ground-motion prediction. One flexible way to represent epistemic uncertainty is to use a GMM with a perturbation added to all predictions (i.e., scaling the model up and down), or to one aspect of the model (e.g., magnitude scaling). Both approaches are used in the right portion of Figure 6.14, labeled "Ground-motion models." The three branches of the left column indicate three distinct models. The three branches in the right column indicate the use of the as-published model ("NGA team gnd," which receives the largest weight), and alternatives where the model is perturbed up or down by a factor. The factor was calibrated to reflect uncertainty in ground-motion prediction that was not captured by the suite of three as-published models (Rezaeian et al., 2014).

The logic-tree branches represent a (discrete representation of a) probability distribution on some model aspect with epistemic uncertainty. As such, it is important to keep in mind that weights are a collective set of probabilities used to represent a complete distribution, rather than a ranking of how well a particular branch agrees with data. One set of guidelines in this regard states that the goal is to specify a set of models and weights to represent the center, body, and range of technically defensible interpretations for these model aspects (SSHAC, 1997; Hanks et al., 2009; NRC, 2012, 2018).

A number of other documents provide more detail and discussion on use of logic trees in PSHA (Bommer and Scherbaum, 2008; Scherbaum and Kuehn, 2011; Musson, 2012b).

Quantifying epistemic uncertainty by specifying alternative models and assigning weights to the models can be a nonintuitive exercise for many analysts, but one that has been studied in seismic hazard and other decision-making contexts. In general, risk analysis and technical decision-making exercises, expert elicitation, and expert ranking systems are often used to quantify lack of knowledge and produce weights on alternative model formulations. Cooke's method (Cooke, 1991; Aspinall, 2010) is one popular approach, which asks experts a set of seed questions (with known answers) to evaluate their proficiency to accurately assess relevant uncertainties. Their proficiency is then used to weight their assessments of (unknown) model aspects. This approach has the benefit of being widely used outside of seismic hazard and risk, so its performance is relatively well understood. A key drawback is that it is quite difficult to develop seed questions that are relevant and informative for issues related to seismic hazards. A more common approach in PSHA is to use a deliberative approach where experts collectively develop epistemic uncertainty characterizations (SSHAC, 1997). With this approach, a team consisting of facilitators, subject-matter experts, peer reviewers, and others work in specific roles to develop the characterization. The interactions among experts risk suppressing uncertainty if not well managed, but the structure of the approach is designed to prevent this. Importantly, this deliberative approach has the advantage of allowing the advancement of science as part of the project (i.e., new models can be developed and integrated, in response to identified gaps in knowledge), rather than considering only existing models. This latter approach is the most common in current PSHA studies with the resources to commission such a team.

A common challenge in building logic trees is a discomfort by analysts with considering all defensible models (including those that are plausible even if unlikely), rather than advocating for a preferred model. Another challenge is that quantifying uncertainties is not an area where many scientists have formal training or intuition. When facing these challenges, it is helpful to keep in mind that epistemic uncertainties exist regardless of whether they are enumerated. Enumerating and parameterizing them allows us to utilize the tools of probability to understand and deal with them, rather than leave them implicit and remain ignorant of their implications (Apostolakis, 1990; Rougier et al., 2013).

6.7.4 Logic Tree Alternatives

Other approaches exist for modeling of epistemic uncertainty in technical evaluations of uncertain phenomena (Rougier et al., 2013). In terms of quantifying alternative models or model parameters, tools used in other fields include ensemble forecasts and Bayesian approaches (e.g., Bauer et al., 2015; Roselli et al., 2016). Moving away from traditional probability representations, imprecise probabilities or Dempster–Shafer approaches can be used to quantify lack of knowledge. However, the reader looking for simpler or more transparent alternatives than logic trees should be cautious in exploring these alternatives. Rougier et al. (2013) note that "the properties of such approaches are less transparent than those of probability, and any type of calculation can rapidly become extremely technical. It is emphatically not the case that a more general uncertainty calculus leads to a more understandable or easier calculation."

Without entering an extended discussion, we note that none of these alternatives is a decidedly superior approach for assessing epistemic uncertainty in PSHA. Logic trees are ubiquitous in real-world applications of PSHA and show no sign at present of becoming obsolete.

6.8 PSHA with Epistemic Uncertainty

With a logic tree specified for a given analysis problem, we can incorporate epistemic uncertainties into a ground-motion hazard calculation. Each terminal node k of the logic tree represents a complete specification of required input models, so an associated ground-motion hazard curve can be computed using Equation 6.1:

$$\lambda^k(IM > im) = \sum_{i=1}^{n_{rup}} P^k(IM > im \mid rup_i, site)\, \lambda^k(rup_i) \qquad (6.37)$$

where $\lambda^k(rup_i)$ quantifies the rate of rupture i, given the SSMs in logic-tree branch k, and $P^k(IM > im \mid rup_i, site)$ quantifies the probability of $IM > im$ (given rup_i and $site$) using the GMM of logic-tree branch k. In general, the rupture scenarios (rup_i), the site definition ($site$), and the number of ruptures (n_{rup}) can vary with k, but we suppress the indexing of these parameters for brevity. This calculation is then repeated for each terminal node of the logic tree.

Further, the set of rate estimates $\lambda^k(IM > im)$ and their associated weights, w^k, specify a probability distribution for the rate of $IM > im$, which can be processed using typical probability tools. Most commonly, the mean rate (mean hazard curve) is computed as follows[4]:

$$E[\lambda(IM > im)] = \sum_{k=1}^{n_{models}} \lambda^k(IM > im)w^k \qquad (6.38)$$

where $E[\ ]$ denotes an expectation, $\lambda^k(IM > im)$ was computed for each logic-tree branch using Equation 6.37, and w^k is the logic-tree weight for branch k.

Hazard curves can also be computed for fractile values of the rates of exceedance. First, define the cumulative distribution function for the rates of exceedance as

$$F_\lambda(l) = \sum_{\substack{k \text{ such that} \\ \lambda^k(IM>im)\le l}} w^k. \qquad (6.39)$$

The fth fractile hazard curve value, $\lambda^f(IM > im)$, can then be computed as

$$\lambda^f(IM > im) = F_\lambda^{-1}(f) \qquad (6.40)$$

where f is the fractile of interest (e.g., $f = 0.5$ for the median hazard curve), and $F_\lambda^{-1}(f)$ is the inverse of the CDF defined by Equation 6.39. In words, the fth fractile exceedance rate for $IM > im$ is the rate λ^* such that a fraction f of the logic-tree weights are associated with branches having rates less than or equal to λ^* (and $1 - f$ have rates greater than λ^*).

The availability of Equations 6.38 and 6.40 means that a choice must be made of how to report hazard results. From a practical perspective, the mean hazard curve of Equation 6.38 is the nearly universal choice for summarizing the results as a single hazard curve. It is also the required result in many codes and regulatory documents. A final practical consideration is that fractiles depend on

---

[4] There is some risk here of ambiguity in terminology. The hazard curve for a particular logic-tree branch, Equation 6.37, computes a mean rate of exceedance, in the sense that it is the long-term, or average, rate of IM exceedances; the actual number of exceedances in a given time period is still random, due to aleatory variability. Equation 6.38 then computes a mean value of those rates when considering epistemic uncertainty, so to be precise, it could be termed the "mean estimate of the mean rate of exceedance." To avoid this cumbersome phrasing we adopt the convention that the long-term average rate of Equation 6.37 is a "rate of exceedance" and Equation 6.38 is a "mean rate of exceedance" (McGuire et al., 2005b).

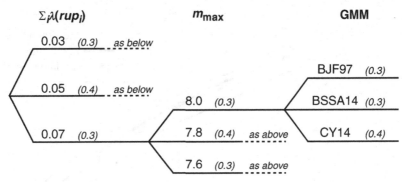

Fig. 6.15 Example logic tree, with branches for overall rupture rates (($\Sigma_i \lambda(rup_i)$)), m_{max}, and the GMM. BJF97 is Boore et al. (1997), BSSA14 is Boore et al. (2014), and CY14 is Chiou and Youngs (2014).

the partitioning of epistemic and aleatory uncertainties (Section 6.6.2), while the mean hazard is less sensitive to this partitioning. Related to this partitioning, for validation exercises (discussed in Chapter 12.3.2), it is necessary to keep aleatory and epistemic contributions distinct, and so fractiles or individual branch hazard curves are needed.

From a theoretical perspective, the above aggregations are done "vertically" in hazard-curve space rather than "horizontally" because the weights serve as probabilities on the logic-tree branches, and hence upon λ values rather than *im* values. We also note that from a decision theory perspective, there are reasons to consider the mean hazard curve as *the* hazard curve. The full suite of hazard curves (represented by fractiles) may help those interested in additional characterization of uncertainties. More detailed discussion of the use and interpretation of mean and fractile hazard curves is provided elsewhere (Abrahamson and Bommer, 2005; McGuire et al., 2005b; Musson, 2005; Bommer, 2012).

6.8.1 Example: Single Source with Gutenberg–Richter Magnitudes

Consider again the example of Section 6.3.3, but now with uncertainties on some of the parameters, as illustrated in Figure 6.15. The overall rate of ruptures ($\Sigma_i \lambda(rup_i)$), m_{max}, and the GMM are considered uncertain.

Figure 6.16 shows the $3 \times 3 \times 3 = 27$ hazard curves associated with the logic tree of Figure 6.15. The mean and several fractile curves are also shown. A few trends are typical of such plots and are discussed here to build intuition. First, note by reference to Equations 6.38 and 6.40 that the means and fractiles are computed with respect to rates and not *im* values. That is, Figure 6.16 shows "vertical" means and fractiles. Second, the median (i.e., 0.5 fractile) hazard curve at a given *im* value is the point at which half of the logic-tree weights are associated with hazard curves above the median curve (and the other half are below). The fractile hazard curves can usually be approximately determined visually, especially when logic-tree weights are somewhat uniform. Because the logic-tree weights are not visually apparent in a plot like Figure 6.16, hazard curves with very uneven weights complicate visual interpretation.

To better visualize the rate distribution, considering weights on the branches, Figure 6.17 shows the cumulative distribution function for $\lambda(SA(1\ s) > 0.4\ g)$, using the same data as Figure 6.16. The rates are also plotted in linear scale here.

An observation from Figures 6.16 and 6.17 is that the mean hazard curve is above the median curve. This is frequently observed because the $\lambda^k(IM > im)$ distribution is skewed toward high

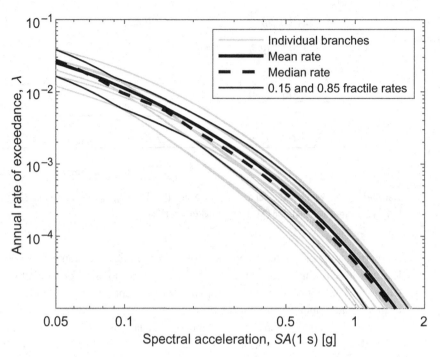

Fig. 6.16 A set of hazard curves associated with branches of a logic tree, with mean and fractile hazard curves noted.

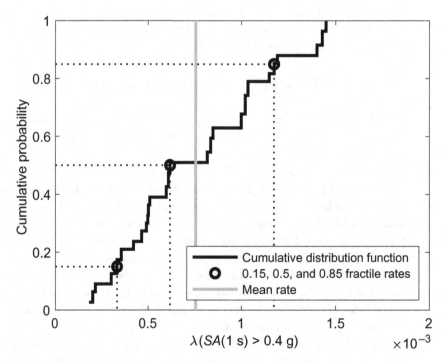

Fig. 6.17 Cumulative distribution function for $\lambda(SA(1\,s) > 0.4\,g)$, using the illustrative data of Figure 6.16.

rates, and for such skewed distributions, the mean is larger than the median. The location of the mean hazard curve may also be in a somewhat nonintuitive location at first glance (e.g., at the right-hand side of Figure 6.16 it is above most individual hazard curves). The visual effect results from plotting the vertical axis of the hazard curve figure in logarithmic scale; this emphasizes the low probabilities of interest to us, but the arithmetic mean of a set of numbers plotted in logarithmic scale is no longer at their visual centroid. Figure 6.17 thus provides a useful complementary visual representation.

6.9 Monte Carlo PSHA

The PSHA equations above are usually computed via direct numerical evaluation, but they are also amenable to Monte Carlo simulation (Ebel and Kafka, 1999; Musson, 2000, 2012a; Assatourians and Atkinson, 2013). In the most basic situation, the steps in this process are as follows:

1. Specify the ground-motion IM of interest.
2. Specify the site properties that help predict ground shaking.
3. Step through all possible rupture scenarios (rup_i) capable of causing ground-motion intensities of interest.
4. For each rup_i, sample (using Monte Carlo simulation) n_{gm} realizations from the IM distribution specified by the GMM.

We can then calculate the IM exceedance rate as

$$\lambda(IM > im) = \sum_{i=1}^{n_{rup}} \left(\frac{1}{n_{gm}} \sum_{k=1}^{n_{gm}} I(im_k > im \mid rup_i, site) \right) \lambda\,(rup_i) \tag{6.41}$$

where im_k is the realization of im from the kth simulation for a given rupture. In comparison with Equation 6.1, Equation 6.41 has replaced the $P(IM > im)$ term with a count of what fraction of IM simulations exceed the target IM.

The first three steps of the above procedure are the same as in the numerical evaluation procedure of Section 6.2, and the SSM and GMMs also need to be specified as in that procedure. The primary difference arises in the tracking of uncertainties in rupture properties and IM values. For Step 4 of the procedure, ground motions can be sampled using an empirical GMM by simulating an ε from a standard normal distribution and using that simulated value (along with the predicted mean and standard deviation of $\ln IM$ for the simulated rup) to compute a simulated im using Equation 6.3. To perform Step 4 with a numerical ground-motion simulation model, all appropriate parameters in the simulation model would be randomized, a simulation with those parameters performed to produce a simulated time series, and the im value computed and used as realization im_k.

In Equation 6.41, the summation is now over all simulations rather than over all rup alternatives. The Monte Carlo simulation of realizations takes the place of the probability distributions previously specified by the $P(IM > im \mid rup_i, site)$ and $\lambda(rup_i)$ terms of Equation 6.1. The two formulations are mathematically equivalent ways of solving the same problem, and the choice between the two involves trade-offs in computational costs and methods of simplifying the problem for numerical tractability.

Epistemic uncertainty can be incorporated into the above procedure in one of two ways. The first option is to sample a terminal branch on the logic tree for each iteration before Step 3, in proportion to the branch weight. In this case, the resulting hazard curve is a mean hazard curve. The second option is to perform the above calculation procedure separately for each branch of interest, and then process the branches as described in Section 6.8. The first option is much easier if the logic tree is complex. The second option preserves the possibility of evaluating hazard curve fractiles, though it can be practically difficult for logic trees with many branches.

The set of ruptures in Step 3 can be produced by stepping through all possible ruptures, as described in the above list. This is the same as in the traditional numerical integration approach. Alternatively, a set of earthquake ruptures can be produced by sampling from the seismic source model distribution. This set of ruptures is sometimes called a synthetic earthquake catalog or a stochastic event set (Ebel and Kafka, 1999; Assatourians and Atkinson, 2013; Silva et al., 2013). The sampling approach is conceptually straightforward within a Monte Carlo framework, but can be inefficient in sampling rare events that are nonetheless important to hazard and risk. Efficiency can be improved through importance sampling or stratified sampling (e.g., Kiremidjian et al., 2007; Jayaram and Baker, 2010b). The sampling approach produces catalogs with temporal randomness included as well, characterized by sampling occurrence times for each rupture. Occurrence times can come from a Poisson process model or some more complex model (Section 3.8). The Monte Carlo approach to sampling ruptures is particularly useful for generating ruptures that have temporal or spatial clustering (e.g., Kagan and Knopoff, 1981; Field et al., 2017b; Zhang et al., 2018).

The Monte Carlo simulation approach has several advantages relative to the numerical calculation approach that was the focus of this chapter. It is very convenient when using physics-based GMMs, where im predictions come naturally in the form of realizations rather than probability distributions (Graves et al., 2011b). As discussed further in Chapter 11, this approach is also beneficial for spatially distributed systems where we need to simulate IM values at multiple locations. In general, it allows flexibility in models—particularly vector IMs and spatially distributed IMs.

The disadvantage of this approach is that many realizations must be simulated to produce a sufficient number of the rare events that are usually of most interest, though importance sampling can substantially mitigate this issue. Another disadvantage is that simulations may entail substantial computational cost and produce significant amounts of data. However, this cost may be worthwhile in order to utilize ground-motion simulations in hazard calculations or to assess risk to distributed systems.

6.10 Discussion

6.10.1 Growing Epistemic Uncertainty Over Time?

In principle, epistemic uncertainty in hazard calculations should reduce over time as hazard analyst knowledge grows. However, a surprising trend is sometimes observed, where uncertainty grows as a seismic hazard study is updated after some time passes (Bommer and Abrahamson, 2006).

A general challenge is that epistemic uncertainty is easy to underestimate, especially in data-poor applications, as observational evidence may not contradict a logic tree that omits rare but important events. Examples include omitting the possibility of very large earthquakes, such as from multiple fault segments rupturing as a single large earthquake (Kagan and Jackson, 2013), or omitting the possibility of significant earthquakes from a background source that reflects unknown seismic sources (Section 2.6). Considering the motivating example of Section 6.6.1, the analogy would be rolling a 7 on the die. Neither model envisions this outcome, so if observed, it would necessitate a reconsideration of candidate models to reflect the greater uncertainty in outcomes of future die roles than had been assumed.

Epistemic uncertainty must be considered carefully to avoid such problems. A fundamental principle is that having limited observational data does not mean that the analyst has limited uncertainty. Conversely, situations with limited observational data should have substantial uncertainty—larger than in data-rich situations. That relative comparison can serve as a guide, where well-studied and data-rich seismic hazard assessments establish a lower bound on epistemic uncertainty compared with other studies.

Issues of appropriate epistemic uncertainty values, and scrutiny of uncertainty assumptions when data are limited, are discussed further in Chapter 12.

6.10.2 Managing Computational Expense

The above formulae are straightforward and easily implemented for the trivial examples shown in this chapter. However, for real-world problems, complex logic trees can produce a vast set of branches: millions or even 10^{26} in one reported case (Bommer, 2012; Bommer et al., 2013). Computational expense also becomes important when maps of seismic hazard results are produced, using analyses at thousands or millions of locations in a region (see Section 7.4). In such cases, complex models can slow down calculations or make it infeasible to perform a complete PSHA calculation. Given the potential complexity of the calculation, several numerical adjustments can be used when performing complex PSHA calculations.

First, the rupture scenarios can be efficiently discretized to reduce the number of terms in the PSHA summation. Small-magnitude and distant ruptures can be omitted if they do not contribute substantially to the exceedance of the IM values of interest. The appropriateness of omissions can be evaluated via sensitivity studies such as that of Section 6.5.2. Further, if performing the calculation using Equation 6.4, the number of ε cases can be reduced by omitting large ε values that are extremely rare and so also do not contribute substantially to hazard (a common choice is to omit $\varepsilon > 3$). However, these are choices of numerical convenience and not a scientific hypothesis that such events are not possible (Section 4.4.3). The $\varepsilon > 3$ omission is sometimes mistakenly framed as scientifically justified, though no such justification exists (Bommer et al., 2004).

Second, complexity in the logic tree can be managed. Because the number of terminal nodes can grow quite large if the tree has many layers of branches, this can be an important issue. Recall that the logic-tree branches are a discretization of some parameter or model with epistemic uncertainty—this discretization should balance complexity and accuracy. A pragmatic way to view the issue is to recognize that a logic tree is a tool for producing the set of hazard curves, with the ultimate usable product being the mean hazard curve and the fractile hazard curves (e.g., Figure 6.16). These metrics can often be produced using a simpler logic tree with a smaller number of terminal nodes. With this goal in mind, it is sometimes useful to run an initial (expensive) analysis with a full logic tree, and then develop a simplified model for subsequent studies that reproduces the initial analysis. For example, if there are 10 parameters of a hazard analysis with epistemic uncertainty, discretizing them each into five branches will produce $5^{10} \approx 10$ million terminal nodes, while discretizing them each into three branches will produce $3^{10} \approx 60,000$ terminal nodes. The simplified model (sometimes termed a "pruned" logic tree) is strongly preferred if it is sufficiently accurate. Another way to address complex logic trees is to sample them via Monte Carlo simulation rather than consider them exhaustively. This is a natural fit with the Monte Carlo–based PSHA calculations of Section 6.9 but can be coupled with a numerical hazard analysis as well. With both pruned and Monte Carlo–sampled logic trees, care should be taken to ensure that branches leading to extreme exceedance rates are sufficiently sampled.

6.10.3 Revisiting Deterministic Hazard Analysis

With PSHA now formally defined, we can revisit the deterministic seismic hazard concept introduced in Section 1.3. For a point of reference, consider the following procedure based on guidelines such as that in Section 21.2.2 of ASCE (2016):

1. Identify the maximum magnitude earthquake rupture that could occur on a given earthquake source.
2. Compute the 84th percentile IM, given the rupture.
3. If there is more than one nearby source, take the maximum of the IM values, as determined above, associated with each source.

Comparing that process to the above PSHA, the deterministic analysis has essentially simplified the SSM to one rup and implicitly assumed that other rups do not contribute to exceedances of the IM value of interest. Further, it has specified a target exceedance probability for the IM intensity but ignored the occurrence rate of the rup. Finally, it has neglected epistemic uncertainty in model components, such as the maximum magnitude a source can produce. Note that this is not a "worst case" calculation and that there is a specific annual frequency of exceedance of a DSHA value (which will be unknown and variable from case to case).

For example, Figure 6.18 shows the probabilistic ground-motion hazard for the example of Section 6.3.3, compared with a deterministic calculation of an $SA(1\text{ s})$ value using the above procedure. The maximum magnitude is 7.9 in the above calculations (considering discretization of magnitudes), and other rupture parameters are as defined in that example. Using the same Boore et al. GMM as in the probabilistic assessment, the median $SA(1\text{ s})$ for this magnitude 7.9 rupture is 0.65 g, and the 84th percentile $SA(1\text{ s})$ is 1.10 g. The 1.10 g value is exceeded at an annual rate of 1.67×10^{-5} per year. Whether this deterministic value is greater than or less than the probabilistic value depends upon what exceedance rates are of interest in the probabilistic assessment. Also, if the source's activity rate were to change, as in Figure 6.9, the probabilistic ground-motion hazard would change proportionally, but the deterministic value would be unaffected.

Finally, note that any reasonable suggestion for a DSHA scenario is also considered in PSHA. The PSHA calculation will have considered the rup associated with a maximum magnitude event. It will also have considered the possibility that the ground motion is 84th percentile (or greater). Whether the probabilistic calculation produces a higher or lower IM value than the deterministic calculation depends on how likely the considered deterministic rup scenario is, how much other rup scenarios contribute to hazard, and what the target return period is in the probabilistic analysis.

While communication and interpretation can benefit from identifying a single rup scenario to consider for a given site, there are better ways to choose that scenario than pre-supposing an earthquake magnitude and IM percentile (Stewart et al., 2020). One useful approach to identifying a key scenario or scenarios—hazard disaggregation—will be a significant focus of the following chapter.

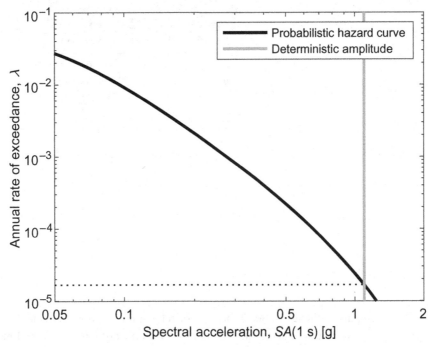

Fig. 6.18 Hazard curve for the example of Section 6.3.3, annotated with the metrics computed in Section 6.4. The plot has been extended to lower $SA(1\text{ s})$ values than other figures in this chapter to illustrate the $\lambda\,(M \geq m_{\min})$ calculation.

Exercises

6.1 Using the ground-motion hazard data in Table 6.7, provide the following:

(a) A plot of the hazard curve in log-log scale.
(b) The $SA(0.5\text{ s})$ value with an annual exceedance rate of 10^{-3}.
(c) The $SA(0.5\text{ s})$ value with an annual exceedance rate of 10^{-4}.
(d) Using Equation 6.31, estimate the slope, k, of the hazard curve between $SA(0.5\text{ s}) = 0.4$ g and 0.6 g. Compute the slope between 1.8 g and 2 g. Superimpose the fitted slopes on the hazard curve, and comment on the results.

6.2 Consider a ground-motion amplitude that will be exceeded with rate 0.01 per year (i.e., the "100-year return-period amplitude" for a site).

(a) What is the probability that one or more ground motions exceeding the 100-year return-period amplitude will occur in a 100-year period?
(b) What is the probability that exactly one ground motion exceeding the 100-year return-period amplitude will occur in a 100-year period?
(c) What is the probability that exactly three ground motions exceeding the 100-year return-period amplitude will occur in a 100-year period?

6.3 Consider the example of Section 6.3.3. Repeat the ground-motion hazard calculations for that example, but increasing or decreasing the Gutenberg–Richter b value by 10% (leaving all other

Table 6.7. Example hazard curve data	
$SA(0.5 \text{ s})$ [g]	$\lambda(SA(0.5 \text{ s}) > x)$
0.2	1.95e-02
0.4	7.40e-03
0.6	3.31e-03
0.8	1.65e-03
1.0	8.92e-04
1.2	5.12e-04
1.4	3.08e-04
1.6	1.92e-04
1.8	1.24e-04
2.0	8.23e-05
2.2	5.59e-05
2.4	3.87e-05
2.6	2.73e-05

inputs unchanged). See Table 3.5 for related information. Provide a plot of the three ground-motion hazard curves, and comment on the effect of b on the ground-motion hazard.

6.4 Consider the example of Section 6.5.3. Instead of using a fixed rate of $M > 5$ earthquakes (from Table 3.6) as was done in that example, consider a source with a fixed moment rate and varying m_{max} (data from Table 3.7). Perform hazard calculations and plot the hazard curves for each case. How do the results vary relative to Figure 6.11? Why?

6.5 Identify whether the following items are epistemic uncertainty or aleatory variability, and explain why:

   (a) The maximum magnitude that a seismic source can produce.
   (b) The magnitude of the next earthquake from a given source.
   (c) The standard deviation of $SA(1 \text{ s})$ for an $M = 6, R = 20$ km, rupture.
   (d) The annual rate of occurrence of $M > 5$ earthquakes from a given source.
   (e) The $SA(1 \text{ s})$ amplitude for an $M = 6, R = 20$ km, rupture.
   (f) The occurrence of an $M > 5$ earthquake from a given source in the next year.

6.6 Without looking up any reference information, make your best estimate of how many $M > 8$ earthquakes will occur in the world in the next calendar year. Estimate an interval that you think is has 90% probability of including the number of $M > 8$ earthquakes that will occur in the world in the next calendar year. Next, look up whatever information you you want for ≤ 30 minutes, and specify a second set of estimates (i.e., best estimate and 90% interval). How do the two relate? Describe your thought process, the information you relied on, and the role of epistemic and aleatory uncertainty in your answers.

6.7 An engineer is designing two structures, and considers a ground-motion amplitude with a return period of 1000 years. However, Structure 1 has a design life of 50 years, while Structure 2 has a design life of 10 years. Assuming that ground motion exceedances follow a Poisson process, what are the probabilities that each structure will see an exceedance of the 1000-year return period ground-motion amplitude over their design lives? If the engineer instead wants a ground-motion amplitude that is exceeded with probability 0.05 of each structure's lifetime, what return periods should they choose?

6.8 For the example of Section 6.3.1, compute the hazard curve derivative, and show that it is the lognormal PDF (with mean and standard deviation of ln IM matching the GMM prediction for the rupture), multiplied by the rate of earthquakes.

6.9 Repeat the example of Section 6.3.2, but considering epistemic uncertainty on the associated rupture magnitudes. Consider rupture 1 to have $M = 6$, 6.5, or 7 and rupture 2 to have $M = 7$, 7.5, or 8. All nine combinations of these rupture magnitudes are equally probable, and the rupture rates and distances are the same as in the example. Compute the mean hazard curve and the 11% and 89% fractiles. State in words why we might consider magnitudes to have epistemic uncertainty in this way, as opposed to considering the magnitudes to have aleatory uncertainty as in the example of Section 6.3.3. If we instead considered the magnitudes to have aleatory uncertainty (and again for all nine magnitude combinations to be equally likely), what would the mean and fractiles of the hazard curve be?

6.10 For the site described in the previous exercise, compute a deterministic $SA(1\ s)$ amplitude by considering the maximum possible magnitude on the fault, and the 84th percentile $SA(1\ s)$ amplitude given that magnitude. What is the return period of that deterministic amplitude, using the mean hazard curve computed in the previous exercise? What are the advantages and disadvantages of this deterministic calculation, relative to a probabilistic calculation?

6.11 Consider a site of interest with an aerial earthquake source surrounding it. Earthquakes are equally probable at all locations, with a rate of 10^{-5} earthquakes/year/km^2. Consider only magnitude 6 strike-slip earthquakes, a site $V_{S,30} = 500$ m/s, and compute PGA hazard using the BJF GMM introduced in Chapter 4. For simplicity, approximate the earthquakes as point sources so that the distance distribution is simple. Compute hazard curves considering the following:

(a) All earthquakes occurring within 20 km of the site.
(b) All earthquakes occurring within 50 km of the site.
(c) All earthquakes occurring within 100 km of the site.

Plot your hazard curves in log-log scale on a single figure. What distance is needed to accurately estimate the PGA amplitude with a 100-year return period? With a 1000-year return period?

7

PSHA Products

Several supplemental calculations can be performed in probabilistic seismic hazard analysis (PSHA), in addition to the basic hazard calculation introduced in Chapter 6. The calculations either generalize the basic hazard calculation or extend it to provide additional insights into ground-motion hazard at a site. Supplemental calculations can also help summarize results from multiple hazard calculations. This chapter aims to introduce these additional calculations and discuss the insights that they can provide.

Learning Objectives

By the end of this chapter, you will be able to do the following:

- Perform a disaggregation calculation.
- Compute a hazard map.
- Compute a uniform hazard spectrum and a conditional mean spectrum.
- Explain the difference between a uniform hazard spectrum and a conditional mean spectrum.
- Describe the connection between vector PSHA and conditional spectrum calculations.
- Describe how aftershocks, or other dependent earthquake sequences, can be incorporated within the PSHA framework.
- List items that should be documented as part of a PSHA study.

7.1 Introduction

A PSHA calculation is typically performed in support of an engineering evaluation or decision. The type of evaluation to be made will inform whether additional analysis or processing of the PSHA results is required. Some typical analysis situations are as follows:

- An equivalent static analysis is performed, requiring a static force that is determined based on a PSHA study (e.g., an IM value with a given exceedance rate, or a response spectral ordinate from a design response spectrum).
- A response spectrum analysis is performed, requiring a response spectrum derived from a hazard study.
- A site response analysis is performed. Depending upon the type of analysis, this may require as input an IM value, a scenario earthquake rupture, a response spectrum, or a set of ground-motion time series consistent with the results from a hazard analysis.
- A response-history analysis of a structure is performed, requiring a suite of ground-motion time series compatible with a hazard study.

For these applications, and in general, the ground-motion hazard curves from Chapter 6 are not sufficient to develop these inputs. Additional information from the PSHA study is needed, or the PSHA calculation outputs must be organized to facilitate other analyses.

One product that supports engineering analyses is identified rupture scenarios that contribute most to the occurrence or exceedance of some IM level. *Disaggregation* analysis, discussed in Section 7.2, is the calculation suited for this identification exercise.

A second product is an aggregation of multiple hazard calculation results into a response spectrum or a map. A response spectrum composed of spectral acceleration values at multiple vibration periods, computed from multiple hazard analyses, is referred to as a *uniform hazard spectrum* and is presented in Section 7.3. A map composed of IM values at multiple locations, computed from multiple hazard analyses, is referred to as a *hazard map* and is presented in Section 7.4.

Additional methods to aggregate hazard calculation information, or extend hazard analysis results, include the following:

- Computation of a *conditional spectrum*, which utilizes disaggregation information to compute the conditional distribution of a set of $SA(T)$ values, conditional on some primary $SA(T^*)$ value computed from a single PSHA calculation (Section 7.5).
- Computation of a joint distribution of multiple IMs, using *vector probabilistic seismic hazard analysis* (Section 7.6).

The calculation of these various products, and a discussion of their utility, is the primary focus of this chapter. In addition, the modification of PSHA to accommodate dependent earthquake sequences is also discussed in Section 7.7.

7.2 Disaggregation

Once the PSHA computations are complete, a natural question is, "Which earthquake rupture is most likely to cause $IM > x$?" The PSHA calculation has aggregated all scenarios together, so the answer is not immediately apparent. In the example PSHA calculations, however, some of the intermediate calculation results indicated the relative contribution of different earthquake sources and magnitudes to the rate of exceedance of a given IM. Here we will formalize those calculations through a process known as disaggregation[1] (McGuire, 1995; Bazzurro and Cornell, 1999).

7.2.1 Motivation: Example with Two Rupture Scenarios

To gain intuition regarding disaggregation, consider the example of Section 6.3.2. Two ruptures influenced this site, so we can compute the contribution of each to the IM exceedance. Consider first the case of $SA(1 \text{ s}) > 0.2$ g, since we previously computed some needed results for that case. From that example, we computed the following results:

$$\lambda(SA(1 \text{ s}) > 0.2 \text{ g}, rup_1) = 0.00378 \tag{7.1}$$

$$\lambda(SA(1 \text{ s}) > 0.2 \text{ g}, rup_2) = 0.00188 \tag{7.2}$$

---

[1] The calculations in this section are known as both *disaggregation* and *deaggregation*. Disaggregation is the only one of the two words found in a dictionary, so we adopt that term here, but deaggregation is also frequently used.

$$\lambda(SA(1 \text{ s}) > 0.2 \text{ g}) = 0.00566. \tag{7.3}$$

Taking the ratio of the exceedance rate from a given rupture to the overall exceedance rate, we can find the probablity that an exceedance is caused by that rupture:

$$P(rup_1 \mid SA(1 \text{ s}) > 0.2 \text{ g}) = 0.668 \tag{7.4}$$

$$P(rup_2 \mid SA(1 \text{ s}) > 0.2 \text{ g}) = 0.332. \tag{7.5}$$

Repeating the same calculations using the results for $SA(1 \text{ s}) > 0.5$ g gives

$$P(rup_1 \mid SA(1 \text{ s}) > 0.5 \text{ g}) = 0.186 \tag{7.6}$$

$$P(rup_2 \mid SA(1 \text{ s}) > 0.5 \text{ g}) = 0.814. \tag{7.7}$$

So, for the relatively lower $SA(1 \text{ s})$ value of 0.2 g, the more active Source 1 has a high probability of being the causal rupture. At the larger $SA(1 \text{ s})$ intensity of 0.5 g, the less active Source 2 has a greater contribution to the exceedance of the $SA(1 \text{ s})$ value, because earthquakes from this source are so much more likely to produce a large $SA(1 \text{ s})$ value. This result is consistent with the qualitative observations made at the end of the original example calculation, using a visualization of the hazard curve (which is reproduced here as Figure 7.1). The probabilities we have computed are exactly proportional to the rates we previously computed for $\lambda(IM > x, rup)$.

The disaggregation calculation is an extension of this concept to more general situations where there are ranges of rupture conditions of interest.

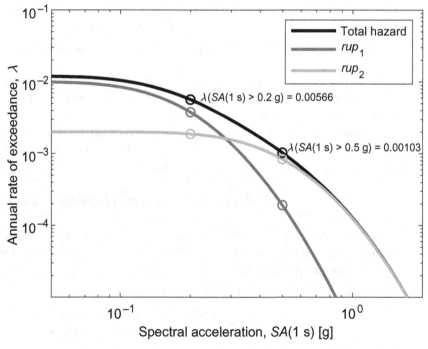

Fig. 7.1 $SA(1 \text{ s})$ hazard curve for the example with two ruptures, which provides the inputs for the example of Section 7.2.1.

7.2.2 Disaggregation Calculations

Consider the annual rate of observing a ground motion at a given *site* with $IM > im$, caused by rupture rup_i. The rate of this event can be computed as the rate of ground motions (from any rupture) with $IM > im$, multiplied by the probability that such a ground motion came specifically from rupture rup_i:

$$\lambda(IM > im, rup_i) = \lambda(IM > im) \times P(rup_i \mid IM > im). \tag{7.8}$$

For a disaggregation calculation, we are interested in the $P(rup_i \mid IM > im)$ term, so let us rearrange to solve for this:

$$P(rup_i \mid IM > im) = \frac{\lambda(IM > im, rup_i)}{\lambda(IM > im)}. \tag{7.9}$$

The denominator of this equation is exactly what we have computed previously in Equation 6.1 (i.e., the primary PSHA equation), and the numerator can be computed using the same inputs.[2] Alternatively, we can decompose the numerator of Equation 7.9 into terms that were used in the original PSHA equation:

$$P(rup_i \mid IM > im) = \frac{P(IM > im \mid rup_i)\lambda(rup_i)}{\lambda(IM > im)}. \tag{7.10}$$

All of the above equations are site-specific, so *site* could be noted explicitly as a conditioning term in all of the above terms, but it is omitted here for brevity. Intuitively, Equations 7.9 and 7.10 show that the probability of rup_i causing $IM > im$ is equal to the rate of rup_i earthquakes that cause $IM > x$, divided by the rate of *all* earthquakes that cause $IM > x$. The left-hand side of these equations always produces a valid probability distribution; that is, the sum over i of $P(rup_i \mid IM > im)$ always equals 1. Additionally, when substituting $im = 0$, such that $P(IM > im \mid rup_i) = 1$ for all rup_i, Equation 7.10 becomes

$$P(rup_i \mid IM > 0) = \frac{\lambda(rup_i)}{\lambda(IM > 0)} \tag{7.11}$$

$$= \frac{\lambda(rup_i)}{\sum_i \lambda(rup_i)}. \tag{7.12}$$

That is, the disaggregation probability for rup_i is the occurrence rate of rup_i, scaled by a constant (reciprocal of the sum of rates for all ruptures). The implication is that the disaggregation distribution tends toward the original seismic source model (SSM) distribution as *im* tends toward zero.

Note from the above equations that all results needed for this calculation are computed when performing the original PSHA calculation. For pedagogical reasons, we have delayed presenting disaggregation until now, but in reality, this calculation is performed simultaneously with the original PSHA calculation to take advantage of these available results.

Consider the example of Section 6.3.3, which had a source producing earthquake magnitudes with a Gutenberg–Richter distribution. We can use disaggregation to find the probability that an $SA(1\text{ s})$ level was exceeded by an earthquake with a given magnitude. Tables 6.3 and 6.4 provide all of the inputs needed to perform these disaggregations.

---

[2] Note also that if the PSHA has been performed using the Monte Carlo procedure of Section 6.9, this ratio can be computed as the fraction of simulations with $IM > im$ that have the specified rupture.

m_i	$P(m_i \mid SA(1\ s) > 0.2\ g)$	$P(m_i \mid SA(1\ s) > 0.5\ g)$
5.1	0.008	0.000
5.3	0.018	0.000
5.5	0.037	0.001
5.7	0.064	0.003
5.9	0.094	0.008
6.1	0.120	0.018
6.3	0.133	0.037
6.5	0.131	0.064
6.7	0.116	0.096
6.9	0.093	0.126
7.1	0.069	0.146
7.3	0.049	0.151
7.5	0.033	0.140
7.7	0.021	0.118
7.9	0.014	0.092
	Sum = 1.000	Sum = 1.000

Table 7.1. Results for the Section 6.3.3 disaggregation example

To start, we will consider disaggregation over magnitude bins that correspond to the rupture definitions used in Section 6.3.3, so that the disaggregation on magnitude corresponds to the disaggregation on ruptures. That is,

$$\lambda(SA(1\ s) > 0.2\ g, m_1) = \lambda(SA(1\ s) > 0.2\ g, rup_1). \tag{7.13}$$

Referring to Table 6.3, the rate of $M = 5.1$ ruptures causing $SA(1\ s) > 0.2\ g$ ground motions is 1.59×10^{-5} (the rightmost column of the top row), and the rates of all ruptures causing $SA(1\ s) > 0.2\ g$ ground motions is 0.00212 (the sum of the rightmost column's values, noted in the bottom row). So the probability of an $SA(1\ s) > 0.2\ g$ ground motion being caused by an earthquake with magnitude 5.1 is

$$P(5 \leq M < 5.2 \mid SA(1\ s) > 0.2\ g) = \frac{\lambda(SA(1\ s) > 0.2\ g, m_1)}{\lambda(SA(1\ s) > 0.2)} = \frac{1.59 \times 10^{-5}}{0.00212} = 0.008. \tag{7.14}$$

This is the first entry in column 2 of Table 7.1, and that table contains results obtained in the same manner for other magnitudes and $SA(1\ s)$ values. These tabulated results are also plotted in Figure 7.2.

Table 7.1 and Figure 7.2 show what we observed qualitatively for this site in Chapter 6. Smaller-magnitude (i.e., $M \cong 6$) earthquakes are likely to cause exceedances of $SA(1\ s) > 0.2\ g$, but are quite unlikely to cause exceedances of $SA(1\ s) > 0.5\ g$. This is because $M \cong 6$ ruptures are relatively likely compared with larger-magnitude ruptures, and also likely to cause smaller-intensity ground motions. However, these $M \cong 6$ ruptures are very unlikely to cause $SA(1\ s) > 0.5\ g$ ground motions, so they contribute little at that higher intensity.

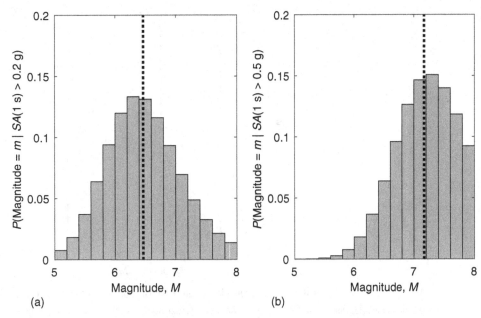

Fig. 7.2 Disaggregation results associated with the example calculation of Section 6.3.3. Mean magnitude values, computed using Equation 7.15, are shown with dashed lines. (a) Magnitude disaggregation, given $SA(1\,s) > 0.2\,g$ ($\overline{M} = 6.47$). (b) Magnitude disaggregation, given $SA(1\,s) > 0.5\,g$ ($\overline{M} = 7.17$).

Means, Medians, and Modes from Disaggregation

Results from the above equations are probability distributions, and those probability distributions can be evaluated using typical stochastic modeling approaches. Most commonly, we are interested in mean, median, and modal values of rupture parameters. These metrics can be computed using typical mean, median, and modal equations, considering the probability distribution of the rupture parameter of interest. For example, the mean magnitude from a disaggregation distribution is

$$\overline{M} = E[M \mid IM > im] = \sum_{rup_i} m_i \times P(rup_i \mid IM > im) \qquad (7.15)$$

where m_i is the magnitude of rup_i. For the data of Table 7.1, the mean magnitude values are 6.47 and 7.17 for the left and right columns, respectively (as the reader can easily verify). Figure 7.2 shows these mean values, and it is apparent that they lie at the bar plots' visual centroids.

The above calculations can be performed for other parameters of interest, such as distance or ε. We will discuss joint disaggregation distributions of magnitude, distance, and ε below, but Equation 7.15 can be correctly evaluated for a single parameter at a time.

A median magnitude can be computed similarly, by summing bins until we reach a 0.5 cumulative probability. Modal values are sometimes computed as well, by selecting the bin with the highest conditional probability. Modal values face the typical problem that they are more sensitive than means and medians to the choice of discretization of ruptures.

With all of the above metrics, challenges can arise with more complex probability distributions, such as Figure 7.3. In this case, the mean magnitude of 6.33 does not correspond to the magnitude of any causal rupture. For reference, the median of this distribution is 5.7, and the mode is 5.5. This type of situation arises when a site is near two distinct sources with differing characteristic rupture

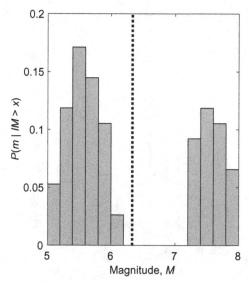

Fig. 7.3 A disaggregation distribution that illustrates potential challenges with interpreting mean disaggregation values. The mean magnitude of 6.33, indicated by the dashed line, has a zero probability of occurrence for this distribution.

properties (Bazzurro and Cornell, 1999). In these cases, rather than modifying these metrics to solve the problem, the analyst should keep in mind that these are merely summary metrics and cannot completely replace the information provided by the full probability distribution.

General Binning of Ruptures

We may also be interested in probabilities associated with a subset of ruptures (e.g., all ruptures with magnitudes between 5 and 5.5). In this case we modify the numerator of Equation 7.10 to incorporate all rupture scenarios of interest:

$$P(rup \in subset \mid IM > im) = \sum_{rup_i \in subset} P(rup_i \mid IM > im) \qquad (7.16)$$

where $rup \in subset$ denotes ruptures in some relevant subset (e.g., all rup with magnitude between 5 and 5.5, or all rup associated with a specific fault source), and the summation is over all ruptures in that subset. The $P(rup_i \mid IM > im)$ term in the summation comes from Equation 7.16.

We can also perform disaggregation over different intervals of rupture properties than those used to compute the initial PSHA summation. Using the above example, we can compute the probability of $5.0 < M < 5.4$, given $SA(1\ \text{s}) > 0.2$ g, as

$$
\begin{aligned}
&P(5 \leq M < 5.4 \mid SA(1\ \text{s}) > 0.2\ \text{g}) \\[4pt]
&= \frac{\lambda(SA(1\ \text{s}) > 0.2\ \text{g}, m_1 = 5.1) + \lambda(SA(1\ \text{s}) > 0.2\ \text{g}, m_2 = 5.3)}{\lambda(SA(1\ \text{s}) > 0.2)} \\[6pt]
&= \frac{1.59 \times 10^{-5} + 3.58 \times 10^{-5}}{0.00212} \\[6pt]
&= 0.0257.
\end{aligned} \qquad (7.17)
$$

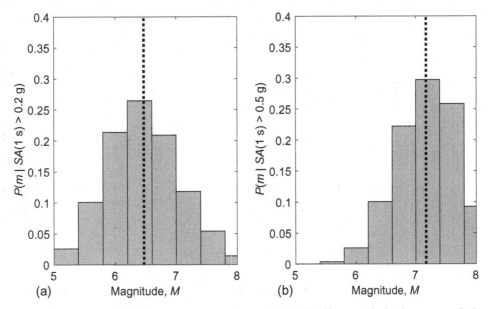

Fig. 7.4 Magnitude disaggregations associated with the example calculation of Section 6.3.3. Mean magnitude values, computed using Equation 7.15, are shown with dashed lines. (a) Results given $SA(1\,s) > 0.2\,g$ ($\overline{M} = 6.47$). (b) Results given $SA(1\,s) > 0.5\,g$ ($\overline{M} = 7.17$).

Similar calculations are performed for other $SA(1\,s)$ values and other magnitude ranges and are plotted in Figure 7.4. Comparing Figure 7.4 with Figure 7.2, the general shapes of the distributions, and the mean magnitude values, are not significantly changed. PSHA software typically uses this approach, as it allows for flexible specification of intervals, independently from the intervals used to compute numerically precise hazard results. A standard PSHA calculation will use a finer discretization for the initial PSHA summation (perhaps a 0.1 magnitude interval), and a coarser discretization for the disaggregation (perhaps 0.2 or 0.5). The inputs to the original summation are never output, and so can be finely discretized to maximize numerical accuracy. However, coarse disaggregation discretization limits the length of output tables. Keep in mind that this example considered only magnitude, but when disaggregating to find the probabilities of combinations of M, R, and ε, the output can quickly become much more substantial.

Disaggregation Conditional on Occurrence Rather than Exceedance

Equations 7.9–7.16 compute probabilities of rup, conditional on IM *exceeding im*. But comparable calculations can be used to find contributions of rup, conditional on IM *equaling im*. This may be appropriate, for example, when using disaggregation to select appropriate ground motions having IM equal to im (McGuire, 1995; Fox et al., 2016). The equality conditioning can be achieved by considering a narrow range of im values:

$$P(rup_i \mid IM = im) \cong P(rup_i \mid im < IM < im + \delta)$$

$$= \frac{\lambda(IM > im, rup_i) - \lambda(IM > im + \delta, rup_i)}{\lambda(IM > im) - \lambda(IM > im + \delta)} \qquad (7.18)$$

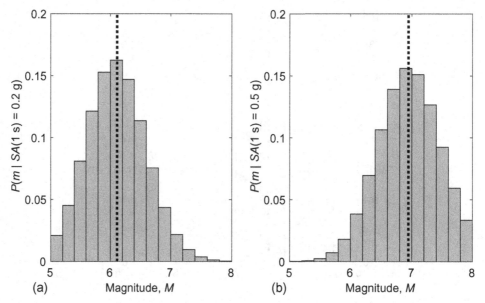

Fig. 7.5 Magnitude disaggregations associated with the example calculation of Section 6.3.3. Mean magnitude values, computed using Equation 7.15, are shown with dashed lines. (a) Results given $SA(1\ s) = 0.2\ g$ ($\overline{M} = 6.11$). (b) Results given $SA(1\ s) = 0.5\ g$ ($\overline{M} = 6.94$).

where δ is some small increment of im, and the result becomes exact in the limit $\delta \rightarrow 0$. The terms in Equation 7.18 are identical to those in Equation 7.9. Equation 7.18 can also be extended to consider the probability of a subset of ruptures causing im, using the same approach as Equation 7.16.

For comparison, Equation 7.18 can be used to perform disaggregation conditional on an occurrence rather than an exceedance of im. Here the equation is calculated using the interval $im < IM < 1.02 \times im$ to represent $IM = im$, though in this case, the results do not vary significantly if the interval is doubled or quadrupled in size. Disaggregation results for $SA(1\ s) = 0.2$ g and $SA(1\ s) = 0.5$ g are shown in Figure 7.5. In comparison to Figure 7.2, we see that the distributions have shifted slightly to the left. That is, $IM = im$ ground motions tend to come from slightly smaller magnitudes than $IM > im$ ground motions. This should be intuitive, as the former condition omits more intense motions that are more likely to come from larger magnitudes. Further, we see that the shift is more significant for the 0.2 g case than for the 0.5 g case (e.g., the mean magnitude decreases by 0.36 for the former and 0.22 for the latter). This differential shift has to do with the steepness of the hazard curve, as observed in Figure 6.7. At $SA(1\ s) = 0.5$ g, the hazard curve is steeper than at 0.2 g, meaning that at 0.5 g, the rate of IM exceedances is dropping more quickly as IM increases. Qualitatively, most of the $SA(1\ s) > 0.5$ g ground motions have $SA(1\ s)$ "close to" 0.5 g, and so the $SA(1\ s) > 0.5$ g condition is not so different from the $SA(1\ s) = 0.5$ g condition (relative to the 0.2 g case). These differences can be significant when selecting ground motions consistent with the hazard results, as will be discussed later in Chapter 10.

7.2.3 Disaggregation of Contributions by ε

In addition to the *rup* definition, it is also often of interest to know the ε value associated with the ground-motion model (GMM) (Equation 4.16) most likely to cause $IM > im$ (or $IM = im$).

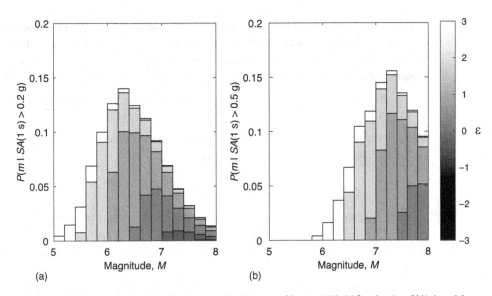

Fig. 7.6   Disaggregation results for magnitude and ε, for the example calculation of Section 6.3.3. (a) Results given SA(1 s) > 0.2 g ($\bar{\varepsilon} = 1.17$). (b) Results given SA(1 s) > 0.5 g ($\bar{\varepsilon} = 1.42$).

Such consideration is useful because ε indicates where the *IM* value is within the predicted probability distribution (see Section 4.6.2).

For disaggregation of ε, we first utilize a PSHA formulation that explicitly includes ε in the summation (Equation 6.4):

$$\lambda(IM > im) = \sum_{i=1}^{n_{rup}} \sum_{k=1}^{n_\varepsilon} I\,(IM > im \mid e_k, rup_i, site)\, f_\varepsilon(e_k)\Delta e\,\lambda\,(rup_i). \tag{7.19}$$

We can then find the disaggregation distribution for ε as above for the *rup* disaggregation

$$P(rup_i, e_k \mid IM > im) = \frac{I\,(IM > im \mid e_k, rup_i, site)\, f_\varepsilon(e_k)\Delta e\,\lambda\,(rup_i)}{\lambda(IM > im)} \tag{7.20}$$

Revisiting the example, Figure 7.6 shows disaggregation results comparable to those from Figure 7.2, but with the bars for each magnitude shaded to reflect the contribution by ε value. Note that the lower-magnitude bars indicate only contributions from positive ε values: for these lower-magnitude ruptures, only larger-than-average (i.e., $\varepsilon > 0$) ground motions can cause exceedances. For larger magnitudes, and the 0.2 g case in particular, negative ε do contribute somewhat. However, positive ε values make much greater contributions than negative. Mean ε values can also be computed for these cases (Equation 7.15), and are $\bar{\varepsilon} = 1.17$ for SA(1 s) > 0.2 g and $\bar{\varepsilon} = 1.42$ for SA(1 s) > 0.5 g. These results are typical of realistic analyses: for ground-motion intensities of engineering interest, ε disaggregation values are typically positive and get larger with increasing ground motion intensity.

7.2.4 Finite Fault Example

We now consider a more realistic example, building on the result of Figure 3.30. We use the probabilities for the magnitudes and distances given on that figure, assume a site $V_{S,30} = 500$ m/s, and use the Chiou and Youngs (2014) GMM to predict SA(1 s) exceedances.

The Figure 3.30 distribution does not specify an overall rate of rupture occurrences. However, it is not needed here because the rate would appear in both the numerator and denominator of Equation 7.10, and so does not influence the disaggregation results.[3] Making this cancellation of earthquake rates, we slightly modify Equation 7.10 to the following equation for this example:

$$P(rup_i \mid IM > im) = \frac{P(IM > im \mid rup_i)P(rup_i)}{P(IM > im)} \tag{7.21}$$

where $P(rup_i)$ is the probability of rup_i, given a rupture of any type (i.e., the probability specified by Figure 3.30), and $P(IM > im)$ is computed as

$$P(IM > im) = \sum_i P(IM > im \mid rup_i)P(rup_i). \tag{7.22}$$

Figure 7.7 shows disaggregation results for $SA(1\,s)$ at two intensities, computed using Equation 7.21. Compared with the SSM distribution, $P(rup_i)$, the disaggregation distribution $P(rup_i \mid IM > im)$ has higher probabilities associated with large magnitudes and small distances, as expected given the discussion above.

Figure 7.8 shows additional disaggregation results using the same SSM and GMM, but this time for $SA(0.1\,s)$ values with similar exceedance probabilities as the values shown in Figure 7.7. Compared with the $SA(1\,s)$ disaggregations, the $SA(0.1\,s)$ disaggregations have higher probabilities associated with small-magnitude and small source-to-site distance ruptures. This is because $SA(0.1\,s)$ values do not vary as strongly with earthquake magnitude as $SA(1\,s)$ values, so it is relatively more likely that the frequent small magnitude/distance ruptures cause exceedances.

The results from this example highlight that there is no uniquely defined "disaggregation for a site." The most important rupture scenarios vary depending upon the type of IM, and the value of that IM. One implication of this issue is the problematic nature of choosing scenario ruptures for deterministic hazard analysis procedures. Even for relatively simple SSMs, the disaggregation distributions generally change with the IM type and value. A second implication is that there may not be a unique rupture scenario to consider for later risk analysis calculations. This is true for selecting scenarios in general and in Chapter 10 when selecting ground motions (and associated ruptures) for input to risk analysis procedures. These implications do not create insurmountable problems, but we note them here in anticipation of later discussion.

7.2.5 Disaggregation Considering Epistemic Uncertainty

Section 6.8 illustrated that PSHA with explicit consideration of epistemic uncertainties results in a multitude of seismic hazard curves. For each logic-tree terminal node (and resulting hazard curve), it is possible to perform disaggregation calculations. The resulting disaggregation distribution, $P(rup_i \mid IM > im)$, will vary from branch to branch because of the different behavior of individual models in the logic-tree combinations, and understanding such differences can be enlightening.

Because of the common use of the mean hazard curve (Equation 6.38 and Figure 6.16), it is convenient to compute the disaggregation of the mean hazard from:

$$E[P(rup_i \mid IM > im)] = \sum_{k=1}^{n_{models}} P^k(rup_i \mid IM > im)\, P^k(im \mid E[\lambda(IM > im)]) \tag{7.23}$$

---

[3] For a case with multiple sources, this cancellation of rates would not apply. The event rates from each source would be needed.

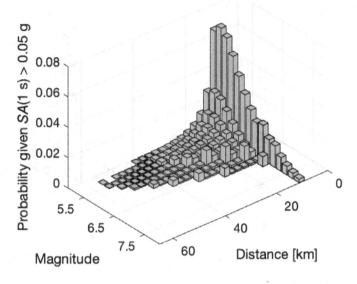

(a)

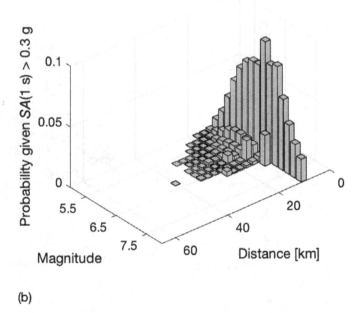

(b)

Fig. 7.7 Disaggregation results using the SSM of Figure 3.30, for (a) $SA(1\,s) > 0.05$ g and (b) $SA(1\,s) > 0.3$ g.

where $E[\ ]$ is the expectation (mean) operator; $P^k(rup_i \mid IM > im)$ is the disaggregation probability for rup_i, obtained in the usual manner (Equation 7.10), for model k; and $P^k(im \mid E[\lambda(IM > im)])$ is the contribution of the kth model to the mean hazard (Lin et al., 2013c; Tarbali et al., 2018):

$$P^k(im \mid E[\lambda(IM > im)]) = \frac{\lambda^k(IM > im)w^k}{E[\lambda(IM > im)]} \qquad (7.24)$$

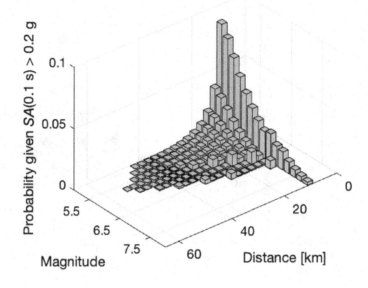

(a)

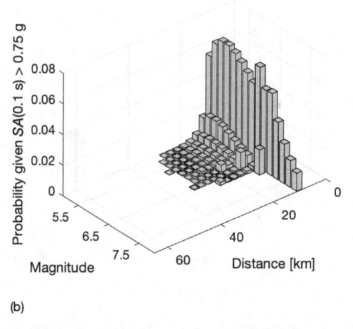

(b)

Fig. 7.8   Disaggregation results using the SSM of Figure 3.30, for (a) $SA(0.1\,s) > 0.2\,g$ and (b) $SA(0.1\,s) > 0.75\,g$.

where all terms have been defined in relation to Equation 6.38. In words, the numerator of this equation is the hazard curve for the kth terminal node, multiplied by that node's weight (i.e., its contribution). The denominator is the mean hazard curve. So the ratio of the numerator to the denominator is the kth terminal node's contribution to exceedances of im. This contribution

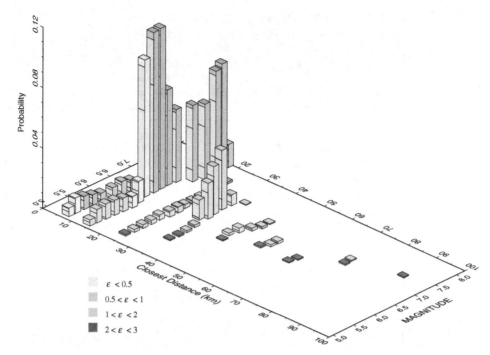

Fig. 7.9 Disaggregation for $SA(1\,s) > 0.2\,g$, the intensity with 50% probability of exceedance in 50 years, at the example Yerba Buena Island site. (Figure adapted from the US Geological Survey).

will be large for a given branch when $\lambda^k(IM > im)$ is large (i.e., that branch predicts a high rate of exceedances) or when w^k is large (i.e., that branch has a high weight in computing the overall hazard).

Equations 7.23 and 7.24 also apply for disaggregation conditional on occurrence, rather than exceedance, as discussed in Section 7.2.2.

7.2.6 Example: Yerba Buena Island

We now investigate a realistic example of hazard disaggregation for the Yerba Buena Island site used throughout the book. Recall from earlier sections that the site is approximately 12 km from the closest point on the Hayward Fault and 18 km from the closest point on the San Andreas fault. The results in this section utilize the SMM, GMMs, and logic tree from the 2008 US Geological Survey hazard calculations. Figure 7.9 shows a disaggregation for the site, conditional on $SA(1\,s) > 0.2\,g$—an intensity exceeded with 50% probability in 50 years. The probabilities are provided for bins of magnitude and distance values (noted via horizontal axis values) and ε values (noted via shading of the bars). In this figure, the largest contributions are from $6 < M < 7.25$ earthquakes at distances of $12 < R_{rup} < 15$ km (consistent with Hayward Fault events) and $7.25 < M < 8.25$ earthquakes at distances of $15 < R_{rup} < 18$ km (consistent with San Andreas events). There are some contributions at other distances (from other nearby faults) and from smaller-magnitude ($M < 6$) earthquakes, indicating the wide range of rupture scenarios in this area that could produce $SA(1\,s) > 0.2\,g$.

The contributions can be grouped by fault (instead of M, R) and plotted on a map. An example is shown in Figure 7.10. The large spikes on the San Andreas and Hayward Faults confirm the

Table 7.2. Table of disaggregation by fault at the example Yerba Buena Island site. Conditional probabilities are provided given $SA(1\ \mathrm{s}) > 0.2$ g (an intensity with 50% probability of exceedance in 50 years) and given $SA(4\ \mathrm{s}) > 0.18$ g (an intensity with 1% probability of exceedance in 50 years). Only faults with greater than 2% contribution are noted. San Andreas Fault segments are as follows: O = offshore, N = north, P = peninsula, and S = south. The rows listing multiple segments refer to ruptures spanning the segments

Fault ID	Given $SA(1\ \mathrm{s}) > 0.2$ g	Given $SA(4\ \mathrm{s}) > 0.18$ g
Hayward–Rodgers Creek	0.37	0
San Andreas (O+N+P+S segments)	0.06	0.47
San Andreas (O+N segments)	0.07	0.22
San Andreas (O+S segments)	0.05	0.11
San Andreas unsegmented	0.03	0.06
San Gregorio	0.03	0.03
Other	0.39	0.11

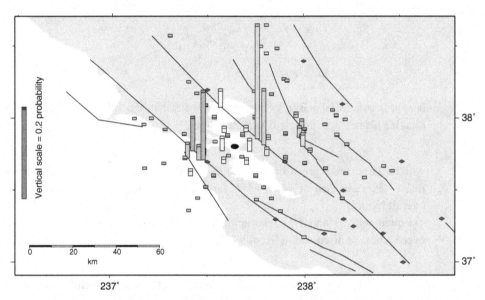

Fig. 7.10   Disaggregation for $SA(1\ \mathrm{s}) > 0.2$ g, the intensity with 50% probability of exceedance in 50 years, at the example Yerba Buena Island site. Bars are located at the associated rupture source. Major regional faults are shown with thin lines. (Figure adapted from the US Geological Survey).

discussion of the previous paragraph. The results can also be tabulated rather than plotted, as shown in Table 7.2. The Figure 7.9 disaggregation format is more common than the Figure 7.10 format for disaggregation plots. The causal seismic sources can often be inferred from magnitude/distance disaggregation, so the map format is generally unnecessary. Furthermore, if we need disaggregation by seismic source, it is generally easier to summarize in a table than a figure. However, the mathematical disaggregation calculation and the principle of grouping contributions by rupture property are the same regardless of the presentation format.

While the numerical results from this example are more complex, reflecting the complexity of sources at a real site, the same general findings from the above examples are confirmed. Figures 7.9 and 7.10 indicate that large-magnitude ruptures on nearby active sources contribute most to exceedances of the high-intensity $SA(1\ s)$ of interest. Table 7.2 shows that, although the San Andreas Fault is a dominant contributor for both conditioning cases shown, the relative contributions from a given fault may vary widely depending upon the IM of interest. For example, the Hayward–Rogers Creek Fault has a 37% chance of causing $SA(1\ s)\ >\ 0.2\ g$, but no chance of causing $SA(4\ s)\ >\ 0.18\ g$ at the same site. This reinforces the statement above that there is no uniquely defined "disaggregation for a site."

7.3 Uniform Hazard Spectrum

A uniform hazard spectrum (UHS) is a convenient way to aggregate results from several hazard calculations performed for a given site. It is developed by first performing the PSHA calculation for spectral accelerations at a range of (oscillator) periods. Second, we identify the spectral acceleration value having the target rate or exceedance at each period. Finally, we plot those spectral acceleration values versus their periods. The figure is a plot of spectral acceleration ordinates versus period, and so is a response spectrum. Since the spectrum ordinates all have the same exceedance rate (i.e., "hazard" level), it is called a *uniform hazard* spectrum.

To illustrate, consider the example site of Section 6.3.3, with ruptures that are 10 km from the site and have a Gutenberg–Richter magnitude distribution. Here we utilize the Chiou and Youngs (2014) GMM to have realistic trends in response spectra as a function of rupture properties (versus the simpler Boore et al. model used above). Figure 7.11 shows hazard curves for the site, for three SA periods, and the SA value at each period corresponding to an annual exceedance rate of 2.1×10^{-3}

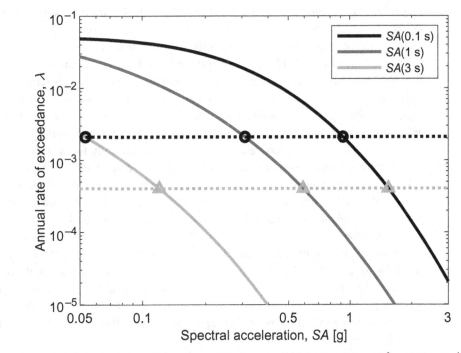

Fig. 7.11 *SA* hazard curves at individual periods, used to generate uniform hazard spectra with $\lambda = 2.1 \times 10^{-3}$ and 4.04×10^{-4} rates of exceedance.

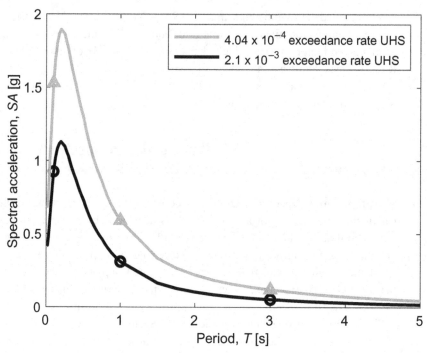

Fig. 7.12 Uniform hazard spectra (UHS) for $\lambda = 2.1 \times 10^{-3}$ and 4.04×10^{-4} rates of exceedance, constructed from the results of Figure 7.11.

or 4.04×10^{-4} (corresponding to $RP = 475$ and 2475 years, or $P = 2\%$ or 10% exceedance probability in 50 years; see Section 6.2.2); these are the hazard curves used to construct the UHS. Figure 7.12 shows those three SA values plotted at their corresponding periods (the circles) along with comparable values at a range of additional periods (the lines).

The spectra of Figure 7.12 are called uniform hazard spectra because every ordinate has an equal (i.e., uniform) rate of being exceeded. Importantly, each spectrum is a set of values from separate PSHA calculations at each period, so the spectral ordinates are in general caused by differing ruptures, rather than representing the spectrum of any single ground motion. To see this, magnitude and ε disaggregation results for the $SA(0.1\ \text{s})$ and $SA(3\ \text{s})$ values from Figure 7.12 are shown in Figure 7.13. The causal ruptures and ε values differ significantly for the two cases. The $SA(0.1\ \text{s}) > 0.92$ g ground motions are most likely to be caused by magnitude $5 < M < 6$ ruptures, while the $SA(3\ \text{s}) > 0.053$ g ground motions are most likely to be caused by magnitude $6 < M < 7$ ruptures.

To build further intuition about this result, consider some additional information about the predicted ground motions. Figure 7.14 shows response spectra associated with the mean values from the disaggregations of Figure 7.13. Specifically, Figure 7.14 shows the $\mu_{\ln SA} + \varepsilon \times \sigma$ response spectra for a rupture with magnitude $\overline{M}$ and other rupture properties as specified for the site (and ε assumed equal for all periods). The $\overline{M} = 5.9$, $\overline{\varepsilon} = 1.9$, case causes larger spectral accelerations at short periods, while the $\overline{M} = 6.5$, $\overline{\varepsilon} = 1.2$, case causes larger spectral accelerations at longer periods. This matches a common trend that IMs associated with high-frequency ground shaking are relatively more likely to be exceeded by small-magnitude motions than IMs associated with low-frequency ground shaking.

Additionally, the probabilities of SA exceedance, for each possible rupture at the site, are provided in Table 7.3 for $SA(0.1\ \text{s})$ and $SA(3\ \text{s})$. The rupture occurrence rates, $\lambda(rup_i)$, are the same for each

Table 7.3. Ruptures, rupture occurrence rates, and probabilities of *SA* exceedance for the *SA*(0.1 s) and *SA*(3 s) hazard calculations used to produce the UHS of Figure 7.12

i	m_i	$\lambda(rup_i)$	$P(SA(0.1\text{ s}) > 1.2\text{ g} \mid rup_i)$	$P(SA(3\text{ s}) > 0.08\text{ g} \mid rup_i)$
1	5.1	0.0185	0.015	0.000
2	5.3	0.0117	0.026	0.002
3	5.5	0.0074	0.041	0.008
4	5.7	0.0046	0.058	0.028
5	5.9	0.0029	0.078	0.077
6	6.1	0.0018	0.100	0.162
7	6.3	0.0012	0.125	0.275
8	6.5	7.35×10^{-4}	0.152	0.397
9	6.7	4.64×10^{-4}	0.190	0.513
10	6.9	2.93×10^{-4}	0.230	0.611
11	7.1	1.85×10^{-4}	0.268	0.690
12	7.3	1.17×10^{-4}	0.304	0.752
13	7.5	7.35×10^{-5}	0.338	0.801
14	7.7	4.64×10^{-5}	0.370	0.840
15	7.9	2.93×10^{-5}	0.401	0.871

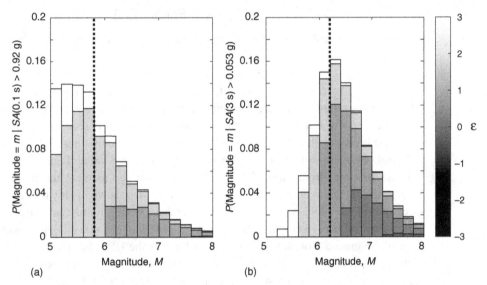

Fig. 7.13 Magnitude and ε disaggregation for the *SA*(0.1 s) and *SA*(3 s) values from the 2.1×10^{-3} rate of exceedance UHS of Figure 7.12. For *SA*(0.1 s), $\overline{M} = 5.9$, $\overline{\varepsilon} = 1.9$. For *SA*(3 s), $\overline{M} = 6.5$, $\overline{\varepsilon} = 1.2$.

SA period. Conditional on a rupture with a given magnitude, however, the probability of exceeding the *SA* value differs for each period. For $SA(0.1\text{ s}) > 0.92$ g, even small magnitude ruptures may cause an exceedance (e.g., $P(SA(0.1\text{ s}) > 0.92\text{ g} \mid m_i = 5.1) = 0.015$); this is important because such ruptures occur much more frequently than large magnitude ruptures, and so these ruptures contribute significantly in the disaggregation of Figure 7.13. Conversely, the smallest-magnitude ruptures have effectively zero probability of causing $SA(3\text{ s}) > 0.053$ g, as also reflected in the disaggregation figure. Thus, disaggregation helps us see that no single earthquake event is the design earthquake for every situation at a given site. The earthquake of interest will depend upon the *IM* type and value.

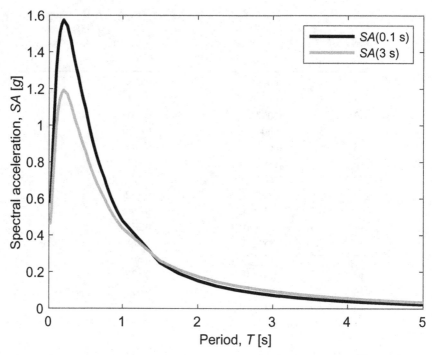

Fig. 7.14 Response spectra corresponding to the mean disaggregation magnitudes and ε values. For $SA(0.1\ \text{s})$, $\overline{M} = 5.9$, $\overline{\varepsilon} = 1.9$, and for $SA(3\ \text{s})$, $\overline{M} = 6.5$, $\overline{\varepsilon} = 1.2$.

This mixing of SA values from different ruptures to create a uniform hazard spectrum has sometimes been used to criticize the entire PSHA procedure. However, the UHS is merely one way to use the output of PSHA. None of the calculations prior to this section required the use of a UHS, and it is entirely possible to use PSHA results without ever computing a UHS.

A final important feature of the UHS has to do with the ε value. Figure 7.15 shows the $SA(1\ \text{s})$ response spectrum from Figure 7.14. Setting aside for the moment the multi-rupture discussion above, consider this as an approximation of the UHS. Also shown in gray in Figure 7.15 are response spectra of 54 recordings in the NGA-West2 database having rupture and site properties similar to those of interest here ($5.6 \leq M \leq 6.2$, $5 \leq R_{rup} \leq 15$ km, and $400 \leq V_{S,30} \leq 600$ m/s). We see that some of the ground-motion intensities are as large as the UHS, but most are not. This is the result of the ε value: if the GMMs used to perform the hazard analysis are well-calibrated, by definition only $\approx 3\%$ of the observed response spectra should exceed the SA value with $\varepsilon = 1.9$. Also shown by a dashed black line in Figure 7.15 is the response spectrum, with $\varepsilon = 0$, predicted by the Chiou and Youngs (2014) GMM. This spectrum is close to the median of the recordings, as expected. This figure serves to illustrate that when $\varepsilon > 0$ from disaggregation, the target spectrum will be larger than the median spectrum associated with the rupture properties. This is helpful to understand in general and will be relevant in Section 7.5 and Chapter 10 when we select ground motions to match target response spectral ordinates (and other IM types).

7.3.1 Utilization of the UHS

Many codified design procedures use the UHS (or some approximation to the UHS) for development of "design" spectra specifying seismic loads (e.g., CEN, 2004; NZS, 2004; American Society of

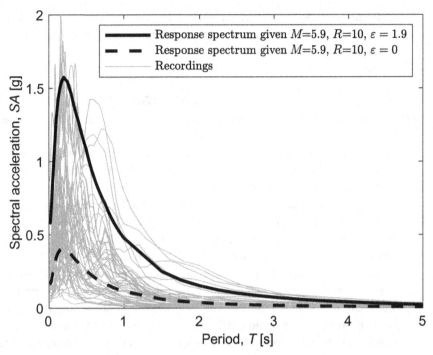

Fig. 7.15 Response spectra associated with the UHS calculation example. The solid black line shows the SA(1 s) response spectrum from Figure 7.14 (the spectrum for $M = 5.9$, $\varepsilon = 1.9$, $R = 10$ km, etc.). The dashed black line shows the spectrum for the same rupture but with $\varepsilon = 0$. The light gray lines show spectra from recordings having similar rupture properties.

Civil Engineers, 2010). Such procedures are conventionally based on the use of an equivalent SDOF system. Therefore only a single "point" on the UHS is utilized (i.e., the SA value for a single vibration period, T). When used for only a single vibration period, the fact that different causal earthquake ruptures dominate the UHS for different vibration periods is not problematic. However, problems arise for methods that simultaneously make use of spectral ordinates at more than a single vibration period, such as modal response spectrum analysis or in the selection of ground-motion time series for seismic response analyses.

Figure 7.15 helps illustrate the issues in using the UHS for problems requiring multiple vibration periods. The majority of the individual ground motions have spectral ordinates below that of the $\varepsilon = 1.9$ case. Ground motions whose SA values exceed this $\varepsilon = 1.9$ spectrum do so for only a small range of vibration periods, and generally have smaller values at other periods. Thus, the use of this $\varepsilon = 1.9$ spectrum (i.e., approximately the UHS) is equivalent to considering a ground motion that is equally severe at all vibration periods—when, for a single ground motion, such large SA ordinates occur for only a limited range of vibration periods.

In summary, for cases such as modal response spectrum analysis or selection of ground-motion time series, the UHS is generally conservative because it does not provide spectral ordinates at multiple vibration periods that are consistent with ground motions from individual earthquake ruptures. Section 7.5 discusses so-called conditional spectra that are mathematically consistent for use in these purposes, and Chapter 10 further generalizes these ideas and discusses their use for ground-motion selection.

The practicality of summarizing IMs with a UHS has led to its ubiquitous use in code-based seismic loading specifications, without separate definitions for seismic loading for use in the other applications alluded to above. Thus, the UHS's pervasiveness indicates its utility, but we should remember how a UHS is constructed to avoid misinterpreting the results as the spectrum from some single ground motion.

7.4 Hazard Maps

For applications such as the specification of seismic loads for building codes, it is useful to perform PSHA calculations for many sites and summarize the outputs in map form. A *hazard map* displays the *IM* value with a specified exceedance rate for each location. An example is shown in Figure 7.16.

Hazard maps are produced by specifying a grid of locations of interest, computing a hazard curve for each location, selecting the value with the specified exceedance rate, and then interpolating and mapping those values for all locations. The hazard calculation itself is identical to that in single-location studies, so nothing new is needed other than the ability to loop over many locations. The input models for map calculations may differ somewhat, however, in terms of their details.

SSMs are conceptually straightforward to handle, as all sources near any part of the mapped region can be specified. For calculations at a given location, only the sources with the potential to produce strong ground motions need to be considered. SSMs for large regions are sometimes more generic than for site-specific studies, as it is not practical (or potentially as relevant) to identify and characterize many small faults that impact only local areas. There is thus a greater tendency to use areal sources with spatially smoothed activity rates in hazard mapping studies.

GMMs can be reused from location to location in local areas.[4] However, care is needed if the study area spans multiple seismic or geologic environments. For example, in the map of Figure 7.16,

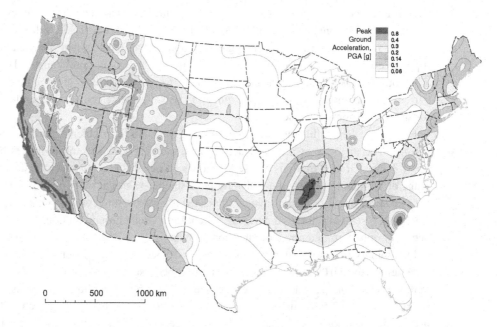

Fig. 7.16 Example hazard map for the continental United States, indicating *PGA* values with 2% probability of exceedance in 50 years (Petersen et al., 2014).

---

[4] This will be revisited later in Chapter 8.

the western portion of the map has active crustal region earthquake sources, the east portion is a stable continental region, and the northwest portion additionally has subduction sources. The appropriate GMMs, therefore, vary throughout the mapped region. An analyst developing a hazard map thus needs to associate relevant GMMs to corresponding sources or geographical regions, and use appropriate models in each circumstance.

It is not feasible to directly measure site condition information for the many locations in a hazard map. Therefore, a fixed site condition is often assumed at all locations, or a proxy such as topographic slope is used to approximately infer site conditions (Wald and Allen, 2007). For example, the map of Figure 7.16 assumes that all locations have $V_{S,30} = 760$ m/s. If the mapped results are used in a location-specific application where more site condition information is known, a site-correction factor is typically used to approximately reflect the anticipated effect of those site conditions on the hazard analysis.

When interpreting hazard maps, it is important to remember that the maps result from a series of independent calculations at each location, so there is no quantification of the simultaneous occurrence of ground motions at multiple locations. In this sense, a hazard map is equivalent to a UHS: both are convenient formats to summarize the results from a set of PSHA calculations, but they do not indicate the response spectrum (in the case of the UHS) or the regional shaking (in the case of a hazard map) that would result from any single earthquake event. Because hazard maps are almost exclusively used for applications where facilities at single locations are considered, there is no need to quantify regional ground motion, and so the hazard map format is not limiting in any way.

If quantification of regional ground motions from individual earthquakes is of interest (e.g., for evaluating risk to portfolios of properties or distributed infrastructure systems), then an alternative calculation is needed to capture spatial dependencies. Chapter 11 provides details on suitable alternatives for calculations in that circumstance.

7.5 Conditional Spectrum

The conditional spectrum calculation takes an alternate approach to the UHS calculation. It first specifies a spectral acceleration at a conditioning period, T^*. Then, conditional on that $SA(T^*)$ having some value, this approach computes the conditional distribution of SA at other periods. This calculation is often used in support of ground-motion selection (Chapter 10). The conditional spectrum is a common output of general PSHA calculations, so it is introduced here and is discussed in further detail in Chapter 10.

7.5.1 Basic Calculation

Consider a simple case where a single *rup* is the sole contributor to exceedances of all $SA(T)$, and consider two periods, denoted T^* and T_i. The SA values at these two periods can be predicted from the GMM (Equation 4.16) as

$$\ln SA(T^*) = \mu_{\ln SA(T^*)}(rup, site) + \sigma_{\ln SA(T^*)}(rup, site)\varepsilon(T^*) \tag{7.25}$$

$$\ln SA(T_i) = \mu_{\ln SA(T_i)}(rup, site) + \sigma_{\ln SA(T_i)}(rup, site)\varepsilon(T_i) \tag{7.26}$$

where $\mu_\bullet$ and $\sigma_\bullet$ are the mean and standard deviation predictions from a GMM, *rup* is the causal rupture event, *site* is given for the particular analysis case, and ε is the variate reflecting the value of SA relative to the mean GMM prediction, as in Section 4.4.

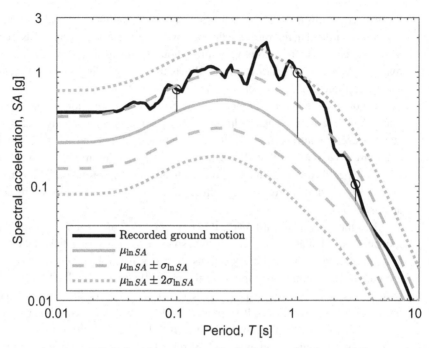

Fig. 7.17   Response spectrum from the 14145 Mulholland Drive recording of the Northridge earthquake, used to illustrate calculation of ε
values at three periods. Mean and standard deviation values are from the Chiou and Youngs (2014) GMM.

The variates $\varepsilon(T^*)$ and $\varepsilon(T_i)$ are not equal to each other, but they are correlated. To illustrate
this, Figure 7.17 shows an example SA_{RotD50} spectrum at three periods. The response spectrum is
from the 14145 Mulholland Drive recording of the Northridge earthquake ($M = 6.7$, $R_{rup} = 17$ km,
$V_{S,30} = 356$ m/s). Circles in the figure highlight spectral values at three example periods (0.1 s, 1 s,
and 3 s), and the ε values at these periods can be estimated graphically by counting the number of
standard deviations away from the $\mu_{\ln SA}$ spectrum they are. For example, the $SA(1\text{ s})$ value is near
the $\mu_{\ln SA} + 2\sigma_{\ln SA}$ line, indicating that $\varepsilon(1\text{ s}) \approx 2$. All three ε values are greater than 0, but they vary
in numerical value because of the "bumpiness" of the spectrum. The precise values are $\varepsilon(0.1\text{ s}) =$
0.85, $\varepsilon(1\text{ s}) = 1.96$, and $\varepsilon(3\text{ s}) = 0.49$. To study this variation more systematically, Figure 7.18
shows scatter plots of residual values from a larger set of ground motions.[5] The correlation in these
residuals varies depending upon the period pair, and can be estimated for each period pair of interest,
as discussed in Appendix B. The following calculations combine those correlations with the base
GMM predictions.

The conditional mean and standard deviation of $\ln SA$ values, given a value of $SA(T^*)$, are
computed with the following equations:

$$\mu_{\ln SA(T_i)|\ln SA(T^*)} = \mu_{\ln SA(T_i)}(rup, site) + \rho\, \varepsilon(T^*)\, \sigma_{\ln SA(T_i)}(rup, site) \qquad (7.27)$$

---

[5] More precisely, these calculations rely on decomposing ε into δW and δB terms, as introduced in Chapter 4. Appendix B
provides additional detail on this decomposition.

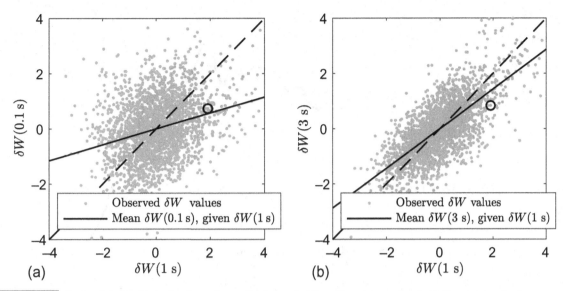

Fig. 7.18 Scatter plots of residual values from a large suite of ground motions. The points associated with the ground motion in Figure 7.17 are highlighted with black circles. (a) $\delta W(1 \text{ s})$ versus $\delta W(0.1 \text{ s})$. (b) $\delta W(1 \text{ s})$ versus $\delta W(3 \text{ s})$.

$$\sigma_{\ln SA(T_i)|\ln SA(T^*)} = \sigma_{\ln SA(T_i)} \sqrt{1 - \rho^2} \tag{7.28}$$

where $\mu_{\ln SA(T_i)|\ln SA(T^*)}$ and $\sigma_{\ln SA(T_i)|\ln SA(T^*)}$ denote the mean and standard deviation, respectively, of $\ln SA(T_i)$, conditioned on $SA(T^*)$ having a specified value. The $\mu_{\ln SA(T_i)}$ and $\sigma_{\ln SA(T_i)}$ terms are the mean and standard deviation, respectively, from the GMM. The ρ term is the correlation coefficient between $\ln SA(T^*)$ and $\ln SA(T_i)$ (i.e., $\rho = \rho_{\ln SA(T_i),\ln SA(T^*)}$). Equation 7.27 is referred to as the conditional mean spectrum (CMS), as it predicts the mean value of the response spectrum, conditional upon the occurrence of $SA(T^*)$. When a complete response spectrum distribution is computed (including the conditional standard deviation of Equation 7.28), the result is referred to as a conditional spectrum (CS; Baker, 2011).

Figure 7.19 shows a plot of the CMS using Equation 7.27. The *site* and *rup* properties from the 14145 Mulholland Drive recording are used to evaluate the GMM. The Chiou and Youngs (2014) GMM is used to compute $\mu_{\ln SA}$ and $\sigma_{\ln SA}$, and the Baker and Bradley (2017) correlation model is used to evaluate $\rho_{\ln SA(T_i),\ln SA(T^*)}$. The conditioning period is $T^* = 1$ s, and $\varepsilon(T^*) = 1.96$ so that the target value equals the recording at T^*. As seen in Figure 7.19, the CMS tracks the shape of the recorded spectrum more closely than the dashed $\mu_{\ln SA} + 1.96\sigma_{\ln SA}$ spectrum; this reflects the fact that $SA(1 \text{ s})$ is a "peak" in the recording's spectrum, rather than a high amplitude seen at all periods.

With this approach, the spectrum represents the amplitude of a real ground motion, rather than the unrepresentative envelope of the UHS. Looking back to Figure 7.15, we also see that individual recorded spectra may get as large as the UHS for a few periods, but those are generally narrow peaks in the spectra. This further illustrates the issue that the CMS solves. Because the CMS more realistically represents the spectrum of a single ground motion, it is now a standard output of PSHA studies. It is utilized in several standards and guidelines for the seismic design of structures (e.g., FEMA, 2012; ASCE, 2016; Moehle et al., 2017). The utilization of this spectrum for ground-motion selection is discussed further in Chapter 10.

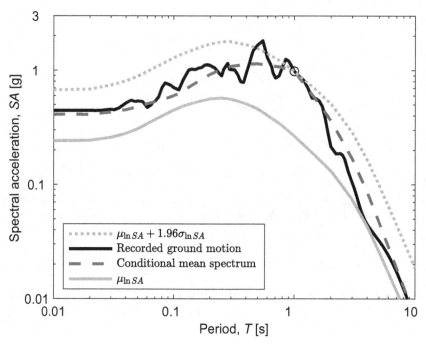

Fig. 7.19 Conditional mean values of spectral acceleration at all periods, given $SA(1\ s)$, and the example 14145 Mulholland Drive ground motion.

7.5.2 Consideration of Multiple Ruptures

Section 7.5.1 considered a single GMM and a single rupture, to provide a basic introduction of the calculation. The calculation can be generalized, however, to incorporate features of realistic hazard analyses: multiple contributing ruptures and epistemic uncertainty from logic trees (Bradley, 2010b; Lin et al., 2013c).

First, we note that when computing the mean spectrum, we take the expectation (with respect to the ruptures and GMMs) of the CMS for each particular rupture and GMM[6]:

$$\mu_{\ln SA(T_i)|\ln SA(T^*)} = \sum_k \sum_j p_{j,k}^d\ \mu_{j,k}(T_i) \tag{7.29}$$

where $\mu_{j,k}(T_i)$ is the CMS for rup_j, GMM_k, at period T_i:

$$\mu_{j,k}(T_i) = \mu_{\ln SA,j,k(T_i)|\ln SA(T^*)} = \mu_{\ln SA,k}(rup_j, T_i) + \rho(T^*, T_i)\varepsilon_j(T^*)\sigma_{\ln SA,k}(T_i) \tag{7.30}$$

and $p_{j,k}^d$ is the probability of rup_j and GMM_k, conditional on occurrence (or exceedance) of the $SA(T^*)$ of interest, as obtained from a disaggregation calculation.

---

[6] Logic trees include branches for uncertain parameters other than GMMs, but this is the only parameter that affects the CMS prediction directly. Uncertainties on SMM parameters affect the weighting of various rupture scenarios.

For the conditional standard deviation, there is now a contribution to variance from the partial $\varepsilon(T)$ correlation, and a contribution from the effect of multiple possible rup and GMM values:

$$\sigma_{\ln SA(T_i)|\ln SA(T^*)} = \sqrt{\sum_k \sum_j p_{j,k}^d \left[\sigma_{\ln SA,j,k(T_i)|\ln SA(T^*)}^2 + \left(\mu_{j,k}(T_i) - \mu_{\ln SA(T_i)|\ln SA(T^*)} \right)^2 \right]} \quad (7.31)$$

where $\sigma_{\ln SA,j,k(T_i)|\ln SA(T^*)}$ is the conditional standard deviation associated with rup_j, GMM_k, at period T_i:

$$\sigma_{\ln SA,j,k(T_i)|\ln SA(T^*)} = \sigma_{\ln SA,k}(T_i)\sqrt{1 - \rho^2(T_i, T^*)}. \quad (7.32)$$

Equation 7.27 is a good approximation of Equation 7.29 in the case where a single rup is a good approximation of the disaggregation, and where the GMMs are in relative agreement in predicting spectra for the rup of interest (Lin et al., 2013c). Equation 7.28 is an underestimate of Equation 7.31 because it omits the contribution to the variance of $\ln SA$ from variation in causal earthquake ruptures and GMMs. Again, the approximation of the simple equation is best when a single rup is a good approximation of the disaggregation, and the GMMs are in relative agreement.

Aggregation Approach

While Equations 7.29 and 7.31 are conceptually straightforward, they require significant output from a PSHA calculation in order to document all rup and GMM combinations, and their weights.

A convenient alternative calculation approach is to calculate, during the original PSHA computation, a CS using Equations 7.29 and 7.31 for each earthquake source and logic-tree terminal node. These sources and terminal nodes are considered in the original PSHA calculation, so the modification is to compute a CS for each at the same time as the exceedance rate associated with that case is considered. These numerous CS results can then be combined using Equations 7.30 and 7.32, but now with j representing all the earthquake sources and k representing all the logic-tree branches. In this case, the $p_{j,k}^d$ come from the PSHA contribution (mean annual exceedance frequency) for each earthquake source and logic-tree branch, normalized by the total aggregated hazard. This approach is straightforward computationally but cannot be postprocessed, and so requires incorporation in PSHA software itself.

7.5.3 Conditional Distributions of Other *IM*s

This subsection focused on response spectra, as SA values are the most common IMs used for these calculations. These concepts can easily be extended to other IMs (Bradley, 2010b). Equations 7.27 and 7.28 are generalized to refer to IMs instead of spectral values:

$$\mu_{\ln IM_i|\ln IM^*}(rup, site) = \mu_{\ln IM_i}(rup, site) + \rho_{\ln IM_i,\ln IM^*}\varepsilon(IM^*)\sigma_{\ln IM_i}(rup, site) \quad (7.33)$$

$$\sigma_{\ln IM_i|\ln IM^*}(rup, site) = \sigma_{\ln IM_i}(rup, site)\sqrt{1 - \rho_{\ln IM_i,\ln IM^*}^2} \quad (7.34)$$

where IM^* is the conditioning IM, and IM_i is another IM of interest.

Even for response spectra, it is useful to note this generality. For example, Gulerce and Abrahamson (2011) apply these concepts to compute a vertical response spectrum, conditional upon a horizontal $SA(T^*)$ value.

7.6 Vector PSHA

Vector PSHA is an alternate hazard calculation where joint occurrences of multiple IM values are considered (in slight contrast to the CS where a primary IM is considered, and conditional distributions of other parameters are calculated). This calculation requires specialized software, but it is occasionally appealing when multiple IMs are relevant for later analysis.

The calculation revises the basic PSHA formula of Equation 6.1 to consider simultaneous occurrences of multiple IM values (Bazzurro and Cornell, 2002). Because characterization of simultaneous exceedances of multiple IM values does not provide a convenient characterization for later analysis, we instead compute a mean rate density of occurrence of the IMs. For the case of two IMs the computation is as follows:

$$MRD_{IM_1,IM_2}(im_1, im_2) = \sum_{i=1}^{n_{rup}} f_{IM_1,IM_2}(im_1, im_2 \mid rup_i) \lambda(rup_i) \qquad (7.35)$$

where $MRD_{IM_1,IM_2}(im_1, im_2)$ is the mean rate density for IM_1 and IM_2, $f_{IM_1,IM_2}(im_1, im_2 \mid rup_i)$ is the joint PDF for IM_1 and IM_2, conditional on rup_i,[7] $\lambda(rup_i)$ is the typical source characterization term, and we sum over all considered ruptures.

The mean rate density can be used to find the rate of occurrence of any range of IM values by integrating over the relevant range:

$$\lambda(a < IM_1 < b, c < IM_2 < d) = \int_a^b \int_c^d MRD_{IM_1,IM_2}(im_1, im_2) \, dim_2 dim_1. \qquad (7.36)$$

To see the connection between this vector result and traditional PSHA, note that the traditional IM-exceedance hazard curve can be obtained by finding the rate of $im < IM_1 < \infty$ and $-\infty < IM_2 < \infty$:

$$\lambda(IM_1 > im) = \int_{im}^{\infty} \int_{-\infty}^{\infty} MRD_{IM_1,IM_2}(im_1, im_2) \, dim_2 dim_1. \qquad (7.37)$$

Evaluation of $f_{IM_1,IM_2}(im_1, im_2 \mid rup_i)$ for a given rupture is performed by assuming that $\ln IM_1$ and $\ln IM_2$ are jointly normal, an assumption shown to be reasonable for spectral values (Jayaram and Baker, 2008). In that case, the jointly normal PDF requires mean values and standard deviations for each IM, and correlation coefficients for the pair of IMs—the same information required for the conditional spectrum calculations in the previous section. So no fundamentally new information is used here, as expected since this is mostly a reformatting of the same information provided by conditional spectra. This calculation can also be performed within a logic-tree structure (for each terminal node), and the results processed to obtain mean hazard and other results of interest.

7.7 Earthquake Sequences in PSHA

The properties of aftershock sequences described in Section 3.3.5 are normally used to identify dependent events while declustering an earthquake catalog. However, events within these clusters

---

[7] This distribution also depends upon the characteristics of the site, i.e., $f_{IM_1,IM_2}(im_1, im_2 \mid rup_i, site)$, but this site dependence is suppressed here for notational simplicity.

can contribute to hazard and risk. Merely removing aftershocks (and foreshocks) to coerce empirical data to conform to a more convenient Poisson model can be hard to justify and may lead to biased results. Indeed, Boyd (2012) highlights the fact that hazard levels can be underestimated if dependent events are removed within the seismicity analysis. The consequences for risk can also be severe, given that damage can accumulate within structures throughout an aftershock sequence. A relatively weak aftershock event can cause significant problems for an already damaged structure. Therefore, this section outlines options to account for these clusters of dependent events within hazard analyses.

There are two main ways in which these dependent sequences can be accounted for in hazard analyses. A direct but challenging approach is to model time-dependent earthquake occurrence explicitly. Several such models have been proposed over many years (e.g., Anagnos and Kiremidjian, 1988), but it is difficult to constrain the parameters of these models using the relatively short periods of observation that are available. The other alternative is to incorporate the effects of these dependent event sequences within our existing framework (Boyd, 2012; Iervolino et al., 2014). This latter approach requires a little more abstraction, but it is generally more convenient to apply and interpret the results. Importantly, this latter approach also has the distinct benefit of maintaining our ability to make forecasts of hazard for generic time windows rather than specific time windows. The effects of dependent events can therefore be incorporated into existing seismic codes if so desired. Both of these general approaches are discussed in the sections that follow.

7.7.1 Aftershock Hazard Analysis

As discussed in Section 3.3.5, aftershock sequences have relatively predictable rates of occurrence and magnitude distributions. Spatial patterns of event densities also show relatively predictable power-law decay behavior, but existing fault networks can complicate this. These properties imply that once a significant event occurs, it becomes possible to make conditional predictions of the time-dependent hazard associated with the aftershock sequence.

Yeo and Cornell (2009) presented a framework for conducting what they referred to as aftershock probabilistic seismic hazard analysis (APSHA). Their framework defines the rate of exceedance of a level of ground-motion intensity *im* over some time interval commencing at time t and lasting Δt (where Δt will typically be days). The framework itself is similar to the traditional hazard integral discussed in Chapter 6, but rather than having a constant average rate of occurrence defining the likelihoods of ruptures, in APSHA, this rate is the time-dependent rate of aftershocks in the particular time window of interest. The overall framework is defined as[8]:

$$\lambda(IM > im, t, \Delta t; rup_{ms}) = \lambda_{as}(t, \Delta t; m_{ms}) \sum_{i=1}^{n_{as}} P\left(IM > im \mid rup_{as(i)}, site\right) P\left(rup_{as(i)} \mid rup_{ms}\right).$$

(7.39)

Here, $\lambda_{as}(t, \Delta t; m_{ms})$ is the time-dependent rate of occurrence of aftershock ruptures that have magnitudes between some lower bound, m_l, and upper bound, m_u, and rup_{ms} represents the mainshock rupture. The term $P\left(rup_{as(i)} \mid rup_{ms}\right)$ defines the probability of aftershock rupture $rup_{as(i)}$ given the

---

[8] In the original article, Yeo and Cornell (2009) worked with basic variables of magnitude, M, and distance, R, and so Equation 7.39 was cast as:

$$\lambda(IM > im, t, \Delta t; m_{ms}) = \lambda_{as}(t, \Delta t; m_{ms}) \int_R \int_{m_l}^{m_u} P(IM > im \mid m, r) f_{M,R}(m, r; m_{ms}) dm dr \qquad (7.38)$$

with magnitudes considered between some lower bound, m_l, and upper bound, m_u.

occurrence of the mainshock rupture rup_{ms}, and a total of n_{as} such aftershocks are considered. To derive $\lambda_{as}(t, \Delta t; m_{ms})$, we first use the modified Omori relationship to describe the time-dependent rate of all events. Assuming that no aftershocks take place with magnitudes between m_u and m_{ms}, we have the expression:

$$\lambda_{as}(t; m_{ms}) = \frac{10^{a+b(m_u-m_l)} - 10^a}{(t+c)^p}. \tag{7.40}$$

The parameter pairs a, b and c, p have previously been shown to relate to the doubly bounded exponential distribution and the modified Omori relationship, respectively. When making forecasts following some mainshock occurrence, the sequence-specific parameters will not yet be available. Therefore, the values of these parameters are instead taken from studies such as that of Reasenberg and Jones (1989, 1994), who derive parameter estimates for California using data from several earthquake sequences.

Equation 7.40 is then integrated to obtain the time-dependent rate of occurrence of events over the forecast period Δt:

$$\begin{aligned}
\lambda_{as}(t, \Delta t; m_{ms}) &= \int_{t}^{t+\Delta t} \lambda_{as}(\tau; m_{ms}) d\tau \\
&= \frac{10^{a+b(m_u-m_l)} - 10^a}{p-1} \left[(t+c)^{1-p} - (t+\Delta t+c)^{1-p} \right].
\end{aligned} \tag{7.41}$$

The conditional probability of occurrence of aftershock ruptures $P(rup_{as(i)} \mid rup_{ms})$ depends upon the magnitude and geometry of the mainshock rupture. However, the magnitude distribution of these conditional ruptures can be described by the doubly bounded exponential distribution. Therefore, if a rupture $rup_{as(i)}$ is regarded as being defined by a set of variables $rup_{as(i)} \equiv \{m_i, \boldsymbol{\theta}_i\}$, where m_i is the magnitude of the rupture, and $\boldsymbol{\theta}_i$ represents all other variables describing the ith rupture, then we can write:

$$P(rup_{as(i)} \mid rup_{ms}) \equiv P(\boldsymbol{\theta}_i \mid m_i)P(m_i; m_{ms}). \tag{7.42}$$

The final component of this expression comes from the magnitude distribution that is defined for a finite range of magnitude from some lower level that might provide ground motions of interest at the site m_l (often $m_l = 5$) up to some maximum magnitude for the aftershock sequence. Yeo and Cornell (2009) used the mainshock magnitude for this upper bound.

7.7.2 Equivalent Homogeneous Hazard Analysis

APSHA proposed by Yeo and Cornell (2009), and described in the previous section, is designed to enable the assessment of hazard over a specific period following the occurrence of a significant event. It does not, however, address the problem of how to represent these clusters of dependent events within a general hazard calculation for an arbitrary time window. Toro and Silva (2001) and Boyd (2012) proposed to work within the traditional framework of PSHA, in which events are a homogeneous Poisson process (i.e., average time-independent rupture rates are calibrated based upon declustered seismicity catalogs). However, each "independent" event is then deemed to represent a mainshock *and* a cluster of associated aftershocks. For each rupture, one can then compute the probability of exceeding some level of *IM* for the entire cluster, rather than just the mainshock.

Iervolino et al. (2014) formalized the concepts in Toro and Silva (2001) and Boyd (2012) and proposed sequence-based PSHA (or SPSHA). The key concept in this extension is that rather than having the term $P(IM > im \mid rup_i, site)$ in our common hazard expression of Equation 6.1, we replace this with

$$P(IM > im \mid rup_i, site) \mapsto P\left[IM(rup_i) > im \cup \boldsymbol{IM}(\boldsymbol{rup}_{as}) > im \mid rup_i, \boldsymbol{rup}_{as}, site\right]. \quad (7.43)$$

That is, we define the probability of exceeding *im* given a mainshock rupture rup_i and all of its associated aftershocks $\boldsymbol{rup}_{as}$ as being the union of the events that *im* is exceeded by the mainshock *or* by the sequence of aftershock ruptures $\boldsymbol{rup}_{as}$.

This expression can also be written in terms of an intersection of nonexceedances. From this viewpoint, the probability of exceeding *im* from either the mainshock or any event in the aftershock cluster is the same as the probability of not exceeding *im* from a mainshock and not exceeding *im* with any aftershock. An assumption is then made that the exceedance of *im* from the aftershock sequence is conditionally independent of the exceedance from the mainshock, given the definition of the mainshock rupture. In addition, the exceedances of *im* from each of the individual aftershock ruptures within the sequence are also assumed independent of one another. In other words, the occurrence of mainshock events is assumed to follow some homogeneous Poisson process, while the events within each aftershock cluster follow a separate nonhomogeneous Poisson process with a time-varying rate governed by the modified Omori law.

These assumptions enable the probability of exceeding *im* to be written as

$$P(IM > im \mid rup_i, site) = 1 - P\left(IM \le im \mid rup_i, site\right) \times P\left(\boldsymbol{IM} \le im \mid \boldsymbol{rup}_{as}, rup_i, site\right). \tag{7.44}$$

The final term of Equation 7.44, $P\left(\boldsymbol{IM} \le im \mid \boldsymbol{rup}_{as}, rup_i, site\right)$, represents the probability that none of the aftershocks causes exceedance of *im*. Equation 7.39 was previously used to define the average rate of exceedance of *im* for an aftershock cluster, and this term can be used again here. For an aftershock cluster having an average rate of exceedance of *im* defined by Equation 7.39, the probability of exceeding *im* over a time period Δt, starting at the time of the mainshock, can be expressed as

$$P\left(\boldsymbol{IM} > im \mid \boldsymbol{rup}_{as}, rup_i, site\right) = 1 - \exp\left[-\lambda(IM > im, t = 0, \Delta t; rup_i)\Delta t\right]. \tag{7.45}$$

The complement of Equation 7.45 is precisely the final term required within Equation 7.44.

Thus far, the presentation has focused on probabilities of exceedance for the cluster of events associated with each mainshock. To include these probabilities within a hazard calculation, we need to combine the probabilities with the rates of occurrence of the mainshock ruptures. Incorporating these rupture rates enables an expression for equivalent homogeneous hazard to be written as

$$\lambda(IM > im) = \lambda(EQ) - \sum_i^{n_{rup}} P(IM \le im \mid rup_i, site) \times e^{-\lambda(IM > im, t=0, \Delta t; rup_i)\Delta t} \lambda(rup_i). \quad (7.46)$$

In Equation 7.46, $\lambda(EQ)$, introduced previously in Section 6.2.1, is the rate of occurrence of all (mainshock) events, i.e., $\lambda(EQ) = \sum_i^{n_{rup}} \lambda(rup_i)$. The term $\lambda(IM > im, t = 0, \Delta t; rup_i)$ comes from Equation 7.39 with each rupture rup_i being regarded as a mainshock rupture rup_{ms} and the time period for each aftershock sequence commencing from the time of the mainshock such that $t = 0$. Equation 7.46 requires that a time period of Δt is defined for each aftershock sequence. In practice, this time period could be made to vary as a function of magnitude, as in the Gardner and Knopoff (1974) declustering algorithm, but it is common to assume a fixed time window for all events.

7.8 Implementation and Documentation of Hazard Studies

While this book focuses on hazard analysis principles rather than guidance on specific models and inputs to the analysis in a given situation, Chapters 6 and 7 point to issues that a real-world hazard study should address.

A hazard study must consider both aleatory variability and epistemic uncertainty. Aleatory variability is considered regularly in hazard studies, at least in the geometry of ruptures, recurrence relationships, and GMM standard deviations. Treatment of epistemic uncertainty, while standard practice in much of the world, is unfortunately not yet universal. Treatment of epistemic uncertainty also requires more careful consideration of issues such as uncertainties in SSM and GMM representations.

A hazard analysis is performed for a particular IM. Thus the hazard analyst and the user of the hazard study must communicate regarding the IM(s) of relevance for a given project. Typically, the user will specify the range of periods for which to compute spectral acceleration hazard, along with other IMs as needed.

Careful documentation of the hazard study is an essential supplement to any numerical results. Items to be documented include the following:

- Geometries, activity rates, and recurrence model parameters for all considered earthquake sources
- Ranges of considered ruptures, magnitudes, and distances
- Considered GMMs
- Ranges of truncation (if any) on ground-motion variability (ε)
- Discretization intervals used for any parameters (e.g., magnitude values, spatial locations for gridded seismicity sources)
- Full documentation of the logic tree used to quantify epistemic uncertainty
- Assumed profile of site conditions (including spatial variation and measurement uncertainty)
- Disaggregation plots and tables
- Findings from any sensitivity studies used to evaluate the impact of model assumptions.

Because a hazard calculation utilizes many input models and data sources, and because the numerical outputs can be sensitive to those inputs, a careful explanation of chosen inputs is crucial. Without clear documentation, it is challenging to understand and interpret results from a study, reproduce a calculation, or critically examine the needed assumptions. Insufficient documentation can also cause ambiguity or confusion when a seismic source model or ground-motion model is developed by one analyst (or team) and transferred to another analyst to perform the PSHA calculation.

In the SSHAC process for hazard analysis, a Hazard Input Document (HID) is developed to report many of the above items (NRC, 2012). It describes all components required to implement the hazard calculations, while justification of model choices and decisions is provided in the main report. The ASCE 7 Standard (ASCE, 2016, Chapter 21) requires many of the above items to be documented for any site-specific hazard study.

Exercises

7.1 Show that Equation 7.10 always produces a valid probability distribution (i.e., that the probabilities over all ruptures will always sum to 1).

7.2 To identify a single representative scenario, defined by a M, R, ε triplet, from a hazard disaggregation we typically choose between the mean $\bar{M}, \bar{R}, \bar{\varepsilon}$ and the mode M^*, R^*, ε^*.

    (a) Discuss the advantages and disadvantages of using each of these.

    (b) Upon choosing a particular M, R, ε triplet, one can insert these parameters back into a GMM. Do you expect this triplet of values to give you a ground-motion amplitude which matches that corresponding to the return period for which the disaggregation was performed? Justify your answer.

7.3 Consider a location 15 km from a fault that produces earthquakes with a Gutenberg–Richter magnitude distribution with $\lambda(M > m_{\min}) = 0.05$, $m_{\min} = 5.0$, $m_{\max} = 8.0$, and $b = 1$. You can check your rupture rate calculations against Table 3.5. Consider the following simple GMM for PGA (Cornell et al., 1979):

$$\mu_{\ln PGA} = -0.152 + 0.859M - 1.803 \ln(R + 25) \qquad (7.47)$$

$$\sigma_{\ln PGA} = 0.57 \qquad (7.48)$$

where PGA is in units of g.

    (a) Plot a ground-motion hazard curve for PGA values between 0.1 and 1 g, and provide tabulated exceedance rates for $PGA = 0.1, 0.2, \ldots, 1$ g.

    (b) Compute magnitude disaggregation distributions, conditional on $PGA > 0.2$ g and $PGA > 1$ g. Provide bar charts and tabulated values for $P(5 < M \leq 5.2 | PGA > x)$, $P(5.2 < M \leq 5.4 | PGA > x)$, etc. Compute the mean magnitude for each distribution. Comment on your results.

    (c) Provide a plot of PGA versus the mean disaggregation magnitude conditional on exceedance of that PGA. Superimpose on the same plot a line showing the magnitude that has 0.5 probability of exceeding each PGA value. State in words why these two lines differ.

    (d) Repeat (c), but for the case where the source has $b = 1.1$. Comment on the impact that this change has.

7.4 For the example of Section 6.3.3, and the Chiou and Youngs (2014) GMM,[9] compute a uniform hazard spectrum with a 1000-year return period. Compute a conditional mean spectrum,

---

[9] The reader should feel free to use an alternate GMM if one is available or preferred. We assume that a numerical code for a relevant GMM is available.

conditional on the $SA(0.2 \text{ s})$ and $SA(2 \text{ s})$ amplitudes with 1000-year return periods. Utilize the mean magnitude values and Equation 7.27 to compute the CMS, and adjust the $\varepsilon(T^*)$ value as needed to ensure that the spectrum matches the target $SA(T^*)$. Provide a plot of the UHS and the two CMSs on one figure, and comment on your results.

7.5 Repeat the calculation from the previous question, but using the complete magnitude disaggregation distribution and Equation 7.29. Superimpose your results on a plot with the result from the previous question.

8

Non-Ergodic Hazard Analysis

Previous chapters have outlined the fundamental inputs, computations, and results of a traditional PSHA. Throughout these earlier chapters, use has been made of generic models derived after pooling data from many different regions. This approach of pooling data has been relevant when discussing source-scaling relationships in Section 2.7.1, deriving the empirical GMMs in Chapter 4 and calibrating components of physics-based GMMs in Chapter 5. Ideally, however, we would use models that are derived explicitly for the location of interest.

It is increasingly common in advanced seismic hazard studies to refine generic models and model components to improve prediction accuracy and precision for the location of interest. Analyses that consider specific features of a particular location are known as *non-ergodic hazard analyses*. The present chapter provides conceptual insight into non-ergodic hazard analyses and examples of the types of refinements to traditional PSHA that are often considered. The specifics of these refinements are not discussed in detail because they depend strongly upon the particular data and information available and the budgetary and temporal constraints. Non-ergodic hazard analysis is an active area of research and development, and best practices are still evolving at this time.

Learning Objectives

By the end of this chapter, you will be able to do the following:

- Describe the ergodic assumption and how it influences PSHA results.
- Determine whether a given set of circumstances would allow the ergodic assumption to be relaxed.
- Transfer apparent aleatory variability into epistemic uncertainty in the context of PSHA.
- Evaluate alternative options for accounting for site-specific source, path, and site effects in a non-ergodic context.

8.1 Introduction

When modeling processes that are not entirely random, it is beneficial to understand what has happened in the past before making predictions about what may happen in the future. Unfortunately, our understanding of what has happened in the past is limited by a lack of empirical data for a specific region or a particular site location. The concepts and approaches discussed in Chapters 2 and 3 can be used to obtain geological evidence of past events that span centuries or more. However, such a period may still include only one or two large events. Similarly, Chapter 4 showed that it is rare to have more than a few decades of empirical ground-motion data for a given instrumented site. Furthermore, our site of interest will generally not have a seismic instrument. To overcome the problem of having a limited observation period at a given site, the *ergodic assumption* is generally employed.

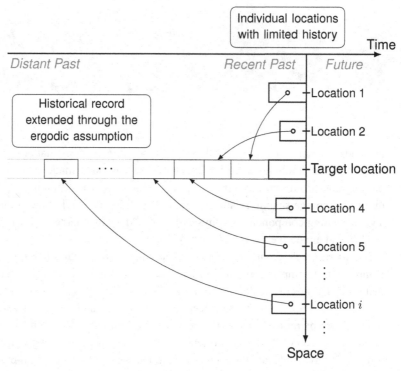

Fig. 8.1   Visualization of the ergodic assumption. The effective historical record at the target location is extended by merging the relatively short historical records (with consequent data) for many other spatial locations thought to be similar to the target location.

In the context of seismic hazard studies, the ergodic assumption means that we trade observations (or a lack thereof) in time with observations in space (Anderson and Brune, 1999). That is, we assume that the observations made at many different spatial locations (potentially distributed worldwide) are representative of what would be observed at a single location (our target location) if it were possible to extend the observation period at this location. This idea is visualized in Figure 8.1, where there are several locations that each have relatively short historical records that cover only the recent past.

Under the ergodic assumption, these multiple, spatially distributed records depicted in Figure 8.1 can be combined and assumed to represent a single long historical record that spans into the distant past. Note that no attempt is made to recreate an actual temporal sequence of observations via the ergodic assumption. Instead, we pool the data to enable estimates of statistics (mean rates of occurrence, maximum magnitudes, expected ground-motions, etc.) to be made using a larger dataset. In practice, the ergodic assumption is most commonly invoked to develop GMMs (particularly, empirical GMMs). In this context, the data compiled from multiple spatial locations are ground-motion recordings from which intensity measures are computed. The current chapter is, therefore, focused primarily upon how the ergodic assumption influences empirical GMMs. However, the concepts presented also apply to aspects of source characterization, as explained in Section 8.9. Source models are always partially non-ergodic to some extent, as they utilize region-specific seismicity analyses. The relative source geometry is also intrinsically coupled to the target location.

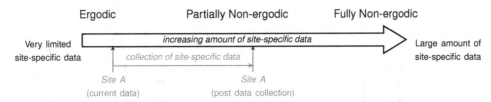

Fig. 8.2 Conceptual illustration of how the extent of available data influences placement on the continuum between ergodic analyses and fully non-ergodic analyses. As indicated for generic Site A, the accumulation of new site-specific data moves one to the right in the figure.

The ergodic assumption is ubiquitous in PSHA (and many other scientific disciplines) and has important implications for the accuracy and precision of PSHA modeling. The critical issue to be discussed herein is that when invoking the ergodic assumption, we pool data from sites or regions that we consider to be *similar*, or *analogous*, to our target site. However, as more data are compiled globally, it is becoming evident that individual earthquake sources, wave propagation paths, and sites have attributes that systematically differ from location to location. For this reason, there has been a recent move toward undertaking non-ergodic hazard analyses. This is not fully feasible in practice, so efforts are focused on removing detrimental effects from the ergodic assumption.

The Ergodic Continuum

The ergodic assumption is not binary in the sense that we choose to invoke it or not. There are many examples of ergodic ground-motion models (it has historically been the default approach). However, an increasing number of hazard studies use non-ergodic or partially non-ergodic approaches to ground-motion modeling. In reality, there are no examples of fully non-ergodic GMMs. Instead, we have an *ergodic continuum* in which a GMM is developed using some combination of generic ergodic components, or ergodic data, and site- or region-specific information. This ergodic continuum is illustrated schematically in Figure 8.2, and shows models for particular locations that are positioned along the continuum according to how much location-specific data is utilized in their development.

Strong parallels could be drawn between using the ergodic continuum in Figure 8.2 and another continuum concept related to epistemic uncertainty. From the perspective of epistemic uncertainty, this uncertainty conceptually reduces as one gathers more data to constrain a model (data carry information, and epistemic uncertainty reflects our lack of information and understanding). So, as one changes position from left to right within Figure 8.2, we should expect that epistemic uncertainty should generally also be reducing. However, as we saw in Figure 2.16 for source models, as more data become available, we tend to adopt more sophisticated modeling approaches that have increased parametric uncertainty. As a result, while the amount of location-specific data increases from left to right in Figure 8.2, the overall uncertainty does not necessarily follow the same trend. Increasing parametric uncertainty trades off against decreasing apparent aleatory variability as more elaborate modeling approaches are adopted.

The more data we have, the more the ergodic models, and their components, can be calibrated by site- or region-specific information, and the further we move to the right in Figure 8.2. However, it is essential to keep these continuum concepts in mind, particularly the link between the epistemic and ergodic continua. Exactly how far one moves to the right on the ergodic continuum should be based upon assessing the overall uncertainty associated with that position, in light of the modeling approach that is simultaneously adopted.

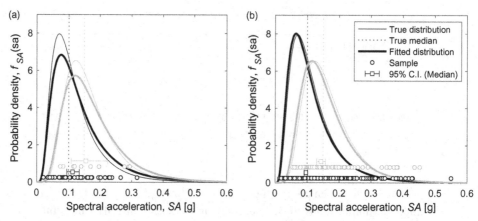

Fig. 8.3 Samples of spectral acceleration, SA, at a particular period representing observations made globally (black curves and points) and for a particular target location (gray curves and points). In panel (a), 50 observations exist from the global dataset, while just 10 exist for the target site. In panel (b), these sample sizes are increased by a factor of 10 (to 500 and 100, respectively), while the parent distributions remain the same. The circular markers represent samples/observations drawn from the true global or target distributions, and while their lateral positions show spectral accelerations, their vertical positions are set for visual purposes. The thin lines correspond to optimal fits made to these samples. Horizontal error bars show 95% confidence intervals (C.I.) in the logarithmic means for each sample.

8.2 Fundamental Concepts

Consider the sets of SA observations for a particular rupture scenario shown in Figure 8.3. In both panels, samples from two different lognormal distributions are shown. The black curves correspond to an ergodic global distribution of spectral ordinates for the rupture scenario and a site that is parametrically equivalent to the target; i.e., it has the same $V_{S,30}$ value. The gray curves correspond to data that has been recorded at the actual target site. In all cases, the heavy lines represent the real underlying distributions from which the data came (in practice, we can never know this underlying distribution). In contrast, the thin lines represent distributions inferred from the observations. Imagine that we wish to define the median SA value (the mean of the logarithmic SA) at the target site for this rupture scenario. We are also interested in the variability for the scenario. The "global" datasets are sampled from a lognormal distribution with a median $\widehat{SA} = 0.1$ g and logarithmic standard deviation of $\sigma_{\ln SA} = 0.6$. The "target" datasets are similarly sampled from a lognormal distribution with $\widehat{SA} = 0.15$ g and $\sigma_{\ln SA} = 0.45$.

In Figure 8.3a, we have a small dataset, with just ten observations made for our target region. Although this is site-specific data, we are reluctant to infer the true median from this sample because the 95% confidence interval for the median is relatively wide at $[0.108, 0.224]$. When we statistically compare the data from the target region with a larger (but still relatively small) global dataset of 50 observations compiled from multiple *similar* regions, we cannot detect any significant differences.[1] In this case, there are two options to proceed:

---

[1] Statistically, the confidence intervals overlap. Hypothesis tests for a common underlying distribution cannot be rejected, and visually it is difficult to be confident about the apparent differences.

- Invoke the ergodic assumption and combine the global data with the data from the target region to have a larger sample size of 60 observations. For this sample with the combined dataset, we would obtain $\widehat{SA} = 0.124$ g and $\sigma_{\ln SA} = 0.631$.[2] Or:
- If a global model already exists, invoke the ergodic assumption and use the global model directly, assuming that it applies to our site. For the current example, this results in a lower median of $\widehat{SA} = 0.113$ g, and a variability of $\sigma_{\ln SA} = 0.634$.[3]

Either option leads to the use of values for subsequent calculations that reflect the global distribution observations more than the actual target region. In particular, the true median for the target region is underestimated, and the variability is overestimated. Carrying this median bias and variance inflation through our subsequent calculations will influence our final hazard estimates.

Given the quantity of available local data, it is impossible to know that these biases exist, although we can gauge the potential for bias through our statistical analysis. Additionally, we would not have been better off working with just the ten observations from the target region. The sample mean and variance are unlikely to represent the underlying distribution.

In Figure 8.3b, the sample sizes are increased 10-fold relative to the example in Figure 8.3a, and now the confidence intervals suggest a statistically significant difference between the data from the target location and the global data. The increased confidence suggests the ergodic assumption could be avoided, or that the global dataset could be refined to remove data seen to differ from the target region's data. Once we recognize that the target location data are significantly different from the global data, we need estimates of the median and standard deviation specific to the target location. In theory, this allows us to remove the bias introduced by merging the data through the ergodic assumption and often reduces the data variance.[4] However, given the much smaller dataset (relative to that globally available), the 95% confidence interval for the median is larger than when using the global data.

If we consider the ergodic dataset as a mixture of the two distributions, the mean for the combined dataset (μ_{ergodic}) can be expressed in terms of the means of the global (μ_G) and target (μ_T) distributions as

$$\mu_{\text{ergodic}} = w_G \mu_G + w_T \mu_T \tag{8.1}$$

where $w_G = n_G / (n_G + n_T)$ and $w_T = n_T / (n_G + n_T)$ are weights reflecting the relative sizes of the datasets, with n_G and n_T being the number of observations for the global and target datasets, respectively. Similarly, the variance of the ergodic dataset, $\sigma^2_{\text{ergodic}}$, is

$$\sigma^2_{\text{ergodic}} = w_G \sigma^2_G + w_T \sigma^2_T + [w_G \mu^2_G + w_T \mu^2_T - (w_G \mu_G + w_T \mu_T)^2] \tag{8.2}$$

where σ^2_G and σ^2_T are the variances of the global and target distributions, respectively. Equation 8.2 indicates that the ergodic variance is a weighted sum of the variances of the two distributions *plus* an additional term in the square brackets (which is always positive) reflecting differences between the means of the two distributions.[5] From Equation 8.1, it can be appreciated that, by invoking the

---

[2] Note that the expected values of the median and logarithmic standard deviation of the underlying distribution are $E\left[\widehat{SA}\right] = 0.107$ g and $E[\sigma_{\ln SA}] = 0.597$ for this example case.

[3] These values are computed from the sample of 50 observations in the global dataset.

[4] In most situations, the non-ergodic variability will be lower, but there are cases where the opposite occurs.

[5] More generally, when an ergodic dataset is compiled from n regions, each with different distributions $f_i(x)$, the resulting distribution will be $f(x) = \sum_i^n w_i f_i(x)$, where w_i is the weight representing the relative contribution of data from region i. With the mean and variance for each region denoted by μ_i and σ^2_i, the corresponding mean and variance for the ergodic mixture are defined using the expressions: $\mu_{\text{ergodic}} = \sum_i^n w_i \mu_i$ and $\sigma^2_{\text{ergodic}} = \sum_i^n w_i \sigma^2_i + \sum_i^n w_i \mu^2_i - (\sum_i^n w_i \mu_i)^2$.

ergodic assumption, a bias in the mean equal to $\Delta_\mu = \mu_{\text{ergodic}} - \mu_T$ is introduced. Similarly, from Equation 8.2, we can state that the variance will differ by an amount $\Delta_{\sigma^2} = \sigma^2_{\text{ergodic}} - \sigma^2_T$.

Invoking the ergodic assumption could, therefore, be viewed as coming with an obvious cost. However, the alternative to this assumption is to directly estimate μ_T and σ_T from whatever limited data are available at the location of interest. Provided that some data exist, estimating μ_T and σ_T from the local data is possible, but the estimates will have significant epistemic uncertainty. This is particularly the case for rarer events, such as ground motions from large-magnitude earthquakes at short source-to-site distances. When conducting a non-ergodic hazard analysis, the primary challenge is to minimize potential modeling biases while also limiting model epistemic uncertainty. The ergodic assumption allows for better-constrained model parameter estimates (low parametric uncertainty), at the potential cost of these estimates being biased. Non-ergodic approaches should theoretically avoid these biases, but lead to much larger parametric uncertainties. Importantly, non-ergodic analyses using insufficient data can have increased parametric uncertainty *and* increased biases compared with using the ergodic assumption because of sampling biases from small observational datasets (Abrahamson et al., 2019). Therefore, it is essential to have sufficient evidence of a statistically significant difference from the ergodic baseline before utilizing a non-ergodic analysis.

8.3 Aleatory Variability versus Epistemic Uncertainty

Figure 8.4 illustrates how an ergodic distribution for a given IM can be viewed as a mixture of many site-specific distributions. The ergodic model is shown with a heavy black line, while the shaded distributions represent 20 other site-specific distributions. The region-specific aleatory variability is assumed constant for all regions, $\sigma^{\text{site},i}_{\ln IM} = \sigma^{\text{site}}_{\ln IM}$ for all i, but the region-specific mean value varies. Assume that this region-specific mean, $\mu^{\text{site},i}_{\ln IM}$, is distributed according to a normal distribution such that

$$\mu^{\text{site}}_{\ln IM} \sim \mathcal{N}\left(\mu^{\text{ergodic}}_{\ln IM}, \sigma^2_{\mu^{\text{site}}_{\ln IM}}\right). \tag{8.3}$$

Equation 8.3 implies that the ergodic mean is an unbiased estimate of the global mean of this IM, and the particular region-specific means are normally distributed around this global mean. In this example, all of the region-specific standard deviations are the same, so we can express the ergodic variance as

$$\left(\sigma^{\text{ergodic}}_{\ln IM}\right)^2 = \left(\sigma_{\mu^{\text{site}}_{\ln IM}}\right)^2 + \left(\sigma^{\text{site}}_{\ln IM}\right)^2. \tag{8.4}$$

Under this conceptual framework, we know that for a given site, i, the IM will be distributed according to

$$\ln IM^{\text{site},i} \sim \mathcal{N}\left[\mu^{\text{site},i}_{\ln IM}, \left(\sigma^{\text{site}}_{\ln IM}\right)^2\right] \tag{8.5}$$

but we will often not know how $\mu^{\text{site},i}_{\ln IM}$ compares with $\mu^{\text{ergodic}}_{\ln IM}$. This lack of knowledge about the value of $\mu^{\text{site},i}_{\ln IM}$ is an epistemic uncertainty. In the typical framework of PSHA, such uncertainties would be treated through a logic tree (previously discussed in Section 6.7).

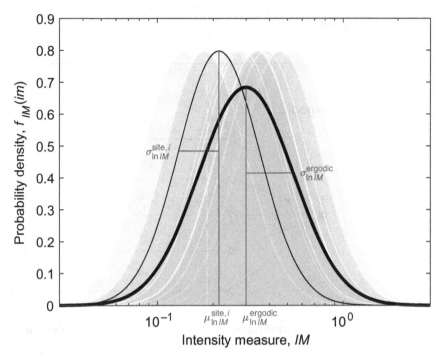

Fig. 8.4 Illustration of the ergodic assumption applied to a generic IM. The thick line shows the ergodic distribution with properties $\mu_{\ln IM}^{\text{ergodic}}$ and $\sigma_{\ln IM}^{\text{ergodic}}$. This distribution represents the composition of data from 20 different regions. The thin black line shows one particular site-specific distribution with properties $\mu_{\ln IM}^{\text{site},i}$ and $\sigma_{\ln IM}^{\text{site},i}$. The distributions for the other 19 regions are shown with lighter gray distributions.

When we do not have any data for the target region, it is reasonable to assume that the region-specific value of $\mu_{\ln IM}^{\text{site}}$ is distributed according to the regional variation observed within the ergodic dataset, i.e., Equation 8.3. For computing ground-motion hazard curves, an ergodic approach would use the ergodic GMM and use the apparent aleatory variability for that model (which includes the region-to-region variation).

Figure 8.5 illustrates the calculation of seismic hazard using ergodic and non-ergodic approaches in the GMM.[6] In the ergodic case, there is a single (ergodic) estimate of $\mu_{\ln IM}$, $\sigma_{\ln IM}$, and the consequent exceedance rate $\lambda(IM > im)^{\text{ergodic}}$. In the non-ergodic case, the uncertainty in the true value of $\mu_{\ln IM}^{\text{site},i}$ is reflected through the depiction of five different values of this mean on logic-tree branches with associated weights. While a discretized representation of $\mu_{\ln IM}^{\text{site},i}$ is useful for this illustration, Figure 8.5 also notes that uncertainty can be equivalently described in continuous or discrete forms.

The calculation of the IM exceedance rate in the non-ergodic case now requires explicit consideration of the unknown value of the mean, $\mu_{\ln IM}^{\text{site},i}$. Using the discretized logic tree in Figure 8.5, we have

$$\lambda(IM > im)^{\text{ergodic}} \equiv \sum_{j=1}^{n} w_j \lambda(IM > im)_j^{\text{site}} \tag{8.6}$$

---

[6] For simplicity, only a single rupture, rup, is shown. However, the result that follows is equally applicable to cases involving many rupture scenarios from a full SSM and a full logic tree for the GMM.

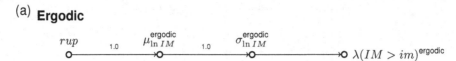

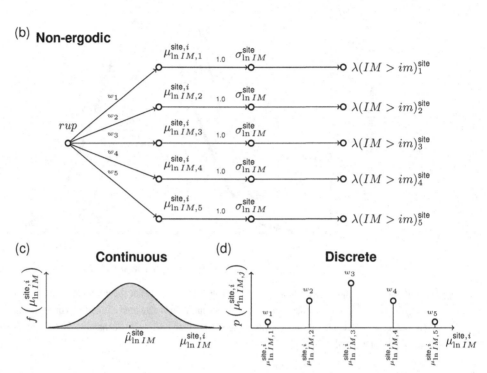

Fig. 8.5 Schematic illustration of an ergodic and non-ergodic approach to deriving a hazard curve. The *ergodic* "logic tree" in panel (a) uses an ergodic model that contains region-to-region variation in the mean logarithmic IM. The *non-ergodic* central logic tree in panel (b) explicitly considers uncertainty in what the site-specific mean logarithmic IM should be. The uncertainty in the value of $\mu_{\ln IM}^{\text{site},j}$ can be represented in *continuous* or *discrete* forms, as noted in the panels (c) and (d), respectively. The representations in panels (c) and (d) can be equivalent if one uses a sufficient number of branches and ensures that the nodes and weights preserve the moments of the distribution (e.g., Keefer and Bodily, 1983; Miller and Rice, 1983).

when the distribution of $\mu_{\ln IM}^{\text{site}}$ is consistent with Equation 8.3.

More generally, we can write Equation 8.6 as

$$\lambda(IM > im) = \lambda(rup) \int \underbrace{P(IM > im \mid \mu_{\ln IM}^{\text{site}}, \sigma_{\ln IM}^{\text{site}})}_{\text{aleatory}} \underbrace{f(\mu_{\ln IM}^{\text{site}} \mid \hat{\mu}_{\ln IM}^{\text{site}}, \hat{\sigma}_{\mu_{\ln IM}^{\text{site}}})}_{\text{epistemic}} \, d\mu_{\ln IM}^{\text{site}} \quad (8.7)$$

where $P(IM > im \mid \mu_{\ln IM}^{\text{site}}, \sigma_{\ln IM}^{\text{site}})$ is the probability of $IM > im$ obtained from a GMM, and $f(\mu_{\ln IM}^{\text{site}} \mid \hat{\mu}_{\ln IM}^{\text{site}}, \hat{\sigma}_{\mu_{\ln IM}^{\text{site}}})$ is the PDF representing the epistemic uncertainty in $\mu_{\ln IM}^{\text{site}}$ that is integrated over. The expression on the right-hand side of Equation 8.6 is effectively the discrete version of Equation 8.7 with the discrete weights playing an equivalent role to the PDF.

When no site-specific data exist, the typical assumption is that $\hat{\mu}_{\ln IM}^{\text{site}} \equiv \mu_{\ln IM}^{\text{ergodic}}$ and $\hat{\sigma}_{\mu_{\ln IM}^{\text{site}}} \equiv \sigma_{\mu_{\ln IM}^{\text{site}}}$ from Equations 8.3 and 8.4. The hazard curves computed from Equations 8.6 and 8.7 are equivalent in this case because we assume the relations in Equations 8.3 and 8.4 hold; i.e., the epistemic uncertainty in $\mu_{\ln IM}^{\text{site}}$ matches the location-to-location variability in this term embedded within the ergodic GMM.

A critical idea within non-ergodic hazard analysis is the relationship of aleatory and epistemic components within Equation 8.7. When there are few or no site-specific data, the epistemic uncertainty represented in Equation 8.7 will nullify the benefits of a site-specific estimate of the parameters ($\mu_{\ln IM}^{\text{site}}$ in this example). This arises because the term $\sigma_{\mu_{\ln IM}^{\text{site}}}$ in Equation 8.4 will be at least as large as the location-to-location variability within an ergodic GMM, and there is no basis for supposing that $\hat{\mu}_{\ln IM}^{\text{site}}$ differs from $\mu_{\ln IM}^{\text{ergodic}}$. However, as the available site-specific data increase, differences between $\hat{\mu}_{\ln IM}^{\text{site}}$ and $\mu_{\ln IM}^{\text{ergodic}}$ are likely to arise, and the estimate $\hat{\sigma}_{\mu_{\ln IM}^{\text{site}}}$ will generally reduce relative to $\sigma_{\mu_{\ln IM}^{\text{site}}}$ from Equation 8.4. Equation 8.7 therefore allows for the prediction of a hazard curve with more appropriate precision and accuracy than the ergodic alternative. These concepts are illustrated explicitly in the following section.

8.3.1 Example of Regional Ground-Motion Effects

Consider an ergodic GMM with the commonly adopted form of Section 4.6.2:

$$\ln im = \mu_{\ln IM} + \delta B + \delta W. \tag{8.8}$$

The between-event term, δB, has a variability of τ_E^2, and the within-event term, δW, has a variability of ϕ^2. The overall ground-motion variability is therefore $\sigma_E^2 = \tau_E^2 + \phi^2$. While δB is considered entirely aleatory variability in this ergodic model, in reality it will have some location dependence. If we consider this location-to-location variablity with the variable δL, which varies among different locations with a variance of τ_L^2, we can express the between-event variance as $\tau_E^2 = \tau_L^2 + \tau_{NE}^2$, where τ_{NE}^2 is the "remaining" non-ergodic between-event variance after removing the ergodic location-to-location effect.

Now consider a non-ergodic approach to this same problem. In this case, we would consider the GMM to have the form:

$$\ln im = \mu_{\ln IM} + \delta L + \delta B_0 + \delta W \tag{8.9}$$

where δL is a systematic location effect, and δB_0 is the between-event term at this particular location. Conceptually, the term δB in Equation 8.8 is equivalent to $\delta L + \delta B_0$ in Equation 8.9. We know there is a source location effect δL (with some unknown value), and we also know the location-to-location variability should not strictly be included in the ground-motion variability for our application because it is an epistemic uncertainty. A specific source location does not have this component of location-to-location variability associated with it. However, at a specific location, there will still be systematic source effects from event to event that are reflected by δB_0.

Without any local recordings, we would assume that $\delta L \sim \mathcal{N}(0, \tau_L^2)$. In a Bayesian framework (Gelman et al., 2013), this would be regarded as our prior distribution for the unknown region effect δL. Imagine now that we observe some new data in our target region. These data consist of several estimates of between-event terms from a suite of recorded events, $\widehat{\delta B_i}$, for $i = 1, \ldots, n$, that enable us to estimate the distribution of δL. For the framework of Equation 8.9, we have $\delta B_0 \sim \mathcal{N}(0, \tau_{NE}^2)$, which is the non-ergodic distribution of δB_0 for our particular location. In contrast, for the ergodic framework in Equation 8.8, we have $\delta B \sim \mathcal{N}(\delta L, \tau_{NE}^2)$ and $\delta L \sim \mathcal{N}(0, \tau_L^2)$. To adopt a non-ergodic approach, we need to estimate $\widehat{\delta L}$ and its uncertainty $\hat{\tau}_L$ using these new data.

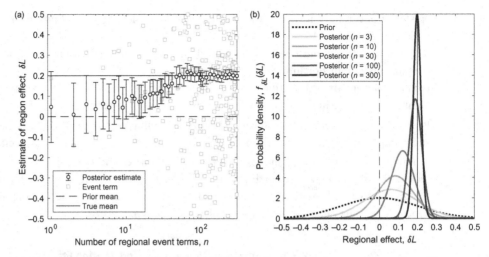

Fig. 8.6 Panel (a) shows the variation of the posterior estimate of δL as increasing numbers of event terms are estimated in a target region. The prior mean, $\delta L_{(0)}$, is shown with a horizontal dashed line, while the true mean δL^{true} is shown by the solid horizontal line. Panel (b) shows the evolution of the posterior distribution of δL as a function of the number of event terms estimated within the region.

As the estimate of the event term for each earthquake, $\widehat{\delta B_i}$, becomes available we can update our prior distribution using the pair of equations:[7]

$$\widehat{\delta L}_{(i+1)} = \frac{\widehat{\delta L}_{(i)}\tau_{NE}^2 + \widehat{\delta B_i}\hat{\tau}_{L(i)}^2}{\hat{\tau}_{L(i)}^2 + \tau_{NE}^2} \tag{8.10}$$

and

$$\hat{\tau}_{L(i+1)} = \frac{\hat{\tau}_{L(i)}\tau_{NE}}{\sqrt{\hat{\tau}_{L(i)}^2 + \tau_{NE}^2}}. \tag{8.11}$$

In Equations 8.10 and 8.11, $\widehat{\delta L}_{(i+1)}$ and $\hat{\tau}_{L(i+1)}$ are the posterior estimates of δL and τ_L, given our prior estimates of $\widehat{\delta L}_{(i)}$ and $\hat{\tau}_{L(i)}$. That is, Equation 8.10 progressively updates the mean estimate of the regional effect, $\widehat{\delta L}$, as each new earthquake is recorded, and Equation 8.11 updates the standard error in the estimate of this regional effect. In these equations, the subscript (i) represents the estimate of the designated terms from earthquake i, enabling an $i + 1$ update of the δL and τ_L. So, when the estimate of $\widehat{\delta B_i}$ arrives, $\hat{\tau}_{L(i)}$ and $\widehat{\delta L}_{(i)}$ are updated to become $\hat{\tau}_{L(i+1)}$ and $\widehat{\delta L}_{(i+1)}$, respectively.

This process is illustrated in Figure 8.6. For this example the true, unknown value of δL for the region is $\delta L^{\text{true}} = 0.2$, and we assume[8] $\tau_L = 0.2$ and $\tau_E = 0.4$, such that $\tau_{NE} = 0.346$, and $\phi = 0.5$. Values of event terms are simulated from the true distribution as $\delta B \sim \mathcal{N}(\delta L^{\text{true}}, \tau_{NE}^2)$ and depicted as gray squares in Figure 8.6a. These event terms are used to progressively update the distribution of δL for our region. The properties of this distribution (the mean and standard error) evolve in Figure 8.6 and approach the true value of δL^{true} as the number of event terms increases.

---

[7] These equations for the posterior mean and standard deviation arise from a Bayesian approach, starting with a conjugate prior for a normal distribution with a known variance.

[8] These are realistic values, similar to those obtained by Stafford (2014a).

Table 8.1. Numerical estimates of the regional effect, δL, and the impact upon non-ergodic variance components corresponding to the example in Section 8.3.1, and illustrated in Figure 8.6. Here, $\tau^2 = \hat{\tau}_L^2 + \tau_{NE}^2$, $\sigma_{NE}^2 = \tau^2 + \phi^2$, and remaining column descriptions are provided in the text

n	$\widehat{\delta L}$	$\hat{\tau}_L$	τ_{NE}	τ	ϕ	σ_{NE}	$e^{\widehat{\delta L}}$	σ_{NE}/σ_E
0	0.000	0.200	0.346	0.400	0.500	0.640	1.000	1.000
3	0.060	0.141	0.346	0.374	0.500	0.624	1.062	0.975
10	0.083	0.096	0.346	0.359	0.500	0.616	1.087	0.962
30	0.122	0.060	0.346	0.352	0.500	0.611	1.130	0.955
100	0.189	0.034	0.346	0.348	0.500	0.609	1.208	0.951
300	0.199	0.020	0.346	0.347	0.500	0.609	1.220	0.950

The evolution of the distribution of δL is shown more explicitly in Figure 8.6b. This figure illustrates that we need a lot of data to obtain good approximations to the true value of δL.[9] However, this result depends upon the relative magnitudes of τ_L and τ_{NE}.

Sample numerical values from this example are shown in Table 8.1. The table shows how the estimates of $\widehat{\delta L}$ and $\hat{\tau}_L$ change as the number of event terms, n, increases. The location-specific between-event variability, τ_{NE}, remains unchanged, but the effective non-ergodic estimate of the between-event variability, τ, gradually reduces as the precision of the $\widehat{\delta L}$ estimate increases. The impact upon the total variability, σ, is relatively small, as it is dominated by the (irreducible) within-event variability ϕ. The final two columns show the change in the median estimate of IM accounting for the updates to δL, and the reduction of the non-ergodic standard deviation relative to the ergodic value. For $n = 0$, we assume $\widehat{\delta L} = 0$ and so $e^{\delta L} = 1$, reflecting that our median predictions match the ergodic GMM. As $n \to \infty$, our estimate $\widehat{\delta L} \to \delta L^{\text{true}}$ and so the median non-ergodic IM value is approximately 22% higher than the ergodic equivalent. Similarly, for $n = 0$ there is no reduction in the total standard deviation, while as $n \to \infty$ the non-ergodic total standard deviation approaches 95% of the ergodic value.

Figure 8.7 shows the associated impact upon seismic hazard curves estimated for this example location. The SSM consists of a single rupture scenario, and is deliberately simple to highlight the impact of the non-ergodic treatment of the GMM. The median spectral acceleration for the scenario, using an ergodic GMM, is 0.1 g, while the true regional effect remains δL^{true}. The heavy gray line shows the true hazard curve. In practice, we cannot know this hazard curve. However, Figure 8.7 demonstrates that with more and more regional data, we can modify the epistemic uncertainty in δL and obtain better estimates of this true hazard. For the example shown here, $\delta L > 0$ and the change in the total standard deviation associated with using the non-ergodic estimate of τ is not very large (see Table 8.1). The differences in the seismic hazard curve are, therefore, driven by changes in the mean of the IM distribution rather than its variance. As a consequence, using the non-ergodic approach leads to higher levels of hazard for this example. However, at longer return periods, the sensitivity to the standard deviation increases (Section 6.5.4). For very low values of the probability of exceedance (outside the range shown in Figure 8.7), the effect of changes in the standard deviation dominate the effect of δL, and the hazard estimates become lower in the non-ergodic case.

---

[9] The statistical error is proportional to $1/\sqrt{n}$, where n is the sample size.

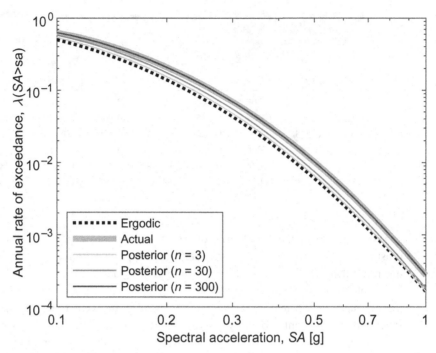

Fig. 8.7 Hazard curves estimated from a single rupture scenario with increasing levels of regional information. Note the evolution from the ergodic hazard curve toward the true non-ergodic hazard curve as the regional information increases.

8.4 When Can Non-Ergodic Approaches Be Applied?

Non-ergodic approaches can be applied when we have enough data to distinguish systematic effects from statistical fluctuations. Referring back to Figure 8.4, if no information exists to constrain the site-specific value of $\mu \equiv \mu_{\ln IM}^{\text{site},i}$, we should treat μ as being an unknown and reflect its uncertainty through $\sigma_\mu \equiv \sigma_{\mu_{\ln IM}^{\text{site},i}}$, i.e., the bottom row of Figure 8.5. This approach explicitly recognizes that σ_μ^2 is epistemic in nature. In contrast, the ergodic approach assumes that the GMM provides an unbiased estimate of μ and treats the site-to-site variability σ_μ^2 as aleatory variability.

When some information exists, an estimate of μ may be obtained. The aleatory variability considered within the hazard integral should not include the effects of σ_μ^2. However, while the aleatory variability is reduced, we must also add in the epistemic uncertainty associated with not knowing μ exactly. For limited location-specific data, the uncertainty in μ can be substantial and may be larger than σ_μ^2 that was removed from the aleatory variability, thus negating any benefit of a non-ergodic approach (Abrahamson et al., 2019). As the amount of site-specific data increases, we can continue to revise (generally reduce) σ_μ^2. This process ultimately allows us to entirely relax the ergodic assumption (as we move to the far right of the continuum shown in Figure 8.2).

The process for deciding whether to adopt a non-ergodic approach is shown in Figure 8.8. The first step is to compare the available data for the site with an ergodic model's predictions. This comparison is made to see whether using the ergodic assumption is likely to lead to biased model predictions and subsequently hazard estimates. Suppose there is no apparent difference between the mean of the

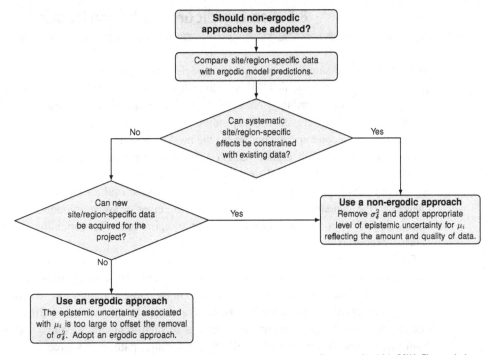

Fig. 8.8 Flowchart showing the typical process used to decide whether to adopt a non-ergodic approach within PSHA. The symbol μ_i represents a location-specific (non-ergodic) estimate of a particular model component. The symbol σ_δ^2 represents the location-to-location variability of this same model component that is ordinarily embedded within the variability of an ergodic model. If the ergodic mean is μ_E, then $\mu_i = \mu_E + \delta$.

site-specific data and the mean predictions of the ergodic model. In that case, this does not determine that ergodic approaches should automatically be applied, because the variance of the ergodic model may still not be appropriate for the target site. Therefore, the second step is to evaluate how well one can constrain the site-specific estimate of the parameter in question. When the site-specific value is strongly supported by data (whether it differs from the ergodic predictions or not), one can modify the ergodic model's variance to make it more site-specific. In the more common case where limited data exist, there will be a nontrivial degree of epistemic uncertainty in the site-specific parameter or model. This epistemic uncertainty must then be represented through logic tree branches, and the benefits of reduced variance via non-ergodic approaches could be offset by the parameter or model uncertainty.

It should be noted that the non-ergodic approach is technically the correct approach in all cases. The conceptual flowchart in Figure 8.8 primarily serves to recognize the practical challenges associated with constraining the epistemic uncertainty in location-specific model components. If data are insufficient to enable a realistic representation of this uncertainty, then the non-ergodic approach can give inaccurate results, even if the underlying framework is more technically defensible.

Figure 8.8 serves as a general guide to the decision process that can be followed but does not provide concrete limits or tests that can be used to reach a decision in practice. The nature of the data available for the project, and project constraints (time, budget, regulatory requirements, etc.) can all influence how one views the decision nodes represented in Figure 8.8. The next section addresses more practical issues that dictate when non-ergodic approaches warrant consideration.

8.5 Non-Ergodic Ground-Motion Models

In Section 8.3, we saw that non-ergodic models more appropriately treat aleatory variability and epistemic uncertainty. Ergodic GMMs include the variability in IM values between regions and sites as apparent aleatory variability. Non-ergodic GMMs require either a bespoke model for the target location or an existing ergodic model modified for the target location. The paucity of empirical data dictates that the second approach is by far the most common.

To modify an existing ergodic GMM for use in a particular application, we adjust both the logarithmic mean and standard deviation of each rupture scenario's distribution. Changes to the logarithmic mean can lead to higher or lower ground-motion values for a given scenario. In principle, changes to the standard deviation can also lead to higher or lower amounts of variability. However, as ergodic GMMs contain a significant amount of region-to-region variability, the standard deviation is usually reduced.

The basis for adjustments to the logarithmic mean follows from the backgrounds presented in Chapters 4 and 5. When sufficient empirical data exist (for the rupture scenarios of interest), the adjustments can be calibrated using these data. In the more common case where empirical data are lacking, theoretical considerations can be used to extrapolate the effects of differences observed from other scenarios (often low-amplitude recordings of small-magnitude events) and derive adjustments from these. Typical approaches to adjustment for site, path, and source effects are subsequently presented in Sections 8.6–8.8.

In contrast, the basis for adjustments to the standard deviation is almost exclusively statistical. It requires an understanding of how regional variations manifest in the observed ground-motion variability. The following section focuses on the components of ground-motion variability and the typical framework to reduce variability in non-ergodic hazard applications.

8.5.1 Components of Ground-Motion Variability

In Section 4.6.2, the aleatory variability in a GMM was represented as a combination of between-event, δB, and within-event, δW, components. This formulation is shown again here in Equation 8.12,[10] but with the rupture being represented by "event" e, and the site being represented by "site" s:

$$\ln im_{es} = \mu_{\ln IM}(e, s) + \delta B_e + \delta W_{es} = \mu_{\ln IM}(e, s) + \eta_e \tau + \varepsilon_{es} \phi. \tag{8.12}$$

The variation of the between-event terms is represented by the between-event standard deviation τ, while the corresponding standard deviation of the within-event components is ϕ. The terms η_e and ε_{es} are standard normal variates that equivalently describe the deviations δB_e and δW_{es} in terms of the number of standard deviations for each component.

In this framework, the apparent variation of ground motions from event to event is based primarily upon recognizing that earthquakes of the same magnitude, occurring on different faults, will have different average ground-motion amplitudes. Therefore, each set of observations from

---

[10] In Section 4.6.2, this equation was defined as $\ln im_{ij} = \mu_{\ln IM}(rup_i, site_j) + \delta B_i + \delta W_{ij}$. However, it is more convenient to introduce subscripts that more clearly relate to source and site effects when extending the framework. Therefore, rup_i is replaced with e for event, $site_j$ is replaced with s for site, and this subsequently allows a particular path connecting the event and site to be denoted as es.

a given earthquake can depart systematically from the global average level for an event of this magnitude. However, what would happen if we could observe multiple ruptures of the same fault? Presumably, there will be subtle changes in the source properties between events that may influence the hypocenter's location and how the rupture propagates. These changes will cause each event's motions to differ, even when they occur on the same segments of the same fault. There will also be source properties that remain very consistent between these ruptures, such as the fault segments' overall geometry. Therefore, it is reasonable to expect that some deviations from the global average will be systematic and repeatable for this particular rupture scenario. While the variations associated with the subtle changes will appear (and may be) genuinely random or aleatory, the systematic and repeatable deviations are, in principle, knowable. Therefore, the variance components of τ and ϕ associated with Equation 8.12 really portray *apparent aleatory variability* and include both genuine aleatory contributions as well as these repeatable epistemic contributions.

Therefore, when decomposing ground-motion variability, it is more appropriate to separate these epistemic terms from the remaining apparent aleatory components (Al Atik et al., 2010; Landwehr et al., 2016). The concepts discussed above for the between-event variability, where repeatable source effects were considered, can also be extended to path and site attributes of ground motions. There are physical reasons why particular source-to-site paths will have repeatable effects. There will also be systematic effects associated with a given recording site. A revised formulation of Equation 8.12 that takes into account these repeatable contributions of source, path, and site is therefore

$$\ln im_{es} = \mu_{\ln IM}(e, s) + \underbrace{\delta_e + \delta_{es} + \delta_s}_{\text{epistemic}} + \underbrace{\psi_e + \psi_{es} + \psi_s}_{\text{aleatory}}. \tag{8.13}$$

Here, the δ_x terms, where $x \in \{e, es, s\}$, represent the systematic deviations that exist for a given rupture scenario, δ_e, source-to-site path, δ_{es}, and a site, δ_s. Also, a given observation will have aleatory contributions ψ_x, which again arise from source, ψ_e, path, ψ_{es}, and site, ψ_s, components. Correlations can exist among these components, but we present the case where all variance components are independent for simplicity. It is common to describe the various deviations in terms of numbers of standard deviations, and hence, similar to Equation 8.12, Equation 8.13 can be expressed as

$$\ln im_{es} = \mu_{\ln IM}(e, s) + \zeta_e \tau_e + \zeta_{es} \tau_{es} + \zeta_s \tau_s + \varepsilon_e \phi_e + \varepsilon_{es} \phi_{es} + \varepsilon_s \phi_s \tag{8.14}$$

where the standard deviations of the epistemic components are all denoted τ_x, while the standard deviations of the aleatory components are denoted by ϕ_x. The use of τ_x and ϕ_x in Equation 8.14 should not be confused with the τ and ϕ from Equation 8.12. The terms in Equation 8.14 relate directly to the distinct epistemic and aleatory components of Equation 8.13. In contrast, the symbols τ and ϕ in Equation 8.12 are adopted for historical reasons, and include both epistemic and aleatory contributions. The standard normal variates are ζ_x and ε_x for the epistemic and aleatory contributions, respectively.

The notation used here departs from that of Al Atik et al. (2010), but there is general consistency between the two frameworks. Importantly, Equation 8.13 highlights the distinction between components of apparent aleatory variability that are actually aleatory, and those that are epistemic and have been entrained into our datasets through the necessary use of the ergodic assumption.

In the framework of Equation 8.13, the true non-ergodic aleatory variance is $\phi^2 = \phi_e^2 + \phi_{es}^2 + \phi_s^2$. Similarly, the total non-ergodic epistemic variance is $\tau^2 = \tau_e^2 + \tau_{es}^2 + \tau_s^2$. However, for a fully ergodic analysis, the systematic deviations in the source, path, and site effects are included within the dataset and become apparent aleatory variability. The total ergodic variance is therefore $\tau^2 + \phi^2$.

Table 8.2. Variance decomposition for different ergodic, non-ergodic, and partially non-ergodic analyses cases. The variance components are all assumed independent of one another, and τ_x and ϕ_x are consistently used to represent epistemic and aleatory components of variability. The terms $\mu \equiv \mu_{\ln IM}$ and $\sigma \equiv \sigma_{\ln IM}$ in all cases, but the subscripts are dropped for notational simplicity. However, subscripts of *FE*, *PNE*, and *FNE* are added to σ to represent full ergodic, partially non-ergodic, and full non-ergodic cases, respectively

Analysis type		Aleatory variability	Epistemic components[a]	Description of available constraint
Full ergodic		$\sigma^2_{FE} = \tau^2 + \phi^2$	μ, σ	No site-specific or region-specific information available
Partially non-ergodic	Site	$\sigma^2_{PNE} = \tau^2_e + \tau^2_{es} + \phi^2$	μ, σ, δ_s	Site-specific empirical data, or numerical/theoretical constraint
	Path	$\sigma^2_{PNE} = \tau^2_e + \tau^2_s + \phi^2$	μ, σ, δ_{es}	Path-specific empirical data, or numerical/theoretical constraint
	Source	$\sigma^2_{PNE} = \tau^2_{es} + \tau^2_s + \phi^2$	μ, σ, δ_e	Source-specific empirical data, or numerical/theoretical constraint
	Path and site	$\sigma^2_{PNE} = \tau^2_e + \phi^2$	$\mu, \sigma, \delta_{es}, \delta_s$	Site-specific constraint, and multiple observations of repeated paths
Full non-ergodic		$\sigma^2_{FNE} = \phi^2$	$\mu, \sigma, \delta_e, \delta_{es}, \delta_s$	Specific constraints upon site, path, and source components

[a] Here, the epistemic uncertainty in σ reflects all of the contributions to the aleatory variability.

This distinction between aleatory and epistemic contributions has implications for the computation of seismic hazard curves within a logic-tree framework. In a fully non-ergodic hazard analysis, the distribution of ground motions for a given rupture scenario has a mean of $\hat{\mu}_{\ln IM} + \hat{\delta}_e + \hat{\delta}_{es} + \hat{\delta}_s$ and a variance of ϕ^2. Each of the systematic $\hat{\delta}$ terms is estimated for the target location, and has epistemic uncertainty associated with it. It is common to account for this uncertainty through additional nodes and branches within the logic tree; e.g., there may be a node with multiple branches for each $\hat{\delta}$ component. Similarly, in a partially non-ergodic hazard analysis, only some components are treated as repeatable. For example, a partially non-ergodic hazard analysis that accounted for site-specific site response (e.g., Rodriguez-Marek et al., 2014) would simply focus upon an adjusted mean of $\mu_{\ln IM} + \hat{\delta}_s$ and a partially reduced variability[11] of $\sigma^2_{\ln IM} = \tau^2_e + \tau^2_{es} + \phi^2_e + \phi^2_{es} + \phi^2_s$ (Villani and Abrahamson, 2015). In this partially non-ergodic case, the uncertainty in estimating δ_s would be reflected in a dedicated node of the logic tree. Importantly, this partially non-ergodic approach does not remove all aspects of site variability from the standard deviation used in the hazard calculations. Even after correcting for the systematic deviations at a given site, the aleatory component ϕ^2_s will remain.

Table 8.2 provides a summary of how variance components are treated within PSHA for common analysis cases. As can be appreciated, the aleatory variability always needs to be included directly within the overall standard deviation. The differences between the partially non-ergodic cases relate to which of the systematic, repeatable components δ_e, δ_{es}, and δ_s can be constrained.

---

[11] Assuming here that all of the variance components are independent.

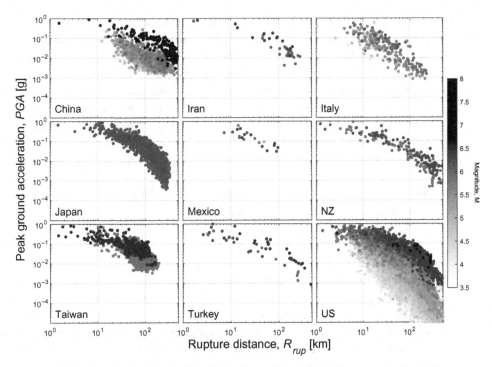

Fig. 8.9 Distance scaling of PGA with data grouped by country of origin. The shading of points indicates the magnitude of the event from which they came.

Example of Partially Non-Ergodic Ground-Motion Modeling

Region-specific anelastic attenuation effects are identified using an ergodic dataset to illustrate a partially non-ergodic approach to ground-motion modeling. The approach follows Stafford (2014a) and uses a random effect on a particular model coefficient to identify these region-specific tendencies. The nine countries with the highest number of records in the NGA-West2 database (Ancheta et al., 2014) are considered. The general distance scaling for each country is shown in Figure 8.9. The shading of points indicates the magnitude of the causative events. Certain countries contribute far more records to the database than others. In particular, the United States contributes the greatest number of recordings, although primarily from relatively small magnitude events.

For this example, a simplified version of the Chiou and Youngs (2014) GMM, which was previously presented in Chapter 4, is adopted. The simplifications involve ignoring nonlinear site response, sediment depth effects, style-of-faulting terms, and finite-fault effects. These simplifications result in the functional form presented in Equation 8.15. The first line of the equation represents the magnitude scaling of the model, the second line the geometric spreading effects, and the third line includes magnitude-dependent anelastic attenuation effects and a term representing linear site response:

$$
\mu_{\ln PGA} = c_1 + c_2(M - 6) + \frac{c_2 - c_3}{c_n} \ln\left(1 + \exp\left[c_n(c_m - M)\right]\right)
$$
$$
+ c_4 \ln\left(R_{rup} + c_5 \cosh[c_6(M - c_{hm})]\right) + (c_{4a} - c_4)\ln\sqrt{R_{rup}^2 + c_{rb}^2}
$$
$$
+ \left(c_{\gamma,1} + \delta_{c_{\gamma,1}} + \frac{c_{\gamma,2}}{\cosh\left[\max(M - c_{\gamma,3}, 0)\right]}\right)R_{rup} + c_v \ln\left(\frac{V_{S,30}}{760}\right) + \delta_e + \delta_s.
$$

(8.15)

In Equation 8.15, M is moment magnitude, R_{rup} is the rupture distance, $V_{S,30}$ is the time-averaged shear-wave velocity over the uppermost 30 m, and all c_x terms are model parameters constrained by theory or determined by regression. The terms δ_e, δ_s, and $\delta_{c_{\gamma,1}}$ represent non-ergodic source, site, and anelastic attenuation terms that will be explained further below.

Of particular interest in Equation 8.15 is the magnitude-dependent anelastic attenuation term. The magnitude dependence of anelastic attenuation arises from the magnitude dependence of the frequency content in the underlying Fourier spectral ordinates that contribute to response spectral ordinates (and PGA) (Stafford et al., 2017). Considering, for the time being, that the term $\delta_{c_{\gamma,1}} = 0$, Equation 8.15 implies that, for very large magnitudes, the anelastic attenuation effects will scale according to $c_{\gamma,1}R_{rup}$, while for very small magnitude events the scaling will be $(c_{\gamma,1}+c_{\gamma,2})R_{rup}$. The functional form includes a smooth transition (using the hyperbolic cosine function, cosh) between these extremes for intermediate magnitudes.

The strength of anelastic attenuation observed in a given region depends upon the crustal properties of that region (see Section 5.4.4). Therefore, to obtain a partially non-ergodic GMM with regional differences in anelastic attenuation rates, one can allow the coefficient $c_{\gamma,1}$ to vary by country. Equation 8.15 includes a random effect $\delta_{c_{\gamma,1}}$ for these regional anelastic attenuation effects[13] along with the more common random effects δ_e and δ_s for systematic earthquake and site effects, respectively. The δ_e terms absorb the effects of ignoring style-of-faulting terms, while the δ_s terms account for site-specific site response effects (including the average degree of nonlinearity experienced at each site). Model coefficients and the random effects for the anelastic attenuation $\delta_{c_{\gamma,1}}$ are obtained using partially crossed and nested mixed-effects regression analysis (Stafford, 2014a) and presented in Figure 8.10.

Figure 8.10 shows that anelastic attenuation rates vary from region to region (where each country represents a single region in this example). The amount of available data differs from region to region and this is reflected in the confidence intervals shown. For example, the confidence intervals for the United States and Mexico have the smallest and largest ranges because they have the most and least data, respectively. The nominal value of $c_{\gamma,1}$ is negative such that ground-motion amplitudes decrease with increasing rupture distance. Therefore, if one were making predictions of ground motions in China, the effective anelastic attenuation coefficient $c_{\gamma,1} + \delta_{c_{\gamma,1}}$ is less negative, representing a lower rate of attenuation. In contrast, the $c_{\gamma,1} + \delta_{c_{\gamma,1}}$ value for Japan is more negative than the ergodic case (where $\delta_{c_{\gamma,1}} = 0$). Therefore, ground-motion amplitudes in Japan are observed to attenuate more rapidly with distance.

In a partially non-ergodic hazard analysis, region-specific ground-motion amplitudes can be modeled by including the relevant value of $\delta_{c_{\gamma,1}}$ in the forward predictions. Simultaneously, the epistemic uncertainty in the estimate of these random effects (indicated by the confidence intervals in Figure 8.10) must be accounted for in the logic tree for median ground-motion predictions.

Using the classification of Table 8.2, the above example is a "partially non-ergodic path" approach, where the term $\delta_{c_{\gamma,1}}$, related to path effects, is considered as non-ergodic. While the random effects of δ_e and δ_s are also included in Equation 8.15, the specific values of these terms do not have utility for forward predictions, unless estimates of regional effects are made, as seen previously in Section 8.3.1, or one of the recording stations coincides with the target location of interest. Despite this classification according to Table 8.2, the reduction in aleatory variability that can be achieved

---

[13] So, the effective value of $c_{\gamma,1}$ in each region is $c_{\gamma,1} + \delta^i_{c_{\gamma,1}}$, where i identifies the relevant region.

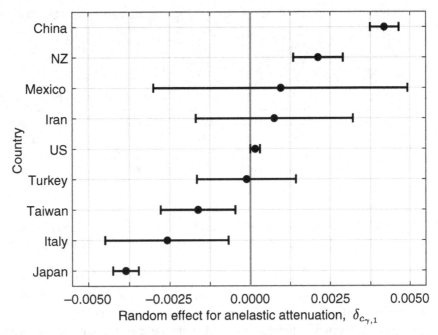

Fig. 8.10 Random effects, $\delta_{c_{\gamma,1}}$, for the anelastic attenuation coefficient $c_{\gamma,1}$. Horizontal lines show the approximate 95% confidence intervals for these random effects. The vertical line at $\delta_{c_{\gamma,1}} = 0$ represents the fully ergodic model. Negative values of $\delta_{c_{\gamma,1}}$ indicate stronger anelastic attenuation in a specific region relative to the ergodic model.

through this partially non-ergodic approach is not as simple to specify as suggested by the table. Partially non-ergodic path effects have been considered, so from Table 8.2 we have total aleatory variability of $\sigma^2 = \tau_e^2 + \tau_s^2 + \phi^2$ and epistemic uncertainty on μ, σ, and δ_{es}. This aleatory variability differs from the fully ergodic case due to removing the systematic path-to-path variability of τ_{es}^2. However, unlike the simple additive models in Equations 8.13 and 8.14, in the example, a parametric approach is adopted with a focus on the coefficient $c_{\gamma,1}$, reflecting the overall crustal Q characteristics for each country. For a given path, the systematic deviation away from the ergodic median is therefore a function of the rupture distance:

$$\hat{\delta}_{es} = \hat{\delta}_{c_{\gamma,1}} R_{rup}. \tag{8.16}$$

The effect of Equation 8.16 is demonstrated in Table 8.3 using the results presented in Figure 8.10 for Japan, the United States, and China. For fixed values of $\hat{\delta}_{c_{\gamma,1}}$, the corresponding $\hat{\delta}_{es}$ terms vary linearly with the rupture distance and cause significant differences in ground-motion amplitudes at greater distances. For example, Table 8.3 shows that PGA values in Japan and China are approximately 50% lower and higher, respectively, than in the United States for a distance of 100 km. At 200 km, these systematic differences in anelastic attenuation translate to factors greater than two in the level of PGA.

If these systematic region-to-region variations had not been considered, then the corresponding apparent aleatory variability that would have been contained in the ergodic GMM would be (based on error propagation using Equation 8.16):

$$\tau_{es}^2 = \left(\frac{\partial \mu_{\ln PGA}}{\partial \delta_{c_{\gamma,1}}}\right)^2 \tau_{c_{\gamma,1}}^2 = (R_{rup})^2 \tau_{c_{\gamma,1}}^2. \tag{8.17}$$

Table 8.3. Numerical example showing the impact of the systematic regional attenuation effects shown in Figure 8.10. Values are computed for $\delta_{c_{\gamma,1}}^{\text{Japan}} = -0.0039$, $\delta_{c_{\gamma,1}}^{\text{US}} = 1.52 \times 10^{-4}$, and $\delta_{c_{\gamma,1}}^{\text{China}} = 0.0042$. The ratio between partially non-ergodic and fully ergodic σ values in the final column are based upon a distance-independent fully ergodic value of $\sigma_{FE} = 0.75$ and $\tau_{c_{\gamma,1}} = 0.0025$

R_{rup} [km]	$\delta_{es}^{\text{Japan}}$	δ_{es}^{US}	$\delta_{es}^{\text{China}}$	$PGA^{\text{Japan}}/PGA^{\text{US}}$	$PGA^{\text{China}}/PGA^{\text{US}}$	σ_{PNE}/σ_{FE}
10	−0.039	0.002	0.042	0.961	1.041	0.999
50	−0.193	0.008	0.210	0.818	1.224	0.986
100	−0.386	0.015	0.419	0.670	1.498	0.943
200	−0.771	0.030	0.839	0.449	2.244	0.745

Therefore, the aleatory variability that should be adopted for the partially non-ergodic analyses will depend upon the rupture distance for each scenario. This variability can be defined as

$$\sigma_{PNE}^2 = \sigma_{FE}^2 - \tau_{es}^2 = \sigma_{FE}^2 - (R_{rup})^2 \tau_{c_{\gamma,1}}^2 \tag{8.18}$$

where the subscripts PNE and FE stand for *partially non-ergodic* and *fully ergodic*, as in Table 8.2. For the particular example shown in Figure 8.10, the variability in anelastic attenuation is $\tau_{c_{\gamma,1}} = 0.0025$. Therefore, when $R_{rup} = 100$ km, the value of τ_{es} that is embedded within ergodic datasets is $\tau_{es} \approx 0.25$. Example ratios of σ_{PNE}/σ_{FE} are included in Table 8.3 for a range of distance values.[14]

Considerations for Physics-Based Ground-Motion Models

Physics-based GMMs (Chapter 5) are inherently partially non-ergodic because they make use of region-specific 1D or 3D crustal models (see Section 5.4.6). The extent of partially non-ergodic path and site representation in such crustal models depends upon how region-specific they are. Although region-specific information is utilized when possible, ergodic modeling components are also inevitably adopted. Common examples include generic relationships between crustal properties (e.g., Brocher, 2005) and models for prescribing slip distributions (e.g., Graves and Pitarka, 2010).

Despite this inherent advantage of physics-based GMMs (Section 5.2.3), the treatment of epistemic uncertainty in such models is challenging because of the need to consider the spatial dependencies among various parameters (see Section 5.6.3). These parameters include those related to crustal properties within the velocity model and those required for specifying slip distributions over a 2D fault surface within kinematic simulations. The assumed correlations among the large numbers of simulation parameters strongly influence the resulting estimated parametric uncertainty (a form of epistemic uncertainty; see Section 5.6.1).

The framework of Equation 8.13 is representative of how non-ergodic models are commonly considered for empirical GMMs. In this framework, the systematic deviations are represented as additive deviations away from the logarithmic mean motions for a given scenario. However, in a physics-based GMM framework, the ergodic assumption is applied to the model parameters calibrated using ergodic datasets. For example, it is common to assume that the velocity of rupture propagation over a fault will be approximately 80% of the shear-wave velocity in the vicinity of the rupture. This is based on observations of events from around the world (Somerville et al., 1999),

---

[14] Note that direct application of Equation 8.18 for very large distances will lead to nonsensical values of σ_{PNE}. However, this is a result of GMMs not currently modeling the distance-dependence in σ_{FE}.

and is therefore ergodic. In relatively simple physics-based approaches (Section 5.5.3), like SMSIM (Boore, 2003), the rupture velocity is often assumed to be a known constant. In more elaborate approaches (e.g., Graves and Pitarka, 2016), it is a derived quantity, obtained from combining a random depth-dependent background rupture velocity with local perturbations that scale with local slip amplitude (Section 5.3.2). However, some faults may generate events that have rupture velocities that differ systematically from the nominal ergodic values (or distributions), so a fault-specific value, or distribution, may be defined. With this non-ergodic approach, a systematic deviation in the rupture velocity would not produce a simple additive deviation from the mean ergodic predictions. Instead, the effect of the systematic deviation in this parameter must be propagated through the model to obtain the net deviation in ground motions.

This thinking applies to all of the physical parameters within a physics-based formulation. When removing the ergodic assumption in a physics-based model component, the effect on the resulting ground motions can be observed only by propagating the systematic deviation through the model. This propagation leads to a more complicated structure of deviations away from the ergodic predictions than is shown in Equation 8.13. One can appreciate the first-order effect on ground motions resulting from parametric deviations using a truncated Taylor-series expansion:

$$\mu_{\ln IM}(e, s \mid \boldsymbol{\beta}_{NE}) \approx \mu_{\ln IM}(e, s \mid \boldsymbol{\beta}_E) + \sum_i^n \frac{\partial \mu_{\ln IM}(e, s \mid \boldsymbol{\beta}_E)}{\partial \beta_{E,i}} (\beta_{NE,i} - \beta_{E,i}). \qquad (8.19)$$

In Equation 8.19, $\boldsymbol{\beta}_E = \{\beta_{E,1}, \ldots, \beta_{E,n}\}$ is the set of ergodic model parameters, while $\beta_{NE,i}$ is a non-ergodic counterpart for parameter i. The equation shows that the first-order effect of a deviation away from the ergodic parameter $\delta\beta_i = \beta_{NE,i} - \beta_{E,i}$ combines with the partial derivative of the model with respect to this parameter (evaluated at the ergodic parameter estimates). The partial derivative in Equation 8.19 represents the sensitivity of the physics-based GMM at predicting $\mu_{\ln IM}$ due to parameter $\beta_{E,i}$. The nature of these derivatives will vary with the details of the specific GMM being considered. However, these derivatives will often have magnitude, distance, and frequency dependence. Furthermore, the epistemic component of the removed apparent variability is related to the square of these partial derivatives.

8.5.2 Impact upon Hazard Calculations

The impact of adopting a non-ergodic approach on seismic hazard curves depends on how different the ergodic and non-ergodic GMMs are. Generally speaking, two factors will influence the degree of impact. The first is the change (almost always a reduction) in the GMM's aleatory variability. This effect can be anticipated from Section 6.5.4, which explored the sensitivity of hazard curves to levels of σ. As the aleatory variability reduces, the slope of the hazard curve will steepen, and large IM values will have smaller exceedance rates (e.g., Bommer and Abrahamson, 2006).

The second effect is the shift in the logarithmic mean, $\Delta\mu$, of the IM for each rupture scenario. Equation 8.19 in Section 8.5.1 showed that this shift could depend upon the rupture scenario itself. However, it could also be reflected by some global adjustment for all rupture scenarios.

Figure 8.11 demonstrates the combined effect of these two factors. The main figure shows an ergodic hazard curve as well as three non-ergodic curves. The ergodic hazard curve is generated using the same parameters as the example in Section 6.5.4[15] and the nominal ergodic aleatory variability

---

[15] The curves are derived using the BJF97 GMM, a Gutenberg–Richter source at a fixed distance of 10 km and with $b = 1.0$, $m_{\min} = 5.0$, $m_{\max} = 8.0$, and $\lambda(M > m_{\min}) = 0.05$. The site has $V_{S,30} = 500$ m/s.

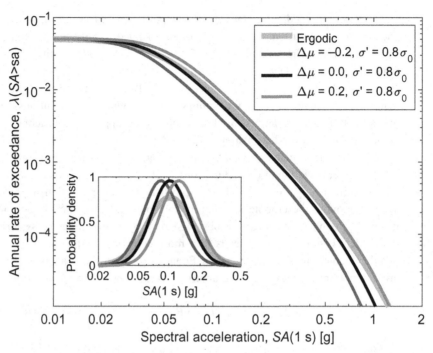

Fig. 8.11 Illustration of the effects of changes to the median amplitudes and aleatory variability upon hazard curves from ergodic to non-ergodic analyses. The main figure shows an ergodic and three non-ergodic hazard curves obtained with reduced aleatory variability and some global shift to the logarithmic mean IM levels $\Delta\mu$. The inset shows example PDFs for a magnitude 6 scenario. Figure based upon Abrahamson et al. (2019).

is denoted by σ_0. The non-ergodic hazard curves are then obtained by using 80% of this ergodic variability, $\sigma' = 0.8\sigma_0$, and shifts to the logarithmic mean IM levels for all rupture scenarios of $\Delta\mu = \pm 0.2$ or zero.

Figure 8.11 indicates that a non-ergodic hazard analysis can either increase or decrease the hazard at a site. While the shifts to the logarithmic mean $\Delta\mu$ increase or decrease hazard, the aleatory variability will almost always be reduced, so hazard is usually decreased, particularly when considering long return periods. That said, the curves shown in Figure 8.11 are generated assuming that $\Delta\mu$ and σ' are known precisely. As stated repeatedly throughout the chapter, there will be significant epistemic uncertainties in these parameters in practice. Consideration of these uncertainties may not significantly impact the overall mean hazard curve, but it will influence the specific locations of fractile hazard curves representing epistemic uncertainty (Section 6.7).

8.5.3 Empirical Constraint upon Source, Path, and Site Components

The PSHA formulations presented in Chapter 6 reflect contributions from numerous rupture scenarios, each of which induces ground shaking at the target site. Each of these modeled contributions has a source component, a path component, and a site component, and the ergodic assumption is used to characterize these components by default. When relaxing the ergodic assumption, it is convenient to consider each component separately.

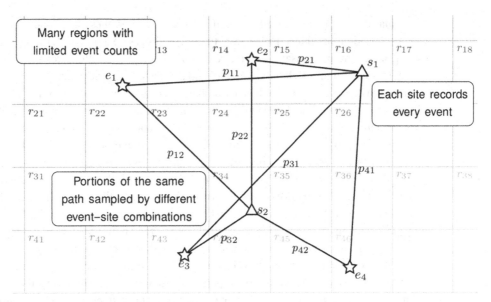

Fig. 8.12   Schematic illustration of how different levels of constraint naturally arise between source, path, and site components within ground-motion datasets. Star symbols denote earthquake events, e_1–e_4, and triangles denote recording stations, s_1 and s_2. Lines connecting events with stations represent travel paths, p_{es}, and the r_{ij} represent discretized cells of the spatial region.

The source component can be further decomposed into static and dynamic attributes. The static attributes relate to the rupture scenario's dimensions, geometry, and rates of occurrence (i.e., the information arising from an ERF). On the other hand, the dynamic attributes relate to source information that influences ground shaking, such as the rupture velocity, the source-time function, or the subfault corner frequency. The source characterization discussed in Chapters 2 and 3 is associated with the static attributes. These attributes will usually be calibrated to the target site or region by identifying specific local sources and using local seismicity data. In a sense, this suggests that the ergodic assumption is not invoked at this point. However, generic components, or assumptions based upon global analogs (such as source-scaling relations or prior m_{max} distributions), are still frequently employed here. The local seismicity data are also often used to calibrate magnitude-frequency distributions that are assumed valid based on observations made in other regions, often involving ergodic datasets. For the dynamic source attributes, and the path and site components of GMMs, the ergodic assumption is strongly utilized. This is the case for both empirical (Chapter 4) and physics-based (Chapter 5) models, although the latter models are normally less reliant upon the ergodic assumption. The main distinction is that the empirical models use ergodic data directly, while physics-based models may use components calibrated from ergodic data. Furthermore, when site- or region-specific physical attributes are identified, such as a crustal velocity model, it is usually easier to account for this within a physics-based approach.

As discussed in Section 8.4, relaxing the ergodic assumption requires sufficient data to discriminate between systematic and statistical deviations from an ergodic model for a given parameter or model component. Therefore, the chance of relaxing the ergodic assumption relates to how much data are available to constrain the source, path, and site components. Figure 8.12 shows that the nature of seismic networks naturally leads to different levels of constraint being available for these components, with the level of constraint decreasing from site to path to the source. In Figure 8.12, the spatial domain is discretized into cells, denoted r_{ij}, and these can conceptually represent areal regions, or individual faults. Either way, there will always be a unique 1:1 mapping of an individual earthquake event to a causative source region or fault r_{ij}. As earthquakes infrequently occur, many spatial regions or faults will have limited to no associated empirical data.

In contrast, when an individual earthquake occurs, it will be recorded at multiple sites. In Figure 8.12, four earthquake events are shown along with two recording sites. While these four events occur in four distinct sources, each station records every event. Therefore, it is possible to accumulate empirical constraint at a given recording station more quickly than for a seismic source.[16]

Between these two end members of source and site, we have constraints for source-to-site travel paths. To truly have repeated observations of a travel path, we need repeated source and site combinations to arise. However, as shown in Figure 8.12, different events may have travel paths to a given site that share common portions. For instance, station s_1 records both events e_1 and e_2 and while the travel paths p_{11} and p_{21} are quite different; they both traverse spatial region r_{15}. Similarly, the travel paths from event e_3 to stations s_1 and s_2 both traverse common regions.[17] Therefore, while repeated observations of unique travel paths are as rare as the events themselves, it is still possible to compile data related to the paths more rapidly than for the sources because of these common portions. Additionally, azimuthal variations can be observed from distinct source–site combinations. That is, paths such as p_{22} and p_{41} in the north-south direction may exhibit attributes that differ from east-west paths like p_{11} and p_{21}.

To summarize, relaxing the ergodic assumption is most easily achieved for linear site response,[18] followed by path effects, and then dynamic source attributes. Partially non-ergodic approaches will also be implemented for static source attributes, with the exception of source-scaling relations and maximum magnitude distributions.

In addition to the above considerations based on passively recorded empirical data, new measurements can also be made. Site-investigation approaches enable geophysicists and geotechnical earthquake engineers to measure, or infer, the dynamic properties of geo-materials influencing linear site response. These approaches typically constrain physical parameters that are otherwise estimated from ergodic datasets, which can then be used to develop site-specific adjustments to ergodic models. Similarly, paleoseismic investigations and targeted geological field surveys can provide site-specific data regarding static source characteristics. In contrast, it is far more challenging to obtain new measurements that constrain the dynamic source attributes or path attributes. The relaxation of the ergodic assumption is less common for these components.

The following three sections discuss the most common approaches to obtaining non-ergodic estimates of source, path, and site effects. The sections' lengths reflect the reality that more data, and hence options, are typically available for constraining site effects than path and source components.

---

[16] Especially when the focus is upon constraining linear site effects, which can be estimated from small-amplitude ground motions that occur more frequently.

[17] Figure 8.12 is simplified in that the regions are in 2D-space compared with the true 3D ray path. This, therefore, assumes any systematic effects in region r_{ij} are (approximately) depth-independent, for the depths of greatest relevance to source and wave propagation.

[18] It is far harder to obtain empirical constraint for nonlinear site response because this requires large-amplitude ground motions to be observed, which are rare, and additionally requires some estimate of the incident ground-motion amplitude at depth.

8.6 Non-Ergodic Site Effects

A site investigation is routinely conducted for a typical engineering design. To design the foundations of structures, or to select the foundation system in the first instance, it is necessary to gather information about the near-surface geo-materials at the site. For static analyses, this information might consist of a model of the soil layering with unit weights, material strengths, and the water table's location, etc. For dynamic analysis, additional information about the strain-dependence and cyclic degradation of the shear modulus (and potentially shear strength) and the strain-dependent level of equivalent viscous damping may be inferred from the constructed stratigraphic profile, or dynamic laboratory testing (e.g., Kramer, 1996). Additionally, geophysical testing may be conducted to more directly assess the site's velocity structure. The information obtained for any given site will depend upon the project's budget and the extent to which the information may enhance the design or assessment decisions. The following sections outline the various approaches that can be employed to utilize this site-specific information when modeling linear impedance and damping and nonlinear response of the near-surface geo-materials.

8.6.1 Options for Incorporation of Site Response

Empirically calibrated GMMs (Chapter 4) usually characterize the effects of site response using the time-averaged shear-wave velocity over the upper 30 m of geo-materials, $V_{S,30}$.[19] Following a site investigation, where a velocity profile for at least 30 m has been obtained,[20] the $V_{S,30}$ can be computed and ground-motion predictions can be made for a given rupture scenario. Using this site-specific $V_{S,30}$ still represents an ergodic approach. The approach assumes that the velocity profiles, three-dimensional material and geometric heterogeneities, and dynamic site properties used during the GMM development are consistent with the current site in question. However, in the ergodic approach, this consistency is typically established only through the $V_{S,30}$ value. Two sites with the same $V_{S,30}$ are assumed to have the same median site response, which implies that all other attributes affecting site response are also common between the two sites.

In the subsections that follow, this ergodic approach is first discussed before refinements that make use of partially non-ergodic site response calculations are presented. These partially non-ergodic refinements either deal specifically with the near-surface dynamic properties or consider the deeper crustal structure that might influence ground motions. The modeled site response is computed based on 1D site response (vertical propagation of waves through a horizontally layered medium) throughout the section. Although this is a simplification, it is commonly employed in practice (e.g., Kwok and Stewart, 2006) and illustrates how the partially non-ergodic approaches are implemented. However, this approach cannot capture all effects that might contribute to site response. In particular, the influence of two- and three-dimensional variations in geometry and material properties cannot be resolved under the 1D assumption (Stewart et al., 2017).

---

[19] It was also noted that this is a modeling choice that reflects the fact that velocity estimates are often available to 30 m depth and that this metric correlates reasonably well with site amplification effects. It does not imply some underlying causality between this particular metric and site response effects, other than representing a degree of change in impedance from the source's crustal properties to those in the near-surface.

[20] Or less than 30 m if correlations are used (Boore, 2004).

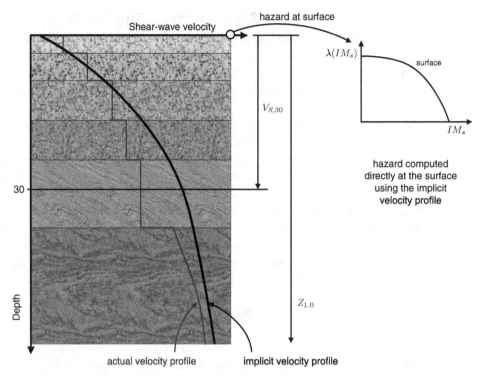

Fig. 8.13 Schematic illustration of how hazard analyses can be conducted directly for the surface as a target horizon. $V_{S,30}$ is computed over the uppermost 30 m only. The parameter $Z_{1.0}$ defines the depth corresponding to the first exceedance of $V_S = 1000$ m/s and therefore carries additional (implicit) information about the profile, which likely differs, to some extent, from the actual velocity profile.

Use of Generic Empirical Site Terms

In Equation 6.1, repeated below as Equation 8.20, the specification of site characteristics was defined generically through the use of the term *site*:

$$\lambda(IM > im) = \sum_{i=1}^{n_{rup}} P(IM > im|rup_i, site)\,\lambda(rup_i). \tag{8.20}$$

The way that this site dependence enters GMMs depends upon the particular model, but for empirical models, the dominant approach is to include dependence upon $V_{S,30}$. Therefore, the most direct option for computing hazard for a given site is to follow the approach schematically illustrated in Figure 8.13. In this figure, a particular site profile has been obtained that consists of several horizontal layers[21] that have different shear moduli and densities (and hence shear-wave velocities[22]). Even though more information about the velocity profile is available regarding the velocities at depth, if a GMM is used that characterizes sites only through $V_{S,30}$, then only the uppermost 30 m of the profile will be considered.

---

[21] This representation of a horizontally layered site profile is necessary under the 1D modeling assumption. The actual crustal structure of the site may be far more complex.

[22] Recall that $V_S = \sqrt{G/\rho}$, with G being the shear modulus and ρ being the density.

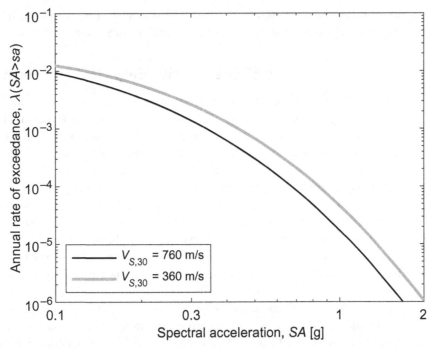

Fig. 8.14 Comparison of hazard curves for $SA(T = 0.01\,\text{s})$ computed directly for average shear-wave velocities of $V_{S,30} = 760$ m/s and $V_{S,30} = 360$ m/s.

On the left of Figure 8.13, two velocity profiles are shown: the site-specific profile and an implicit profile. The site-specific profile is used to compute the $V_{S,30}$ value for the site, but is not used for any other purpose. The implicit profile is a hypothetical profile implicitly reflected by the site response terms within a GMM. It can loosely be thought of as the mean velocity profile of the sites with a $V_{S,30}$ similar to the site of interest in the database used to develop the GMM. The site effects predicted by an ergodic GMM will be associated with this implicit velocity profile. In the figure, the actual velocities in the uppermost layers are slightly stiffer (higher V_S) than in the implicit profile, while the deeper layers are softer. However, the time-averaging used in the $V_{S,30}$ calculation means that the profiles have the same $V_{S,30}$.

Figure 8.14 shows the hazard curves obtained using the single point-source model, with a Gutenberg–Richter magnitude-frequency distribution, presented in Section 6.3.3 and the Chiou and Youngs (2014) GMM for two different values of $V_{S,30}$ (360 and 760 m/s). As expected, the softer site with $V_{S,30} = 360$ m/s has higher rates of exceedance than the $V_{S,30} = 760$ m/s case because the motions from most rupture scenarios are amplified more at the softer site.

Obtaining the hazard results in this manner is straightforward. However, it can be appreciated from Figure 8.13 that much information about the site-specific velocity profile, and potentially the more complex underlying site model, has been ignored. Ignoring this information leads to biased hazard estimates. Although the $V_{S,30}$ values for the actual and implicit profiles are equal, the site response predicted by the ergodic GMM will not accurately reflect the real site response. This impacts the median IM predicted for each considered rupture scenario. Also, given that we know the actual velocity profile, the GMM variability should not include profile-to-profile variability for a given $V_{S,30}$ that is present in an ergodic model. Instead, we should use a smaller value for the standard deviation in the hazard calculations, and this will lead to individual hazard curves becoming "steeper,"—as previously shown in Figure 8.11, and more generally in Section 6.5.4.

Use of Site-Specific Empirical Site Terms

The generic empirical site terms described in the previous section are derived from observations at many different sites. Therefore, the ergodic GMM standard deviation will include the effect of site-to-site variability. In the most common case, where no ground motions have been recorded at the target site, options to develop site-specific empirical site terms become very limited. However, a recording station may be nearby (≈ 0.5–1.0 km), and geophysical investigations at the target site and recording station may suggest similar site characteristics exist at these locations. In such cases, Stewart et al. (2017) suggest that empirical data from the recording station may be used to develop site terms for the target site. However, the epistemic uncertainty associated with assuming the validity of these data must be accounted for. If no nearby recordings are available, the only remaining option is to analytically or numerically model the site-specific response. That said, simplified analytical and numerical techniques usually account only for impedance effects and the nonlinear behavior of the soil (Stewart et al., 2017). Other contributors to surface ground motions, such as topographic effects and three-dimensional heterogeneity in the sub-surface, are ignored.

In Section 8.5.1, we saw that a particular site may exhibit systematic deviations away from an ergodic model and that these are represented as $\delta_s \equiv \zeta_s \tau_s$ in Equations 8.13 and 8.14. The deviation δ_s is the product of a site-specific standard normal variate ζ_s and the site-to-site variability τ_s. When a reasonable number of ground-motion recordings exist at a site, one can compute an empirical estimate, $\hat{\delta}_s$, of what δ_s is for the target site as well as an estimate of the associated uncertainty.[23] The predictions of an ergodic GMM can then be adjusted so that we have a site-specific mean prediction defined by $\mu_{\ln IM}(e, s) + \hat{\delta}_s$. The standard deviation can also be adjusted, as discussed in Sections 8.3.1 and 8.5.1, for the hazard calculations. After adjusting the GMM, hazard computation is the same as when generic empirical site effects are used (Equation 8.20).

This empirical approach is not often used in practice for two main reasons. First, it is rare to have a sufficient number of recordings at a site of interest to constrain $\hat{\delta}_s$ adequately. Furthermore, it is often necessary to consider recordings from relatively small earthquakes to obtain sufficient data to constrain δ_s. However, Stafford et al. (2017) have shown that estimates of site effects from small magnitude events can be biased due to inadequate treatment of damping effects within existing GMMs. Second, the empirical correction is likely to capture only linear site effects, whereas a hazard analysis will often need to consider the nonlinear response of the ground induced through strong shaking. Stewart et al. (2017) presented an approach whereby a site-specific site response model can be obtained by combining empirically constrained linear site response with nonlinear site response from an ergodic model or from numerical site response analyses. However, there can often be a disconnect between the linear and nonlinear components. The empirical motions include site response effects associated with linear impedance and damping, but also inherently capture the effects of topography and other three-dimensional effects associated with the near-surface stratigraphy. When such a disconnect is observed, Stewart et al. (2017) recommend adopting the empirical estimates of the linear site response and then anchoring the nonlinear site response function

---

[23] The statistical approach is analogous to that shown previously in Section 8.3.1.

to the amplitude of the linear site response. This approach assumes that site effects that manifest empirically, but that the numerical modeling does not capture, will persist when the site responds nonlinearly.

Surface Hazard from the Convolution Integral

While the previous two options make use of Equation 8.20 and either an ergodic or site-adjusted GMM, the two remaining options discussed hereafter require modifications to our hazard formulation. These approaches require the development of a site-specific amplification function that describes the change in some IM at a particular reference velocity horizon (usually at depth), IM_r, to the surface, IM_s. The reference horizon should generally be at a strong impedance contrast within the site profile such that the medium below the reference horizon can be regarded as an elastic half-space.[24] This amplification function, describing the *amplification factor* $AF = IM_s/IM_r$, either is combined with a hazard curve obtained for the reference velocity horizon (the convolution approach in this section) or is embedded within the hazard calculations for each rupture scenario (the following section).

The convolution approach (Bazzurro and Cornell, 2004b) to obtaining the surface hazard, $\lambda(IM_s > im_s)$, is defined by

$$\lambda(IM_s > im_s) = \int_0^\infty P\left(AF > \frac{im_s}{im_r} \mid im_r\right) |d\lambda(IM_r > im_r)| \qquad (8.21)$$

where $\lambda(IM_r > im_r)$ is the hazard curve derived for some predefined reference horizon.

In Equation 8.21, the probability that the amplification exceeds some level im_s/im_r can be written as[25]

$$P\left(AF > \frac{im_s}{im_r} \mid im_r\right) = 1 - \Phi\left[\frac{\ln(im_s/im_r) - \mu_{\ln AF}(im_r)}{\sigma_{\ln AF}(im_r)}\right] \qquad (8.22)$$

where Φ denotes the cumulative distribution function of the standard normal distribution. Equation 8.22 assumes that AF is described by a lognormal distribution ($\ln AF$ is normally distributed) and that the mean and standard deviation are a function of im_r in order to account for nonlinear effects on AF. Site response effects are commonly represented in this way so that the response $AF(im_r)$ is a function of intensity levels at some reference horizon (Bazzurro and Cornell, 2004a; Choi and Stewart, 2005). The remaining term $|d\lambda(IM_r > im_r)|$ in Equation 8.21 can be interpreted as the rate of *occurrence* of im_r (see Section 6.4.2).

The convolution integral in Equation 8.21 can thus be regarded as sweeping over each value of ground motion represented by the hazard curve for the reference horizon and computing the conditional probability of exceeding some value of surface motion im_s. These probabilities are combined in a "weighted" manner, with the "weights" (the rates of occurrence) representing the likelihood of each im_r value. The site amplification is assumed to be conditionally independent of the rupture characteristics, given the amplitude $IM_r = im_r$. Therefore, the convolution integral is not strictly correct in cases where the site amplification depends upon attributes of the underlying rupture scenario (Stafford et al., 2017). If this scenario dependence is a concern, one can either adopt the approach discussed in the following section or otherwise derive the site-specific amplification

---

[24] This is not a strict requirement, but it makes the site response calculations easier as we can ignore reflected waves from deeper impedance contrasts.

[25] Note that the left-hand side of Equation 8.22 is equivalent to $P(IM_s > im_s \mid im_r)$, as $im_s = AF \times im_r$.

using records that have been carefully selected to reflect the scenarios responsible for the hazard at each level of im_r. Such approaches are described in Chapter 10.

Surface Hazard from Embedded Site Amplification

The final and most rigorous way to incorporate site amplification is to embed the amplification function within the hazard integral directly. This treatment is shown mathematically as

$$\lambda(IM_s > im_s) = \sum_{i=1}^{n_{rup}} \left[\int_0^\infty P\left(AF > \frac{im_s}{im_r} \mid im_r\right) f_{IM_r}(im_r \mid rup_i, site) \, dim_r \right] \lambda(rup_i). \quad (8.23)$$

Here, $f_{IM_r}(im_r \mid rup_i, site)$ is the PDF of the IM at the reference horizon, IM_r, conditioned upon the rupture scenario and site conditions.[26] The integral term can also be thought of as the limit of a discrete sum. In this sum, each possible value of reference motion for a given rupture scenario is considered. The probability of each value is then combined with the probability that the amplification required to exceed im_s is realized.

Note that in Equation 8.23, there is no need for conditional independence of the site amplification with respect to the rupture scenario. That is, the site amplification can be expressed as

$$P\left(AF > \frac{im_s}{im_r} \mid im_r\right) = P\left(AF > \frac{im_s}{im_r} \mid im_r, rup_i, site\right). \quad (8.24)$$

Scenario-dependence in the amplification function is therefore readily accounted for. However, this requires the amplification function to be derived to reflect each rupture scenario's dependency.

The other main difference between embedding the site amplification within the hazard integral (Equation 8.23) and using the convolution approach (Equation 8.21) is how fractiles of the surface hazard are obtained when logic-tree formulations are employed. When the site amplification is embedded directly within the hazard integral, the logic tree for all aspects of the rupture scenarios, ground-motion predictions to the reference horizon, and the site response will be considered. As a result, the hazard fractiles obtained through embedding the site response will reflect the full distribution of hazard values for each level of im_s. However, when using the convolution approach, the logic tree could be split into two smaller trees: one relevant for defining the hazard at the reference horizon, and another that just focused upon epistemic uncertainty in the site response. Each hazard curve arising from the logic tree for the reference horizon must be convolved with each site amplification function represented by the site response logic tree to capture the full uncertainty and to obtain equivalent results between the two cases. Suppose the convolution approach is implemented where only the mean hazard curve for the reference horizon interacts with the logic tree for the site response (and associated amplification functions). In that case, the corresponding fractiles will underestimate the full uncertainty at each level of im_s.

8.6.2 Site-Specific Site Response

It is well known that two sites with the same values of $V_{S,30}$ can have site effects that differ considerably (e.g., Papaspiliou et al., 2012), for both linear and nonlinear site response. The specific details of the velocity model influence the locations and amplitudes of resonant peaks in the site

---

[26] These are the site conditions *below* the reference velocity horizon as this PDF describes the likelihoods of levels of motion entering the near-surface deposits from below.

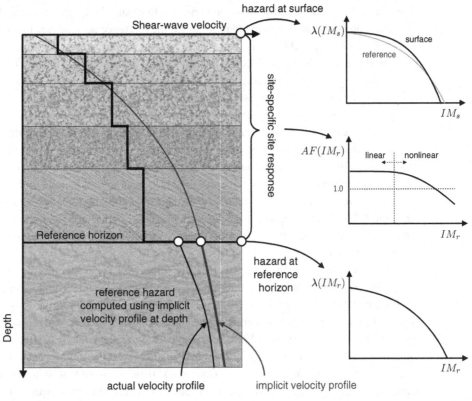

Fig. 8.15 Schematic illustration of how surface hazard for an intensity measure can be computed using an initial hazard analysis for some reference horizon combined with site-specific site response.

transfer function, while the constitutive behavior of the various layers influences their nonlinear response. To account for site-specific information, it is common practice to conduct site-specific site response analysis to reflect details of the near-surface stratigraphy and combine these analyses with hazard computations for motions entering the near-surface materials from below. A partially non-ergodic framework must account for both the near-surface and deeper velocity profile, as explained in the following sections.

Adjustments for Near-Surface Response

Figure 8.15 shows a schematic illustration of how a site-specific profile can be used to obtain a hazard curve at the surface. In contrast to Figure 8.13, where the site-specific velocity profile was used to compute only the average shear-wave velocity, $V_{S,30}$, now the entire profile is used. This profile is first used to identify a horizon above which site effects are likely to be important. Often this involves identifying a strong impedance contrast below which the medium can be regarded as being an elastic half-space. For the example shown in Figure 8.15, this horizon is located below the 30 m depth indicated in Figure 8.13, but this need not be the case. With this reference velocity horizon identified, a hazard analysis is then conducted using a GMM set to have its $V_{S,30}$ value consistent with the assumed elastic half-space. Figure 8.15 shows that a hazard curve for this reference horizon $\lambda(IM_r)$ is thus obtained.

A site-specific amplification function is needed to obtain the hazard curve at the surface horizon. This function maps values of the input motion at the reference horizon im_r to values of motion at the surface im_s, i.e., $im_s = AF(im_r) \times im_r$. Therefore, the amplification function needs to cover the full range of linear and nonlinear response that the site may experience (Stewart et al., 2017).

The function itself is derived using fully nonlinear, or equivalent linear (when appropriate), site response analysis to compute the surface response associated with specified input motions prescribed at the reference horizon. These analyses could involve response-history analyses for the site profile, but can also be computed using random vibration theory (e.g., Kottke and Rathje, 2013). The details of such methods are beyond this book's scope, but key concepts are discussed in Kramer (1996). Upon performing many such analyses, a set of $\{im_r, im_s\}$ combinations will be obtained, for a broad range of im_r values, and regression analysis can be conducted to obtain both $\mu_{\ln AF}(im_r)$ and $\sigma_{\ln AF}(im_r)$, where, as is typical, we assume that the amplification is lognormal (similar to Equation 8.22). If the amplification depends upon the underlying scenario that gave rise to the $\{im_r, im_s\}$ combinations, the input motions need careful specification via ground-motion selection methods (Chapter 10).

It is important to note that the primary objective is to define the appropriate level of $\mu_{\ln AF}(im_r, rup)$. As shown in Section 8.5.1, only the systematic component of site response δ_s is accounted for in a partially non-ergodic framework, and we want to ensure that $\mu_{\ln AF}(im_r)$ represents this site-specific systematic response. The random component associated with the site response (ψ_s from Equation 8.13) is conceptually equivalent to $\sigma_{\ln AF}(im_r, rup)$. However, it is extremely challenging to decompose the apparent aleatory variability to constrain the site component ψ_s. If we assumed that $\psi_s \equiv \sigma_{\ln AF}(im_r, rup)$, then we would need to be able to remove the contribution of ψ_s from the reference motions to ensure that these random site effects are not double-counted. Unfortunately, we do not currently know how to isolate the appropriate value of ψ_s, and current best practice assumes that this random site variability is already contained within the aleatory variability used to compute the reference hazard. The term $P(AF > im_s/im_r \mid im_r)$ in Equation 8.24 is therefore effectively a Heaviside function that filters the reference motions to retain only those that lead to exceedance of im_s from the site-specific amplification function.

Correction for Deeper Velocity Horizon

A GMM developed using only recordings from a given geographical region can use a metric like $V_{S,30}$ to distinguish between sites within that region. The deeper velocity structure is likely common to, or at least similar for, all sites in the region. In this context, the $V_{S,30}$ value is not merely characterizing the uppermost 30 m of the site. Instead, it acts as a proxy for the overall velocity profile going from the surface down into some common velocity structure at depth.

Consider now a GMM developed using data from one region but then applied in another region with differing deep geological structure. In this case, it will still be appropriate to consider the site-specific nature of the near-surface materials. However, it is now also necessary to correct for the differences between the actual deep velocity profile in the target region and the implicit deep velocity profile associated with the imported GMM. Figure 8.16 illustrates this situation and mirrors Figure 8.15. Figure 8.16 highlights that the deeper velocity profile below the reference horizon may be different from the implicit profile associated with the GMM. Now, to compute the hazard curve for the reference horizon, a correction must be made to the GMM to reflect the target region's actual velocity profile at depth.

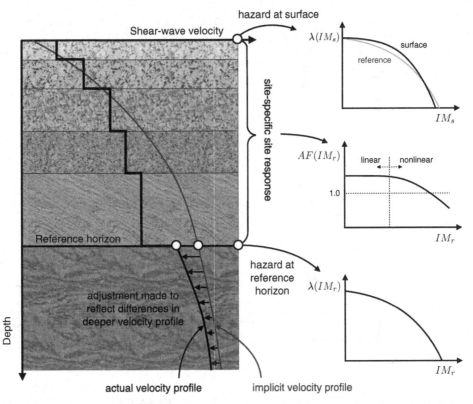

Fig. 8.16 Schematic illustration of adjustments for the deeper velocity horizon when undertaking site-specific site response calculations to obtain surface hazard. The hazard at the reference horizon accounts for the region-specific velocity profile below the reference horizon.

How such corrections are made is an area of active research. However, proposed approaches aim to account for differing seismic impedance between the implicit and actual velocity profiles, and differing damping (referred to as κ or as a quality factor Q; see Section 5.5.3). Such differences can be accounted for most easily within a Fourier spectrum-based representation of ground motions (Bora et al., 2014, 2016). Most of the proposed adjustment methods use this thinking and can be considered variants of the hybrid empirical method (Campbell, 2003).

8.6.3 Example Site-Specific Hazard Calculations

To demonstrate site-specific hazard calculations using the convolution approach, three example sites are considered that all have the same value of $V_{S,30} = 360$ m/s, and reference horizons of $V_{S,30} = 760$ m/s. Figure 8.17 shows these three profiles in its upper row. The thin vertical line indicates the time-averaged $V_{S,30}$ value of 360 m/s, while the heavy lines indicate the actual profiles. For each of these profiles, 500 equivalent-linear site response calculations have been conducted, and the response spectral ratios obtained are shown in the central row of Figure 8.17. The equivalent-linear approach to site response calculation is very commonly employed in practice. It is described in Kramer (1996) and is implemented in freely available software packages like SHAKE (Schnabel et al., 1972) and STRATA (Kottke and Rathje, 2009) (with STRATA used for this example). The thin

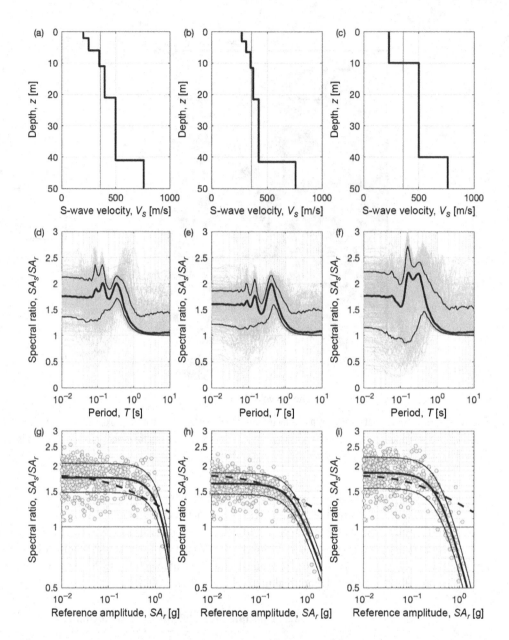

Fig. 8.17 Comparison of site response for three profiles with the same $V_{S,30}$. The upper row shows the shear-wave velocity profiles. The central row shows the response spectral ratios obtained from 500 equivalent-linear site response analyses for each profile. The lower row shows the response spectral ratios plotted against the strength of the input motion for $SA(T = 0.01\,\text{s})$. The dashed lines in the lower row show the expected spectral ratio from the Chiou and Youngs (2014) GMM. Hereafter, the profiles from left to right are referred to as "Profile 1," "Profile 2," and "Profile 3."

gray lines in these central panels show each of the 500 individual results, while the heavy black lines show the 10th, 50th, and 90th percentiles of amplification in each case. The central row illustrates that amplification levels and the periods of the resonant peaks differ significantly for the three considered cases, despite having the same $V_{S,30}$ and using the same 500 ground motions.

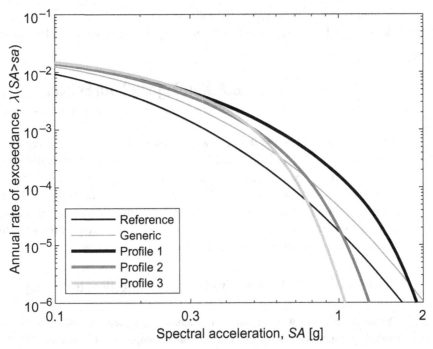

Fig. 8.18 Comparison of hazard curves using direct predictions to the surface, or via the convolution integral for each of the three site-specific profiles shown in Figure 8.17. Hazard is computed for the same single point-source scenario used in Section 6.3.3 and for Figure 8.14.

Spectral ratios obtained at a response period of 0.01 seconds were extracted and plotted against the reference level of the 0.01-second response spectral ordinate at the reference horizon. These ratios are then used to develop site-specific nonlinear amplification functions via regression analysis. The amplification functions shown in the lower row of Figure 8.17 are developed for each of the three cases using the same functional form. The $\pm\sigma_{\ln AF}(im_r)$ amplifications are shown in these panels in addition to the amplification corresponding to $\mu_{\ln AF}(im_r)$. This lower row of the figure also includes the ergodic nonlinear amplification function associated with the Chiou and Youngs (2014) GMM. For the site-specific examples shown here, the onset of nonlinear response is "delayed" relative to the Chiou and Youngs (2014) model. Additionally, the decrease in amplification, and tendency to deamplify motions, for high reference amplitudes is greater.

The comparison between the ergodic nonlinear response function of Chiou and Youngs (2014) and the profile-specific amplification functions highlights that average shear-wave velocity is only a simple proxy measure that reflects the approximate stiffness of the near-surface layers. When more detailed information exists regarding these near-surface layers, it makes sense to utilize this for more than merely estimating the $V_{S,30}$ to insert into a GMM.

Figure 8.18 shows the impact of using site-specific amplification functions rather than the ergodic functions within GMMs. Figure 8.18 includes the curves previously shown in Figure 8.14 obtained for shear-wave velocities of 760 and 360 m/s with the Chiou and Youngs (2014) model. Hazard curves computed using the convolution approach are also included. While there are differences in linear amplification levels between the three profiles (Figure 8.17), the hazard curves for low levels of spectral acceleration are reasonably similar. However, as the SA values increase, nonlinear behavior differences play a more influential role, and the hazard curves diverge quite markedly for small annual rates of exceedance. This sentiment is generally observed and not unique to this example.

8.7 Non-Ergodic Path Effects

Path effects, as evident via the distance scaling observed in a given region, depend upon the local crustal structure and spatial variation of seismic impedance. Partially non-ergodic analyses that capture systematic path effects use a local crustal model (primarily describing spatial variations in velocity and density) within physics-based approaches, or make empirical or parametric adjustments to ergodic GMMs. Differences in crustal structure manifest as differences in the apparent rates of geometric spreading and the rate of anelastic attenuation. Several approaches to identify and constrain these differences are briefly described in the following sections.

8.7.1 Data-Driven Empirical Corrections to Existing Models

The most straightforward approach to identifying systematic path effects is to compare an ergodic model's predictions with an empirical dataset comprised of data from the target region. Ideally, the data available for the region will be relatively rich such that the various source-to-site travel paths that are of interest for the hazard study are represented. In this case, the residuals computed using the local data can be inspected in order to identify any trends against distance (e.g., Scasserra et al., 2009). A model fit to any observed residual trends can then be used to adjust the ergodic model predictions for application in the target region. This is sometimes referred to as the *referenced empirical approach* (Atkinson, 2008). This relatively simple approach acts to primarily capture systematic deviations associated with a given region, rather than effects associated with specific source-to-site travel paths.

A more elaborate approach that can target individual travel paths involves constraining the various systematic source, path, and site effects from the available empirical data (Anderson and Uchiyama, 2011). Once the repeatable source and site effects have been identified, the computed systematic path residuals can be spatially smoothed to enable general prediction of path deviations for source-to-site paths not explicitly covered by the data. Anderson and Uchiyama (2011) propose one smoothing scheme, but many alternatives are available (including approaches discussed in the next section).

Anderson and Uchiyama (2011) showed that accounting for systematic site and path effects can reduce the GMM's apparent aleatory standard deviation by a factor of two or more. However, the extent of this reduction will depend upon how densely instrumented the study region is and how spatially variable the regional crustal structure is.

8.7.2 Advanced Model Calibration to Empirical Data

The data-driven empirical corrections discussed in the previous section derive corrections to an existing model. An alternative is to derive a new empirical model to limit potential contamination of the ergodic assumption from the outset. This requires advanced model-fitting approaches, such as the example presented earlier in Section 8.5.1. However, it is also possible to adopt some of these advanced model-fitting approaches under a referenced empirical approach (e.g., Kuehn et al., 2019).

Mixed-effects regression formulations such as those employed by Chen and Tsai (2002) or Stafford (2014a) first classify particular paths a priori and then compute estimates of systematic deviations associated with these paths. In these approaches, the regression analysis objective is to obtain an unbiased model representing the overall region covered by the empirical data. However, a side effect of these approaches is that potential causes of systematic bias need to be identified and represented by *random effects*. Subsequently, path-specific predictions can be made using the appropriate random effects.

A more general implementation that does not rely upon predefined paths has been introduced by Landwehr et al. (2016). In this approach, entire sets of model parameters, or just path-specific parameters, vary spatially according to some background distribution (a spatial Gaussian process). A Bayesian framework is adopted in which individual paths sampled by the data provide local constraints upon the level of this spatial field. In contrast, levels tend to a prior distribution elsewhere. This is a more rigorous version of the approach of Anderson and Uchiyama (2011), with the advantage that the posterior distribution of the Gaussian field can be used to inform estimates of epistemic uncertainty.

Another use of Bayesian methods was proposed by Kuehn et al. (2019) to compute path correction terms within a Bayesian hierarchical framework. In their approach, a region is discretized into spatial cells with properties reflecting path attributes (these are essentially attenuation rates assigned to discrete spatial cells). A Gaussian spatial process is then used to ensure smooth variations of these properties over nearby cells. Non-ergodic path corrections can then be obtained by aggregating the effects of path attenuation over the discrete cells that are sampled by specific source-to-site travel paths. Again, particular paths' epistemic uncertainty can be quantified from the posterior distribution of cell parameters.

8.7.3 Parametric Adjustments

The final approach applies primarily to physics-based GMMs where physical parameters reflect the path's attributes. In simplified simulation methods (see Section 5.5.3), these parameters will typically be rates of geometric spreading and the regional quality factor $Q(f)$. Custom parameters result in region-specific predictions but do not capture the features of particular source-to-site travel paths.

In more elaborate physics-based methods (see Section 5.5.1), where a detailed velocity model is used to prediction the Green's functions between source and site, one can obtain bespoke predictions of path scaling for individual source-to-site paths. For such predictions to be accurate, the spatial resolution of the velocity model must be high enough to capture the critical impedance contrasts and crustal heterogeneities that influence the trajectories of seismic rays. Typically, the resolution available in practice reasonably constrains low-frequency ground-motion IMs (e.g., spectral ordinates at response periods longer than one second). Simplified approaches are required to constrain the scaling of IMs reflecting high-frequency content (see Section 5.5.2). It is also possible to use high-resolution velocity models to compute numerical predictions of path scaling effects and to use these as constraints within simplified physics-based modeling approaches (e.g., Edwards et al., 2019). Similarly, empirical Green's functions (Hutchings, 1994) could be used to represent specific path scaling. In this case, small events that occurred in virtually the same location as the source of interest and recorded at the target site can be assumed to have captured the actual crustal structure empirically.

A key challenge when working with parametric adjustments, rather than direct empirical approaches, is that epistemic uncertainties relate to the model's parameters rather than to the predicted motions themselves. This presents a challenge to obtain reasonable estimates of the epistemic uncertainty in predicted motions. Both the uncertainties in the individual parameters and the correlation structure among these parameters must be well-modeled. Characterizing this correlation structure is difficult, particularly given that parameter estimates are often constrained from observations of relatively small events representing point-to-point paths, whereas larger events will involve the interaction of many such paths.

8.8 Non-Ergodic Source Effects

As discussed in Section 8.4, the chances of empirically constraining non-ergodic source effects are far lower than for path and site effects. However, partially non-ergodic source effects are frequently considered within high-level hazard analyses using a parametric approach. The most common method is to derive region-specific adjustment factors applied to ergodic GMMs to reflect perceived differences in source properties. In particular, a number of studies (e.g., Bommer et al., 2015, 2017) identify region-specific estimates of the Brune stress parameter (Section 5.5.3) and then either derive hybrid empirical adjustments (Campbell, 2003) to ergodic models (e.g., Bommer et al., 2015) or derive bespoke models for these stress parameter values using simplified physics-based methods (e.g., Bommer et al., 2017; Edwards et al., 2019). Differences in source properties are usually inferred from comparisons between sets of parameters obtained from the inversion of Fourier spectral amplitudes. That is, Fourier amplitude spectra that are assumed to be representative of the ergodic dataset are compiled. A simple physical model (Section 5.5.3) is then fit using this data to obtain a set of parameter estimates (such as stress parameter $\Delta\sigma$, anelastic attenuation properties $Q(f)$, and site kappa κ_0, along with geometric spreading rates). The process is then repeated using region-specific data to obtain a second set of parameter estimates. These parameter sets are then used to develop adjustment factors following the hybrid empirical method (Campbell, 2003).

When empirical approaches are considered, they rely upon the identification of spatial patterns within systematic source effects. For example, Kuehn et al. (2016) derive spatially varying systematic deviations from residuals computed from an ergodic ground-motion model for spectral ordinates. Similarly, Lee et al. (2020) use a physics-based approach to determine systematic deviations from predictions obtained using generic source parameters assumed to be representative for a given region.

The idea, previously discussed in Section 8.3.1, is that the between-event component of apparent aleatory variability, δB, can be represented as a function of location, $\boldsymbol{x}$:

$$\delta B(\boldsymbol{x}) = \delta B_0 + \delta L(\boldsymbol{x}). \tag{8.25}$$

At a given location we can then consider that some component of this event-specific deviation is genuinely aleatory, δB_0, while the remaining component, $\delta L(\boldsymbol{x})$, is a systematic deviation varying with spatial location.

As recurrence intervals for large magnitude events are relatively long compared with our observation period, multiple similar events at a particular location are likely to have small magnitudes. These repeated events can be used to determine $\delta L(\boldsymbol{x})$ at various locations throughout a study region, and from these, a location-to-location variance τ_{L2L}^2 can be computed. The ergodic between-event variance can then be represented as

$$\tau^2 = \tau_0^2 + \tau_{L2L}^2 \qquad (8.26)$$

with τ_0^2 being the variance of δB_0 and τ_{L2L}^2 being the variance of the $\delta L(x)$ terms. To consider non-ergodic source effects we need to estimate the magnitude of τ_{L2L}^2 so that it can be removed from the apparent aleatory between-event variability, and we also need an estimate of $\delta L(x)$ such that x coincides with a seismic source of interest. The estimates $\hat{\tau}_{L2L}^2$ and $\widehat{\delta L}(x)$ will have epistemic uncertainty that also needs to be specified.

With this framework, a critical practical challenge is relating the estimates of $\delta L(x)$ from the small, essentially point-source, events to larger ruptures spanning tens of kilometers or more. In particular, an important question to address is whether the $\delta L(x)$ values computed for small events rupturing portions of a larger fault retain their systematic deviation from an ergodic model when these same portions rupture as part of a larger event. Another challenge is associated with considering time-dependent effects. As previously discussed in Section 3.8, physical conditions in the vicinity of faults vary in time, particularly following large rupture events. There is some expectation that $\delta L(x)$ should, therefore, vary in time and be $\delta L(x, t)$ to allow for fluctuations in crustal fluid pressures, redistribution of regional stress via aftershock events, and frictional healing of rupture surfaces. This refinement is beyond the reach of current applications for large tectonic events that are most important for typical hazard and risk studies but will likely be addressed in time.

8.9 Non-Ergodic Components in Seismic-Source Models

A SSM includes many components whose calibration may utilize the ergodic assumption. The present section's objective is not to identify all such components, but rather to explain the general concepts in the context of selected examples related to an ERF. Two examples will consider (1) activity rates within areal sources and (2) a source-scaling relation to predict magnitude as a function of logarithmic rupture area. In both cases, the examples illustrate how partially non-ergodic approaches are implemented by understanding components of variability within the hazard model inputs. The first step is to recognize that the apparent aleatory variability from an ergodic dataset is comprised of actual aleatory variability and other systematic epistemic deviations. Once the epistemic components are identified, we must estimate their expected deviations for a given site or region and move the epistemic uncertainty out of the apparent aleatory variability and into the logic tree.

8.9.1 Components of Variability within Activity Rates

Area sources have constant earthquake activity rates throughout the source (see Section 2.6). However, the available seismicity data for a given area source will always show spatial fluctuations. Under the ergodic assumption, these spatial fluctuations are assumed to be statistical and reflect the historical sampling from the source's underlying uniform distribution of activity. When ruptures are generated for hazard calculations, they arise from discrete points within the area source. The rate assigned to events at each point is a fraction of the total source rate, computed based on the fractional area associated with the point (Section 3.6.1). With this approach, the rate of occurrence assigned to each point is ergodic. The activity rate at each specific point is based on observed events at many spatial locations throughout the source, and is assumed constant at all other points.

In reality, the observed spatial fluctuations may be systematic, rather than statistical, in nature. If the activity rate at a given spatial location x is defined as $\lambda(M \geq m_{\min}|x)$, then we can describe this local rate in terms of an ergodic rate $\lambda_0(M \geq m_{\min})$ and systematic deviations from this:

$$\lambda(M \geq m_{\min}|x) = \lambda_0(M \geq m_{\min}) + \delta\lambda(M \geq m_{\min}|x) + \varepsilon(x). \qquad (8.27)$$

In Equation 8.27, the ergodic rate $\lambda_0(M \geq m_{\min})$ does not vary with position and represents the average rate of occurrence over the entire area source (when data from all locations within the source are pooled together). The term $\delta\lambda(M \geq m_{\min}|x)$ represents a systematic deviation away from this ergodic rate and is specific to location x, while $\varepsilon(x)$ represents inherent variability in the activity rate that is relevant for this same location. Both of these terms have zero mean and standard deviations over all spatial positions in the source of $\tau_{\delta\lambda(M \geq m_{\min})}$ and ϕ_ε, respectively.

When using the ergodic assumption for each point within the source, we recognize the possible existence of systematic deviations like $\delta\lambda(M \geq m_{\min}|x)$ but assume they cannot be constrained. These deviations therefore appear as apparent inherent variability indistinguishable from $\varepsilon(x)$. For our modeling, we use $\lambda_0(M \geq m_{\min})$ to define the average rate of occurrence at every point within the source. We then define the variability in this rate, at every point, to be

$$\sigma^2_{\lambda(M \geq m_{\min})} = \tau^2_{\delta\lambda(M \geq m_{\min})} + \phi^2_\varepsilon. \qquad (8.28)$$

To use a partially non-ergodic approach, for each point within the area source, we need to estimate the systematic deviation $\widetilde{\delta\lambda}(M \geq m_{\min}|x)$. The rate assigned to a given location would then be $\lambda_0(M \geq m_{\min}) + \widehat{\delta\lambda}(M \geq m_{\min}|x)$ and the variability at each point would be $\sigma^2_{\lambda(M \geq m_{\min}|x)} = \phi^2_\varepsilon$.

For the ergodic case, we have epistemic uncertainty in the values of $\lambda_0(M \geq m_{\min})$ and $\sigma_{\lambda(M \geq m_{\min})}$. However, as we have estimated these by pooling all of the source data, these uncertainties are lower than for the non-ergodic case. For the partially non-ergodic case, the epistemic uncertainty must relate to $\lambda_0(M \geq m_{\min}) + \widehat{\delta\lambda}(M \geq m_{\min}|x)$, which is spatially varying, and ϕ_ε, which could also be spatially varying. For the locations with more historical data, the epistemic uncertainty will be lower than for locations with fewer observations. However, in all cases, the epistemic uncertainty will be greater than for the ergodic model because the data available to constrain $\widehat{\delta\lambda}(M \geq m_{\min}|x)$ are fewer than that available to constrain $\lambda_0(M \geq m_{\min})$.

When using partially non-ergodic approaches in this example, we are breaking the overall area source into smaller sources where the rates are allowed to differ. Whether or not partially non-ergodic approaches will lead to different hazard results depends upon whether spatial patterns exist among the $\delta\lambda(M \geq m_{\min}|x)$ terms. When spatial clustering of relatively high or low deviations exists, an ergodic model is more likely to produce differing hazard estimates.

There are many parallels between the approach of spatially smoothing activity rates presented in Section 3.5.10 and the discussion presented in this section. Spatially smoothing activity rates is a well-established approach and not ordinarily regarded as partially non-ergodic. The critical difference between the method of Section 3.5.10 and Equation 8.27 is that under the partially non-ergodic paradigm we explicitly distinguish between systematic and aleatory variations in the observed activity rates.

8.9.2 Components of Variability in Source-Scaling Relations

In many analyses, we need a model to predict moment magnitude M as a function of logarithmic rupture area, i.e., $M(\ln A)$. Imagine that we need to predict magnitudes for ruptures on the Hayward Fault as part of a seismic hazard study for Yerba Buena Island in San Francisco Bay. There are too few historical data to develop a model using only ruptures on the Hayward Fault and so the ergodic assumption needs to be invoked. Therefore, we need a dataset of rupture areas and magnitudes from other similar faults. To start, we could define *similar* faults to be other strike-slip faults within the San Francisco Bay Area that have a similar stress environment and rupture history. If insufficient data still exist, the stress environment and rupture history constraints can be relaxed, so we would still look for strike-slip faults in the San Francisco Bay Area, but would not worry about their orientation or slip rate, etc. Subsequently, *similar* faults could be defined as strike-slip faults within California, North America, or even globally, until a sufficiently large database could be compiled to constrain $M(\ln A)$.

As the conditions defining similarity are relaxed, more data become available. However, at the same time, more variability and potential bias enter the dataset as even small differences in fault geometry, slip rate, stress regime, rupture history, etc., all contribute characteristics to the dataset that are not strictly appropriate for the Hayward Fault. With increasing data, the statistical robustness of the model should improve. However, this robustness reflects the model's ability to represent ruptures within all of the regions from which that data originated; it does not ensure the robustness of predictions for the Hayward Fault. Therefore, it remains necessary to quantify the extent to which ruptures on the Hayward Fault differ from the derived model's predictions, which is an example of epistemic uncertainty.

An ergodic source-scaling model, $\mu_M(\ln A; \boldsymbol{\theta})$ (e.g., Section 2.7.1), will be a function of logarithmic rupture area $\ln A$, but could depend upon other parameters, such as fault slip rate (Anderson et al., 1996), encapsulated within $\boldsymbol{\theta}$. These additional parameters could explain some of the apparent variability that exists if the model were only a function of the rupture area. Without loss of generality, we will ignore these secondary effects and assume that the ergodic model has the linear form $\mu_M(\ln A) = \alpha + \beta \ln A$. In this case, the ergodic model has two fixed parameters α and β, and ergodic variability represented as $\sigma^2_{M(\ln A)}$.

In a non-ergodic approach, we could recognize that the parameters α and β may vary systematically from region to region, and from fault to fault within a given region. A non-ergodic model for the moment magnitude associated with some rupture could then be written as

$$M(\ln A) = (\alpha + \delta\alpha_r + \delta\alpha_{r,f}) + (\beta + \delta\beta_r + \delta\beta_{r,j})\ln A + \varepsilon. \tag{8.29}$$

This formulation implies that the parameter α in the ergodic model is really composed of some global average value plus systematic deviations from region to region, $\delta\alpha_r$, as well as nested deviations from fault to fault within a given region, $\delta\alpha_{r,f}$. A similar situation exists for the slope parameter β.

The ergodic variability $\sigma^2_{M(\ln A)}$ is therefore comprised of several components that reflect the intrinsic variability of source-scaling relations for the rupture area as well as the systematic deviations among regions and faults that appear as variability within the ergodic framework. Assuming that all of these systematic deviations are uncorrelated, the total ergodic variability can be written as

$$\sigma^2_{M(\ln A)} = \tau^2_{\delta\alpha_r} + \tau^2_{\delta\alpha_{r,f}} + (\ln A)^2 \, (\tau^2_{\delta\beta_r} + \tau^2_{\delta\beta_{r,f}}) + \phi^2. \tag{8.30}$$

In Equation 8.30, the various $\tau^2_{\delta x}$ terms correspond to the region-to-region or fault-to-fault (within a region) variance of the zero-mean δx terms of Equation 8.29.

In a fully non-ergodic analysis, we would seek source-scaling parameters tuned explicitly to the Hayward Fault. In this case we need estimates of $\widehat{\delta\alpha}_r$, $\widehat{\delta\alpha}_{r,f}$, $\widehat{\delta\beta}_r$, and $\widehat{\delta\beta}_{r,f}$. The mean magnitude for a particular Hayward Fault rupture would then be $(\alpha + \widehat{\delta\alpha}_r + \widehat{\delta\alpha}_{r,f}) + (\beta + \widehat{\delta\beta}_r + \widehat{\delta\beta}_{r,f})\ln A$,

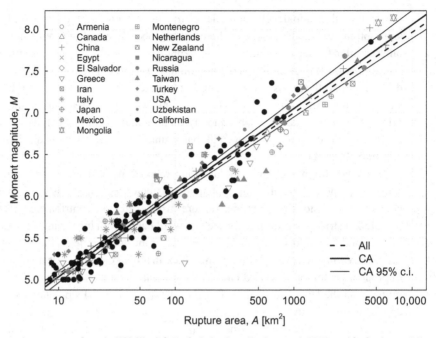

Fig. 8.19 Magnitudes and rupture areas from the NGA-West2 finite-fault database (Ancheta et al., 2014) used for for the partially non-ergodic source-scaling example. The dashed line shows the best global fit to all data, while the solid lines represent the fit to Californian data only.

with r and f set to identify the Hayward Fault, and the variance would just be ϕ^2. We would also need to represent the epistemic uncertainty in each estimated $\widehat{\delta x}$ value.

In a less ambitious partially non-ergodic analysis, we could relax the ergodic assumption to account for the region-to-region variability. In this case the appropriate model would be $(\alpha + \widehat{\delta \alpha_r}) + (\beta + \widehat{\delta \beta_r}) \ln A$ and we would need a larger apparent aleatory variance of $\tau^2_{\delta \alpha_{r,f}} + (\ln A)^2 \, \tau^2_{\delta \beta_{r,f}} + \phi^2$. Again, we would need to define the epistemic uncertainty in the two $\widehat{\delta x}$ values, but this should be more straightforward than in the fully non-ergodic case.

Example of Partially Non-Ergodic Source Scaling

Figure 8.19 presents a dataset of moment magnitude and rupture area for a broad range of well-documented earthquakes from around the world. The data here come from the NGA-West2 finite-fault database (Ancheta et al., 2014), and combines all style-of-faulting events. The objective is to define a relationship between moment magnitude and logarithmic rupture area to infer the size of events on a specific fault. Although the data used in the example are real, the example itself is somewhat contrived as we would not usually mix data from all styles of faulting. We should expect (at least) a bilinear relation between magnitude and rupture area (e.g., Hanks and Bakun, 2008; Shaw, 2009), with the location of the break in scaling being a function of the style of faulting (Stafford, 2014b). However, this aggregated data help to demonstrate the issues raised in the previous section.

In Figure 8.19, two lines are plotted that show fits to this data using different techniques. In all cases, the model being fit is of the form $M_i = \alpha + \beta \log_{10} A_i + \varepsilon_i$. For the case where all data are

Table 8.4. Coefficients, and associated standard errors (s.e.), of linear source-scaling relations between magnitude and logarithmic rupture area. The first three lines relate to events of all mechanism, while the fourth is for strike-slip events in California (CA) only. The last three columns give estimated mean magnitude for three different areas (in units of km²)

Region	n_{obs}	$\hat{\alpha}$	s.e.(α)	$\hat{\beta}$	s.e.(β)	$\hat{M}(1 \times 10^3)$	$\hat{M}(3 \times 10^3)$	$\hat{M}(1 \times 10^4)$
All	234	4.096	0.058	0.968	0.028	7.00	7.46	7.97
USA	106	4.017	0.058	1.014	0.030	7.06	7.54	8.07
CA	95	4.024	0.068	1.009	0.037	7.05	7.53	8.06
CA (SS)	68	4.043	0.080	0.989	0.044	7.01	7.48	8.00

considered, the objective is to obtain the best estimate of a fully ergodic model. For that reason, a mixed-effects regression analysis is conducted so that the impacts of individual countries do not bias the estimate. The line for California (CA), on the other hand, specifically targets this region. So a fixed-effects regression analysis is performed using only data from California. Although the fitted lines are similar over a broad range of rupture areas, there are visible differences for larger rupture areas.

In addition to Figure 8.19, Table 8.4 presents the model parameters obtained for the above-mentioned cases, as well as an additional fit that is made just to data from the United States. In each case, the best estimates of the two coefficients ($\hat{\alpha}$ and $\hat{\beta}$) are provided along with the associated standard errors in these estimates. The table shows that as one progressively refines the dataset to be more specific, the coefficients' standard errors tend to increase. This reflects the trade-off between the number of available observations and the precision with which estimates of the model parameters can be made.

While the coefficient differences shown in Table 8.4 are not *statistically significant*,[27] they have important impacts for a source model. For example, the estimated magnitudes shown in Table 8.4 correspond to seismic moment estimates that differ by 19%, 28%, and 36% as one moves through rupture areas of 1×10^3, 3×10^3, and 1×10^4 km², respectively. Suppose one is conserving seismic moment release rates. In that case, these differences in moment release from these relatively large events can greatly impact the implied rates of occurrence of smaller, more frequent events (see Section 3.5.5).

The estimates of uncertainty shown in Table 8.4 represent only the parametric uncertainty *given* the particular datasets used in this example. In practice, one also needs to consider the additional uncertainty associated with the database itself. That is, if we observed a new set of 95 earthquakes in California, to what extent would the model parameters differ from those shown here? The confidence interval shown in Figure 8.19 also grows as one moves away from the bulk of the data toward larger rupture areas. Most of the Californian data are also observed at relatively small magnitudes. Therefore, if we are interested in focusing upon source-scaling relations for larger events (which is often the case), the relative contribution of Californian data reduces considerably. For example, Californian events comprise 44% of ruptures with areas less than 1000 km² (approximately $M7$), but less than 10% of the data above this limit.

---

[27] This can be seen in Figure 8.19, as the confidence interval for the Californian predictions contains the global prediction for all data.

8.9.3 Practical Issues Common to Both SSM Examples

In the previous examples of Sections 8.9.1 and 8.9.2, we have stated which components need to be estimated and have shown how the variance components are included or removed from the expressions for the ergodic variability. However, the details of decomposing this variance and estimating the systematic deviations can be quite involved. While the approaches in this section are valid, they are rarely used due to insufficient data and high amounts of inherent parameter variability. This makes it challenging to identify systematic deviations and leads toward the use of ergodic models.

Exercises

8.1   What is the ergodic assumption? Give an example of how this assumption is used in the development of an empirical GMM.

8.2   Consider three different empirical GMMs published in each decade from 1980 to 2020 (e.g., see Douglas, 2019) and comment on their relative position on the ergodic continuum of Figure 8.1 and classification of Table 8.2.

8.3   Section 4.6.1 presents example calculations using the BJF97 GMM for $SA(1 \text{ s})$, an average shear-wave velocity $V_{S,30} = 500$ m/s, and a magnitude 6.5 earthquake. In those calculations, it is assumed that the total (ergodic) aleatory variabilty is $\sigma_{\ln SA(1 \text{ s})} = 0.52$. For the present exercise, assume that the non-ergodic within-event standard deviation is $\phi = 0.4$ and that the ergodic between-event standard deviation is $\tau = 0.33$. Further assume that this ergodic between-event standard deviation reflects a "within-region" between-event standard deviation of $\tau_0 = 0.25$.

   (a) What is the "region-to-region" standard deviation τ_L?
   (b) If an adjustment were made to the BJF97 GMM to reflect a regional effect of $\delta L = 0.2$, but no modification is made to the ergodic variability, compute the probabilities of exceeding levels of $SA(1 \text{ s})$ of 0.5 g and 1.2 g.
   (c) If the same location adjustment of $\delta L = 0.2$ was applied, and the appropriate reduction to the aleatory variability was also made, compute revised estimates of the probabilities of exceeding the levels of $SA(1 \text{ s})$ defined in (b).
   (d) Comment upon the relative roles of δL and τ_L in the exceedance probability calculations of (c) and (d).

8.4   The example in Section 6.3.3 computes a hazard curve for a site influenced by a single source with a Gutenberg–Richter magnitude-frequency distribution. Repeat the hazard curve calculations in that example for the following conditions:

   (a) Assume that the site-to-site variability within the Boore et al. GMM is equal to 0.25 and that the site-specific site response is perfectly represented by the GMM, i.e., $\delta_s = 0$.
   (b) Assume that the estimated site-specific site response remains at $\delta_s = 0$, but that there is significant epistemic uncertainty associated with this estimate. Specifically, the uncertainty in δ_s is assumed to be equal to the site-to-site variability of 0.25. Compute the mean hazard curve using the expression of Equation 8.7, and compare this with the result from (a).

(c) Repeat (b) using a logic-tree approach with explicit branches for δ_s. Use three branches corresponding to quantiles of 10%, 50%, and 90% for the δ_s distribution, along with weights of 0.3, 0.4, and 0.3, following Keefer and Bodily (1983).

(d) Compare and comment upon the results obtained in each of the three previous parts.

8.5 Discuss the non-ergodic nature of physics-based ground-motion models in the context of their treatment of source, path, and site effects.

8.6 Repeat the example presented in Section 8.3.1, but assume that $\delta L^{\text{true}} = -0.15$ and that all other parameters remain as in the example. How many earthquake events need to be observed to obtain an estimate $\widehat{\delta L}$ within ± 0.05 of the true value, at a confidence level of 95%?

8.7 Demonstrate that treating site-specific site response via the convolution integral or through embedding the site response within the hazard integral are equivalent when the site response is purely a function of the reference intensity level.

8.8 What are the four potential methods for treatment of site response in the context of ergodic or non-ergodic PSHA? Discuss the advantages and disadvantages of each method and the conditions in which it would be appropriate and/or pragmatic to adopt each option.

8.9 Ground motions are commonly considered in the context of source, path, and site contributions. Discuss the relative ability of observational data to constrain each of these three effects.

8.10 Non-ergodic features can be considered through data from observations or from the results of numerical simulations. Discuss the pros and cons of each of these two data sources in the context of PSHA predictions for typical problems of interest.

8.11 Observed ground-motion residuals can be partitioned according to Equation 8.13 as

$$\ln im = \mu_{\ln IM}(rup, site) + \delta_e + \delta_{es} + \delta_s + \varepsilon.$$

However, Equation 8.16, and the associated example, explained that there are reasons why the amplitude of the "path effect" δ_{es} should be a function of distance, i.e., $\delta_{es} \equiv \delta_Q R$, where R is a distance measure and δ_Q is a random effect reflecting differences in the crustal attenuation model. Imagine that the true value of δ_Q for this particular region is $\delta_Q = -0.0025$, that $\tau_{\delta_Q} = 0.0025$ for a general ergodic model, and that the standard deviation of the δ_e and δ_s terms are $\tau_e = 0.3$ and $\tau_s = 0.3$, respectively. The standard deviation of ε is $\phi = 0.4$.

(a) Consider two sites located distances of $R_1 = 50$ km and $R_2 = 150$ km from an active seismic source. Assume that the event and site effects at each site are perfectly known, and further assume that only observations from a single site are considered at a time. How many events would need to be observed at each site in order to recognize that $\delta_Q \neq 0$ for this region, at a 95% confidence level?

(b) How would the numbers computed in (a) change if the site effects were unknown at both stations?

(c) A detailed site investigation is conducted at the most distant site, such that it is known that $\delta_{s,2} = 0.25$, exactly. At the closest site, no detailed investigations are conducted, but similarities between the two locations lead to an estimate of correlation between the site effects to be assessed as $\rho_{\delta_{s,1}, \delta_{s,2}} = 0.7$. How does this information influence the numbers of events required to demonstrate that $\delta_Q \neq 0$ using recordings from the closest site only? Assume that the observations are not used to estimate $\delta_{s,1}$.

8.12 Interperiod correlations between response spectral ordinates can be found from the residuals of ground-motion models under the framework of Equation 8.12, as explained in Section B.7.

(a) Assuming that systematic site effects are also considered, within the following framework:

$$\ln SA_{es} = \mu_{\ln SA}(e, s) + \delta B_e + \delta W_{es} + \delta S_s,$$

derive an expression for the overall interperiod correlation among $\ln SA_{es}$ ordinates. Note that δB and δS are random effects for the event and site, respectively, while δW_{es} is the event and site corrected within-event residual.

(b) Within a non-ergodic analysis, explain how the expression for the correlation in (a) would be modified.

(c) Explain under what circumstances the non-ergodic correlation would be greater or smaller than the ergodic correlation.

8.13 Section 5.5.3 explained that the Fourier spectral amplitudes at the ground surface can be represented by the product of source, path, and site contributions. For a soil deposit of depth H_s, with shear-wave velocity V_s, density ρ_s, and damping ratio ζ_s overlying an elastic medium with shear-wave velocity V_r, density ρ_r, and damping ratio ζ_r, the site transfer function is defined by

$$|S(f)| = \left| \frac{1}{\cos \beta^* + i\alpha^* \sin \beta^*} \right|$$

where $\beta^* = 2\pi f H_s / V_{s,s}^*$, $\alpha^* = (\rho_s V_{s,s}^*)/(\rho_r V_{r,s}^*)$, and $V_{s,s}^* = V_s(1 + i\zeta_s)$ (and similar for $V_{s,r}^*$) (Kramer, 1996).

A series of seven recording stations are located with a 1-km spacing on horizontal ground, at coordinates of $x = \{-3, -2, -1, 0, 1, 2, 3\}$ km. The ground profile below each site consists of a single layer of soil overlying elastic bedrock. The shear-wave velocity, density, and damping ratio for the soil are $V_s = 300$ m/s, $\rho_s = 1.8$ t/m³, and $\zeta_s = 2\%$, respectively, while the corresponding parameters for the bedrock are $V_r = 1000$ m/s, $\rho_r = 2.0$ t/m³, and $\zeta_r = 0.1\%$. The depth to bedrock can be defined as

$$H_s(x) = 30 + 15 [1 - \tanh(x)].$$

Assume that the stations record ground motions from a series of distant earthquakes, such that site response remains linear and source-to-site distances can be assumed equal for all sites.

(a) If a GMM for logarithmic Fourier amplitudes is derived from the data at these sites, but uses $V_{S,30}$ to characterize site effects, how much apparent aleatory variability can be attributed to the imperfect characterization of site effects at frequencies of 1 Hz and 3 Hz?

(b) If the target site is the location with coordinate $x = 0$, what adjustment should be made to the generic model to achieve an accurate site-specific characterization?

(c) How many records would be required to empirically demonstrate that the site response for Fourier amplitudes at 1 Hz and 3 Hz at coordinate $x = -1$ is different from the average over this seven-station network? Assume a 95% confidence level is desired, and that the between-event variability is $\tau = 0.3$ and the event and site-corrected within-event variability is $\phi_{SS} = 0.4$.

(d) Revisit (c) after reading Chapter 11. How would spatial correlations influence the required numbers of records?

8.14 A source-scaling relation for computing moment magnitude as a function of rupture area is defined as $M = \log A + 4.0$, with $\sigma_M = 0.25$. This relationship is derived from an ergodic database. Equation 2.12 suggested that scaling relations of this form implicitly reflect static stress drop effects.

(a) For a rupture area of $A = 1000$ km^2, and the ergodic source-scaling model above, what is the probability of observing a magnitude of at least $M7.2$?

(b) Assume that the logarithmic stress drop variability is $\sigma_{\log \Delta \sigma} = 0.3$ for the ergodic database. If a target region is known to have the same stress drop as the mean of the ergodic database, what is the probability of observing a magnitude of at least $M7.2$ for a 1000-km^2 rupture area?

(c) Reconsider (b) under the circumstances that stress drop in the target region is known to be 70% of the mean stress drop in the ergodic database.

PART III

RISK

9

Seismic Risk

Until now, this book has focused on quantifying the hazard associated with ground motions, but it is the consequence of ground motions that is usually of concern when designing or assessing facilities. In this chapter, we develop tools for *risk analysis*—a quantitative probabilistic evaluation of what adverse consequences can occur from an earthquake, and their likelihoods of occurrence. The potential for these adverse consequences depends upon the earthquake ground-motion *hazard* (the focus of the preceding chapters), the *exposure* of assets to those hazards, and the *vulnerability* of the exposed assets to ground motions and other earthquake effects. This chapter is concerned with the mathematical evaluation of risk while considering these factors. We will introduce the concepts of risk analysis for seismic problems utilizing hazard analysis tools, while also maintaining consistency with broader risk analysis principles.

Learning Objectives

By the end of this chapter, you will be able to do the following:

- Define fragility and vulnerability functions to characterize the impact of ground motions.
- List common types of consequences considered in risk analyses.
- Compute several metrics to measure the risk to buildings and other facilities.
- Evaluate how the shape of a fragility curve affects the implied risk to a structure.
- Describe the advantages and disadvantages of three general options for developing and calibrating fragility and vulnerability functions.

9.1 Introduction

A seismic risk analysis aims to predict adverse consequences from earthquakes, considering all relevant uncertainties (in earthquake occurrence, resulting ground motion, structural response, and consequences to the structure). The predictions are formulated in terms of the probability of an adverse consequence, or the average consequence in a given exposure period (e.g., the expected financial loss per year due to earthquake damage). Such analyses are utilized widely in assessment of facilities and activities subject to a variety of perils, not just for earthquakes (e.g., NRC, 1975; NUREG, 1983; Grossi et al., 2005).

Several decisions can be supported by this information, such as the following:

- Direct evaluation of whether a facility is sufficiently safe
- Calibration of design standards that produce structures with an acceptably low probability of failure

- Pricing of insurance contracts, such that the insurance company generates revenue sufficient to cover possible future losses, and the building owner pays a fair price for transferring their risk
- Evaluation of potential losses to a range of existing structural types, to identify primary drivers of risk and efficiently devote societal resources to mitigating those risks.

The above decisions require a probabilistic assessment of consequences for efficient decisions to be identified. Further, the above decisions all require an understanding of the consequences of damage, rather than just ground-motion hazard, so the results from the previous chapters alone are insufficient for addressing these problems.

This chapter will discuss how consequences can be formulated probabilistically and illustrate how those formulations can be incorporated with hazard calculations to gain insights and support decision making. The development and calibration of models to predict consequences could quickly fill an entire book on its own, so this chapter focuses on the narrower question of how to utilize such models once they are available. Specific models are available in other publications (e.g., Yepes-Estrada et al., 2016; Porter, 2020).

9.2 Fragility and Vulnerability Functions

To link with results from previous chapters, risk calculations require a connection between ground-motion intensity and the resulting impacts. These links come in many forms, but broadly speaking can be categorized into fragility and vulnerability (loss) functions, depending upon whether the outcome of interest is binary or continuous. As with the earlier treatment of hazard analysis, our primary focus is on the impact of ground motions, and so the functions below take IM as an input. However, recall that other phenomena such as liquefaction, tsunami, or fire can also cause damage, and so inputs to the following functions could also use metrics to describe the severity of those phenomena.

This section will discuss the definition and use of such functions,[1] and subsequently, Section 9.3 will discuss how these functions are developed and calibrated.

9.2.1 Fragility Functions

A fragility function provides a prediction of a binary outcome, F (failure or nonfailure), as a function of ground-motion intensity. Failure is defined with respect to some outcome of interest, such as the examples given in Table 9.1. That is, failure is not necessarily a structural collapse.

A lognormal cumulative distribution function (Section A.5.2) is often used to define a fragility function:[2]

$$P(F \mid IM = x) = \Phi\left(\frac{\ln(x/\theta)}{\beta}\right) \tag{9.1}$$

---

[1] The terminology for these functions varies substantially in other literature. Fragility and vulnerability are used by some interchangeably or in other ways, and other terms such as loss functions are also used. We have chosen terminology here that is relatively common in the literature, but readers should be aware that it is not universal.

[2] This equation is similar in form to Equation 4.23, with $\ln\theta = \mu_{\ln IM}$ and $\beta = \sigma_{\ln IM}$. The change in notation has been made to reflect typical conventions in the respective ground-motion model and fragility fields, so that the two equations will look similar to those in related literature.

Table 9.1. Examples of binary failure criteria and continuous consequence metrics. Entries in a given row are not necessarily related

Failure criteria	Consequence metrics
Material yielding	Repair cost
Cracking of windows	Time to reopen a building
Reduction of x% capacity in an element	Time to repair a component
Exceedance of y floor acceleration	Number of fatalities
Structural collapse	Number of displaced people
Breakage in a pipe	Number of injuries
Breach of a levee	Amount of levee settlement
Soil liquefaction triggering	

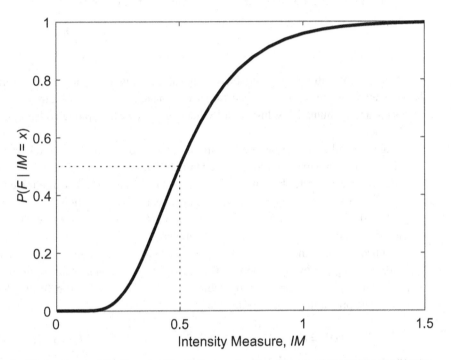

Fig. 9.1   Illustrative fragility function providing the probability of failure as a function of the ground-motion intensity, *IM*, using Equation 9.1 with $\theta = 0.5$ and $\beta = 0.4$.

where $P(F \mid IM = x)$ is the probability that a ground motion with $IM = x$ will cause failure to occur, $\Phi()$ is the standard normal cumulative distribution function, θ is the median of the fragility function (the IM level with a 50% probability of failure), and β is the standard deviation of the $\ln IM$ level at which failure will occur (sometimes referred to as the *dispersion* of IM). An illustrative fragility function, using Equation 9.1, is shown in Figure 9.1. The value of $\theta = 0.5$ results in the fragility function giving a probability of 0.5 when $IM = 0.5$, and $\beta = 0.4$ controls the slope of the function (with larger β values giving a flatter slope of the fragility function at its median).

One way to view this function is that the capacity to resist failure (in terms of the level of IM it can sustain) is a random variable (Cornell et al., 2002; Gardoni et al., 2002). Equation 9.1 implies that this capacity is lognormally distributed, the median capacity is θ, and the standard deviation of the $\ln IM$ value causing failure is β. The assumption of lognormality dates back to the first formulation of fragility functions for evaluating risk for nuclear power plants (Kennedy et al., 1980; Kennedy and Ravindra, 1984). The lognormal model has been confirmed as reasonable in a number of cases (e.g., Ibarra and Krawinkler, 2005; Porter et al., 2007; Bradley et al., 2010; Ghafory-Ashtiany et al., 2011; Eads et al., 2013), and it is advantageous because it requires only two parameters. Despite the above factors, a lognormal function is not required, and other assumptions can easily be used. The defining aspect of a fragility function is that it provides a probability of failure as a function of IM.

Equation 9.1 characterizes the probability associated with a binary failure or nonfailure outcome. This formulation can be generalized to consider a sequential set of outcomes of increasing severity. Most commonly, we consider a discrete set of damage states (DS), and specify fragility functions for the probability of a structure reaching that damage state or worse:

$$P(DS \geq ds_i \mid IM = x) = \Phi\left(\frac{\ln(x/\theta_i)}{\beta_i}\right) \tag{9.2}$$

where ds_i is the ith damage state, increasing values of i indicate more severe damage, and the fragility parameters θ_i and β_i are specified for each damage state. The multiple damage states are typically assumed to be mutually exclusive and collectively exhaustive, though other assumptions can be made if needed.

Figure 9.2 illustrates structural fragility functions from HAZUS, a widely used model for assessing risk to a range of structure types in the United States (FEMA, 2015). Results are shown for a high-code low-rise concrete shear-wall building with retail occupancy, using the equivalent PGA results from that report. These fragility functions are specified by Equation 9.2 with $IM = PGA$ in units of g, $\theta = (0.24, 0.45, 0.9, 1.55)$, and $\beta = (0.64, 0.64, 0.64, 0.64)$, corresponding to damage states of "slight," "moderate," "extensive," and "complete."

With multiple damage states of sequentially increasing severity, and damage state exceedance probabilities given by Equation 9.2, the probability of damage state occurrence (i.e., of being in a particular ds) can be computed by subtracting the exceedance probabilities for sequential damage states. The probability of observing exactly ds_i is

$$P(DS = ds_i \mid IM = x) = P(DS \geq ds_i \mid IM = x) - P(DS \geq ds_{i+1} \mid IM = x) \tag{9.3}$$

where the right-hand side terms are defined by Equation 9.2 and $P(DS \geq ds_{i+1} \mid IM = x) = 0$ if ds_i is the most severe damage state. Using the specific example values above, and considering $PGA = 0.5$ g, we can compute:

$$P(DS \geq ds_1 \mid PGA = 0.5 \text{ g}) = \Phi\left(\frac{\ln(0.5/0.24)}{0.64}\right) = 0.87 \tag{9.4}$$

$$P(DS \geq ds_2 \mid PGA = 0.5 \text{ g}) = \Phi\left(\frac{\ln(0.5/0.45)}{0.64}\right) = 0.57 \tag{9.5}$$

$$P(DS = ds_1 \mid PGA = 0.5 \text{ g}) = 0.87 - 0.57 = 0.3. \tag{9.6}$$

These values are annotated in Figure 9.2, to illustrate their relationship to the visual fragility functions.

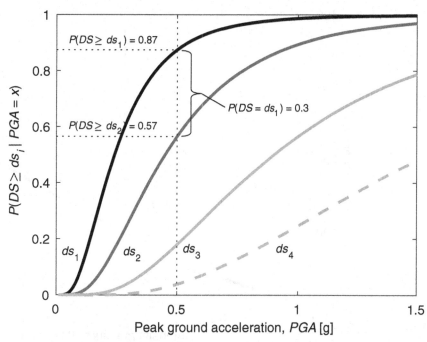

Fig. 9.2 HAZUS structural fragility functions for a high-code low-rise concrete shear-wall building with retail occupancy, using the "equivalent *PGA*" functions of HAZUS Section 5.4.4 for simplicity. For brevity, the "| *PGA* = 0.5 g" conditioning has been suppressed in the labels.

9.2.2 Vulnerability Functions

A vulnerability function is used to quantify outcomes when the consequence of interest is a continuous outcome, rather than a binary "failure" or "nonfailure." See Table 9.1 for examples of such consequence metrics. A vulnerability function (also sometimes referred to as a damage function, a loss function, or a consequence function) predicts the probability distribution of consequences as a function of IM. A general equation for this type of function is as follows:

$$P(C > c \mid IM = x) = 1 - F(c \mid x) \tag{9.7}$$

where C is the consequence metric of interest (e.g., Table 9.1), and $F(c \mid x)$ is a cumulative distribution function for the consequence C, evaluated at c and dependent on the IM amplitude x. There are several ways in which Equation 9.7 can be parameterized in practice. In the subsections that follow, we consider several common forms of $F(c \mid x)$, based on lognormal and beta distributions, and a nonparametric form. Following those subsections, a model focused only on mean consequence is introduced.

Lognormal Model

In the case where the consequence C is unbounded (e.g., a structural displacement), a lognormal CDF is often used in Equation 9.7. One common model in this situation is that the structural response distribution at a given IM level takes the form

$$[C \mid IM = x] = ax^b \epsilon \tag{9.8}$$

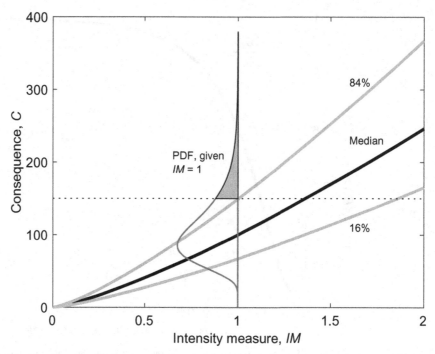

Fig. 9.3 Illustrative vulnerability function, using Equation 9.10 with parameter values $a = 100$, $b = 1.3$, and $\beta = 0.4$. A lognormal PDF is shown for $IM = 1$, and fractiles of the lognormal distributions are shown for other IM levels. The shaded region of the lognormal PDF indicates $P(C > 150 \mid IM = 1)$.

or, if logarithms are taken of both sides,

$$[\ln C \mid IM = x] = \ln a + b \ln x + \ln \epsilon \tag{9.9}$$

where a and b are constants, and ϵ is a lognormal random variable with median equal to 1 (so that $E[\ln \epsilon] = 0$) and logarithmic standard deviation of $\sigma_{\ln \epsilon} = \beta$ (Cornell et al., 2002). With this model, the vulnerability function takes the form

$$P(C > c \mid IM = x) = 1 - \Phi\left(\frac{\ln(c/ax^b)}{\beta}\right). \tag{9.10}$$

An illustration of this model and vulnerability function is shown in Figure 9.3. This figure was generated using the numerical values $a = 100$, $b = 1.3$, and $\beta = 0.4$. The heavy black line in the figure is the median consequence, $100x^{1.3}$. From Equation 9.7, the 84th percentile consequence is associated with the 84th percentile value of ϵ, which is $e^\beta = 1.49$. Thus the line $100x^{1.3} \times 1.49$ is the 84th percentile consequence given IM and is labeled as such in the figure. A similar exercise can be used to determine that the 16th percentile is $100x^{1.3}/1.49$. At any particular IM level, these parameters can be used to determine a complete distribution of consequences as well. A probability density function (PDF) given $IM = 1$ is annotated on Figure 9.3: a lognormal PDF with a median of $100 \times (1)^{1.3}$ and logarithmic standard deviation of 0.4. Finally, Equation 9.10 can be used to compute probabilities of exceeding a given consequence level. For $IM = 1$ and $c = 150$,

$$P(C > 150 \mid IM = 1) = 1 - \Phi\left(\frac{\ln\left(150/(100 \times 1^{1.3})\right)}{0.4}\right) = 0.155. \tag{9.11}$$

This probability is indicated graphically by the shaded portion of the PDF shown in Figure 9.3.

If there is an upper bound to the consequence (e.g., if the consequence is structural demand and there are failures that produce nonnumerical demands), the lognormal model can be generalized. The typical generalized model is to provide a probability of failure, $P(F \mid IM = x)$ (for which the consequence level of interest is assumed to be exceeded), and a lognormal distribution for the nonfailure consequences. In this case, because failure and nonfailure are mutually exclusive and collectively exhaustive, Equation 9.10 is modified to

$$P(C > c \mid IM = x) = P(F \mid IM = x) + [1 - P(F \mid IM = x)] \left[1 - \Phi \left(\frac{\ln \left(c/ax^b \right)}{\beta} \right) \right]. \qquad (9.12)$$

Other adjustments of this model can be made as well, such as replacing the lognormal CDF term in this equation with a truncated lognormal CDF (e.g., Shome and Cornell, 2000; Stoica et al., 2007).

Beta Model

Another consequence metric of common interest is the repair cost for the structure divided by the replacement cost.[3] This metric is typically referred to as a loss ratio, and sometimes by other names such as damage factor (Rojahn and Sharpe, 1985) or damage ratio (FEMA, 2015). The value of this metric is typically considered bounded between zero and one. In such cases, a beta distribution is often used due to its flexibility in modeling bounded distributions (Rojahn and Sharpe, 1985; Lallemant and Kiremidjian, 2015).

The beta distribution PDF is given by

$$f_C(c) = \frac{(c - a)^{q-1} (b - c)^{r-1}}{B(q,r)(b - a)^{q+r-1}}, \qquad a \leq c \leq b \qquad (9.13)$$

where a and b are constants that specify the distribution bounds, r and q are positive constants that specify the shape of the distribution, and $B(q,r)$ is the beta function[4] (e.g., Benjamin and Cornell, 2014).

For use with loss ratios bounded between zero and one, $a = 0$, $b = 1$, and the PDF simplifies to

$$f_C(c) = \frac{(c)^{q-1} (1 - c)^{r-1}}{B(q,r)}, \qquad 0 \leq c \leq 1. \qquad (9.14)$$

Figure 9.4 illustrates this PDF for several values of r and q, to illustrate the flexible shape of the distribution.

The beta distribution alone is not flexible enough, however, to model real loss data that have substantial fractions of outcomes with either no loss or total loss. To address this, the above distribution can be extended to consider discrete probabilities of consequences equaling either zero or one (Shome et al., 2012). The resulting distribution is

$$f_C(c) = p_0 \, \delta(c) + (1 - p_0 - p_1) \times f_X(c) + p_1 \, \delta(1 - c) \qquad (9.15)$$

where $\delta()$ is a Dirac delta function (Equation A.26) and $f_X(c)$ is the beta distribution from Equation 9.14 with support over $0 < c < 1$ (note that X has the basic beta distribution while C has the

---

[3] A similar metric for insurance applications is the insured loss divided by the maximum insured loss.

[4] The beta function, $B(q,r)$, is a special function defined explicitly as $B(q,r) = \int_0^1 t^{q-1} (1 - t)^{r-1} \, dt$, and usually evaluated via the gamma function as $B(q,r) = \Gamma(q)\Gamma(r)/\Gamma(q + r)$.

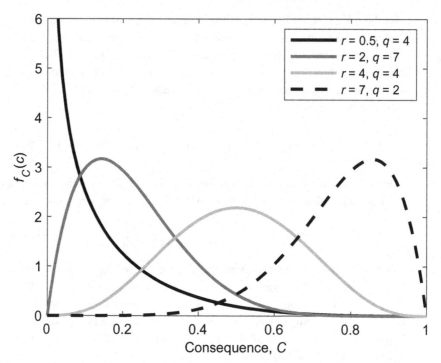

Fig. 9.4 Beta distribution PDF of Equation 9.14, bounded between zero and one and plotted for several values of r and q.

extended distribution). The four parameters that define this distribution are p_0 and p_1 (to specify the probabilities of no loss or complete loss), plus r and q (to specify the shape of the beta distribution). An example for $p_0 = 0.2$, $p_1 = 0.1$, $r = 2$, $q = 7$ is shown in Figure 9.5, where the PDF has been discretized into discrete intervals of width 0.05 (Equation A.20). In general, p_0, p_1, r, and q would all be functions of IM.

Nonparametric Model

The consequence distribution need not be specified by a parameterized distribution, such as the lognormal or beta distributions above. This situation often arises when the consequences are specified as conditional on a damage state, rather than being conditional on the IM directly. In such a case, the consequence distribution is given by

$$P(C > c \mid IM = x) = \sum_{i=1}^{n} P(C > c \mid DS = ds_i)P(DS = ds_i \mid IM = x) \qquad (9.16)$$

where $ds_1, ds_2, \ldots, ds_n$ are potential damage states for the structure, with probabilities $P(DS = ds_i \mid IM = x)$ given by fragility functions (Section 9.2.1), and $P(C > c \mid DS = ds_i)$ are consequence functions conditional on the damage state, rather than the IM.

For example, HAZUS specifies fragility functions for various damage states (such as in Figure 9.2) and loss ratios conditional on each of those damage states. In such a case, Equation 9.16 can be used to compute a loss distribution conditional on IM.

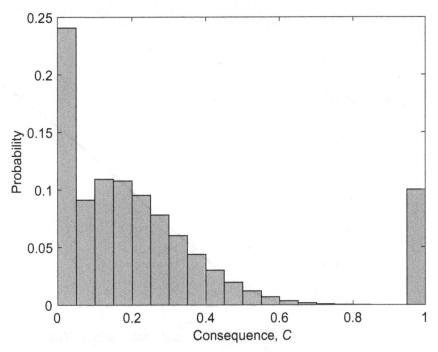

Fig. 9.5 Four parameter beta distribution with discrete probabilities (Equation 9.15), plotted for parameter values $p_0 = 0.2, p_1 = 0.1$, $r = 2, q = 7$. Probabilities are presented for discrete consequence intervals of width 0.05.

Mean Vulnerability

Sometimes the mean consequence is of primary importance in the vulnerability function, and uncertainty in consequences for a given IM is of lesser importance. The mean loss might be specified directly as a function of IM or might be defined via a nonparametric model. In the nonparametric case, an equation similar in form to Equation 9.16 can be used:

$$E[C \mid IM = x] = \sum_{i=0}^{n} E[C \mid DS = ds_i]P(DS = ds_i \mid IM = x) \tag{9.17}$$

where mean values of consequences, $E[C \mid DS = ds_i]$, are considered rather than probability distributions.

For example, the HAZUS damage states associated with the fragility functions in Figure 9.2 have mean associated loss ratios of $(0.01, 0.1, 0.41, 1)$ for the four damage states.[5] Taking the $PGA = 0.5$ g case considered above in Section 9.2.1, the occurrence probabilities are 0.13 for no damage, and 0.3, 0.39, 0.14, and 0.04 for damage states 1, 2, 3, and 4, respectively. Substituting into Equation 9.17, the mean loss ratio given $PGA = 0.5$ g is

$$E[C \mid PGA = 0.5 \text{ g}] = 0(0.13) + 0.01(0.3) + 0.1(0.39) + 0.41(0.14) + 1(0.04) \tag{9.18}$$

$$= 0.14. \tag{9.19}$$

---

[5] HAZUS separates losses into structural, nonstructural drift-sensitive, and nonstructural acceleration-sensitive losses. Here the three types of loss are summed and associated with the structural fragility functions from above, for simplicity of illustration.

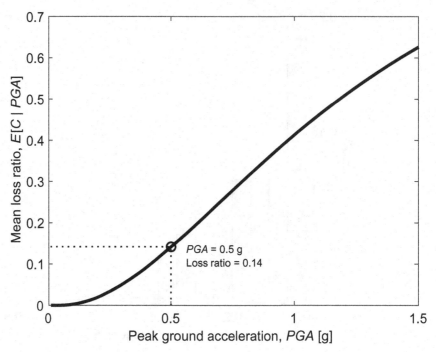

PGA = 0.5 g
Loss ratio = 0.14

Fig. 9.6 Mean loss ratio given *PGA*, using HAZUS fragility functions and loss ratios, for a high-code low-rise concrete shear-wall building with retail occupancy.

If the "no-damage" case (ds_0) has no consequences, it need not be included as its term in the summation will be zero. Nonetheless, the no-damage case was included in Equation 9.18 simply to highlight that the probabilities of all possible damage state outcomes sum to one. More generally, the ds_0 case could have consequences associated with inspections or other impacts. For a portfolio of structures the cost of these inspections could be nontrivial, so the $i = 0$ term is retained in Equation 9.17 as a reminder that such consequences could be included.

This calculation can be repeated at other *PGA* levels. The probabilities of each damage state will change, per the specified fragility functions, but the mean loss ratios *given* the damage state will not. Figure 9.6 plots the mean loss ratio for a range of *PGA* values for this case, with the example number from Equation 9.18 annotated. While the curve in Figure 9.6 is smoothly varying, there is no parametric equation that defines it.

9.3 Calibrating Fragility and Vulnerability Functions

There are several ways in which models for fragility and vulnerability are calibrated, and the general features of these approaches are discussed in this section. Three approaches are (1) to observe damage in the field after earthquakes, (2) to use analytical models to predict consequences, and (3) to use the judgment of domain experts.[6] All three of these approaches have strengths and weaknesses,

---

[6] Experimental testing is another possible method, but it is difficult to perform experiments over the broad set of circumstances needed to constrain these functions. As such, in this context, experiments are primarily used to calibrate and validate

meaning that none is clearly superior over all circumstances (though ideally, they should agree in their predictions when more than one procedure is suitable). The following subsections discuss each of these methods.

9.3.1 Empirical Approach

The empirical approach is the most straightforward, conceptually and procedurally. Direct observations are made of earthquake damage, and then used to calibrate predictive functions for the observed consequences (see Section B.8 for details on the numerical calibration of functions). Ideally, paired observations of failures (F) or consequences (C), and the associated ground-motion IMs, are obtained and used to build a regression model for a fragility or vulnerability function. Data typically come from post-earthquake field surveys, in combination with estimates of ground motion or structural deformations (Calvi et al., 2006; Crowley and Pinho, 2011; Jaiswal et al., 2011b; Rossetto and Ioannou, 2018). This approach works well for common classes of buildings where there are significant numbers of observations to use for calibration; examples include single-family houses built from masonry or light wood framing. It is also the most common approach for estimating consequences such as casualties and fatalities, which depend upon a complex set of factors including highly nonlinear structural response, damage, and human behavior (Coburn and Spence, 2002; Jaiswal et al., 2011a; So and Pomonis, 2012; Kwak et al., 2016a). A primary advantage of the approach is that it is directly validated by observational evidence, and so requires few modeling assumptions.

There are, however, several challenges with the empirical approach. First, it requires careful collection of data, relative to what might be collected for other purposes. Both damaged and undamaged structures must be surveyed randomly, to obtain an unbiased estimate of the probability that a building subjected to a given level of ground motion will be damaged. For most other post-earthquake data collection efforts, the focus is primarily damaged buildings (rather than undamaged ones), so data collected with a focus on damaged buildings will produce overestimates of damage probability if not properly adjusted. Additionally, the state of each building must be classified consistently; this requires careful training and unambiguous classification guidelines.

A second challenge is that even if the structures are surveyed appropriately, obtaining the parameter values of interest is difficult. There are rarely ground-motion recordings at the site of the structure of interest, so the IM associated with a damage observation can be quite uncertain. If the fragility or vulnerability function is calibrated using an estimated IM, then the IM uncertainty manifests as apparent uncertainty in consequence. For example, buildings with the same best-estimate IM values actually have varying IM values, and that variation in IM can cause extra apparent variation in observed consequences. Further, the consequence of interest may be difficult to obtain via observation. For consequences such as structural collapse, post-earthquake observations are sufficient. For transient outcomes, such as exceedance of a displacement amplitude, it may be impossible to collect direct observations (except in the rare cases where the structure has instrumentation). And, for consequence metrics such as repair costs, physical observation of the structure is insufficient to obtain this information. When repair costs are reported (typically as part of insurance claims), there are also reasons to be cautious about data quality. Owners may overstate the cost of repairs as part of insurance negotiations, while insurers may understate costs

computer models that can then be used to calibrate fragility and vulnerability functions, rather than being directly used for this task.

(unless reinsurance will cover claims, in which case they may more willingly pay out claims). For all of the above reasons, empirical loss data are difficult to aquire.

A third challenge is that this approach is limited to events that have been observed. Therefore, it is most relevant for predicting minor rather than severe damage states, because the former are much more commonly observed.[7] This also means that the method is more useful for assessing common building types, but is of little use for special situations like high-rise or base-isolated buildings, because of the negligible observational data for those buildings in earthquakes. Grouping structures into broad classes can increase the amount of available data but, conversely, can suppress features within the class that cause differing consequences.

For the case of fragility functions for structural components rather than full structures (e.g., Gardoni et al., 2002; Pagni and Lowes, 2006), empirical data can be gathered from laboratory experiments rather than field data. In such cases, the above challenges can be mitigated because the cases to be studied can be proactively chosen, and data acquisition is more straightforward. However, for full structural fragility functions, collecting experimental data is generally difficult or impossible.

9.3.2 Analytical Approach

The analytical approach is the most flexible and conceptually is also straightforward. Computer models (or, in rare cases, an analytical model) are developed and used to simulate consequences under a variety of ground-motion intensities. By incorporating relevant uncertainties in the model, the uncertainty in consequences should also be reflected in this procedure. This approach is most suitable for incorporating asset-specific features into the consequence predictions, because computer models can be adjusted to account for unique structural properties. It also allows exploration of a wide range of circumstances, because models are relatively inexpensive to run. It is also an ideal approach for developing vulnerability functions for structural classes with little observational data, such as modern tall buildings (e.g., Jayaram et al., 2012). Because of the flexibility and potential generality of using numerical models, this approach is by far the most active of the three in terms of ongoing development and refinement.

Nonlinear response-history analysis is the most direct structural modeling method for analytical fragility calibration. These models use input ground motions for which the IM can be determined, and directly predict dynamic response of the structure (e.g., Singhal and Kiremidjian, 1996; Ibarra and Krawinkler, 2005; Ellingwood and Kinali, 2009; D'Ayala et al., 2014; Silva et al., 2019). Alternatively, analytical fragility models have also sometimes been developed using simplified approaches such as pushover or capacity spectrum analysis (e.g., Kircher et al., 1997; Crowley et al., 2004; Rossetto and Elnashai, 2005; Silva et al., 2014). Those simplified approaches were especially useful before response-history analysis became common, but they require significant assumptions to relate computed static response to the real dynamic response. For these reasons, response-history analysis is increasingly the preferred approach for analytical calibration.

The numerical structural models can directly predict outcomes of interest, such as collapse, as long as the models are calibrated to predict the relevant phenomenon. For consequence metrics such as

---

[7] A related issue is that minor damage has its own complications in being predicted, because preexisting damage and deterioration may look like earthquake-induced damage, and there is typically no pre-earthquake survey to help resolve this. Earthquake sequences and aftershocks can also complicate the determination that a particular earthquake caused the observed damage.

building repair costs, structural models can be coupled with models for component-level damage and consequences to produce a building-level consequence prediction; this approach will be discussed further in Section 9.5. The resulting failure or consequence predictions can then be used to calibrate a fragility or vulnerability function (Section B.8).

The major limitation of this approach is that it requires the analytical consequence predictions to represent reality. Issues such as the role of soil flexibility at the structural foundation level, damage to floor diaphragms, the impact of environmental deterioration, and the role of nonmodeled features such as "nonstructural components" in buildings require careful consideration. Furthermore, extreme consequences such as collapse require refined modeling of behavior under severe nonlinearity. The analysis results also depend strongly on the ground motions used to perform the analysis; care is needed to utilize appropriate input ground motions (Chapter 10). These issues can often be ignored or simplified in traditional structural analysis to verify code compliance, but the importance of realistic response predictions for risk analysis means that much more model refinement is needed for this type of analysis than for traditional structural design applications.

Uncertainties associated with modeling choices and model parameters also need to be accounted for, to reflect uncertainty due to modeling limitations. If the resulting predictions are intended to be applied to a class of similar structures rather than a single structure, uncertainties are also needed to account for the variability in consequences observed in the real world for the range of structures associated with the prediction. Uncertainties can be incorporated directly into the modeling exercise, by varying input ground motions, model parameters, and modeling approaches (Dolsek, 2009; Liel et al., 2009; Bradley, 2013b). Alternatively, they can be incorporated by postprocessing the structural analysis results to inflate the variability observed in the initial structural modeling. Direct modeling is appealing because it allows for direct assessment of the impact of particular model uncertainties on consequence uncertainties Not all uncertainties are easily modeled, however, and so postprocessing approaches are also needed and are widely used (FEMA, 2009). Further, because the results are predictions rather than observations, model validation is required to gauge the suitability of the results.

9.3.3 Expert Opinion

In the absence of empirical observations or results from analytical models, expert opinion can be used to develop consequence functions. Despite the significant efforts devoted to empirical and analytical calibration over several decades, there remain many circumstances where those methods have not yet provided models. As in other technical fields, expert opinion can be used to fill these gaps. This approach consists of synthesizing a model based on responses from a group of domain experts.

Expert opinion is often advantageous when developing general sets of fragility or vulnerability functions for a wide range of structural types. The advantages arise because empirical data and analytical studies inevitably have gaps in the set of cases considered. This often leads to expert-based judgments of how a case in one of these gaps would perform, relative to cases that are better-constrained by empirical or analytical data. An important, and still-used, example of this method is the Applied Technology Council's ATC-13 report (Rojahn and Sharpe, 1985). This study used opinions from 85 experts, aggregated systematically, to provide estimates of damage for 78 classes of buildings and infrastructure (defined by lateral force resisting system and material).

Challenges with the method are that it requires care in selecting experts, eliciting unbiased judgments, and combining opinions. It is also well-known that experts tend to be overconfident in their predictions, requiring care to manage (Russo and Schoemaker, 1992). Jaiswal et al. (2012)

discuss some alternative approaches and their features, considering the specific application of calibrating fragility and vulnerability functions.

9.4 Risk Metrics

The previous sections of this chapter defined models for quantifying consequences conditional on IM. This section discusses several ways in which those fragility and vulnerability functions can be combined with ground-motion hazard curves to obtain information about consequences that are not conditional on the occurrence of a single IM level.

9.4.1 Failure Rate

A metric of common interest is the annual failure rate. As this metric quantifies the level of safety of a structure as a whole, or one of its components, it is utilized to calibrate design guidelines, safety-critical facility licensing, and other industrial decision-making contexts.

This metric is computed by combining the ground-motion hazard curve (quantifying the likelihoods of various IM levels) with the fragility function (quantifying the consequence of the IM) as follows:[8]

$$\lambda(F) = \int_0^\infty P(F \mid IM = x) \left| \frac{d\lambda(IM > x)}{dx} \right| dx \tag{9.20}$$

where $\lambda(F)$ denotes the annual rate of failure, F, $P(F \mid IM = x)$ is the fragility function for the failure limit state from Equation 9.1, and $\lambda(IM > x)$ is the ground-motion hazard curve from Equation 6.1. The ground-motion hazard curve derivative used here was discussed in Section 6.4.2.

Canceling the dx terms in Equation 9.20 produces a simplified presentation:

$$\lambda(F) = \int_0^\infty P(F \mid IM = x) |d\lambda(IM > x)|. \tag{9.21}$$

Alternatively, the failure rate can be computed using the derivative of the fragility function instead of the hazard curve (Kennedy and Short, 1994):

$$\lambda(F) = \int_0^\infty \frac{dP(F \mid IM = x)}{dx} \lambda(IM > x)\, dx. \tag{9.22}$$

It is left to the interested reader to confirm the equivalence of these equations.

All of the risk metrics in this section are typically solved by numerical techniques, as analytical solutions rarely exist. When the fragility function is lognormal, and the hazard curve is log-linear with the functional form of Equation 6.31, however, the above failure rate integral can be computed analytically. In this special case, the analytical solution of Equation 9.21 is

$$\lambda(F) = k_0 \theta e^{-k + 0.5(k\beta)^2} \tag{9.23}$$

where θ and β are the fragility function parameters from Equation 9.1, and k_0 and k are the hazard curve parameters from Equation 6.31.

---

[8] Note that this calculation is similar to the site response calculation of Equation 8.21. That calculation considered a surface ground-motion amplitude, conditional on ground-motion amplitude at depth, while this calculation considers a system failure conditioned on ground-motion amplitude, but the calculation procedure is the same in both cases.

Because lognormal fragility functions often appear to represent real-world data, and because hazard curves can at least locally approximate Equation 6.31 (Figure 6.7), this analytical solution has been used in several applications. In these cases, probabilistic risk concepts can be used without requiring users to evaluate integrals, or even be aware of the underlying basis for the relevant formula (e.g., Kennedy and Short, 1994; Cornell et al., 2002; Kennedy, 2011). Additionally, the analytical formulation provides useful qualitative insight into the interaction of various inputs to the calculation (e.g., how the β term influences failure rates). Nonetheless, the hazard curve approximation often results in appreciable error (Bradley and Dhakal, 2008), so numerical integration is often preferred when numerical accuracy is important.

When analytical solutions are not appropriate, Equation 9.21 is typically evaluated numerically using a discrete summation instead of integration. To do this, we take the differences of the hazard curve at discrete IM levels, as was previously introduced in Equation 6.30:

$$\Delta\lambda_i = \lambda(IM > x_i) - \lambda(IM > x_{i+1}) \tag{9.24}$$

where $x_1, x_2, \ldots, x_n$ are the discrete IM amplitudes of interest, ordered from smallest to largest, and we define $\lambda(IM > x_{n+1}) = 0$ so that the difference can be evaluated for x_n.

With this definition of $\Delta\lambda_i$, Equation 9.21 can be evaluated in a discrete form:

$$\lambda(F) \approx \sum_{i=1}^{n} P(F \mid IM = x_i)\Delta\lambda_i. \tag{9.25}$$

This discrete summation converges to the continuous integral of Equation 9.21 as the discretization interval gets small. The requirement for convergence is that the fragility function $P(F \mid IM = x)$ does not change appreciably in value over the discrete im intervals considered. As the discretization interval goes to 0 in the limit, Equation 9.25 is exactly the Reimann sum that defines the integral of Equation 9.21.

Example Failure Rate Calculation

Consider the mean ground-motion hazard curve of Section 6.8.1, and a fragility function with $\theta = 0.5$ g and $\beta = 0.4$ (as shown in Figure 9.1). This hazard curve and fragility function are shown together in Figure 9.7.

Summary data for the calculation are provided in Table 9.2. Column 1 provides the IM levels of interest here (which usually would be more finely discretized, but are provided coarsely for convenience of tabulation). Column 2 provides the $\lambda(SA(1 \text{ s}) > x)$ values, and the $\Delta\lambda_i$ values of Equation 9.24 are given in column 3. Column 4 provides the $P(F \mid SA(1 \text{ s}) = x)$ values obtained using Equation 9.1. The product inside the summation of Equation 9.25 is shown in column 5, and the summation is shown at the bottom of the column. For this example, the annual rate of failure is $\lambda(F) = 4.82 \times 10^{-5}$ (corresponding to a return period for failure of approximately 20,000 years).

Column 5 of Table 9.2 provides insight into which IM levels contribute most to failure. The lowest IM levels lead to a near-zero probability of failure, while the highest IM levels have low rates of occurrence. To see this visually, Figure 9.8 shows the $P(F \mid IM = x_i)\Delta\lambda_i$ values by IM level. This is analogous to the disaggregation plots of Chapter 7. Reviewing Figures 9.7 and 9.8, it is apparent that IM values of 0.3–0.6 contribute most to the rate of failure, as this is where the fragility function indicates nonnegligible probabilities of failure, while the hazard curve indicates nonnegligible exceedance rates.

Table 9.2. Intermediate calculations used to compute $\lambda(F)$ for the example of Section 9.4.1

x_i [g]	$\lambda(IM > x_i)$	$\Delta\lambda_i$	$P(F \mid IM = x_i)$	$P(F \mid IM = x_i)\Delta\lambda_i$
0.001	0.0252	0.0193	0	0
0.01	0.00585	0.00528	0	0
0.1	0.000568	0.000351	2.9×10^{-5}	1.01×10^{-8}
0.2	0.000217	0.000102	0.011	1.12×10^{-6}
0.3	0.000115	4.44×10^{-5}	0.10	4.48×10^{-6}
0.4	7.08×10^{-5}	2.32×10^{-5}	0.29	6.68×10^{-6}
0.5	4.77×10^{-5}	1.36×10^{-5}	0.50	6.79×10^{-6}
0.6	3.41×10^{-5}	8.71×10^{-6}	0.68	5.88×10^{-6}
0.7	2.54×10^{-5}	5.91×10^{-6}	0.80	4.73×10^{-6}
0.8	1.95×10^{-5}	4.07×10^{-6}	0.88	3.58×10^{-6}
0.9	1.54×10^{-5}	3.05×10^{-6}	0.93	2.83×10^{-6}
1.0	1.24×10^{-5}	2.21×10^{-6}	0.96	2.12×10^{-6}
1.1	1.01×10^{-5}	1.71×10^{-6}	0.98	1.67×10^{-6}
1.2	8.43×10^{-6}	1.34×10^{-6}	0.99	1.32×10^{-6}
1.3	7.09×10^{-6}	1.07×10^{-6}	0.99	1.06×10^{-6}
1.4	6.03×10^{-6}	8.61×10^{-7}	0.99	8.57×10^{-7}
1.5	5.17×10^{-6}	5.17×10^{-6}	1.00	5.15×10^{-6}

$Sum = 4.82 \times 10^{-5}$

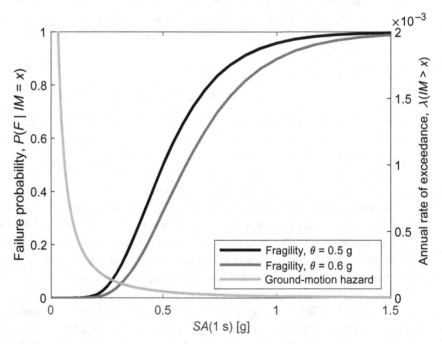

Fig. 9.7 Hazard curve and fragility curves for the example of Section 9.4.1. The left axis refers to fragility values, and the right axis refers to ground-motion hazard values.

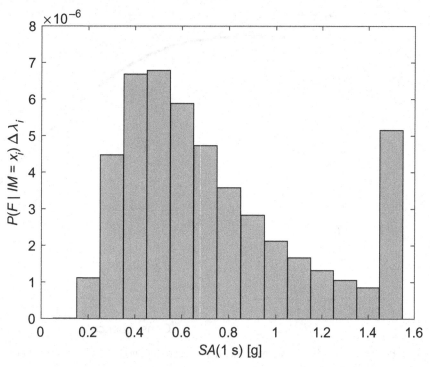

Fig. 9.8 Contributions to $\lambda(F)$ by *IM* level for the example of Section 9.4.1.

The large contribution of $IM = 1.5$ g in these results may be surprising, as it deviates from the pattern of adjacent *IM* values. This spike in contribution is because the summation was truncated at 1.5 g, with all $IM > 1.5$ g values being associated with $IM = 1.5$ g, as defined in the discrete difference calculation of Equation 9.24. This is an accurate treatment of the calculation in this case, because $P(F \mid IM = 1.5 \text{ g}) = 1$, as can be seen in the last row of column 4 of Table 9.2, so all larger *IM*s cause equivalent failure probabilities. Coarse discretization of the large *IM* levels is thus a convenient approach to shorten the summation (and shorten the table and figure limits), without omitting any *IM* levels from consideration.

Repeating this calculation, but with θ increased to 0.6 g and β held constant at 0.4, we obtain $\lambda(F) = 3.59 \times 10^{-5}$ (or return period of ~28,000 years). This alternate fragility function is also shown in Figure 9.7. In this case, increasing the fragility function median by 20% results in a 25% decrease in the failure rate. This nonlinear relationship is due to the steep drop-off of the hazard curve; moving the fragility function to the right shifts it to much less likely *IM* values. The fragility parameter β also plays an important role in the numerical results, and it is left to an exercise for the reader to examine its role.

9.4.2 Loss Exceedance Curve

A *loss exceedance curve* provides the rates of exceeding various levels of losses (i.e., consequences) by combining a ground-motion hazard curve with a vulnerability function. The exceedance rate is computed for a particular loss level as

$$\lambda(C > c) = \int_0^\infty P(C > c \mid IM = x) |d\lambda(IM > x)| \tag{9.26}$$

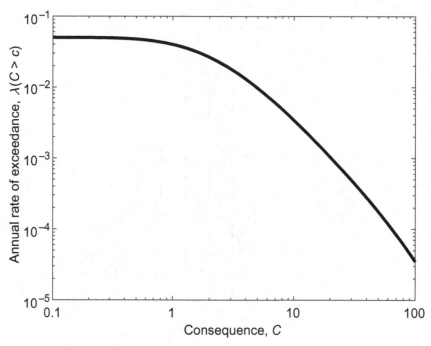

Fig. 9.9 Example loss exceedance curve, computed using the vulnerability model from Section 9.2.2, the ground-motion hazard curve from Section 6.3.3, and Equation 9.21.

where $\lambda(C > c)$ is the annual rate of consequence metric C exceeding threshold c, $P(C > c \mid IM)$ is a vulnerability function (Section 9.2.2), and $\lambda(IM > x)$ is again the ground-motion hazard curve. As with the failure rate calculation of Equation 9.21, this equation can be discretized for numerical solution:

$$\lambda(C > c) = \sum_{i=1}^{n} P(C > c \mid IM = x_i)\, \Delta\lambda_i \tag{9.27}$$

where the x_i are a discrete set of IM values, and $\Delta\lambda_i$ is defined in Equation 9.24. The above equations are then evaluated for a range of c values to develop a curve that can be plotted or tabulated.

Using the lognormal vulnerability model of Section 9.2.2, and the ground-motion hazard curve from Section 6.3.3, Equation 9.27 can be used to compute an exceedance curve. The 0.155 probability from Equation 9.11 is an example of the first term in the summation, and the hazard curve differences are the same as computed in Table 9.2. Example results for this case are shown in Figure 9.9. The figure shows that as the loss level tends toward zero, the exceedance rate tends toward the rate of $IM > 0$, and as the loss levels get very large, the exceedance rate tends toward zero.

Exceedance Probability Curve

An *exceedance probability* (EP) curve (sometimes abbreviated as an 'EP Curve') provides the probability of exceeding various levels of loss during a specified window of time (often 1 year). This connects the loss exceedance rates to the probability of an event occurring over some period of interest.

The above exceedance rates can be used to compute probabilities of exceedance over some time period t, assuming that the exceedances are Poissonian in nature and using Equation 6.14:

$$P(C > c) = 1 - \exp\left[-\lambda(C > c)\,t\right]. \tag{9.28}$$

Alternatively, if the probabilities of interest are all small, the IM rates can be approximated as probabilities (Equation 6.16), and probabilities can be computed directly:

$$P(C > c) \approx \sum_{x_i} P(C > c \mid IM = x_i)P(IM = x_i) \tag{9.29}$$

where the IM values have been discretized, and $P(IM = x_i)$ is the probability of occurrence of x_i over the time period of interest.

Example: Exceedance Probability Approximation

To illustrate the approximation with Equation 9.29, consider two example calculations based on the results of Section 1.4.1, to explore the conditions under which the calculation is reasonable.

Case 1: Consider IM levels of 0.1, 0.2, and 0.3, with exceedance rates[9] of 0.06, 0.016, and 0.003, respectively. Assume that the IM levels of 0.1, 0.2, and 0.3 cause $P(C > c) = 0.1$, 0.4, and 0.6, respectively (for some arbitrary value of c), and assume that we are interested in a time period of 1 year ($t = 1$).

For these values, $\Delta\lambda_i = 0.044$, 0.013, 0.003, and we can compute the annual rate of exceedance using Equation 9.27:

$$\lambda(C > c) = (0.1)0.044 + (0.4)0.013 + (0.6)0.003 = 0.0114 \tag{9.30}$$

and the exceedance probability using Equation 9.28:

$$P(C > c) = 1 - e^{-0.0114(1)} = 0.0113. \tag{9.31}$$

The annual rate of exceedance (0.0114) and the annual probability of exceedance (0.0113) are almost identical in this case, because the probability of the event in the time frame of interest is small, as discussed in Section 6.2.2. Because of this, we can anticipate that the approximate formulation of Equation 9.29 will be accurate. Indeed, the annual probabilities of IM occurrences can be computed using Equation 6.14 as $P(IM = x_i) = 0.043$, 0.0129, 0.0030 for IM levels of 0.1, 0.2, and 0.3. These values can then be substituted into Equation 9.29 to obtain $P(C > c) \approx 0.0113$. So the approximate result is accurate to three significant figures in this example.

Case 2: To illustrate where the approximation breaks down, we repeat the above example, but with the rates multiplied[10] by 20. The $\Delta\lambda_i$ are then 0.88, 0.26, 0.06. Equation 9.27 then evaluates to

$$\lambda(C > c) = (0.1)0.88 + (0.4)0.26 + (0.6)0.06 = 0.228 \tag{9.32}$$

and the exceedance probability using Equation 9.28 is

$$P(C > c) = 1 - e^{-0.228(1)} = 0.204. \tag{9.33}$$

---

[9] Here we consider rates rather than probabilities as in the original example, having seen in Chapter 6 that rates are the more fundamental quantity.

[10] This could represent either a case where the annual earthquake rate λ is increased by a factor of 20 or a case where we consider a time span of $t = 20$ years instead of $t = 1$ year.

The exceedance rate has increased by a factor of exactly 20 relative to the previous example (Equation 9.30), but the exceedance probability did not. Alternatively, Equation 6.14 gives $P(IM = x_i) = 0.5852, 0.2289, 0.0582$ for IM levels of 0.1, 0.2, and 0.3, and Equation 9.29 gives $P(C > c) \approx 0.185$ (approximately 10% less than the exact answer). The breakdown of the approximation comes because of the high event rates under consideration. To be specific, $IM = 0.1$ occurs 0.88 times per year on average, but it only has a 0.585 probability of happening in a given year (because some years there will be more than one occurrence of $IM = 0.1$, balancing other years where there are zero occurrences); in such cases, using rates and probabilities interchangeably is a poor approximation.

A general insight from these calculations is that when the rate of events of interest is large (i.e., greater than ≈ 0.1), one should work with rates exclusively throughout the calculation, and convert to probabilities only at the final step. Note also that the above discussion and calculations all assumed Poissonion occurrences of events; relaxing that assumption leads to another set of considerations around rates versus probabilities (Section 3.8).

9.4.3 Average Annual Loss

The average annual loss (AAL) measures the expected amount of loss experienced per year.[11] This metric is of interest for insurance transactions, as the annual cost of an insurance policy is influenced by the average payouts expected under the policy. It is also useful in evaluating risk reduction actions, as the cost of the action can be compared with the expected reduction of loss produced by the action.

The calculation uses a mathematical framework similar to Equation 9.26, but with expected consequences replacing the consequence exceedance probabilities:

$$E[C] = \int_0^\infty E[C \mid IM = x] |d\lambda(IM > x)| \tag{9.34}$$

where $E[C]$ is the expected loss (consequence) per unit time. Since $\lambda(IM > x)$ is typically an annual rate, these units persist and $E[C]$ is an expected annual loss.

An alternate method to obtain the AAL is to compute the area under the loss exceedance curve:

$$E[C] = \int_0^\infty \lambda(C > c) \, dc. \tag{9.35}$$

Equations 9.35 and 9.34 can be shown to be equivalent as follows:

$$\int_0^\infty \lambda(C > c) \, dc = \int_0^\infty \int_0^\infty P(C > c \mid IM = x) \left| \frac{d\lambda(IM > x)}{dx} \right| dx \, dc \tag{9.36}$$

$$= \int_0^\infty \int_0^\infty \left[\int_c^\infty f_{C|IM}(y \mid x) \, dy \right] \left| \frac{d\lambda(IM > x)}{dx} \right| dx \, dc \tag{9.37}$$

$$= \int_0^\infty \left[\int_0^\infty \left\{ \int_0^y f_{C|IM}(y \mid x) \, dc \right\} dy \right] \left| \frac{d\lambda(IM > x)}{dx} \right| dx \tag{9.38}$$

$$= \int_0^\infty \left[\int_0^\infty y f_{C|IM}(y \mid x) \, dy \right] \left| \frac{d\lambda(IM > x)}{dx} \right| dx \tag{9.39}$$

---

[11] This is sometimes also termed an expected annual loss (EAL).

$$= \int_0^\infty E[C \mid IM = x] \left| \frac{d\lambda(IM > x)}{dx} \right| dx \qquad (9.40)$$

$$= \int_0^\infty E[C \mid IM = x] \left| d\lambda(IM > x) \right|. \qquad (9.41)$$

Equation 9.36 expands the loss exceedance curve term using the definition of Equation 9.26, and uses the more explicit dx notation from earlier in order to provide clarity with regard to the multiple integrals. Equation 9.37 computes the probability of $C > c$ by integrating the PDF of C from c to ∞ (and uses the notation of Section A.3.2 to denote that the PDF of C is conditional on IM). Equation 9.38 reverses the order of integration of y and c, while still integrating over all values of $y > c$. Equation 9.39 evaluates the integral with respect to c; all terms in the integral are constant with respect to c, so it simply produces a y associated with the upper limit of integration. Then, noting that the remaining term inside the [ ] is the expected value of C given IM, Equation 9.40 substitutes the expected value. Finally, Equation 9.41 simplifies the dx notation to show equivalence to Equation 9.34.

While Equation 9.35 is mathematically correct, it has two practical limitations. First, it is less convenient to compute than Equation 9.34 (because it requires two integrals) unless the $\lambda(C > c)$ result is already available. Second, this approach can be more sensitive to numerical integration errors resulting from the discretization of IM levels. While both methods of computing $E[C]$ are sensitive to numerical discretization (discussed further below), this approach is particularly sensitive, as the $\lambda(C > c)$ curve is generally heavy-tailed. Qualitatively, the losses consist of frequent small losses and rare large losses. Depending upon how the loss levels are discretized, and what rates they are associated with, large errors in $E[C]$ can result.

Example

Consider again the case used to produce Figure 9.9: combining the vulnerability model of Section 9.2.2 and the ground-motion hazard curve from Section 6.3.3. Table 6.5 provided exceedance rates for the ground-motion hazard curve. Those rates are repeated in Table 9.3, along with additional results for this calculation. The third column provides discrete hazard curve derivative values, $\Delta\lambda_i$, as defined by Equation 9.24. The fourth column provides the mean values of the consequence distribution for each IM level. The mean consequence as defined by Equation 9.8 is

$$E[C \mid IM = x] = E[ax^b \epsilon] \qquad (9.42)$$

$$= ax^b E[\epsilon] \qquad (9.43)$$

$$= ax^b e^{0.5\beta^2} \qquad (9.44)$$

where the final line comes from Equation A.70. To limit errors from discretization, we compute mean losses for the midpoint of each IM interval. For example, when $x = 0.1$, and using the parameter values as above, $E[C \mid IM = x] = (100)0.15^{1.3} e^{(0.5)0.4^2} = 9.2$ (where the 0.15 in the formula is the midpoint of the IM interval from 0.1 to 0.2). The fifth column is the product of $E[C \mid IM = x]$ and $\Delta\lambda_i$, and the sum of that column is the AAL as defined by Equation 9.34.

While Table 9.3 illustrates the AAL calculation, the coarse discretization of IM values leads to substantial numerical errors in this case. A more finely discretized calculation, with IM intervals of 0.01, leads to an AAL of 0.195, versus the 0.122 of Table 9.3. While the finely discretized results cannot be easily tabulated, they are shown visually in Figure 9.10. Figure 9.10a shows

x_i	$\lambda(IM > x_i)$	$\Delta\lambda_i$	$E[C \mid IM = x_i]$	$E[C \mid IM = x_i]\Delta\lambda_i$
0.1	0.00920	0.00708	9.2	0.0652
0.2	0.00212	0.00130	17.9	0.0232
0.3	0.00082	0.00042	27.7	0.0116
0.4	0.00040	0.00018	38.4	0.00688
0.5	0.00022	8.96×10^{-5}	49.8	0.00446
0.6	0.00013	4.92×10^{-5}	61.9	0.00304
0.7	8.15×10^{-5}	2.88×10^{-5}	74.5	0.00215
0.8	5.27×10^{-5}	1.77×10^{-5}	87.7	0.00155
0.9	3.50×10^{-5}	1.12×10^{-5}	101.0	0.00114
1.0	2.38×10^{-5}	2.38×10^{-5}	115.0	0.00275

Table 9.3. Ground-motion hazard and mean loss results for the example of Section 9.4.3

Sum = 0.122

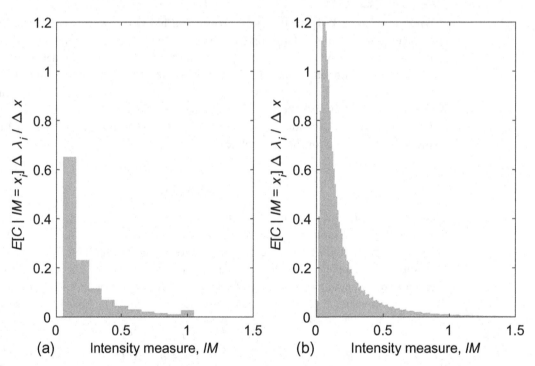

Fig. 9.10   Contributions by *IM* to average annual loss (AAL), for the example of Section 9.4.3. (a) Results from Table 9.3, with *IM* intervals of 0.1. (b) Results for the same calculation repeated using *IM* intervals of 0.01.

the results from Table 6.5, while Figure 9.10b shows the more finely discretized results. To ease comparison, the vertical axes are $E[C \mid IM = x_i]\Delta\lambda_i$ divided by the discretization interval (Δx), so that the area under the bars represents the AAL. Figure 9.10b shows that both small and large *IM* levels contribute to AAL, and the rapid variations in value at small *IM* levels indicate that the *IM* levels need to be discretized finely in order for the calculation to be precise. The AAL calculation, even more than some others in this book, is sensitive to discretization. So if AAL is of interest, and hazard or consequence results are available at only a few *IM* levels, the results should be interpolated and extrapolated carefully before being discretized, and preferably results should be obtained over as many *IM* levels as possible.

Another observation from this example is that the AAL (0.195) lies at the lower end of the loss distribution. Figure 9.9 shows that consequence values of 100 are exceeded with a rate of 4×10^{-5} per year—a consequence more than 500 times larger than the average loss in a given year. So while AAL is a useful metric for some financial calculations and decision-making, it does not convey the heavy tail of large consequences that are often associated with seismic risk assessments.

9.4.4 Limitations

There are three major assumptions inherent in the above formulations that can limit the generality of these calculations.

First, all of the above equations assume that the ground-motion IM is *sufficient* to predict consequences (Luco and Cornell, 2007). That is, given IM, the consequence is not dependent upon other properties of the earthquake or resulting ground motion (e.g., the magnitude of the earthquake). This *Markovian dependence* means that the consequence assessment can be decoupled from hazard calculations. The assumption has significant practical benefits in terms of calibrating fragility and vulnerability functions, and in terms of performing the above calculations. When this assumption is not appropriate, more complexity in the loading can be introduced by using ground-motion time series as inputs to structural models used to calibrate fragility and loss functions. Calibration of fragility and loss functions is discussed briefly in the following section, and the use of ground-motion time series is discussed in Chapter 10.

Second, in the above calculations, there is only a single location of interest, and so a scalar IM at that location is sufficient. For spatially distributed infrastructure systems or portfolios of insured properties, the IM value will vary for components at differing locations. In such situations, an IM parameter will need to be specified at every location to predict consequences, so additional complications arise when evaluating risk using hazard calculations. Methods for addressing this situation will be discussed further in Chapter 11.

Third, there is an implicit assumption of ergodicity when we combine fragility and vulnerability functions and ground-motion hazard curves in this way (Der Kiureghian, 2005). This will be explored further in Section 9.6.

9.5 PEER Framework

A special case of the analytical consequence function approach discussed in Section 9.3.2 is the so-called PEER framework (Cornell and Krawinkler, 2000; Porter et al., 2001; Krawinkler and Miranda, 2004; Moehle and Deierlein, 2004). While the framework has broad-ranging applications, in the context of this book, it is relevant as a tool for producing vulnerability predictions using analytical modeling. Specifically, it is a tool for linking predictions of structural and geotechnical response to financial and other consequence metrics.

Mathematically, the framework is well known by its integral formulation:

$$\lambda(DV > x)$$
$$= \int_{DM} \int_{EDP} \int_{IM} P(DV > x \mid DM) f(DM \mid EDP) f(EDP \mid IM) \, |d\lambda(IM)| \, dEDP \, dDM$$

(9.45)

where DV is a decision variable equivalent to the consequence, C, used above in this chapter,[12] and $\lambda(DV > x)$ is the rate of DV exceeding threshold x. The calculation relies on several intermediate variables, with boldface notation indicating variables that are vectors. $\boldsymbol{DM}$ is a vector damage measure indicating the discrete damage states of each component in the building, $\boldsymbol{EDP}$ is a vector of engineering demand parameters such as story drift ratios and peak floor accelerations, and IM is a ground-motion intensity measure.[13] The $d\lambda(IM)$ term is the ground-motion hazard curve derivative as defined in Section 9.4.2. The $f(a \mid b)$ terms are conditional PDFs for a, given b.

The $\boldsymbol{EDP}$ vector may consist of dozens of peak story drifts and peak floor accelerations at each story and direction. The $\boldsymbol{DM}$ vector may consist of a damage state for each of the hundreds, or thousands, of components present in a building. Because these variables are high-dimensional, the integrals of Equation 9.45 are very difficult to evaluate analytically. This problem can be addressed by separating the equation into two sets of integrals, and using Monte Carlo simulation for the high-dimensional set.

With this approach, the calculations become

$$P(DV > x \mid IM) = \int_{\boldsymbol{DM}} \int_{\boldsymbol{EDP}} P(DV > x \mid \boldsymbol{DM}) f(\boldsymbol{DM} \mid \boldsymbol{EDP}) \, f(\boldsymbol{EDP} \mid IM) \, d\boldsymbol{EDP} \, d\boldsymbol{DM}$$

(9.46)

and

$$\lambda(DV > x) = \int_{IM} P(DV > x \mid IM) \, |d\lambda(IM)|.$$ (9.47)

Equation 9.46 contains the high-dimensional terms. The approximate workflow is as follows: specify the IM; sample $\boldsymbol{EDP}$ from the appropriate distribution, conditional on the specified IM; sample $\boldsymbol{DM}$ from the appropriate distribution, conditional on a simulated $\boldsymbol{EDP}$; and sample a DV based on the sampled $\boldsymbol{DM}$ values. This sequence produces a single realization of DV, and repeating these simulations many times produces a set of realizations that can be used to estimate $P(DV > x \mid IM)$. Equation 9.47 uses the output of Equation 9.46 and is a traditional loss exceedance curve calculation that can be solved numerically (Section 9.4.2).

This framework has found significant uptake in the last decade. The systematic framing of ground-motion hazard, structural response, and consequences has enabled significant progress in refining each of these components and integrating them to produce risk metrics. Note that the framework assumes conditional independence at each stage of the calculation (e.g., DV is independent of $\boldsymbol{EDP}$ when $\boldsymbol{DM}$ is known). This conditional independence is likely not strictly true, but is a reasonable model in many cases, and the assumption greatly simplifies the required modeling and calculation effort.

The Federal Emergency Management Agency (FEMA) P-58 procedure (FEMA, 2012), one standardized form of this PEER framework, utilizes Monte Carlo simulation to evaluate Equation 9.46, and numerical integration to evaluate Equation 9.47, as described above. In the P-58 procedure, the $P(DV > x \mid IM)$ result is referred to as an intensity-based assessment (as it is an assessment of loss conditional on an intensity); note that this is equivalent to the vulnerability function defined in Section 9.2.2. When the $P(DV > x \mid IM)$ result is combined with a ground-motion hazard curve

---

[12] Here we have used the DV notation for consistency with the PEER framework literature.

[13] IM is typically scalar, though with vector hazard (Section 7.6) an $\boldsymbol{IM}$ vector would be used.

using Equation 9.47, the result is referred to as a time-based assessment (as it is an assessment of the frequency of losses over time). Note that this is equivalent to the loss exceedance rate calculation of Section 9.4.2.

9.6 Epistemic Uncertainty

As with hazard models, fragility and vulnerability models are subject to lack-of-knowledge uncertainty. With empirically calibrated consequence models, there are typically uncertainties about the degree to which the empirical data represent some forward-prediction application and the degree to which the empirical data represent a single type of structure versus being a mixture of multiple structure types with differing behavior (a parallel being the ergodic and non-ergodic ground-motion predictions of Section 8.1). There are also uncertainties associated with using limited empirical data to calibrate a model, and often the model has some degree of extrapolation at high consequence levels.

With analytically derived consequence models, there is uncertainty about the models and model parameters chosen to predict structural consequences. This includes uncertainty in structural modeling approaches, uncertainty in structural component behavior, and uncertainty in the consequences of a predicted level of structural response (Bradley, 2013b). Model parameter uncertainty, in particular, is most amenable to systematic study.

With expert opinion–based consequence models, there is inherent uncertainty in the judgments underlying the models, as well as issues related to equivalence of expertise.

Another consideration in consequence modeling is that the magnitude of epistemic uncertainty can be quite case-specific. For a given structure, the properties of that structure may be well known or not; when modeling risk to a portfolio of structures, even the most basic properties may not be known or may be cost-prohibitive to obtain. The failure of the structure may be well-understood and relatively predictable (e.g., the onset of cracking in a regular concrete wall), or may be poorly understood (e.g., progressive structural collapse after damage to a column).

Explicit consideration of epistemic uncertainties in fragility and vulnerability models dates back to the early years of probabilistic seismic risk analysis (e.g., Kennedy et al., 1980; Veneziano et al., 1983; Kennedy and Ravindra, 1984; Schotanus et al., 2004; Dolsek, 2009; Liel et al., 2009), though broad consideration of the issue is not as widespread as in seismic hazard studies. Nonetheless, guidance on quantifying consequence model uncertainties is provided in some consensus documents (FEMA, 2009, 2012).

Treatment of epistemic uncertainties can be handled in several ways, depending upon the desired output of the risk analysis. The most straightforward approach is to include epistemic and aleatory uncertainty into a single representation of total uncertainty in the consequence function. An example of this approach is the "double-lognormal" fragility model presented in the following section.

If epistemic uncertainty in the final loss metrics is of concern, however, then uncertainties must be tracked separately. In this case, extending the logic tree approach of Chapter 6 to consider consequences is a straightforward concept, though it can require substantial effort to implement. In this situation, it is important that all analysis stages have consistent detail in the treatment of epistemic uncertainty. While the decoupling of hazard and risk calculations, as described in this chapter, has significant practical advantages, one drawback is that different groups will lead the two stages of analysis, and epistemic uncertainties are not always treated comparably by the groups. In practice at present, many projects treat epistemic uncertainties in hazard analysis quite rigorously, while the consequence models have much more crude consideration of epistemic uncertainties. This often has the effect of limiting insights into the final impacts of epistemic uncertainty.

Double-Lognormal Fragility Model

One popular, and mathematically tractable, way to represent epistemic uncertainty in fragility functions is to treat the median of the fragility function, θ, as a random variable (Kennedy et al., 1980; Kennedy and Short, 1994; Shinozuka et al., 2000). This allows the probability of failure to be computed as

$$P(F \mid IM = x) = \int_0^\infty \Phi\left(\frac{\ln(x/s)}{\beta_R}\right) f_\theta(s)ds \qquad (9.48)$$

where $f_\theta(s)$ is a lognormal PDF with median θ^* and logarithmic standard deviation β_U. With this formulation, β_R represents the aleatory uncertainty in the fragility function, and is constant (i.e., there is no epistemic uncertainty considered in β_R). Pragmatically, β_R can be inflated to represent additional uncertainty in other factors while maintaining this tractable formulation. Additionally, this formulation assumes independence of aleatory and epistemic uncertainty.

Using the above formulation, the mean probability of failure from Equation 9.48 is given by

$$P(F \mid IM = x) = \Phi\left(\frac{\ln(x/\theta^*)}{\beta_T}\right) \qquad (9.49)$$

where

$$\beta_T = \sqrt{\beta_R^2 + \beta_U^2}. \qquad (9.50)$$

That is, the fragility function is lognormal with logarithmic standard deviation of β_T, which combines β_R and β_U in a square-root-sum-of-squares manner. A graphical illustration of this formulation is given in Figure 9.11.

Example

To consider the impact of epistemic uncertainty, the example of Section 6.8.1 is expanded here to consider epistemic uncertainty in fragility. The uncertain source and ground-motion model parameters are retained, and here we also consider an uncertain fragility function. The fragility function is assumed to have $\beta = 0.4$, and an unknown median with an equal probability of $\theta = 0.4$ or $\theta = 0.6$. The IM for the fragility function is $SA(1 \text{ s})$, for consistency with the hazard curves. Figure 9.12 shows the logic tree of Figure 6.15, expanded with one more node to consider the fragility uncertainty.

For this example, $\lambda(F)$ is the parameter of interest. There are now 54 terminal nodes on the logic tree, each representing a plausible combination of source, ground-motion, and fragility models. Equation 9.21 is evaluated for each of these terminal nodes, to produce 54 estimates of $\lambda(F)$, along with associated weights.

Figure 9.13 shows an empirical CDF of the 54 $\lambda(F)$ values, plotted by the cumulative weight associated with each $\lambda(F)$ interval. Figure 9.14 shows a histogram of the same data. The values of $\lambda(F)$ range from 0.0001 (for the branch with low earthquake rates, small m_{max}, weak ground-motion predictions, and a large fragility function median) to 0.0022 (for the opposite case): a factor

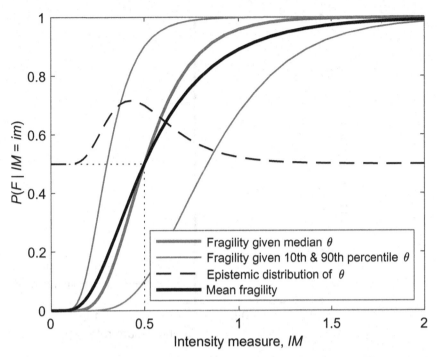

Fig. 9.11 Illustration of fragility curve with aleatory and epistemic uncertainty defined by Equation 9.48, $\theta^* = 0.5$, $\beta_R = 0.4$, and $\beta_U = 0.4$.

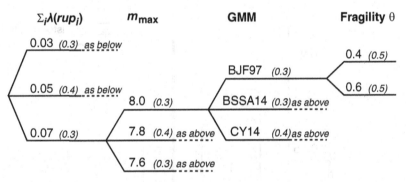

Fig. 9.12 Logic tree for the example of Section 9.6.

of 20 difference. While the particular logic tree parameters are arbitrary in this example, order-of-magnitude differences in risk among logic-tree branches are plausible.

Along with the distributions of $\lambda(F)$ shown in these figures, a mean value of $\lambda(F)$ is shown with a black vertical line. This mean value is computed in the same manner as was done to compute the mean rate of exceeding IM values in Section 6.8, and Equation 6.38 specifically:

$$E[\lambda(F)] = \sum_{k=1}^{n_{models}} \lambda_k(F)\, w_k \tag{9.51}$$

where $\lambda_k(F)$ denotes the failure rate associated with the kth terminal node, and w_k denotes the terminal node weight. Alternatively, the mean $\lambda(F)$ can be obtained by computing the failure rate

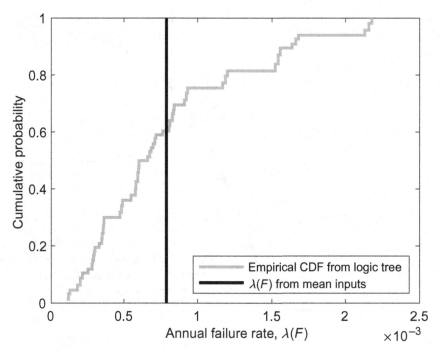

Fig. 9.13 Empirical CDF of failure rates for the example of Section 9.6.

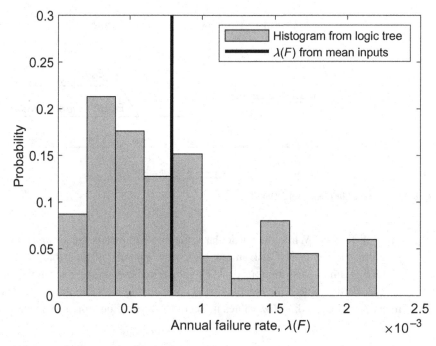

Fig. 9.14 Histogram of failure rates for the example of Section 9.6.

with Equation 9.21, using the mean ground-motion hazard curve from Section 6.8.1, and the mean probability of failure over the fragility function logic tree computed using the following equation:

$$E[P(F \mid IM = x)] = \sum_{j=1}^{n_{frag.models}} P_j(F \mid IM = x)w_j \tag{9.52}$$

where the summation is only over the fragility model alternatives (i.e., the branches dealing with ground-motion hazard can be ignored for this step). If the ground-motion hazard uncertainties and the fragility function uncertainties are independent (as would typically be the case), then the mean of the logic tree failure rates is identical to the failure rate obtained from the mean ground-motion hazard and mean fragility. For this example, the mean failure rate is $E[\lambda(F)] = 7.93 \times 10^{-4}$, and this is the value marked with the black vertical line on the figures.

One convenient representation of epistemic uncertainty in calculations like this is to treat the fragility median uncertainty as a lognormal random variable, so that the double-lognormal model above can be used. To maintain a comparable uncertainty representation to the above calculations, we can find the median and standard deviation of $\ln \theta$ associated with the logic tree of Figure 9.12:

$$\mu_{\ln \theta} = \frac{\ln 0.4 + \ln 0.6}{2} = -0.714 \tag{9.53}$$

$$\beta_U = \sqrt{\frac{(\ln 0.4 - \mu_{\ln \theta})^2 + (\ln 0.6 - \mu_{\ln \theta})^2}{2}} = 0.203. \tag{9.54}$$

With these parameters, we can evaluate Equation 9.49 with $\theta = e^{-0.714} = 0.49$ and $\beta_T = \sqrt{0.4^2 + 0.203^2} = 0.448$. Figure 9.15 shows the two fragility function branches from the logic tree

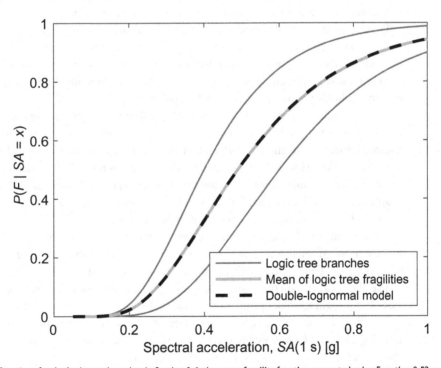

Fig. 9.15 Fragility functions for the logic-tree branches in Section 9.6, the mean fragility function computed using Equation 9.52, and an equivalent lognormal fragility function.

for this example, plus the mean fragilities computed using Equation 9.52 and this double-lognormal model. Interestingly, while the two mean fragility models come from somewhat different epistemic uncertainty representations (one with discrete θ values and the other with a lognormal distribution, albeit both having the same mean and standard deviation), the overall mean fragility functions are nearly indistinguishable. This is somewhat due to the epistemic uncertainty being low relative to the aleatory uncertainty ($\beta_U < \beta_R$), but the correspondence is generally good over a range of cases. And on the other hand, the representation of θ by a discrete distribution with only two values is somewhat extreme—should there be considerable variation in possible θ values, a realistic logic tree would likely have more than two branches for this parameter.

Given the good correspondence of the two models for mean fragility, it should not be a surprise that the mean failure rates are in excellent agreement: $E[\lambda(F)] = 7.94 \times 10^{-4}$ for the double-lognormal fragility, versus $E[\lambda(F)] = 7.93 \times 10^{-4}$ for the above calculations. On the other hand, if only a "median" fragility function with $\theta = e^{-0.714} = 0.49$ and $\beta = 0.4$ is considered (i.e., epistemic uncertainty in fragility is neglected), the failure rate using the mean hazard curve is $\lambda(F) = 7.24 \times 10^{-4}$, an approximately 10% reduction. This example thus suggests the following: (1) to obtain a mean failure rate (considering epistemic uncertainty), it is sufficient to use the mean ground-motion hazard and mean fragility; and (2) the precise representation of the fragility epistemic uncertainty (i.e., via a discrete distribution or lognormal distribution) is not critical, but it is important to include epistemic uncertainty in some form. The above trends are often seen in other cases.

Discussion

Considering the above example, and revisiting the discussion of Section 6.8, a few practical implications of the epistemic uncertainty calculation are apparent.

First, while the example considered the calculation of $\lambda(F)$, the findings hold for other risk metrics such as those related to vulnerability functions. Because the form of the calculation for $\lambda(C > c)$ is nearly identical to that for $\lambda(F)$, the mean $\lambda(C > c)$ estimate can be obtained from the mean ground-motion hazard curve and the mean vulnerability function $P(C > c \mid IM = im)$. The same conclusion holds for computation of AAL and other metrics.

Second, because the mean ground-motion hazard curve is sufficient for computing the mean values of all of these risk metrics, the mean hazard curve is almost always the reported ground-motion hazard metric. Specifically, the mean rather than the median hazard curve is the needed input to all of these calculations (McGuire et al., 2005a). Recall the Section 6.8 discussion of mean versus fractile hazard curves; the motivation for using the mean curves is now more apparent in the context of this chapter. Additionally, the distribution of hazard curves is not needed *if* the goal is to obtain the mean rate of failure, the mean rate of exceeding some consequence, or the mean average annual loss. If, on the other hand, it is desired to partition aleatory and epistemic uncertainties, then it will be necessary to track results from distinct branches of hazard and risk logic trees. In either case, it is still necessary to consider epistemic uncertainties in the underlying calculations.

The above discussion then raises the question of whether mean estimates of these risk metrics are sufficient, or whether the epistemic probability distribution is also of interest (e.g., in the example, is it sufficient to know the mean failure rate of 7.93×10^{-4}, or is the distribution of failure rates from Figure 9.14 of additional interest?). Individual applications vary, so it is not possible to provide absolute guidance, but a few observations can be made.

Concerning the interpretation of risk results, there are several insights offered by the characterization of epistemic uncertainty. Quantifying epistemic uncertainty provides insight into the sensitivity of the model to uncertain inputs, which ultimately helps the analyst to understand where uncertainty reduction efforts can most effectively be targeted. Additionally, quantification of epistemic uncertainty can guard against overconfidence in analysis results: given the large uncertainties in these calculations, we should not mistake numerical calculation precision for lack of uncertainty. Results like those in Figure 9.14 provide a direct reminder that, depending upon modeling approaches or model parameter choices, the implied performance of a system can vary dramatically.

Concerning the quantitative evaluation of acceptable risk, the value of explicitly considering epistemic uncertainty is less clear. First, most current regulatory regimes are based on mean probabilities of failure rather than evaluations of distributions—though there are exceptions to this, such as the requirement that there is a high confidence of low probability of failure (HCLPF) in US nuclear power plant evaluations (Budnitz et al., 1985). Second, we note the implications of possibly considering the full distribution of failure rates. In either case, we must accept that occasional failures are inevitable (though hopefully rare). The question is whether we are more averse to a failure caused because reality was associated with a logic tree branch with a high failure rate, or a failure that happened because our particular interval of time just happened to have a ground motion that caused a failure (regardless of the long-term failure rate). If we do care about this distinction, then we are "risk averse" per decision theory, and we do need to maintain the epistemic distributions. On the other hand, if all failures are equivalently undesirable, then mean failure rates are sufficient for decision-making. Finally, we note (as in Section 6.8) that obtaining a probability distribution of the metric rather than the mean estimate of the metric requires maintaining more careful distinctions between epistemic and aleatory uncertainty.

In the end, practical considerations lead us to generally consider mean ground-motion hazard and mean consequence probabilities when performing a risk assessment. Estimation of epistemic uncertainties may be desired in some cases for risk-averse decision-makers, but this is not standard in most circumstances that readers are likely to encounter. Additionally, epistemic uncertainties may need to be tracked separately for validation studies, as will be discussed in Chapter 12.

9.7 Risk-Targeted Ground-Motion Intensity

In Chapters 2–8, ground-motion hazard was computed without consideration of structural performance, and only in this chapter has the structural performance been incorporated. This decoupling of hazard calculations and risk calculations has practical benefits, as seen with the above calculation procedures. One challenge that arises with this approach, however, is that ground-motion hazard maps presented in Section 7.4 are developed based on hazard metrics alone: they map the ground-motion amplitude with a specified exceedance rate. The ultimate goal of these maps, however, is to produce structures that achieve some target risk level (i.e., to limit the rate of building collapse to some tolerably low rate). Because the structural risk depends upon a full ground-motion hazard curve rather than the amplitude at a single exceedance rate, two locations with the same "hazard map value" and the same structure may have different risk results.

To consider target risks while still decoupling ground-motion hazard from structural analysis, so-called risk-targeted hazard maps have been proposed (Luco et al., 2007, 2015), following concepts developed earlier in the nuclear industry (Kennedy and Short, 1994). With this approach, a map of design values can be produced that has a direct linkage to targeted building risk. Specifically, the

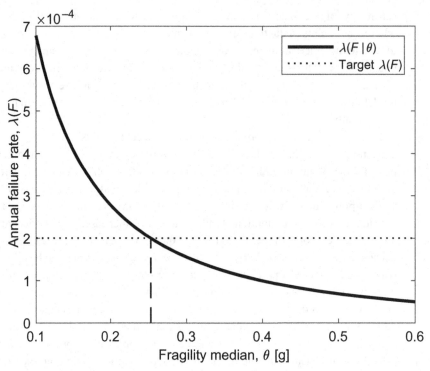

Fig. 9.16 Collapse risk (Equation 9.20) as a function of the fragility function median θ, where the fragility function is defined by Equation 9.2.

maps show the ground-motion amplitude that would cause a specified probability of collapse, based on a fragility function of a structure with a given collapse rate. The fragility function for this hypothetical building has a prespecified β, and a θ value that is varied to achieve the target collapse rate (where both parameters are as defined in Equation 9.1).

To illustrate, consider the ground-motion hazard curve from Section 6.8.1, a target failure rate of $\lambda(F) = 2 \times 10^{-4}$, and a fragility function defined by Equation 9.48 with $\beta = 0.6$. For a given value of θ, Equation 9.21 can be used to compute the annual failure rate of the structure. Depending upon the value of θ, the failure rate may be higher or lower than the target rate, and so θ can be adjusted until the target is reached; Figure 9.16 illustrates how $\lambda(F)$ varies with θ for this example, and indicates that $\theta = 0.25$ effectively gives the target failure rate. Thus a structure at the given site (i.e., with the given ground-motion hazard), and with a fragility function having $\theta = 0.25$ and $\beta = 0.6$, would have the target rate of failure. Further, structures with fragilities "to the right" (i.e., with generally lower probabilities of failure) would have lower rates of failure, and fragilities "to the left" would have higher rates of failure and thus would be unsatisfactory.

This process can be repeated for other values of β, to identify the fragilities of structures with β values that would satisfy the target collapse rate. Figure 9.17 shows such fragility functions for three values of β, all using the same ground-motion hazard and all having the same target failure rate. Notably, these fragility functions all have similar probabilities of failure in the lower tail (i.e., for probabilities of failure between approximately 0.1 and 0.25). This stability of lower-tail fragilities is true more generally and is one motivation for developing risk-targeted hazard maps that map the IM amplitude with $P(F|IM = x) = 0.1$. The 0.1 probability of failure level is relatively stable for typical hazard curve shapes and β ranges (Kennedy and Short, 1994). On the other hand, Kennedy and Short

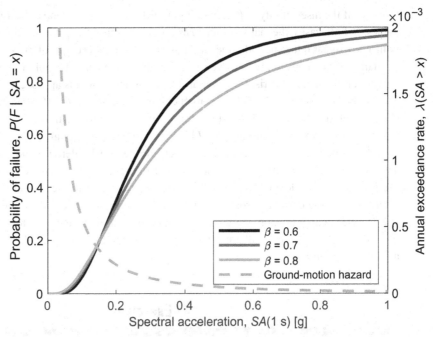

Fig. 9.17   Fragility functions with three values of β, and with corresponding fragility function median (θ) values chosen to obtain the target collapse rate of $\lambda(F) = 2 \times 10^{-4}$.

(1994) recommended against use of small $P(F|IM = x)$ values (e.g., $P(F|IM = x) = 0.01$) because the design value is then unstable as β varies, and in reality β is not known with precision. It is also apparent from Figure 9.17 that the median values of the fragilities vary substantially with β.

To produce a risk-targeted map of ground-motion amplitudes useful for structural assessment, the following steps are used:

1. Compute ground-motion hazard curves for each site of interest.
2. Specify a reference β, a target $\lambda(F)$ for consideration, and a target $P(F|IM = x)$ value to consider.
3. For each site, compute $\lambda(F)$ for a range of θ values, and determine the θ value that produces the target $\lambda(F)$ (Figure 9.16).
4. For each site, given the θ value, compute the x value that gives the target $P(F|IM = x)$.
5. Map the x values determined in Step 4.

With a map produced in this manner, a designer or analyst need only determine that a structure under consideration has $P(F|IM = x)$ less than the target probability to infer that the structure has a satisfactory failure rate. More precisely, current building codes aim to produce designs that have $P(F|IM = x)$ less than the target probability when used in conjunction with the risk-targeted map, even if the designer never explicitly checks this probability.

The first adoption of this approach in the United States used parameter $\beta = 0.8$, targeted a structural collapse rate of $\lambda(F) = 2 \times 10^{-4}$ (i.e., 1% probability of failure in 50 years), and mapped the IM amplitudes with $P(F \mid IM = x) = 0.1$ (BSSC, 2009). This β was larger than in earlier derivations, but was seen to produce stable numerical results. Subsequent provisions revised β to 0.6, while retaining the other parameter values (American Society of Civil Engineers, 2010; BSSC, 2015).

Because of the insensitivity to β seen in Figure 9.17, this revision had a negligible impact on mapped values. In both cases, the maps produce IM amplitudes that are generally similar to the IM exceeded with a 2475-year return period—the target used in previous versions of these documents. The general similarity of the maps to previous versions, but the refinement of amplitudes up and down locally to achieve uniform collapse risk, was a motivation for adopting this approach.

A study for France used $\beta = 0.5$, a target collapse rate of $\lambda(F) = 10^{-5}$, and $P(F \mid IM = x) = 10^{-5}$ (Douglas et al., 2013). A related study for all of Europe proposed to use $\beta = 0.6$, a target collapse rate of $\lambda(F) = 5 \times 10^{-5}$, and a $P(F \mid IM = x) = 0.001$ (Silva et al., 2016). The lower $\lambda(F)$ values than the United States were deemed to reflect European risk tolerance and current design standards. The very low $P(F \mid IM = x)$ values chosen in these studies were selected so that, on average, the IM amplitudes determined from this process are similar to the IM amplitudes with a 475-year return period (the current European standard). While these choices illustrate the adaptability of the risk-targeted mapping approach depending upon design goals, the very low $P(F \mid IM = x)$ values produce numerical results that are extremely sensitive to the assumed value of β, as discussed above and illustrated in Figure 9.17.

Exercises

9.1   You are trying to calibrate a fragility function with the functional form of Equation 9.1. You know that $P(F) = 0.1$ when $IM = 0.3$, but do not yet have information about the rest of the function. Provide an expression for the θ and β values that could satisfy this constraint. Use this expression to determine the required θ values if $\beta = 0.3$ and if $\beta = 0.5$.

9.2   A particular building has four damage stages that can result from ground shaking. The fragility functions have $\theta = 0.3, 0.5, 0.8,$ and 1.2 g (defined in terms of $SA(1 \text{ s})$); all have $\beta = 0.5$. The mean loss ratios given each damage state are $0.01, 0.05, 0.3,$ and 1. Plot the four fragility functions on a single figure. Then for $SA(1 \text{ s}) = 0.2$ g, 0.5 g, and 1 g, compute the following:

(a) The probability of the building being in each damage state.
(b) The mean loss ratio for the building.

9.3   Repeat the exercise used to produce Table 9.2, for the following cases, and comment on the impact on λ_F and on the IM levels contributing most to failure.

(a) $\theta = 0.6$, $\beta = 0.4$ (the additional case discussed in the text of the example section).
(b) $\theta = 0.5$, $\beta = 0.5$.

The ground-motion hazard values can be reused from the table as-is. Produce a table like Table 9.2 for each case, and produce a single figure plotting the fragility functions for the above two cases plus the original $\theta = 0.5$, $\beta = 0.4$, case, and comment on the results and the trends relative to the analytical solution of Equation 9.23.

9.4   Consider a risk problem with $\lambda(IM > x) = 10^{-5}x^{-3.5}$, and consequences given by $[C|IM = x] = 10x^2\epsilon$ with $E[\ln \epsilon] = 0$ and $\sigma_{\ln \epsilon} = \beta$ (following Equation 9.8). Compute the loss exceedance curve and AAL using both Equations 9.34 and 9.35, for the following three cases and comment on your results:

(a) *IM* values are discretized into intervals of 0.1.

(b) *IM* values are discretized into intervals of 0.01.

(c) *IM* values are discretized into intervals of 0.001.

9.5 Consider a site with $\lambda(IM > x) = 10^{-4}x^{-2.5}$ and a structure with a fragility function having epistemic uncertainty. The fragility function has $\beta = 0.3$, and θ equals either 0.15 or 0.4 (you think both are equally likely).

(a) What is the mean annual failure rate of structure?

(b) What is the probability that the structure fails during its 50-year lifespan, if *IM* exceedances are a Poisson process? (Hint: First consider the probabilities that the structure fails during its lifespan, if it has $\theta = 0.15$ or $\theta = 0.4$, and then compute the total probability by considering that those two cases each have a 0.5 probability of being true.)

(c) If you used the mean annual failure rate to compute a failure probability over 50 years, what do you get? Explain why the answer differs from the answer for (b).

9.6 Consider a site with a ground-motion hazard curve represented by $\lambda(IM > x) = (5\times10^{-6})x^{-3}$, and a structure with a target failure probability of $\lambda_F = 2 \times 10^{-4}$. Find the risk-targeted *IM* value (x) for all permutations of the following cases: $\beta = 0.4, 0.5, 0.6$, and $P(F|IM = x) = 0.001, 0.01, 0.1, 0.5$. Provide your results in a table, with one column for each β value, and one row for each $P(F|IM = x)$ value. For which value of $P(F|IM = x)$ are the targeted values most stable across the β cases?

10

Ground-Motion Selection

The most sophisticated approaches for assessing the seismic performance of structures and other infrastructure use response-history analysis, which requires solving the equation of motion for the structure when excited by a ground-motion time series.[1] The seismic response in such analyses depends strongly upon the specific ground motions used, therefore impacting seismic risk calculations (Chapter 9). In contrast, seismic hazard analysis outputs in Chapters 6 and 7—hazard curves, disaggregation, uniform hazard spectra, and conditional response spectra—are all in the form of ground-motion intensity measures (IMs). This chapter, therefore, focuses on the processes and procedures to obtain an ensemble of ground-motion time series that is consistent with the IM-based seismic hazard at the site.

Learning Objectives

By the end of this chapter, you will be able to do the following:

- Describe the principles of hazard-consistent ground-motion selection.
- Compute conditional distributions of ground-motion IMs from seismic hazard and disaggregation information.
- Understand ground-motion selection as a constrained optimization problem and perform ground-motion selection with appropriate constraints.
- Evaluate potential errors in response-history analysis associated with hazard-consistent ground-motion selection.
- Comprehend conventional code requirements for ground-motion selection and their consistency, or otherwise, with first principles.
- Understand the key differences between risk-based and intensity-based assessments, and how each influences ground-motion selection.

10.1 Introduction

In this chapter, we describe the calculations necessary to select a *hazard-consistent* ensemble of ground motions for use in response-history analysis. Hazard-consistent ground motions are defined

---

[1] In this chapter the phrase "ground motion," or "motion" for short, is used. Existing literature also frequently uses "record" as a synonym, reflecting traditional use of ground-motion recordings. However, we avoid this term because it does not generalize to a ground-motion simulation.

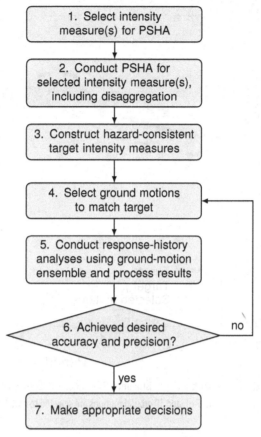

Fig. 10.1 Overall seismic analysis and decision-making process, with ground-motion selection embedded. Steps 3, 4, and 6 are the primary focus in this chapter.

as motions whose IM values are consistent with PSHA calculations for the site being considered. There are several ways in which this consistency can be evaluated, as discussed below.

Figure 10.1 illustrates the overall seismic analysis and decision-making process involving ground-motion selection and response-history analyses. Step 1 involves selecting the IMs used to characterize the seismic hazard, which should be specific to the type of problem considered (Section 10.6.2). Step 2 involves performing PSHA and obtaining its derivative products, as discussed in Chapters 6 and 7. The present chapter focuses upon Steps 3, 4, and 6, starting by explaining how to define target IMs that are consistent with results from seismic hazard analysis (Step 3; Section 10.3), then explaining the main concepts associated with selecting ground motions to match this target (Step 4; Section 10.4), and evaluating the resulting seismic responses in this context (Step 6; Section 10.5). Step 5 is not explicitly discussed here, and readers are referred to textbooks on structural dynamics and finite element analysis for seismic problems (e.g., Cook et al., 1989; Chopra, 2016).

Figure 10.2 provides an example of the result from Steps 3 and 4 in Figure 10.1. Based on a seismic hazard analysis for $SA(0.5 \text{ s})$, a conditional spectrum (Section 7.5) is computed to define the hazard-consistent target IM distribution, which is depicted in terms of the target mean and standard

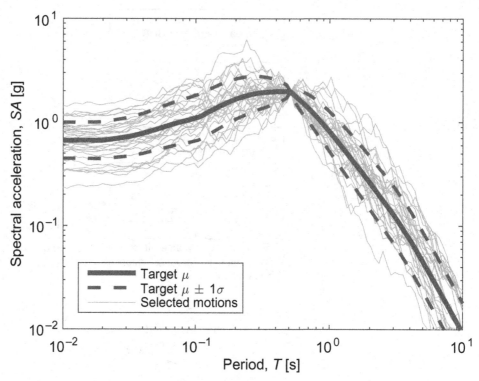

Fig. 10.2 Ground-motion selection to match a target distribution. The target distribution is defined as a conditional spectrum (conditioned on $SA(0.5\text{ s}) = 1.97$ g), and 30 ground motions are selected to match this distribution.

deviation range. Thirty ground motions are then selected from a ground-motion database such that their SA values have a distribution that is statistically consistent with the target conditional spectrum. These ground motions can then be used to compute the seismic response of the structure of interest (i.e., Steps 5–7 of Figure 10.1).

The large number of approaches and heuristics for ground-motion selection that have been proposed in technical literature (e.g., Katsanos et al., 2010) can make this field appear intimidating, and sometimes contradictory. Consequently, this chapter is primarily concerned with how to select ground motions in a hazard-consistent manner from first principles. This is reflected in the layout of the material covered, beginning with underpinning principles and mathematics (Section 10.2), and then addressing various simplifications or pragmatic alternatives that can be made to obtain multiple approaches for constructing the target IM distribution and selecting ground motions to match this target (Sections 10.3 and 10.4). It is also acknowledged that in many practical situations, the approach to selecting ground motions may be dictated or constrained by the requirements of relevant guidelines or codes of practice. Section 10.7 briefly addresses code requirements and highlights some issues that practitioners should be aware of. A reader narrowly focused on the practical application of ground-motion selection methods may prefer to skip the mathematical details of Section 10.2 in the first instance.

Ultimately, the objective of this chapter is to outline the key concepts that enable appropriate decisions, not to provide formulaic guidance for every conceivable analysis case.

10.2 Principles of Hazard-Consistent Ground-Motion Selection

Ground-motion selection provides the link between seismic hazard and response analyses. It is therefore critical that selected ground-motion ensembles are consistent with hazard analysis results. From an observational perspective, the most appropriate ground motions would be those observed at the site over a substantial duration of time. Such motions would intrinsically capture the key features of the seismic sources that influence the hazard at the site, the effects of specific ray paths from the various sources to the site, and the surficial soil response of the site itself. However, acquiring ground motions in this manner requires a very long period of observation and is not practical (Section 12.4.1). Ground-motion time series must, therefore, be obtained from recorded ground-motion databases, or through simulations.

When empirical ground-motion models are used in PSHA (Chapter 4), the hazard is provided solely in the form of ground-motion IMs, not ground-motion time series. It is therefore necessary to select ground-motion ensembles that are consistent with the IMs from PSHA to achieve hazard consistency. Furthermore, when considering recorded ground motions from databases containing different tectonic, geographic, and site conditions, care is needed to ensure that selected ground motions are consistent with the hazard at the specific site of interest.

Seismic hazard analysis using simulation-based ground-motion methods (Chapter 5) could, in concept, provide a numerical equivalent to a significant time duration of observations. However, at present, such simulation methods are generally not considered sufficiently realistic to replace the conventional use of recorded ground motions. Also, even as simulated ground motions become increasingly used, a limited number of ground motions will be typically desired for response-history analysis, with a focus on larger ground-motion intensities to assess undesirable outcomes. The need to select a small number of ground motions from a large set of simulations, and the greater importance of higher-amplitude motions, indicate that ground-motion selection methods are still required for simulation-based seismic hazard analyses (Bradley et al., 2015).

10.2.1 Coupled Approach

Figure 10.3 illustrates two general approaches, referred to as *coupled* and *uncoupled*, that could be adopted to assess the seismic performance of a structure. In the coupled approach, depicted in the upper half of the figure, a seismic source model (SSM; Chapter 3) is used to provide an earthquake rupture forecast specifying all possible ruptures that can occur within the modeled domain along with their likelihood. For each of these ruptures, consistent ground motions are obtained, and response-history analyses are performed to assess the system's performance. This coupled approach is referred to as such because the ground motions used for the response-history analysis are directly coupled to the source model.

The ground motions representing each rupture scenario should reflect *future* motions. Therefore, an ensemble of motions is required for each rupture scenario to reflect the significant conditional ground-motion variability (Chapter 4). In the common case, where individual ruptures in a source model are defined by a magnitude and rupture geometry, an ensemble of ground motions will result from different possible kinematic ruptures (e.g., Figure 5.4) with this geometry and magnitude. The large number of rupture scenarios from a SSM, and the subsequent ensemble of possible ground

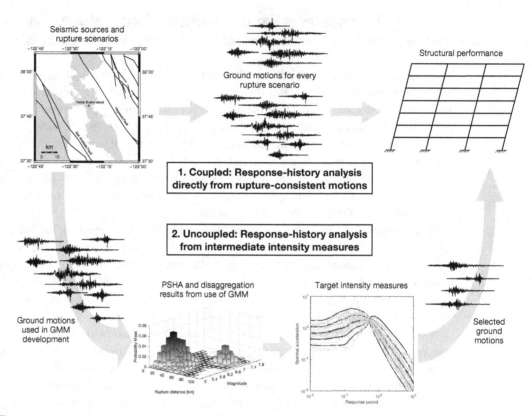

Seismic sources and rupture scenarios

Ground motions for every rupture scenario

Structural performance

1. Coupled: Response-history analysis directly from rupture-consistent motions

2. Uncoupled: Response-history analysis from intermediate intensity measures

Ground motions used in GMM development

PSHA and disaggregation results from use of GMM

Target intensity measures

Selected ground motions

Fig. 10.3 Schematic illustration of the two options for conducting response-history analysis. In the coupled approach (upper half of figure), ground motions for every rupture scenario arising from the seismic source model are applied directly to the structural model. In the uncoupled approach (lower half of figure), ground-motion models are used to perform PSHA and then obtain intermediate IM targets based on the rupture scenarios that influence the seismic hazard analysis. A relatively small number of ground motions are then selected to be consistent with the intermediate IMs and applied to the structural model.

motions for each rupture, produce a huge number of motions (e.g., Hancock et al., 2008) and the computational demands are typically prohibitive. Furthermore, this conceptual approach assumes that suitable motions can be obtained, which is challenging given available empirical ground-motion databases with insufficient large-amplitude observations (Chapter 4) and nascent physics-based ground-motion simulations capabilities (Chapter 5).

10.2.2 Uncoupled Approach

Given the practical limitations of the coupled approach, the overall process of assessing seismic performance through response-history analysis is therefore typically *uncoupled*[2] into two key steps, shown in the bottom half of Figure 10.3. In Figure 10.3, the results of the hazard analysis are represented by intermediate "target" IMs that are consistent with the hazard analysis results. An ensemble of motions is then chosen to reflect the characteristics of these IMs and to obtain the conditional structural response via response-history analysis. This uncoupled approach allows several

---

[2] This approach is also utilized for the seismic risk calculations of Chapter 9.

benefits, including requiring far fewer response-history analyses than in the coupled approach, and a more transparent separation of the uncertainties associated with the hazard and consequent demand.

It is worth keeping Figure 10.3 in mind when considering the requirements for ground-motion selection. In principle, if both coupled and uncoupled approaches could feasibly be implemented with the same level of accuracy and precision, their results would be equivalent (Bradley, 2012c). For the uncoupled approach, this requires that the selected ground motions reflect the attributes of the motions that would have been used in the coupled approach. The mathematical basis of the uncoupled approach is discussed in the next subsection.

10.2.3 Mathematical Basis

The most common approach in ground-motion selection is to make use of a conditioning IM to link the seismic hazard for a specific rate of exceedance with the distribution of seismic response (i.e., "uncoupled" in Figure 10.3). The IM provides a means to uncouple the seismic hazard and response analyses. The IM can consist of either a single scalar quantity or a vector of quantities, $\boldsymbol{IM}$. For generality, we will begin with a vector of IMs here (denoted with boldface notation), and later note the special conditions that permit the use of a single scalar IM.

The vector of IMs, $\boldsymbol{IM}$, is initially assumed to comprise all characteristics of the ground motion. That is, we can view this exhaustive vector $\boldsymbol{IM}$ as containing the same information as a complete ground motion.[3] If the mean annual rate of exceedance of this vector is denoted as $\lambda_{\boldsymbol{IM}}(\boldsymbol{im})$, then the mean annual frequency of exceeding a measure of seismic demand, denoted as an engineering demand parameter (EDP),[4] is given by[5]

$$\lambda(EDP > edp) = \int_{\boldsymbol{IM}} P(EDP > edp \mid \boldsymbol{im}) \, MRD_{\boldsymbol{IM}}(\boldsymbol{im}) d\boldsymbol{IM} \tag{10.1}$$

where $P(EDP > edp \mid \boldsymbol{im})$ is the probability of $EDP > edp$ given $\boldsymbol{IM} = \boldsymbol{im}$; $MRD_{\boldsymbol{IM}}(\boldsymbol{im})$ is the mean rate density (see Section 7.6) of $\boldsymbol{IM} = \boldsymbol{im}$; and $d\boldsymbol{IM}$ is shorthand notation for the product $dIM_1 dIM_2 \ldots dIM_N$. From Equation 7.35, it follows that $MRD_{\boldsymbol{IM}}(\boldsymbol{im})$ can be expressed as

$$MRD_{\boldsymbol{IM}}(\boldsymbol{im}) = \sum_{i=1}^{n_{rup}} f_{\boldsymbol{IM}}(\boldsymbol{im}|rup_i)\lambda(rup_i) \tag{10.2}$$

where $f_{\boldsymbol{IM}}(\boldsymbol{im}|rup_i)$ is the multivariate PDF of $\boldsymbol{IM}$, conditional on rup_i, and $\lambda(rup_i)$ is the mean annual occurrence rate of rup_i.

In the idealized case, when $\boldsymbol{im}$ completely describes the ground-motion time series, there is a unique mapping from $\boldsymbol{im} \rightarrow edp$; i.e., the deterministic equation of motion will give one seismic response result for a given ground-motion loading. In this case, the distribution $P(EDP > edp \mid \boldsymbol{im})$ is equivalent to a Heaviside step function equal to 1 for $EDP \geq edp$ and 0 otherwise. The vector $\boldsymbol{IM}$ in this ideal case is perfectly *efficient* (Luco and Cornell, 2007).

---

[3] Or even a complete ground-motion field for spatially distributed structures (Chapter 11).
[4] The case of a vector of engineering demand parameters, $\boldsymbol{EDP} = \{EDP_1, EDP_2, \ldots\}$, is an extension of the scalar EDP case presented here.
[5] Equation 10.1 is equivalent to the loss exceedance curve equation in Section 9.4.2. Note that Equation 10.1 represents the uncoupled approach. For the coupled approach, the corresponding expression is:

$$\lambda(EDP > edp) = \sum_{i=1}^{n_{rup}} P(EDP > edp \mid rup_i)\lambda(rup_i)$$

In the more realistic case, where the vector IM does not completely describe the ground motion, the incomplete representation through im will lead to motion-to-motion variability in the response, which must be reflected in the distribution $P(EDP > edp \mid im)$. A vector of IMs that results in smaller motion-to-motion variability reflected in $P(EDP > edp \mid im)$ is more efficient (Luco and Cornell, 2007). That is, *efficiency* is a measure of the conditional variability of EDP given some $IM = im$, with lower variability corresponding to greater efficiency.

We cannot reduce the dimension of IM arbitrarily. It is important to ensure that the EDP of interest systematically depends only upon the selected IM vector, and not upon other characteristics of the underlying ruptures. That is, the distribution $P(EDP > edp \mid im)$ is not, practically speaking, a function of any other variables. This requirement is known as *sufficiency* and is necessary to permit the full uncoupling between the hazard and demand analyses (Luco and Cornell, 2007). If the condition of sufficiency is not met, then it is not possible to condition the response solely upon the rate of occurrence of im, as done in Equation 10.3.

When the dimension of the vector IM is small, it is feasible (in principle) to conduct vector PSHA (Section 7.6). That said, even when vector PSHA is performed, the computational demands associated with accurately obtaining $P(EDP > edp \mid im)$ can remain very high. To obtain this distribution, large numbers of response-history analyses are required, especially as the dimensionality of im increases,[6] which is typically prohibitive. As a consequence, the conventional approach to obtaining $\lambda(EDP > edp)$ is to instead work with a single conditioning IM (Shome et al., 1998; Baker, 2007; Bradley, 2012c). We will denote this conditioning IM as IM_1 and express the overall vector of IMs in a partitioned form as $IM = [IM_1, IM_{2+}]$. The "2+" subscript denotes that this vector contains the IMs with indices of 2 or greater. With the full vector of IMs partitioned in this way, we can then express the demand hazard integral from Equation 10.1 as

$$\lambda(EDP > edp)$$

$$= \int\limits_{IM_1} \left[\int\limits_{IM_{2+}} P(EDP > edp \mid im_1, im_{2+}) \, f_{IM_{2+}|IM_1}(im_{2+} \mid im_1) \, dim_{2+} \right] |d\lambda(IM_1 > im_1)|.$$

$$(10.3)$$

For the term $P(EDP > edp \mid im_1, im_{2+})$—which is equivalent to $P(EDP > edp \mid im)$ in Equation 10.1—we have simply partitioned out IM_1 from the vector IM and left all remaining IMs in the reduced vector IM_{2+}. The consequence of this partitioning of IM is that the integral[7] can now be viewed as a univariate integral over IM_1, since $|d\lambda(IM > im)|$ from Equation 10.1 is now represented by the product $f_{IM_{2+}|IM_1}(im_{2+} \mid im_1) \, dim_{2+} \, |d\lambda(IM_1 > im_1)|$. This is the rate of occurrence of IM_1 multiplied by the conditional joint-density function of the remaining IMs given $IM_1 = im_1$.

The inner integral of Equation 10.3 is equivalent to the conditional probability of exceedance of edp given $IM_1 = im_1$:

$$P(EDP > edp \mid im_1) = \int\limits_{IM_{2+}} P(EDP > edp \mid im_1, im_{2+}) \, f_{IM_{2+}|IM_1}(im_{2+} \mid im_1) \, dim_{2+}. \quad (10.4)$$

Equations 10.3 and 10.4 effectively state that you can work with a single conditioning IM, IM_1, provided that you derive the response distribution in a manner that accounts for the conditional

---

[6] Referred to as the "curse of dimensionality" (Hastie et al., 2001).
[7] Note that dim_{2+} in this expression is equivalent to $dim_2 \, dim_3 \, dim_4 \ldots dim_n$.

distribution of the other IMs not explicitly being considered. If a set of n_{gm} ground motions were selected such that all motions had $IM_1 = im_1$, and this set of motions had an empirical distribution $f_{IM_{2+}|IM_1}(im_{2+} \mid im_1)$ that is statistically equivalent[8] to the corresponding distribution in Equation 10.4, then the n_{gm} ground motions can be viewed as a set of realizations from this distribution (i.e., an ensemble) (Shome and Cornell, 1999; Baker, 2007; Bradley, 2012c). The edp values obtained from response-history analyses with the n_{gm} ground motions can then be used to determine $P(EDP > edp \mid im_1)$. The desired result of $\lambda(EDP > edp)$ can then be obtained by substituting Equation 10.4 into Equation 10.3 to obtain

$$\lambda(EDP > edp) = \int_{IM_1} P(EDP > edp \mid im_1) \; |d\lambda(IM_1 > im_1)| \, . \qquad (10.5)$$

10.2.4 Practical Implementation Concepts

Based on the theory in the previous subsection, the key steps to obtain an unbiased estimate of $\lambda(EDP > edp)$ through Equation 10.5 can be summarized with reference to the steps in Figure 10.1 as follows (Bradley, 2012c):

1. Identify a vector of IMs, $\boldsymbol{IM}$, that sufficiently represents the characteristics of the ground motions that influence EDP (Step 1).
2. Select a scalar conditioning IM, IM_1, from $\boldsymbol{IM}$, and perform PSHA for IM_1 (Step 2).
3. Select ensembles of ground motions that are consistent with $f_{IM_{2+}|IM_1}(im_{2+} \mid im_1)$, for the range of IM_1 values of interest (Steps 3 and 4).
4. Perform response-history analyses and compute $P(EDP > edp \mid im_1)$ (Step 5).
5. Verify that the selected ground motions provide sufficient accuracy and precision in estimating the $EDP \mid im_1$ distribution (Step 6).
6. Use Equation 10.5 to compute $\lambda(EDP > edp)$ (Step 7).

Figure 10.4 illustrates how Equation 10.5 is computed in practice. For each of 10 IM_1 values of interest, 20 ground motions are selected, and response-history analyses are performed. Graphically, for particular values of IM_1, the response estimates are plotted in vertical "stripes."[9] Using these 20 EDP values, a distribution of $EDP \mid IM_1 = im_1$ is obtained, and can be evaluated to determine $P(EDP > edp \mid im_1)$.

The key point to note is that the EDP variability in Figure 10.4 needs to represent the contributions of the IMs that were not explicitly considered in the hazard analysis. That is, if one were to examine the ground motions selected for a particular value of $IM = IM_1$, their distribution of $\boldsymbol{IM}_{2+}$ should be consistent with those of the conditional distribution $f_{IM_{2+}|IM_1}(im_{2+} \mid im_1)$. Application-specific decisions associated with the selection of the conditioning IM, IM_1, and other IMs to consider (i.e., $\boldsymbol{IM}_{2+}$) are discussed later in Section 10.6.

To obtain a single point on the demand hazard curve, Equation 10.5 shows that the rate of occurrence of each IM value $|d\lambda(IM_1 > im_1)|$ must be combined with the conditional probability of

---

[8] That is, one could not reject the hypothesis that the motions were drawn from an underlying population described by the true $f_{IM_{2+}|IM_1}(im_{2+} \mid im_1)$.

[9] This is approach is often referred to as the *multiple-stripe* method to estimate the $EDP \mid IM$ relationship, as compared with *cloud* or *incremental dynamic analysis* methods (Vamvatsikos and Cornell, 2002; Jalayer, 2003; Mackie and Stojadinovic, 2005).

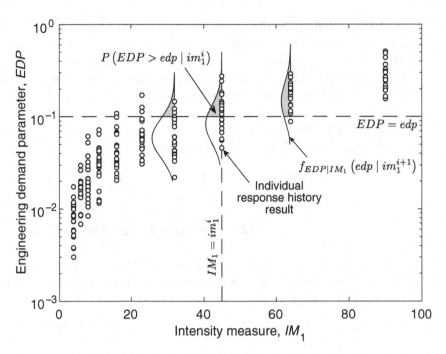

Fig. 10.4 Illustration of how response-history analysis results are used to obtain the distribution of a scalar engineering demand parameter, *EDP*, as a function of IM_1, $f_{EDP|IM_1}(edp \mid im_1)$, in order to compute the demand hazard via Equation 10.5. Individual response-history results are shown with markers.

$EDP > edp$ given $IM_1 = im_1$. That is, the shaded regions of the distributions of $EDP \mid IM_1$ shown in Figure 10.4 are combined across all levels of IM_1 to obtain the total rate of exceeding *edp*.

The need for consistency in selecting ground motions according to $f_{IM_{2+}|IM_1}(\boldsymbol{im}_{2+} \mid im_1)$ has not always being well appreciated. For example, Shome et al. (1998) assumed that first-mode spectral acceleration was a good predictor of various EDPs (for which they considered both energy and peak-response parameters) and that secondary IMs did not have a significant influence upon the prediction of the EDP (i.e., $SA(T_1)$ was sufficient). That concept was popular for several years. More refined studies later demonstrated that a scalar IM is often not sufficient, and careful ground-motion selection is needed to represent other parameters (e.g., Baker, 2007; Bradley et al., 2010).

As a prelude to Sections 10.3 and 10.4, Figure 10.5 illustrates the comparison of a selected ensemble of ground motions against several target IM distributions obtained as a derivative product from PSHA. The target IM distributions include response spectra in the form of a conditional spectrum in Figure 10.5a (see Section 7.5), as well as the conditional distributions of other IMs in Figure 10.5b-d (Arias intensity and significant durations from 5–75% to 5–95%). These distributions are based on a conditioning IM_1 of spectral acceleration for $T = 0.5$ s.

The ground-motion selection process therefore aims to identify an ensemble of ground motions whose statistics match those of the conditional distribution $f_{IM_{2+}|IM_1}(\boldsymbol{im}_{2+} \mid im_1)$.[10] In Figure 10.5,

---

[10] Relevant IMs to consider in $\boldsymbol{IM}_{2+}$ are discussed in Section 10.6.2.

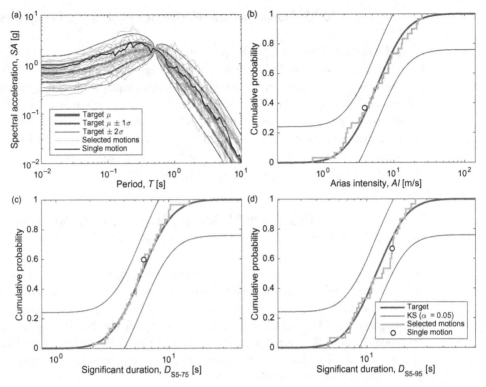

Fig. 10.5 Example of conditional IM targets and the fit of 30 ground motions selected to be consistent with the targets. Panel (a) shows a conditional mean spectrum along with fractiles corresponding to $\pm\sigma$ and $\pm 2\sigma$ (where σ is the conditional standard deviation). The remaining panels share the common legend of panel (d) and depict the empirical cumulative distribution function for the 30 ground motions along with the conditional target distribution and Kolmogorov–Smirnov (KS) bounds that quantify the degree of match to the target. In all cases the IMs corresponding to one particular ground motion are also highlighted.

the spectra and empirical CDF results for the selected ground-motion ensemble are shown for comparison with the target IM distributions. For further reference, the IM values corresponding to a single ground motion are also depicted. For the Arias intensity and significant duration IMs, Kolmogorov–Smirnov (KS) bounds are shown around the conditional target CDF. Provided that the empirical CDF of the selected ground motions is contained within these bounds, the hypothesis that the selected ground motions are consistent with the target distribution cannot be rejected (Section 10.4.3). For the conditional spectra, no such KS bounds are shown, but similar approaches can be applied to evaluate the extent to which the selected ground motions are compatible with the (multivariate) target distribution.

The above concept of an appropriate hazard-consistent multivariate distribution of IMs is the most important point to grasp in this chapter. Once we have the target IM distributions, we can then find corresponding ground motions. Modern ground-motion selection procedures rely on the above thinking regarding hazard consistency. The main differences between various proposals that have been made come from how conditional distributions are approximated (Section 10.3) and how ground motions are judged to be representative of the hazard-consistent target distribution (Section 10.4).

10.2.5 Ground-Motion Selection for Single Scenario Ruptures

Ground-motion selection is typically performed in the context of PSHA, as previously discussed. However, it may occasionally be of interest to select ground motions for a specific scenario rupture, *rup*. The general sentiments expressed throughout this chapter apply equally to this situation, so we will not belabor discussions by separately presenting this situation in each section of the chapter. The IM-based approach for ground-motion selection discussed in this section is still the preferred approach, but ground motions are selected based on $IM \mid rup$ rather than $IM_{2+} \mid im_1$. Additional details and examples of ground-motion selection for scenario ruptures are discussed elsewhere (Baker et al., 2011; Wang, 2011; Tarbali and Bradley, 2015).

10.3 Target Intensity Measure Distributions

Section 10.2 emphasized the importance of selecting ground motions in a manner consistent with seismic hazard analysis. In particular, when hazard curves from PSHA are computed for a particular IM_1, the target distribution of other IMs, $f_{IM_{2+}|IM_1}(im_{2+} \mid im_1)$, is conditioned upon a value of this particular IM_1 (see Equation 10.3). These target IM distributions, or simply *IM targets*, are fundamental to the overall ground-motion selection process and are the focus of this section. The various selection algorithms that are subsequently summarized in Section 10.4 identify ground motions that conform to these IM targets.

The basic concepts for computing IM targets were covered in Section 7.5.1 in the context of conditional response spectra. They required models to predict the unconditional distribution of each IM of interest, as well as the correlation between these different IMs. In this section, these ideas are further generalized, and two potential convenient approximations are considered:

1. Whether the conditional distribution of $IM_{2+} \mid IM_1$ is evaluated in a nonparametric fashion or approximated via its first two statistical moments
2. Whether the entire disaggregation distribution is used to compute the IM targets or simply a single "effective" (typically, mean) rupture scenario.

These choices lead to different specific mathematical forms for defining these IM targets, and this section examines the differences and the circumstances under which the approximations are appropriate. Ultimately, however, the concepts underlying the different forms remain the same.

10.3.1 Computing Conditional Distributions

Because PSHA considers the aggregation of hazard from all potential seismic sources (i.e., the summation in Equation 6.1), calculation of the conditional distribution $f_{IM_{2+}|IM_1}(im_{2+} \mid im_1)$ requires the use of the total probability theorem (Section A.2.2). Given $IM_1 = im_1$, the conditional distribution is obtained from (Bradley, 2010b):

$$f_{IM_{2+}|IM_1}(im_{2+} \mid im_1) = \sum_{i=1}^{n_{rup}} f_{IM_{2+}|IM_1,Rup}(im_{2+} \mid im_1, rup_i) P(rup_i \mid im_1) \tag{10.6}$$

where $f_{IM_{2+}|IM_1,Rup}(im_{2+} \mid im_1, rup_i)$ is the PDF of IM_{2+} conditional on $IM_1 = im_1$ and $Rup = rup_i$, and $P(rup_i \mid im_1)$ is the probability of $Rup = rup_i$ given $IM_1 = im_1$. $P(rup_i \mid im_1)$ can be directly obtained from seismic hazard disaggregation, based on the *occurrence* of $IM_1 = im_1$ (not exceedance; see Section 7.2.2). Note that the conditioning on the *site* of interest has been omitted for brevity in this equation.

$IM \mid Rup$ is the distribution of ground-motion IMs for a given rupture. From the typical assumption that $\ln IM \mid Rup$ has a multivariate normal distribution,[11] it follows that $\ln IM_{2+} \mid IM_1, Rup$ also has a conditional multivariate normal distribution (see Section A.5.1). Equations A.64 and A.65 provide the complete mathematical description of the mean and covariance of this distribution.[12] Of particular interest in the ensuing discussion is the conditional distribution for each individual IM from the vector IM_{2+} (denoted, in nonbold, as IM_2), which is a normal distribution with mean and standard deviation of (see Equations A.66 and A.67):

$$\mu_{\ln IM_2 \mid im_1, rup_i} = \mu_{\ln IM_2 \mid rup_i} + \rho_{\ln IM_2, \ln IM_1} \sigma_{\ln IM_2 \mid rup_i} \varepsilon_{im_1} \tag{10.7}$$

$$\sigma_{\ln IM_2 \mid im_1, rup_i} = \sigma_{\ln IM_2 \mid rup_i} \sqrt{1 - \rho_{\ln IM_2, \ln IM_1}^2} \tag{10.8}$$

where $\mu_{\ln IM_2 \mid rup_i}$ and $\sigma_{\ln IM_2 \mid rup_i}$ are the mean and standard deviation of $\ln IM_2$ for $Rup = rup_i$, $\varepsilon_{im_1} = (\ln im_1 - \mu_{IM_1 \mid rup_i})/\sigma_{IM_1 \mid rup_i}$ is the *epsilon* value (normalized residual) of the observation im_1 relative to the predicted distribution of $IM_1 \mid rup_i$ (Equation 4.16), and $\rho_{\ln IM_2, \ln IM_1}$ is the correlation between the two IMs *conditional on* rup_i.[13] Comparison of Equations 10.7 and 10.8 with those of the conditional spectra discussed in Section 7.5 illustrates the equivalence of the two, as alluded to in Section 7.5.3.

Equations 10.6–10.8, in combination with the seismic hazard (including disaggregation), provide the necessary information to compute the conditional IM targets for ground-motion selection. From these equations we can see that the following model information is required:[14]

1. The conditioning IM, its conditioning value, $IM_1 = im_1$, and seismic hazard disaggregation, $P(rup_i \mid im_1)$
2. Ground-motion models (GMMs) for each of the IMs in IM_{2+} to enable computation of $\mu_{\ln IM_2 \mid rup_i}$ and $\sigma_{\ln IM_2 \mid rup_i}$
3. Models for the correlation between each of the different IM combinations ($\rho_{\ln IM_2, \ln IM_1}$).

Empirical ground-motion models exist for many IMs (discussed in Section 4.2), and models for correlations among these IMs have also been developed (e.g., Baker and Jayaram, 2008; Bradley, 2015). These correlation models are obtained from statistical analysis of residuals (see Section B.7), and therefore the derivation of such models requires GMMs for these same IMs to be available. While the correlation models are derived through the use of particular GMMs, the resulting correlations are relatively insensitive to the choice of these models (Baker and Jayaram, 2008; Baker and Bradley, 2017).

10.3.2 Mean and Standard Deviation of the Target IM Distributions

Equation 10.6 provides the basis to compute the distribution of $f_{IM_{2+} \mid IM_1}$, but it is additionally useful to compute the mean and standard deviation of this distribution. The mean and standard deviation can be used as summary statistics or directly as targets for ground-motion selection (see Section 10.4).

---

[11] This assumption is not strictly required, but is the basis for the specific mathematics that follow.
[12] Substituting $X_a = IM_{2+} \mid Rup$ and $X_b = IM_1 \mid Rup$.
[13] That is, the correlation between the prediction residuals of $\ln IM_1$ and $\ln IM_2$, not the values themselves.
[14] The specifics differ for the case of simulation-based ground-motion prediction, as discussed in Bradley et al. (2015).

From Equation 10.6 it follows that, for each IM within IM_{2+}, the mean and standard deviation of $f_{IM_{2+}|IM_1}$ are (Section A.4; Lin et al., 2013c; Carlton and Abrahamson, 2014):

$$\mu_{\ln IM_2|IM_1} = \sum_i^{n_{rup}} \mu_{\ln IM_2|IM_1,Rup}(rup_i)\, P(rup_i \mid im_1) \tag{10.9}$$

$$\sigma^2_{\ln IM_2|IM_1} = \sum_i^{n_{rup}} \left[\sigma^2_{\ln IM_2|IM_1,Rup}(rup_i) + \left(\mu_{\ln IM_2|IM_1,Rup}(rup_i) - \mu_{\ln IM_2|IM_1}\right)^2 \right] P(rup_i \mid im_1) \tag{10.10}$$

where all terms have been previously defined. Equations 10.9 and 10.10 are equivalent to Equations 7.29 and 7.31, presented in Section 7.5.2, for the case of only one GMM (the consideration of epistemic uncertainties is discussed subsequently in Section 10.3.4).

It is useful to intuitively understand how multiple ruptures contribute to the mean and standard deviation of the conditional IM target distributions. In Equation 10.9, $\mu_{\ln IM_2|IM_1,Rup}(rup_i)$ is the conditional mean for rupture scenario rup_i, and in Equation 10.10 the term $\sigma_{\ln IM_2|IM_1,Rup}(rup_i)$ is the conditional standard deviation for this same IM and rupture scenario. Therefore, when the contribution of one rupture is very large, i.e., $P(rup_i \mid im_1) \gg P(rup_j \mid im_1)$ for all $j \neq i$, we can see that $\mu_{\ln IM_2|IM_1} \approx \mu_{\ln IM_2|IM_1,Rup}(rup_i)$ and $\sigma_{\ln IM_2|IM_1} \approx \sigma_{\ln IM_2|IM_1,Rup}(rup_i)$. However, when we have a strongly bimodal distribution, Equation 10.9 will define a conditional mean that lies somewhere in between those associated with the two modes (as illustrated subsequently via example). Similarly, Equation 10.10 will inflate the conditional standard deviations that correspond to each of these main modes through the term representing the difference between the conditional means for these modes and the overall conditional mean defined in Equation 10.9.

Second-Moment Approximation of the Conditional Distribution

Equations 10.9 and 10.10 enable calculation of the mean and standard deviation of the IM targets for each IM in IM_{2+}. Using the calculated mean and standard deviation, it is possible to make the second-moment approximation of the distribution by assuming a normal distribution for $\ln IM_2 \mid IM_1$. That is, Equation 10.6 provides the true target distribution, $f_{IM_{2+}|IM_1}(im_{2+} \mid im_1)$, and Equations 10.9 and 10.10 for all the IMs define a multivariate normal distribution[15] that approximates this true target.

To investigate the performance of this second-moment approximation, as well as to provide an example for use in subsequent sections, Figure 10.6 shows the disaggregation distribution for a fictitious site that has a strong bimodality.

Figure 10.7 shows a comparison between the exact probability density function obtained through the use of Equation 10.6 and the distributions obtained using the second-moment approximations of Equations 10.9 and 10.10. While the second-moment representation is not perfect, the main features of the true distributions are well matched. Because the distribution of $\ln IM_2$ for each rupture is normal, then the assumption that $\ln IM_2 \mid IM_1$ is also normal will be most appropriate when one rupture dominates this quantity, in the same manner as discussed with respect to Equations 10.9 and 10.10.

---

[15] In addition to the standard deviations from Equation 10.10, conditional covariance terms are also needed to define the multivariate distribution.

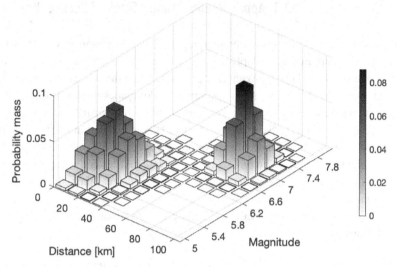

Fig. 10.6   Example of a strongly bimodal disaggregation distribution. A significant contribution to the hazard comes from events with magnitude just under 6 and at close distances, while another major contribution comes from larger events of around magnitude 7 at 80 km.

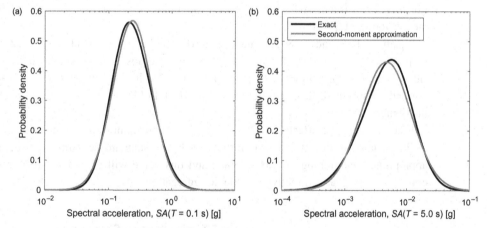

Fig. 10.7   Accuracy of the second-moment approximation for two example distributions. The exact probability density function is given by Equation 10.6, and the second-moment approximation from Equations 10.9 and 10.10. Panel (a) shows the distribution for $T = 0.1$ s, and panel (b) $T = 5.0$ s. In both cases, the conditioning period is $T^* = 0.5$ s.

The second-moment representation is often adopted for convenience. It defines the target distribution in a simple parametric form and transparently extends past practices that used a single representative scenario. Furthermore, in many situations, it will provide a good approximation to the true target distribution.

10.3.3 Approximation Using a Single "Effective" Rupture Scenario

Another approximation to obtain $f_{IM_{2+}|IM_1}$ from Equation 10.6 is to avoid summation over all contributing ruptures, and simply use a single "effective" rupture scenario. We will denote this effective rupture as $\overline{rup}$. This approach is appropriate when the seismic hazard disaggregation indicates that a single rupture scenario, say, rup_j, strongly dominates the hazard, such that $\sum_i P(rup_i) \approx P(rup_j)$. Alternatively, many ruptures may contribute to the hazard, but for simplicity, an effective rupture scenario ($\overline{rup}$) may be used as a first-order representation of the full disaggregation distribution. Furthermore, such a simplification may be necessary if only dissagregation summary statistics are available, rather than the full distribution. Note that "effective rupture scenario" is usually interpreted as the "mean rupture scenario" (Section 7.2.2), but other alternatives are possible as discussed subsequently.

With the approximation of a single effective rupture scenario, Equations 10.9, 10.10, and 10.6 become, respectively:

$$\mu_{\ln IM_2|IM_1} \approx \mu_{\ln IM_2|IM_1,Rup}(\overline{rup}) \tag{10.11}$$

$$\sigma_{\ln IM_2|IM_1} \approx \sigma_{\ln IM_2|IM_1,Rup}(\overline{rup}) \tag{10.12}$$

$$f_{IM_{2+}|IM_1}(im_{2+} \mid im_1) \approx f_{IM_{2+}|IM_1,Rup}(im_{2+} \mid im_1, \overline{rup}). \tag{10.13}$$

Equation 10.11 is unbiased when $\overline{rup}$ is the mean rupture scenario and the functional dependence is linear in $\overline{rup}$ (see Equation A.48), which is usually a reasonable assumption. In contrast, Equation 10.12 is a biased estimate (underestimation) of the true standard deviation, due to ignoring the $(\cdot)^2$ term in Equation 10.10 (Lin et al., 2013c).

Equations 10.7 and 10.8 can then be used with Equations 10.11 and 10.12 to the obtain the (approximate) conditional distributions. Note that because only a single rupture scenario is considered in these approximations, and because $\ln IM_2 \mid \ln IM_1, Rup$ is conventionally considered as normally distributed, then it follows that $\ln IM_2 \mid \ln IM_1$ is also implicitly assumed as normally distributed.

When making the effective rupture scenario approximation, the value of ε_{im_1} in Equation 10.7 is also usually recalculated based on these mean and standard deviation approximations, to ensure appropriate conditioning on $IM_1 = im_1$, and this value will, therefore, deviate slightly from the corresponding ε value directly from the disaggregation.

Choice of Effective Rupture Scenario

Many sites will not have their hazard dominated by a single rupture scenario. For example, it is quite common for disaggregation to show significant contributions from both relatively small-magnitude nearby scenarios and larger-magnitude more distant scenarios (Section 7.2), as illustrated in Figure 10.6. Despite this, in the interests of computational efficiency, it was common for some time to use a single representative scenario that could be used for constructing IM targets and selecting ground-motion ensembles. The mean and the mode of the dissagregation distribution are the two most common options to identify this scenario (Section 7.2.2).

Consider again the disaggregation plot in Figure 10.6 with two clear contributions arising from distinct magnitude-distance regions. While one can readily compute the mean magnitude and source-to-site distance for the distribution (which are magnitude 6.3, distance 40 km), from inspection this

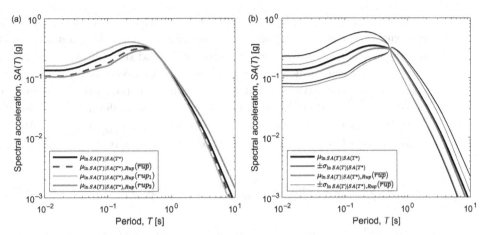

Fig. 10.8   Target spectra derived from the disaggregation distribution shown in Figure 10.6. Panel (a) compares the exact CMS obtained from Equation 10.9 along with the approximate CMS obtained using Equation 10.11 and two local CMS corresponding to the two modal scenarios shown in that figure. Panel (b) shows the exact conditional spectra using Equations 10.9 and 10.10 and the approximation obtained for the mean rupture scenario using Equations 10.11 and 10.12. The conditioning IM is $SA(0.5\text{ s})$ and has a target value of 0.3 g in all cases. The spectra are computed using the response spectral model of Chiou and Youngs (2014).

"centroid" lies in between the two main scenarios that dominate the hazard. Likewise, a modal scenario can represent only one of the two dominant scenarios.

Figure 10.8 illustrates the influence of effective rupture scenario choice on response spectra IM targets for the disaggregation plot in Figure 10.6. Figure 10.8a illustrates the difference in mean SA ordinates of the target IM distribution for the exact, mean, and two dominant modal scenarios, while Figure 10.8b additionally compares the standard deviation range for the exact and mean-based approximation.

It is important to appreciate that the aim of choosing the single effective scenario is to make the approximations in Equations 10.11 and 10.12 as accurate as possible. First-order statistical theory (Section A.4.1) suggests that using the mean of the disaggregation distribution is the most accurate approximation. Practically, this requires the computation of the mean values of all of the causal parameters such as magnitude, distance, etc. This process is straightforward for continuous variables, but it is not possible for discrete parameters such as style of faulting (e.g., normal, strike-slip) or tectonic type (e.g., active shallow crustal, subduction interface). In these cases, a discrete choice needs to be made, which would usually be the discrete parameter value with the highest contribution. When alternative discrete parameter values have similar contributions, it is prudent to use the full disaggregation distribution and compute the conditional distributions exactly.

10.3.4 Consideration of Epistemic Uncertainties

The target IM distributions presented so far in this chapter have not addressed epistemic uncertainties. Following the logic-tree approach (Chapter 6) for the treatment of epistemic uncertainties, it is possible to develop target IM distributions for each possible outcome from the logic-tree structure. Given the intent of using these target IM distributions for ground-motion selection, different ground-motion ensembles can be selected for each of the corresponding target IM distributions. Naturally, for a large number of logic-tree combinations, the total number of target distributions and consequent ground-motion ensembles becomes practically unmanageable.

Tarbali et al. (2018) and Lin et al. (2013c) discuss the computation of target IM distributions when epistemic uncertainties in the hazard are explicitly accounted for. Tarbali et al. (2018) demonstrate that the results from different target distributions (and subsequent ground-motion ensembles) for each logic-tree combination can be sufficiently approximated simply via the mean (with respect to epistemic uncertainties) estimate of the target IM distribution, and one ensemble of ground motions based on this single IM target. This mean hazard-estimate of the target IM distribution is obtained via multiplying the target IM distribution from model k, $f^k_{IM_{2+}|IM_1}$ (Equation 10.6), by the contribution of the kth model to the mean hazard, $P^k(im_1 \mid E[\lambda(IM_1 > im_1)])$ (Equation 7.24):

$$E[f_{IM_{2+}|IM_1}(\boldsymbol{im}_{2+} \mid im_1)] = \sum_{k=1}^{n_{models}} f^k_{IM_{2+}|IM_1}(\boldsymbol{im}_{2+} \mid im_1) \, P^k(im_1 \mid E[\lambda(IM_1 > im_1)]). \quad (10.14)$$

In a similar fashion, the first and second statistical moments of the target IM distributions (Equations 10.9 and 10.10) for the different n_{models} can be obtained from:

$$E\left[\mu_{\ln IM_2|IM_1}\right] = \sum_{k}^{n_{models}} \sum_{i}^{n_{rup}} \mu^k_{\ln IM_2|IM_1}(rup_i) \, P^k(rup_i \mid im_1) \, P^k(im_1 \mid E[\lambda(IM_1 > im_1)])$$

$$(10.15)$$

$$E\left[\sigma^2_{\ln IM_2|IM_1}\right] = \sum_{k}^{n_{models}} \sum_{i}^{n_{rup}} \left[\left(\sigma^k_{\ln IM_2|IM_1}(rup_i)\right)^2 + \left(\mu^k_{\ln IM_2|IM_1}(rup_i) - \mu^k_{\ln IM_2|IM_1}\right)^2\right] \quad (10.16)$$

$$\times P^k(rup_i \mid im_1) P^k(im_1 \mid E[\lambda(IM_1 > im_1)]).$$

10.3.5 Challenges Defining Target Distributions in Non-Ergodic Studies

Several additional challenges arise when computing conditional distributions for non-ergodic hazard applications (see Section 8.1). Specifically, while the conditioning GMM for IM_1 will have been defined with considerable care within the non-ergodic framework, the additional GMMs for other IMs used within the target for ground-motion selection (i.e., $\boldsymbol{IM}_{2+}$) will often not have received this same level of attention. Also, available IM correlation models make use of ergodic GMMs.

It is expected that non-ergodic correlations between IMs will often be "weaker" than their ergodic counterparts because, in the ergodic case, a portion of the apparent correlation will arise from systematic biases. An effect similar to this case can be seen in the spatial correlation model for response spectral ordinates of Jayaram and Baker (2009). This model predicts differing levels of correlation depending upon whether the region contains clustered or random site conditions. The effect of clustered site conditions combined with a GMM that imperfectly represents these site conditions leads to systematic biases that appear as stronger spatial correlations. Another example of this effect can be seen when accounting for site-specific effects within interfrequency correlations for Fourier spectra (Stafford, 2017) or interperiod correlations for response spectra (Kotha et al., 2017). In these cases, the removal of systematic site effects leads to the reduction of the within-event correlations among spectral ordinates.

Based on the sentiments above, and the growing emphasis on non-ergodic modeling approaches (Chapter 8), partially non-ergodic correlation models among general combinations of IMs will no doubt arise in the future.

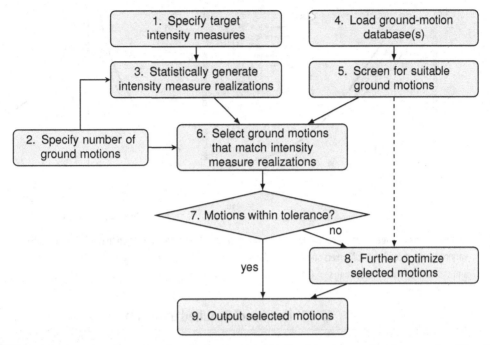

Fig. 10.9 High-level overview of ground-motion selection algorithms. This flowchart can be considered an expansion of Steps 4 and 5 from Figure 10.1.

10.4 Selection Algorithms

Figure 10.1 illustrated how ground-motion selection fits as part of a seismic response assessment, and ground-motion selection algorithms relate to Steps 4 and 5 of that process. Ground-motion selection algorithms also involve several common steps (regardless of algorithm-specific details), and these are depicted in Figure 10.9.

All ground-motion selection algorithms require three types of information: (1) target IMs, (2) database(s) of ground motions to select from, and (3) miscellaneous additional details (some algorithm-specific). The target IM distributions (Step 1) were discussed in Section 10.3. Associated with the definition of the target is the specification of how many ground motions should be identified from the selection process, n_{gm} (Step 2). A database(s) of recorded or simulated candidate ground motions must be available (Step 4). Databases may contain a very large number of motions, and therefore some degree of preselection, or screening, will typically be applied by imposing bounds upon the ground-motion causal parameters (Step 5). With the IM target and prospective ground motion database(s) defined, particular algorithms will identify motions that match the target in some manner (Steps 6 and 8), and the quality of this match is assessed using one or more evaluation criteria (Step 7).

Having discussed the definition of the target IM distribution in detail in the previous section, we now provide details on each of the remaining steps illustrated in Figure 10.9. While a large number of ground-motion selection algorithms have been proposed, those with a strong theoretical basis share common attributes that are the focus here.

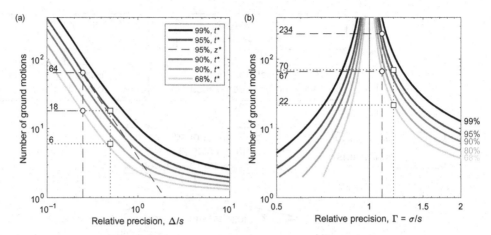

Fig. 10.10 Number of ground motions required to constrain the (a) mean and (b) standard deviation to within some relative precision at varying levels of confidence. Markers and annotated numbers correspond to factors of $\gamma \approx 1.1$ and $\gamma \approx 1.2$ around the mean and standard deviation (assuming $s = 0.4$).

10.4.1 Number of Required Ground Motions

A finite number of ground motions, n_{gm}, is desired to empirically represent the target IM distribution for use in response-history analysis. The larger the number of ground motions utilized, the smaller the statistical uncertainty in the estimated seismic response will be. Our objective is to obtain reliable estimates of the response quantities of interest, using the smallest possible number of ground motions to avoid unnecessary computations. However, motion-to-motion variability[16] in seismic responses (e.g., Figure 10.4) is a real phenomenon, and we must ensure that we have a sufficient number of analyses to obtain reliable information.

Statistically speaking, for a sample of n_{gm} values of EDP that result from our response-history analyses, we can compute "point" statistics such as the sample mean, $m_{\ln EDP}$, and sample standard deviation, $s_{\ln EDP}$, as approximations of the true, but unknown, population values – $\mu_{\ln EDP}$ and $\sigma_{\ln EDP}$, respectively. These point statistics are uncertain estimates of their population counterparts because of the finite number of n_{gm} values that they have been estimated from. Section B.1 provides the mathematical details[17] for estimating the confidence intervals for the mean and standard deviation, and the number of ground motions required to estimate these quantities with a specific precision and confidence level. In doing so, we make the common assumption that response quantities are lognormally distributed (Chapter 9).

Figure 10.10 illustrates how the required number of ground motions n_{gm} varies with the desired relative precision and confidence level (from 68 to 99%) for the mean and standard deviation, based on the normalized variables (relative precision) $\Delta/s = |\mu_{\ln EDP} - m_{\ln EDP}|/s_{\ln EDP}$ and $\Gamma = \sigma_{\ln EDP}/s_{\ln EDP}$. In addition, Tables 10.1 and 10.2 present specific values of n_{gm} required to constrain the mean as a function of the relative precision (Δ/s and Γ) for two different confidence levels. For the relative precision in the mean, Figure 10.10a also illustrates the commonly used normal distribution approximation (Equation B.10) as discussed by Shome et al. (1998), among others.

[16] Also commonly referred to as "record-to-record" variability.
[17] Using the variable substitution $X = \ln EDP$ and $n = n_{gm}$.

This approximation assumes that the variance is known and suggests the requirement of a smaller number of ground motions than is strictly appropriate, and is therefore not recommended.

The typical context for the results depicted in Figure 10.10 is that an allowable percentage error in the distribution parameters will be specified, and the required n_{gm} is then determined. Because of the fact that EDP is lognormally distributed (see Section A.5.2), then an allowable error of 10% is equivalent to a factor of $\gamma = 1.1$ around the true median $e^{\mu_{\ln EDP}}$ while the same 10% error for the standard deviation leads to a relative precision of $\Gamma = [\gamma^{-1}, \gamma]$. The factor γ corresponds to Δ via $\Delta = \ln(\gamma)$, so $\Delta = \ln(1.1) \approx 0.1$ for the 10% case. Values of s (needed to determine the relative precision Δ/s in the mean) are not known in advance for a given structure and EDP of interest, but reasonable estimates can be made from similar past studies, and a representative value of $s = 0.4$ will be used here for discussion (Hancock et al., 2008).

Consider an allowable relative error of 10% in the mean and standard deviation estimates. Based on $s = 0.4$, this corresponds to $\Delta/s = 0.25$, which from Figure 10.10a implies that 64 ground motions are required to estimate the mean at the 95% confidence level, and 18 at the 68% confidence level. To achieve a relative precision of 10% in the estimated standard deviation (i.e., $\Gamma = [1/1.1, 1.1]$), the required number of ground motions is 234 at the 95% confidence level, and 67 at the 68% level. To further illustrate these ideas, now consider a desired relative error of approximately 20% ($\Delta/s = 0.5$ and $\Gamma = [1/1.2, 1.2]$). At the 95% confidence level, 18 motions are required for the mean constraint, and 70 motions of the standard deviation. At the 68% confidence level, the corresponding numbers are 6 and 22, respectively.

The above numerical examples highlight several important features about the number of required ground motions. First, greater response variability and a greater desired level of confidence or precision will require a larger number of ground motions. The example calculations discussed above show that moving from a 95% confidence level to a 68% confidence level has a similar effect to working with half the level of relative precision. Second, for typical levels of response variability ($s \approx 0.4$) a larger number of ground motions is required to estimate the standard deviation to a desired level of precision as for the mean. The numbers presented here suggest that we need on the order of three times as many ground motions to constrain the standard deviation compared with the mean.

As discussed further in Section 10.7, design codes and standards have historically required small n_{gm} values (e.g., minimum values of $n_{gm} = 3$ or 7) on the basis of computational considerations. Table 10.1 indicates that the point-estimated median response from three ground motions has an error of 50% at the 68% confidence interval. Using seven ground motions reduces the error to slightly less than 20% at the 68% confidence interval, but still 50% at the 95% confidence interval. These large errors are the basis behind the impetus to increase the minimum number of n_{gm}. The larger n_{gm} values required to estimate the response standard deviation also explain why code-based documents principally avoid its estimation, considering only the mean response.

10.4.2 Screening for Suitable Ground Motions

The previous section demonstrated that using more ground motions for response-history analysis should produce more accurate estimates of the response mean and standard deviation. In practice, this possibility is limited by the finite number of historical ground motions we have to choose from.[18] We have emphasized that ground-motion selection should be performed in a hazard-consistent manner,

---

[18] This problem is alleviated through simulated ground motions (Chapter 5), which is a major reason for their increasing utilization.

Table 10.1. Number of ground motions n_{gm} required to achieve a given level of relative precision for two different confidence levels. Approximate factors (rounded to the nearest 0.05) γ are also provided for a representative value of $s = 0.4$

Factor $\gamma (s = 0.4)$	Relative precision Δ/s	68% confidence $n_{gm}(t*)$	95% confidence $n_{gm}(t*)$
1.10	0.25	18	64
1.20	0.50	6	18
1.35	0.75	4	10
1.50	1.00	3	7

Table 10.2. Number of ground motions n_{gm} required to achieve a given level of relative precision on the standard deviation. Only the upper bound values are provided as these always govern (see Equation B.8)

Factor γ	Relative precision Γ_{upper}	Upper bound 68% confidence	Upper bound 95% confidence
1.10	1.10	67	234
1.20	1.20	22	70
1.30	1.30	13	37
1.40	1.40	9	25
1.50	1.50	7	18

and this is reflected through both the disaggregation of the seismic hazard (Section 7.2) and also the conditional IM target distributions (Section 10.3). The disaggregation results identify the *implicit* causal parameters (rupture magnitude, source-to-site distance, etc.) that result in the ground-motion hazard but are not measures of the ground motions themselves. The IM targets provide an *explicit* description of the characteristics of the ground motions that result in the ground-motion hazard via the vector of IMs.

The distinction between implicit causal parameters and explicit IMs is a fundamental concept in ground-motion selection. Historically, the emphasis was placed on implicit causal parameters in ground-motion selection. However, it is now widely appreciated that a focus on explicit IMs is more important (noting that implicit causal parameters from disaggregation affect the IM targets via Equation 10.6). This appreciation is important because the databases of historical ground motions are underrepresented for the implicit causal parameters that are often of interest, as discussed below.

General Constraints from Ground-Motion Databases

As noted in Section 10.2, recorded ground-motion databases will generally be used as a source of prospective ground motions. However, as discussed in the context of empirical ground-motion prediction (Section 4.3.3), there is a paucity of recorded ground motions for causal parameter combinations that are often of interest in seismic design and assessment—and thus desired for ground-motion time series. Figure 4.13, in particular, illustrated the distribution of ground motions with magnitude and source-to-site distance within the NGA-West2 database. It is clear from this figure that observed ground motions are very unevenly distributed in magnitude-distance space, with many magnitude-distance combinations having few observations.

Large-magnitude and short-distance recordings are particularly important but rarely recorded. Ground motion features also vary as a function of site conditions, geographical regions, and rupture tectonic type, among others, but these conditions may also not be well-represented in recordings. Thus, recorded ground-motion databases do not sufficiently sample the multidimensional parameter space to make ground-motion selection straightforward. Although ground-motion instrumentation networks continue to increase in density, this undersampling problem is likely to persist into the future.

Formal Constraints on Causative Rupture Scenarios

An obvious way to improve the available number of ground motions is to relax constraints on implicit causal parameters. Indeed, it is common to constrain causal rupture parameters before applying a selection algorithm. For example, if the effective scenario identified from hazard disaggregation is M^*, then one may consider only ground motions from events with $M \in [M^* \pm \Delta M]$, where ΔM may take values such as 0.2 (e.g., Bommer and Acevedo, 2004) or 0.25 (e.g., Stewart et al., 2001).

These constraints were historically based upon the conceptual thinking outlined in the previous section. However, particular values of ΔM (for instance) were often chosen based upon intuition and assumptions about what influences the scaling of IMs, such as response spectra and duration. Importantly, these limits did not take into consideration how many ground motions were available to select from, nor did they account for the actual disaggregation distribution of causal parameters. That historical thinking is now outdated. Rather than excessively focusing upon the parameters defining the dominant rupture scenario, and indirectly trying to capture the properties of other IMs, it is now appreciated that it is better to focus directly upon the IMs for each ground motion based on the conditional IM targets from Section 10.3.

Tarbali and Bradley (2016) discuss the effect of limiting the ranges of causal parameters before selecting ground motions for response-history analysis. As the target IM distributions directly account for the attributes of other IMs not considered in the hazard disaggregation, the focus shifts to capturing the range of scenarios that contribute to the target IM distribution. In particular, they define the causal bounds based on the marginal disaggregation distributions of the causal parameters (magnitude and distance, primarily), but also monitor how the bounds impact the numbers of available ground motions. They find, counter to historical guidance, that it is generally better to have relatively wide bounds on the causal parameters to permit both a larger pool of ground motions to be selected from and to ensure that the selected ground motions provide a better representation of the distribution of rupture scenarios implied by the hazard disaggregation.

The recommendations of Tarbali and Bradley (2016) are presented in Table 10.3 and are visualized for an example in Figure 10.11. For magnitude and distance, bounds are the relatively extreme 1% or 99% percentiles, or some extension from the 10% or 90% percentiles. The most extreme of the resulting bounds are then adopted. These ranges may extend past the values in the disaggregation distributions, because ground motions in these ranges may still match the target IM distributions.

It will still often be the case that insufficient numbers of recorded ground motions are available within allowable causal parameter ranges, keeping in mind the required numbers suggested in Figure 10.10. We, therefore, need approaches to supplement the numbers of available ground

Table 10.3. Recommended bounds to be imposed upon implicit causal parameters prior to ground-motion selection

Causal parameter	Lower bound	Upper bound
Magnitude, M	$\min(M^{1\%}, M^{10\%} - 0.5)$	$\max(M^{99\%}, M^{90\%} + 0.5)$
Rupture distance, R_{rup}	$\min(R_{rup}^{1\%}, 0.5R_{rup}^{10\%})$	$\max(R_{rup}^{99\%}, 1.5R_{rup}^{90\%})$
Average shear-wave velocity, $V_{S,30}$	$0.5V_{S,30}$	$1.5V_{S,30}$

Note: $X^{y\%}$ represents the yth percentile of the marginal disaggregation distribution for causal parameter X.

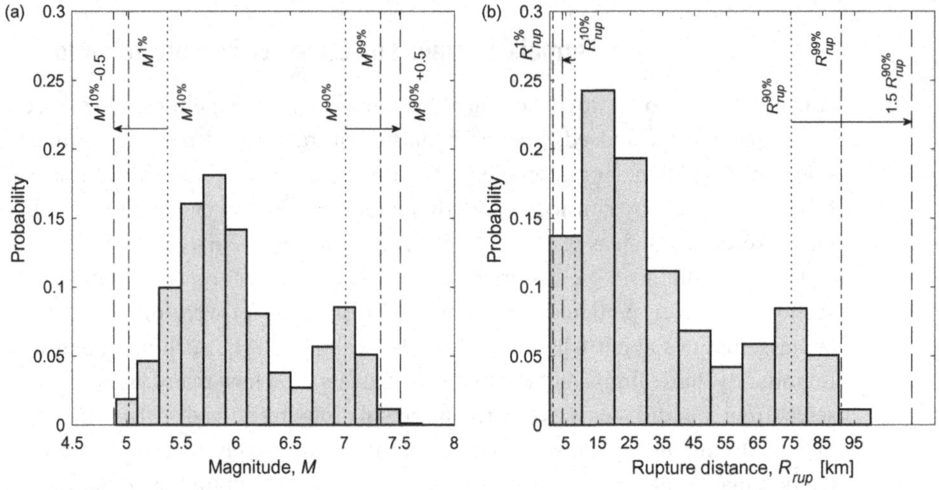

Fig. 10.11 Example application of the causal bounds of Table 10.3. The two panels collectively show a disaggregation distribution with respect to magnitude and distance. Panel (a) shows the distribution of magnitude and the associated causal bounds, while panel (b) corresponds to distance.

motions. Two main options exist: scaling or modifying recorded ground motions, and using simulated ground motions. These options are discussed in the following subsections.

Scaling or Modifying Ground Motions

As ground-motion selections are made conditional on $IM_1 = im_1$, recorded motions are typically "scaled," by multiplying their acceleration amplitudes, such that they are consistent with this conditioning. Similar to causal parameter bounds, the literature also contains many suggestions for limits on ground-motion scaling. In most cases, these recommendations are also based upon intuition rather than any quantitative analysis (e.g., see the discussions in Watson-Lamprey and Abrahamson, 2006; Luco and Bazzurro, 2007). Scaling a ground motion can produce combinations of IM values that are not naturally seen in a ground motion with the target IM_1. However, if only ground motions with reasonable IM combinations are selected, or if the response metric of interest is not sensitive to the IMs that have been distorted, then scaling is unlikely to introduce any significant biases (Luco and Bazzurro, 2007; Bradley, 2010b).

When a scalar factor is applied to a ground-motion time series, it does not result in a proportional change to all IMs. To appreciate this, consider a scaled ground motion, $a_s(t) = \alpha a_r(t)$, obtained from multiplying a raw ground motion, $a_r(t)$, by a factor α. From the definition of response spectral ordinates presented in Chapter 4, $SA_s(T) = \alpha SA_r(T)$, where the subscripts r and s denote the raw and scaled spectral ordinates, respectively. However, consider the Arias intensity (Equation 4.10) of the raw and scaled ground motions. In this case, the relationship between the values from the scaled and unscaled motions is $AI_s = \alpha^2 AI_r$. Furthermore, the significant duration (Section 4.2.4) is entirely unaffected by scaling, i.e., $D_{S,s} = D_{S,r}$.

In general, the relationship between IMs from raw and scaled ground motions can often be expressed as[19]

$$IM_s = IM_r \alpha^\gamma \tag{10.17}$$

where α is the scale factor applied to the ground motion, and γ is an IM-specific exponent. Thus, $\gamma = 1$ for peak ground acceleration or spectral acceleration, $\gamma = 2$ for Arias intensity, and $\gamma = 0$ for significant duration. Therefore, scaling a ground motion can distort the natural characteristics of $im \mid rup$ for the scaled ground motion.

When considering the impact of amplitude scaling, it is useful to consider the large IM aleatory variability associated with a given rupture scenario. In Chapter 4, we saw that ergodic ground-motion models for response spectral ordinates often have total aleatory standard deviations on the order of 0.6–0.7 natural logarithmic units. These values imply that around 30% of spectral amplitudes for a particular rupture scenario will be a factor of two higher or lower than the median value for the scenario. Therefore, while scaling a ground motion changes the relative values of its IMs, significant scaling is often required to produce problematic distortion, given the inherent aleatory variability in IM values.

Potential problems from scaling are largely circumvented by selecting ground motions based on a *generalized* target IM distribution that considers more than response spectral ordinates (Bradley, 2010b, 2012b). As scaling was historically done with a focus upon matching response spectral amplitudes, scaling could influence duration- and energy-related IMs. Including these IMs in the target IM distribution prevents significant distortions. The reason is that relatively large scale factors are required to introduce perceptible distortion, and ground motions that require this level of scaling to match an amplitude-based IM are unlikely to match the target distribution for the other IMs. Hence using a generalized target IM distribution effectively limits the extent to which ground motions can be scaled. There is no remaining reason to impose specific constraints upon amplitude scale factors, if the target distribution includes IMs with as many distinct attributes as possible (i.e., a mix of amplitude, cumulative, duration, and energy-based IMs).

In addition to traditional amplitude-scaling of ground motions, various methods exist to modify ground motions in nonlinear ways. The most common example of these is spectral-matching, which selectively modifies parts of a ground motion rather than scaling the entire signal by a constant. Recognizing that any signal may be represented by its Fourier or wavelet decomposition, it is possible to determine scaling factors for individual Fourier components or wavelets that gives rise to a near-perfect match to a target response spectrum.

There are two circumstances where spectral matching is appealing. The first is when an analyst wishes to take recorded ground motions from softer sites (which are relatively plentiful) and modify them to have richer high-frequency content expected for a stiffer site. Spectral matching can be

---

[19] There are some exceptions, such as bracketed duration and CAV_5 (Kramer and Mitchell, 2006), for which there is no parametric relationship between the IM and amplitude scale factor. In these cases the required scale factor, α, to achieve the desired value of IM_s must be obtained via iteration.

applied in such cases, and when this is done, the spectral variability in the ground motions is usually retained. The second usage case for spectral matching is where one wishes to suppress spectral variability. This is common in code-based contexts, as discussed in Section 10.7. Suppression might also be desirable in a research context to obtain ensembles of ground motions that share common spectral amplitudes but possess different duration- or energy-related properties (e.g., Hancock and Bommer, 2007). These modification techniques are, in some sense, simulating a new ground motion, and are discussed further below.

Use of Simulated Motions

While a generalized target distribution largely solves issues associated with the scaling of recorded ground motions, it also limits the available number of prospective ground motions. The methods for simulating ground motions discussed within Chapter 5 become prime candidates to supplement databases of recorded ground motions in this case.

Recorded ground motions (even if scaled) have historically been preferred to simulations. This was primarily due to concern that simulated motions did not adequately produce the full range of real ground-motion characteristics. This concern remains valid, but the quality of ground-motion simulation approaches has improved drastically in recent years, and simulated motions are becoming more indistinguishable from recorded motions.

Importantly, simulated motions do not need to possess all of the attributes of real ground motions before they can be considered for use in response-history analyses. Consider the example of a structure whose relevant EDPs are sensitive to response spectra, but are entirely insensitive to other IMs. In such cases, whether or not a simulated motion has the correct duration or energy content is irrelevant because they have no bearing upon the response. Whether or not simulated motions can be used for response-history analyses then depends upon the nature of the application.

On the other hand, recorded ground motions are also susceptible to having inappropriate characteristics, due to impacts of ground-motion scaling and the use of recordings from sites other than the specific site of interest.

Whether using recorded or simulated ground motions, we need to ensure that the selected set of ground motions matches the target distribution $f_{IM_{2+}|IM_1}(im_{2+} \mid im_1)$, where IM_{2+} contains all IMs relevant for the given application. Douglas and Aochi (2008) provide a survey of potential simulation methods, of which the following three broad classes could be considered to meet this objective:

- **Physics-based simulations**, discussed in detail in Chapter 5, generate time series from physically motivated parameters related to source, path, and site processes. These approaches include comprehensive wave-propagation simulations or more simplified, but still physics-based, processes. With these methods, $f_{IM|Rup}(im \mid rup)$ arises implicitly from the combination of the physical models and modeling assumptions with the distributions of any input parameters. They do not generally allow the prespecification of target IM values.

- **Empirically calibrated simulations** generate time series using concepts from random vibrations and signal processing (e.g., Gasparini and Vanmarcke, 1976; Rezaeian and Der Kiureghian, 2008, 2011; Sgobba et al., 2011). The properties of the resulting time series are empirically calibrated and can vary as a function of source, path, and site parameters. The user specifies a rupture scenario and target site conditions and uses this information to obtain estimates (and, potentially, distributions) of the stochastic process's parameters. Simulated time series can then be produced. In some cases, it is also possible to condition the simulated motions upon certain IMs, like Arias intensity or duration (e.g., Stafford et al., 2009; Rezaeian and Der Kiureghian, 2011).

- **Target-based simulations** generate time series with a target response spectrum, or take a seed ground motion and then modify it to match the target spectrum. The first of these options, while permitted by some codes (e.g., CEN, 2004), should be avoided as it leads to unrealistic ground motions (Bommer and Acevedo, 2004). The second option, which includes approaches based upon wavelet adjustments (e.g., Al Atik and Abrahamson, 2010) as well as discrete Volterra series (e.g., Alexander et al., 2014), is preferred. With these methods, if the seed ground motion has duration and energy-based IMs that are broadly consistent with the target distribution, then these properties are largely retained after the spectral matching. Applications of these approaches focus upon matching smooth target spectra and so the resulting ground motions have unrealistically low spectral variability that can lead to biased response estimates (Carballo, 2000; Bazzurro and Luco, 2006; Iervolino et al., 2010; Seifried and Baker, 2016).

Of the three general classes of options outlined above, physics-based simulations are garnering increasing attention and credibility. Physics-based simulations have developed considerably in recent years due to the combination of incremental understanding of how to model the physics of the problem and significant increases in computational resources being applied to this problem. Also, these approaches are increasingly focused on replicating natural ground-motion characteristics over a broader range of IMs. However, physics-based simulations currently still possess characteristics that are known to lead to biases in structural response (e.g., Bayless and Abrahamson, 2018).

When simulated ground motions are used in ground-motion selection, they can be treated in the same manner as recorded motions. Note the use of simulations for this task does not require that the simulations be fully validated for ground-motion prediction, as in this task it is only required that some simulations have the desired $f_{IM_{2+}|IM_1}(im_{2+} \mid im_1)$, and not that they have the correct distribution of $IM \mid rup$ (Bradley et al., 2015).

10.4.3 Evaluation of Match to Target IM Distributions

This subsection addresses the central task in selecting ground motions that match the target IM distribution. This is Step 6 in Figure 10.9, and also relates to Steps 3, 7, and 8.

A number of methods been proposed in the literature for evaluating the match to a target, based on the distribution mean (e.g., Baker and Cornell, 2006b; Beyer and Bommer, 2007; Buratti et al., 2011), variance (e.g., Kottke and Rathje, 2008; Shi and Stafford, 2018), or some metric of distribution mismatch (e.g., Bradley, 2012b). A key consideration is that assessing the match to a target mean can be made on a motion-by-motion basis, whereas matching the distribution requires considering the entire ensemble of motions. These particular cases are discussed below.

Match to Target Intensity Measures

Despite the probability distribution of IM being the correct target for ground-motion selection, it is still commonplace to adopt a target comprised only of deterministic IM values. Examples include code-based ground-motion selection procedures discussed in Section 10.7. Matching a set of deterministic IM values is a useful intermediate step before considering a target IM distribution.

Matching a set of target IM values requires finding a set of ground motions with a mean that is as "close" as possible to the target. This is normally achieved by searching for the individual ground motions that are closest to the target mean.[20] However, what constitutes "closeness" depends upon the IMs that are being considered. The following three issues need to be addressed when evaluating the match to a target mean:

1. How does one compare differences in IMs that have different units?
2. How likely are we to observe a difference Δim_i between a ground motion and a target IM, given that some IMs are inherently more variable than others?
3. How does this difference in IM values between a ground motion and its target (Δim_i) influence the EDP?

For item 1, consider a target vector of IMs that includes spectral acceleration ordinates and Arias intensity. How should a 0.1-g difference in spectral acceleration be compared with a 0.1-m/s difference in Arias intensity? The solution is to work with ratios between a candidate ground motion and the target mean, as this allows the units of each IM to cancel out. Equivalently, we can compute the difference between the logarithms of the two IMs.

For item 2, consider whether a ratio of 1.1 in spectral acceleration is equivalent to a ratio of 1.1 in Arias intensity, if the inherent variability of Arias intensity is significantly greater than that for spectral acceleration. This issue can be addressed by considering a variate where differences in logarithmic IM are normalized by their logarithmic standard deviations. We define the normalized variate as

$$z_{\ln im_i}^{(s)} = \frac{\ln im_i^{(s)} - \ln im_i^{target}}{\sigma_{\ln im_i}} = \frac{\ln\left(im_i^{(s)}/im_i^{target}\right)}{\sigma_{\ln im_i}}. \tag{10.18}$$

Here $z_{\ln im_i}^{(s)}$ is the normalized variate for ground motion s evaluated for IM i, $\ln im_i^{(s)}$ is the logarithmic IM value for ground motion s and IM_i, $\ln im^{target}$ is the logarithmic target IM value, and $\sigma_{\ln IM_i}$ is the logarithmic standard deviation for IM_i. The rightmost expression of Equation 10.18 explicitly illustrates that the numerator is the logarithm of the ratio of the IM values.

Finally, item 3 is typically addressed by applying a weight w_i to each IM that reflects how sensitive the EDP is to perturbations of each IM,[21] and therefore how much emphasis should be given to matching it accurately. The match between ground motion s at the target IM values is then defined in terms of a weighted sum-of-squared-errors (SSE):

$$SSE(s) = \sum_{i=1}^{n_{IM}} w_i \left(z_{\ln im_i}^{(s)}\right)^2 \tag{10.19}$$

$$= \sum_{i=1}^{n_{IM}} w_i \left(\frac{\ln im_i^{(s)} - \ln im_i^{target}}{\sigma_{\ln im_i}}\right)^2. \tag{10.20}$$

Note that the evaluation of SSE for ground motion s is independent of all other ground motions, and thus Equation 10.20 can be evaluated on a motion-by-motion basis. As a result, to select n_{gm} ground motions that match a target, we evaluate $SSE(s)$ for all candidate ground motions, and select those with the n_{gm} smallest values.

---

[20] This is not guaranteed to result in the optimal set because we are implicitly assuming that the errors in approximating the target from each ground motion are entirely random and cancel out over the ensemble. This assumption will not always be valid, but the approach is straightforward and will typically give good results.

[21] Such sensitivity is not always known in advance; see Section 10.5.

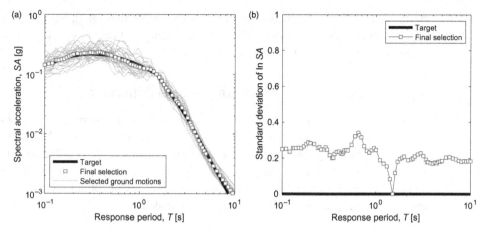

Fig. 10.12 Ground-motion selection for a target conditional mean spectrum based on $SA(1.5\,\text{s})$. The conditional mean spectrum is obtained from the disaggregation distribution of Figure 10.11 using Equation 10.10, and the ground-motion match, SSE, obtained using $w_i = 1$ and $\sigma_{\ln im_i} = 1$ for all spectral ordinates. Panel (a) shows the target spectrum, 30 individual response spectra corresponding to each selected ground motion, and the median of the 30 motions at each response period. Panel (b) shows the standard deviation of the logarithmic spectral acceleration values. The target standard deviation when trying to match the conditional mean spectrum is $\sigma_{\ln SA,i} = 0$ for all i as we wish to closely match the target. The actual variability obtained is $\sigma_{\ln SA,i} \approx 0.2$ for response periods other than $T = 1.5\,\text{s}$.

In the common case where ground-motion scaling is allowed, Equation 10.20 is evaluated for the scaled ground motion. Equation 10.17, and associated text, discussed how this could be obtained from the unscaled (raw) ground motion. Substitution of Equation 10.17 into Equation 10.18 means that Equation 10.20 is now a function of the amplitude scale factor, which can be solved to determine the scale factor that gives the minimum SSE value. Alternatively, the amplitude scale factor may be fixed by enforcing that the scaled ground motion has the desired value of the conditioning IM, $IM_1 = im_1$.

To illustrate the above concepts, an example of a set of 30 ground motions chosen to match a target conditional mean spectrum (CMS) is shown in Figure 10.12. In this example, the target spectrum is constructed to be consistent with the disaggregation distribution previously shown in Figure 10.11. Each spectral ordinate is weighted equally, i.e., $w_i = 1/n_{IM}$ in Equation 10.20. The depicted 30 ground motions are selected from the NGA-West2 database (Ancheta et al., 2014).

Application of Equation 10.20 provides a measure of agreement to the target, and will produce a unique set of ground motions from any given dataset. If the match to the target is unsatisfactory, the pool of available ground motions may need to be enlarged. This may be achieved by relaxing the causal parameter ranges, relaxing limits on scale factors, or supplementing a database of recorded ground motions with simulations.

Figure 10.12 also provides a basis to discuss two analyst decisions pertaining to the assignment of weights w_i, and the IM standard deviation $\sigma_{\ln IM_i}$. First, weights on IMs allow the analyst to consider a larger number of IMs, while retaining discretion regarding their relative importance. If weights are not explicitly considered, it is equivalent to considering equal weighting for all IMs in the target IM vector. Second, if $\sigma_{\ln IM_i}$ is not explicitly considered as a normalization factor, the implication is that all IMs are considered to have an approximately equal level of variability. These decisions may not be influential when considering only spectral acceleration ordinates (which have similar levels of $\sigma_{\ln IM_i}$ at different oscillator vibration periods; see Section 4.4.3), but may be problematic if the $\sigma_{\ln IM_i}$ values vary significantly. The general framing of Equations 10.18 and 10.20 is, therefore, useful when considering a general vector of IMs.

Match to a Target Distribution

In contrast to the previous section, when examining a ground-motion ensemble's distributional properties, a single motion's suitability cannot be assessed in isolation. That is, we cannot evaluate whether a particular ground motion is suitable for matching an IM distribution without also knowing the other ground motions in the ensemble. The direct evaluation of all possible ensembles, arising from all potential combinations of motions, is computationally prohibitive for typical database sizes and values of n_{gm}, so an alternative approach is needed.

Statistically Generate Intensity Measure Realizations

Rather than directly solve the computationally demanding problem of identifying a ground-motion ensemble that is consistent with the full IM distribution, Jayaram et al. (2011) proposed that it can be indirectly solved by simply generating random realizations of IM values from the target IM distribution (Step 3 in Figure 10.9), and finding individual ground motions that most closely resemble each realization. Because the realizations are drawn directly from the target IM distribution, they will be statistically consistent with it. If individual ground motions can be found that are "close" to each realization, then the ensemble of selected ground motions will also be consistent with the target IM distribution. Section 10.3 provided the target multivariate IM distribution from which the random realizations can be generated. When the target IM distribution is approximated as multivariate lognormal (Section 10.3.2), then random realizations are generated from the multivariate lognormal distribution (Jayaram et al., 2011). Alternatively, if the target IM distribution is considered in an exact form (Equation 10.6), then the random realizations are generated in a two-step fashion: (1) generating a random rupture from the disaggregation distribution, then (2) generating $\boldsymbol{IM}$ values from the (lognormal) target IM distribution for this specific rupture (Bradley, 2012b). If $\boldsymbol{im}^m$ is the mth random realization of the IM vector considered (and im_i^m is the ith element in the vector), then these are used as im_i^{target} values in Equation 10.20 to identify the ground motion with the lowest misfit.

Misfit of the Ensemble Mean and Standard Deviation

Once an ensemble of ground motions is obtained using the process described above, its match to the target IM distribution is evaluated. This match is most commonly quantified based either (1) on the mean and standard deviation or (2) directly on the distribution. The examples below illustrate this further.

Given an ensemble of n_{gm} ground motions, the sample mean $m_{\ln IM_i}$ and standard deviation $s_{\ln IM_i}$ can be computed for each IM_i (see Section B.1) and compared with the target IM distribution values of $\mu_{\ln IM_i}$ and $\sigma_{\ln IM_i}$, respectively. In the same manner that the misfit for an individual ground motion is computed from Equation 10.20, the misfit of the ground-motion ensemble mean and standard deviation can be computed as

$$SSE_\mu = \sum_{i=1}^{n_{IM}} w_i (m_{\ln IM_i} - \mu_{\ln IM_i})^2 \tag{10.21}$$

$$SSE_\sigma = \sum_{i=1}^{n_{IM}} w_i (s_{\ln IM_i} - \sigma_{\ln IM_i})^2. \tag{10.22}$$

Given the interest in minimizing the misfit in both the mean and standard deviation, SSE_μ and SSE_σ can be combined in a weighted fashion to compute a "total" misfit (Jayaram et al., 2011):

$$SSE_T = w_\mu SSE_\mu + w_\sigma SSE_\sigma \tag{10.23}$$

where SSE_T is the "total" (or combined) sum-squared-error, and w_μ and w_σ are the relative weights given to the misfit in the mean and standard deviation, respectively.

Other metrics can be considered for describing the misfit between the mean and standard deviation. For example, in addition to Equation 10.23, Baker and Lee (2018) also compute maximum percentage differences between the sample and target medians and standard deviations.

Figure 10.13 shows an example of ground-motion selection for a target distribution of response spectral ordinates that is constructed to be consistent with the disaggregation distribution shown in Figure 10.11. Figure 10.13a illustrates the generated individual realizations of response spectral amplitudes from the IM distribution. Then Figure 10.13b illustrates the resulting ground motions that individually best match each of these response spectra realizations. This initial selection has median

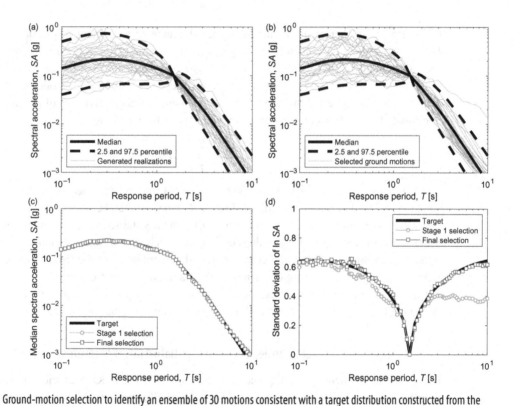

Fig. 10.13 Ground-motion selection to identify an ensemble of 30 motions consistent with a target distribution constructed from the disaggregation distribution of Figure 10.11. The mean magnitude and distance are $\bar{M} = 6.0$ and $\bar{R} = 34$ km, and each spectral ordinate carries the same weight. (a) Percentiles from the target distribution along with generated response spectra realizations that represent this target. (b) The same percentiles along with the spectra of the selected ground motions. (c) The target (conditional mean) spectrum and the median spectral ordinates of the 30 selected ground motions (before and after the optimization step). (d) The target conditional standard deviation and the standard deviation of the 30 selected ground motions.

spectral acceleration values that match the target very well, but the variability is slightly underestimated at periods below the conditioning period of $T = 1.5$ s, and more significantly underestimated at longer periods above approximately $T = 3$ s (the "Stage 1 selection" results Figure 10.13d). The evaluation of the overall performance of the ground-motion ensemble is made using Equation 10.23. A subsequent step will be applied in Section 10.4.3 to further optimize the results.

Misfit of the Ensemble Distribution

An alternative approach to comparing estimates of the first and second moments of the target distribution is to adopt the Kolmogorov–Smirnov (KS) measure of difference.[22] In the KS approach, the CDF of an IM from the target distribution and the corresponding empirical CDF from the selected ground motions are computed. The maximum difference between the two CDFs is then obtained for various values of an IM. Bradley (2010b, 2012b) adopted the KS statistic to assess the degree of agreement between the empirical CDF, $F_{n_{gm}}(im_i)$, for the IMs associated with a set of n_{gm} ground motions, and the target IM distribution. For each IM i, the conditional distribution of the IM_i given IM_1 is used as the target CDF, $F_{IM_i|IM_1}(im_i \mid im_1)$. The KS statistic is then defined as

$$D_{KS,i} = \max_{IM_i} \left| F_{IM_i|IM_1}(im_i \mid im_1) - F_{n_{gm}}(im_i) \right|. \tag{10.24}$$

To determine whether or not a sufficiently good degree of agreement has been obtained between the selected ground motions and the target, a critical value of the KS statistic corresponding to a particular confidence level can be defined. This approach treats each of the IMs individually and so multiple values of the $D_{KS,i}$ will be obtained depending upon the number of IMs considered. Bradley (2012b) suggested that the relative performance of alternative sets of ground motions could be evaluated on the basis of computing an overall statistic according to:

$$R = \sum_{i=1}^{n_{IM}} w_i D_{KS,i}^2 \tag{10.25}$$

where w_i are weights that are applied to the IMs to reflect their importance (similarly to Equation 10.20). The ensemble with the lowest R value can then be selected.

For the ground motion example in Figure 10.13, the KS statistics are shown in Figure 10.14. This figure shows the individual $D_{KS,i}$ values evaluated at each response period, and the critical KS value. If the $D_{KS,i}$ values are below the critical value, the distribution of the IMs for the selected ground motions are regarded as being consistent with a random sample from the target. The figure shows that at all but the longest response periods, this consistency is achieved. The overall measure of agreement is shown using $\sqrt{R}$, with all w_i being the same in Equation 10.25.

Similarities between Alternative Misfit Measures

The errors in moments (i.e., Equations 10.21 and 10.22) and KS statistics (Equation 10.24) of a ground-motion ensemble are related. Figure 10.15 illustrates how a difference in the mean or standard deviation of a distribution translates into a KS statistic value. In Figure 10.15a the

---

[22] Or alternative metrics for assessing differences between distributions such as the Anderson–Darling test (Tarbali and Bradley, 2015), the Kullback–Leibler divergence (Kullback and Leibler, 1951), or the multivariate energy check in Shi and Stafford (2018).

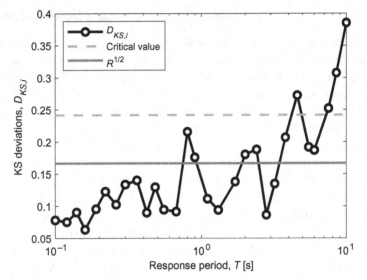

Fig. 10.14 Kolmogorov–Smirnov (KS) deviations and statistics as a function of response spectral vibration period for the ground-motion selection example shown in Figure 10.13. The solid gray line shows $\sqrt{R}$ in order to compare all metrics on a common scale.

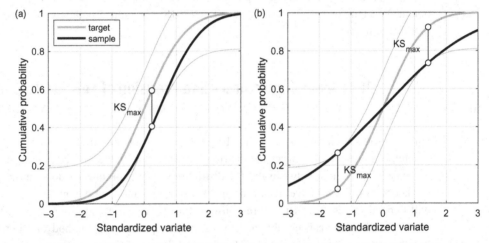

Fig. 10.15 Illustration of the connection between a difference in the distribution mean and variance, and the KS statistic. (a) Example where the sample mean deviates from the target mean. (b) Example where the sample variance deviates from the target variance. In both cases, the deviation causes the KS statistic to coincide with the critical value for a sample size of 50 and a confidence level of 95%. These critical KS values are shown using the thin gray lines.

sample[23] distribution has a different mean than the target distribution, but the same variance, whereas in Figure 10.15b the sample distribution has the same mean, but different variance. Therefore, a change to either the mean or standard deviation leads to a change in the CDF that results in a vertical difference between the target and sample distributions, the maximum of which is the KS statistic.

---

[23] In reality, the KS statistic is computed with the empirical CDF of the sample distribution, which is shown as a smooth line here for conceptual simplicity.

Optimizing Selected Ground-Motion Ensembles

Because the ground-motion ensemble is selected based on random realizations of the target IM distribution, different ensembles will be selected for different sets of random realizations. Furthermore, a given set of random realizations may have properties that differ from the underlying target IM distribution. For relatively large ensembles (e.g., $n_{gm} > 25$), this effect will likely be unimportant. However, it can be problematic for smaller ensemble sizes (Jayaram et al., 2011; Bradley, 2012b), and thus further optimization of the selected motions may be desired as depicted in Step 7 of Figure 10.9.

A brute-force solution is to simply perform ground-motion selection multiple times, determining the ensemble misfit each time (via Equation 10.23 or 10.25), and then selecting the ensemble with the lowest misfit. Bradley (2012b) and Tarbali and Bradley (2016) (based on Equation 10.25) represent examples of this approach. Alternatively, a local optimization approach can be used to "improve" the ground-motion ensemble in a more computationally efficient manner (Jayaram et al., 2011). After a ground-motion ensemble has been initially obtained, the analyst can consider the substitution of other ground motions from the candidate databases to "locally" improve the misfit metric (Equation 10.23 or 10.25). An example of the application of this "greedy" optimization was shown previously in the "Final selection" results of Figure 10.13. Computationally efficient implementations of the greedy optimization algorithm enable ground-motion ensembles to be obtained in tens of seconds (Baker and Lee, 2018) for typical databases and numbers of ground motions of interest.

10.5 Assessing Accuracy and Precision of Seismic Responses

Once a ground-motion ensemble has been obtained (Section 10.4), we will know how well this ensemble represents the target IM distribution. Because of the limited available ground motions, we expect differences between the IM distribution of our ground-motion ensemble $\hat{f}_{IM}(\boldsymbol{im})$ and the target $f_{IM}(\boldsymbol{im})$. However, we do not know in advance the extent to which these differences will impact the resulting seismic response-history analyses. An understanding of the likely impact of such differences can be obtained by using the approximate method of Bradley (2010b). The distribution of seismic response (EDP) obtained directly from the response-history analysis results can be expressed as

$$\hat{f}_{EDP|IM_1}(edp \mid im_1) = \int f_{EDP|\boldsymbol{IM}_{2+},IM_1}(edp \mid \boldsymbol{im}_{2+}, im_1)\hat{f}_{\boldsymbol{IM}_{2+}|IM_1}(\boldsymbol{im}_{2+} \mid im_1)d\boldsymbol{IM}_{2+}$$

(10.26)

where $\hat{f}_{\boldsymbol{IM}_{2+}|IM_1}(\boldsymbol{im}_{2+} \mid im_1)$ represents the empirical distribution of the ground-motion ensemble used for the response-history analyses, and $f_{EDP|\boldsymbol{IM}_{2+},IM_1}(edp \mid \boldsymbol{im}_{2+}, im_1)$ explicitly reflects how the PDF of EDP depends on $\boldsymbol{IM}_{2+}$. The hat on $\hat{f}_{EDP|IM_1}(edp \mid im_1)$ is used to denote that it is an approximation of the true $EDP \mid IM_1$ distribution as a result of the ground-motion ensemble being an imperfect approximation of $f_{\boldsymbol{IM}_{2+}|IM_1}$.

Following the response-history analyses, it becomes possible to explore the sensitivity of EDP with respect to the various IM_i that were considered in the specification of the ground-motion target.

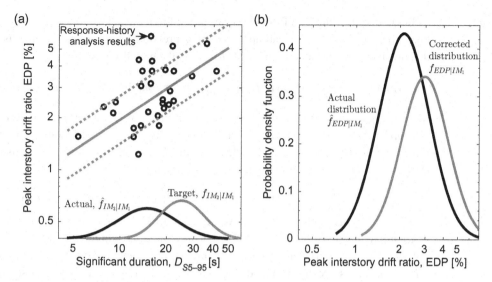

Fig. 10.16 Assessing bias in seismic response as a result of the selected ground-motion ensemble. (a) Examination of the seismic response dependence of 5–95% significant duration, D_{S5-95}. (b) Approximation of the effect of the selected ground-motion ensemble on EDP bias through comparison of the difference between the actual and corrected EDP distributions.

If $\ln IM$ and $\ln EDP$ are linearly related, one can perform a simple linear regression between the obtained edp values from response-history analysis and the corresponding values of im_i. The mean of this regression can be expressed as

$$\mu_{\ln EDP|IM_i, IM_1} = \beta_0 + \beta_1 \ln IM_i \tag{10.27}$$

where β_0 and β_1 are regression parameters. The standard deviation resulting from this regression is denoted as $\sigma_{\ln EDP|IM_i, IM_1}$. It follows that if a particular IM_i is responsible for a statistically significant difference between $\hat{f}_{IM}(im)$ and $f_{IM}(im)$, and the coefficient β_1 is statistically different from zero, a bias in the estimated distribution of EDP will exist.

As an example, consider the 30 ground motions selected in Figure 10.13 applied in the response-history analysis of a multistory structure, with the EDP of interest being the peak interstory drift ratio. Figure 10.16a illustrates the 30 obtained values of EDP, which have a median of $EDP_{50} = 2.5\%$ ($\mu_{\ln EDP|IM_1} = \ln(EDP_{50}) = -3.69$) and lognormal standard deviation of $\sigma_{\ln EDP|IM_1} = 0.4$. The ground-motion ensemble in Figure 10.13 was selected based on an IM target distribution comprising response spectra, but no non-SA IMs. Figure 10.16a also illustrates the target distribution of 5–95% significant duration D_{S5-95} (denoted as $f_{IM_2|IM_1}$), compared with the actual D_{S5-95} distribution of the 30 selected ground motions, $\hat{f}_{IM_2|IM_1}$. The target D_{S5-95} distribution has a median of 25 s and lognormal standard deviation of 0.4, whereas the actual D_{S5-95} distribution of the 30 selected ground motions has a median of 15 s and lognormal standard deviation of 0.3. Finally, Figure 10.16a illustrates that the seismic response does exhibit a dependence on D_{S5-95}, as indicated by the $\mu \pm \sigma$ trend lines. The specific values of the regression parameters[24] from Equation 10.27 are $\beta_0 = -5.31$ and $\beta_1 = 0.6$, with a residual standard deviation of $\sigma_{\ln EDP|IM_i, IM_1} = 0.32$.

---

[24] These parameter values give a much stronger dependence on duration than is seen in typical structural analyses, but the strong dependence is used here to make the conceptual illustration clear.

To assess the impact of the bias in ground-motion selection, such as in the example of Figure 10.16a, we introduce an approximate expression[25] for the "corrected" distribution of EDP that involves only the conditioning IM, IM_1, and the biased IM_i:

$$f_{EDP|IM_1}(edp \mid im_1) \approx \int f_{EDP|IM_i,IM_1}(edp \mid im_i, im_1) f_{IM_i|IM_1}(im_i \mid im_1) dIM_i \qquad (10.28)$$

where $f_{IM_i|IM_1}(im_i \mid im_1)$ is the theoretical distribution of IM_i given IM_1, and $f_{EDP|IM_i,IM_1}(edp \mid im_i, im_1)$ is the distribution of EDP as a function of IM_i obtained from the regression analyses. Further assuming that both $EDP \mid IM_i, IM_1$ and $IM_i \mid IM_1$ are lognormally distributed, it can be shown that $EDP \mid IM_1$ is also lognormal according to

$$EDP \mid IM_1 \sim \mathcal{LN}\left(\beta_0 + \beta_1 \mu_{\ln IM_i|IM_1}, \sigma^2_{\ln EDP|IM_i,IM_1} + \beta_1^2 \sigma^2_{\ln IM_i|IM_1}\right). \qquad (10.29)$$

The corrected distribution obtained via Equation 10.29 can then be compared with the actual distribution obtained directly from the n_{gm} values of edp to gauge the extent of any bias.

Returning to the previous example, Figure 10.16b compares the actual and corrected PDFs of $EDP \mid IM_1$. As previously noted, the actual distribution has a mean and standard deviation of $\mu_{\ln EDP|IM_1} = -3.69$ and $\sigma_{\ln EDP|IM_1} = 0.4$, respectively. From Equation 10.29, the mean and variance of the corrected distribution are estimated as

$$\mu_{\ln EDP|IM_1} = -5.31 + 0.6 \times \ln(25) = -3.38 \qquad (10.30)$$

$$\sigma^2_{\ln EDP|IM_1} = 0.32 + 0.6^2 \times 0.3^2 = 0.103. \qquad (10.31)$$

These correspond to median and standard deviation values of 3.4% and 0.37, respectively. This corrected median EDP is 3.7/2.5 = 1.36 times greater than the uncorrected median (i.e., a 36% underestimate). The EDP bias in this example is significant, and would likely require repeating the ground-motion selection with explicit consideration of D_{S5-95}. This dependence on D_{S5-95} is specific to the structural system considered, but will also be a function of ground-motion intensity. For example, Chandramohan et al. (2016) illustrate that EDP dependence on significant duration is generally present only for large seismic demands that cause near-collapse responses in structural systems.

In considering the above theory and example, it is worth reiterating that bias will only be introduced if both (1) the ground-motion ensemble selected is biased with respect to a particular IM_i (i.e., $\hat{f}_{IM_i|IM_1}(im_i \mid im_1)$ differs from $f_{IM_i|IM_1}(im_i \mid im_1)$) and (2) the seismic response measure EDP is sensitive to IM_i (i.e., $\beta_1 \neq 0$). Further, these differences must produce a substantial effect relative to the EDP motion-to-motion variability, and the estimation uncertainty inherent from using a finite ensemble of ground motions. This dual requirement guides the emphasis given to different IMs in the ground-motion selection process (Section 10.6.2). During ground-motion selection it may, therefore, be acceptable to ignore a particular IM_i (by setting $w_i = 0$ in the ground-motion selection via Equation 10.20). For example, all non-SA IMs had $w_i = 0$ in Figure 10.13. Ignoring these IMs will, however, potentially result in the ground-motion ensemble distribution being inconsistent with the target distribution. The implications for the seismic response analysis for EDP can then be assessed via Equation 10.29. If potential bias is found, then ground-motion selection should be

---

[25] The approximation arises from treating the apparent bias in IM_i in isolation, whereas it will likely be correlated with other elements of IM_{2+}.

repeated with consideration given to this IM_i. Prior experience will guide the identification of important IMs before seismic response analyses are performed, so experienced analysts can generally avoid iterative calculations.

10.6 Application-Specific Decisions

Earlier sections have addressed computing conditional IM distributions (Section 10.3) and consequent ground-motion selection (Section 10.4). Performing these tasks requires the analyst to make several decisions. Two of the most important decisions are: Which IM to use as the conditioning IM, and which IMs to consider in ground-motion selection? These two considerations are discussed in the subsections below.

10.6.1 Choice of Conditioning Intensity Measure

Selection of the conditioning IM, IM_1, has implications for the characteristics of the target distribution used for ground-motion selection. In this section, we make a few notes regarding this topic. There is an important distinction to be made between cases where multiple im_1 values are considered for the purpose of computing a seismic "risk-based assessment" (e.g., Equation 10.5), compared with cases where a single im_1 value is considered for a code-based seismic design "intensity-based assessment" (Bradley, 2013a; Lin et al., 2013a,b).

Risk-Based Assessments

When considering risk applications, expressions such as Equation 10.3 involve integration over all values of the considered IMs. For these applications, the choice of the conditioning IM should actually have no practical impact upon the evaluation of $\lambda(EDP > edp)$ (Bradley, 2012c; National Institute of Standards and Technology, 2012) as long as ground motions are correctly selected (see Section 10.5).

To appreciate why this is the case, consider the definition of $\lambda(EDP > edp)$ in terms of two IMs: IM_1 and IM_2. Simplifying Equation 10.1 for this two-variable case gives

$$\lambda(EDP > edp) = \int_{IM_1} \int_{IM_2} P(EDP > edp \mid im_1, im_2) MRD_{IM_1, IM_2}(im_1, im_2) \, dIM_2 \, dIM_1.$$

(10.32)

Similarly, the two-variable case of Equation 10.2 is equal to Equation 7.35, which is repeated here:

$$MRD_{IM_1, IM_2}(im_1, im_2) = \sum_{i=1}^{n_{rup}} f_{IM_1, IM_2}(im_1, im_2 \mid rup_i) \lambda(rup_i).$$

(10.33)

However, this same expression can also be represented as either

$$MRD_{IM_1, IM_2}(im_1, im_2) = \sum_{i=1}^{n_{rup}} f_{IM_2 \mid IM_1}(im_2 \mid im_1, rup_i) f_{IM_1}(im_1 \mid rup_i) \lambda(rup_i)$$

(10.34)

or

$$MRD_{IM_1,IM_2}(im_1, im_2) = \sum_{i=1}^{n_{rup}} f_{IM_1|IM_2}(im_1 \mid im_2, rup_i) f_{IM_2}(im_2 \mid rup_i) \lambda(rup_i). \quad (10.35)$$

Substituting the first of these expressions into Equation 10.32, and noting that

$$|d\lambda(IM_1 > im_1)| = \sum_{i=1}^{n_{rup}} f_{IM_1}(im_1 \mid rup_i) \lambda(rup_i) \, dim_1 \quad (10.36)$$

illustrates that Equation 10.32 is in fact equivalent to Equation 10.3 in the case of a scalar IM_2 term.[26] Furthermore, as Equations 10.34 and 10.35 are equivalent, it implies that the exact same value of $\lambda(EDP > edp)$ should be obtained regardless of whether one conditions upon IM_1 or IM_2. That said, there are still practical aspects of estimating EDP distributions from finite numbers of analyses, which make it preferable to use an efficient IM_1. Bradley (2012c) and Lin et al. (2013a) present examples that illustrate the practical independence of $\lambda(EDP > edp)$ on the choice of the conditioning IM with proper ground-motion selection.

Intensity-Based Assessments

An intensity-based assessment involves the determination of $EDP \mid IM_1$ for a single value of $IM_1 = im_1$, and is commonly adopted in seismic design guidelines and prescriptions. The primary issue in intensity-based assessments is that the choice of IM_1 directly influences other IMs' values. Consider two different conditioning IMs, IM_a and IM_b, and conditioning values such that $\lambda(IM_a > im_a) = \lambda(IM_b > im_b) = y$; i.e., the conditioning IMs have the same rate of exceedance computed via PSHA. Despite this, the two target IM distributions of other IMs of interest, $\boldsymbol{IM}_{2+} \mid IM_a = im_a$ and $\boldsymbol{IM}_{2+} \mid IM_b = im_b$, will not be the same.

To demonstrate this point, Figure 10.17 shows conditional mean spectra for three choices of the conditioning period T^*. In each case, the value of the target spectrum at the conditioning period has the same probability of exceedance, as illustrated via the UHS. The conditional mean spectra vary in shape quite considerably, and ground motions selected to match each of these conditional mean spectra would produce different response-history analysis results. In practical applications, the choice of the conditioning IM will, therefore, have to be made on a case-by-case basis (in the same way that the choice of IM of interest has to be made for any PSHA calculation).

There is an important implication underlying the relationship between IM_1 and the shape of the conditional target distribution: it is not possible to select a single conditional target that represents an equivalent hazard level for all IMs (in this instance, response spectral periods), while also maintaining a target shape representative of real ground motions (see Section 7.5). This poses practical issues when developing a target spectrum (spectra, or general IM targets) for design based on single ground-motion IM values. The UHS (and its equivalent for non-SA IMs) can be a desirable tool in some cases because it is invariant to the IMs being considered, with the trade-off that it is conservative in enveloping design values across all IMs. Analysts are thus faced with a trade-off between the convenient but conservative results obtained using the UHS or the elimination of conservatism at the expense of additional required analyses when using multiple conditional IM target distributions (Baker and Cornell, 2006b). Alternatively, analysts could depart from code-based

---

[26] That is, in Equation 10.3 the $\boldsymbol{IM}_{2+}$ term simply becomes $\boldsymbol{IM}_2$.

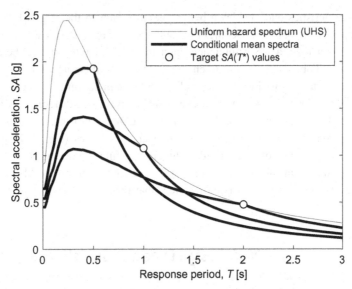

Fig. 10.17 Conditional mean spectra using $IM_1 = SA(T^*)$, obtained based on $SA(T^*)$ with a fixed rate of exceedance and with $T^* = 0.5$ s, $T^* = 1.0$ s, and $T^* = 2.0$ s. For a given response period, the three conditional mean spectra have different SA values, and therefore would be expected to produce different seismic demand distributions despite having the same rate of $SA(T*)$ exceedance (as indicated by the UHS).

prescriptions requiring intensity-based assessments and instead measure performance using the more theoretically consistent seismic demand hazard via risk-based assessments (Bradley, 2013c).

10.6.2 Intensity Measures to Consider in Ground-Motion Selection

Specifying the number of IMs (Step 1 in Figure 10.1), as well as their associated weights (Equation 10.20) is an important part of the ground-motion selection process. Equations 10.20–10.25 utilize weights to favor particular IMs during the ground-motion selection. Conceptually, this approach is straightforward, but two challenges commonly arise with specifying weights for a given application:

1. Which IMs are most important for a given EDP?
2. Which IMs are most important when multiple EDPs are of interest?

The first point is challenging to address as it is not easy to know the demands that will be imposed upon a structure and how it will behave nonlinearly until response-history analyses have been conducted. Therefore, we have to assume which IMs will be most important before conducting response-history analyses to enable this assumption to be tested (Section 10.5). Bradley (2012b) and Tarbali and Bradley (2015) evaluate how the choice of weights influences the match of the selected ground-motion ensemble to the underlying target distribution. They find that including IMs that are weakly correlated with each other leads to greater independent constraints on ground-motion characteristics and enables a better overall match to the target. However, as more constraints are imposed, the harder it becomes to obtain a precise match to the target for individual IM_i distributions. IMs that have a weak influence upon the EDP of interest can therefore be down-weighted. The principal objective remains to ensure that the IMs of greatest relevance are consistent with the target mean or target distribution (depending on the application).

This leads to the second challenge listed above, since different EDPs are sensitive to different IMs (e.g., Bradley et al., 2010). When multiple EDPs are of interest, as is generally the case, restricting attention to a subset of ground-motion characteristics (e.g., only spectral ordinates) can lead to biased estimates of some EDPs. Therefore, while the actual values of the weights that should be applied remain subjective, it is generally true that as more weakly correlated IMs are considered in the selection process, it is more likely that response distributions for any EDP of interest will be unbiased.

10.7 Design Code and Guideline Requirements

The primary focus of this chapter has been ground-motion selection in the context of seismic risk analysis. As noted in Section 10.6.1, for these types of analyses, the risk results should be somewhat independent of the choice of the conditioning IM. However, when an intensity-based assessment is undertaken, this ceases to be the case (see Section 10.6.1). Seismic design codes generally use intensity-based assessments, with loading typically characterized by a response spectrum that approximates a UHS or CMS. Many other specific details vary by individual code (e.g., Katsanos et al., 2010; NIST, 2011).

Generally speaking, codes require the analyst to ensure that the mean of the selected ensemble of ground motions is consistent with the target UHS or CMS. Usually, the mean response spectrum of the (relatively small set of) ground motions must be comparable to the design spectrum, over some period range defined relative to the fundamental period of the structure. The rationale for basing the selection upon the mean of the ensemble is discussed next in Section 10.7.1. It is not always easy to find suitable ground motions to match the design spectrum, because the codified spectra are typically parameterized to enable them to work for many conditions, at the cost of being unrealistic relative to observed ground motions. This is particularly the case when one considers combined horizontal and vertical spectral targets, as the approximate rules to define vertical spectra generally ignore the scenario dependence of vertical-to-horizontal spectral ratios (Bozorgnia and Campbell, 2004).

In some code-based applications, analysts are advised to account for the attributes of the ruptures that dominate the hazard at the site, or to account for the appropriate ground-motion duration. By accounting for these other attributes, codes are suggesting that ground-motion characteristics in addition to response spectra may influence estimated EDPs. This suggestion is consistent in concept, though differs in execution, with the hazard-consistent approaches outlined earlier in this chapter that explicitly consider response spectra and other IMs.

The intensity-based assessment context of codes also raises the issue of what conditioning IM(s) should be adopted for a particular structure. As noted in Section 10.6.1, the conditioning IM adopted will influence the intensity-based assessment results. In many situations, $SA(T_1)$ (where T_1 is the structure's elastic first-mode period) is suggested as the conditioning IM. This will be appropriate for EDPs that are strongly correlated to $SA(T_1)$, e.g., peak displacements. However, care is needed to ensure that EDPs, which are primarily affected by other IMs, e.g., peak floor accelerations or shear forces, are appropriately estimated.

Because codes define loadings in terms of response spectral ordinates, it is often not feasible to adopt the approaches addressed in this chapter when site-specific PSHA information is unavailable. However, in cases where codes do encourage the use of additional information about the causative earthquake scenarios, the analyst should use as much of this information as possible. The use of this additional information will help to reduce the dependence upon the choice of the conditioning IM (e.g., Tarbali and Bradley, 2015).

While ultimately being required to satisfy the prescriptions of the relevant code of practice, the practicing analyst should explore their approach to response-history analysis in detail, particularly asking how might one conduct these analyses if unrestricted by the code requirements. The advantage of this thought experiment is to appreciate where the code prescriptions may be conservative or otherwise, and where sensitivity analyses could help to develop a greater understanding of the likely performance of their structure of interest.

10.7.1 Rationale for the Mean IM Target

To estimate $\lambda(EDP > edp)$ in seismic risk analyses, significant emphasis in this chapter is placed on ground-motion selection that considers the full distribution of other IMs beyond the conditioning IM. In other applications, the objective may not be to derive $\lambda(EDP > edp)$. For example, in a design situation, an engineer is often interested only in the expected response given some code-specified value of ground motion, $E[EDP \mid im_1]$. From Equation A.33, it follows that

$$E[EDP \mid im] = \int_{EDP} edp\, f_{EDP|IM}(edp \mid im)\, dedp. \tag{10.37}$$

Noting that $f_{EDP|IM_1}(edp \mid im_1)$ is the derivative of Equation 10.4, it can be substituted into the above equation to obtain

$$E[EDP \mid im_1]$$
$$= \int_{IM_{2+}} \left[\int_{EDP} edp\, f_{EDP|IM_1,IM_{2+}}(edp \mid im_{2+}, im_1) f_{IM_{2+}|IM_1}(im_{2+} \mid im_1)\, dedp \right] dIM_{2+}. \tag{10.38}$$

Equation 10.38 can be reinterpreted by considering the seismic response edp, conditioned on im_1, as a function of the ground motions as represented through the other IMs im_{2+}. That is, $edp = g(im_{2+})$, where $g()$ indicates the functional dependency. The first-order second-moment method (Section A.4.1) then provides an approach to approximate the mean response via

$$E[EDP \mid im_1] \cong g(\mu_{IM_{2+}}) = g(\mu_{IM_2}, \mu_{IM_3}, \ldots). \tag{10.39}$$

Conceptually, this approximation suggests that the mean seismic response can be estimated from a ground motion with IM values equal to their conditional mean values. This approximation is used in many code-based prescriptions to motivate the selection of ground motions to match a mean-only target.

With a focus on estimating the mean response, a remaining question is the required number of ground motions to enable accurate estimation (Section 10.4.1). The number of ground motions required to reliably estimate the mean is notably less than the number required to estimate the standard deviation, which is why code-based guidance has historically allowed for very small numbers of ground motions, e.g., $n_{gm} = 3$, 7, or more recently 11 ground motions (ASCE, 2016). Figure 10.10 indicates that for typical levels of response variability, and code-required numbers of ground motions, code-based estimates of response mean have appreciable uncertainty. Because the required ground motions presented in Section 10.4.1 depend upon the variability of the response, an effective way to reduce the uncertainty in the estimated mean response is to suppress the variability in the input motions. Some seismic codes permit this suppression of spectral variability through spectral matching. However, spectral matching may lead to biased estimates of response (Seifried and Baker, 2016) if other factors in the code are not appropriately calibrated.

10.8 Documentation

Careful documentation of ground-motion selection for seismic response analysis is an essential supplement to any numerical results. Items to be documented include the following:

- The seismic hazard analysis that the ground-motion selection is based upon (whether a site-specific PSHA or a code-based prescriptive hazard)
- The methods and algorithms used to perform the ground-motion selection, as well as the constituent input parameters (e.g., the number of ground motions in the ensemble, ground-motion database(s) considered and causal parameter constraints, the considered IMs and relative weighting in assessing the match to the target IM distribution)
- Properties of the selected ground motions in terms of IMs and causal parameters, and their match to the target IM distributions
- A statement of the intended use for the ground motions (e.g., the site and structure being analyzed) and the associated reasoning for the adopted ground-motion selection details.

Because ground-motion selection can consider a wide variety of algorithmic variations and parameter values, and because the numerical outputs can be sensitive to those inputs, a careful explanation of chosen inputs is crucial. Without clear documentation, it is challenging to understand and interpret results from a study, reproduce a calculation, or critically examine the assumptions that have been made.

Exercises

10.1 Consider the ground motion from an $M6$ strike-slip earthquake observed at a site located at a distance of $R = 20$ km with $V_{S,30} = 300$ m/s.

    (a) What is the mean and standard deviation of the distribution of $SA(0.2$ s$)$ and $SA(1.0$ s$)$ predicted by the BJF97 model?

    (b) Using the Baker and Jayaram (2008) interperiod correlation model, what is the correlation between the $SA(0.2$ s$)$ and $SA(1.0$ s$)$?

    (c) The observed ground motion has a value of $SA(0.2$s$) = 0.4$ g. Compute the conditional mean and standard deviation of the distribution of $SA(1.0$ s$)$ | $SA(0.2$ s$) = 0.4$ g, and comment on its difference to the (unconditional) distribution computed in (a).

10.2 In probabilistic seismic hazard analysis, we keep track of rupture scenarios that contribute to the hazard as well as levels of ε that describe the relative strength of a ground motion with respect to the mean for its causative scenario. By making reference to the conditional mean spectrum, explain why consideration of ε is important for the purposes of ground-motion selection following a seismic hazard analysis.

10.3 Explain how a uniform hazard spectrum (UHS) is constructed, and discuss how suitable this spectrum is when used as a target spectrum for the selection of ground motions for response-history analyses.

10.4 Consider ground-motion selection for a 10-story structure, for a single hazard level conditioned on $SA(T_1)$, in which we wish to predict the maximum interstory drift ratio (MIDR) and peak floor acceleration (PFA) over all floors.

(a) How many ground motions would be required to predict the mean of the MIDR distribution to within 10% of the true value if $\sigma_{\ln MIDR|SA(T_1)} = 0.3$?

(b) How many ground motions would be required if the required precision was relaxed to 20%?

(c) The prediction of PFA has a higher variability of $\sigma_{\ln PFA|SA(T_1)} = 0.5$. How many ground motions are required to predict the mean of the PFA distribution to within 20% of its true value? Compare your answer to that of (a).

(d) How many ground motions are required to predict the distribution standard deviation to the same 10% precision as the mean $MIDR$?

10.5 A design code allows you to design for the median response obtained from the results of $n_{gm} = 11$ response-history analyses.

(a) If the response metric of interest is EDP_1 (with a variability of $\sigma_{EDP_1|IM_1} = 0.4$), what level of precision can be obtained?

(b) What is the probability that we underdesign by a factor of more than $x = 1.2$?

10.6 The consideration of appropriate bounds on causal parameters for ground-motion selection must balance the variation in IM value due to these causal parameters with the inherently aleatory variability that exists for a given set of causal parameters. In order to explore this:

(a) Use the BJF97 GMM to compute the median and lognormal standard deviation of $SA(1.0 \text{ s})$ for a scenario with $M = 7$, $R = 30$ km, and $V_{S,30} = 500$ m/s.

(b) Determine the range of PGA that is associated with epsilon values of $\varepsilon = \pm 0.5$ (which is a 38% confidence interval)

(c) Keeping all parameters except one constant, determine the individual ranges of M, R, and $V_{S,30}$ for which the median of $SA(1.0 \text{ s})$ from the BJF97 model is equal to the range computed in (b).

(d) Comment on the implications of (c) for the use of stringent causal parameter bounds in ground-motion selection.

10.7 A recorded ground motion has $PGA = 0.31$ g, $SA(1 \text{ s}) = 0.22$ g, $FAS(f = 5\text{Hz}) = 0.8$ g-s, $AI = 0.13$ m/s, and $D_{S5-95} = 14.2$ s. Ground-motion selection is undertaken conditioned on a seismic hazard of $SA(1\text{s}) = 0.63$ g.

(a) What is the required amplitude scale factor for this ground motion?

(b) What are the values of the other four IMs stated above after scaling this ground motion?

(c) The unscaled ground motion uniform duration, D_U, for a threshold acceleration of 0.1 g was 4.1 s. How would the D_U value of the scaled ground motion be computed?

10.8 The probability of exceedance of the maximum interstory drift ratio $MIDR$ of a five-story structure is represented by a function of spectral accelerations at the first, $SA(T_1 = 1.0 \text{ s})$, and second, $SA(T_2 = 0.5 \text{ s})$, fundamental periods of the structure. The function is given by

$$P[MIDR > x|SA(T_1), SA(T_2)] = 1 - \Phi\left[\frac{x - (\alpha_0 + \alpha_1 \ln SA(T_1) + \alpha_2 \ln SA(T_2))}{\sigma}\right]$$

where SA is in units of g, $[\alpha_0, \alpha_1, \alpha_2] = [0.1, 0.7, 0.35]$, and $\sigma = 0.4$. Assume that the structure is situated in California on a hard rock site ($V_{S,30} = 1000$ m/s) located 10 km from a vertical strike-slip fault that ruptures characteristically with an $M7$, surface rupturing event. The recurrence interval for the characteristic scenario is 100 years. Use the Chiou and Youngs (2014) ground-motion model and the Baker and Jayaram (2008) interperiod correlation model.[27]

(a) Demonstrate that the average rate of exceeding a maximum interstory drift ratio of 0.01 is the same in a risk-based assessment, regardless of whether the conditioning period is T_1 or T_2.

(b) Compare the results obtained to computations based upon vector-valued PSHA.

(c) Now consider an intensity-based assessment where we are interested in a return period of 500 years. Compare the probability of exceeding MIDR of x if scalar hazard is conducted for $SA(T_1)$ or $SA(T_2)$.

(d) For a return period of 500 years, determine how conservative the use of the UHS would be in comparison to the results of the previous part.

---

[27] The reader should feel free to use alternate models if preferred. We assume that relevant numerical codes are available.

11

Spatially Distributed Systems

Until this point, we have considered calculations of seismic hazard and risk at individual locations. Most structures have a small enough spatial extent that we can consider ground motions or other seismic effects to be effectively uniform across its dimensions. However, there are some circumstances where we analyze structures or systems with differing inputs at distinct locations. Stakeholders interested in such analyses include governments, infrastructure system operators, and insurers of property portfolios.

Of primary interest are regional-scale consequence analyses where structures are separated by several kilometers or more. At that spatial scale, the ground motion and consequent IM that each structure experiences will differ. The other situation of interest is the analysis of spatially extensive structures, such as bridges, where the ground-motion time series at each support are not entirely coherent (even if they do have the same IM). Tools for generalizing our prior calculations for those two situations are presented in this chapter.

Learning Objectives

By the end of this chapter, you will be able to do the following:

- Describe when it is important to consider spatially variable ground motions for seismic risk calculations.
- Compute the joint distribution of IM values at multiple sites, conditional upon a rupture.
- Describe how spatial correlations in ground shaking are quantified in empirical- and physics-based ground-motion models.
- Describe the variability in ground-motion time series using the coherency parameter.

11.1 Introduction

A ground-motion hazard curve (Chapter 6) is used to quantify the exceedance rates of IM values at any single site, but risk assessment of spatially distributed systems requires calculations of IM values at multiple locations. The typical analysis process is illustrated in Figure 11.1. Ground shaking IMs are needed at all locations of interest. Those IM values are then used to predict damage to components of the system (e.g., pipes, bridges, buildings). Then, the consequences of the component damage are modeled to quantify performance for the overall system.

When analyzing these systems, the IM that each structure experiences will differ. This variability results from the differing distances from each considered earthquake rupture to the locations of interest, differing site conditions at each location, and other effects such as rupture heterogeneity and wave propagation and scattering that are difficult to predict in a deterministic manner. In this

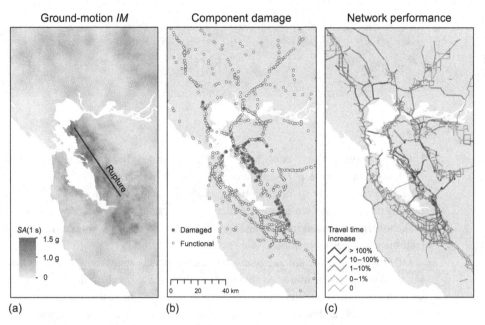

Ground-motion *IM*   Component damage   Network performance

(a)   (b)   (c)

Fig. 11.1 Typical steps in the assessment of a spatially distributed system, illustrated for a transportation network: (a) sample *IM* values at all locations of interest, (b) sample component damage, and (c) compute aggregate system consequences.

situation, a distinct *IM* value is needed for each site, necessitating a revision of the basic single-location seismic hazard calculation of the previous chapters. Additionally, the consequences of shaking to each system component may have dependencies that should be considered in a risk analysis. Finally, the need to consider many locations in the analysis leads to the ubiquitous use of Monte Carlo simulation for these problems, as opposed to the numerical calculations that were more common in previous chapters. All of these issues will be discussed in this chapter, but primary emphasis will be on characterization of *IM*s and hazard analysis.

11.1.1 Limitation of Seismic Hazard Maps

Traditional seismic hazard maps (Section 7.4) provide *IM* values over a spatial region for a specific exceedance rate. These results are inappropriate for assessing the risk to regional-scale systems because they do not represent *IM* values from a single earthquake.[1] Recall that the hazard maps were produced by computing individual hazard curves for each location and *IM* metric. The map serves only to summarize the results from these individual analyses. Figure 11.2 helps illustrate this limitation. There are three example city locations labeled on the map. Berkeley and San Leandro are located near the Hayward Fault, and hazard disaggregations indicate that the high *IM* values at these locations will likely come from a Hayward Fault rupture. Daly City can also experience high-amplitude ground motions, but disaggregation shows that the high amplitudes at that location

---

[1] When the risk metric of interest is average annual loss (Section 9.4.3), and the consequence of interest is the summation of consequences at individual locations, then hazard maps can be used to extract the needed ground-motion information. However, for all other cases, the joint distributions of *IM* and consequences at all locations are required.

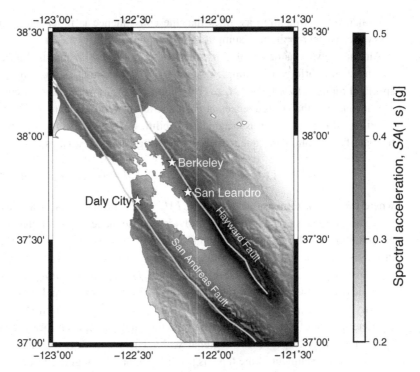

Fig. 11.2 Hazard map for $SA(1\ s)$ with a 475-year return period, assuming $V_{S,30} = 760$ m/s at all sites. Three example locations and the Hayward and San Andreas Faults are indicated. Hazard data from Petersen et al. (2014).

are likely to come from a San Andreas Fault rupture.[2] It is quite improbable, however, that the large Hayward and San Andreas Fault events will occur simultaneously, so it is also unlikely that the three locations will all see these mapped IM values simultaneously. When aiming to understand the simultaneous occurrence of ground motions at these sites, we need additional analyses, as described below.

11.1.2 Single-Location versus Regional-Scale Risk Analysis

Consider the consequences of shaking for an asset at a single location i, C_i, given an earthquake rupture, rup_k. We can compute the probability of exceeding some consequence threshold, given rup_k, by multiplying the vulnerability function for the site by the probability density function (PDF) for $IM \mid rup_k$, and integrating over all possible values of IM:

$$P(C_i > c \mid rup_k) = \int_{IM_i} P(C_i > c \mid IM = im_i)\, f_{IM_i|Rup}(im_i \mid rup_k)\, dim_i \qquad (11.1)$$

where $P(C_i > c \mid IM = im_i)$ comes from a vulnerability model (Section 9.2.2) and $f_{IM_i|Rup}(im_i \mid rup_k)$ comes from a ground-motion model (GMM, Chapters 4 and 5). The i index on IM is used to denote the location, and is effectively used in lieu of explicit conditioning upon $site_i$ in the GMM.

---

[2] The actual disaggregation results are not provided here, but are available from the US Geological Survey. The results should be intuitive from looking at the map and noting that all three locations are very near to the high slip-rate faults mentioned above.

Note that this formulation assumes that the consequence is independent of the rupture, if IM is known, consistent with the assumptions of Chapter 9.

Alternatively, consider total consequences associated with multiple assets at multiple locations (the most straightforward example being $C = \sum_i C_i$ where C_i is the consequence at site i). Given a specified rupture (rup_k), the IM_i over the multiple locations are dependent. The dependence of the IM_i means that the consequences at each site are also dependent. All possible values of IM at all locations, along with their associated probabilities of simultaneous occurrence, must be considered to quantify portfolio consequences. The following equation satisfies this requirement:

$$P(C > c \mid rup_k) = \int_{IM} P(C > c \mid IM = im)\, f_{IM|Rup}(im \mid rup_k)\, dim \qquad (11.2)$$

where im is an n-dimensional vector of IM_i values at all n sites of interest (and boldface denotes a vector or matrix). This equation provides consequences conditional on a single rupture, but by summing results over multiple ruptures, the approach can be extended to the general cases discussed in Chapter 9. In addition to the marginal distributions of IM_i at each site (the focus of previous chapters), the joint PDF calculation of $f_{IM|Rup}(im \mid rup_k)$ now requires specification of dependencies between IMs at multiple sites. The consideration and impact of IM dependency is a primary focus of this chapter. Additionally, the vulnerability function $P(C > c \mid IM = im)$ requires characterization of the joint distribution of consequences at each location and how they impact the aggregate consequences at the regional scale. These consequence models are system-specific, so only a general treatment of the topic is provided. Finally, integration over the vector of IM values is nontrivial, so numerical implementation issues are discussed in Section 11.4.

Spatial dependence among ground motions is known to impact predicted consequences to spatially distributed systems (e.g., Wesson and Perkins, 2001; Adachi and Ellingwood, 2007; Lee and Kiremidjian, 2007; Park et al., 2007; Shiraki et al., 2007; Sokolov and Wenzel, 2011). As a numerical example, Figure 11.3 shows risk curves for the travel time delay in a regional transportation

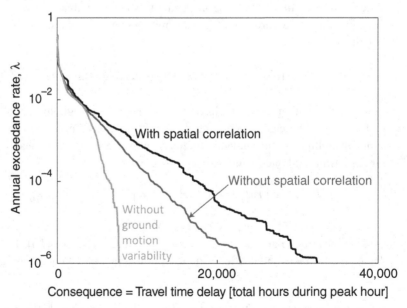

Fig. 11.3 Effect of ground-motion spatial correlation on estimated risk of travel time disruption (C) over a transportation network (Data from Jayaram and Baker, 2010b).

network (i.e., a measure of consequences), under different models for spatial dependence of ground motions. The "with spatial correlation" result considers spatial dependence in $f_{IM|Rup}(im \mid rup_k)$ in Equation 11.2, and is considered here as the most accurate estimate. The "without spatial correlation" analysis case treats the marginal distributions of $IM_i \mid rup_k$ in Equation 11.2 as mutually independent, and underestimates the rate of severe disruptions to the network. This underestimation results because the "with spatial correlations" model allows for the possibility that IMs in a cluster of locations could all be larger than the median for a given rupture (an outcome that can produce severe consequences). Conversely, the 'without spatial correlation' model incorrectly implies that larger-than-median motions would occur at individual locations independently. Finally, the "without ground-motion variability" result treats $f_{IM|Rup}(im \mid rup_k)$ in Equation 11.2 as deterministic (i.e., with no aleatory variability). This case dramatically underestimates the rate of severe disruption because larger-than-median IM values are never considered.

11.2 Parameterization Using Empirical Ground-Motion Models

11.2.1 Model Formulation

Consideration of spatially distributed systems requires an extension of the calculations from previous chapters to account for spatial correlation in ground motions and efficiently characterize hazard when the IM of interest is a vector of IMs at hundreds or thousands of locations. For implementation with empirical GMMs, spatially correlated residuals are introduced to parameterize spatially correlated IM values.

Consider the situation shown in Figure 11.4, where we are interested in IM values[3] at two sites near a given rupture, indexed as $j = 1$ and $j = 2$. The GMM formulation of Equation 4.31, evaluated for two sites, is

$$\ln im_{i,1} = \mu_{\ln IM}(rup_i, site_1) + \delta B_i + \delta W_{i,1}$$
$$\ln im_{i,2} = \mu_{\ln IM}(rup_i, site_2) + \delta B_i + \delta W_{i,2} \tag{11.3}$$

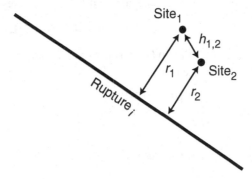

Fig. 11.4 Map view of the geometry of a vertical fault rupture and two sites of interest separated by distance $h_{1,2}$.

---

[3] For simplicity, we assume that the same IM metric is of interest at both sites. The approach can be extended to consider multiple IM metrics, as will be briefly discussed below.

where all terms are as defined in Section 4.6.2. The mean values of $\ln IM_{i,1}$ and $\ln IM_{i,2}$ are deterministic functions given by GMMs, as before. The between-event residual, δB_i, is common for the two sites because it depends upon the rupture and not the specific site. The uncharacterized portion of this model is the dependence of $\delta W_{i,1}$ and $\delta W_{i,2}$—that is, the model formulation requires a characterization of how the within-event residuals vary at sites near each other.

As before (Figure 4.15), we make the standard assumption that $\ln IM_{i,1}$ and $\ln IM_{i,2}$ have marginally normal distributions. We also assume that $\ln \boldsymbol{IM}$ at multiple sites have a multivariate normal distribution (Jayaram and Baker, 2008). Under that assumption, the joint distribution of $\ln IM_{i,1}$ and $\ln IM_{i,2}$ is bivariate normal, with marginal mean values and standard deviations given by the GMM, and with covariance

$$\sigma_{\ln IM_{i,1}, \ln IM_{i,2}} = \rho_{\ln IM_{i,1}, \ln IM_{i,2}} \sigma_{\ln IM_{i,1}} \sigma_{\ln IM_{i,2}} \tag{11.4}$$

$$= \rho_{\delta W_{i,1}, \delta W_{i,2}} \sigma_{\delta W_{i,1}} \sigma_{\delta W_{i,2}} + \rho_{\delta B_i, \delta B_i} \sigma_{\delta B_i} \sigma_{\delta B_i} \tag{11.5}$$

$$= \rho_{\delta W_{i,1}, \delta W_{i,2}} \sigma_{\delta W_{i,1}} \sigma_{\delta W_{i,2}} + \sigma_{\delta B_i}^2. \tag{11.6}$$

Equation 11.6 comes from the fact that $\rho_{\delta B_i, \delta B_i} = 1$. The above equations can also be further generalized to cases where the between-event residuals vary from site to site.

In the case where the between-event and within-event residuals have constant standard deviations at all sites ($\sigma_{\delta W_{i,j}} = \phi$ and $\sigma_{\delta B_i} = \tau$), Equation 11.6 simplifies to

$$\sigma_{\ln IM_{i,1}, \ln IM_{i,2}} = \rho_{\delta W_1, \delta W_2} \phi^2 + \tau^2 \tag{11.7}$$

and the correlation coefficient between $\ln IM$ values is:

$$\rho_{\ln IM_{i,1}, \ln IM_{i,2}} = \frac{\rho_{\delta W_1, \delta W_2} \phi^2 + \tau^2}{\sigma^2} \tag{11.8}$$

where τ, ϕ, and $\sigma = \sqrt{\tau^2 + \phi^2}$ are the between-event, within-event, and total standard deviations for $\ln IM$, respectively (specified by the GMM; see Equation 4.29), and $\rho_{\delta W_1, \delta W_2}$ is the correlation coefficient for the within-event residuals. We note that Equation 11.8 is a correlation coefficient for $\ln IM_1$ and $\ln IM_2$, conditional on rup_i and $site_j$, and so formally should be denoted $\rho_{\ln IM_{i,1}, \ln IM_{i,2}|rup,site}$, but the conditioning notation is suppressed for brevity.

Generalizing the above formulae to the case of multiple sites, we can write that the distribution of $\ln \boldsymbol{IM}$, conditional upon a rupture and on the sites of interest, is multivariate normal:

$$\ln \boldsymbol{IM} \sim \mathcal{N}(\boldsymbol{M}, \boldsymbol{\Sigma}) \tag{11.9}$$

where $\sim \mathcal{N}(\,)$ denotes that $\ln \boldsymbol{IM}$ has multivariate normal distribution, parameterized by the mean vector $\boldsymbol{M}$ and covariance matrix $\boldsymbol{\Sigma}$ as follows:

$$\boldsymbol{M} = \begin{bmatrix} \mu_{\ln IM}(rup_i, site_1) \\ \mu_{\ln IM}(rup_i, site_2) \\ \vdots \\ \mu_{\ln IM}(rup_i, site_n) \end{bmatrix} \tag{11.10}$$

$$\boldsymbol{\Sigma} = \begin{bmatrix} \sigma_{\ln IM_{i,1}}^2 & \sigma_{\ln IM_{i,1}, \ln IM_{i,2}} & \cdots & \sigma_{\ln IM_{i,1}, \ln IM_{i,n}} \\ \sigma_{\ln IM_{i,2}, \ln IM_{i,1}} & \sigma_{\ln IM_{i,2}}^2 & & \\ \vdots & & \ddots & \\ \sigma_{\ln IM_{i,n}, \ln IM_{i,1}} & & & \sigma_{\ln IM_{i,n}}^2 \end{bmatrix} \tag{11.11}$$

and where $\sigma_{\ln IM_{i,j}, \ln IM_{i,k}}$ is the covariance between $\ln IM$ at sites j and k, from Equation 11.6. When the conditions for Equation 11.7 hold, the covariance matrix can be written

$$\mathbf{\Sigma} = \tau^2 \mathbf{1} + \phi^2 \mathbf{R} \tag{11.12}$$

where $\mathbf{1}$ is an $n \times n$ matrix of ones and $\mathbf{R}$ is the within-event correlation matrix:

$$\mathbf{R} = \begin{bmatrix} 1 & \rho_{\delta W_{i,1}, \delta W_{i,2}} & \cdots & \rho_{\delta W_{i,1}, \delta W_{i,n}} \\ \rho_{\delta W_{i,2}, \delta W_{i,1}} & 1 & & \\ \vdots & & \ddots & \\ \rho_{\delta W_{i,n}, \delta W_{i,1}} & & & 1 \end{bmatrix}. \tag{11.13}$$

Note that in any of the above series of equations, the within-event standard deviation ϕ can vary from site to site (as is the case with some GMMs). In this case, the ϕ terms will need a site index in the above equations, and Equation 11.12 will become

$$\mathbf{\Sigma} = \tau^2 \mathbf{1} + diag\,(\boldsymbol{\phi})\,\mathbf{R}\,diag\,(\boldsymbol{\phi}) \tag{11.14}$$

where $\boldsymbol{\phi}$ is an $n \times 1$ vector of standard deviation values for the n sites, and $diag\,(\boldsymbol{\phi})$ is a diagonal $n \times n$ matrix with $\boldsymbol{\phi}$ as the diagonal entries.

11.2.2 Causes of Correlation

Several phenomena associated with earthquake rupture and wave propagation can produce spatial correlation in the within-event residuals. First, the source description in GMMs is simple and does not account for commonalities in adjacent locations over the rupture. Ruptures have spatial variation in slip (Section 5.3), and sites near high-slip asperities tend to have larger ground motions than sites near low-slip patches of the rupture. Adjacent sites will tend to be located near rupture segments with similar slip, inducing correlation in their resulting IM amplitudes. Second, as waves propagate from the rupture through the crust, geologic structures such as sedimentary basins will tend to focus and trap waves (Section 5.4). Locations near such geologic structures will thus tend to have common amplification or deamplification of waves. Finally, surficial geology impacts IM values in complex ways that are only approximated by simple predictive parameters such as $V_{S,30}$. Because these site effects are only partially captured by empirical GMMs, and because the surficial geology causing these effects tends to be spatially continuous, the unexplained site effects will also introduce spatial correlation in observed ground-motion residuals.

Empirical evidence of these effects is present in observations from well-recorded historical earthquakes. To illustrate, Figure 11.5 shows observations and median predictions of $SA(1\text{ s})$ values from the 1999 Chi-Chi, Taiwan, earthquake. Figure 11.6 shows within-event residuals from these data, computed using Equation 11.3 and the Chiou and Youngs (2014) GMM. The patches of similarly shaded marks indicate that there are groups of adjacent stations with similar values of residuals. For example, Figure 11.5 shows that at the western central portion of Taiwan, the observed $SA(1\text{ s})$ values were smaller in amplitude than the median predictions, while observed $SA(1\text{ s})$ values at the northern end of the island were larger than median predictions. These deviations of amplitudes and predictions result in clusters of negative residuals in the west of the island and positive residuals in the north (Figure 11.6). With this qualitative understanding of the causes of correlations, we next proceed to quantify this effect.

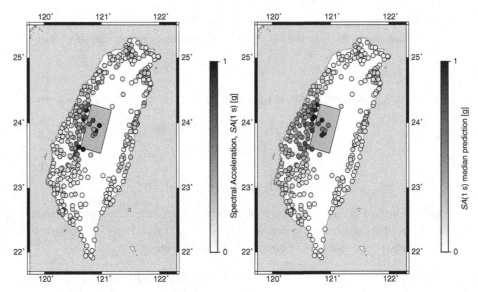

Fig. 11.5 (a) Observed $SA(1\,s)$ values from the 1999 Chi-Chi, Taiwan, earthquake. (b) Predicted median $SA(1\,s)$ values, computed using the Chiou and Youngs (2014) GMM. The approximate surface projection of the rupture is shown with a gray rectangle, and the shaded circles indicate the $SA(1\,s)$ amplitude observed or predicted at each location.

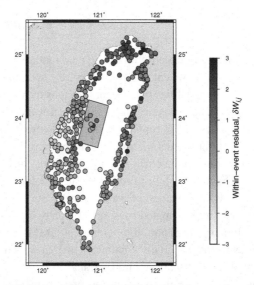

Fig. 11.6 Within-event residuals ($\delta W_{i,j}$) for $SA(1\,s)$ values observed in the 1999 Chi-Chi, Taiwan, earthquake, as computed using the Chiou and Youngs (2014) GMM.

11.2.3 Estimating Correlations

The spatial correlations of $\delta W_{i,j}$ can be estimated from observations such as those in Figure 11.6. Ideally, we would use a traditional empirical correlation calculation to estimate these correlations, using pairs of observed residuals from multiple earthquakes recorded at the sites of interest:

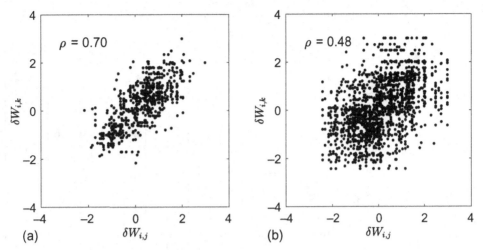

Fig. 11.7   Paired within-event residuals ($\delta W_{i,j}$) for $SA(1\text{ s})$ values observed in the 1999 Chi-Chi, Taiwan, earthquake at distances of (a) 0–3 km, and (b) 30–33 km. Sample correlation coefficients are reported on each figure.

$$\hat{\rho}_{\delta W_{i,j},\delta W_{i,k}} = \frac{\sum_{i=1}^{n}(\delta W_{i,j} - \delta\bar{W}_{i,j})(\delta W_{i,k} - \delta\bar{W}_{i,k})}{\sqrt{\sum_{i=1}^{n}(\delta W_{i,j} - \delta\bar{W}_{i,j})^2}\sqrt{\sum_{i=1}^{n}(\delta W_{i,k} - \delta\bar{W}_{i,k})^2}} \tag{11.15}$$

$$= \frac{\sum_{i=1}^{n}\delta W_{i,j}\delta W_{i,k}}{\phi^2} \tag{11.16}$$

where $\delta\bar{W}_{i,j}$ denotes a sample mean, and the simplifications of the second line of the equation rely on the property that $E[\delta W] = 0$ and $\phi_i = \phi_j = \phi$ (see Equation 11.7). This calculation requires n observations of δW available at the given pair of sites. It is difficult to obtain a sufficient number (typically $n \geq 10$) of observations because pairs of stations with recordings from multiple strong earthquakes are rare.

To circumvent this limitation, we typically assume that the correlations depend only upon the separation distance between the sites, and not on the specific locations or orientations of the sites themselves (that is, the correlations are *stationary* and *isotropic*). These assumptions are related to the assumption of ergodicity in empirical ground-motion modeling (Section 8.1). With these assumptions, all pairs of residual values at a given separation distance can be pooled and used to estimate correlations at that distance. To illustrate this concept, Figure 11.7 shows pooled pairs of residuals from the Chi-Chi earthquake, at separation distances of 0–3 km and 30–33 km. The decreased correlation in residuals for the larger separation distance is apparent in these data. The stationarity assumption creates a significant number of data pairs for consideration at a given distance (650 and 1632 pairs in Figure 11.7, for the smaller and larger distances, respectively). These pairs result in part from the reuse of data: the horizontal and vertical "stripes" of data in this figure result from an individual recording (and associated $\delta W_{i,j}$) being paired with multiple other records, each producing a point on the figure. Figure 11.8 shows the number of paired residuals available from this earthquake, using distance bins that are 3 km wide. Having hundreds or thousands of paired observations available at each distance is why it is so advantageous to make the assumptions of stationarity and isotropy when estimating spatial correlations.

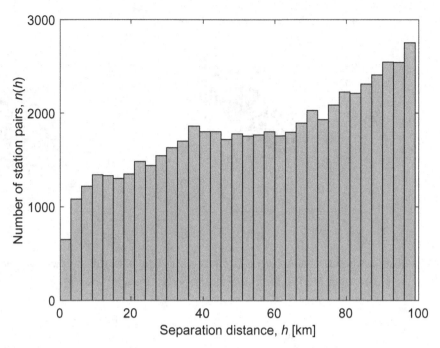

Fig. 11.8 Number of paired observations of within-event residuals ($\delta W_{i,j}$) for SA(1 s) values observed in the 1999 Chi-Chi, Taiwan, earthquake, at each considered separation distance.

To turn data like that of Figure 11.6 into a model for correlations, a semivariogram is often used. This popular geostatistics calculation measures dissimilarity in two observations (Journel and Huijbregts, 1978; Goovaerts, 1997). The semivariogram can be related to a correlation coefficient, as will be discussed later. However, this form is more popular than the correlation coefficient in the spatial statistics literature, so we consider it for now. Under the stationarity and isotropy assumptions introduced above, the semivariogram is defined as

$$\gamma(h) = \frac{1}{2} E\left[(Z_j - Z_k)^2\right] \tag{11.17}$$

where $E[\ ]$ denotes an expected value, Z_j and Z_k are paired variables of interest at a separation distance of h, and $\gamma(h)$ is the semivariogram value. Substituting $Z_j = \delta W_{i,j}$, our parameter of interest, this equation becomes

$$\gamma(h) = \frac{1}{2} E\left[(\delta W_{i,j} - \delta W_{i,k})^2\right]. \tag{11.18}$$

The semivariogram can be empirically estimated from observed data using the following equation:

$$\hat{\gamma}(h) = \frac{1}{2n(h)} \sum_{d_{j,k}=h} (\delta W_{i,j} - \delta W_{i,k})^2 \tag{11.19}$$

where $d_{j,k}$ is the distance between sites j and k, the summation is over all j and k with separation distance h (within a user-specified tolerance, i.e., $|d_{j,k}-h| \leq \Delta h$), and $n(h)$ is the number of observed station pairs with separation distance h (i.e., the counts shown in Figure 11.8). Figure 11.9 shows the empirical semivariogram estimated from the Chi-Chi SA(1 s) residuals, using Equation 11.19 with distance increments of 5 km and a tolerance of $\Delta h = 1.5$ km.

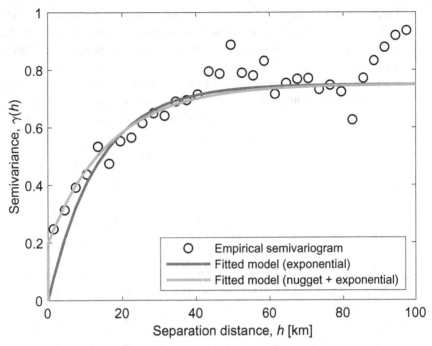

Fig. 11.9 Semivariogram for within-event residuals (δW_{ij}) for SA(1 s) values observed in the 1999 Chi-Chi, Taiwan, earthquake. The "Fitted model (exponential)" semivariogram uses the functional form of Equation 11.20, and the "Fitted model (nugget + exponential)" uses Equation 11.24.

11.2.4 Fitted Semivariograms

The estimated values from Equation 11.19 can then be used to create a model that predicts semivariogram values as a continuous function of separation distance. A common functional form for ground-motion residuals is the so-called exponential model:

$$\tilde{\gamma}(h) = s \left[1 - \exp\left(\frac{-3h}{r} \right) \right] \tag{11.20}$$

where ~ refers to the value from a model. This model has two parameters to be fit from data: the sill (s) and the range (r). The sill is equal to the variance of $Z = \delta W$. For the exponential semivariogram, s represents the value to which $\tilde{\gamma}(h)$ asymptotically converges as h tends to infinity. The range is the separation distance h at which $\tilde{\gamma}(h)$ is equal to 95% of the exponential semivariogram sill (i.e., the distance at which 95% of the correlation is lost). Figure 11.9 shows an example fitted semivariogram superimposed on the Chi-Chi data, using the functional form of Equation 11.20 and estimated parameter values $s = 0.75$ and $r = 40$ km.

Because the empirical semivariogram calculation of Equation 11.19 involves reuse of each observed station and is dependent upon the geometries of the specific stations recording a particular earthquake, there is no general solution for optimally estimating model parameters from an empirical semivariogram. Further, the smaller-distance semivariogram values are of more practical importance than the larger-distance values. For the above two reasons, popular automated techniques such as least-squares regression are not recommended for estimating sill and range parameters (Journel and

Huijbregts, 1978). Instead, best practice in estimating parameters is to fit the model by eye (focusing on the distance values of most importance) or to use a heuristic fitting technique.

There are several functional forms besides that of Equation 11.20 that can be used to develop a fitted semivariogram. The critical constraint is that the model must produce positive definite covariance matrices for residuals at any possible configuration of sites, for reasons discussed below. There are a few "admissible" functional forms that are proven to ensure this property (Journel and Huijbregts, 1978), such as the exponential function of Equation 11.20, the Gaussian model

$$\tilde{\gamma}(h) = s \left[1 - \exp\left(\frac{-3h^2}{r^2} \right) \right],$$ (11.21)

the spherical model

$$\tilde{\gamma}(h) = \begin{cases} s \left[\frac{3}{2} \left(\frac{h}{r} \right) - \frac{1}{2} \left(\frac{h}{r} \right)^2 \right] & \text{if } h \leq r \\ s & \text{otherwise,} \end{cases}$$ (11.22)

which are special cases of the Matérn model (Stein, 2012), and the nugget effect model, which indicates total loss of similarity at separation distances greater than zero:

$$\tilde{\gamma}(h) = s \left[I(h > 0) \right]$$ (11.23)

where $I(h > 0)$ is an indicator variable, equal to 1 when $h > 0$ and equal to 0 otherwise.

Any linear combination of the above models is also admissible. One popular choice is to combine a nugget model with an exponential model. This combination works well because the data often have a rapid drop of the semivariogram away from zero separation distance followed by an exponential decay at greater distances (e.g., Figure 11.9). To illustrate, Figure 11.9 also shows a fitted semivariogram with the following functional form, which is a combination of the above nugget and exponential models:

$$\tilde{\gamma}(h) = 0.2 \left[I(h > 0) \right] + 0.55 \left[1 - \exp\left(\frac{-3h}{50} \right) \right].$$ (11.24)

The Gaussian and spherical functional forms are rarely, if ever, used for IM data.

After choosing the appropriate semivariogram equation (or a combination thereof) from above, and determining the corresponding fitted parameter values, we can then return to correlation calculations. Recall from Equations 11.8–11.11 specification of these joint distributions requires a model for correlations ($\rho_{\ln IM_{i,1}, \ln IM_{i,2}}$) in addition to the mean and variance terms for each individual site that are provided by the empirical GMMs. The semivariogram is directly related to a covariance function, by the following relationship:

$$C(h) = C(0) - \gamma(h)$$ (11.25)

where $C(h)$ is the covariance of $\delta W_{i,j}$ values at separation distance h. Similarly, the correlation coefficient of $\delta W_{i,j}$ values at separation distance h is

$$\rho(h) = C(h)/C(0).$$ (11.26)

While one could directly study covariance or correlation instead of semivariance, the semivariogram is often preferred in geostatistical practice as it does not require a prior estimation of the mean or standard deviation of the considered parameter.

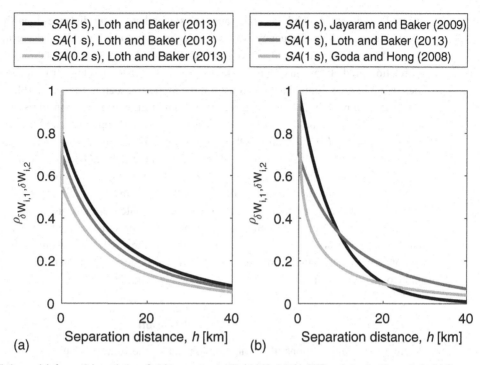

Fig. 11.10   Predictive models for spatial correlation of within-event variability. (a) Models for $SA(T)$ at three vibration periods. (b) Three models for spatial correlations of $SA(1\ s)$ spectral acceleration. The Goda and Hong (2008) model depicts their version based on California ground motion data.

11.2.5 Available Models

The preceding section described major steps in developing a spatial correlation model, but in many applications, the analyst can utilize an existing model. A number of models for spatial correlations have been developed (e.g., Boore et al., 2003; Wang and Takada, 2005; Goda and Hong, 2008; Jayaram and Baker, 2009; Goda and Atkinson, 2010; Esposito and Iervolino, 2011; Goda, 2011; Foulser-Piggott and Stafford, 2012; Loth and Baker, 2013; Sokolov and Wenzel, 2013a; Schiappapietra and Douglas, 2020). At present, these correlation models are developed separately from published GMMs. Some studies have attempted to simultaneously fit a GMM and a spatial correlation model because they both rely on the same data for calibration, but those studies generally use simplified GMM functions (e.g., Jayaram and Baker, 2010a; Ming et al., 2019). If a correlation model were included along with a GMM, Equations 11.8–11.11 can be completely evaluated from a single reference document. But typically, the mean and variance terms for these equations come from a GMM, and the correlation terms come from a separate spatial correlation model.

Figure 11.10 shows several example correlation results, and illustrates typically observed trends. First, correlation coefficients in $\delta W_{i,j}$ are relatively large[4] (> 0.6) at separation distances of a few kilometers, and are small (< 0.1) at distances greater than 40 km. Second, the degree of predicted correlation depends upon the IM parameter of interest. Figure 11.10a shows that long-period

---

[4] Strictly, the correlation coefficients are equal to one at a separation distance of zero (i.e., the correlation of an IM with itself is perfect) and tend to zero at infinite separation distance.

spectral accelerations exhibit correlations over somewhat greater distances than short-period spectral accelerations, as high-frequency waves lose coherence more quickly and are more influenced by fine-scale rupture and crustal heterogeneities that tend to reduce correlation (Section 5.4). Third, correlation predictions vary from model to model. This is particularly true at short distances, where there are relatively limited data to constrain the model, as seen in Figure 11.10b.

Correlation models can also be generalized so that Equation 11.3 considers differing IM parameters at spatially separated sites. In such a case, both spatial separation and the differing IM metrics will influence the IM correlation. Observations of two differing IMs with a separation distance of h will have a lower correlation than two observations of the same IM with the same separation distance of h. The overall correlation for differing IMs can be approximated by multiplying a correlation term for the spatial correlation (as above) by a correlation term for the differing IMs (Goda and Hong, 2008). Alternatively, the total correlation can be estimated directly using observations of the IM types of interest from ground-motion recordings with appropriate separation distances. With direct estimation, additional care is needed to ensure that the final correlation model produces positive definite covariance matrices, but several approaches are available to address this requirement (Goda and Hong, 2008; Loth and Baker, 2013; Markhvida et al., 2018).

Discussion

There are two implications of the above models that merit discussion: they assume stationarity and isotropy, and are dependent upon the reference GMM.

First, these models all make stationarity and isotropy assumptions. These assumptions are essential to make the estimation exercise tractable, and the limited amount of available data makes alternative assumptions challenging to utilize. Spatial correlation estimation (more so than the estimation of IM values at a single location) requires relatively dense ground-motion observations from individual earthquakes. At present, there are only dozens of historical earthquakes with a sufficient number of recordings for this type of calculation (Goda, 2011; Loth and Baker, 2013; Heresi and Miranda, 2019). There are even fewer instances of sites where there are multiple strong earthquakes that have been recorded by the same dense set of instruments. It is thus difficult to quantify the impact of various earthquake rupture and site conditions on observed correlations. Some researchers report variations regionally or from earthquake to earthquake: the two are difficult to disentangle since we do not always have multiple large earthquakes in a given region (Goda and Hong, 2008; Foulser-Piggott and Stafford, 2012; Schiappapietra and Douglas, 2020). Others have reported impacts of variability in near-surface geology (Jayaram and Baker, 2009; Sokolov et al., 2012) or earthquake magnitude (Sokolov and Wenzel, 2013b; Foulser-Piggott and Goda, 2015). Uncertainty in estimating correlations, and apparent variability in correlations from earthquake to earthquake, are issues undergoing active study (Heresi and Miranda, 2019; Baker and Chen, 2020; Schiappapietra and Douglas, 2020). Given the limited available data, the typical practice is to use generic models of the type shown in Figure 11.10 and apply them globally. The continued growth of ground-motion databases and increased use of dense ground-motion arrays is slowly changing this situation. In future years we may have the data to develop more situation-specific correlation models.

Second, spatial correlation models are dependent upon a reference GMM. At a numerical level, this is because the spatial models are computing residuals of observations with respect to the reference model. The choice of reference model does not always have a practical impact, because in many cases the GMMs do not have differences in means or standard deviations that appreciably change

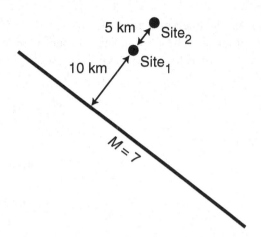

Fig. 11.11   Map view of fault rupture and two sites for the example calculation.

the computed residuals, and some potential effect of the GMM is removed when calculating the δB event term. When a non-ergodic GMM is used (Chapter 8), however, this issue may be more critical. A non-ergodic model will presumably account for unique site conditions or path effects in its mean IM predictions, and those effects are the cause of some spatial correlations. If these effects are removed from the residuals, then the new residuals may have a different and presumably weaker spatial correlation structure (Stafford et al., 2019). Furthermore, once GMM parameters are introduced that vary with location, a range of model generalizations is possible, such as including between-event residuals that have a spatial pattern due to geometric variations in faulting properties (Kuehn and Abrahamson, 2020). Analyses including both regional scale analysis (requiring spatial correlations) and a non-ergodic GMM (requiring careful study of one individual site) are presently rare. However, some applications do exist, and this combination highlights how correlations connect with a reference empirical GMM.

11.2.6 Example: Simple Rupture Model

To illustrate the above calculations and discuss numerical results, we compute the joint distribution of $SA(1 \text{ s})$ amplitudes for the situation shown in Figure 11.11. The rupture has a strike-slip mechanism, the sites both have $V_{S,30} = 500$ m/s, and we use the Boore et al. (1997) GMM presented in Chapter 4. We adopt the following model for correlations (Jayaram and Baker, 2009):

$$\rho_{\delta W_{i,1}, \delta W_{i,2}} = \exp\left(\frac{-3\,h_{1,2}}{25.7}\right) \tag{11.27}$$

where $h_{1,2}$ is the distance between the sites 1 and 2 in units of km, and 25.7 is an empirically calibrated coefficient for $SA(1 \text{ s})$ (with units of km), which controls the rate of decay in correlation as a function of distance.

Evaluating the GMM for the given *rup* and *site* conditions provides the following mean terms (Equation 11.10):

$$\boldsymbol{M} = \begin{bmatrix} -1.277 \\ -1.583 \end{bmatrix} \tag{11.28}$$

and the total standard deviation for both sites is $\sigma = 0.5201$ (and, equivalently, $\sigma^2 = 0.2705$). Additionally, this total standard deviation is decomposed into between- and within-event standard deviations, $\tau = 0.214$ and $\phi = 0.474$. Given the separation distance between the sites, the within-event correlation is

$$\rho_{\delta w_{i,1}, \delta w_{i,2}} = \exp\left(\frac{-3}{25.7}5\right) = 0.558 \tag{11.29}$$

and from Equation 11.8

$$\rho_{\ln IM_{i,1}, \ln IM_{i,2}} = \frac{\tau^2 + \rho_{\delta w_{i,1}, \delta w_{i,2}} \phi^2}{\sigma^2} \tag{11.30}$$

$$= \frac{0.214^2 + 0.558(0.474^2)}{0.5201^2} \tag{11.31}$$

$$= 0.6327. \tag{11.32}$$

This correlation can be used to compute a covariance:

$$\sigma_{\ln IM_{i,1}, \ln IM_{i,2}} = 0.6327(0.5201)^2 = 0.1711. \tag{11.33}$$

The covariance and site standard deviations can then be combined to develop the covariance matrix for $\ln IM$ at the two sites (Equation 11.11):

$$\Sigma = \begin{bmatrix} 0.2705 & 0.1711 \\ 0.1711 & 0.2705 \end{bmatrix}. \tag{11.34}$$

Note that this covariance matrix needs to be positive definite in order for the joint distribution to be valid. This requirement motivates the use of admissible spatial correlation functions as discussed in Section 11.2.4. These results fully specify the multivariate normal distribution for $\ln SA(1\ s)$, given the rupture.

Contours of the distribution in Figure 11.12 show a few relevant features. The Site 1 amplitudes tend to be larger (with a median of $e^{-1.277} = 0.28$ g, versus $e^{-1.583} = 0.21$ g for Site 2), but there is still a significant chance that Site 2 has a larger amplitude due to the variability in predictions (indicated by PDF contour values below the diagonal 1:1 line on the figure). The positive correlation of the $\ln IM$ values leads the contours to be elliptical, with a positive slope of the major axis, indicating a greater likelihood that both IM values tend to be large or both tend to be small, relative to their mean values.

11.2.7 Example: San Francisco Bay Area

To further illustrate the above calculations in a more realistic circumstance, this section presents numerical results for $SA(1\ s)$ values in the San Francisco Bay Area and three example sites shown in Figure 11.13. Site A is Yerba Buena Island, Site B is Berkeley, and Site C is Richmond. The letter names are used below for brevity and because the focus here is on procedures rather than specific locations. To facilitate the statistical sampling of realistic rupture scenarios, the OpenSHA event set simulator (Field et al., 2003) is used to sample ruptures from the UCERF2 rupture model (Field et al., 2009), with the Chiou and Youngs (2008) GMM and the Jayaram and Baker (2009) correlation model.

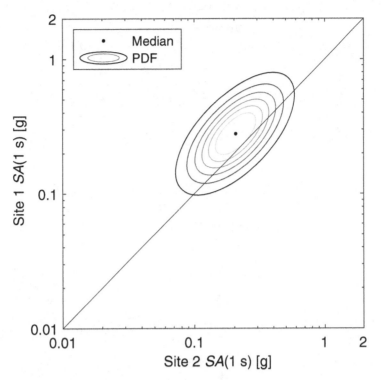

Fig. 11.12 Contours of probability density for SA(1 s), for the example of Figure 11.11.

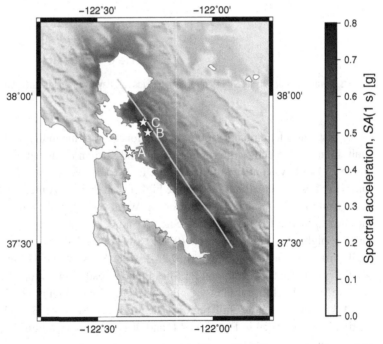

Fig. 11.13 Median predicted IM values, given a magnitude 7.05 rupture of the Hayward Fault. The rupture's surface projection and three example locations of interest are annotated.

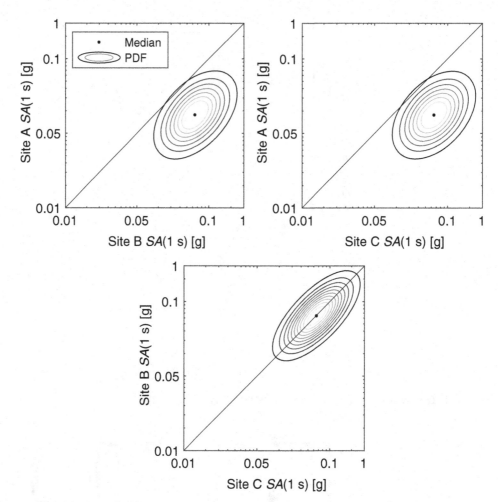

Fig. 11.14   Contours of probability density for SA(1 s), given a magnitude 7.05 Hayward Fault rupture, at the three example sites.

We first consider ground motions conditional on a magnitude 7.05 rupture of the Hayward Fault. Figure 11.13 shows the rupture extent and median predicted ground-motion amplitudes, and Figure 11.14 shows contours of the joint PDF for each pair of sites. We see that the median SA(1 s) values at Sites B and C (both 0.65 g) are higher than the Site A median, so the IM values at Sites B and C tend to be larger than those for Site A. We also see that the correlations between pairs of sites vary due to their differing distances. Sites B and C have the smallest separation distance (4.5 km) and the highest $\ln IM$ correlation (0.59). Sites A and C have the largest separation distance (12.4 km) and smallest correlation (0.23), while Sites A and B are intermediate ($h = 10.6$ km, $\rho = 0.29$). This higher correlation for Sites B and C leads their probability contours in Figure 11.2 to be more strongly elliptical than the other pairs' contours. Figure 11.15 then shows one realization of SA(1 s) values for the full region. The values at Sites A, B, and C were sampled from the distributions illustrated in Figure 11.14, and distributions for all other locations were obtained in a similar manner.

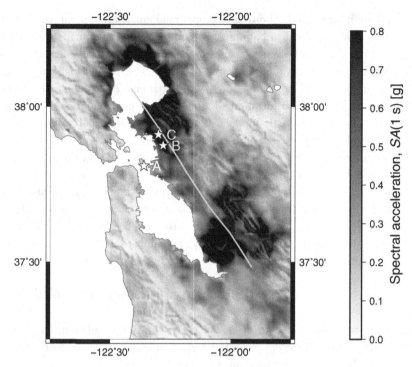

Fig. 11.15 Statistically sampled SA(1 s) values, using an empirical GMM and correlation model. The amplitudes are for a magnitude 7.05 rupture of the Hayward Fault. The rupture's surface projection and three example locations of interest are annotated.

11.3 Parameterization Using Physics-Based Simulations

When we use physics-based simulations (Chapter 5) in place of empirical GMMs, spatial correlations may be inherently present in the simulations, depending upon the nature of the simulation approach. Physics-based ground-motion simulations do not have the parameterization of empirical GMMs (Equation 11.3), with a separate model for variability, and instead require uncertainty propagation through introducing uncertainty into the simulation inputs (Section 5.6). This section discusses how correlations may arise in simulated ground motions.

Some physics-based simulations explicitly consider and model the factors discussed above as causing correlations in empirical GMM residuals. For example, Figure 5.4 illustrated that rupture models can have spatial dependencies that produce spatial correlations in resulting IM values. So, to the extent that simulations adequately model spatial dependencies in rupture characteristics, crustal velocity models, and near-surface geotechnical models, the simulated ground-motion amplitudes should have spatial variations consistent with those observed in recordings. In such cases, simulations could be used to capture spatial variations by merely running an ensemble of simulations and observing differences in amplitudes at the sites of interest. This approach is similar to the discussion of Figure 5.22, regarding the potential for ensembles of simulations to produce appropriate IM

variability if the simulations have proper randomization of input parameters. Here the interpretation is extended to joint distributions of IMs at two sites, rather than the distribution at a single site.

While the use of physics-based simulations is conceptually appealing because of the potential to quantify, and directly model, the physical sources of spatial dependencies, not all physics-based models provide useful predictions of spatial variations. Point-source stochastic models (Section 5.5.3) have no explicit consideration of finite-rupture parameters, and some stochastic models do not have physically realistic models of wave propagation that would produce spatial correlation in amplitudes. Similarly, the high-frequency portions of hybrid-broadband simulations (Section 5.5.2) do not rely on the physically parameterized model, and so generally lack appropriate spatial correlations. Simulations that do not consider processes causing spatial correlation are inappropriate for the considerations of this chapter. However, simulations that utilize finite rupture models and three-dimensional crustal velocity models may produce appropriate spatial correlations. Simulations producing appropriate correlations are also our best hope for calibrating partially non-ergodic correlation models, due to the difficulty of obtaining suitable observational data for empirical models. The following section explores this issue using example data.

11.3.1 Observed Correlations in Simulations

To illustrate the presence of spatial correlations in simulations, we consider $SA(3\text{ s})$ amplitudes observed in CyberShake ground-motion simulations.[5] The results in this section come from a more in-depth study by Chen and Baker (2019).

Here we consider 355 simulations of magnitude 8.05 Southern San Andreas Fault ruptures, from CyberShake Study 15.12 (Graves et al., 2011b). The simulations all have the same total rupture extent, but differing hypocenter locations and slip distributions along the rupture plane. The crustal velocity model used for the wave propagation calculations is three-dimensional, including the effects of geologic features that vary spatially throughout the study region. Each simulation consists of time series at 336 sites throughout Southern California, and $RotD50\ SA$ values for each time series are obtained. For each of the 355 simulations, the rupture extent and site parameters were fixed, so with respect to Equation 11.3 we can consider the $\mu_{\ln IM}(rup_i, site_j)$ term constant for a given site. We thus compute the within-event residual using Equation 11.3 by estimating $\mu_{\ln IM}(rup_{i,j}, site_j)$ as the average (over all 355 simulations) of the observed $\ln IM$ values. The between-event residual is estimated for each simulation as the average (over all 336 sites) of the total residuals. With these within-event residuals for each simulation and site, we compute correlations for each pair of sites using Equation 11.15.

Figure 11.16 shows results from these calculations. Each panel shows correlations in δW values, with respect to a particular reference site indicated with a triangle symbol. The approximate extent of the Los Angeles Basin, a deep sedimentary basin in the study area, is also shown with a black line in each panel. Figure 11.16a shows correlations from the CyberShake simulations for a site within the basin. This figure indicates that some positive correlations in $\ln SA(3\text{ s})$ values persist over large distances between the reference site and the rupture (indicated by the black lines near the top of the map).

---

[5] The physics-based portions of the simulations do not entirely govern the $SA(1\text{ s})$ values that were considered in previous sections, so we shift here to a longer period that is governed by the physics of the model.

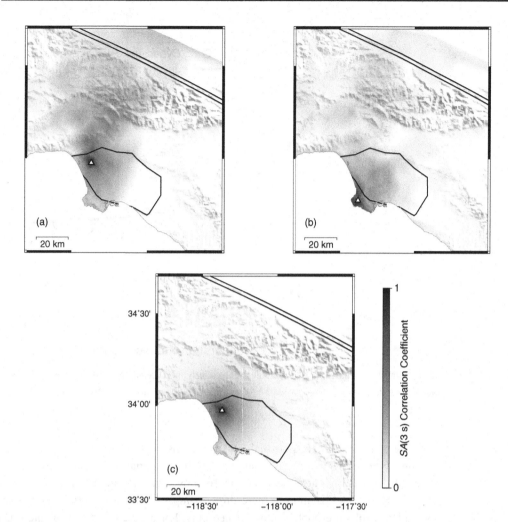

Fig. 11.16 Maps of within-event residual correlation with respect to a reference site (marked with a white triangle), given a magnitude 8.05 Southern San Andreas Fault rupture (surface projection marked with lines to the top of the map). The contour of locations with 0.5 km depth to the 1 km/s shear wave velocity layer is shown with lines near the center of the map and indicates the approximate extent of the Los Angeles Basin. (a) Measurements from CyberShake, for a reference in the Los Angeles Basin. (b) Measurements from CyberShake, for a reference site located to the south of the Los Angeles Basin. (c) Predictions from the empirical model of Jayaram and Baker (2009).

Figure 11.16b shows correlations from the CyberShake simulations, using a second reference site located to the south of the sedimentary basin. This figure indicates that sites near the reference site but outside the basin have high correlations and that the correlations drop for sites in the basin.

Figure 11.16c shows correlations from Jayaram and Baker (2009), for comparison with the simulation-based results. With the empirical model, the correlations depend only on the distance to the reference site. In this case, the sedimentary basin and the geometry of the causative rupture have no bearing on predicted correlations. This simple prediction is not because the empirical data have proven a lack of dependence of correlations on geologic or rupture properties. Instead, it is because the empirical data are insufficient to detect any dependence if it exists.

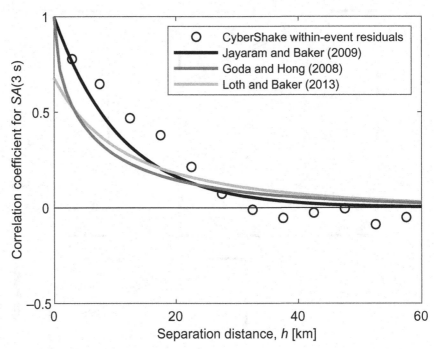

Fig. 11.17 Average correlation coefficients of $SA(3\ \text{s})$ within-event residuals from CyberShake simulations of magnitude 8.05 San Andreas Fault ruptures, relative to the empirical GMM mean as a reference. Three empirical correlation models are superimposed for reference.

Figure 11.16 shows the potential of simulations to produce spatial correlations, but it does not directly indicate whether the simulations have spatial correlations that are consistent with observed data. This is because Figure 11.16 shows correlations for specific pairs of sites, rather than general trends. To allow a more direct comparison with empirical models, the CyberShake correlation results are pooled into distance bins. For a given distance of interest, the correlation coefficients for all combinations of station pairs with that distance (± 2.5 km) are averaged. This operation provides a spatially averaged correlation coefficient analogous to the empirical estimates from observed ground motions. Figure 11.17 shows the resulting correlation estimates, as well as estimates from three empirical models discussed in earlier sections. There is a general agreement between the spatial correlation of the CyberShake residuals and those of the empirical models, with the CyberShake residuals having somewhat larger correlations than the empirical models at distances less than 20 km. To a reasonable degree, however, these data indicate that the CyberShake simulations have spatial correlations that are, on average, consistent with observed ground-motion data (at least for $SA(3\ \text{s})$). Similar findings have been observed using other simulated datasets that use source and path simulation techniques that potentially preserve spatial correlations (Jayaram et al., 2010).

Collectively, the results in Figure 11.16 suggest that repeated ground motions produced by simulation ensembles have the potential to reveal complex patterns in spatial correlations due to geologic features and wave propagation paths from the rupture to the sites of interest. Such patterns are generally not detectable from empirical observations due to the sparse nature of available data.

Further, the simulations can produce these complex patterns of correlations while at the same time maintaining average levels of correlation that are consistent with empirical observations (Figure 11.17). The limitation of this approach is that the accuracy of the simulation-based correlations depends on the quality of the rupture and crustal velocity models used to produce the simulations. And even for high-quality input models, the physics-based simulation results can only, at present, resolve long-period response spectral values. As with the ground-motion predictions discussed in Chapter 5, progress in this field is likely to come from greater use of simulations while also continuing to benchmark the results against empirical data where they are available.

11.4 Numerical Implementation

In concept, a hazard calculation for multiple sites is similar to calculations discussed earlier for single-location cases. The seismic source model (SSM) inputs are identical, and the GMM inputs are similar, but with revisions discussed in the prior sections. However, the multisite nature of the problem changes the numerical calculation.

Because there are often many sites of interest, it is not feasible to evaluate the high-dimensional joint distribution of IM over all values in the integral of Equation 11.2. Instead, we usually use Monte Carlo simulation to sample rupture scenarios and then sample from the joint IM distribution, conditional upon each rupture. The subsections below provide further details on this calculation.

11.4.1 Site Information

For single-location hazard analysis, it is feasible to perform geotechnical investigations to measure near-surface site conditions and infer related parameters such as $V_{S,30}$. For regional analysis, however, it is usually not possible to directly measure site conditions at all sites of interest, and so alternative methods are often used to estimate $V_{S,30}$. These alternate inference methods use topographic slope and/or surficial geology to estimate $V_{S,30}$, since these parameters are more readily observable on a regional scale (Wald and Allen, 2007; Wills et al., 2015) and are observed to be correlated with $V_{S,30}$ (which is itself only indirectly related to the ground-motion effects of interest here). These alternate metrics are termed $V_{S,30}$ *proxies*. While proxies are convenient, there is some evidence for regional differences in the relationships between these proxies and $V_{S,30}$, and even when well-calibrated, they are imperfect indicators relative to direct estimates of $V_{S,30}$ from site-specific measurements. So, while these inference methods are a necessary tool for numerical evaluation of regional risks, they are imperfect and should be used with appropriate caution.

Figure 11.18 illustrates $V_{S,30}$ values inferred for the San Francisco Bay area, using the topographic-slope-based model of Wald and Allen (2007). The flat, low-lying areas near the bay and the Central Valley at the northeast of the figure have low inferred $V_{S,30}$ values, while the hillier areas have higher inferred $V_{S,30}$ values. Some of the example calculations below use these $V_{S,30}$ values.

11.4.2 Hazard Quantification Procedure

With the above models, we can characterize IM hazard as follows.

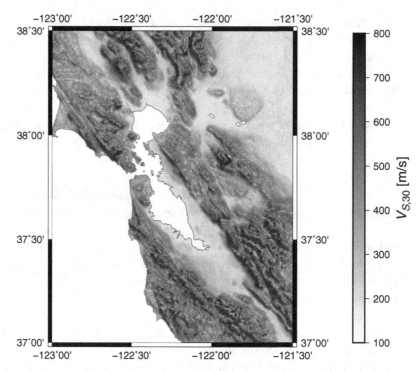

Fig. 11.18 $V_{S,30}$ values for the San Francisco Bay area, inferred using the topographic slope–based model of Wald and Allen (2007).

1. Compute *site* parameters for all sites of interest.
2. Sample a rup_i from the SSM, and evaluate $\mu_{\ln IM}(rup_i, site_j)$ and $\sigma_{\ln IM}(rup_i, site_j)$ for all sites of interest. Also, evaluate the mean occurrence rate of the rupture, $\lambda(rup_i)$.
3. Sample a between-event residual, δB_i, from the distribution specified by the GMM.
4. Sample within-event residuals, $\delta W_{i,j}$, for all sites of interest, from the joint distribution specified by the GMM and the correlation model, as described in Section 11.2.
5. Combine the variables sampled in the above steps, using Equation 11.3, to obtain the resulting vector of $\boldsymbol{IM}$ at each site of interest.

Steps 2–5 are repeated as many times as needed to obtain robust estimates of the probabilistic information regarding $\boldsymbol{IM}$ occurrences. The result is a "stochastic catalog" or "stochastic event set" of $\boldsymbol{IM}$ realizations having the desired multivariate distribution, as depicted schematically in Figure 11.19a. Some software packages such as OpenSHA and OpenQuake provide outputs in this form (Field et al., 2003; Pagani et al., 2014).

Note that we can recover single-site hazard curves by looking only at one location. The hazard curve is computed by

$$\lambda(IM_j > im) = \sum_{i=1}^{n_{sim}} I(im_{i,j} > im)\, \lambda(rup_i) \tag{11.35}$$

where IM_j is the IM at site j, and $im_{i,j}$ is the ith simulation of IM at site j, produced using the steps above. Figure 11.19b shows an example hazard curve computed in this way for the location marked in Figure 11.19a.

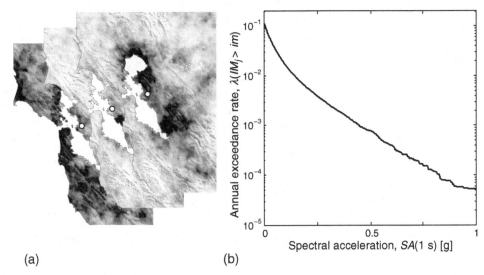

(a)   (b)

Fig. 11.19 (a) Three example simulations of IM values throughout the study region, with an example location marked with a circle. (b) Ground-motion hazard curve for the example location, using the simulations of (a) and Equation 11.35.

In addition to the single-station hazard curves, the paired observations at multiple locations contain information about the joint distribution of IM. The following section discusses these distributions.

11.4.3 Impact of Rupture Scenarios

Equations 11.3 and 11.8 characterized the distributions of IM, conditional upon a rupture. However, the magnitudes and locations of future earthquake ruptures are also random and will contribute to correlation in IM values. Over the set of potential future earthquake ruptures, the $\mu_{\ln IM}(rup_i, site_j)$ terms will vary in a way that produces dependence in the values at sites $j = 1$ and $j = 2$. In a given rupture, the common rupture properties create dependence in IM values at the two sites (loosely speaking, the fact that the two sites will both experience either a high magnitude or low magnitude in a given earthquake will tend to make the resulting IM values similar). The distances from the rupture to the two sites will, in general, differ, and the degree of difference will depend upon the orientation of the earthquake sources relative to the two sites. Finally, the $site_j$ values for the sites may differ, although they may have some dependence due to commonalities in geologic conditions at nearby sites (Thompson et al., 2007). Additional predictor parameters, such as hanging-wall effects or depth to bedrock, may also produce spatial dependence (Figure 4.25). The induced correlation caused by commonality in earthquake events and site conditions is here termed "correlation in means" as it manifests itself in correlation of the mean $\mu_{\ln IM}(rup_i, site_j)$ term when empirical GMMs are used to model IM distributions.

To illustrate this dependence, the $\mu_{\ln IM}(rup_i, site_j)$ values, for the three sites in Figure 11.15, are shown in Figure 11.20. Results are shown for all UCERF2 ruptures within 200 km of the study area, using $SA(1\ s)$ as the IM. This figure illustrates that the mean predictions are highly correlated among all three sites—much more so than the correlation in residuals discussed above. Further, Sites B and C have almost identical mean predictions for most ruptures, as the sites are nearly equidistant from

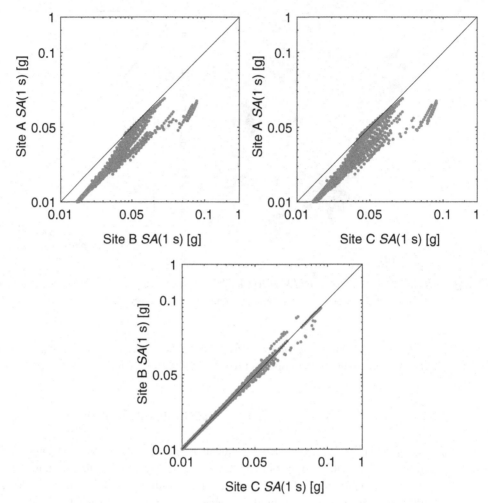

Fig. 11.20 Scatterplots of mean predicted ln SA(1 s) values (exponentiated in the figure for ease of interpretation) at pairs of sites from Figure 11.15. One point is plotted for each considered rupture in the region.

any possible rupture (Hayward Fault, San Andreas Fault, or any other source). On the other hand, Site A is more distant from the Hayward Fault, and closer to the San Andreas Fault, than Sites B and C. So, the joint distribution of SA(1 s) values for Site A with Site B or C show somewhat more scatter.

Finally, Figure 11.21 shows scatterplots of SA(1 s) values produced using the procedure of Section 11.4.2. The $\mu_{\ln IM}(rup_i, site_j)$ values of Figure 11.20 are used for Steps 1 and 2 of the procedure, and the joint distributions of Figure 11.14 are used to sample the residuals for Steps 3 and 4. The final results of Step 5 (Figure 11.21) show the combined effect of the random ruptures and the random but spatially correlated residuals at the sites of interest. These simulated SA(1 s) values can then be used as inputs to consequence predictions for risk analysis, as will be discussed in Section 11.6.

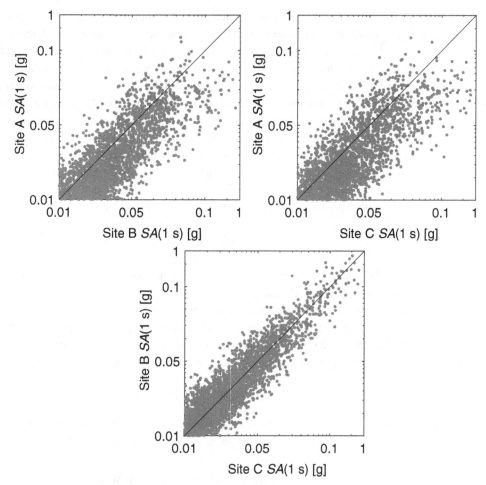

Fig. 11.21   Scatterplots of *IM* values at pairs of sites from Figure 11.15, including GMM residuals. One point is plotted for each considered rupture in the region.

11.5 Coherency

The prior sections in this chapter have focused on cases where the *IM* varies from location to location. A second situation in which ground motions vary spatially is for massive structures analyzed using response-history analysis. The input ground motions vary spatially in the time domain, causing differential displacements that can cause consequences for the structure. The quintessential example is a bridge with differing excitations at each support, which often increases the demands on connecting decks and seismic gaps (Harichandran et al., 1996). Other structures where coherency may be relevant include pipelines, dams, and tunnels.

The typical assumption in these cases is that *IM* amplitude is not varying across the supports, in contrast with the sections above. We assume that the *IM* is fixed (either at the surface or at a given reference depth), and the associated ground-motion time series are incoherent in the temporal domain.[6] In this case, the traditional PSHA calculation of Chapter 6 is sufficient for determining the *IM* level(s) of interest. The extra task required here is to incorporate coherency in the time series.

Several effects cause variations in ground-motion time series (Abrahamson, 1993). Seismic waves travel at finite speeds, so they arrive at stations at differing times (the *wave passage effect*). Seismic sources have finite dimensions, and the rupture propagates at a finite speed, so the timing of wave arrivals varies spatially (the *extended source effect*). Finally, as waves propagate through the crust, they are scattered by heterogeneities, meaning that different locations receive differing sets of seismic wave arrivals (the *scattering effect*). At larger spatial separation distances, the extended source and scattering effects also cause spatial differences in *IM* values, so the distinction between *IM* correlation and time-series coherency is not absolute. Still, it is a useful categorization, as typically one or the other phenomenon will dominate for a given problem.

11.5.1 Coherency Models

As with *IM* correlations, observed ground motions can be used to calibrate coherency models.[7] In this case, the required data come from arrays of closely spaced instruments. For example, Figure 11.22 shows observations of ground motions from the SMART-1 array in Taiwan.

With paired ground-motion observations like these, it is possible to estimate coherency parameters. The parameterization is typically performed using a cross-spectral density function (Zerva, 2009). We begin by taking a Fourier transform[8] of the ground-motion time series, $a(t)$:

$$a(\omega) = \frac{1}{2\pi} \int_{-\infty}^{\infty} a(t) e^{-i\omega t} dt. \tag{11.36}$$

The above equation is for a continuous signal of infinite length, and in discrete form for a finite time series (which is relevant for recorded ground motions) can be computed as

$$L_j(\omega_n) = \sqrt{\frac{\Delta t}{2\pi N_T}} \sum_{l=0}^{N_T-1} a_j(l\Delta t) e^{-i\omega_n l\Delta t} \tag{11.37}$$

where $a_j(t)$ is the jth ground motion with time step Δt, and N_T is the number of time steps in the ground motion.

A power spectral density can be computed from this result as

$$\hat{S}_{jj}(\omega_n) = |L_j(\omega_n)|^2. \tag{11.38}$$

Then, for two ground motions, $a_j(t)$ and $a_k(t)$, a cross-spectral density can be estimated as

$$\hat{S}_{jk}(\omega_n) = L_j^*(\omega_n) L_k(\omega_n) \tag{11.39}$$

where * denotes a complex conjugate.

---

[6] Using a fixed *IM* at a reference depth, coupled with a site response model, can address cases where supports are on differing ground conditions that thus have differing surface *IM* amplitudes.

[7] In principle, physics-based ground-motion simulations could also directly represent coherency, if they include the above-mentioned physical phenomena in their formulations, and consider a spectral domain sufficient to characterize high-frequency incoherence.

[8] Similar calculations were presented previously in Section 4.2.3. For consistency with other literature, here we present the Fourier transform in terms of angular frequency (ω) instead of hertz ($f = \omega/2\pi$) as used in Section 4.2.3. This leads to some differences by a factor of 2π in the two sets of equations.

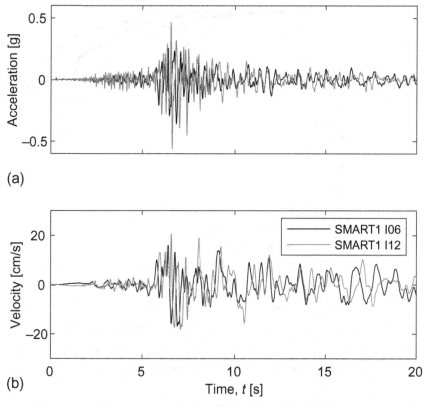

Fig. 11.22 (a) Acceleration and (b) velocity time series recorded at a separation distance of 400 m. Data are from a 1981 *M*5.9 earthquake recorded by the I06 and I12 stations of the SMART1, Taiwan, array located approximately 30 km from the rupture.

Finally, the coherency of two ground motions is computed as the cross-spectral density normalized by the power spectral densities of the ground motions:

$$\bar{\gamma}_{ij}(\omega_n) = \frac{\hat{S}_{jk}(\omega_n)}{\sqrt{\hat{S}_{jj}(\omega_n)\hat{S}_{kk}(\omega_n)}}. \tag{11.40}$$

The coherence $\bar{\gamma}_{ij}(\omega_n)$ is a complex number. Its absolute value is termed the lagged coherency:

$$\left|\bar{\gamma}_{ij}(\omega_n)\right| = \frac{\left|\hat{S}_{jk}(\omega_n)\right|}{\sqrt{\hat{S}_{jj}(\omega_n)\hat{S}_{kk}(\omega_n)}}. \tag{11.41}$$

The amplitude term preserved here measures the similarity of ground motions, without quantifying a phase shift between the signals (i.e., the imaginary term) that would be associated with the wave passage effect. The Fourier amplitudes themselves also vary on small spatial scales (Zerva and Zhang, 1996).

The spectral densities are typically smoothed during the estimation process, to reduce estimation variability, and this smoothing influences the coherency result of Equation 11.41. Readers interested in quantitative assessment of coherency from data should read further about stochastic process estimation (e.g., Abrahamson et al., 1991; Lutes and Sarkani, 2004).

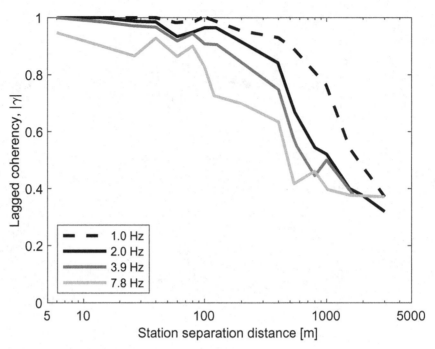

Fig. 11.23 Estimated lagged coherency for varying frequencies and varying separation distances between the ground motions. Data are from several earthquakes recorded by dense arrays in Taiwan (Abrahamson et al., 1991).

The above calculations use strong assumptions regarding the properties of the ground motions—specifically that they are stationary and ergodic (Zerva, 2009). While these assumptions are not strictly valid for ground-motion data, they can be circumvented using techniques such as considering windows of the time series for which the assumptions are more reasonable.

The lagged coherency of Equation 11.41 is a measure of the similarity of the two ground motions $a_j(t)$ and $a_k(t)$. It is frequency-dependent so that a pair of ground motions have differing lagged coherency at each frequency. It can, in general, also vary as a function of earthquake magnitude, source-to-site distance, and near-surface soil conditions (Somerville et al., 1988, 1991). Figure 11.23 shows estimates of lagged coherency from an earthquake recorded by Taiwanese arrays, showing variation in results as a function of both frequency and the separation between the stations. We see that the coherency decreases as the separation distances increase, as would be expected intuitively. We also see that the coherency decreases with increasing frequency for a given separation distance.

The lagged coherency measured in Figure 11.23 is near 1 for separation distances of less than 100 m, and drops to approximately 0.4 for separation distances of 1–3 km. There are few structures large enough and stiff enough to be sensitive to differential excitations at distances of a kilometer or more. We thus see that coherency models are relevant at spatial scales of hundreds of meters, while the spatial IM correlations discussed earlier are relevant at spatial scales of kilometers. For that reason, we rarely need to consider incoherence and spatial IM correlations simultaneously.

11.5.2 Applications

Many stochastic simulation models have been proposed to generate time series with phase differences that produce appropriate coherency properties (e.g., Zerva, 2009). These approaches are, to some degree, compatible with the time series generation techniques of Chapter 5. Random vibrations–based analyses of structures also utilize these coherence models, but those approaches usually require strong assumptions (linearity of structural response and excitations with stationary amplitude and frequency content over time), which are usually not appropriate for the seismic risk analysis topics of interest here. Several other studies have considered the effect of coherency on ground strains and other metrics of engineering interest (e.g., Hashash et al., 2001; Ancheta et al., 2010). Some design tools also offer approximate approaches for quantifying the effect of coherency (FEMA, 2005; NIST, 2012).

When the near-surface site conditions vary across the input locations (a common occurrence for bridges crossing riverbeds with abutments in more competent material, or for structures with inputs occurring at differing depths), the impact of site conditions must also be considered. Some spatial incoherency models provide cross-spectral density functions for such conditions (Zerva, 2009). Alternatively, ground motions can be defined at a reference depth and then separately propagated toward the surface using site response analysis. The input bedrock motions can include appropriate incoherency as needed, and the output motions will then incorporate both incoherency and the effects of surficial soils.

11.6 Risk

Given that the numerical implementation issues of Section 11.4 lead to near-universal use of Monte Carlo simulation for **IM** values, Monte Carlo simulation is also usually the appropriate approach for computing risk metrics.

To compute an exceedance probability curve (e.g., Equation 9.26), rather than integrating over a high-dimensional vector IM, we can sum over the rupture simulations:

$$\lambda(C > c) = \sum_{i=1}^{n_{sim}} I(c_i > c) \, \lambda_i \tag{11.42}$$

where c_i is a statistical sample of the consequence associated with the ith rupture simulation (and associated im_i simulation), $I(\ )$ is an indicator function, and λ_i is the annual rate of occurrence associated with the ith simulation.

The consequence associated with an im_i simulation is typically computed by using fragility or vulnerability functions to sample consequences for all relevant components of the system. The consideration of multiple consequence functions means that a joint distribution of consequence outcomes (conditional on IM) must be specified for the multiple components. The dependence of IM values at multiple sites can be accounted for using the previous section's approaches. The additional consideration here is whether there are commonalities in construction characteristics, modeling uncertainties, or other factors that could lead to correlations in consequence outcomes.

It is challenging to measure such correlations from empirical or analytical data: Shome et al. (2012) is a rare example of an empirical study of these correlations. Loss ratios from the Northridge earthquake were considered, and nonzero correlations were seen at distances of less than 15 km, though the study was limited to single-family houses, and the locations of the properties were known to only limited precision. In other studies, the typical assumption is that consequences to multiple components are independent, conditional on IM (Wesson et al., 2009), though some studies assume an idealized correlation model for the purpose of evaluating sensitivity to this assumption (Bazzurro and Luco, 2005; Lee and Kiremidjian, 2007; Kwak et al., 2016b; Silva, 2019).

Once component consequence outcomes have been simulated, they can then be used to evaluate an aggregate system consequence, usually either a summation of component consequences or some other more complex relationship, as discussed in the following subsections.

11.6.1 Summed Consequences

The first category of consequences is where the total consequences are the sum of consequences computed for each site of interest. A typical example is the computation of insured financial losses, where insured losses are computed for each property, and then summed over all properties in a portfolio (e.g., Grossi et al., 2005; Park et al., 2007; Sokolov and Wenzel, 2011; Weatherill et al., 2015). In this case, it may be possible (though cumbersome) to provide an equation relating the consequence, C, to the IM values at all sites of interest. More typically, evaluating C for a given im_j realization involves generating samples of losses to individual properties using the consequence functions described in Chapter 9, and then summing the losses for each property.

11.6.2 General Regional Performance

The second category of consequences is the general case where consequences are a more complex function of impacts at individual locations. A typical example is the computation of impacts to infrastructure systems, where component damage at individual locations relates to the ability of the system to provide services (e.g., Lee and Kiremidjian, 2007; Adachi and Ellingwood, 2009; Jayaram and Baker, 2010b; Sokolov and Wenzel, 2014). Another example is the computation of regional economic impacts, where the output depends in a complex way on the damage throughout the region (e.g., Dorra et al., 2013).

In this case, no explicit equation exists to relate simulated IM values to consequences. Instead, a specialized simulation or computation is performed for each realization of im_j values (e.g., Figure 11.1). The consequence metric of interest may be the number of customers who have lost service, or the difference between customer demands and supplied service. Typically, the damage to individual components (e.g., bridges or pipe segments) is sampled with probabilities specified by fragility functions. Then the system performance is computed by performing a network analysis accounting for the impact of damaged components (which may result in the rerouting of traffic or flow due to leakage from pipes) and then assessing the consequence metric of interest.

In this case, the computational cost of evaluating consequences may be substantial, limiting the number of simulations (i.e., limiting n_{sim} in Equation 11.42). Even if a consequence calculation takes a few minutes for a given damage realization, performing these calculations for thousands of damage realizations can be burdensome. Moreover, for complex analysis situations, the consequence calculation can easily take hours. In such situations, it is common to develop a reduced and reweighted set of im_j samples that maintain the same probability distribution as the original full set of samples. This reduced set of samples is sometimes termed a *hazard consistent* or *stochastically representative* stochastic event set. Techniques for developing the reduced set include clustering or optimization (Chang et al., 2000; Jayaram and Baker, 2010b; Han and Davidson, 2012). The extra effort required to implement these procedures is often justified when the time to reduce the set is much less than the time to simulate consequences for a full stochastic catalog of im_j simulations.

Table 11.1. Example risk calculation data. These results are plotted in Figure 11.24. Individual columns of the Table are described in the text							
i	$SA_1(1\text{ s})$	$SA_2(1\text{ s})$	$p_{f,1}$	$p_{f,2}$	$Failed_1$	$Failed_2$	n_{Failed}
1	0.47	0.20	0.27	0.00	0	0	0
2	0.42	0.25	0.19	0.01	0	0	0
3	0.24	0.16	0.01	0.00	0	0	0
4	0.67	0.16	0.60	0.00	1	0	1
5	0.66	0.37	0.59	0.12	0	1	1
6	0.93	0.46	0.86	0.25	1	1	2
7	0.46	0.33	0.25	0.07	0	0	0
8	0.37	0.33	0.12	0.07	0	0	0
9	0.26	0.13	0.02	0.00	0	0	0

Example

To illustrate the numerical process of evaluating risk to distributed systems, consider here a simple example. For the ground-motion hazard, we consider the rupture example of Section 11.2.6. For the consequences, we assume that there is one structure at each of the two locations in that example. Each structure has a lognormal fragility function for failure with a median of $\theta = 0.6$ g and log standard deviation of $\beta = 0.4$ (Equation 9.1). The failure or nonfailure of the two structures, conditional upon the IM values at the sites, is independent. The consequence metric of interest will be the number of collapsed structures, given the rupture.

Table 11.1 provides nine example Monte Carlo simulations for this example, and Figure 11.24 plots 200 simulations of the same data. This table and figure can be used to discuss the steps in the risk calculations. Referring to Table 11.1, the first column (i) is simply an index to the simulation number, to facilitate referencing. The second and third columns ($SA_1(1\text{ s})$ and $SA_2(1\text{ s})$) are simulations of spectral acceleration values for the two sites, obtained using Monte Carlo simulation, with a multivariate normal distribution for $\ln SA$ and mean and covariance matrices from Equations 11.28 and 11.34. Because the covariance matrix includes correlation between the SA values at the two locations, the SA values tend to be both high or both low; this can be seen anecdotally in the table, and in the pattern of points in Figure 11.24 lying along the diagonal. Note also that the points in Figure 11.24 fall in the same location as the contours of Figure 11.12. This correspondence is by design, as the Monte Carlo simulations are coming directly from the probability distribution plotted in Figure 11.12.

Next, for a given simulation of $SA(1\text{ s})$, a probability of failure (p_f) can be computed using the specified fragility function. For example, considering the $SA_1(1\text{ s})$ value simulated in row 1 of Table 11.1, and the fragility function of Equation 9.1 with the parameters above, gives

$$p_{f,1} = \Phi\left(\frac{\ln(0.47/0.6)}{0.4}\right) = 0.27. \tag{11.43}$$

This value is shown in column 4 of Table 11.1, and all other values in columns 4 and 5 are computed in the same way. Columns 6 and 7 are generated using additional Monte Carlo samples. For example, considering column 6 of row 1, the failure of structure 1 is sampled as a "1" (failed) with probability 0.27, and a "0" (not failed) with probability 0.73. In this particular case, the sample was a "0." The final column sums the number of failed structures in that particular simulation: either zero,

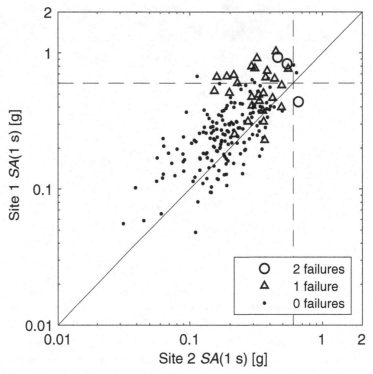

Fig. 11.24 Two hundred samples of $SA(1\,s)$ at the two sites of the example of Figure 11.11. The dashed lines show the $\theta = 0.6\,g$ value, indicating the amplitude with a 0.5 probability of failure of the structure at the given site. The sampled values are formatted to indicate whether the simulation produced zero, one, or two structural failures.

one, or two. Looking at Figure 11.24, we see that the cases with at least one failure tend to be near the top or right of the figure, and the cases with two failures tend to be in the top and right, though there is some randomness due to the probabilistic nature of the fragility function. Using 10,000 simulations of IM values and failures, and counting the fraction of simulations with a given number of failures, we can estimate that $P(0\text{ Failed}) = 0.846$, $P(1\text{ Failed}) = 0.135$, and $P(2\text{ Failed}) = 0.019$. Note that these probabilities are conditional upon the rupture. To compute the exceedance rate for these consequences (Equation 11.42), these probabilities should be multiplied by the occurrence rate of the rupture.

With a slight modification to the problem, we can measure the effect of the SA correlations on these risk results. To do this, we resimulate the SA values from alternate joint distributions. In all cases, we keep the means and standard deviations of the $\ln SA$ values fixed, and only alternate the correlation coefficient between the two sites. Table 11.2 shows results for three choices of $\rho_{\ln SA_{i,1}, \ln SA_{i,2}}$: perfect correlation, $\rho = 0.63$ (the value from Equation 11.32), and zero correlation. Columns 2–4 of the table show the probabilities of zero, one, or two failures in each case. The probability of two failures decreases with decreasing correlation, as there is a decreasing likelihood of observing the extreme high-amplitude shaking at both sites in a given simulation; this is the same phenomenon shown in Figure 11.3 at the start of the chapter. The probability of a single failure is lowest for the $\rho = 1$ case, as both SA values will be high or both will be low, so the only way to get a single failure is due to the randomness in the failure of each structure for a given SA amplitude.

Table 11.2. Effect of $\rho_{\ln IM_{i,1}, \ln IM_{i,2}}$ on the number of failed structures, for the example calculation

$\rho_{\ln SA_{i,1}, \ln SA_{i,2}}$	$P(0\ \text{Failed})$	$P(1\ \text{Failed})$	$P(2\ \text{Failed})$
1.00	0.872	0.109	0.020
0.63	0.846	0.135	0.019
0.00	0.855	0.140	0.005

This example, and the results of Table 11.2, highlight one other subtle, but important, aspect of computing risk to distributed systems. This example made the typical assumption that the damage states of individual structures were independent, conditional on the IM values at each site. The IM correlation does produce some dependence in resulting damage states, but for a given simulation of the IM values, the collapses or non-collapses of the individual structures are independent. If the structures have commonalities not captured by the fragility functions, there may be dependence in damage given IM, and this dependence would influence system risk calculations (Wesson and Perkins, 2001; Lee and Kiremidjian, 2007; Bocchini and Frangopol, 2011; DeBock et al., 2014). Measuring and calibrating this dependence using observational data is much more difficult than measuring IM dependence. So, at present, assessments typically assume independence of damage states, conditional on IM.

Exercises

11.1 Use the definition of the semivariogram (Equation 11.17) to show that an exponential semivariogram (i.e., Equation 11.20) is equivalent to a correlation function of the form:

$$\rho(h) = \exp\left(\frac{-3h}{r}\right) \tag{11.44}$$

where h is the separation distance, and r is the range parameter.

11.2 Consider the exponential model for within-event correlations given by Equation 11.44.

(a) If the range parameter for $SA(T)$ is a function of the vibration period T according to the expression $r = 22 + 3.7T$, for $T \geq 1$ s, then compute the correlation of the within-event residuals for $SA(1\ \text{s})$ separation distances of $h = 2$ km and $h = 10$ km.

(b) Repeat (a), but for $SA(5\ \text{s})$.

(c) If the between- and within-event standard deviations are $\tau = 0.3$ and $\phi = 0.5$, respectively, determine the correlation of the *total* residuals for the two separation distances in (a) and (b).

11.3 Consider two locations, A and B, near a magnitude 6.5 vertical strike-slip rupture. The distances to each location from the rupture are denoted r_A and r_B, and the distance between the two locations is denoted h. Both locations have $V_{S,30} = 400$ m/s. For each of the cases listed below, compute $P(SA_A(2\ \text{s}) > 0.3\ \text{g})$, $P(SA_B(2\ \text{s}) > 0.3\ \text{g})$, and $P(\text{Both } SA(2\ \text{s}) > 0.3\ \text{g})$. Discuss your results.

(a) $r_A = 10$ km, $r_B = 10$ km, $h = 100$ km.

(b) $r_A = 10$ km, $r_B = 10$ km, $h = 10$ km.

(c) $r_A = 10$ km, $r_B = 10$ km, $h = 5$ km.

(d) $r_A = 10$ km, $r_B = 10$ km, $h = 0.01$ km.

(e) $r_A = 10$ km, $r_B = 5$ km, $h = 10$ km.

(d) $r_A = 10$ km, $r_B = 5$ km, $h = 5$ km.

You can solve this problem either using Monte Carlo simulation or numerically using a multivariate normal cumulative distribution function in a computer tool.

11.4 Consider the conditions from the previous problem, but now with a structure located at each location. The probability of failure of each structure is given by a lognormal fragility function with a median of $\theta = 0.3$ g and a log standard deviation of $\beta = 0.4$. Assume that, conditional on the $SA(2$ s) values, the failures are independent. For each case, compute the probability of the structure at location A failing, the structure at location B failing, and both structures failing. Discuss your results.

11.5 Discuss how the modeled loss exceedance curve for an insurance property portfolio would vary if you assume the following for ground motion variability: (1) no spatial correlation in GMM residuals, (2) perfect spatial correlation in GMM residuals, and (3) partial spatial correlation in GMM residuals, using an empirical model. Assume the portfolio is comprised of 100 houses at 100 distinct locations. Is it appropriate to also consider correlations among the consequence functions, conditional on IM? If so, what features do you think would affect correlations?

Validation

Probabilistic seismic hazard and risk methods discussed in the preceding chapters provide the scientific basis for the seismic design and assessment of societal infrastructure. As a scientific approach, they are subject to the scientific method and observational data used to examine their validity.

While simple in concept, such validation is complicated by the fact that most hazard and risk analysis predictions are for events that are rare and thus infrequently observed. This chapter examines these challenges, approaches to validate the seismic hazard curve distribution, and their underlying constituent models' predictive capabilities.

Learning Objectives

By the end of this chapter, you will be able to do the following:

- Understand verification and validation as a means to develop confidence in the predictive capability of scientific models.
- Compare observational data with seismic hazard and risk models and infer the validity of these models.
- Understand the link between assessing predictive capability and the quantification of epistemic uncertainty, particularly in the context of model extrapolation.
- Describe the utility of seismic hazard and risk calculations for decision-making, despite the difficulties in comprehensive validation.
- Explain common errors in using observational data to (in)validate seismic hazard analysis results.

12.1 Introduction

Probabilistic seismic hazard and risk analysis (PSHA and PSRA, respectively) predict future seismic hazards and their impacts. The modeling components—seismic source models (SSMs) and ground-motion models (GMMs) in PSHA (Chapter 6), and additionally fragility and vulnerability functions in PSRA (Chapter 9) – are developed utilizing observational data from historical earthquakes, among other datasets.

Given the limited observational data that usually exist during initial model development, additional observational data provide an opportunity to validate models and identify potential improvements. This chapter considers how observations of earthquakes, ground motions, and damage (or lack thereof) can be used to validate (or invalidate) PSHA and PSRA results and their underlying model components.

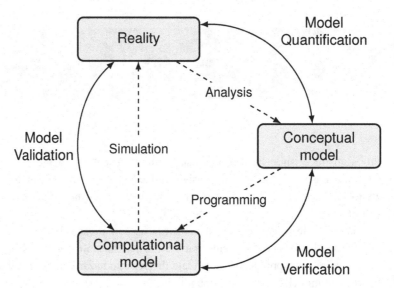

Fig. 12.1 Phases in computational modeling and simulation and the role of verification and validation in connecting conceptual and computational models to reality. Solid arrows indicate the pivotal steps, and dashed lines the principle tecnhiques required. Adapted from Oberkampf et al. (2004).

In the following sections, we begin by distinguishing between verification and validation in computational modeling. We next present, by example, a common misconception in comparing PSHA results with observed ground motions. Appropriate approaches for validating the results of PSHA and PSRA, and their constituent models, are then addressed. Finally, we discuss the use of PSHA and PSRA for decision-making, despite imperfect models and validation challenges.

12.2 Verification and Validation

Confidence in computational model results can be developed by demonstrating the model's fidelity, including the validity of input models and parameters for the considered problem, and documentation of its implementation. Figure 12.1 provides a schematic illustration of the three general phases in computational modeling and simulation, and an overview of verification and validation concepts is provided in the subsections below to contextualize validation in the context of seismic hazard and risk. A detailed discussion of verification and validation as a formal process for developing predictive capability in computational modeling is provided in Oberkampf et al. (2004).

12.2.1 Verification

As shown in Figure 12.1, software programming and numerical methods are used to take a conceptual (i.e., mathematical) model and implement the model for computational analysis. In its simplest form, verification assesses the accuracy of the result of the computational model. Verification is needed to ensure that there are no programmatic errors (i.e., bugs) in the code that implements the methodology, and that the numerical methods are suitable (e.g., they converge) for the problem being considered. Importantly, verification does not tell us anything about whether the conceptual model is an appropriate representation of reality, just that it is implemented correctly.

An obvious means by which to verify a computational algorithm is to make comparisons with known analytical solutions. This process is useful for verifying scientific software and for verification against simple problems with analytical solutions. However, since PSHA and PSRA software are often utilized for solving problems with no analytical solutions, verification is often obtained by comparing results from multiple programs and approaches developed and maintained by different entities (e.g., Bielak et al., 2010; Chaljub et al., 2010; Thomas et al., 2010; Hale et al., 2018)

Verification is critical when implementing existing programs on new computational resources and developing new numerical solution methods. However, for verification problems without analytical solutions, complete asymptotic convergence of numerical solutions from different methods or algorithms is unnecessary. Specifically, prior comparisons of independently developed computational codes have illustrated that the intercode differences are often small relative to the differences between the models and actual observations (e.g., Maufroy et al., 2015). That is, the verification problem is more easily "solved" than the validation problem.

12.2.2 Validation

As shown in Figure 12.1, a (verified) computational model is used to provide a prediction of reality. Validation assesses whether the computational implementation of the conceptual model is representative of reality as measured using experimental observations. Unlike verification, which is a computer science and mathematical problem, validation is a physics problem:[1] Does the conceptual model adequately represent reality?

Because probabilistic seismic hazard analysis calculations and their resulting outputs are determined through mathematically combining models for earthquake occurrence and ground-motion shaking (i.e., Equation 6.1), validation of PSHA outputs can be examined from two perspectives. We can attempt to directly validate the results of the PSHA calculation (Section 12.4). Alternatively, recognizing that direct validation is generally impossible for typical ground-motion intensities of interest (Section 1.4.2), we can instead validate model components (Section 12.5). The second approach recognizes that complex phenomena (i.e., earthquakes, ground motions, and infrastructure damage) involve a multifaceted array of physical processes. Therefore, model validation in a hierarchical fashion can ultimately build predictive confidence for a situation that has not been directly observed—prediction is inherently extrapolation (Oberkampf et al., 2004).

Validation can consider predictions made before, or following, the occurrence of the observations in question. *Prospective validation* involves predictions made before the observations occur, whereas *retrospective validation* involves the prediction of observations that have occurred in the past. Prospective validation provides the truest test of predictive capability. In the context of seismic hazard, a simple example of prospective validation would be to predict the locations and characteristics (e.g., magnitude) of earthquakes in a given region over the next five years (via an earthquake rupture forecast, ERF). We would then observe earthquakes over this period and compare prediction and observation. In both prospective and retrospective settings, a guiding principle is that a model cannot be validated by data used in its construction.

---

[1] In the context of vulnerability and fragility models used for seismic risk calculations, validation may also be a socioeconomic problem.

Despite its conceptual benefits, prospective validation can be hindered by an ill-posed problem definition. For example, to specifically validate GMMs, it is necessary to know the features of the earthquake that caused the recorded ground motions. Since such information is known only once the earthquake occurs, then true prospective validation of GMMs is not possible. Instead, GMMs can be validated in a *pseudo-prospective* fashion (Section 12.5.2), via predictions that use automatically determined characteristics of the earthquake. This process has the benefit of requiring no analyst intervention, but imperfections in the automated determination of earthquake characteristics will be mapped into consequent errors in the GMM prediction.

In retrospective validation, the passage of time can allow for more accurate determination of the necessary inputs to specific models,[2] enabling direct evaluation of a model's predictive capability conditional on the assumption that the inputs are correct. While model inputs will never be known with certainty,[3] the uncertainty can be small enough that it is not consequential, or validation methods can explicitly consider input uncertainties through appropriate statistical methods. The limitation of retrospective validation is that the prediction is not truly "blind" (i.e., a model proponent may have a conscious or unconscious tendency to specify the model inputs in a manner that increases prediction performance). Determining model inputs in an objective and reproducible manner is therefore essential to understanding model validity and its link to epistemic uncertainty (Section 1.7).

A final consideration for validation is the overlap in conditions for which validation is possible and those for which prediction is desired. Oberkampf et al. (2004, Figure 5) consider two dimensions of this situation. The first dimension reflects inputs to models, such as the range of observed earthquake magnitudes used in validation, compared with those of interest for prediction. The second dimension reflects the salient physics that the models are trying to represent. A disconnect in the second dimension could occur, for example, when validating a GMM that predicts nonlinear site response for large-amplitude ground motions. If observational data include only small-amplitude ground motions, the model's nonlinear site response aspect cannot be validated. Where overlap in these two dimensions is not possible using observations available at the site or region of interest, global analog observations may be useful. As a result, seismic hazard problems can generally be considered as having a "partial degree of overlap" for validation (Oberkampf et al., 2004), which makes their validation challenging, but possible.

12.3 Validation from Limited Observations

Before examining appropriate validation approaches, it is instructive to understand common, but improper, attempts to compare observational data with seismic hazard results. The most frequent comparison is that of a seismic design spectrum (derived from a PSHA) with ground motions from a recent earthquake. Such an example is used in this section to illustrate problems with comparing probabilistic quantities and individual observations.

---

[2] For example, earthquake characteristics needed for GMM predictions or ground-motion estimates at a specific location of damage used in fragility function development.

[3] Some model inputs may even be the outputs of other models themselves, rather than actual observations.

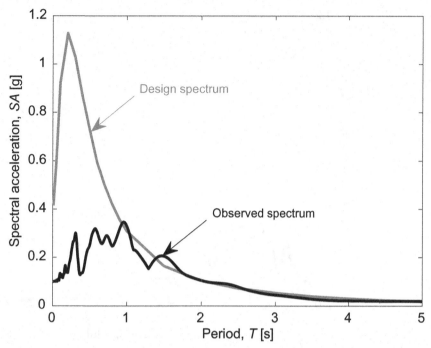

Fig. 12.2 Comparison of a design response spectrum for a specific location and the response spectrum of an observed ground motion at the same location. In this example, the design spectrum is the UHS for an annual exceedance rate of 2.1×10^{-3} (10% probability of exceedance in 50 years), and the observed ground motion is one horizontal component of the Treasure Island record from the 1989 Loma Prieta earthquake.

12.3.1 Comparison of Seismic Design Spectra with Observed Ground Motions

Following a damaging earthquake, it is common to evaluate seismic design codes, guidelines, and underpinning methodologies. For structural and geotechnical engineers, this may involve comparing the intensity of one or more observed ground motions with the design ground-motion intensity and, subsequently, comparing the observed and expected performance of structures.

Consider the situation depicted in Figure 12.2, where an observed ground motion at a specific location is compared with a seismic design spectrum for that same location. The observed response spectrum is seen to exceed the design spectrum at several vibration periods. This exceedance may lead an engineer to conclude that the design seismic intensity is flawed, resulting from flawed inputs or a flawed PSHA methodology (see Section 12.6).

Reconciling observations with seismic design values requires an appreciation of how seismic design values are determined—particularly their probabilistic nature. As is common in contemporary seismic design standards and guidelines, the example seismic design spectrum in Figure 12.2 is based on a UHS (Section 7.3.1). This example design spectrum is the UHS with a 10% probability of exceedance in 50 years from Section 7.3,[4] and we use the Treasure Island ground motion from Chapter 4 as the example observation.

---

[4] In practice, the UHS would correspond to the actual site of interest, but the link to the results from the hypothetical site in Section 7.3 is made for familiarity.

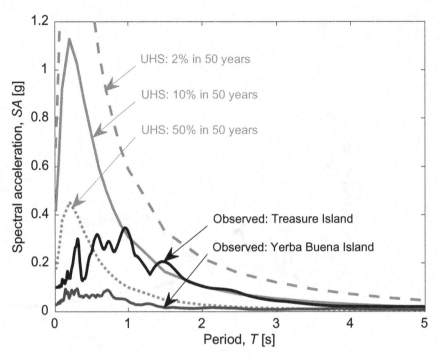

Fig. 12.3 Comparison of UHS for multiple exceedance probabilities compared with multiple observed response spectra.

Figure 12.3 extends on Figure 12.2 by plotting the UHS for 2, 10, and 50% exceedance probabilities in 50 years. Displaying the UHS for different exceedance probabilities conveys the idea that the PSHA result is a probabilistic quantity and not a single deterministic number. The question therefore is not "Is the observed ground motion higher or lower than the PSHA result?" but "What exceedance probability does the observed ground motion correspond to?" Figure 12.3 also illustrates that even this second question depends on the considerd ground-motion IM. For example, at a vibration period of $T = 1.5$ s the ground-motion intensity exceeds the 10% in 50-year UHS. The $SA(0.5$ s) amplitude is approximately equal to the 50% in 50-year UHS, and the SA amplitudes for $T < 0.5$ s are all smaller than the 50% in 50-year UHS.

Figure 12.3 also presents the observed spectra at both Treasure Island and Yerba Buena Island from the 1989 Loma Prieta earthquake (Section 4.2). That is, both of these observations are from the same earthquake. Assuming, for discussion, that the UHS for these two (nearby) locations are equal,[5] the lower SA amplitudes at Yerba Buena Island indicate that this specific observation corresponds to a ground-motion intensity with larger exceedance probabilities than that for the Treasure Island observation. For example, at $T = 1.5$ s, the Treasure Island amplitudes exceed the 10% in 50-year UHS, while the Yerba Buena Island amplitudes are below the 50% in 50-year UHS. Hence, not only does the exceedance probability of a ground-motion amplitude change as a function of the IM considered (e.g., SA for different T), but the IM amplitudes of different ground motions from the same earthquake will also have different exceedance probabilities.

---

[5] As discussed in Chapter 4, these two observations differ in amplitude due to near-surface site-effects. So the observations should be compared with their own (differing) uniform hazard spectrum. However, here for illustration, we compare them with a generic set of UHS.

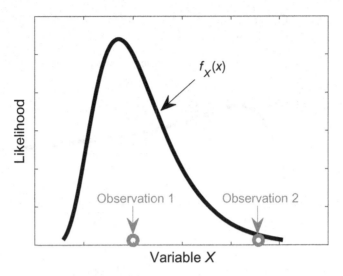

Fig. 12.4 Two observations of a quantity of interest, X, compared with their presumed underlying probability distribution, $f_X(x)$. While Observation 2 has a lower likelihood of being observed than Observation 1, both are possible.

It is invalid to state, "That earthquake corresponded to a $Z\%$ in 50-year event." Instead, it is necessary to specify the ground-motion location and considered IM. That is, we can state, "The ground motion at location X had an intensity measure IM_Y value corresponding to a $Z\%$ in 50-year exceedance probability."

12.3.2 Comparing Single Observations with Probabilistic Predictions

The presentation of observed ground motions in the form of Figure 12.3 reinforces that PSHA provides a probability distribution for a specific IM value, and a single observation cannot validate or invalidate a distribution. Figure 12.4 illustrates this situation in a more general setting. A variable, X, has a probability density function, $f_X(x)$, and a single observation x_i is made. The observation may be near the center of the distribution (Observation 1) or its tail (Observation 2). While an observation near the center of the distribution is more likely, tail observations are possible. Therefore, based on either one of these observations alone, it is impossible to validate or invalidate the adequacy of $f_X(x)$.[6] With any single observation, as shown in Figure 12.4, the only valid statement is that the observation represents a specific percentile of the assumed underlying distribution. When this distribution is a seismic hazard curve,[7] this is equivalent to saying the observed IM value has a specific exceedance probability, as in the previous section.

Given that it is impossible to draw any conclusions on the validity of $f_X(x)$ given one observation, there are two options to make progress. The first is to obtain a greater number of observations, $x = \{x_1, \ldots, x_N\}$—this is the focus of Section 12.4. The second is to instead focus on validating constituent models for earthquake occurrence, ground-motion shaking, and fragility or vulnerability. This is addressed in Section 12.5.

---

[6] The only exception is if the distribution is bounded and a single observation occurs outside this bounded range. An example is an earthquake event larger than the maximum magnitude in the seismic source model (Section 12.5.1). This point is less relevant for a seismic hazard curve or empirical GMM, which are generally unbounded.

[7] The seismic hazard curve is the complementary cumulative distribution function (CCDF) of the IM of interest multiplied by the total rate of earthquakes (Section 6.2.2)

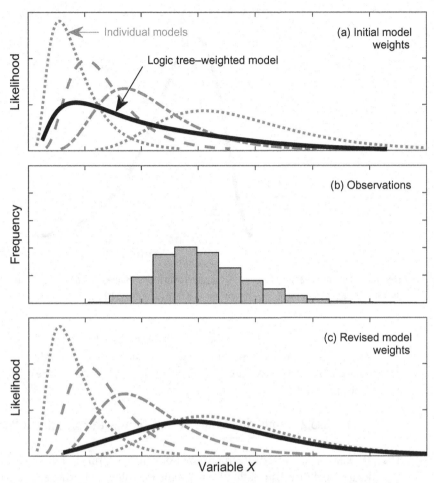

Fig. 12.5 Illustration of validation of individual models representing epistemic uncertainty in a logic tree. The validity of model prediction should occur in the context of individual models and observations and used to update the weights assigned to the underlying models. (a) Four models with initially equal weights of 0.25, (b) new observational data, and (c) revised model weights for the individual models (which, from left to right, are [0.01, 0.04, 0.2, 0.75]) and a new logic tree–weighted mean model.

12.3.3 Separation of Epistemic and Apparent Aleatory Uncertainties

Separating a logic tree–weighted mean prediction from the underlying model predictions for each branch of the logic tree (see Section 6.8) is important for validation, as illustrated in the following example. Figure 12.5 extends the problem posed in Figure 12.4, but with multiple models and observational data. Figure 12.5a illustrates PDFs from four individual models that are combined in a logic tree, with equal weights, to obtain a logic tree–weighted mean model prediction. Figure 12.5b illustrates a histogram reflecting multiple observations. Epistemic uncertainty is associated with the notion that there is only one true, but unknown, representation of reality. Thus, observations indicate the validity of each individual model, over the parameter space consistent with the data. The weights can be revised based on this new information about their validity. Figure 12.5c illustrates the same four individual models, and a new mean model based on revised weights of $[0.01, 0.05, 0.2, 0.75]$ that are more consistent with the observations.

Conversely, it is not informative to directly compare the observations in Figure 12.5b with the logic tree–weighted model in Figure 12.5a, as it is not expected that observational data should exactly match that distribution.

The extent to which model weights are revised in light of new observations will depend on the ratio of the information provided by the new observations relative to the information that produced the original weights. Where the information from new observations is small relative to existing information, the weights should not materially change. However, if the new observations provide important information or contradict an individual model, substantial reweighting may be needed. "Information" in this regard relates to the quantity and relevance of the observations. Quantity is relatively easy to understand because it helps establish significant deviations from natural variations in small samples. Relevance relates to the observations' underlying causal parameters compared with the causal parameters important for the problem of interest. For example, observations of ground motions from small-magnitude earthquakes at large source-to-site distances are abundant (i.e., Figure 4.13), yet the prediction of large-magnitude earthquakes at near-source distances is often of principal concern, so such observations may provide little information. A second example is that observed earthquake activity over a short time period provides little information for an earthquake rupture forecast of rare events, due to limited data and possible sample bias due to time-dependent earthquake activity (e.g., Figure 3.6). In these examples, numerous data of low relevance may lead to insignificant changes in model weights. The need to consider the relevance for the specific forward prediction of interest makes updating model weights inherently subjective. Techniques such as Bayesian updating can guide model updating but are generally not used blindly without considering the model's intended application.

In the above discussion, the focus was on validating individual models and adjustments in model weights. The individual models themselves can also be adjusted based on the observational data, as discussed further in Section 12.5.

12.4 Direct Validation of Seismic Hazard Curves

"Direct" comparison of a site's seismic hazard curve to the observed rate of ground-motion intensity at the site is the most explicit means to validate it. The challenges with this approach are two-fold: (1) it is uncommon to have direct observations at the location of interest, and (2) a long observation period is needed to constrain the hazard at exceedance probabilities of interest. While having instruments at a specific location of interest for an extended period of time is very uncommon, such settings do exist at generic locations where seismometers are installed as part of instrumentation networks. Therefore, a direct comparison can be performed in limited instances, enabling examination in a research context. The issues associated with using direct observations to understand seismic hazard were introduced in Section 1.4.2, and similar challenges exist here. We will first examine the problem via an illustrative calculation in the next subsection, and then subsequently examine the potential uses of direct validation in the face of such challenges.

12.4.1 Required Observation Period

Assume that some ground-motion intensity level z is exceeded with probability p in 1 year at a given location. How many years do we need to observe ground motions to obtain a reasonable empirical (frequentist) estimate of p?

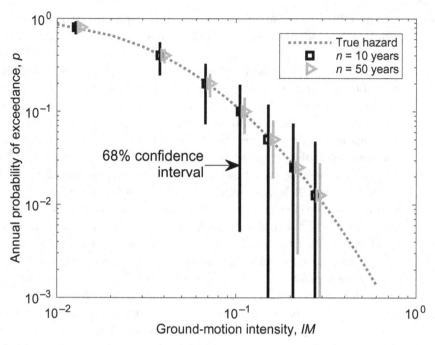

Fig. 12.6 Illustration of direct validation of a seismic hazard curve based on observations. At a given IM level, the true hazard provides the p, and the $n = 10$ and 50 years results indicate the mean (in squares) and 68% confidence interval (in lines) of the $\hat{p}$ estimates that would result from the given amount of observational data.

Let n denote the number of years of observations, X denote the number of years when we observe a ground motion with amplitude greater than z, and $\hat{p} = X/n$ be our frequentist estimate of p. Assuming exceedances of z are independent from year to year, then X is a binomial random variable with mean np and standard deviation $\sqrt{np(1-p)}$ (Equations A.74 and A.75). Then $\hat{p} = X/n$ has mean and standard deviation

$$\mu_{\hat{p}} = p \tag{12.1}$$

$$\sigma_{\hat{p}} = \sqrt{\frac{p(1-p)}{n}}. \tag{12.2}$$

Figure 12.6 illustrates the effect of observation period and exceedance probability on our ability to validate the hazard curve with any reasonable precision.

For a given observation period (e.g., $n = 10$ years), as p decreases, the 68% confidence interval[8] increases significantly. Similarly, for a given p, the confidence interval is appreciably larger for $n = 10$ than $n = 50$ years.

Noting that we are interested in small values of p (such that $1 - p \cong 1$), the coefficient of variation of our estimate is

$$\delta_{\hat{p}} = \frac{\sigma_{\hat{p}}}{\mu_{\hat{p}}} \cong \frac{1}{\sqrt{np}}. \tag{12.3}$$

---

[8] Based on asymptotic convergence to a normal distribution via the central limit theorem, this confidence interval is approximately equal to $[\mu_{\hat{p}} \pm \sigma_{\hat{p}}]$.

The form of this expression should make intuitive sense. Both larger n (longer observation window) and larger p (higher probability of observation) reduce the coefficient of variation because they increase the expected number of observed events used to generate the estimate.

If we want an estimate of p with a standard deviation of one-third of the true value of p (i.e., $\delta_{\hat{p}} = 1/3$), we can substitute into Equation 12.3 and rearrange to obtain

$$n = \frac{1}{p\delta_{\hat{p}}^2} = \frac{9}{p} \approx \frac{10}{p}. \tag{12.4}$$

So, if we would like to estimate a probability $p = 1/50$ with this desired level of precision, we need to observe 450 years of ground motions. Given this equation, a useful rule of thumb is that we should observe 10 times the length of the return period of interest ($1/p$) to get a "reasonable" direct constraint on that probability. Furthermore, this rule of thumb gives us only a relatively imprecise estimate (i.e., a coefficient of variation of 1/3). If we desired $\delta_{\hat{p}} = 1/10$, then we would need an observation period of approximately 5000 years for this $p = 1/50$ case. Beauval et al. (2008) provide further discussion on the challenges of direct hazard curve validation.

12.4.2 Effective Direct Hazard Validation

The results of Figure 12.6 illustrate that such validation is likely to be useful only (1) for small ground-motion levels that occur very frequently, or (2) if the seismic hazard model is grossly erroneous. Nonetheless, the direct validation of seismic hazard estimates has been performed as a means to assess the overall validity of seismic hazard predictions (e.g., Ordaz and Reyes, 1999; Mak and Schorlemmer, 2016; Petersen et al., 2017). In most such studies, it is common to increase the power of statistical significance testing by considering multiple locations, which increases the observational data for a given time period. The logic is that while the deviation at a single site of the predicted probability, p, from the observed probability, $\hat{p}$, might not have sufficient statistical significance, a consistent deviation across many sites may have. The principal conclusion of such studies to date has been that if PSHA calculations use seismicity catalogs that were declustered to remove aftershocks, the resulting seismic hazard curves tend to underestimate the observed exceedance rate of small ground-motion intensities, which are dominated by aftershock sequences (which is retrospectively unsurprising).

Another indirect means to validate over longer observational periods is to use noninstrumental measures of ground-motion shaking. Macro-seismic intensity observations from written records of large past earthquakes are available over hundreds of years in some regions of the world (e.g., Miyazawa and Mori, 2009). Similarly, fragile geologic features such as unstable rock formations that topple when subject to certain levels of shaking (Brune, 1999; Hanks and Abrahamson, 2008) can also indicate ground-motion intensity thresholds that have not been exceeded over tens to hundreds of thousands of years. These noninstrumental measures of ground shaking greatly increase the observational period, n, which allows examination of the hazard curve validity at smaller exceedance probabilities. However, the link between macro-seismic intensity or fragile geologic feature toppling capacity (including its time-dependence) and instrumental ground-motion intensity is uncertain. These uncertainties are significant and make it challenging to validate seismic hazard results unless there are order-of-magnitude differences between the model and data (e.g., Baker et al., 2013). Such differences have, however, been observed in seismic hazard predictions at very low exceedance probabilities (typically, $p < 1 \times 10^{-4}$) (Hanks and Abrahamson, 2008).

12.5 Validation of Model Components

As explained in the previous section, because seismic hazard analysis principally focuses on low annual probabilities of exceedance, direct validation of the seismic hazard curve is generally not practical on a site-specific basis. However, significant progress toward validation can occur by recognizing that the seismic hazard calculation (Equation 6.1) is a mathematical statement of the total probability theorem. Therefore, rather than validate the final result, it is sufficient to validate every one of its constituent components. At the highest level, these constituents are the SSM and GMM. Seismic risk calculations (i.e., Equations 9.20 and 9.26) also require the validation of fragility or vulnerability functions. The validation of each of these three types of constituent models is addressed in the subsections to follow.

A general consideration in the validation of model components, which applies to all of the following subsections, is the relevance of potential revisions. For example, observational data used in validation may indicate that a SSM underpredicts the occurrence rate of $M < 3$ earthquakes (or that a GMM overpredicts ground motions for such $M < 3$ events), but not for $M > 3$ earthquakes. However, at exceedance rates of interest for a specific problem, the seismic hazard may be dominated by events with $M > 6$. Therefore, the apparent model invalidity associated with $M < 3$ events is not relevant, and the effort required to undertake model revision is not warranted. Because relevance will vary between applications, such processes and decisions should be adequately documented to ensure the seismic hazard model results are not used in instances of known invalidity.

12.5.1 Validation of Seismic Source Models

The SSM provides the locations, characteristics, and frequencies of rupture scenarios in a particular region. The output of an SSM for a specific time period is referred to as an earthquake rupture forecast (ERF) (Chapter 3). Validation of an SSM, therefore, requires the comparison of an ERF with observed earthquakes over the same forecast time period, and should consider the following two basic questions:

1. Are the locations and characteristics of the earthquake observations physically permissible according to the model?
2. Is the distribution of the earthquake observations consistent with the probabilistic model description?

The first question determines whether the observations are possible, according to the model. If they are not, then the observations invalidate the model. If the observations are permitted by the model, the second question considers whether they are probabilistically consistent with the model. The following two subsections address these two questions. A third subsection discusses indirect SSM validation through the continuous assimilation of data that are relevant in the SSM development itself, particularly for rare large-magnitude events.

In the following discussion, note that the two questions ask whether the model is consistent with the data, and not how informative it is. Very simple and uninformative models can satisfy the two questions, but may not usefully forecast the ruptures that dominate the seismic hazard. A more informative model that is also consistent with the data is thus often preferred.

Plausibility of Observed Earthquakes

Seismic sources are defined either as fault sources or distributed sources, as discussed in Section 2.6. Therefore, it is first necessary to determine whether each observed earthquake should be associated with the fault or distributed source components of the model. Figure 3.11 provides the specific approach and illustration of this process. While the resulting source assignment is probabilistic, a likely source will be assigned, from which the validation process can continue.[9]

After associating observed earthquakes with modeled seismic sources, the observations' characteristics can be compared with the seismic sources. As discussed in Chapter 2, earthquake sources have many characteristics, including tectonic type, magnitude, rake/dip/strike, fault length/width, and depth, among others. For each of these characteristics it can be asked whether the observation[10] is permissible according to the SSM. The magnitude of an observed earthquake is the most notable characteristic and therefore discussed further below.

A maximum magnitude for a distributed source is usually assigned in its magnitude-frequency distribution, such as the doubly bounded exponential or the characteristic distribution (Section 3.5). Hence, observed earthquakes exceeding such maximum magnitudes would invalidate the SSM or lead to reweighting of logic-tree branches (Figure 12.5). This potential for invalidation provides a tendency to increase the maximum magnitude assigned to distributed sources in SSMs over time.[11] Increasing this maximum magnitude does not always result in a practically-significant difference in seismic hazard results, however, because the rate of such events may be low relative to those of fault sources in the region.

The maximum magnitude for fault sources is typically determined from source-scaling relationships (Section 2.7.1). Therefore, if an observed earthquake magnitude exceeds the maximum source magnitude, it is likely because the geometric extent of the source, or its static stress drop, was significantly underestimated. The subducting Pacific Plate off the east coast of Japan is an example where the SSM for the subduction interface was comprised of many different segments. The 2011 Tohoku earthquake ruptured multiple such segments (which was not permitted by the model; Headquarters for Earthquake Research Promotion, 2005; Simons et al., 2011), resulting in a much larger earthquake magnitude than that specified for the individual possible segmented ruptures (Kagan and Jackson, 2013). The persistent observations that earthquakes do not rupture with consistent "segment boundaries" have led to an increased motivation to adopt SSM methods in which fault sources do not require segmentation assumptions (Section 3.6.2).

Spatial, Temporal, and Magnitude Distributions

If the observed earthquakes are plausible according to the SSM, the next question is whether they are consistent with the probabilistic description. Because it is not possible to use single observations to validate or invalidate probabilistic models (Section 12.3.2), addressing the second question requires multiple observations.

---

[9] If the largest two (or more) potential sources have similar probabilities, then the subsequent validation can occur for each relevant source.

[10] Many aspects of an earthquake "observation" are not directly observable, but are inferred based on inverse modeling. Surface rupture deformations can be measured, but rupture subsurface geometry is inferred from source inversion models. Validation is possible with inferred characteristics, but their uncertainty should be considered.

[11] For example, in Northern California, the maximum magnitude in consensus models increased from 7.25 ± 0.25 in 2002 (Working Group on California Earthquake Probabilities, 2003) to 7.6 ± 0.3 in 2014 (Field et al., 2014).

The Collaboratory for the Study of Earthquake Predictability (CSEP) (Jordan, 2006; Schorlemmer et al., 2018) is a global effort to advance formalized statistical testing of ERFs obtained from SSMs, in both retrospective and prospective settings. It has developed a coherent set of validation tests and nomenclature. Examples include the following:

1. Number of events (N-test): For a given time period, spatial region, and magnitude range, determine the consistency of the number of observed and forecasted earthquakes.
2. Information gain per event (T-test): For a given time period, spatial region, and magnitude range, compare the (statistical) information gain per earthquake of one model relative to a reference model (Rhoades et al., 2014).

The utility of the above tests is a function of whether they are applied in a retrospective or prospective manner, and the exceedance probabilities that are of interest. Figure 3.21, for example, compares a modeled magnitude-frequency distribution with the observed distribution. In that (retrospective) example, the model was directly developed using the same data, hence the validation tests should indicate the model is consistent with that data. Nonetheless, Figure 3.21 is useful to illustrate the issue that despite using decades of observational data (the specific time period being a function of the magnitude completeness level; see Section 3.3.5), the number of observations at larger magnitudes of interest is often insufficient to identify deficiencies. This is particularly difficult with prospective testing, which may span time periods of days, months, or years (Schorlemmer et al., 2018). Over such short time durations, only small-magnitude earthquakes (and consequent ground-motion observations) occur frequently enough for statistically significant prospective validation (Section 12.4.1). Thus, CSEP-type research efforts have focused on small-magnitude events (Schorlemmer et al., 2018).

Small-magnitude validation is necessary, but not sufficient, for seismic hazard analysis, which tends to be concerned with larger-magnitude events that can cause appreciable damage to engineered structures. For example, Field (2015) highlights that the best model in a 5-year forecast of California seismicity was based on smoothed seismicity, which produces significantly lower seismic hazard estimates than models utilized in practice that consider additional geologic and paleoseismic information about the San Andreas Fault, for example. Field (2015) concludes that such validation tests are therefore not currently sufficient to determine model usefulness because they could be misleading at the larger magnitudes that dominate seismic risk.

Fortunately, the same methods for prospective validation can also be applied in a retrospective sense. Retrospective validation can consider longer time periods to enable tests at larger magnitudes, and it also enables testing of models with time-dependent seismicity. For example, if the annual probability of an $M7$ event in a particular region is 0.01, Equation 12.4 implies that an observation period of 1000 years is required for meaningful validation inferences. Rhoades et al. (2016) provides an example of the significant time-varying changes in seismicity in Canterbury, New Zealand, associated with the 2010–2011 Canterbury earthquake sequence.

Given the above challenges in the direct validation of ERFs with observed earthquakes, principally for the larger magnitudes of interest, alternative validation testing approaches are necessary. One approach is to use the SSM to generate synthetic catalogs of earthquakes in the region of interest and then examine their statistical properties relative to features of observed seismicity. For example, Musson and Winter (2012) examined multiple realizations of synthetic catalogs and determined whether the observed catalog was an outlier (indicating a flaw in the underlying SSM used to generate the synthetic catalogs). Similarly, Page and van der Elst (2018) present an approach for examining synthetic California catalogs with spatial and temporal clustering from the UCERF3-ETAS forecast, compared with a 28-year historical catalog.

Assimilation of New Information

Validating a model by comparing its prediction with observations is the most direct approach to assess predictive capability. However, the previous subsections have highlighted the challenges with SSM validation in this direct fashion.

A supplemental approach is the continuous assimilation of new data that form the subcomponents and parameters that underpin an SSM. Specifically, Chapter 3 (e.g., Figures 3.1 and 3.2) illustrates how seismological observations and geological, geospatial, and geophysical data are synthesized in order to develop a SSM. Therefore, new geological, geospatial, and geophysical data can be used to reexamine the validity of SSM model parameters and assumptions. For example:

1. Identification of a "new" fault: Does the background source adequately represent this new fault, or is it necessary to represent it as a fault source explicitly?
2. Revised characterization of fault geometry: Does the new understanding of source geometry significantly affect maximum rupture areas and the likelihood of multifault ruptures?
3. New slip rate data: Do the new data imply significant differences in the mean slip rate or slip rate variability?
4. New paleoseismic data: Do the new data on historical rupture size and timing lead to significant differences in the magnitude-frequency distribution (and time since the last rupture in the context of time-dependent forecasts) of the relevant sources?
5. Analog region data: Have earthquakes in other analog regions led to, for example, revised magnitude source-scaling relations and maximum permissible magnitudes that deviate significantly from that assumed in the present SSM?

The above list illustrates data that can be used to improve the components and parameters of a SSM. Items 1 and 2 are associated with fault representation, while items 3 and 4 are specific to individual faults. Items 1–4 are all specific to the region in question, whereas item 5 suggests how observations in other regions might provide insights for the site of interest.

12.5.2   Validation of Ground-Motion Models

Ground motions measured with seismic instruments provide the most direct means to validate GMMs. Ground-motion prediction is also conditional on the (unknown) future earthquake's properties, which creates complications. Similar to SSMs, validation of GMMs requires consideration of whether observed ground motions are both physically plausible and probabilistically consistent with the GMM. Challenges in validation also differ between empirical and physics-based GMMs (Chapters 4 and 5), and separate discussion is given to these two approaches below.

Pseudo-Prospective Validation

Prospective validation requires predictions to be made in advance of observations. As alluded to in Section 12.2.2, because the precise nature of future earthquakes is unknown, prospective validation of GMMs in the strictest sense is generally not informative.[12] Pseudo-prospective validation is an alternative approach in which earthquake source information is obtained in an automated fashion, in near real time, and the subsequent prediction from a GMM is also obtained automatically. Thus, while near-real-time prediction occurs following the causative earthquake event, because it is automated (i.e., there is no human "in the loop"), it avoids the primary issue with retrospective validation—that the analyst knows the "answer" before making the prediction.

The specific features that are required for near-real-time rupture characterization can vary depending on the GMM examined. For empirical GMMs, they would include the rupture area geometry, magnitude, average rake angle, and (potentially) hypocenter location. The rupture geometry is necessary to determine the fault dip angle, depth of the top of rupture, and to compute source-to-site distances. The hypocenter location is necessary if rupture directivity is explicitly considered. For physics-based GMMs, the required inputs are those of the utilized kinematic rupture generator (Section 5.3.2). Some such models (e.g., Graves and Pitarka, 2010) can use as little information as those for empirical GMMs, with the remaining kinematic rupture generation parameters adopting default values.

While the above sentiments may imply that pseudo-prospective and prospective GMM validation are equivalent for all intents and purposes, there is one major assumption. Pseudo-prospective GMM prediction inherently assumes that the (automated) representation of the causative earthquake rupture is correct. This becomes more important as the magnitude of the causative earthquake increases. For events with $M < 5$, a point-source representation (see Section 2.3.2) is generally sufficient, and the earthquake location, magnitude, and average rake can be easily obtained from a centroid moment tensor (CMT) solution. For these small magnitudes, the effect of finite-fault geometry on ground motions is small, so an approximate finite-fault geometry can be estimated, with the fault length and width approximated from a source-scaling relationship (Section 2.7.1). For large magnitude ($M > 7$) events, the uncertainties in the rupture geometry, magnitude, and hypocenter can have a major effect on predicted ground motions. Examples of significant differences between CMT solutions and alternative finite-fault source inversions in recent earthquakes include the 2010 Darfield (Gledhill et al., 2011) and 2016 Kaikoura (Berryman et al., 2018) earthquakes, as a result of multiple fault segments with differing rake and dip directions. As an extreme example, Hayes et al. (2011) provides an evolution of earthquake rupture representation, and subsequent ground-motion prediction, for the $M9$ 2011 Tohoku, Japan, earthquake during the first several days following the event.

In concept, it is possible to represent uncertainties in the characterization of the earthquake rupture probabilistically, allowing ground-motion prediction to be conditioned on this distribution, rather than a single "best" rupture model. However, uncertainties in the description of complex ruptures are often incomplete (due to an insufficient examination of the solution space of potential ruptures). Large source uncertainties can also overwhelm that associated with the GMM itself, and therefore limit validation inferences.

In contrast, retrospective prediction allows progress in understanding past earthquake ruptures to be included in GMM validation, albeit with its own potential pitfalls. These pitfalls can be avoided through transparent processes, including open-source software implementations and community-driven centralized testing infrastructure, such as CSEP (Michael and Werner, 2018). Hence, there

---

[12] In concept, the modeled event set from the ERF could be used to identify the "closest event" to that which occurred, and the ground-motion prediction for this closest event could be compared with the observations. However, differences between the closest-modeled and observed events will lead to biases in the ground-motion prediction that typically prohibit validation inferences.

is a natural synergy in using pseudo-prospective and retrospective GMM validation, as long as the limitations associated with each are adequately borne in mind.

Plausibility of Observed Ground-Motion Intensity Measures

A GMM provides a probabilistic prediction of ground motion resulting from a specific earthquake rupture. The nature of the prediction is specific to whether an empirical or physics-based approach is adopted. An empirical GMM (Chapter 4) provides a prediction of one or more ground-motion intensity measures (IMs), whereas a physics-based GMM (Chapter 5) provides a prediction of ground-motion time series, from which IMs of interest can be computed.

Empirical GMMs have prediction bounds of $[0, \infty)$. The IM of an observed ground motion is always within this range, and hence always physically plausible. Statistical tests of the assumed lognormal distribution (e.g., Jayaram and Baker, 2008; Strasser et al., 2009) indicate its validity for response spectral ordinates to at least $\varepsilon = 3$ standard deviations from the mean. Any proposed alternative distributions are also unbounded (e.g., Bullock, 2019).

Physics-based GMMs produce ground-motion time series that can be considered as a high-dimensional representation of the ground motion. No simulated time series will exactly match an observed time series,[13] but it is not practically important to achieve such a match. Physics-based GMMs are therefore also most appropriately validated through a comparison of the IMs (i.e., summary statistics) of observed and simulated ground-motion time series. Unlike empirical GMMs, the prediction distribution from physics-based GMMs can be considered in parametric or nonparametric forms. If an (unbounded) parametric distribution is adopted, then, similar to empirical GMMs, an observed ground motion is always plausible. If a nonparametric distribution is adopted from the ensemble of N simulations (Section 5.6.2; Figure 5.22), then many simulations are needed to quantify the tails of the distribution. If each of the N simulations has equal probability (i.e., $1/N$), and if the model is correct, there is a $2/(N + 1)$ probability that an observation will lie outside the (min,max) bounds of the ensemble. Representing a distribution to two standard deviations from the mean (a 95.5% prediction interval) would, therefore, require an ensemble of $N \approx 40$, whereas to three standard deviations (a 99.7% prediction interval) would require an ensemble of $N \approx 700$. The ensemble of simulations should also reflect appropriate model uncertainty, as discussed in Section 5.6.2. Finding a model implausible in this manner is therefore nontrivial.

Probabilistic Consistency

Similar to SSMs, examining the probabilistic consistency of GMMs requires comparison with multiple ground-motion observations. Section 4.6.2 previously illustrated the comparison of observations to a GMM, including partitioning the (total) residual between the observation and mean prediction into between- and within-event terms.[14] One important concept illustrated in Section 4.6.2 is that 82 observations from one earthquake led to 82 within-event residuals and only one between-event residual. Therefore, validation of the between-event component requires observations from multiple earthquakes.

---

[13] Considering that a $t = 100$ s time series with a time increment of $\Delta t = 0.005$ s is a vector of 20,000 values, even extensive uncertainty consideration is extremely unlikely to yield a simulation that matches this 20,000-dimension vector.

[14] This same approach can be used for physics-based GMMs, as alluded to in Figure 5.21. Multiple IMs should be used to examine the physics-based GMM comprehensively.

The between- and within-event residual partitioning can be generalized further to include additional residual terms for systematic path and site effects (i.e., Table 8.2). In the context of GMM validation, the observational data should ideally sample all of the features of the model (e.g., Equation 8.13) that are relevant in the seismic hazard calculation. This means that ground-motion data should ideally be from multiple earthquakes, that traverse different wave propagation paths, to the one or more sites of interest. Furthermore, the multiple earthquakes should represent the range of source-related parameters (e.g., tectonic type, magnitude, depth, focal mechanism) that are relevant for the subsequent seismic hazard and risk calculation.[15]

With suitable validation data, it is possible to evaluate the predictive capability of GMMs. Trends in prediction-observation residuals can be used to examine the model's mean, variance, and scaling with predictor variables. For instance, the example in Section 4.6.2 examined the mean and standard deviation of the within-event residuals, and its dependence on R and $V_{S,30}$. It is also possible to validate the assumed parametric distribution for ground-motion uncertainty (Section 4.4.3), although this will require a large number of observations.

Site- and Region-Specific Validation

Ground-motion prediction, and consequently its validation, is inherently of interest for a specific site, or region, in question. There is a trade-off between the number of available ground-motion observations for validation and the size of the considered region. Ergodic models lie at one end of this spectrum, using many observations, but with the downside that the models represent global effects rather than those of a specific site. Fully non-ergodic models (Chapter 8) reside at the other end of the spectrum, providing a region- or site-specific prediction, but with the challenge of needing sufficient observational data in their region or site for validation. This trade-off leads to a variation in the use of the ergodic assumption as a function of the amount of site-specific data available (Figure 8.2).

Comparing site- or region-specific observations with ergodic empirical GMMs can show whether the ergodic assumption is appropriate, or whether there are sufficient data to consider a partially non-ergodic GMM (Chapter 8). If observations are compared with an empirical GMM that has already relaxed the ergodic assumption, then the same features previously discussed in Section 4.6.2 are of interest. It is important to keep in mind that the variability in the observation residuals for a particular site or region should normally be less than the ergodic variability of the GMM. This makes it challenging to draw strong inferences as to the appropriateness of the ergodic GMM variability.

Physics-based GMMs, principally those solving the 3D wave equation (Section 5.5.1), have further validation challenges based on the significant spatial variation in their predicted ground motions as a result of 3D crustal models. Region-specific 3D crustal models are inherently non-ergodic. Bradley et al. (2017) discuss generic, region-, and site-specific approaches for validating physics-based GMMs. They also emphasize that model validity can vary significantly on a site-by-site basis, and that the spatial density of ground-motion observations is typically insufficient to scrutinize some of these aspects at present.

A challenge that exists with both empirical and physics-based GMMs is the "extrapolation" of their predictive performance against available observations compared with the seismic scenarios that principally influence seismic hazard and risk calculations (e.g., Musson, 2010). Beauval et al. (2012), for example, illustrate the differences in the predictive performance of alternative empirical GMMs

---

[15] This criterion will generally not be satisfied due to the lower likelihood of observing larger-magnitude earthquakes, particularly for region-specific validation.

for observed ground motions from earthquakes over different magnitude ranges. Importantly, the differences in magnitude scaling of the GMMs can lead to higher or lower performance against the ground motions from smaller magnitude events than that associated with higher magnitude events. Validation of extrapolation for empirical GMMs is a question of whether the model uses an appropriate parametric functional form. In contrast, the inherent physical basis of physics-based GMMs provides a rationale for extrapolation beyond observations. For example, available small-magnitude earthquakes can be used to validate a 3D crustal model (Bradley et al., 2017). Such events will not provide the ability to validate nonlinear site response associated with large amplitude ground motions, but these effects can be examined using other means (Section 8.6). The same is true for any specific features of large-magnitude sources, as discussed in Section 8.8.

Caution in extrapolating predictive performance of GMMs (and SSMs) should also apply to ranking models and assigning degrees of belief. Figure 12.5 illustrated how observational data could be used to reweight alternative models. This reweighting is valid for the specific conditions that the observational data represent. It not certain, however, that the same weights are valid when the models are extrapolated beyond the validation data.

12.5.3 Validation of Fragility and Vulnerability Functions

Methods for the calibration of seismic fragility and vulnerability functions were presented in Section 9.3. Validation of such functions using observational data is similar to the empirical calibration approach in Section 9.3.1. The only difference is that during calibration, empirical observations are used to estimate the parameters of the function, whereas, during validation, goodness-of-fit tests (Ang and Tang, 2007) are used to ascertain whether the observations are consistent with the function. As noted in Section 9.3, the principal challenges in the development and validation of fragility and vulnerability functions using observational data are (1) the acquisition of damage and loss estimates using consistent methods and (2) obtaining accurate values of the damage, loss, and the causative ground-motion intensity.

One way to obtain validation observations is through laboratory testing. In laboratory settings, uncertainty in the causative ground-motion intensity can be eliminated through direct measurement, and damage can be measured in great detail. The losses associated with damage can be estimated from a suitably trained quantity surveyor or loss adjuster. To account for apparent variability in the ground motion and fragility of the component (or system) considered, multiple laboratory tests can be performed. The principal challenge with laboratory-based data is the degree to which it represents reality—in particular, the ground-motion loading applied, and the extent to which the construction of the laboratory specimen represents the boundary conditions in the field (Bradley, 2010a). Fragility and vulnerability functions for individual assets can be aggregated to define loss models for distributed infrastructure systems (e.g., Figure 11.1) or insurance portfolios. Crowley et al. (2008) considered whether observed losses from an earthquake can be used to validate such models, and concluded that without a dense array of seismic instruments, uncertainty in the causative ground motion prevents assessment of loss model validity.

12.6 Do Failures of Past Calculations Invalidate the PSHA Methodology?

There are instances where SSM and GMM assumptions can be demonstrated to be invalid and to produce incorrect results as a consequence. Extending this train of thought further, is it possible that failures of past PSHA calculations can invalidate the PSHA methodology itself? If you have arrived here having read the preceding chapters, then hopefully your intuition at this point is a resounding "No!" As discussed when introducing PSHA intuitively in Chapter 1, and then more formally in Chapter 6, PSHA is merely an application of the total probability theorem—effectively an accounting or bookkeeping tool—to consistently combine models for earthquake occurrence and consequent ground-motion intensity. Hence, it follows mathematically that if the two key ingredients of PSHA— an SSM and GMM—are valid, then the resulting PSHA result will also be valid. Conversely, if the PSHA result is shown to be invalid, it is due to flaws in either the adopted SSM, GMM, or both.

Despite the above sentiments, damaging earthquakes often grant skeptics an opportunity to question the use of PSHA. Stein et al. (2011) and Geller (2011) represent examples following the 2011 $M9$ Tohoku, Japan, earthquake, which Hanks et al. (2012) rebut by pointing out the main misunderstandings of PSHA results: (1) individual observations cannot be used to invalidate probabilistic statements (see Section 12.3); (2) spatial patterns of observations over short time periods cannot be used to draw inferences on relative seismic hazard between different regions; and (3) low (relative) seismic hazard does not equal no seismic hazard. Proposed alternatives to PSHA offered by detractors invariably represent some form of deterministic approach and, as discussed in Section 1.3, have significant and self-evident deficiencies.

Another, more concrete, concern in PSHA is the treatment of variability and uncertainties. Bommer and Abrahamson (2006) discuss instances where seismic hazard estimates at the same location increased over time as PSHA studies were repeated (see also Section 6.10.1). While science (and consequent PSHA results) will continually evolve, analysts should strive to ensure that at any point in time, PSHA results are unbiased (neither conservative nor unconservative). The principal problem raised by Bommer and Abrahamson (2006) is that the applications of PSHA in earlier decades commonly ignored or grossly underestimated ground-motion variability. As shown in Section 6.5.4, underestimating uncertainty leads to an underestimation of the seismic hazard.

In the present day, appropriate treatment of apparent aleatory variability is commonplace. However, appropriate consideration of epistemic uncertainties sometimes remains a challenge. Logic trees (Section 6.8) should represent the true uncertainty in seismic hazard analysis inputs. In practice, however, they frequently reflect simply the range of available alternative scientific models (Abrahamson, 2006). Consequently, using the few available models for a site with little data may produce small epistemic uncertainty, when the lack of data and models should in fact suggest considerable uncertainty. A general solution to this problem is challenging (Marzocchi and Jordan, 2017). A useful approach in these cases would be to establish epistemic uncertainties for well-studied regions and using them as lower bounds for poorly-studied regions (e.g., Bradley, 2009; Douglas et al., 2014). This would ensure that future studies (when epistemic uncertainties are more rigorously considered) are not systematically different, as was noted above for apparent aleatory variability.

12.7 Seismic Hazard and Risk Analysis for Decision-Making

In the end, if hazard and risk calculations cannot be definitively validated, are they still of value? This question can be answered by considering their ultimate purpose—to support effective decision-making—and the context that all models are wrong, but some are useful (Box, 1980; Field, 2015). A wide range of societal decisions utilize, or even require, analogous risk calculations that also can be partially, but not completely, validated (Pate-Cornell, 1994). Genetically modified crops, industrial operations, and sources of pollution are all regulated based on assessments of the likelihood and consequences of events that are difficult to observe or have not yet happened (e.g., NRC, 1975; Anderson, 1983; Krayer von Krauss et al., 2004). In such cases, the likelihoods of various models being correct are considered using tools analogous to logic trees presented here, and probabilities of initiating events (analogous to earthquake ruptures) are assessed from empirical observations or other data. The outputs of interest in such problem domains are the probabilities of adverse outcomes, so that they can be weighed against the costs of regulations that could decrease risks.

Seismic hazard and risk analyses are ultimately bookkeeping tools that provide a transparent method to understand the implications of our scientific models (National Research Council, 1996). They are not magical procedures that can somehow produce good decisions despite bad input models. Continued attention is needed to refine and improve the input seismic source, ground motion, and consequence models, as our understanding of these phenomena grows, and validate them to the extent possible. But the mathematical procedures are sound, and conceptually equivalent procedures are widely adopted in other fields. With careful framing and high-quality input models, they are an invaluable tool to support decision-making in domains such as calibration of building codes, pricing of insurance, or private owners' decisions about upgrading their facilities.

Exercises

12.1 Compare and contrast verification and validation in the context of seismic hazard and risk analysis, discussing, in particular, the role of theory and observational data.

12.2 Realistic model representations of seismic sources, ground motions, and the resulting seismic hazard are complex to the point where they cannot be expressed analytically. In this context, discuss how the implementations of such models can be verified.

12.3 Discuss the benefits and drawbacks of retrospective validation, relative to prospective validation.

12.4 The seismic hazard for PGA at a site is given by $\lambda(PGA > pga) = k_0[pga]^{-k}$ (see Equation 6.31), with $k_0 = 2 \times 10^{-6}$ and $k = 3$.

   (a) What is the PGA value corresponding to the "design" intensity with a mean annual exceedance rate of $\lambda = 1/500$?

   (b) An earthquake occurs in the region and produces a ground motion with $PGA = 0.15$ g at this site. What was the probability that this PGA value, or greater, was observed in the $T = 100$ year lifetime of the structure?

   (c) How does this probability determined in (b) compare with the probability that the "design" level PGA value was observed?

12.5 Consider the required seismic hazard loading for a new building. A critical component of the proposed construction method is temporary scaffolding that has a capacity of $PGA = 0.05$ g, and to satisfy the construction design criteria, must have a probability of failure (which occurs if $PGA > 0.05$ g) less than 10% during the construction phase of 1 year. The site is located beside a seismic instrument that has been installed for 25 years, which has recorded two events with a ground motion exceeding $PGA = 0.05$ g, with the largest recording of $PGA = 0.18$ g. A site-specific PSHA estimates the two ground-motion levels for the seismic performance assessment of the structure as being $PGA = 0.15$ g and 0.35 g, with mean annual exceedance rates of $\lambda = 0.02$ and 0.002, respectively.

(a) What is the mean estimate of the probability of scaffolding failure based on direct observational data from the seismic instrument?

(b) What is the coefficient of variation of this likelihood in (a)?

(c) What is the mean and coefficient of variation of the estimated probabilities for the two different ground-motion levels for the design of the structure itself? Are the observations consistent with the PSHA results?

(d) How long would you estimate the seismic instrument needs to be present in order to confirm the appropriateness of the PSHA result for $PGA = 0.15$ g and 0.35 g with a coefficient of variation of 0.5?

12.6 Seismic hazard and risk analysis results usually depend strongly on predictions about scenarios that have been observed in limited quantities (if at all). Comment on the importance of validation for understanding the predictive capability of models for extrapolation beyond their observational data, and how predictive performance against observed data should be considered in weighting alternative models that are considered to reflect epistemic uncertainty.

12.7 Chapters 4 and 5 described empirical and physics-based methods of ground-motion characterization, including their relative strengths and weaknesses. Considering the availability of predictions from multiple empirical and physics-based models for a particular PSHA project, discuss how you would consider the validity of these alternative methods, and how to weight them.

Appendix A Basics of Probability

Probability is so fundamental to probabilistic seismic hazard and risk analysis that the word appears in its name. Thus, the calculations in this book rely heavily on the use of probability concepts and notation, to facilitate calculations that account for uncertainty. This appendix provides a brief overview of several important concepts. Readers desiring more details will benefit from reviewing a textbook dedicated to engineering applications of probability concepts (e.g., Ang and Tang, 2007; Benjamin and Cornell, 2014).

A.1 Random Events

The basic building block of probability calculations is the *random event*: an event having more than one possible outcome. The *sample space* (denoted S) is the collection of all possible outcomes of a random event. Any subset of the sample space is called an *event* and denoted E. Sample spaces and events are often illustrated graphically using Venn diagrams, as illustrated in Figure A.1.

As an example, the number obtained from rolling a die is a random event. The sample space is $S = \{1, 2, 3, 4, 5, 6\}$. The outcomes in the event that the number is odd are $E_1 = \{1, 3, 5\}$, and that the number is greater than three are $E_2 = \{4, 5, 6\}$

We are commonly interested in two operations on events. The first is the *union* of E_1 and E_2, denoted as $E_1 \cup E_2$, which is the event that contains all outcomes in either E_1 or E_2. The second is the intersection, denoted as $E_1 E_2$ (or $E_1 \cap E_2$), which is the event that contains all outcomes in both E_1 and E_2. Continuing the die example from above, $E_1 \cup E_2 = \{1, 3, 4, 5, 6\}$ and $E_1 \cap E_2 = \{5\}$.

The following concepts are often useful for probability calculations:

1. Events E_1 and E_2 are *mutually exclusive* when they have no common outcomes (i.e., $E_1 E_2 = \phi$, where ϕ is the *null event*).
2. Events $E_1, E_2 \ldots E_n$ are *collectively exhaustive* when their union contains every possible outcome of the random event (i.e., $E_1 \cup E_2 \cup, \ldots, \cup E_n = S$).
3. The *complementary event*, $\overline{E_1}$, of an event E_1, contains all outcomes in the sample space that are not in event E_1. By this definition, $\overline{E_1} \cup E_1 = S$ and $\overline{E_1} E_1 = \phi$. That is, $\overline{E_1}$ and E_1 are mutually exclusive and collectively exhaustive.

We will be interested in the probabilities of occurrence of various events. These probabilities must follow three axioms of probability:

$$0 \leq P(E) \leq 1, \tag{A.1}$$

$$P(S) = 1, \tag{A.2}$$

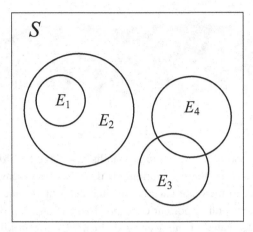

Venn diagram illustrating a sample space and four events, E_1 to E_4.

and, for mutually exclusive events E_1 and E_2,

$$P(E_1 \cup E_2) = P(E_1) + P(E_2). \tag{A.3}$$

These axioms form the building blocks of all other probability calculations. It is easy to derive additional laws using these axioms and the previously defined events. For example,

$$P(\bar{E}) = 1 - P(E) \tag{A.4}$$

$$P(\phi) = 0 \tag{A.5}$$

$$P(E_1 \cup E_2) = P(E_1) + P(E_2) - P(E_1 E_2). \tag{A.6}$$

A.2 Conditional Probability

The probability of the event E_1 may depend upon the occurrence of another event E_2. The conditional probability $P(E_1|E_2)$ is defined as the probability that event E_1 occurs, given that event E_2 has occurred. That is, we are computing the probability of E_1, if we restrict the sample space to only those outcomes in event E_2, as depicted in Figure A.2.

We can deduce the following from Figure A.2:

$$P(E_1|E_2) = \begin{cases} \frac{P(E_1 E_2)}{P(E_2)} & \text{if } P(E_2) > 0 \\ 0 & \text{if } P(E_2) = 0. \end{cases} \tag{A.7}$$

Rearranging this equation, for the nontrivial case of $P(E_2) > 0$, gives

$$P(E_1 E_2) = P(E_1|E_2)P(E_2). \tag{A.8}$$

A.2.1 Independence

Conditional probabilities give rise to the concept of independence. We say that two events are stochastically *independent* if they are not related probabilistically in any way. More precisely, we say that events E_1 and E_2 are independent if

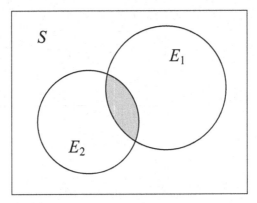

Fig. A.2 Schematic illustration of the events E_1 and E_2. The shaded region depicts the area corresponding to event $E_1 E_2$.

$$P(E_1|E_2) = P(E_1).\tag{A.9}$$

That is, the probability of E_1 is not in any way affected by knowledge of the occurrence of E_2. Substituting Equation A.9 into Equation A.8 gives

$$P(E_1 E_2) = P(E_1)P(E_2),\tag{A.10}$$

which is an equivalent way of stating that E_1 and E_2 are independent. Note that equations A.9 and A.10 are true *if and only if* E_1 and E_2 are independent.

A.2.2 Total Probability Theorem

Consider an event A and a set of mutually exclusive and collectively exhaustive events $E_1, E_2, \ldots, E_n$. The Total Probability Theorem states that

$$P(A) = \sum_{i=1}^{n} P(A|E_i)P(E_i).\tag{A.11}$$

In words, this tells us that we can compute the probability of A if we know the probabilities of the E_i events, and know the probability of A, given each E_i. The schematic illustration in Figure A.3 may help us to understand what is being computed. This calculation is valuable when the probability of A is challenging to determine directly, but the problem can be broken down into several pieces whose probabilities can be computed.

PSHA is a direct application of the Total Probability Theorem. The probabilities of earthquake occurrences are studied independently of the conditional distribution of resulting ground-motion intensity, and this probabilistic framework allows us to correctly recombine the various sources of information.

A.2.3 Bayes' Rule

Consider an event A and a set of mutually exclusive and collectively exhaustive events $E_1, E_2, \ldots, E_n$. Using Equation A.9, we can write

$$P(AE_j) = P(E_j|A)P(A) = P(A|E_j)P(E_j).\tag{A.12}$$

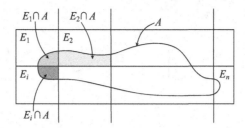

Fig. A.3 Venn diagram illustrating the Total Probability Theorem.

Rearranging the last two terms gives

$$P(E_j|A) = \frac{P(A|E_j)P(E_j)}{P(A)}. \tag{A.13}$$

This equation is known as Bayes' Rule. An alternate form substitutes Equation A.11 for $P(A)$:

$$P(E_j|A) = \frac{P(A|E_j)P(E_j)}{\sum\limits_{i=1}^{n} P(A|E_i)P(E_i)}. \tag{A.14}$$

Like the Total Probability Theorem, Bayes' Rule provides a calculation approach for combining pieces of information to compute a probability that may be difficult to determine directly. The utility of Bayes' Rule lies in its ability to compute conditional probabilities when available information has conditioning in the reverse order of what is desired. That is, you would like to compute $P(A|B)$ but know only $P(B|A)$.

A.3 Random Variables

A *random variable* is a numerical variable whose specific value cannot be predicted with certainty before the occurrence of an "event." Examples of random variables relevant to seismic hazard and risk are the time to the next earthquake in a region, the magnitude of a future earthquake, the distance from a future earthquake to a site, ground-shaking intensity at a site, etc.

We need a notation to refer to the random variable itself, and to numerical values that the random variable might take. The standard convention denotes a random variable with an uppercase letter and denotes the values it can take by the same letter in lowercase. That is, $x_1, x_2, x_3, \ldots$ denote possible numerical outcomes of the random variable X. We can then talk about probabilities of the random variable taking those outcomes (i.e., $P(X = x_1)$ is the probability of X taking value x_1).

In general, we can treat all random variables using the same tools, except for distinguishing between discrete and continuous random variables. If the number of values a random variable can take are countable, the random variable is called *discrete*. An example of a discrete random variable is the number of earthquakes occurring in a region in a specified period of time. The probability distribution for a discrete random variable can be quantified by a probability mass function (PMF), defined as

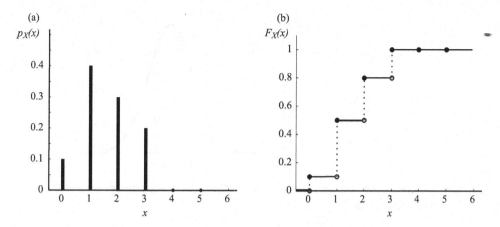

Example descriptions of a discrete random variable. (a) Probability mass function (PMF). (b) Cumulative distribution function (CDF).

$$p_X(x) = P(X = x). \tag{A.15}$$

The cumulative distribution function (CDF) is defined as the probability of the event that the random variable takes a value less than or equal to the value of the argument:

$$F_X(x) = P(X \leq x). \tag{A.16}$$

The PMF and CDF have a one-to-one relationship:

$$F_X(a) = \sum_{\text{all } x_i \leq a} p_X(x_i). \tag{A.17}$$

Examples of the PMF and CDF of a discrete random variable are shown in Figure A.4.

In many cases, we are interested in the probability of $X > x$, rather than the $X \leq x$ addressed by the CDF. Noting that those two outcomes are mutually exclusive and collectively exhaustive events; we can use the previous axioms of probability to see that $P(X > x) = 1 - P(X \leq x)$. Note that the $>$ and $\leq$ signs are used so that the $X = x$ case is not considered in both of the mutually exclusive events, but pragmatically this is not important because $P(X = x) = 0$.

In contrast to discrete random variables, *continuous* random variables can take any value on the real axis. Because there is an infinite number of possible realizations, the probability that a continuous random variable X will take any single value x is zero. This forces us to use an alternate approach for calculating probabilities. We define the probability density function (PDF) as

$$f_X(x)\, dx = P(x < X \leq x + dx) \tag{A.18}$$

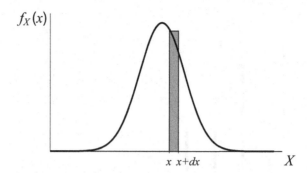

Plot of a continuous PDF. The area of the shaded rectangle, $f_X(x)\, dx$, represents the probability of the random variable X taking values between x and $x + dx$.

where dx is a differential element of infinitesimal length. An illustration of the PDF and related probability calculation is given in Figure A.5. We can compute the probability that the outcome of X is between a and b by "summing" (integrating) the PDF over the interval of interest:

$$P(a < X \leq b) = \int\limits_a^b f_X(x)\, dx. \tag{A.19}$$

In this book, we sometimes approximate continuous random variables by discrete random variables, for ease of numerical integration. In those cases, we replace the infinitesimal dx by a finite Δx, so that Equation A.18 becomes:

$$p_{\widetilde{X}}(x) = f_X(x)\,\Delta x = P(x < X \leq x + \Delta x) \tag{A.20}$$

where $p_{\widetilde{X}}(x)$ is the PMF for $\widetilde{X}$, the discretized version of the continuous random variable X. Reference to Figure A.5 should help the reader understand that the probabilities of any outcome between x and $x + \Delta x$ are assigned to the discrete value of x.

Another way to describe a continuous random variable is with its CDF:

$$F_X(x) = P(X \leq x). \tag{A.21}$$

The PDF and CDF are related by

$$F_X(x) = P(X \leq x) = \int\limits_{-\infty}^x f_X(u)\, du \tag{A.22}$$

$$f_X(x) = \frac{d}{dx} F_X(x). \tag{A.23}$$

Note that the CDF of a continuous and a discrete random variable has the same definition. This is because probabilities of outcomes within an interval are identically defined for discrete and continuous outcomes.

A related function quantifying random variables is the inverse CDF. If a random variable X has CDF such that $F_X(x) = u$, then the inverse CDF is defined by

$$F_X^{-1}(u) = x. \tag{A.24}$$

This is the same definition that applies to any inverse function. It is useful in applications such as random number generation (Section A.6).

In cases where both discrete and continuous outcomes may be possible, we can formulate a *mixed distribution*. For example, if there is a probability p that no earthquake will occur (and thus ground-shaking amplitude is 0), and probability $1 - p$ that an earthquake will occur (and ground-shaking amplitude will have a continuous distribution $f_X(x)$), then the following PDF can be formulated for Y, representing ground-shaking amplitude:

$$f_Y(y) = p\,\delta(0) + (1 - p)\,f_X(y) \tag{A.25}$$

where $\delta(x)$ is the Dirac delta function, which has the following properties:

$$\delta(x) = \begin{cases} \infty & \text{if } x = 0 \\ 0 & \text{otherwise} \end{cases} \tag{A.26}$$

$$\int_{-\infty}^{\infty} \delta(x)\,dx = 1. \tag{A.27}$$

That is, $\delta(x)$ is a function that takes value ∞ at $x = 0$, takes value 0 elsewhere, and has area = 1.

A.3.1 Notation

This PMF/PDF/CDF notation allows us to compactly and precisely describe probabilities of outcomes of random variables. Note that the following conventions have been used:

1. The initial letter indicates the type of probability being described (i.e., "p" for PMFs, "f" for PDFs, and "F" for CDFs).
2. The subscript denotes the random variable (e.g., X), and thus is always a capital letter.
3. The argument in parentheses indicates the numerical value being considered (e.g., x), and is thus either a lowercase letter or a numeric value (e.g., $F_X(2) = P(X \leq 2)$).

These conventions are not chosen arbitrarily here or unique to PSHA. They are used almost universally in probability papers and books, regardless of the field of application.

A.3.2 Conditional Distributions

We are often interested in conditional probability distributions of random variables. We can adopt all of the results from Section A.2 if we recognize that the random variable X exceeding some value x is an event. So we can adapt Equation A.8, for example, to write

$$\begin{aligned} f_{X|Y}(x|y)dx &= P(x < X \leq x + dx \,|\, y < Y \leq y + dy) \\ &= \frac{P(x < X \leq x + dx \cap y < Y \leq y + dy)}{P(y < Y \leq y + dy)} \end{aligned} \tag{A.28}$$

where the notation $f_{X|Y}(x|y)$ is introduced to denote the conditional PDF of X, given that the random variable Y has taken value y. If we further introduce the following notation for the joint PDF of X and Y:

$$f_{X,Y}(x, y)\,dx\,dy = P(x < X \leq x + dx \cap y < Y \leq y + dy), \tag{A.29}$$

then Equation A.29 becomes

$$f_{X|Y}(x|y) = \frac{f_{X,Y}(x, y)}{f_Y(y)}. \tag{A.30}$$

Similarly, Equation A.9 can be used to show that random variables X and Y are independent if and only if

$$f_{X,Y}(x, y) = f_X(x)f_Y(y). \tag{A.31}$$

These types of manipulations, which are only briefly introduced here, are useful for computing probabilities of outcomes of random variables, conditional upon knowledge of other probabilistically dependent random variables.

A.4 Expectations and Moments

A random variable is completely defined by its PMF or PDF (for discrete and continuous random variables, respectively). Sometimes, however, it is convenient to use measures that describe general features of the distribution, such as its "average" value, breadth of feasible values, and whether it has a heavier tail to the left or right. We can measure these properties using moments of a random variable.

The *mean* is the most common moment, and is used to describe the central location of a random variable. The mean of X is denoted μ_X or $E[X]$. It can be calculated for a discrete random variable as

$$\mu_X = \sum_{\text{all } i} x_i \, p_X(x_i) \tag{A.32}$$

and for a continuous random variable as

$$\mu_X = \int_{-\infty}^{\infty} x \, f_X(x) \, dx. \tag{A.33}$$

The equations may be recognizable to some readers as computing the centroid of the PMF or PDF. Because the mean is computed using summation or integration, and both operations are linear functions, the mean value is also a linear operator. As a result, several useful formulas can be derived for mean value computations. For instance, if $Y = aX$, where a is a constant, then $\mu_Y = a\mu_X$; and if $Y = X_1 + X_2$, then $\mu_Y = \mu_{X_1} + \mu_{X_2}$.

The variation of values from a random variable can be measured using the *variance*, denoted σ_X^2 or $Var[X]$. It is calculated for a discrete random variable as

$$\sigma_X^2 = \sum_{\text{all } i} (x_i - \mu_x)^2 p_X(x_i) \tag{A.34}$$

and for a continuous random variable as

$$\sigma_X^2 = \int_{-\infty}^{\infty} (x - \mu_x)^2 \, f_X(x) \, dx. \tag{A.35}$$

This is the moment of inertia of the PDF (or PMF) about the mean. Again, several useful formulas can be derived from this operation, but they have slightly different forms due to the squared x term in the integral. For instance, if $Y = aX$, where a is a constant, then $\sigma_Y^2 = a^2\sigma_X^2$.

The square root of the variance is known as the *standard deviation*, and is denoted σ_X. It is often preferred to the variance when reporting a description of a random variable because it has the same units as the random variable itself (unlike the variance, whose units are the original units squared).

Means and variances are special cases of the expectation operation. The expectation of $g(X)$ is defined for discrete random variables as

$$E[g(X)] = \sum_{\text{all } i} g(x_i) \, p_X(x_i) \tag{A.36}$$

and for continuous random variables as

$$E[g(X)] = \int_{-\infty}^{\infty} g(x) \, f_X(x) \, dx. \tag{A.37}$$

A.4.1 First-Order Second-Moment Analysis

Using the expectation and moment calculations, and a Taylor series expansion, we can derive several useful formulas related to functions of random variables.

Consider the random variable Y, which is a linear function of random variables $X = [X_1, X_2, \ldots, X_n]$:

$$Y = a_0 + \sum_{i=1}^{n} a_i X_i. \tag{A.38}$$

The mean and variance of Y can be determined as a function of the moments of $\mathbf{X}$, using the expectation operations from earlier in the section. A few intermediate steps are shown here to illustrate the process:

$$\mu_Y = E\left[a_0 + \sum_{i=1}^{n} a_i X_i \right] \tag{A.39}$$

$$= a_0 + \sum_{i=1}^{n} a_i \mu_{X_i} \tag{A.40}$$

$$\sigma_Y^2 = E\left[(Y - \mu_Y)^2 \right] \tag{A.41}$$

$$= E\left[\sum_{i=1}^{n} \sum_{j=1}^{n} a_i a_j \left(X_i - \mu_{X_i} \right) \left(X_j - \mu_{X_j} \right) \right] \tag{A.42}$$

$$= \sum_{i=1}^{n} \sum_{j=1}^{n} a_i a_j \, \rho_{X_i, X_j} \sigma_{X_i} \sigma_{X_j}. \tag{A.43}$$

Next, let $g(\mathbf{X})$ be a nonlinear function of $X_1, X_2, \ldots, X_n$. The Taylor series expansion of $Y = g(\mathbf{X})$ about the point $\tilde{\mathbf{x}}$ is

$$Y = g(\tilde{\mathbf{x}}) + \sum_{i=1}^{n} (X_i - \tilde{x}_i) \left. \frac{\partial g}{\partial x_i} \right|_{\mathbf{x} = \tilde{\mathbf{x}}} + \frac{1}{2} \sum_{i=1}^{n} \sum_{j=1}^{n} (X_i - \tilde{x}_i)(X_j - \tilde{x}_j) \left. \frac{\partial^2 g}{\partial x_i \partial x_j} \right|_{\mathbf{x} = \tilde{\mathbf{x}}} + \cdots \tag{A.44}$$

where $\partial g / \partial x_i |_{\mathbf{x} = \tilde{\mathbf{x}}}$ denotes the partial derivative of $g(\ )$ with respect to x_i, evaluated at $\mathbf{x} = \tilde{\mathbf{x}}$.

If we keep only the first two terms, we have the first-order approximation of Y:

$$Y \cong g(\tilde{\mathbf{x}}) + \sum_{i=1}^{n} (X_i - \tilde{x}_i) \left. \frac{\partial g}{\partial x_i} \right|_{\mathbf{x}=\tilde{\mathbf{x}}}. \tag{A.45}$$

The approximate function is linear and of the same form Equation A.38, so the expectation operation produces similar (but now approximate) results:

$$\mu_Y \cong g(\tilde{\mathbf{x}}) + \sum_{i=1}^{n} \left(\mu_{X_i} - \tilde{x}_i \right) \left. \frac{\partial g}{\partial x_i} \right|_{\mathbf{x}=\tilde{\mathbf{x}}} \tag{A.46}$$

$$\sigma_Y^2 \cong \sum_{i=1}^{n} \sum_{j=1}^{n} \left. \frac{\partial g}{\partial x_i} \frac{\partial g}{\partial x_j} \right|_{\mathbf{x}=\tilde{\mathbf{x}}} \rho_{X_i,X_j} \sigma_{X_i} \sigma_{X_j}. \tag{A.47}$$

Setting the linearization point $\tilde{\mathbf{x}}$ equal to the mean of $\mathbf{X}$, $(\tilde{\mathbf{x}} = \boldsymbol{\mu}_X = [\mu_{X_1}, \mu_{X_2}, \ldots, \mu_{X_n}])$ we obtain the first-order, mean-centered approximations to the mean and variance of Y:

$$\mu_Y \cong g(\mu_{X_1}, \mu_{X_2}, \ldots, \mu_{X_n}) \tag{A.48}$$

$$\sigma_Y^2 \cong \sum_{i=1}^{n} \sum_{j=1}^{n} \left. \frac{\partial g}{\partial x_i} \frac{\partial g}{\partial x_j} \right|_{\mathbf{x}=\mu_X} \rho_{X_i,X_j} \sigma_{X_i} \sigma_{X_j}. \tag{A.49}$$

The choice to linearize the functions at the mean values of the input random variables is typically a good one in this situation, as the linear approximation is most accurate close to $\tilde{\mathbf{x}}$, so this approach ensures a good approximation for X values near the centers of the individual variable distributions.

In words, Equation A.48 indicates that the mean of Y can be approximated by using the mean values of the input random variables and evaluating the function $g(\ )$. Equation A.49 indicates that the variance of Y can be approximated by a function that depends only on the partial derivatives of $g(\ )$, the standard deviations of the X_i terms, and the correlation coefficients between each X_i pair. These are approximations due to the linearization of the $g(\ )$ function. The accuracy of the approximation is dependent upon how linear the function is over the range of $\mathbf{X}$ values of interest. This approach is termed a first-order second-moment analysis, as it computes first and second moments of Y, using a first-order approximation of $g(\ )$.

A.5 Common Probability Distributions

In the following subsections, some useful basic results for commonly used distributions are presented.

A.5.1 The Normal Distribution

A random variable is said to be *normal* (or *Gaussian*) if it has the following PDF:

$$f_X(x) = \frac{1}{\sigma_X \sqrt{2\pi}} \exp\left(-\frac{1}{2} \left(\frac{x - \mu_X}{\sigma_X} \right)^2 \right) \qquad -\infty \leq x \leq \infty \tag{A.50}$$

where μ_X and σ_X denote the mean value and standard deviation, respectively, of X. The shorthand notation, $X \sim \mathcal{N}\left(\mu_X, \sigma_X^2\right)$, is also often used. This PDF forms the familiar bell curve seen above in

Figure A.5. This is one of the most common distributions. It has been found to accurately describe the distribution of the logarithm of ground-motion intensity associated with a given earthquake magnitude and distance. There is no analytic solution to the normal distribution CDF, but it is widely available as a built-in function in numerical software tools.

A normal random variable, X, can be transformed into a standard normal random variable as

$$U = \frac{X - \mu_X}{\sigma_X} \tag{A.51}$$

where U is a standard normal random variable. This standardization is useful for several reasons, one of which is that its values can easily be tabulated or computed using simple numerical functions. Because of the widespread use of the standard normal PDF and CDF, it is given unique notations $\phi(\ )$ and $\Phi(\ )$, respectively.

Using the transformation of Equation A.51 and this notation, we can write the CDF for general normal random variable as

$$P(X \le x) = \Phi\left(\frac{X - \mu_X}{\sigma_X}\right). \tag{A.52}$$

Bivariate Normal Distribution

The normal distribution can be generalized to the case of more than one random variable. Two random variables are said to have a *bivariate normal distribution* if they have the following joint PDF:

$$f_{X,Y}(x, y) = \frac{1}{2\pi\sigma_X\sigma_Y\sqrt{1 - \rho_{X,Y}^2}} \exp\left\{-\frac{z}{2(1 - \rho_{X,Y}^2)}\right\} \qquad -\infty \le x, y \le \infty \tag{A.53}$$

where $\rho_{X,Y}$ is the correlation coefficient between X and Y, and

$$z = \frac{(x - \mu_X)^2}{\sigma_X^2} - \frac{2\rho_{X,Y}(x - \mu_X)(y - \mu_Y)}{\sigma_X\sigma_Y} + \frac{(y - \mu_Y)^2}{\sigma_Y^2}. \tag{A.54}$$

A useful property of random variables having this distribution is that if X and Y are jointly normal, then their marginal distributions ($f_X(x)$ and $f_Y(y)$) are normal, and their conditional distributions are also normal. Specifically, the distribution of X given $Y = y$ has conditional mean

$$\mu_{X|Y=y} = \mu_X + \rho_{X,Y}\,\sigma_X\left(\frac{y - \mu_Y}{\sigma_Y}\right) \tag{A.55}$$

and conditional standard deviation

$$\sigma_{X|Y=y} = \sigma_X\sqrt{1 - \rho_{X,Y}^2}. \tag{A.56}$$

These properties are convenient when computing joint distributions of ground-motion parameters.

Multivariate Normal Distribution

Extending the bivariate case further, a vector of random variables is said to be *multivariate normal* if it has the following joint PDF:

$$f_X(x) = \frac{1}{(2\pi)^{n/2}(\det\Sigma)^{1/2}} \exp\left[-\frac{1}{2}(x - M)^T\Sigma^{-1}(x - M)\right] \tag{A.57}$$

where boldface denotes a matrix, X is the vector of random variables:

$$X = \left\{ \begin{array}{c} X_1 \\ X_2 \\ \vdots \\ X_n \end{array} \right\} \tag{A.58}$$

with mean M

$$M = \left\{ \begin{array}{c} \mu_{X_1} \\ \mu_{X_2} \\ \vdots \\ \mu_{X_n} \end{array} \right\} \tag{A.59}$$

and covariance matrix Σ given by

$$\Sigma = \begin{bmatrix} \sigma_{X_1}^2 & \sigma_{X_1,X_2} & \cdots & \sigma_{X_1,X_n} \\ \sigma_{X_2,X_1} & \sigma_{X_2}^2 & & \vdots \\ \vdots & & \ddots & \\ \sigma_{X_n,X_1} & \cdots & & \sigma_{X_n}^2 \end{bmatrix}. \tag{A.60}$$

Conditional Multivariate Normal Distribution

Through combining the concepts from Sections A.3.2 and A.5.1, consider a multivariate normal distribution for X, which is partitioned into two vectors of cumulative length n, as

$$X = \left\{ \begin{array}{c} X_a \\ X_b \end{array} \right\}. \tag{A.61}$$

The mean and covariance matrices can then be expressed in a partitioned form as

$$M = \left\{ \begin{array}{c} \mu_{X_a} \\ \mu_{X_b} \end{array} \right\} \tag{A.62}$$

$$\Sigma = \begin{bmatrix} \Sigma_{aa} & \Sigma_{ab} \\ \Sigma_{ba} & \Sigma_{bb} \end{bmatrix}. \tag{A.63}$$

The distribution of X_a conditional on $X_b = x_b$ also has a multivariate normal distribution with mean and covariance:

$$\mu_{X_a|X_b} = \mu_{X_a} + \Sigma_{ab}\Sigma_{bb}^{-1}(x_b - \mu_{X_b}) \tag{A.64}$$

$$\Sigma_{X_a|X_b} = \Sigma_{aa} - \Sigma_{ab}\Sigma_{bb}^{-1}\Sigma_{ba}. \tag{A.65}$$

In the special case for which X_b is a scalar (i.e., X_a is length $n-1$), then the further simplification of these equations is of interest. Noting that the (i,j) element of the covariance matrix is $\Sigma_{ij} = \sigma_{X_i,X_j} = \rho_{X_i,X_j}\sigma_{X_i}\sigma_{X_j}$ (Equation A.60), and that $\varepsilon_{X_b} = (x_b - \mu_{X_b})/\sigma_{X_b}$, the conditional mean for the ith index of X_a is

$$\mu_{X_i|X_b=x_b} = \mu_{X_i} + \rho_{X_i,X_b}\sigma_{X_i}\varepsilon_{X_b}. \tag{A.66}$$

Similarly, the conditional standard deviation for the ith index of X_a can be expressed as

$$\sigma_{X_i|X_b=x_b} = \sigma_{X_i}\sqrt{1 - \rho_{X_i,X_b}^2}.$$ (A.67)

A.5.2 The Lognormal Distribution

A random variable Y is said to have a *lognormal distribution* if its logarithm, $X = \ln Y$, has a normal distribution. Using this transformation, and the normal distribution equations above, we can determine that the PDF and CDF of Y are given by

$$f_Y(y) = \frac{1}{y\sigma_{\ln Y}\sqrt{2\pi}} \exp\left(-\frac{1}{2}\left(\frac{\ln y - \mu_{\ln Y}}{\sigma_{\ln Y}}\right)^2\right) \quad 0 \le y \le \infty$$ (A.68)

$$F_Y(y) = \Phi\left(\frac{\ln y - \mu_{\ln Y}}{\sigma_{\ln Y}}\right) \quad 0 \le y \le \infty$$ (A.69)

where $\mu_{\ln Y}$ and $\sigma_{\ln Y}$ are the mean and standard deviation of $\ln Y$. The distribution and parameters are often indicated by the notation $Y \sim \mathcal{LN}(\mu_{\ln Y}, \sigma_{\ln Y}^2)$. These parameters are related to the mean and standard deviation of Y by the following equations:

$$\mu_Y = e^{\mu_{\ln Y}} e^{\frac{1}{2}\sigma_{\ln Y}^2}$$ (A.70)

$$\sigma_Y = \mu_Y\sqrt{e^{\sigma_{\ln Y}^2} - 1}.$$ (A.71)

The relationship between the median of Y, y_{50}, and $\mu_{\ln Y}$ can be determined by setting the CDF of Equation A.69 equal to 0.5 when y equals the median, y_{50}:

$$0.5 = \Phi\left(\frac{\ln y_{50} - \mu_{\ln Y}}{\sigma_{\ln Y}}\right) \quad \rightarrow \quad y_{50} = e^{\mu_{\ln Y}}.$$ (A.72)

The equivalence of $\ln y_{50}$ and $\mu_{\ln Y}$ can be stated in words as "the log of the median is equal to the logarithmic mean."

A.5.3 The Bernoulli Sequence

A *Bernoulli sequence* is a series of discrete trials having the following properties:

1. In each trial, *success* and *failure* are the only two possible outcomes.
2. The probability of success, p, in each trial is constant.
3. The trials are all stochastically independent.

The number of successes observed in n trials of a Bernoulli sequence has a *binomial distribution*. Specifically, let X denote the number of successes in n trials. Then X has PMF

$$p_X(x) = \binom{n}{x}p^x(1-p)^{n-x}, \quad x = 0, 1, 2, \ldots, n$$ (A.73)

and mean and standard deviation

$$\mu_X = np$$ (A.74)

$$\sigma_X = \sqrt{np(1-p)}.$$ (A.75)

A.5.4 The Poisson Process

A *Poisson process* is a sequence of discrete events having the following properties:

1. Stationarity: the probability of an event in a short interval from time t to $t + h$ is approximately λh, for any t.
2. Nonmultiplicity: the probability of two or more events in a short time interval is negligible compared with λh.
3. Independence: the number of events in any interval of time is independent of the number of events in any other (nonoverlapping) interval of time.

This process is an extension of a Bernoulli sequence to the case where there are infinite trials. The number of events observed in time t from a Poisson process has a *Poisson distribution*. Specifically, let X denote the number of successes in time t, from a process with a mean rate of events λ. Then X has a Poisson PMF:

$$p_X(x) = \frac{(\lambda t)^x}{x!} \exp(-\lambda t), \qquad x = 0, 1, 2, \ldots \tag{A.76}$$

with mean and standard deviation

$$\mu_X = \lambda t \tag{A.77}$$

$$\sigma_X = \sqrt{\lambda t}. \tag{A.78}$$

A.6 Random Number Generation

Some analysis in this book requires the use of randomly generated numbers for Monte Carlo analysis. These procedures require an algorithm to produce a sequence of numbers having properties consistent with a target probability distribution. Typical applications use *pseudo-random numbers* (produced by a deterministic algorithm) rather than truly random numbers because the former are faster to produce and can be reproduced. Most numerical computing environments have a built-in function to produce pseudo-random numbers with a *uniform distribution* between 0 and 1. These functions (typically named rand) are usually suitable for the applications envisioned in this book.

If the computing environment does not have a built-in function to produce pseudo-random numbers with a given arbitrary CDF $F_X(x)$, they can easily be produced using the following *inverse transform method* algorithm:

1. Use the built-in rand function to sample a uniform random number between 0 and 1. Call the ith sampled number u_i.
2. Generate a sampled number from the target distribution by computing $x_i = F_X^{-1}(u_i)$, where $F_X^{-1}()$ is the inverse CDF of X.

While the above algorithm is simple, it is limited to univariate probability distributions. There are also faster algorithms available for special cases if computational time is an issue. Simulation of samples from multivariate distributions and other cases is discussed in many textbooks devoted to the topic (e.g., Gentle, 2003).

Appendix B Basics of Statistics for Model Calibration

This appendix provides basic overviews of statistical procedures commonly encountered when calibrating models for hazard and risk calculations. Unlike Appendix A, most of these sections are written with specific applications in mind, because the algorithms are problem-specific and the insight provided by general statistical discussion would be limited.

B.1 Confidence Intervals for the Sample Mean and Standard Deviation

Consider a set of n observations,[1] $\{x_1, x_2, \ldots, x_n\}$, assumed to come from a probability distribution $f_X(x)$ with mean μ_X and standard deviation σ_X. Given the data, the point-estimates of the mean and standard deviation are

$$\bar{x} = \frac{1}{n} \sum_{i=1}^{n} x_i \tag{B.1}$$

$$s_x = \sqrt{\frac{1}{n-1} \sum_{i=1}^{n} (x_i - \bar{x})^2}. \tag{B.2}$$

There is "finite-sample uncertainty" in these estimates (i.e., a different set of n observations would yield different point-estimates). Specifically, the sample mean can be shown to have the following mean and standard deviation:

$$\mu_{\bar{x}} = \mu_X \tag{B.3}$$

$$\sigma_{\bar{x}} = \frac{\sigma_X}{\sqrt{n}}. \tag{B.4}$$

Further, when n is large, $\bar{x}$ is approximately normally distributed, by the central limit theorem. Given the above, $(\bar{x} - \mu_{\bar{x}})/\sigma_{\bar{x}}$ has a standard normal distribution.

We can thus use normal distribution properties to evaluate how far the sample mean may vary from the true mean, given a sample of a given size. For example, there is 0.95 probability of a normal random variable taking a value within ± 1.96 standard deviations of its mean, so

$$P\left(\mu_X - 1.96\frac{\sigma_X}{\sqrt{n}} \leq \bar{x} \leq \mu_X + 1.96\frac{\sigma_X}{\sqrt{n}}\right) = 0.95. \tag{B.5}$$

---

[1] Given that many distributions in this book are lognormal, it is often natural and more effective to take logarithms of the data and then proceed with these calculations. In such a case, $x_i = \ln y_i$, where y_i is an observation from the lognormal distribution.

More generally, we can state that the $100(1 - \alpha)\%$ *confidence interval* for μ_X is

$$\overline{x} \pm z_{(1-\alpha/2)} \frac{\sigma_X}{\sqrt{n}} \tag{B.6}$$

where $z_{(1-\alpha/2)}$ is the normal statistic, or quantile, for a cumulative probability of $1 - \alpha/2$ (e.g., 1.96 for $\alpha = 0.05$). Note that there is a subtle difference between Equations B.5 and B.6. Equation B.5 is a statement about the expected range of the sample mean $\overline{x}$, when μ_X and σ_X are known. In contrast, Equation B.6 assumes that σ_X is known, but then constructs an interval around the sample mean $\overline{x}$. The interpretation of this confidence interval is that such an interval will contain true mean μ_X, $100(1 - \alpha)\%$ of the time. That is, there is no guarantee that the true mean is actually contained within the interval defined by Equation B.6.

The above formula requires σ_X to be known, which is often not the case in practice. We can estimate σ_X by the sample standard deviation of Equation B.2, but $(\overline{x} - \mu_{\overline{x}})/(s_x/\sqrt{n})$ has a Student's t-distribution and the resulting confidence interval for μ_X becomes

$$\overline{x} \pm t_{(1-\alpha/2, n-1)} \frac{s_x}{\sqrt{n}} \tag{B.7}$$

where $t_{(1-\alpha/2, n-1)}$ is the Student's t-statistic for a cumulative probability of $1 - \alpha/2$ and $n - 1$ degrees of freedom.

Similarly, the confidence interval for the standard deviation σ_X can be constructed in terms of the sample standard deviation s_x as

$$s_x \sqrt{\frac{n-1}{\chi^2_{(\alpha/2, n-1)}}} \leq \sigma_X \leq s_x \sqrt{\frac{n-1}{\chi^2_{(1-\alpha/2, n-1)}}} \tag{B.8}$$

where $\chi^2_{(\alpha/2, n-1)}$ is the chi-squared statistic for a cumulative probability of $\alpha/2$ and $n - 1$ degrees of freedom.

In order to assess the finite sample uncertainties in the sample mean and standard deviation, it is useful to define two normalized variables: the relative precision, $\Delta/s_x = |\mu_X - \overline{x}|/s_x$, and the relative variability, $\Gamma = \sigma_X/s_x$. When the number of observations is controllable by the analyst (e.g., the number of response-history analyses to perform), then it can be useful to invert these confidence interval expressions to identify the number of observations, n, that are required to achieve a certain relative precision or variability.

In the case of the mean, Equation B.7 can be rearranged to the form

$$n \leq \left(\frac{t_{(1-\alpha/2, n-1)}}{\Delta/s_x} \right)^2. \tag{B.9}$$

This equation is used to define the number of observations required to construct confidence intervals that will contain the true mean $100(1 - \alpha)\%$ of the time. The number of observations needs to be an integer, so we adopt the smallest integer satisfying Equation B.9. The Student's t-statistic is a function of the sample size (hence this equation is implicit in n and requires iterative solution), but for large samples ($n > 30$) it is approximately equal to the standard normal variate equivalent, $z_{(1-\alpha/2)}$. Combined with the fact that $s_x \approx \sigma_X$ for such large samples, Equation B.9 becomes:

$$n = \left\lceil \left(z_{(1-\alpha/2)} \frac{\sigma_X}{\Delta} \right)^2 \right\rceil \tag{B.10}$$

where $\lceil \rceil$ represents rounding up to an integer value.

While this equation leads to an explicit expression for n, and therefore is easy to apply, Equation B.9 should be preferred for smaller sample sizes ($n < 30$) that often arise in practical applications.

Unlike the mean, the confidence interval for the standard deviation (Equation B.8) is asymmetric, and therefore the number of observations, n, to achieve a desired level of precision is obtained in an iterative fashion.

B.2 Hypothesis Testing for Statistical Significance

In many facets of PSHA, one wishes to evaluate the consistency, or otherwise, of observed data with respect to a probabilistic model. Examples throughout this book include:

1. Observed seismicity data with the magnitude-frequency distribution from a seismic source model (SSM).
2. Observed ground-motion data with a ground-motion model (GMM).
3. Appropriateness of the lognormal distribution for modeling ground motion or fragility functions.
4. Selected ground motions compared with the hazard-consistent target distribution.
5. Bias in response-history analysis results due to improper ground-motion selection.

A hypothesis test evaluates whether an observed deviation is a plausible result of small-sample variability, or if it is unlikely to have been due to sampling variability (and so possibly due to a true deviation).

With reference to the calculations of Section B.1, consider comparing a model for $f_X(x)$, with mean μ_X and standard deviation σ_X, to a set of observed data $\{x_1, x_2, \ldots, x_n\}$. If the sample mean $\overline{x}$ differs from μ_X, how can we determine whether this is due to small-sample variability, or is a "significant" difference?

A hypothesis test starts with an assumption, e.g., "the data come from a distribution with mean μ_X," and then uses data to evaluate this hypothesis. This initial assumption is termed a *null hypothesis*, which in this case is

$$H_0 : \overline{x} = \mu_X. \tag{B.11}$$

Rejection of that hypothesis would imply acceptance of the *alternative hypothesis*. In this case, that is

$$H_1 : \overline{x} \neq \mu_X. \tag{B.12}$$

Under the null hypothesis, the sample mean, $\overline{x}$, has a confidence interval given by Equation B.7. We can thus use that interval to evaluate whether the difference between a particular value of $\overline{x}$ and the assumed underlying mean μ_X is greater than would be expected from sampling variability.

We define the *p-value* as the probability that a sample statistic would arise as a result of sampling variability. In the above example, the sample statistic is defined as $t_{sample} = |\mu_X - \overline{x}| \sqrt{n}/s_x$, using Equation B.7. The probability of observing such a statistic is evaluated from the CDF of the Student's t-distribution with $n - 1$ degrees of freedom—this is the p-value. As the above example relates to a "two-tailed" hypothesis test (we care only about the difference between μ_X and $\overline{x}$, not whether $\overline{x}$ is higher or lower than μ_X), the sample statistic t_{sample} is compared with the test statistic $t_{(1-\alpha/2, n-1)}$, where α is specified to be some threshold value. If the p-value is sufficiently small, then the null hypothesis can be rejected in favor of the alternative hypothesis. Typical threshold values are 0.05 and 0.01, and these are specified as the threshold α values to define the test statistic.

For example, consider an assumed distribution with $\mu_X = 5$, $\sigma_X = 2$, and $n = 25$ observations from the distribution. If we set $p = 0.05$, then the confidence interval is $5 \pm 2.06(2)/\sqrt{25} = [4.17, 5.83]$. We would expect to observe an $\bar{x}$ value outside that interval only 5% of the time, and so an observation outside that interval may be taken as evidence that the null hypothesis is possibly wrong, and to thus reject it. References to *statistically significant* differences are made in this context. That is, if a sample mean was outside the above interval, one could assert that there is a statistically significant difference between μ_X and $\bar{x}$ at the 95% confidence level.

A second common example arises when inferring the dependence between two quantities X and Y. Linear regression of the observed (x_i, y_i) values yields the expression

$$\hat{y}(x) = \hat{\beta}_0 + \hat{\beta}_1 x \tag{B.13}$$

where $\hat{\beta}_0$ and $\hat{\beta}_1$ are estimated parameters, and where the predictive equation has an estimation standard deviation of $\hat{\sigma}_{Y|X}$. The ^ symbols are used to denote point estimates of values that have uncertainty as a result of the finite number of observations used in their estimation. If X and Y are linearly dependent, then the linear regression equation will have a nonzero slope, i.e., $\beta_1 \neq 0$. Therefore, we wish to use hypothesis testing to examine the null hypothesis:

$$H_0 : \beta_1 = 0 \tag{B.14}$$

versus the alternative hypothesis:

$$H_1 : \beta_1 \neq 0. \tag{B.15}$$

As with the prior example, to test the null hypothesis, we determine whether the point-estimate of the slope, $\hat{\beta}_1$ is sufficiently far from zero. Because of the finite number of observations, there is a standard error in the estimate of $\hat{\beta}_1$, referred to as $SE(\hat{\beta}_1)$, and obtained as an output from linear regression (e.g., Ang and Tang, 2007). The observed test statistic in this example also has a Student's t-distribution given by

$$t_{obs} = \frac{\hat{\beta}_1 - 0}{SE(\hat{\beta}_1)}. \tag{B.16}$$

From this observed test statistic, a p-value can be determined, and rejection of the null hypothesis considered based on the critical rejection threshold.

The above hypothesis tests, and similar tests of other statistics, are appealing in that they account for finite sample variability in a rigorous way, and thus eliminate concerns about apparent features that are likely due to small sample sizes. They are also appealing in that they provide a straightforward procedure with a binary outcome: accept or reject.

However, these tests have a number of limitations that should be considered before deciding how to interpret (or even whether to perform) a hypothesis test. First, the p-value from a hypothesis test does not provide the probability that either hypothesis is correct, but rather the probability that the observed test statistic would occur if the null hypothesis is correct. The latter is a much more limited statement, although results from these hypothesis tests are sometimes (incorrectly) interpreted with regard to the former statement. Second, these tests are strongly dependent upon the number of observations, n. With a large enough dataset, even small deviations from the null hypothesis can be detected. Whether the small deviation is of practical significance must still be determined by the analyst. In particular, it becomes essential to ensure that the data reflect an unbiased sample of the underlying distribution one is attempting to model. This situation is quite important in seismic hazard models: we may have a large number of observations from small-magnitude earthquakes that suggest some minor deviation from an assumed model, but our application of the model is for predicting a feature of larger-magnitude earthquakes. Conversely, especially in a limited data environment, failing to reject a null hypothesis does not indicate that the hypothesis is true. A judgment and the overall validity and usefulness of a model requires a more holistic evaluation than this single hypothesis test. Third, due to the nature of the hypothesis test, we expect to fail the test some percentage of the time, even if the null hypothesis is correct. That is, 5% of the time, we will obtain data that fail the hypothesis test when using a *p*-value of 0.05. This is particularly important when testing multiple models and multiple datasets. If, for example, we are testing whether earthquake rates in a gridded area source are the same at all grid points, 5% of grid points may fail a hypothesis test simply due to this *multiple testing* issue. With these limitations in mind, some academic journals have even eliminated the use of hypothesis tests and required more appropriate types of data interpretation, some being less formalized and others utilizing Bayesian and other frameworks (Halsey et al., 2015; Trafimow and Marks, 2015; Amrhein et al., 2019).

As this discussion indicates, there is a wide range of methods for comparing observed data with models, and a number of classical hypothesis tests that are relevant for other circumstances and other data features than illustrated here. Interested readers should consult textbooks dedicated to the topic (e.g., Kutner et al., 2004; Gelman et al., 2013).

B.3 Statistical Estimation of $m_{\max}$

This section describes the procedure that can be used to derive Equation 3.32 for the maximum magnitude from a seismic source. For a seismicity dataset containing n events that has been declustered (such that all events can be considered independent of one another) and sorted (such that M_1 represents the smallest event and M_n the largest), an expression for the probability distribution of the largest observed earthquake, $M_n \equiv m_{\max}^{obs}$, can be written as

$$F_{M_n}(m) = P(M_1 \leq m \cap M_2 \leq m \cap \cdots \cap M_n \leq m) = \prod_{i}^{n} P(M_i \leq m). \qquad (B.17)$$

Here, the condition of independence is used to write the product form. Equation B.17 defines the probability that all events in our catalogue would have a magnitude below m.

This expression can also be written in terms of the CDF of the magnitude-frequency distribution (Kijko, 2004):

$$F_{M_n}(m) = \begin{cases} 0 & \text{for } m < m_{\min} \\ [F_M(m)]^n & \text{for } m_{\min} \leq M \leq m_{\max} \\ 1 & \text{for } m > m_{\max}. \end{cases} \qquad (B.18)$$

The parameter that we are interested in estimating, $m_{\max}$, appears in Equation B.18 only within the inequalities.

The expected value of the maximum observed magnitude can then be found from the distribution defined by Equation B.18. This expectation is shown as

$$E(M_n) = \int_{m_{\min}}^{m_{\max}} m \, dF_{M_n}(m) = \int_{m_{\min}}^{m_{\max}} m f_{M_n}(m) \, dm. \tag{B.19}$$

Using integration by parts, and the limiting values of $F_{M_n}(m_{\min}) \equiv 0$ and $F_{M_n}(m_{\max}) \equiv 1$ allows Equation B.19 to be written as

$$E(M_n) = m_{\max} - \int_{m_{\min}}^{m_{\max}} F_{M_n}(m) \, dm. \tag{B.20}$$

Now, as the maximum likelihood estimate of the expected value of the largest magnitude in a given time period is the maximum observed magnitude over that period, i.e., $E(M_n) \equiv m_{\max}^{obs}$ (Kijko, 2004), we can rearrange the above equation to obtain

$$m_{\max} = m_{\max}^{obs} + \int_{m_{\min}}^{m_{\max}} F_{M_n}(m) \, dm \equiv m_{\max}^{obs} + \Delta m_{\max}. \tag{B.21}$$

Equation B.21 shows $m_{\max}$ appearing on both the left-hand side and in the upper limit of the integral on the right-hand side. Therefore, an iterative procedure is required to determine the value of $m_{\max}$ (starting by using $m_{\max} \equiv m_{\max}^{obs}$ within the integral). However, it should be noted that there is no guarantee of convergence under this procedure. For the GR distribution, it has been shown that convergence will not occur when $m_{\max}^{obs} > m_{\max,crit}^{obs}$ where $m_{\max,crit}^{obs} = m_{\min} + (\gamma + \ln n)/\beta$, $\gamma = 0.577\ldots$ is Euler's constant, and β is the parameter of the GR distribution (Haarala and Orosco, 2016). Furthermore, even when the procedure does converge, the estimated variance in Equation 3.33 can be unrealistically large.

B.4 Bayesian Estimation of $m_{\max}$

This section describes the procedure that can be used to derive Equation 3.35 for Bayesian estimation of the maximum magnitude from a seismic source. Given a magnitude-frequency distribution represented by its PDF as $f_M(m; \boldsymbol{\theta})$, where $\boldsymbol{\theta}$ are the parameters of the distribution, the likelihood for a given set of n observed magnitudes $\boldsymbol{m} = \{m_1, m_2, \ldots, m_n\}$ can be defined as

$$\mathcal{L}(\boldsymbol{\theta} \mid \boldsymbol{m}) = \prod_{i=1}^{n} f_M(m_i; \boldsymbol{\theta}). \tag{B.22}$$

The optimal set of parameters, $\hat{\boldsymbol{\theta}}$, is found by maximizing the likelihood function.

When working with the doubly bounded exponential (GR) distribution, we have two parameters $\boldsymbol{\theta} = \{\beta, m_{\max}\}$, once we select a value for the minimum magnitude $m_{\min}$. The likelihood function is therefore

$$\mathcal{L}(\beta, m_{\max} \mid \boldsymbol{m}) = \prod_{i=1}^{n} \frac{\beta \exp\left[-\beta(m_i - m_{\min})\right]}{1 - \exp\left[-\beta(m_{\max} - m_{\min})\right]} \tag{B.23}$$

which is equivalent to

$$\mathcal{L}(\beta, m_{\max} \mid \boldsymbol{m}) = \left(\frac{1}{1 - \exp\left[-\beta(m_{\max} - m_{\min})\right]}\right)^n \prod_{i=1}^{n} \beta \exp\left[-\beta(m_i - m_{\min})\right]. \tag{B.24}$$

While this expression is maximized by manipulating both β and m_{max}, it is reasonable to assume a value for β (which is usually well constrained) in order to make the likelihood solely a function of m_{max}. In this case, the likelihood function becomes

$$\mathcal{L}(m_{max} \mid \boldsymbol{m}, \beta) = \alpha(\boldsymbol{m}, \beta)\left(\frac{1}{1 - \exp\left[-\beta(m_{max} - m_{min})\right]}\right)^n \tag{B.25}$$

where $\alpha(\boldsymbol{m}, \beta)$ is a constant equal to the product term in Equation B.24. In the context of Bayes' theorem, this constant will be absorbed into the denominator and so can be omitted from the likelihood expression. The likelihood function used within the Bayesian approach for estimating m_{max} is therefore

$$\mathcal{L}(m_{max} \mid \boldsymbol{m}, \beta) \equiv \begin{cases} \left(\frac{1}{1-\exp\left[-\beta(m_{max}-m_{min})\right]}\right)^n & \text{for } m_{max} \geq m_{max}^{obs} \\ 0 & \text{for } m_{max} < m_{max}^{obs}. \end{cases} \tag{B.26}$$

The second condition arises because the true maximum magnitude must be at least as large as the maximum already observed (ignoring the effects of uncertainties in the estimate of this magnitude).

For any other magnitude-frequency distribution the same approach can be adopted to derive the corresponding likelihood expression.

B.5 Maximum Likelihood Estimation of Seismicity Parameters

Maximum likelihood estimation (MLE) is an approach to calibrate the parameters of a distribution. We start by assuming that a dataset $\boldsymbol{x}$ has arisen from some underlying distribution whose form is known, but whose parameters need to be estimated. The likelihood function, $\mathcal{L}(\boldsymbol{\theta} \mid \boldsymbol{x})$, defines the likelihood that the distribution has the parameters $\boldsymbol{\theta}$, given the data $\boldsymbol{x}$. This is equivalent to defining the probability of $\boldsymbol{x}$, given the parameters $\boldsymbol{\theta}$:

$$\mathcal{L}(\boldsymbol{\theta} \mid \boldsymbol{x}) = P(\boldsymbol{x} \mid \boldsymbol{\theta}). \tag{B.27}$$

The "maximum" in MLE comes from defining the optimal parameters of the distribution, $\hat{\boldsymbol{\theta}}$, to be those maximizing the likelihood function.

For independent observations, the joint probability of the observations is defined in terms of the product of the likelihoods (probability mass values for discrete distributions or probability densities for continuous distributions) of the observations, given the parameters $\boldsymbol{\theta}$:

$$\mathcal{L}(\boldsymbol{\theta} \mid \boldsymbol{x}) = \prod_i P(x_i \mid \boldsymbol{\theta}). \tag{B.28}$$

When determining the optimal values of $\boldsymbol{\theta} \equiv \hat{\boldsymbol{\theta}}$, it is common to work with the log-likelihood for numerical reasons:

$$\ln \mathcal{L}(\boldsymbol{\theta} \mid \boldsymbol{x}) = \sum_i \ln P(x_i \mid \boldsymbol{\theta}). \tag{B.29}$$

Here, we demonstrate this approach to obtain the activity rate and the b-value of the doubly bounded exponential distribution.

A seismic hazard study will often need seismicity parameters for large area sources. An issue that commonly arises is that the catalog completeness may not be uniform over the region, due to the nonuniform distribution of recording stations. If we assume a single level of completeness for the source, we may be forced to adopt an unnecessarily high magnitude of completeness—removing events from our limited datasets. For this reason, a region may be subdivided such that periods of completeness vary spatially, even within the same area source. The presentation below accounts for this approach, but can easily be applied for simpler applications also.

Define k_{ij} as the number of earthquakes observed to have a magnitude in discrete magnitude bin i and occurring in subregion j. Although magnitude is continuous, it is common to aggregate event counts in discrete magnitude bins. For example, the total range of magnitudes would be subdivided into discrete bins such that the ith magnitude represents all events in the range $m_{i-1} \leq M < m_i$. In most cases, the difference $m_i - m_{i-1}$ will be some constant interval Δm, but the presentation below does not require this.

The log-likelihood function for the doubly bounded exponential distribution, given some predefined values of m_{min} and m_{max}, is

$$\ln \mathcal{L}\left(\{\alpha, \beta\} \mid k\right) = \sum_{j=1}^{J} \sum_{i=1}^{I} \ln P(K = k_{ij} \mid \{\alpha, \beta\}) \qquad (B.30)$$

where α relates to the activity rate and β is linked to the b-value through $\beta = \ln(10)b$. The expression above is generic aside from the explicit dependence upon α and β parameters, which depends on the form of the magnitude-frequency distribution.

Under the assumption that the observed earthquakes follows a Poisson process, the probability of observing a count k_{ij} given the recurrence parameters is

$$P[K = k_{ij}] = \frac{(\lambda_{ij} t_{ij})^{k_{ij}} \exp[-\lambda_{ij} t_{ij}]}{k_{ij}!} \qquad (B.31)$$

where λ_{ij} is the annual rate of occurrence of earthquakes with the ith magnitude in sub-region j, and t_{ij} is the complete period of observation for these events. For the doubly bounded exponential distribution, the average annual rate of events within magnitude bin i and within subregion j is given by

$$\lambda_{ij}(\alpha, \beta) = \alpha A_j \frac{\exp[\beta \delta_i] - \exp[\beta \delta_{i-1}]}{\exp[\beta \delta_{max}] - 1} \qquad (B.32)$$

with $\alpha \equiv \alpha(M_{min})$ representing the average rate per unit area of events of any magnitude within the overall region being considered and A_j representing the area of the jth sub-region. The terms δ_i, δ_{i-1}, and δ_{max} are simply shorthand expressions to make the equations more compact. They respectively represent $\delta_i = m_{min} - m_i$, $\delta_{i-1} = m_{min} - m_{i-1}$, and $\delta_{max} = m_{min} - m_{max}$. Expanding these terms, Equation B.32 is equivalent to

$$\lambda_{ij}(\alpha, \beta) = \alpha A_j P[m_{i-1} \leq M < m_i] = \alpha A_j \left[F_M(m_i; \beta) - F_M(m_{i-1}; \beta)\right]. \qquad (B.33)$$

Upon inserting the above rates into the log-likelihood function we obtain

$$\ln \mathcal{L}\left(\{\alpha, \beta\} \mid k\right) = \sum_{j=1}^{J} \sum_{i=1}^{I} \left\{k_{ij} \ln[t_{ij}] + k_{ij} \ln[\lambda_{ij}(\alpha, \beta)] - t_{ij} \lambda_{ij}(\alpha, \beta) - \ln[k_{ij}!]\right\}. \qquad (B.34)$$

Equation B.34 can then be maximized to find the optimal estimates of the parameters $\{\alpha, \beta\}$. The covariance matrix for these parameters can be computed using the inverse of the Fisher information matrix, $\mathcal{I}\left(\{\alpha, \beta\}\right)$ (which is the Hessian of the log-likelihood function evaluated at the optimal parameter estimates). The Fisher information matrix is

$$\mathcal{I}\left(\{\alpha, \beta\}\right) = \begin{bmatrix} -\dfrac{\partial^2 \ln \mathcal{L}(\{\alpha,\beta\})}{\partial \alpha^2} & -\dfrac{\partial^2 \ln \mathcal{L}(\{\alpha,\beta\})}{\partial \alpha \partial \beta} \\ -\dfrac{\partial^2 \ln \mathcal{L}(\{\alpha,\beta\})}{\partial \alpha \partial \beta} & -\dfrac{\partial^2 \ln \mathcal{L}(\{\alpha,\beta\})}{\partial \beta^2} \end{bmatrix}. \tag{B.35}$$

For the doubly bounded exponential distribution the partial derivatives can be computed analytically. For arbitrary magnitude-frequency distributions it is often easiest to evaluate the derivatives using algorithmic differentiation. The covariance matrix is defined in Equation B.36, and the standard errors are obtained from the square-root of the diagonal elements of this matrix:

$$\Sigma(\{\alpha, \beta\}) = \mathcal{I}\left(\{\alpha, \beta\}\right)^{-1}. \tag{B.36}$$

B.6 Empirical GMM Calibration

This section describes some statistical concepts used in calibrating the empirical GMMs discussed in Chapter 4. For each IM of interest, mixed effects residuals are computed using the following formulation (Al Atik et al., 2010):

$$\ln im_{i,j} = \mu_{\ln IM}(rup_i, site_j) + \delta B_i + \delta W_{i,j} \tag{B.37}$$

where $\ln im_{i,j}$ is the natural logarithm of the intensity measure of interest from the observed ground motion, and $\mu_{\ln IM}(rup_i, site_j)$ is the prediction (from a GMM) of the mean $\ln IM$ value, as a function of rup and $site$ parameters as discussed in Chapter 4. The subscripts indicate the jth ground motion from the ith earthquake, δB_i is the between-event residual (or random effect) for the ith earthquake, and $\delta W_{i,j}$ is the within-event residual for the jth observation from the ith earthquake. The GMM also specifies the standard deviations of δB_i and $\delta W_{i,j}$, denoted τ and ϕ, respectively.[2] Hundreds of models have been calibrated with this general type of formulation (Douglas and Edwards, 2016).

Calibrating a GMM requires specifying a functional form for $\mu_{\ln IM}(rup_i, site_j)$, estimating the parameters (coefficients) of the functional form, and estimating the standard deviations τ and ϕ. Development of the functional form is documented in individual GMM publications. Here we will discuss statistical aspects of estimating model parameters.

The estimation process is more complex than basic least-squares regression, because some of the variability in observed ground motion intensities results from variation in source properties and is observed in all ground motions from a given earthquake. This necessitates the use of two residual terms in Equation B.37, rather than the single residual common in basic regression formulations.[3]

---

[2] Some GMMs predict that τ and ϕ vary with rup and $site$ parameters as well, in which case there are additional model coefficients to estimate. We will address the case of constant τ and ϕ here to simplify the discussion.

[3] There is no requirement or physical evidence that between-event residuals must be due to source properties only, but that is a common explanation to for adopting this model formulation.

Brillinger and Preisler (1984, 1985) first proposed regressing a GMM as a *mixed-effects* model. A mixed-effects model distinguishes between *fixed effects* and *random effects*. Fixed effects are assumed to be the parameters of the underlying population distribution, and the primary objective is to obtain these parameters. Random effects are deviations away from the population model attributed to a particular effect, and we take these into account to ensure unbiased estimates of the fixed effects. In GMM models, random effects are often attributed to specific features of particular earthquakes or sites within the dataset used to calibrate a model.

A simple numerical algorithm to calibrate a mixed effects model is that of Abrahamson and Youngs (1992). This model considers random effects only for earthquake effects (and homoskedastic variance), but has been used extensively by GMM developers. The algorithm alternates between standard nonlinear least-squares regression analyses and an expectation maximization step, but ultimately finds the fixed effects and variance components that maximize the log-likelihood function:

$$\ln \mathcal{L} = -\frac{N}{2}\ln(2\pi) - \frac{1}{2}\ln|\mathbf{\Sigma}| - \frac{1}{2}\mathbf{\Delta}^T\mathbf{\Sigma}^{-1}\mathbf{\Delta} \tag{B.38}$$

where N is the number of observed ground motions, and $\mathbf{\Sigma}$ is the covariance matrix for $\mathbf{\Delta}$. The vector $\mathbf{\Delta}$ represents the total residuals and is defined from:

$$\Delta_{i,j} = \delta B_i + \delta W_{i,j} = \ln im_{i,j} - \mu_{\ln IM}(rup_i, site_j). \tag{B.39}$$

With the standard estimation approach, it is assumed that within-event residuals at multiple locations are mutually independent, but the observations from the same event are correlated, with correlation $\tau^2/(\tau^2 + \phi^2)$. The covariance matrix can then be written:

$$\mathbf{\Sigma} = \oplus_{i=1}^M \mathbf{\Sigma}_i \tag{B.40}$$

with the covariance matrix for event i, $\mathbf{\Sigma}_i$ being defined by

$$\mathbf{\Sigma}_i = \tau^2 \mathbf{1}_{n_i \times 1} \otimes \mathbf{1}_{n_i \times 1} + \phi^2 \mathbf{I}_{n_i \times n_i} \tag{B.41}$$

where $\oplus$ is the direct sum operator used to construct a block diagonal matrix $\mathbf{\Sigma}$ from the M event-specific block matrices $\mathbf{\Sigma}_i$, and $\otimes$ is the Kronecker product. The term $\mathbf{1}_{n_i \times 1}$ is a vector of length n_i of ones, while $\mathbf{I}_{n_i \times n_i}$ is an $n_i \times n_i$ identity matrix. A given earthquake event contributes n_i observations to the overall dataset.

The Abrahamson and Youngs algorithm starts by assuming that the random-effects terms $\delta B = \{\delta B_1, \delta B_2, \ldots, \delta B_M\}$ are all zero. A fixed effects regression (nonlinear least squares) is then performed on $\ln im_{i,j} = \mu_i' + \delta W_{i,j}$, where $\mu_i' \equiv \mu_{\ln IM}(rup_i, site_j) + \delta B_i$. Total residuals are then obtained from Equation B.39, and estimates of τ and ϕ are found from maximizing Equation B.38. New estimates of δB are then obtained from

$$\delta B_i = \frac{\tau^2 \sum_{j=1}^{n_i} \Delta_{i,j}}{n_i \tau^2 + \phi^2}. \tag{B.42}$$

With the new δB terms computed, the algorithm returns to the fixed effects regression step with the revised μ_i' reflecting the new δB terms, and the algorithm loops through this procedure until convergence.

The above approach was commonly adopted for many years, but a number of extensions have since become relevant. For example, extensions have been made to include multiple random effects reflecting different phenomena (Stafford, 2014a), to account for effects of spatial correlations (Jayaram and Baker, 2010a), to accommodate nonlinear site effects (Stafford, 2015), and to consider variable uncertainties (Kuehn and Abrahamson, 2018). Some of these effects can be accommodated through relatively minor extensions to the Abrahamson and Youngs algorithm above, but most require the use of different numerical approaches. Bayesian approaches making use of Markov-chain Monte Carlo methods have proven to be a flexible tool that enables consideration of very complex (hierarchical) mixed-effects models (Stafford, 2014a; Kuehn and Abrahamson, 2018; Stafford, 2019).

B.7 Estimation of *IM* Correlations from GMMs

This section describes estimation of correlations that are required for calculations of target *IM* values and ground-motion selection (Sections 7.5 and 10.3). A ground-motion time series, for a given site and rupture, can be quantified via a number of *IM* metrics, as discussed in Chapter 4. GMMs predict the probability distribution of any single metric, but some situations require a joint distribution for multiple *IM* metrics. In the common case that the $\ln IM$ metrics are Gaussian, it is common to assume that a vector of metrics is multivariate Gaussian (Section A.5.1). Characterization of this multivariate distribution requires specification of correlations between $\ln IM$ metrics, given the site and rupture.

To empirically estimate these correlations, the *IM* metrics of interest are computed from ground-motion recordings, and the within- and between-event residuals are computed using a ground-motion model. Applying the definition of the correlation coefficient to Equation B.37, the following equation can be derived:

$$\rho_{\ln IM_1, \ln IM_2} = \frac{\rho_{\delta B_1, \delta B_2} \tau_1 \tau_2 + \rho_{\delta W_1, \delta W_2} \phi_1 \phi_2}{\sigma_1 \sigma_2} \tag{B.43}$$

where τ_k, ϕ_k, and $\sigma_k = \sqrt{\tau_k^2 + \phi_k^2}$ are the between-event, within-event, and total standard deviations for IM_k, respectively (specified by the GMM), and $\rho_{\delta B_1, \delta B_2}$ and $\rho_{\delta W_1, \delta W_2}$ are the estimated correlation coefficients from the mixed effects data. Equation B.43 is obtained from the residuals of Equation B.37, so it is a correlation coefficient for $\ln IM_1$ and $\ln IM_2$, conditional on *rup* and *site*, but the conditioning notation is suppressed for brevity.

Example values of correlation coefficients for some *IM* pairs are shown in Figure B.1. Correlation coefficients in spectral acceleration residuals are generally highest for pairs of periods that are similar to each other, though short-period and long-period spectral values can also be highly correlated due to properties of frequency response functions of oscillators (Bora et al., 2016). Correlation coefficients for *IM* pairs that include parameters other than spectral accelerations depend upon how much the *IM* parameters are sensitive to similar features in the ground motion.

A number of researchers have estimated correlation coefficients for various *IM* pairs (e.g., Inoue and Cornell, 1990; Akkar et al., 2014; Azarbakht et al., 2014; Bradley, 2015, 2011). These studies have generally found that correlations given a *rup* are not strongly dependent on rupture properties such as magnitude and distance (Baker and Cornell, 2006a; Baker and Bradley, 2017; Stafford, 2017; Bayless and Abrahamson, 2019). For multicomponent ground motions, response spectra correlations are the same for single-component spectra, and for the Sa_{RotD50} and $Sa_{RotD100}$ spectral definitions discussed in Section 4.2.6 (Baker and Bradley, 2017). Finally, the correlation coefficients for a number of *IM* metrics, as estimated from heterogeneous data, may not produce a positive definite correlation matrix. This is problematic for some calculations requiring a valid multivariate distribution, but can be fixed via minor adjustments to the correlation matrix (Qi and Sun, 2006).

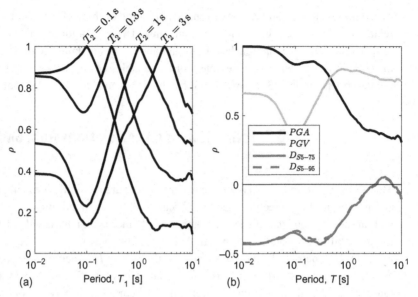

Fig. B.1 Example correlation coefficients for *IM* pairs, estimated from shallow crustal earthquake ground motions (adapted from Baker and Bradley, 2017). (a) ln $SA(T_1)$ and ln $SA(T_2)$, for four values of T_2. (b) *PGA*, *PGV*, D_{S5-75}, and D_{S5-95} versus ln $SA(T)$ at various periods.

B.8 Fragility Function Fitting

Fragility functions (Equation 9.1) often need to be calibrated from data. This requires estimating the θ and β (denoted $\hat{\theta}$ and $\hat{\beta}$) that are most consistent with the data. Estimation of these parameter values depends on way in which the data are obtained. The following sections provide two examples, following a presentation by Baker (2015).

B.8.1 Incremental Dynamic Analysis Data

Incremental dynamic analysis involves scaling each ground motion in a suite until it causes failure of the structure (Vamvatsikos and Cornell, 2002). This process produces a set of *IM* values associated with the onset of failure. Fragility function parameters can be estimated from these data by taking logarithms of each ground motion's *IM* value associated with onset of failure, and computing their mean and standard deviation:

$$\ln \hat{\theta} = \frac{1}{n} \sum_{i=1}^{n} \ln im_i \tag{B.44}$$

$$\hat{\beta} = \sqrt{\frac{1}{n-1} \sum_{i=1}^{n} \left(\ln(im_i/\hat{\theta})\right)^2} \tag{B.45}$$

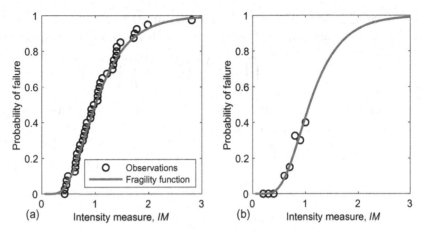

(a) Example incremental dynamic analyses data and a fragility function estimated using Equations B.44 and B.45. (b) Example multiple stripe analysis data and a fragility function estimated using Equation B.49.

where n is the number of ground motions considered, and im_i is the IM value associated with onset of failure for the ith ground motion. This is a method of moments estimator, as $\ln\theta$ and β are the mean and standard deviation, respectively, of the normal distribution representing the $\ln IM$ values. Example data of this type, and a fitted fragility function, are shown in Figure B.2a.

B.8.2 Empirical or Multiple Stripe Analysis Data

The fitting approach in this section is appropriate when the data for calibration take the form of IM values, with associated observations of "failure" or "no failure." This arises from multiple stripe analysis, where response-history analysis is performed at discrete IM levels and observations of failure or nonfailure are observed. This also applies to field observations of failures, where a failure or nonfailure is recorded and associated with the IM value from the earthquake at that site. The method of maximum likelihood can be used in this case (Shinozuka et al., 2000).

At a given intensity level $IM = x_j$, there are z_j observed failures out of n_j total observations. Assuming that each observation of failure or no-failure is independent, the probability of observing z_j failures is given by the binomial distribution (Equation A.73)

$$P(z_j \text{ failures in } n_j \text{ ground motions}) = \binom{n_j}{z_j} p_j^{z_j} (1 - p_j)^{n_j - z_j} \tag{B.46}$$

where p_j is the probability that a failure will be observed when $IM = x_j$.

The fragility function predicts p_j, and the maximum likelihood approach identifies the fragility function parameters that give the highest likelihood of producing the observed failure data. With data from multiple IM levels, we take the product of the binomial probabilities (from Equation B.46) at each IM level to get the likelihood for the entire dataset:

$$\mathcal{L} = \prod_{j=1}^{m} \binom{n_j}{z_j} p_j^{z_j} (1 - p_j)^{n_j - z_j} \tag{B.47}$$

where $\mathcal{L}$ is the likelihood, m is the number of IM levels, and $\prod$ denotes a product over all j. We then substitute Equation 9.1 for p_j, so the fragility parameters are explicit in the likelihood function:

$$\mathcal{L} = \prod_{j=1}^{m} \binom{n_j}{z_j} \Phi\left(\frac{\ln(x_j/\theta)}{\beta}\right)^{z_j} \left(1 - \Phi\left(\frac{\ln(x_j/\theta)}{\beta}\right)\right)^{n_j - z_j}. \tag{B.48}$$

Parameter estimates are obtained by maximizing this likelihood function, though it is equivalent and numerically easier to maximize the logarithm of the likelihood function:

$$\{\hat{\theta}, \hat{\beta}\} = \underset{\theta, \beta}{\arg\max} \sum_{j=1}^{m} \left\{ \ln\binom{n_j}{z_j} + z_j \ln \Phi\left(\frac{\ln(x_j/\theta)}{\beta}\right) + \left(n_j - z_j\right) \ln\left(1 - \Phi\left(\frac{\ln(x_j/\theta)}{\beta}\right)\right) \right\}. \tag{B.49}$$

Example data of this type, and a fitted fragility function using this approach, are shown in Figure B.2b. Generalized linear regression with a Probit link function is an equivalent alternative to Equation B.49.

References

Aagaard, B. T., Brocher, T. M., Dolenc, D., Dreger, D., Graves, R. W., Harmsen, S., Hartzell, S., Larsen, S., and Zoback, M. L. 2008. Ground-Motion Modeling of the 1906 San Francisco Earthquake, Part I: Validation Using the 1989 Loma Prieta Earthquake. *Bulletin of the Seismological Society of America*, **98**(2), 989–1011.

Aagaard, B. T., Hall, J. F., and Heaton, T. H. 2001. Characterization of Near-Source Ground Motions with Earthquake Simulations. *Earthquake Spectra*, **17**(2), 177–207.

Abrahamson, N. A. 1993. Spatial Variation of Multiple Support Inputs. In: *Proc of 1st US Seminar on Seismic Evaluation and Retrofit of Steel Bridges*.

Abrahamson, N. A. 2006. Seismic Hazard Assessment: Problems with Current Practice and Future Developments. In: *First European Conference on Earthquake Engineering and Seismology*.

Abrahamson, N., Atkinson, G., Boore, D., Bozorgnia, Y., Campbell, K., Chiou, B., Idriss, I. M., Silva, W., and Youngs, R. 2008. Comparisons of the NGA Ground-Motion Relations. *Earthquake Spectra*, **24**(1), 45–66.

Abrahamson, N. A., and Bommer, J. J. 2005. Probability and Uncertainty in Seismic Hazard Analysis. *Earthquake Spectra*, **21**(2), 603–607.

Abrahamson, N. A., Kuehn, N. M., Walling, M., and Landwehr, N. 2019. Probabilistic Seismic Hazard Analysis in California Using Nonergodic Ground-Motion Models. *Bulletin of the Seismological Society of America*, **109**(4), 1235–1249.

Abrahamson, N., Schneider, J. F., and Stepp, J. C. 1991. Spatial Coherency of Shear Waves from the Lotung, Taiwan Large-Scale Seismic Test. *Structural Safety*, **10**(1), 145–162.

Abrahamson, N., and Silva, W. 2008. Summary of the Abrahamson & Silva NGA Ground-Motion Relations. *Earthquake Spectra*, **24**(1), 67–97.

Abrahamson, N. A., Silva, W. J., and Kamai, R. 2014. Summary of the ASK14 Ground Motion Relation for Active Crustal Regions. *Earthquake Spectra*, **30**(3), 1025–1055.

Abrahamson, N., and Youngs, R. R. 1992. A Stable Algorithm for Regression Analysis Using the Random Effects Model. *Bulletin of the Seismological Society of America*, **82**(1), 505–510.

Adachi, T., and Ellingwood, B. 2007. Impact of Infrastructure Interdependency and Spatial Correlation of Seismic Intensities on Performance Assessment of a Water Distribution System. In: *Proceedings of the 10th International Conference on Applications of Statistics and Probability in Civil Engineering*.

Adachi, T., and Ellingwood, B. R. 2009. Serviceability Assessment of a Municipal Water System under Spatially Correlated Seismic Intensities. *Computer-Aided Civil and Infrastructure Engineering*, **24**(4), 237–248.

Aki, K. 1980. Scattering and Attenuation of Shear Waves in the Lithosphere. *Journal of Geophysical Research: Solid Earth*, **85**(B11), 6496–6504.

Aki, K. 2003. A Perspective on the History of Strong Motion Seismology. *Physics of the Earth and Planetary Interiors*, **137**(May), 5–11.

Aki, K., and Richards, P. G. 2002. *Quantitative Seismology*. University Science Books.

Akkar, S., and Bommer, J. J. 2006. Influence of Long-Period Filter Cut-off on Elastic Spectral Displacements. *Earthquake Engineering & Structural Dynamics*, **35**(9), 1145–1165.

Akkar, S., Sandıkkaya, M. A., and Ay, B. Ö. 2014. Compatible Ground-Motion Prediction Equations for Damping Scaling Factors and Vertical-to-Horizontal Spectral Amplitude Ratios for the Broader Europe Region. *Bulletin of Earthquake Engineering*, **12**(1), 517–547.

Al Atik, L., and Abrahamson, N. 2010. An Improved Method for Nonstationary Spectral Matching. *Earthquake Spectra*, **26**(3), 601–617.

Al Atik, L., Abrahamson, N., Bommer, J. J., Scherbaum, F., Cotton, F., and Kuehn, N. 2010. The Variability of Ground-Motion Prediction Models and Its Components. *Seismological Research Letters*, **81**(5), 794–801.

Al Atik, L., and Youngs, R. R. 2014. Epistemic Uncertainty for NGA-West2 Models. *Earthquake Spectra*, **30**(3), 1301–1318.

Alavi, B., and Krawinkler, H. 2004. Behavior of Moment-Resisting Frame Structures Subjected to Near-Fault Ground Motions. *Earthquake Engineering & Structural Dynamics*, **33**(6), 687–706.

Aldor-Noiman, S., Brown, L. D., Buja, A., Rolke, W., and Stine, R. A. 2013. The Power to See: A New Graphical Test of Normality. *The American Statistician*, **67**(4), 249–260.

Alexander, N. A., Chanerley, A. A., Crewe, A. J., and Bhattacharya, S. 2014. Obtaining Spectrum Matching Time Series Using a Reweighted Volterra Series Algorithm (RVSA). *Bulletin of the Seismological Society of America*, **104**(4), 1663–1673.

Allen, T. I., and Hayes, G. P. 2017. Alternative Rupture-Scaling Relationships for Subduction Interface and Other Offshore Environments. *Bulletin of the Seismological Society of America*, **107**(3), 1240–1253.

Ameri, G., Gallovic, F., Pacor, F., and Emolo, A. 2009. Uncertainties in Strong Ground-Motion Prediction with Finite-Fault Synthetic Seismograms: An Application to the 1984 M 5.7 Gubbio, Central Italy, Earthquake. *Bulletin of the Seismological Society of America*, **99**(2A), 647–663.

American Society of Civil Engineers. 2010. *Minimum Design Loads for Buildings and Other Structures, ASCE 7-10*. American Society of Civil Engineers/Structural Engineering Institute.

Amrhein, V., Greenland, S., and McShane, B. 2019. Scientists Rise up against Statistical Significance. *Nature*, **567**(7748), 305–307.

Anagnos, T., and Kiremidjian, A. S. 1984. Stochastic Time-Predictable Model for Earthquake Occurrences. *Bulletin of the Seismological Society of America*, **74**(6), 2593–2611.

Anagnos, T., and Kiremidjian, A. S. 1988. A Review of Earthquake Occurrence Models for Seismic Hazard Analysis. *Probabilistic Engineering Mechanics*, **3**(1), 3–11.

Ancheta, T. D., Darragh, R. B., Stewart, J. P., Seyhan, E., Silva, W. J., Chiou, B. S.-J., Wooddell, K. E., Graves, R. W., Kottke, A. R., Boore, D. M., Kishida, T., and Donahue, J. L. 2014. NGA-West2 Database. *Earthquake Spectra*, **30**(3), 989–1005.

Ancheta, T. D., Stewart, J. P., and Abrahamson, N. A. 2010. Engineering Characterization of Spatially Variable Ground Motions. PhD Thesis, University of California, Los Angeles.

Anderson, E. L. 1983. Quantitative Approaches in Use to Assess Cancer Risk. *Risk Analysis*, **3**(4), 277–295.

Anderson, J. G., and Brune, J. N. 1999. Probabilistic Seismic Hazard Analysis without the Ergodic Assumption. *Seismological Research Letters*, **70**(1), 19–28.

Anderson, J. G., and Hough, S. E. 1984. A Model for the Shape of the Fourier Amplitude Spectrum of Acceleration at High Frequencies. *Bulletin of the Seismological Society of America*, **74**(5), 1969–1993.

Anderson, J. G., and Uchiyama, Y. 2011. A Methodology to Improve Ground-Motion Prediction Equations by Including Path Corrections. *Bulletin of the Seismological Society of America*, **101**(4), 1822–1846.

Anderson, J. G., Wesnousky, S. G., and Stirling, M. W. 1996. Earthquake Size as a Function of Fault Slip Rate. *Bulletin of the Seismological Society of America*, **86**(3), 683–690.

Andrews, D. J., Hanks, T. C., and Whitney, J. W. 2007. Physical Limits on Ground Motion at Yucca Mountain. *Bulletin of the Seismological Society of America*, **97**(6), 1771–1792.

Andrews, D. J., and Schwerer, E. 2000. Probability of Rupture of Multiple Fault Segments. *Bulletin of the Seismological Society of America*, **90**(6), 1498–1506.

Ang, A. H.-S., and Tang, W. H. 2007. *Probability Concepts in Engineering Emphasis on Applications in Civil & Environmental Engineering*. New York: Wiley.

Apostolakis, G. 1990. The Concept of Probability in Safety Assessments of Technological Systems. *Science*, **250**(4986), 1359–1364.

Arias, A. 1970. A Measure of Earthquake Intensity. Pages 438–483 of: Hansen, R. (ed.), *Seismic Design for Nuclear Power Plants*. Cambridge, MA: MIT Press.

Arroyo, D., and Ordaz, M. 2011. On the Forecasting of Ground-Motion Parameters for Probabilistic Seismic Hazard Analysis. *Earthquake Spectra*, **27**(1), 1–21.

ASCE. 2016. *Minimum Design Loads for Buildings and Other Structures, ASCE 7-16*. Reston, VA: American Society of Civil Engineers/Structural Engineering Institute.

Aspinall, W. 2010. A Route to More Tractable Expert Advice. *Nature*, **463**(7279), 294–295.

Assatourians, K., and Atkinson, G. M. 2013. EqHaz: An Open-Source Probabilistic Seismic-Hazard Code Based on the Monte Carlo Simulation Approach. *Seismological Research Letters*, **84**(3), 516–524.

Atkinson, G. M. 2008. Ground-Motion Prediction Equations for Eastern North America from a Referenced Empirical Approach: Implications for Epistemic Uncertainty. *Bulletin of the Seismological Society of America*, **98**(3), 1304–1318.

Atkinson, G. M., and Beresnev, I. 1997. Don't Call It Stress Drop. *Seismological Research Letters*, **68**(1), 3–4.

Atkinson, G. M., Bommer, J. J., and Abrahamson, N. A. 2014. Alternative Approaches to Modeling Epistemic Uncertainty in Ground Motions in Probabilistic Seismic-Hazard Analysis. *Seismological Research Letters*, **85**(6), 1141–1144.

Atkinson, G. M., and Silva, W. 2000. Stochastic Modeling of California Ground Motions. *Bulletin of the Seismological Society of America*, **90**(2), 255–274.

Atwater, B. F., Tuttle, M. P., Schweig, E. S., Rubin, C. M., Yamaguchi, D. K., and Hemphill-Haley, E. 2003. Earthquake Recurrence Inferred from Paleoseismology. Pages 331–350 of: *Developments in Quaternary Sciences*. The Quaternary Period in the United States, vol. 1. Elsevier.

Azarbakht, A., Mousavi, M., Nourizadeh, M., and Shahri, M. 2014. Dependence of Correlations between Spectral Accelerations at Multiple Periods on Magnitude and Distance. *Earthquake Engineering & Structural Dynamics*, **43**(8), 1193–1204.

Baker, J. W. 2007. Probabilistic Structural Response Assessment Using Vector-Valued Intensity Measures. *Earthquake Engineering & Structural Dynamics*, **36**(13), 1861–1883.

Baker, J. W. 2011. Conditional Mean Spectrum: Tool for Ground Motion Selection. *Journal of Structural Engineering*, **137**(3), 322–331.

Baker, J. W. 2015. Efficient Analytical Fragility Function Fitting Using Dynamic Structural Analysis. *Earthquake Spectra*, **31**(1), 579–599.

Baker, J. W., Abrahamson, N. A., Whitney, J. W., Board, M. P., and Hanks, T. C. 2013. Use of Fragile Geologic Structures as Indicators of Unexceeded Ground Motions and Direct Constraints on Probabilistic Seismic Hazard Analysis. *Bulletin of the Seismological Society of America*, **103**(3), 1898–1911.

Baker, J. W., and Bradley, B. A. 2017. Intensity Measure Correlations Observed in the NGA-West2 Database, and Dependence of Correlations on Rupture and Site Parameters. *Earthquake Spectra*, **33**(1), 145–156.

Baker, J. W., and Chen, Y. 2020. Ground Motion Spatial Correlation Fitting Methods and Estimation Uncertainty. *Earthquake Engineering & Structural Dynamics*, **49**(15), 1662–1681.

Baker, J. W., and Cornell, C. A. 2006a. Correlation of Response Spectral Values for Multi-component Ground Motions. *Bulletin of the Seismological Society of America*, **96**(1), 215–227.

Baker, J. W., and Cornell, C. A. 2006b. Spectral Shape, Epsilon and Record Selection. *Earthquake Engineering & Structural Dynamics*, **35**(9), 1077–1095.

Baker, J. W., and Cornell, C. A. 2006c. Which Spectral Acceleration Are You Using? *Earthquake Spectra*, **22**(2), 293–312.

Baker, J. W., and Jayaram, N. 2008. Correlation of Spectral Acceleration Values from NGA Ground Motion Models. *Earthquake Spectra*, **24**(1), 299–317.

Baker, J. W., and Lee, C. 2018. An Improved Algorithm for Selecting Ground Motions to Match a Conditional Spectrum. *Journal of Earthquake Engineering*, **22**(4), 708–723.

Baker, J. W., Lin, T., Shahi, S. K., and Jayaram, N. 2011. *New Ground Motion Selection Procedures and Selected Motions for the PEER Transportation Research Program*. PEER Technical Report 2011/03.

Bao, H., Bielak, J., Ghattas, O., Kallivokas, L. F., O'Hallaron, D. R., Shewchuk, J. R., and Xu, J. 1998. Large-Scale Simulation of Elastic Wave Propagation in Heterogeneous Media on Parallel Computers. *Computer Methods in Applied Mechanics and Engineering*, **152**(1), 85–102.

Båth, M. 1965. Lateral Inhomogeneities of the Upper Mantle. *Tectonophysics*, **2**(6), 483–514.

Bauer, P., Thorpe, A., and Brunet, G. 2015. The Quiet Revolution of Numerical Weather Prediction. *Nature*, **525**(7567), 47–55.

Bayless, J., and Abrahamson, N. A. 2018. Evaluation of the Interperiod Correlation of Ground-Motion Simulations. *Bulletin of the Seismological Society of America*, **108**(6), 3413–3430.

Bayless, J., and Abrahamson, N. A. 2019. An Empirical Model for the Interfrequency Correlation of Epsilon for Fourier Amplitude Spectra. *Bulletin of the Seismological Society of America*, **109**(3), 1058–1070.

Bazzurro, P., and Cornell, C. A. 1999. Disaggregation of Seismic Hazard. *Bulletin of the Seismological Society of America*, **89**(2), 501–520.

Bazzurro, P., and Cornell, C. A. 2002. Vector-Valued Probabilistic Seismic Hazard Analysis. In: *7th U.S. National Conference on Earthquake Engineering*. Boston: Earthquake Engineering Research Institute.

Bazzurro, P., and Cornell, C. A. 2004a. Ground-Motion Amplification in Nonlinear Soil Sites with Uncertain Properties. *Bulletin of the Seismological Society of America*, **94**(6), 2090–2109.

Bazzurro, P., and Cornell, C. A. 2004b. Nonlinear Soil-Site Effects in Probabilistic Seismic-Hazard Analysis. *Bulletin of the Seismological Society of America*, **94**(6), 2110–2123.

Bazzurro, P., and Luco, N. 2005. Accounting for Uncertainty and Correlation in Earthquake Loss Estimation. In: *9th International Conference on Structural Safety and Reliability (ICOSSAR05)*.

Bazzurro, P., and Luco, N. 2006. Do Scaled and Spectrum-Matched Near-Source Records Produce Biased Nonlinear Structural Responses? In: *Proceedings, 8th National Conference on Earthquake Engineering*.

Beauval, C., Bard, P. Y., Hainzl, S., and Gueguen, P. 2008. Can Strong-Motion Observations Be Used to Constrain Probabilistic Seismic-Hazard Estimates? *Bulletin of the Seismological Society of America*, **98**(2), 509–520.

Beauval, C., Honoré, L., and Courboulex, F. 2009. Ground-Motion Variability and Implementation of a Probabilistic–Deterministic Hazard Method. *Bulletin of the Seismological Society of America*, **99**(5), 2992–3002.

Beauval, C., Tasan, H., Laurendeau, A., Delavaud, E., Cotton, F., Guéguen, P., and Kuehn, N. 2012. On the Testing of Ground-Motion Prediction Equations against Small-Magnitude Data. *Bulletin of the Seismological Society of America*, **102**(5), 1994–2007.

Bedford, T., and Cooke, R. 2001. *Probabilistic Risk Analysis: Foundations and Methods*. Cambridge University Press.

Bender, B. 1983. Maximum-Likelihood Estimation of b-Values for Magnitude Grouped Data. *Bulletin of the Seismological Society of America*, **73**(3), 831–851.

Benjamin, J. R., and Cornell, C. A. 2014. *Probability, Statistics, and Decision for Civil Engineers*. Mineola, NY: Dover Publications.

Berrill, J. B. 1988. Diversion of Faulting by Hills. *Quarterly Journal of Engineering Geology and Hydrogeology*, **21**(4), 371–374.

Berryman, K., Hamling, I., Kaiser, A., and Stahl, T. 2018. Introduction to the Special Issue on the 2016 Kaikōura Earthquake. *Bulletin of the Seismological Society of America*, **108**(3B), 1491–1495.

Beyer, K., and Bommer, J. J. 2007. Selection and Scaling of Real Accelerograms for Bi-directional Loading: A Review of Current Practice and Code Provisions. *Journal of Earthquake Engineering*, **11**(suppl. 1), 13–45.

Biasi, G. P. 2013. *Appendix H: Maximum Likelihood Recurrence Intervals for California Paleoseismic Sites*. Open File Report 2013-1165. United States Geological Survey.

Biasi, G. P., and Weldon, R. J. I. 2006. Estimating Surface Rupture Length and Magnitude of Paleoearthquakes from Point Measurements of Rupture Displacement. *Bulletin of the Seismological Society of America*, **96**(5), 1612–1623.

Biasi, G. P., and Weldon, R. J. 2009. San Andreas Fault Rupture Scenarios from Multiple Paleoseismic Records: Stringing Pearls. *Bulletin of the Seismological Society of America*, **99**(2A), 471–498.

Biasi, G. P., Weldon, R. J. I., and Dawson, T. E. 2013. *Appendix F: Distribution of Slip in Ruptures*. Open File Report 2013-1165. United States Geological Survey.

Biasi, G. P., Weldon, R. J., Fumal, T. E., and Seitz, G. G. 2002. Paleoseismic Event Dating and the Conditional Probability of Large Earthquakes on the Southern San Andreas Fault, California. *Bulletin of the Seismological Society of America*, **92**(7), 2761–2781.

Bielak, J., Graves, R. W., Olsen, K. B., Taborda, R., Ramírez-Guzmán, L., Day, S. M., Ely, G. P., Roten, D., Jordan, T. H., Maechling, P. J., Urbanic, J., Cui, Y., and Juve, G. 2010. The ShakeOut Earthquake Scenario: Verification of Three Simulation Sets. *Geophysical Journal International*, **180**(1), 375–404.

Bielak, J., Karaoglu, H., and Taborda, R. 2011. Memory-Efficient Displacement-Based Internal Friction for Wave Propagation Simulation. *Geophysics*, **76**(6), T131–T145.

Bielak, J., Loukakis, K., Hisada, Y., and Yoshimura, C. 2003. Domain Reduction Method for Three-Dimensional Earthquake Modeling in Localized Regions, Part I: Theory. *Bulletin of the Seismological Society of America*, **93**(2), 817–824.

Bird, P. 2003. An Updated Digital Model of Plate Boundaries. *Geochemistry Geophysics Geosystems*, **4**(3), 1027.

Bizzarri, A. 2011. On the Deterministic Description of Earthquakes. *Reviews of Geophysics*, **49**(RG3002).

Bocchini, P., and Frangopol, D. M. 2011. Generalized Bridge Network Performance Analysis with Correlation and Time-Variant Reliability. *Structural Safety*, **33**(2), 155–164.

Bommer, J., Abrahamson, N., Strasser, F., A, P., Bard, P., Bungum, H., Cotton, F., Fah, D., Sabetta, F., Scherbaum, F., and Studer, J. 2004. The Challenge of Defining Upper Bounds on Earthquake Ground Motions. *Seismological Research Letters*, **75**, 82–95.

Bommer, J., Douglas, J., and Strasser, F. 2003. Style-of-Faulting in Ground-Motion Prediction Equations. *Bulletin of Earthquake Engineering*, **1**(2), 171–203.

Bommer, J. J. 2012. Challenges of Building Logic Trees for Probabilistic Seismic Hazard Analysis. *Earthquake Spectra*, **28**(4), 1723–1735.

Bommer, J. J., and Abrahamson, N. A. 2006. Why Do Modern Probabilistic Seismic-Hazard Analyses Often Lead to Increased Hazard Estimates? *Bulletin of the Seismological Society of America*, **96**(6), 1967–1977.

Bommer, J. J., and Acevedo, A. B. 2004. The Use of Real Earthquake Accelerograms as Input to Dynamic Analysis. *Journal of Earthquake Engineering*, **8**(Special Issue 1), 43–91.

Bommer, J. J., and Akkar, S. 2012. Consistent Source-to-Site Distance Metrics in Ground-Motion Prediction Equations and Seismic Source Models for PSHA. *Earthquake Spectra*, **28**(1), 1–15.

Bommer, J. J., Coppersmith, K. J., Coppersmith, R. T., Hanson, K. L., Mangongolo, A., Neveling, J., Rathje, E. M., Rodriguez-Marek, A., Scherbaum, F., Shelembe, R., Stafford, P. J., and Strasser, F. O. 2015. A SSHAC Level 3 Probabilistic Seismic Hazard Analysis for a New-Build Nuclear Site in South Africa. *Earthquake Spectra*, **31**(2), 661–698.

Bommer, J. J., and Crowley, H. 2017. The Purpose and Definition of the Minimum Magnitude Limit in PSHA Calculations. *Seismological Research Letters*, **88**(4), 1097–1106.

Bommer, J. J., Douglas, J., Scherbaum, F., Cotton, F., Bungum, H., and Fah, D. 2010. On the Selection of Ground-Motion Prediction Equations for Seismic Hazard Analysis. *Seismological Research Letters*, **81**(5), 783–793.

Bommer, J. J., and Martinez-Pereira, A. 1999. The Effective Duration of Earthquake Strong Motion. *Journal of Earthquake Engineering*, **3**(2), 127–172.

Bommer, J. J., and Scherbaum, F. 2008. The Use and Misuse of Logic Trees in Probabilistic Seismic Hazard Analysis. *Earthquake Spectra*, **24**(4), 997–1009.

Bommer, J. J., and Stafford, P. J. 2016. Chapter 2: Seismic Hazard and Earthquake Actions. Pages 7–40 of: *Seismic Design of Buildings to Eurocode 8*. CRC Press.

Bommer, J. J., and Stafford, P. J. 2020. Selecting Ground-Motion Models for Site-Specific PSHA: Adaptability versus Applicability. *Bulletin of the Seismological Society of America*, **110**(6), 2801–2815.

Bommer, J. J., Stafford, P. J., and Alarcon, J. E. 2009. Empirical Equations for the Prediction of the Significant, Bracketed, and Uniform Duration of Earthquake Ground Motion. *Bulletin of the Seismological Society of America*, **99**(6), 3217–3233.

Bommer, J. J., Stafford, P. J., Alarcon, J. E., and Akkar, S. 2007. The Influence of Magnitude Range on Empirical Ground-Motion Prediction. *Bulletin of the Seismological Society of America*, **97**(6), 2152.

Bommer, J. J., Stafford, P. J., Edwards, B., Dost, B., van Dedem, E., Rodriguez-Marek, A., Kruiver, P., van Elk, J., Doornhof, D., and Ntinalexis, M. 2017. Framework for a Ground-Motion Model for Induced Seismic Hazard and Risk Analysis in the Groningen Gas Field, the Netherlands. *Earthquake Spectra*, **33**(2), 481–498.

Bommer, J. J., Strasser, F. O., Pagani, M., and Monelli, D. 2013. Quality Assurance for Logic-Tree Implementation in Probabilistic Seismic-Hazard Analysis for Nuclear Applications: A Practical Example. *Seismological Research Letters*, **84**(6), 938–945.

Boore, D. M. 2003. Simulation of Ground Motion Using the Stochastic Method. *Pure and Applied Geophysics*, **160**, 635–676.

Boore, D. M. 2004. Estimating $\bar{V}_s(30)$ (or NEHRP Site Classes) from Shallow Velocity Models (Depths < 30 m). *Bulletin of the Seismological Society of America*, **94**(2), 591–597.

Boore, D. M. 2009. Comparing Stochastic Point-Source and Finite-Source Ground-Motion Simulations: SMSIM and EXSIM. *Bulletin of the Seismological Society of America*, **99**(6), 3202–3216.

Boore, D. M. 2010. Orientation-Independent, Nongeometric-Mean Measures of Seismic Intensity from Two Horizontal Components of Motion. *Bulletin of the Seismological Society of America*, **100**(4), 1830–1835.

Boore, D. M. 2013. The Uses and Limitations of the Square-Root-Impedance Method for Computing Site Amplification. *Bulletin of the Seismological Society of America*, **103**(4), 2356–2368.

Boore, D. M. 2016. Determining Generic Velocity and Density Models for Crustal Amplification Calculations, with an Update of the Boore and Joyner (1997) Generic Site Amplification for VS(Z) = 760 m/s. *Bulletin of the Seismological Society of America*, **106**(1), 313–317.

Boore, D. M., and Boatwright, J. 1984. Average Body-Wave Radiation Coefficients. *Bulletin of the Seismological Society of America*, **74**(5), 1615–1621.

Boore, D. M., and Bommer, J. J. 2005. Processing of Strong-Motion Accelerograms: Needs, Options and Consequences. *Soil Dynamics and Earthquake Engineering*, **25**, 93–115.

Boore, D. M., Gibbs, J. F., Joyner, W. B., Tinsley, J. C., and Ponti, D. J. 2003. Estimated Ground Motion from the 1994 Northridge, California, Earthquake at the Site of the Interstate 10 and La Cienega Boulevard Bridge Collapse, West Los Angeles, California. *Bulletin of the Seismological Society of America*, **93**(6), 2737–2751.

Boore, D. M., and Joyner, W. B. 1997. Site Amplifications for Generic Rock Sites. *Bulletin of the Seismological Society of America*, **87**(2), 327–341.

Boore, D. M., Joyner, W. B., and Fumal, T. E. 1997. Equations for Estimating Horizontal Response Spectra and Peak Acceleration from Western North American Earthquakes: A Summary of Recent Work. *Seismological Research Letters*, **68**(1), 128–153.

Boore, D. M., Joyner, W. B., and Wennerberg, L. 1992. Fitting the Stochastic ω^{-2} Source Model to Observed Response Spectra in Western North America: Trade-offs between $\Delta\sigma$ and κ. *Bulletin of the Seismological Society of America*, **82**(4), 1956–1963.

Boore, D. M., Stewart, J. P., Seyhan, E., and Atkinson, G. M. 2014. NGA-West2 Equations for Predicting PGA, PGV, and 5% Damped PSA for Shallow Crustal Earthquakes. *Earthquake Spectra*, **30**(3), 1057–1085.

Boore, D. M., and Thompson, E. M. 2014. Path Durations for Use in the Stochastic-Method Simulation of Ground Motions. *Bulletin of the Seismological Society of America*, **104**(5), 2541–2552.

Boore, D. M., and Thompson, E. M. 2015. Revisions to Some Parameters Used in Stochastic-Method Simulations of Ground Motion. *Bulletin of the Seismological Society of America*, **105**(2A), 1029–1041.

Boore, D., Watson-Lamprey, J., and Abrahamson, N. 2006. Orientation-Independent Measures of Ground Motion. *Bulletin of the Seismological Society of America*, **96**(4A), 1502–1511.

Bora, S. S., Scherbaum, F., Kuehn, N., and Stafford, P. J. 2014. Fourier Spectral- and Duration Models for the Generation of Response Spectra Adjustable to Different Source-, Propagation-, and Site Conditions. *Bulletin of Earthquake Engineering*, **12**, 467–493.

Bora, S. S., Scherbaum, F., Kuehn, N., and Stafford, P. J. 2016. On the Relationship between Fourier and Response Spectra: Implications for the Adjustment of Empirical Ground-Motion Prediction Equations (GMPEs). *Bulletin of the Seismological Society of America*, **106**(3), 1235–1253.

Bourne, S. J., Oates, S. J., and van Elk, J. 2018. The Exponential Rise of Induced Seismicity with Increasing Stress Levels in the Groningen Gas Field and Its Implications for Controlling Seismic Risk. *Geophysical Journal International*, **213**(3), 1693–1700.

Box, G. E. P. 1980. Sampling and Bayes' Inference in Scientific Modelling and Robustness. *Journal of the Royal Statistical Society. Series A (General)*, **143**(4), 383–430.

Boyd, O. S. 2012. Including Foreshocks and Aftershocks in Time-Independent Probabilistic Seismic-Hazard Analyses. *Bulletin of the Seismological Society of America*, **102**(3), 909–917.

Bozorgnia, Y., and Campbell, K. 2004. The Vertical-to-Horizontal Response Spectral Ratio and Tentative Procedures for Developing Simplified V/H and Vertical Design Spectra. *Journal of Earthquake Engineering*, **8**(2), 175–207.

Bradley, B. 2009. Seismic Hazard Epistemic Uncertainty in the San Francisco Bay Area and Its Role in Performance-Based Assessment. *Earthquake Spectra*, **25**(4), 733–753.

Bradley, B. 2010a. Epistemic Uncertainty in Component Fragility Functions. *Earthquake Spectra*, **26**(1), 41–62.

Bradley, B. A. 2010b. A Generalized Conditional Intensity Measure Approach and Holistic Ground-Motion Selection. *Earthquake Engineering & Structural Dynamics*, **39**(12), 1321–1342.

Bradley, B. A. 2011. Correlation of Significant Duration with Amplitude and Cumulative Intensity Measures and Its Use in Ground Motion Selection. *Journal of Earthquake Engineering*, **15**(6), 809–832.

Bradley, B. A. 2012a. Empirical Correlations between Cumulative Absolute Velocity and Amplitude-Based Ground Motion Intensity Measures. *Earthquake Spectra*, **28**(1), 37–54.

Bradley, B. A. 2012b. A Ground Motion Selection Algorithm Based on the Generalized Conditional Intensity Measure Approach. *Soil Dynamics and Earthquake Engineering*, **40**, 48–61.

Bradley, B. A. 2012c. The Seismic Demand Hazard and Importance of the Conditioning Intensity Measure. *Earthquake Engineering & Structural Dynamics*, **41**(11), 1417–1437.

Bradley, B. A. 2013a. A Comparison of Intensity-Based Demand Distributions and the Seismic Demand Hazard for Seismic Performance Assessment. *Earthquake Engineering & Structural Dynamics*, **42**(15), 2235–2253.

Bradley, B. A. 2013b. A Critical Examination of Seismic Response Uncertainty Analysis in Earthquake Engineering. *Earthquake Engineering & Structural Dynamics*, **42**(11), 1717–1729.

Bradley, B. A. 2013c. Practice-Oriented Estimation of the Seismic Demand Hazard Using Ground Motions at Few Intensity Levels. *Earthquake Engineering & Structural Dynamics*, **42**(14), 2167–2185.

Bradley, B. A. 2015. Correlation of Arias Intensity with Amplitude, Duration and Cumulative Intensity Measures. *Soil Dynamics and Earthquake Engineering*, **78**, 89–98.

Bradley, B. A., Burks, L. S., and Baker, J. W. 2015. Ground Motion Selection for Simulation-Based Seismic Hazard and Structural Reliability Assessment. *Earthquake Engineering & Structural Dynamics*, **44**(13), 2321–2340.

Bradley, B. A., and Dhakal, R. P. 2008. Error Estimation of Closed-Form Solution for Annual Rate of Structural Collapse. *Earthquake Engineering & Structural Dynamics*, **37**(15), 1721–1737.

Bradley, B. A., Dhakal, R. P., MacRae, G. A., and Cubrinovski, M. 2010. Prediction of Spatially Distributed Seismic Demands in Specific Structures: Ground Motion and Structural Response. *Earthquake Engineering & Structural Dynamics*, **39**(2), 501–520.

Bradley, B. A., Pettinga, D., Baker, J. W., and Fraser, J. 2017. Guidance on the Utilization of Earthquake-Induced Ground Motion Simulations in Engineering Practice. *Earthquake Spectra*, **33**(3), 809–835.

Bray, J. D., and Travasarou, T. 2007. Simplified Procedure for Estimating Earthquake-Induced Deviatoric Slope Displacements. *Journal of Geotechnical and Geoenvironmental Engineering*, **133**(4), 381–392.

Brillinger, D. R., and Preisler, H. K. 1984. An Exploratory Analysis of the Joyner–Boore Attenuation Data. *Bulletin of the Seismological Society of America*, **74**(4), 1441–1450.

Brillinger, D. R., and Preisler, H. K. 1985. Further Analysis of the Joyner–Boore Attenuation Data. *Bulletin of the Seismological Society of America*, **75**(2), 611–614.

Brocher, T. M. 2005. Empirical Relations between Elastic Wavespeeds and Density in the Earth's Crust. *Bulletin of the Seismological Society of America*, **95**(6), 2081–2092.

Brune, J. 1970. Tectonic Stress and the Spectra of Seismic Shear Waves from Earthquakes. *Journal of Geophysical Research*, **75**(26), 4997–5009.

Brune, J. 1971. Correction. *Journal of Geophysical Research*, **76**(20), 5002.

Brune, J. 1999. Precarious Rocks along the Mojave Section of the San Andreas Fault, California: Constraints on Ground Motion from Great Earthquakes. *Seismological Research Letters*, **70**(1), 29–33.

BSSC. 2009. *NEHRP Recommended Seismic Provisions for New Buildings and Other Structures*. FEMA P-750. Building Seismic Safety Council, Washington, DC.

BSSC. 2015. *NEHRP Recommended Seismic Provisions for New Buildings and Other Structures*. FEMA P-1050. Building Seismic Safety Council, Washington, DC.

Budnitz, R. J., Amico, P. J., Cornell, C. A., Hall, W. J., Kennedy, R. P., Reed, J. W., and Shinozuka, M. 1985. *An Approach to the Quantification of Seismic Margins in Nuclear Power Plants*. NUREG/CR 4334. US Nuclear Regulatory Commission, Washington, DC.

Bullock, Z. 2019. Log-Logistic Uncertainty Is More Durable than Lognormal Uncertainty in Ground-Motion Prediction Equations. *Bulletin of the Seismological Society of America*, **109**(2), 567–574.

Buratti, N., Stafford, P. J., and Bommer, J. J. 2011. Earthquake Accelerogram Selection and Scaling Procedures for Estimating the Distribution of Drift Response. *Journal of Structural Engineering*, **137**(3), 345–357.

Burbank, D. W., and Anderson, R. S. 2011. *Tectonic Geomorphology*, 2nd ed. Chichester, UK: John Wiley & Sons.

Burridge, R., and Knopoff, L. 1964. Body Force Equivalents for Seismic Dislocations. *Bulletin of the Seismological Society of America*, **54**(6), 1875–1888.

Burridge, R., and Knopoff, L. 1967. Model and Theoretical Seismicity. *Bulletin of the Seismological Society of America*, **57**(3), 341–371.

Calvi, G. M., Pinho, R., Magenes, G., Bommer, J. J., Restrepo-Vélez, L. F., and Crowley, H. 2006. Development of Seismic Vulnerability Assessment Methodologies over the Past 30 Years. *ISET Journal of Earthquake Technology*, **43**(3), 75–104.

Campbell, K. W. 2003. Prediction of Strong Ground Motion Using the Hybrid Empirical Method and Its Use in the Development of Ground-Motion (Attenuation) Relations in Eastern North America. *Bulletin of the Seismological Society of America*, **93**(3), 1012–1033.

Campbell, K. W., and Bozorgnia, Y. 2014. NGA-West2 Ground Motion Model for the Average Horizontal Components of PGA, PGV, and 5% Damped Linear Acceleration Response Spectra. *Earthquake Spectra*, **30**(3), 1087–1115.

Carballo, J. E. 2000. *Probabilistic Seismic Demand Analysis: Spectrum Matching and Design*. Dept. of Civil and Environmental Engineering. Stanford, CA: Stanford University.

Carlson, J. M., and Langer, J. S. 1989. Mechanical Model of an Earthquake Fault. *Physical Review A*, **40**(11), 6470–6484.

Carlson, J. M., Langer, J. S., Shaw, B. E., and Tang, C. 1991. Intrinsic Properties of a Burridge–Knopoff Model of an Earthquake Fault. *Physical Review A*, **44**(2), 884–897.

Carlton, B., and Abrahamson, N. 2014. Issues and Approaches for Implementing Conditional Mean Spectra in Practice. *Bulletin of the Seismological Society of America*, **104**(1), 503–512.

CEN. 2004. *Eurocode 8: Design of Structures for Earthquake Resistance—Part 1: General Rules, Seismic Actions and Rules for Buildings*. Comité Européen de Normalisation, European Standard EN 1998-1:2004: E. Brussels.

Cesca, S., Zhang, Y., Mouslopoulou, V., Wang, R., Saul, J., Savage, M., Heimann, S., Kufner, S. K., Oncken, O., and Dahm, T. 2017. Complex Rupture Process of the Mw 7.8, 2016, Kaikoura Earthquake, New Zealand, and Its Aftershock Sequence. *Earth and Planetary Science Letters*, **478**, 110–120.

Chaljub, E., Maufroy, E., Moczo, P., Kristek, J., Hollender, F., Bard, P.-Y., Priolo, E., Klin, P., de Martin, F., Zhang, Z., Zhang, W., and Chen, X. 2015. 3-D Numerical Simulations of Earthquake Ground Motion in Sedimentary Basins: Testing Accuracy through Stringent Models. *Geophysical Journal International*, **201**(1), 90–111.

Chaljub, E., Moczo, P., Tsuno, S., Bard, P.-Y., Kristek, J., Käser, M., Stupazzini, M., and Kristekova, M. 2010. Quantitative Comparison of Four Numerical Predictions of 3D Ground Motion in the Grenoble Valley, France. *Bulletin of the Seismological Society of America*, **100**(4), 1427–1455.

Chandramohan, R., Baker, J. W., and Deierlein, G. G. 2016. Impact of Hazard-Consistent Ground Motion Duration in Structural Collapse Risk Assessment. *Earthquake Engineering & Structural Dynamics*, **45**(8), 1357–1379.

Chang, S. E., Shinozuka, M., and Moore, J. E. 2000. Probabilistic Earthquake Scenarios: Extending Risk Analysis Methodologies to Spatially Distributed Systems. *Earthquake Spectra*, **16**(3), 557–572.

Chen, Y., and Baker, J. W. 2019. Spatial Correlations in CyberShake Physics-Based Ground-Motion Simulations. *Bulletin of the Seismological Society of America*, **109**(6), 2447–2458.

Chen, P., Zhao, L., and Jordan, T. H. 2007. Full 3D Tomography for the Crustal Structure of the Los Angeles Region. *Bulletin of the Seismological Society of America*, **97**(4), 1094–1120.

Chen, Y.-H., and Tsai, C.-C. P. 2002. A New Method for Estimation of the Attenuation Relationship with Variance Components. *Bulletin of the Seismological Society of America*, **92**(5), 1984–1991.

Chiou, B., Darragh, R., Gregor, N., and Silva, W. 2008. NGA Project Strong-Motion Database. *Earthquake Spectra*, **24**(1), 23–44.

Chiou, B., Youngs, R., Abrahamson, N., and Addo, K. 2010. Ground-Motion Attenuation Model for Small-to-Moderate Shallow Crustal Earthquakes in California and Its Implications on Regionalization of Ground-Motion Prediction Models. *Earthquake Spectra*, **26**(4), 907–926.

Chiou, B. S.-J., and Youngs, R. R. 2008. An NGA Model for the Average Horizontal Component of Peak Ground Motion and Response Spectra. *Earthquake Spectra*, **24**(1), 173–215.

Chiou, B. S.-J., and Youngs, R. R. 2014. Update of the Chiou and Youngs NGA Model for the Average Horizontal Component of Peak Ground Motion and Response Spectra. *Earthquake Spectra*, **30**(3), 1117–1153.

Choi, Y., and Stewart, J. P. 2005. Nonlinear Site Amplification as Function of 30 m Shear Wave Velocity. *Earthquake Spectra*, **21**(1), 1–30.

Choi, Y., Stewart, J. P., and Graves, R. W. 2005. Empirical Model for Basin Effects Accounts for Basin Depth and Source Location. *Bulletin of the Seismological Society of America*, **95**(4), 1412–1427.

Chopra, A. 2016. *Dynamics of Structures*, 5th ed. Hoboken, NJ: Pearson.

Clough, R., and Penzien, J. 1975. *Dynamics of Structures*. New York: McGraw-Hill.

Coburn, A., and Spence, R. 2002. *Earthquake Protection*. John Wiley & Sons.

Cochran, W. G. 1977. *Sampling Techniques*, 3rd ed. New York: John Wiley & Sons.

Console, R., and Murru, M. 2001. A Simple and Testable Model for Earthquake Clustering. *Journal of Geophysical Research*, **106**(B5), 8699–8711.

Cook, R., Malkus, D., and Plesha, M. 1989. *Concepts and Applications of Finite Element Analysis*, 3rd ed. New York: John Wiley and Sons.

Cooke, R. 1991. *Experts in Uncertainty: Opinion and Subjective Probability in Science*. Oxford University Press on Demand.

Cordova, P. P., Deierlein, G. G., Mehanny, S. S., and Cornell, C. 2001. Development of a Two-Parameter Seismic Intensity Measure and Probabilistic Assessment Procedure. Pages 187–206 of: *The Second U.S.–Japan Workshop on Performance-Based Earthquake Engineering Methodology for Reinforced Concrete Building Structures*.

Cornell, C. A. 1968. Engineering Seismic Risk Analysis. *Bulletin of the Seismological Society of America*, **58**(5), 1583–1606.

Cornell, C. A. 1994. Statistical Analysis of Maximum Magnitudes. In: *The Earthquakes of Stable Continential Regions*. EPRI Report, nos. TR–102261s–V1–V5. Electric Power Research Institute.

Cornell, C. A., Banon, H., and Shakal, A. F. 1979. Seismic Motion and Response Prediction Alternatives. *Earthquake Engineering & Structural Dynamics*, **7**(4), 295–315.

Cornell, C. A., Jalayer, F., Hamburger, R. O., and Foutch, D. A. 2002. Probabilistic Basis for 2000 SAC Federal Emergency Management Agency Steel Moment Frame Guidelines. *Journal of Structural Engineering*, **128**(4), 526–533.

Cornell, C. A., and Krawinkler, H. 2000. Progress and Challenges in Seismic Performance Assessment. *PEER Center News*, **3**(2).

Cornell, C. A., and Vanmarcke, E. H. 1969. The Major Influences on Seismic Risk. Pages 69–83 of: *Proceedings of the Fourth World Conference on Earthquake Engineering*, vol. 1.

Cornell, C. A., and Winterstein, S. R. 1988. Temporal and Magnitude Dependence in Earthquake Recurrence Models. *Bulletin of the Seismological Society of America*, **78**(4), 1522–1537.

Cotton, F., Scherbaum, F., Bommer, J., and Bungum, H. 2006. Criteria for Selecting and Adjusting Ground-Motion Models for Specific Target Regions: Application to Central Europe and Rock Sites. *Journal of Seismology*, **10**, 137–156.

Crempien, J. G. F., and Archuleta, R. J. 2015. UCSB Method for Simulation of Broadband Ground Motion from Kinematic Earthquake Sources. *Seismological Research Letters*, **86**(1), 61–67.

Crowley, H., and Pinho, R. 2011. Global Earthquake Model: Community-Based Seismic Risk Assessment. Pages 3–19 of: *Protection of Built Environment against Earthquakes*. Springer.

Crowley, H., Pinho, R., and Bommer, J. J. 2004. A Probabilistic Displacement-Based Vulnerability Assessment Procedure for Earthquake Loss Estimation. *Bulletin of Earthquake Engineering*, **2**(2), 173–219.

Crowley, H., Stafford, P. J., and Bommer, J. J. 2008. Can Earthquake Loss Models Be Validated Using Field Observations? *Journal of Earthquake Engineering*, **12**(7), 1078–1104.

Daniell, J. E., Khazai, B., Wenzel, F., and Vervaeck, A. 2011. The CATDAT Damaging Earthquakes Database. *Natural Hazards and Earth System Sciences*, **11**(8), 2235.

Day, S. M., and Bradley, C. R. 2001. Memory-Efficient Simulation of Anelastic Wave Propagation. *Bulletin of the Seismological Society of America*, **91**(3), 520–531.

Day, S. M., Graves, R., Bielak, J., Dreger, D., Larsen, S., Olsen, K. B., Pitarka, A., and Ramirez-Guzman, L. 2008. Model for Basin Effects on Long-Period Response Spectra in Southern California. *Earthquake Spectra*, **24**(1), 257–277.

D'Ayala, D., Meslem, A., Vamvatsikos, D., Porter, K., Rossetto, T., Crowley, H., and Silva, V. 2014. *Guidelines for Analytical Vulnerability Assessment of Low-to Mid-Rise Buildings—Methodology*. GEM Technical Report 2014-12 V1.0.0. Global Earthquake Model.

de la Puente, J., Dumbser, M., Käser, M., and Igel, H. 2008. Discontinuous Galerkin Methods for Wave Propagation in Poroelastic Media. *Geophysics*, **73**(5), T77–T97.

de la Torre, C. A., Bradley, B. A., and Lee, R. L. 2020. Modeling Nonlinear Site Effects in Physics-Based Ground Motion Simulations of the 2010–2011 Canterbury Earthquake Sequence. *Earthquake Spectra*, **36**(2), 856–879.

DeBock, D. J., Garrison, J. W., Kim, K. Y., and Liel, A. B. 2014. Incorporation of Spatial Correlations between Building Response Parameters in Regional Seismic Loss Assessment. *Bulletin of the Seismological Society of America*, **104**(1), 214–228.

DeMets, C., Gordon, R. G., Argus, D. F., and Stein, S. 1990. Current Plate Motions. *Geophysical Journal International*, **101**(2), 425–478.

Der Kiureghian, A. 2005. Non-Ergodicity and PEER's Framework Formula. *Earthquake Engineering and Structural Dynamics*, **34**(13), 1643–1652.

Der Kiureghian, A., and Ditlevsen, O. 2008. Aleatory or Epistemic? Does It Matter? *Structural Safety*, **31**(2), 105–112.

Detweiler, S. T., and Wein, A. M. 2017. *The HayWired Earthquake Scenario*. USGS Numbered Series 2017-5013. US Geological Survey, Reston, VA.

Dieterich, J. H. 1979. Modeling of Rock Friction: 1. Experimental Results and Constitutive Equations. *Journal of Geophysical Research*, **84**(B5), 2161–2168.

Dolsek, M. 2009. Incremental Dynamic Analysis with Consideration of Modeling Uncertainties. *Earthquake Engineering & Structural Dynamics*, **38**(6), 805–825.

Donahue, J. L., and Abrahamson, N. A. 2014. Simulation-Based Hanging Wall Effects. *Earthquake Spectra*, **30**(3), 1269–1284.

Dorra, E. M., Stafford, P. J., and Elghazouli, A. Y. 2013. Earthquake Loss Estimation for Greater Cairo and the National Economic Implications. *Bulletin of Earthquake Engineering*, **11**(4), 1217–1257.

Douglas, J. 2003. Earthquake Ground Motion Estimation Using Strong Motion Records: A Review of Equations for the Estimation of Peak Ground Acceleration and Response Spectral Ordinates. *Earth Science Reviews*, **61**, 43–104.

Douglas, J. 2019. *Ground Motion Prediction Equations 1964–2018*. Online Report www.gmpe.org.uk. University of Strathclyde, Glasgow.

Douglas, J., and Aochi, H. 2008. A Survey of Techniques for Predicting Earthquake Ground Motions for Engineering Purposes. *Surveys in Geophysics*, **29**(3), 187–220.

Douglas, J., and Aochi, H. 2016. Assessing Components of Ground-Motion Variability from Simulations for the Marmara Sea Region (Turkey). *Bulletin of the Seismological Society of America*, **106**(1), 300–306.

Douglas, J., and Boore, D. 2011. High-Frequency Filtering of Strong-Motion Records. *Bulletin of Earthquake Engineering*, **9**(2), 395–409.

Douglas, J., and Edwards, B. 2016. Recent and Future Developments in Earthquake Ground Motion Estimation. *Earth-Science Reviews*, **160**(Sept.), 203–219.

Douglas, J., Ulrich, T., Bertil, D., and Rey, J. 2014. Comparison of the Ranges of Uncertainty Captured in Different Seismic-Hazard Studies. *Seismological Research Letters*, **85**(5), 977–985.

Douglas, J., Ulrich, T., and Negulescu, C. 2013. Risk-Targeted Seismic Design Maps for Mainland France. *Natural Hazards*, **65**(3), 1999–2013.

Dreger, D. S., and Jordan, T. H. 2015. Introduction to the Focus Section on Validation of the SCEC Broadband Platform V14.3 Simulation Methods. *Seismological Research Letters*, **86**(1), 15–16.

Eads, L., Miranda, E., Krawinkler, H., and Lignos, D. G. 2013. An Efficient Method for Estimating the Collapse Risk of Structures in Seismic Regions. *Earthquake Engineering & Structural Dynamics*, **42**(1), 25–41.

Eaton, J. P. 1992. Determination of Amplitude and Duration Magnitudes and Site Residual from Short-Period Seismographs in Northern California. *Bulletin of the Seismological Society of America*, **82**(2), 533–579.

Ebel, J. E., and Kafka, A. L. 1999. A Monte Carlo Approach to Seismic Hazard Analysis. *Bulletin of the Seismological Society of America*, **89**(4), 854–866.

Eberly, D. 2020. *Distance between Point and Triangle in 3D*. Geometric Tools, Online Report. www.geometrictools.com/Documentation/DistancePoint3Triangle3.pdf.

Edwards, B., and Fäh, D. 2013. A Stochastic Ground Motion Model for Switzerland. *Bulletin of the Seismological Society of America*, **103**(1), 78–98.

Edwards, B., Zurek, B., van Dedem, E., Stafford, P. J., Oates, S., van Elk, J., deMartin, B., and Bommer, J. J. 2019. Simulations for the Development of a Ground Motion Model for Induced Seismicity in the Groningen Gas Field, the Netherlands. *Bulletin of Earthquake Engineering*, **17**, 4441–4456.

Ellingwood, B. R., and Kinali, K. 2009. Quantifying and Communicating Uncertainty in Seismic Risk Assessment. *Structural Safety*, **31**(2), 179–187.

Ellsworth, W. L., Matthews, M. V., Nadeau, R. M., Nishenko, S. P., Reasenberg, P. A., and Simpson, R. W. 1999. *A Physically Based Earthquake Recurrence Model for Estimation of Long-Term Earthquake Probabilities*. Open File Report 99-522.

Elms, D. G. 2004. Structural Safety: Issues and Progress. *Progress in Structural Engineering and Materials*, **6**(2), 116–126.

EPRI. 1991. *Standardization of the Cumulative Absolute Velocity*. EPRI TR-100082-T2. Electric Power Research Institute, Palo Alto, CA.

Esposito, S., and Iervolino, I. 2011. PGA and PGV Spatial Correlation Models Based on European Multievent Datasets. *Bulletin of the Seismological Society of America*, **101**(5), 2532–2541.

Evison, F. F., and Rhoades, D. A. 2004. Demarcation and Scaling of Long-Term Seismogenesis. *Pure and Applied Geophysics*, **161**(1), 21–45.

Felzer, K. R. 2013a. *Appendix K: The UCERF3 Earthquake Catalog*. Open File Report 2013-1165. United States Geological Survey.

Felzer, K. R. 2013b. *Appendix L: Estimate of the Seismicity Rate and Magnitude-Frequency Distribution of Earthquakes in California from 1850 to 2011*. Open File Report 2013-1165. United States Geological Survey.

Felzer, K. R., and Brodsky, E. E. 2005. Testing the Stress Shadow Hypothesis. *Journal of Geophysical Research: Solid Earth*, **110**(B5), 702.

Felzer, K. R., and Brodsky, E. E. 2006. Decay of Aftershock Density with Distance Indicates Triggering by Dynamic Stress. *Nature*, **441**(7094), 735–738.

FEMA. 2005. *Improvement of Nonlinear Static Seismic Procedures*. FEMA-440. Federal Emergency Management Agency, FEMA, Washington, DC.

FEMA. 2009. *Quantification of Building Seismic Performance Factors (FEMA P695, ATC-63)*. FEMA P695. Federal Emergency Management Agency, prepared by the Applied Technology Council.

FEMA. 2012. *Seismic Performance Assessment of Buildings*. FEMA P-58. Prepared by Applied Technology Council for the Federal Emergency Management Agency.

FEMA. 2015. *Hazus-MH 2.1*. Multi-hazard Loss Estimation Methodology Technical and User Manuals. Federal Emergency Management Agency.

Field, E. H. 2015. "All Models Are Wrong, but Some Are Useful." *Seismological Research Letters*, **86**(2A), 291–293.

Field, E. H., Arrowsmith, R. J., Biasi, G. P., Bird, P., Dawson, T. E., Felzer, K. R., Jackson, D. D., Johnson, K. M., Jordan, T. H., Madden, C., Michael, A. J., Milner, K. R., Page, M. T., Parsons, T., Powers, P. M., Shaw, B. E., Thatcher, W. R., Weldon, R. J., and Zeng, Y. 2014. Uniform California Earthquake Rupture Forecast, Version 3 (UCERF3)—The Time-Independent Model. *Bulletin of the Seismological Society of America*, **104**(3), 1122–1180.

Field, E. H., Biasi, G. P., Bird, P., Dawson, T. E., Felzer, K. R., Jackson, D. D., Johnson, K. M., Jordan, T. H., Madden, C., Michael, A. J., Milner, K. R., Page, M. T., Parsons, T., Powers, P. M., Shaw, B. E., Thatcher, W. R., Weldon, R. J. I., and Zeng, Y. 2013. *Uniform California Earthquake Rupture Forecast, Version 3 (UCERF3): The Time-Independent Model*. Open File Report 2013-1165. United States Geological Survey.

Field, E. H., Biasi, G. P., Bird, P., Dawson, T. E., Felzer, K. R., Jackson, D. D., Johnson, K. M., Jordan, T. H., Madden, C., Michael, A. J., Milner, K. R., Page, M. T., Parsons, T., Powers, P. M., Shaw, B. E., Thatcher, W. R., Weldon, R. J., and Zeng, Y. 2015. Long-Term Time-Dependent Probabilities for the Third Uniform California Earthquake Rupture Forecast (UCERF3). *Bulletin of the Seismological Society of America*, **105**(2A), 511–543.

Field, E. H., Dawson, T. E., Felzer, K. R., Frankel, A. D., Gupta, V., Jordan, T. H., Parsons, T., Petersen, M. D., Stein, R. S., Weldon, R. J., and Wills, C. J. 2009. Uniform California Earthquake Rupture Forecast, Version 2 (UCERF 2). *Bulletin of the Seismological Society of America*, **99**(4), 2053–2107.

Field, E. H., and Jordan, T. H. 2015. Time-Dependent Renewal-Model Probabilities When Date of Last Earthquake Is Unknown. *Bulletin of the Seismological Society of America*, **105**(1), 459–463.

Field, E. H., Jordan, T. H., and Cornell, C. A. 2003. OpenSHA: A Developing Community-Modeling Environment for Seismic Hazard Analysis. *Seismological Research Letters*, **74**(4), 406–419.

Field, E. H., Jordan, T. H., Page, M. T., Milner, K. R., Shaw, B. E., Dawson, T. E., Biasi, G. P., Parsons, T., Hardebeck, J. L., Michael, A. J., Weldon, R. J. I., Powers, P. M., Johnson, K. M., Zeng, Y., Felzer, K. R., van der Elst, N., Madden, C., Arrowsmith, R., Werner, M. J., and Thatcher, W. R.

2017a. A Synoptic View of the Third Uniform California Earthquake Rupture Forecast (UCERF3). *Bulletin of the Seismological Society of America*, **88**(5), 1259–1267.

Field, E. H., Milner, K. R., Hardebeck, J. L., Page, M. T., van der Elst, N., Jordan, T. H., Michael, A. J., Shaw, B. E., and Werner, M. J. 2017b. A Spatiotemporal Clustering Model for the Third Uniform California Earthquake Rupture Forecast (UCERF3-ETAS): Toward an Operational Earthquake Forecast. *Bulletin of the Seismological Society of America*, **107**(3), 1049–1081.

Field, E. H., and Page, M. T. 2011. Estimating Earthquake-Rupture Rates on a Fault or Fault System. *Bulletin of the Seismological Society of America*, **101**(1), 79–92.

Finn, W. D. L., Ventura, C. E., and Wu, G. 1993. Analysis of Ground Motions at Treasure Island Site during the 1989 Loma Prieta Earthquake. *Soil Dynamics and Earthquake Engineering*, **12**(7), 383–390.

Foulser-Piggott, R., and Goda, K. 2015. Ground-Motion Prediction Models for Arias Intensity and Cumulative Absolute Velocity for Japanese Earthquakes Considering Single-Station Sigma and Within-Event Spatial Correlation. *Bulletin of the Seismological Society of America*, **105**(4), 1903–1918.

Foulser-Piggott, R., and Stafford, P. J. 2012. A Predictive Model for Arias Intensity at Multiple Sites and Consideration of Spatial Correlations. *Earthquake Engineering & Structural Dynamics*, **41**(3), 431–451.

Fox, M. J., Stafford, P. J., and Sullivan, T. J. 2016. Seismic Hazard Disaggregation in Performance-Based Earthquake Engineering: Occurrence or Exceedance? *Earthquake Engineering & Structural Dynamics*, **45**(5), 835–842.

Frankel, A. 1995. Simulating Strong Motions of Large Earthquakes Using Recordings of Small Earthquakes: The Loma Prieta Mainshock as a Test Case. *Bulletin of the Seismological Society of America*, **85**(4), 1144–1160.

Frankel, A., and Vidale, J. 1992. A Three-Dimensional Simulation of Seismic Waves in the Santa Clara Valley, California, from a Loma Prieta Aftershock. *Bulletin of the Seismological Society of America*, **82**(5), 2045–2074.

Frankel, A., Wirth, E., Marafi, N., Vidale, J., and Stephenson, W. 2018. Broadband Synthetic Seismograms for Magnitude 9 Earthquakes on the Cascadia Megathrust Based on 3D Simulations and Stochastic Synthetics, Part 1: Methodology and Overall Results. *Bulletin of the Seismological Society of America*, **108**(5A), 2347–2369.

Fukushima, Y. 1996. Scaling Relations for Strong Ground Motion Prediction Models with M2 Terms. *Bulletin of the Seismological Society of America*, **86**(2), 329–336.

Gardner, J. K., and Knopoff, L. 1974. Is the Sequence of Earthquakes in Southern California, with Aftershocks Removed, Poissonian? *Bulletin of the Seismological Society of America*, **64**(5), 1363–1367.

Gardoni, P., Der Kiureghian, A., and Mosalam, K. M. 2002. Probabilistic Capacity Models and Fragility Estimates for Reinforced Concrete Columns Based on Experimental Observations. *Journal of Engineering Mechanics*, **128**(10), 1024–1038.

Gasparini, D., and Vanmarcke, E. H. 1976. *SIMQKE: A Program for Artificial Motion Generation*. MIT Department of Civil Engineering, Cambridge, MA.

Gatz, D. F., and Smith, L. 1995. The Standard Error of a Weighted Mean Concentration—I. Bootstrapping vs Other Methods. *Atmospheric Environment*, **29**(11), 1185–1193.

Geller, R. J. 2011. Shake-up Time for Japanese Seismology. *Nature*, **472**(Apr.), 407.

Gelman, A., Carlin, J. B., Stern, H. S., Dunson, D. B., Vehtari, A., and Rubin, D. B. 2013. *Bayesian Data Analysis*, 3rd ed. Chapman and Hall/CRC.

Gentle, J. E. 2003. *Random Number Generation and Monte Carlo Methods*, 2nd ed. New York: Springer-Verlag.

Gerstenberger, M. C., Rhoades, D. A., and McVerry, G. H. 2016. A Hybrid Time-Dependent Probabilistic Seismic-Hazard Model for Canterbury, New Zealand. *Seismological Research Letters*, **87**(6), 1311–1318.

Ghafory-Ashtiany, M., Mousavi, M., and Azarbakht, A. 2011. Strong Ground Motion Record Selection for the Reliable Prediction of the Mean Seismic Collapse Capacity of a Structure Group. *Earthquake Engineering & Structural Dynamics*, **40**(6), 691–708.

Gledhill, K., Ristau, J., Reyners, M., Fry, B., and Holden, C. 2011. The Darfield (Canterbury, New Zealand) Mw 7.1 Earthquake of September 2010: A Preliminary Seismological Report. *Seismological Research Letters*, **82**(3), 378–386.

Goda, K. 2011. Interevent Variability of Spatial Correlation of Peak Ground Motions and Response Spectra. *Bulletin of the Seismological Society of America*, **101**(5), 2522–2531.

Goda, K., and Atkinson, G. M. 2010. Intraevent Spatial Correlation of Ground-Motion Parameters Using SK-Net Data. *Bulletin of the Seismological Society of America*, **100**(6), 3055–3067.

Goda, K., and Atkinson, G. M. 2014. Variation of Source-to-Site Distance for Megathrust Subduction Earthquakes: Effects on Ground Motion Prediction Equations. *Earthquake Spectra*, **30**(2), 845–866.

Goda, K., and Hong, H. P. 2008. Spatial Correlation of Peak Ground Motions and Response Spectra. *Bulletin of the Seismological Society of America*, **98**(1), 354–365.

González Ortega, J. A., González García, J. J., and Sandwell, D. T. 2018. Interseismic Velocity Field and Seismic Moment Release in Northern Baja California, Mexico. *Seismological Research Letters*, **89**(2A), 526–533.

Goovaerts, P. 1997. *Geostatistics for Natural Resources Evaluation*. Applied Geostatistics Series. New York: Oxford University Press.

Gospodinov, D. 2017. Spatio-Temporal Evolution of Aftershock Energy Release Following the 1989, $M_w 6.9$, Loma Prieta Earthquake in California. *Acta Geophysica*, **65**(3), 565–573.

Graves, R. 1993. Modeling Three-Dimensional Site Response Effects in the Marina District Basin, San Francisco, California. *Bulletin of the Seismological Society of America*, **83**(4), 1042–1063.

Graves, R., Jordan, T. H., Callaghan, S., Deelman, E., Field, E., Juve, G., Kesselman, C., Maechling, P., Mehta, G., Milner, K., et al. 2011a. CyberShake: A Physics-Based Seismic Hazard Model for Southern California. *Pure and Applied Geophysics*, **168**(3–4), 367–381.

Graves, R., Jordan, T., Callaghan, S., Deelman, E., Field, E., Juve, G., Kesselman, C., Maechling, P., Mehta, G., Milner, K., Okaya, D., Small, P., and Vahi, K. 2011b. CyberShake: A Physics-Based Seismic Hazard Model for Southern California. *Pure and Applied Geophysics*, **168**(3), 367–381.

Graves, R., and Pitarka, A. 2015. Refinements to the Graves and Pitarka (2010) Broadband Ground-Motion Simulation Method. *Seismological Research Letters*, **86**(1), 75–80.

Graves, R., and Pitarka, A. 2016. Kinematic Ground-Motion Simulations on Rough Faults Including Effects of 3D Stochastic Velocity Perturbations. *Bulletin of the Seismological Society of America*, **106**(5), 2136–2153.

Graves, R. W. 1996. Simulating Seismic Wave Propagation in 3D Elastic Media Using Staggered-Grid Finite Differences. *Bulletin of the Seismological Society of America*, **86**(4), 1091–1106.

Graves, R. W., and Pitarka, A. 2010. Broadband Ground-Motion Simulation Using a Hybrid Approach. *Bulletin of the Seismological Society of America*, **100**(5A), 2095–2123.

Gregor, N., Abrahamson, N. A., Atkinson, G. M., Boore, D. M., Bozorgnia, Y., Campbell, K. W., Chiou, B. S.-J., Idriss, I. M., Kamai, R., Seyhan, E., Silva, W., Stewart, J. P., and Youngs, R. 2014. Comparison of NGA-West2 GMPEs. *Earthquake Spectra*, **30**(3), 1179–1197.

Grossi, P., Kunreuther, H., and Patel, C. C. 2005. *Catastrophe Modeling: A New Approach to Managing Risk*. Springer.

Guatteri, M., Mai, P. M., and Beroza, G. C. 2004. A Pseudo-Dynamic Approximation to Dynamic Rupture Models for Strong Ground Motion Prediction. *Bulletin of the Seismological Society of America*, **94**(6), 2051–2063.

Gulerce, Z., and Abrahamson, N. A. 2011. Site-Specific Design Spectra for Vertical Ground Motion. *Earthquake Spectra*, **27**(4), 1023–1047.

Gutenberg, B., and Richter, C. F. 1944. Frequency of Earthquakes in California. *Bulletin of the Seismological Society of America*, **34**(4), 185–188.

Haarala, M., and Orosco, L. 2016. Analysis of Gutenberg-Richter b-Value and M_{max}. Part I: Exact Solution of Kijko–Sellevoll Estimator of M_{max}. *Cuadernos de Ingeniería, Nueva Serie*, **9**, 51–77.

Hale, C., Abrahamson, N., and Bozorgnia, Y. 2018. *Probabilistic Seismic Hazard Analysis Code Verification*. PEER Report 2018/03. Berkeley, CA.

Halsey, L. G., Curran-Everett, D., Vowler, S. L., and Drummond, G. B. 2015. The Fickle P Value Generates Irreproducible Results. *Nature Methods*, **12**(3), 179–185.

Han, Y., and Davidson, R. A. 2012. Probabilistic Seismic Hazard Analysis for Spatially Distributed Infrastructure. *Earthquake Engineering & Structural Dynamics*, **41**(15), 2141–2158.

Hancock, J., and Bommer, J. J. 2006. A State-of-Knowledge Review of the Influence of Strong-Motion Duration on Structural Damage. *Earthquake Spectra*, **22**, 827.

Hancock, J., and Bommer, J. J. 2007. Using Spectral Matched Records to Explore the Influence of Strong-Motion Duration on Inelastic Structural Response. *Soil Dynamics and Earthquake Engineering*, **27**(4), 291–299.

Hancock, J., Bommer, J. J., and Stafford, P. J. 2008. Numbers of Scaled and Matched Accelerograms Required for Inelastic Dynamic Analyses. *Earthquake Engineering & Structural Dynamics*, **37**(14), 1585–1607.

Hanks, T., and McGuire, R. 1981. The Character of High-Frequency Strong Ground Motion. *Bulletin of the Seismological Society of America*, **71**(6), 2071–2095.

Hanks, T. C. 1977. Earthquake Stress Drops, Ambient Tectonic Stresses and Stresses That Drive Plate Motions. *Pure and Applied Geophysics*, **115**, 441–458.

Hanks, T. C., and Abrahamson, N. 2008. A Brief History of Extreme Ground Motions. *Seismological Research Letters*, **79**, 282–283.

Hanks, T. C., Abrahamson, N. A., Boore, D. M., Coppersmith, K. J., and Knepprath, N. E. 2009. *Implementation of the SSHAC Guidelines for Level 3 and 4 PSHAs—Experience Gained from Actual Applications*. Open-File Report 2009-1093. US Geological Survey.

Hanks, T. C., and Bakun, W. H. 2002. A Bilinear Source-Scaling Model for M-logA Observations of Continental Earthquakes. *Bulletin of the Seismological Society of America*, **92**(5), 1841–1846.

Hanks, T. C., and Bakun, W. H. 2008. M-logA Observations for Recent Large Earthquakes. *Bulletin of the Seismological Society of America*, **98**(1), 490–494.

Hanks, T. C., Beroza, G. C., and Toda, S. 2012. Have Recent Earthquakes Exposed Flaws in or Misunderstandings of Probabilistic Seismic Hazard Analysis? *Seismological Research Letters*, **83**(5), 759–764.

Hanks, T. C., and Kanamori, H. 1979. A Moment Magnitude Scale. *Journal of Geophysical Research*, **84**(B5), 2348.

Harichandran, R. S., Hawwari, A., and Sweidan, B. N. 1996. Response of Long-Span Bridges to Spatially Varying Ground Motion. *Journal of Structural Engineering*, **122**(5), 476–484.

Hartford, D. N. D. 2009. Legal Framework Considerations in the Development of Risk Acceptance Criteria. *Structural Safety*, **31**(2), 118–123.

Hartzell, S., Frankel, A., Liu, P., Zeng, Y., and Rahman, S. 2011. Model and Parametric Uncertainty in Source-Based Kinematic Models of Earthquake Ground Motion. *Bulletin of the Seismological Society of America*, **101**(5), 2431–2452.

Hartzell, S., Harmsen, S., and Frankel, A. 2010. Effects of 3D Random Correlated Velocity Perturbations on Predicted Ground Motions. *Bulletin of the Seismological Society of America*, **100**(4), 1415–1426.

Hartzell, S., Harmsen, S., Frankel, A., and Larsen, S. 1999. Calculation of Broadband Time Histories of Ground Motion: Comparison of Methods and Validation Using Strong-Ground Motion from the 1994 Northridge Earthquake. *Bulletin of the Seismological Society of America*, **89**(6), 1484–1504.

Hashash, Y. M., Hook, J. J., Schmidt, B., John, I., and Yao, C. 2001. Seismic Design and Analysis of Underground Structures. *Tunnelling and Underground Space Technology*, **16**(4), 247–293.

Hastie, T., Tibshirani, R., and Friedman, J. 2001. *The Elements of Statistical Learning: Data Mining, Inference, and Prediction*. New York: Springer.

Hawkes, A. G., and Adamopoulos, L. 1973. Cluster Models for Earthquakes: Regional Comparisons. *Bulletin of the International Statistical Institute*, **45**(3), 454–461.

Hayes, G. P., Earle, P. S., Benz, H. M., Wald, D. J., Briggs, R. W., and the USGS/NEIC Earthquake Response Team. 2011. 88 Hours: The U.S. Geological Survey National Earthquake Information Center Response to the 11 March 2011 Mw 9.0 Tohoku Earthquake. *Seismological Research Letters*, **82**(4), 481–493.

Headquarters for Earthquake Research Promotion. 2005. *National Seismic Hazard Maps for Japan*. www.jishin.go.jp/main/index-e.html.

Heaton, T. H. 1990. Evidence for and Implications of Self-Healing Pulses of Slip in Earthquake Rupture. *Physics of the Earth and Planetary Interiors*, **64**(1), 1–20.

Heresi, P., and Miranda, E. 2019. Uncertainty in Intraevent Spatial Correlation of Elastic Pseudo-Acceleration Spectral Ordinates. *Bulletin of Earthquake Engineering*, **17**(3), 1099–1115.

Holt, W. E., and Haines, A. J. 1995. The Kinematics of Northern South Island, New Zealand, Determined from Geologic Strain Rates. *Journal of Geophysical Research: Solid Earth*, **100**(B9), 17991–18010.

Hough, S. E., and Anderson, J. G. 1988. High-Frequency Spectra Observed at Anza, California: Implications for Q Structure. *Bulletin of the Seismological Society of America*, **78**(2), 692–707.

Husen, S., and Hardebeck, J. 2010. *Earthquake Location Accuracy*. Community Online Resource for Statistical Seismicity Analysis, CORSSA, www.corssa.org.

Hutchings, L. 1994. Kinematic Earthquake Models and Synthesized Ground Motion Using Empirical Green's Functions. *Bulletin of the Seismological Society of America*, **84**(4), 1028–1050.

Ibarra, L. F., and Krawinkler, H. 2005. *Global Collapse of Frame Structures under Seismic Excitations*. 152. John A. Blume Earthquake Engineering Center, Stanford, CA.

Idriss, I., and Boulanger, R. 2008. *Soil Liquefaction during Earthquakes*. Earthquake Engineering Research Institute.

Idriss, I., and Seed, H. 1968. Seismic Response of Horizontal Soil Layers. *Journal of Soil Mechanics and Foundations (ASCE)*, **94**(SM4), 1003–1031.

Iervolino, I., De Luca, F., and Cosenza, E. 2010. Spectral Shape-Based Assessment of SDOF Nonlinear Response to Real, Adjusted and Artificial Accelerograms. *Engineering Structures*, **32**(9), 2776–2792.

Iervolino, I., Giorgio, M., and Polidoro, B. 2014. Sequence-Based Probabilistic Seismic Hazard Analysis. *Bulletin of the Seismological Society of America*, **104**(2), 1006–1012.

Imperatori, W., and Mai, P. M. 2013. Broad-band Near-Field Ground Motion Simulations in 3-Dimensional Scattering Media. *Geophysical Journal International*, **192**(2), 725–744.

Imperatori, W., and Mai, P. M. 2015. The Role of Topography and Lateral Velocity Heterogeneities on Near-Source Scattering and Ground-Motion Variability. *Geophysical Journal International*, **202**(3), 2163–2181.

Imtiaz, A., Causse, M., Chaljub, E., and Cotton, F. 2015. Is Ground-Motion Variability Distance Dependent? Insight from Finite-Source Rupture Simulations. *Bulletin of the Seismological Society of America*, **105**(2A), 950–962.

Inoue, T., and Cornell, C. A. 1990. *Seismic Hazard Analysis of Multi-Degree-of-Freedom Structures*. RMS-8. Reliability of Marine Structures, Stanford, CA.

Ishihara, K., and Cubrinovski, M. 2005. Characteristics of Ground Motion in Liquefied Deposits during Earthquakes. *Journal of Earthquake Engineering*, **9**(S1), 1–15.

Jaeger, J. C., Cook, N. G. W., and Zimmerman, R. W. 2007. *Fundamentals of Rock Mechanics*. Blackwell Publishing.

Jaiswal, K., Wald, D., and D'Ayala, D. 2011b. Developing Empirical Collapse Fragility Functions for Global Building Types. *Earthquake Spectra*, **27**(3), 775–795.

Jaiswal, K. S., Aspinall, W., Perkins, D., Wald, D., and Porter, K. A. 2012. Use of Expert Judgment Elicitation to Estimate Seismic Vulnerability of Selected Building Types. In: *15th World Conference on Earthquake Engineering*.

Jaiswal, K. S., Wald, D. J., Earle, P. S., Porter, K. A., and Hearne, M. 2011a. Earthquake Casualty Models within the USGS Prompt Assessment of Global Earthquakes for Response (PAGER) System. Pages 83–94 of: Spence, R., So, E., and Scawthorn, C. (eds.), *Human Casualties in Earthquakes: Progress in Modelling and Mitigation*. Advances in Natural and Technological Hazards Research. Dordrecht: Springer Netherlands.

Jalayer, F. 2003. Direct Probabilistic Seismic Analysis: Implementing Non-linear Dynamic Assessments. PhD thesis, Dept. of Civil and Environmental Engineering, Stanford University.

Jayaram, N., and Baker, J. W. 2008. Statistical Tests of the Joint Distribution of Spectral Acceleration Values. *Bulletin of the Seismological Society of America*, **98**(5), 2231–2243.

Jayaram, N., and Baker, J. W. 2009. Correlation Model for Spatially Distributed Ground-Motion Intensities. *Earthquake Engineering & Structural Dynamics*, **38**(15), 1687–1708.

Jayaram, N., and Baker, J. W. 2010a. Considering Spatial Correlation in Mixed-Effects Regression, and Impact on Ground-Motion Models. *Bulletin of the Seismological Society of America*, **100**(6), 3295–3303.

Jayaram, N., and Baker, J. W. 2010b. Efficient Sampling and Data Reduction Techniques for Probabilistic Seismic Lifeline Risk Assessment. *Earthquake Engineering & Structural Dynamics*, **39**(10), 1109–1131.

Jayaram, N., Lin, T., and Baker, J. W. 2011. A Computationally Efficient Ground-Motion Selection Algorithm for Matching a Target Response Spectrum Mean and Variance. *Earthquake Spectra*, **27**(3), 797–815.

Jayaram, N., Park, J., Bazzurro, P., and Tothong, P. 2010. Estimation of Spatial Correlation between Spectral Accelerations Using Simulated Ground-Motion Time Histories. In: *9th US National and 10th Canadian Conference on Earthquake Engineering*.

Jayaram, N., Shome, N., and Rahnama, M. 2012. Development of Earthquake Vulnerability Functions for Tall Buildings. *Earthquake Engineering & Structural Dynamics*, **41**(11), 1495–1514.

Jeon, S.-S., and O'Rourke, T. D. 2005. Northridge Earthquake Effects on Pipelines and Residential Buildings. *Bulletin of the Seismological Society of America*, **95**(1), 294–318.

Johnston, A. C., Coppersmith, K. J., Kanter, L. R., and Cornell, C. A. 1994. *The Earthquakes of Stable Continental Regions*. EPRI Report TR-102261s-V1-V5. Electric Power Research Institute, Palo Alto, CA.

Jordan, T. H. 2006. Earthquake Predictability, Brick by Brick. *Seismological Research Letters*, **77**(1), 3–6.

Journel, A. G., and Huijbregts, C. J. 1978. *Mining Geostatistics*. London: Academic Press.

Kagan, Y. Y. 2006. Why Does Theoretical Physics Fail to Explain and Predict Earthquake Occurrence? Pages 303–362 of: Bhattacharyya, P., and Chakrabarti, B. K. (eds.), *Modelling Critical and Catastrophic Phenomena in Geoscience*. Springer.

Kagan, Y. Y., and Knopoff, L. 1981. Stochastic Synthesis of Earthquake Catalogs. *Journal of Geophysical Research: Solid Earth*, **86**(B4), 2853–2862.

Kagan, Y. Y., and Jackson, D. D. 2013. Tohoku Earthquake: A Surprise? *Bulletin of the Seismological Society of America*, **103**(2B), 1181–1194.

Kamai, R., Abrahamson, N. A., and Silva, W. J. 2014. Nonlinear Horizontal Site Amplification for Constraining the NGA-West2 GMPEs. *Earthquake Spectra*, **30**(3), 1223–1240.

Kanamori, H., Mori, J., Hauksson, E., Heaton, T. H., Hutton, L. K., and Jones, L. M. 1993. Determination of Earthquake Energy Release and M_L Using Terrascope. *Bulletin of the Seismological Society of America*, **83**(2), 330–346.

Katsanos, E. I., Sextos, A. G., and Manolis, G. D. 2010. Selection of Earthquake Ground Motion Records: A State-of-the-Art Review from a Structural Engineering Perspective. *Soil Dynamics and Earthquake Engineering*, **30**(4), 157–169.

Kawase, H. 1996. The Cause of the Damage Belt in Kobe: "The Basin-Edge Effect," Constructive Interference of the Direct S-Wave with the Basin-Induced Diffracted/Rayleigh Waves. *Seismological Research Letters*, **67**(5), 25–34.

Keefer, D. L., and Bodily, S. E. 1983. Three-Point Approximations for Continuous Random Variables. *Management Science*, **29**(5), 595–609.

Kennedy, R. P. 2011. Performance-Goal Based (Risk Informed) Approach for Establishing the SSE Site Specific Response Spectrum for Future Nuclear Power Plants. *Nuclear Engineering and Design*, **241**(3), 648–656.

Kennedy, R. P., Cornell, C. A., Campbell, R. D., Kaplan, S., and Perla, H. F. 1980. Probabilistic Seismic Safety Study of an Existing Nuclear Power Plant. *Nuclear Engineering and Design*, **59**(2), 315–338.

Kennedy, R. P., and Ravindra, M. K. 1984. Seismic Fragilities for Nuclear Power Plant Risk Studies. *Nuclear Engineering and Design*, **79**(1), 47–68.

Kennedy, R., and Short, S. 1994. *Basis for Seismic Provisions of DOE-STD-1020. UCRL-CR-111478 and BNL-52418*. Lawrence Livermore National Laboratory and Brookhaven National Laboratory.

Kijko, A. 2004. Estimation of the Maximum Earthquake Magnitude, M_{max}. *Pure and Applied Geophysics*, **161**, 1655–1681.

Kijko, A., and Graham, G. 1998. Parametric-Historic Procedure for Probabilistic Seismic Hazard Analysis. Part I: Estimation of Maximum Regional Magnitude M_{max}. *Pure and Applied Geophysics*, **152**, 413–442.

Kijko, A., and Sellevoll, M. A. 1989. Estimation of Earthquake Hazard Parameters from Incomplete Data Files. Part I. Utilization of Extreme and Complete Catalog with Different Threshold Magnitudes. *Bulletin of the Seismological Society of America*, **79**(3), 645–654.

King, G. C. P., Stein, R. S., and Lin, J. 1994. Static Stress Changes and the Triggering of Earthquakes. *Bulletin of the Seismological Society of America*, **84**(3), 935–953.

Kircher, C. A., Nassar, A. A., Kustu, O., and Holmes, W. T. 1997. Development of Building Damage Functions for Earthquake Loss Estimation. *Earthquake Spectra*, **13**(4), 663–682.

Kircher, C. A., Seligson, H. A., Bouabid, J., and Morrow, G. C. 2006. When the Big One Strikes Again: Estimated Losses Due to a Repeat of the 1906 San Francisco Earthquake. *Earthquake Spectra*, **22**(S2), 297–339.

Kiremidjian, A. S., Stergiou, E., and Lee, R. 2007. Issues in Seismic Risk Assessment of Transportation Networks. Pages 461–480 of: *Earthquake Geotechnical Engineering*. Springer.

Klein, F. 2006. *Y2000 Shadow Format & NCSN Data Codes*. Online Report www.ncedc.org/ftp/pub/doc/ncsn/shadow2000.pdf. Northern California Earthquake Data Center.

Klinger, Y., Sieh, K., Altunel, E., Akoglu, A., Barka, A., Dawson, T., Gonzalez, T., Meltzner, A., and Rockwell, T. 2003. Paleoseismic Evidence of Characteristic Slip on the Western Segment of the North Anatolian Fault, Turkey. *Bulletin of the Seismological Society of America*, **93**(6), 2317–2332.

Komatitsch, D., and Tromp, J. 1999. Introduction to the Spectral Element Method for Three-Dimensional Seismic Wave Propagation. *Geophysical Journal International*, **139**(3), 806–822.

Konno, K., and Ohmachi, T. 1998. Ground-Motion Characteristics Estimated From Spectral Ratio between Horizontal and Vertical Components of Microtremor. *Bulletin of the Seismological Society of America*, **88**(1), 228–241.

Kotha, S. R., Bindi, D., and Cotton, F. 2017. Site-Corrected Magnitude- and Region-Dependent Correlations of Horizontal Peak Spectral Amplitudes. *Earthquake Spectra*, **33**(4), 1415–1432.

Kottke, A., and Rathje, E. M. 2008. A Semi-Automated Procedure for Selecting and Scaling Recorded Earthquake Motions for Dynamic Analysis. *Earthquake Spectra*, **24**(4), 911–932.

Kottke, A. R., and Rathje, E. M. 2009. *Technical Manual for Strata*. PEER Report 2008/10. Pacific Earthquake Engineering Research Center.

Kottke, A. R., and Rathje, E. M. 2013. Comparison of Time Series and Random-Vibration Theory Site-Response Methods. *Bulletin of the Seismological Society of America*, **103**(3), 2111–2127.

Kramer, S. L. 1996. *Geotechnical Earthquake Engineering*. Prentice Hall.

Kramer, S. L., and Mitchell, R. A. 2006. Ground Motion Intensity Measures for Liquefaction Hazard Evaluation. *Earthquake Spectra*, **22**(2), 413–438.

Krawinkler, H., and Miranda, E. 2004. Performance-Based Earthquake Engineering. In: Bozorgnia, Y., and Bertero, V. V. (eds.), *Earthquake Engineering: From Engineering Seismology to Performance-Based Engineering*. Boca Raton, FL: CRC Press.

Krayer von Krauss, M. P., Casman, E. A., and Small, M. J. 2004. Elicitation of Expert Judgments of Uncertainty in the Risk Assessment of Herbicide-Tolerant Oilseed Crops. *Risk Analysis: An International Journal*, **24**(6), 1515–1527.

Ktenidou, O.-J., Cotton, F., Abrahamson, N. A., and Anderson, J. G. 2014. Taxonomy of κ: A Review of Definitions and Estimation Approaches Targeted to Applications. *Seismological Research Letters*, **85**(1), 135–146.

Kuehn, N. M., and Abrahamson, N. A. 2018. The Effect of Uncertainty in Predictor Variables on the Estimation of Ground-Motion Prediction Equations. *Bulletin of the Seismological Society of America*, **108**(1), 358–370.

Kuehn, N. M., and Abrahamson, N. A. 2020. Spatial Correlations of Ground Motion for Non-ergodic Seismic Hazard Analysis. *Earthquake Engineering & Structural Dynamics*, **49**(1), 4–23.

Kuehn, N. M., Abrahamson, N. A., and Baltay, A. 2016. Estimating Spatial Correlations between Earthquake Source, Path and Site Effects for Non-ergodic Seismic Hazard Analysis. In: *Annual Meeting of the Seismological Society of America*.

Kuehn, N. M., Abrahamson, N. A., and Walling, M. 2019. Incorporating Non-ergodic Path Effects into the NGA West 2 Ground-Motion Prediction Equations. *Bulletin of the Seismological Society of America*, **109**(2), 575–585.

Kulkarni, R. B., Youngs, R. R., and Coppersmith, K. J. 1984. Assessment of Confidence Intervals for Results of Seismic Hazard Analysis. Pages 263–270 of: *Proceedings of the Eighth World Conference on Earthquake Engineering*, vol. 1.

Kullback, S., and Leibler, R. A. 1951. On Information and Sufficiency. *The Annals of Mathematical Statistics*, **22**(1), 79–86.

Kutner, M. H., Nachtsheim, C., and Neter, J. 2004. *Applied Linear Regression Models*, 4th ed. Boston: McGraw-Hill/Irwin.

Kwak, D. Y., Stewart, J. P., Brandenberg, S. J., and Mikami, A. 2016a. Characterization of Seismic Levee Fragility Using Field Performance Data. *Earthquake Spectra*, **32**(1), 193–215.

Kwak, D. Y., Stewart, J. P., Brandenberg, S. J., and Mikami, A. 2016b. Seismic Levee System Fragility Considering Spatial Correlation of Demands and Component Fragilities. *Earthquake Spectra*, **32**(4), 2207–2228.

Kwok, A. O., and Stewart, J. P. 2006. Evaluation of the Effectiveness of Theoretical 1D Amplification Factors for Earthquake Ground-Motion Prediction. *Bulletin of the Seismological Society of America*, **96**(4A), 1422–1436.

Lallemant, D., and Kiremidjian, A. 2015. A Beta Distribution Model for Characterizing Earthquake Damage State Distribution. *Earthquake Spectra*, **31**(3), 1337–1352.

Landwehr, N., Kuehn, N. M., Scheffer, T., and Abrahamson, N. 2016. A Nonergodic Ground-Motion Model for California with Spatially Varying Coefficients. *Bulletin of the Seismological Society of America*, **106**(6), 2574–2583.

Lay, T., and Wallace, T. C. 1995. *Modern Global Seismology*. Academic Press.

Lee, R., and Kiremidjian, A. 2007. Uncertainty and Correlation for Loss Assessment of Spatially Distributed Systems. *Earthquake Spectra*, **23**(4), 753–770.

Lee, R. L., Bradley, B. A., Ghisetti, F. C., and Thomson, E. M. 2017. Development of a 3D Velocity Model of the Canterbury, New Zealand, Region for Broadband Ground-Motion Simulation. *Bulletin of the Seismological Society of America*, **107**(5), 2131–2150.

Lee, R. L., Bradley, B. A., Stafford, P. J., Graves, R. W., and Rodriguez-Marek, A. 2020. Hybrid Broadband Ground Motion Simulation Validation of Small Magnitude Earthquakes in Canterbury, New Zealand. *Earthquake Spectra*, **36**(2), 673–699.

Leonard, M. 2010. Earthquake Fault Scaling: Self-Consistent Relating of Rupture Length, Width, Average Displacement, and Moment Release. *Bulletin of the Seismological Society of America*, **100**(5A), 1971–1988.

Leonard, M. 2014. Self-Consistent Earthquake Fault-Scaling Relations: Update and Extension to Stable Continental Strike-Slip Faults. *Bulletin of the Seismological Society of America*, **104**(6), 2953–2965.

Liel, A., Haselton, C., Deierlein, G. G., and Baker, J. W. 2009. Incorporating Modeling Uncertainties in the Assessment of Seismic Collapse Risk of Buildings. *Structural Safety*, **31**(2), 197–211.

Lienkaemper, J. J. 2002. A Record of Large Earthquakes on the Southern Hayward Fault for the Past 500 Years. *Bulletin of the Seismological Society of America*, **92**(7), 2637–2658.

Lienkaemper, J. J., and Borchardt, G. 1996. Holocene Slip Rate of the Hayward Fault at Union City, California. *Journal of Geophysical Research: Solid Earth*, **101**(B3), 6099–6108.

Lienkaemper, J. J., and Williams, P. L. 2007. A Record of Large Earthquakes on the Southern Hayward Fault for the Past 1800 Years. *Bulletin of the Seismological Society of America*, **97**(6), 1803–1819.

Lin, T., Harmsen, S. C., Baker, J. W., and Luco, N. 2013c. Conditional Spectrum Computation Incorporating Multiple Causal Earthquakes and Ground-Motion Prediction Models. *Bulletin of the Seismological Society of America*, **103**(2A), 1103–1116.

Lin, T., Haselton, C. B., and Baker, J. W. 2013a. Conditional Spectrum-Based Ground Motion Selection. Part I: Hazard Consistency for Risk-Based Assessments. *Earthquake Engineering & Structural Dynamics*, **42**(12), 1847–1865.

Lin, T., Haselton, C. B., and Baker, J. W. 2013b. Conditional spectrum-based ground motion selection. Part II: Intensity-based assessments and evaluation of alternative target spectra. *Earthquake Engineering & Structural Dynamics*, **42**(12), 1867–1884.

Liu, P., Archuleta, R. J., and Hartzell, S. H. 2006. Prediction of Broadband Ground-Motion Time Histories: Hybrid Low/High-Frequency Method with Correlated Random Source Parameters. *Bulletin of the Seismological Society of America*, **96**(6), 2118–2130.

Loth, C., and Baker, J. W. 2013. A Spatial Cross-Correlation Model for Ground Motion Spectral Accelerations at Multiple Periods. *Earthquake Engineering & Structural Dynamics*, **42**(3), 397–417.

Luco, N., Bachman, R. E., Crouse, C. B., Harris, J. R., Hooper, J. D., Kircher, C. A., Caldwell, P. J., and Rukstales, K. S. 2015. Updates to Building-Code Maps for the 2015 NEHRP Recommended Seismic Provisions. *Earthquake Spectra*, **31**(S1), S245–S271.

Luco, N., and Bazzurro, P. 2007. Does Amplitude Scaling of Ground Motion Records Result in Biased Nonlinear Structural Drift Responses? *Earthquake Engineering & Structural Dynamics*, **36**(13), 1813–1835.

Luco, N., and Cornell, C. 2007. Structure-Specific Scalar Intensity Measures for Near-Source and Ordinary Earthquake Ground Motions. *Earthquake Spectra*, **23**(2), 357–392.

Luco, N., Ellingwood, B. R., Hamburger, R. O., Hooper, J. D., Kimball, J. K., and Kircher, C. A. 2007. Risk-Targeted versus Current Seismic Design Maps for the Conterminous United States. In: *Proceedings of the 2007 Structural Engineers Association of California (SEAOC) Convention*.

Lutes, L. D., and Sarkani, S. 2004. *Random Vibrations: Analysis of Structural and Mechanical Systems*. Burlington, MA: Butterworth-Heinemann.

Macedo, J., Abrahamson, N., and Bray, J. D. 2019. Arias Intensity Conditional Scaling Ground-Motion Models for Subduction Zones. *Bulletin of the Seismological Society of America*, **109**(4), 1343–1357.

Mackie, K., and Stojadinovic, B. 2005. Comparison of Incremental Dynamic, Cloud, and Stripe Methods for Computing Probabilistic Seismic Demand Models. In: *ASCE Structures Congress*.

Madariaga, R. 1977. Implications of Stress-Drop Models of Earthquakes for the Inversion of Stress Drop from Seismic Observations. *Pure and Applied Geophysics*, **115**, 301–316.

Magistrale, H., and Day, S. 1999. 3D Simulations of Multi-segment Thrust Fault Rupture. *Geophysical Research Letters*, **26**(14), 2093–2096.

Mai, P. M., and Beroza, G. C. 2002. A Spatial Random Field Model to Characterize Complexity in Earthquake Slip. *Journal of Geophysical Research*, **107**(10.1029), 2001.

Mai, P. M., and Beroza, G. C. 2003. A Hybrid Method for Calculating Near-Source, Broadband Seismograms: Application to Strong Motion Prediction. *Physics of the Earth and Planetary Interiors*, **137**(1), 18.

Mai, P. M., Galis, M., Thingbaijam, K. K. S., Vyas, J. C., and Dunham, E. M. 2017. Accounting for Fault Roughness in Pseudo-Dynamic Ground-Motion Simulations. *Pure and Applied Geophysics*, **174**(9), 3419–3450.

Mai, P. M., Imperatori, W., and Olsen, K. B. 2010. Hybrid Broadband Ground-Motion Simulations: Combining Long-Period Deterministic Synthetics with High-Frequency Multiple S-to-S Backscattering. *Bulletin of the Seismological Society of America*, **100**(5A), 2124–2142.

Mai, P. M., Spudich, P., and Boatwrigth, J. 2005. Hypocenter Locations in Finite-Source Rupture Models. *Bulletin of the Seismological Society of America*, **95**(3), 965–980.

Mak, S., and Schorlemmer, D. 2016. A Comparison between the Forecast by the United States National Seismic Hazard Maps with Recent Ground-Motion Records. *Bulletin of the Seismological Society of America*, **106**(4), 1817–1831.

Mancini, S., Segou, M., Werner, M. J., and Parsons, T. 2020. The Predictive Skills of Elastic Coulomb Rate-and-State Aftershock Forecasts during the 2019 Ridgecrest, California, Earthquake Sequence. *Bulletin of the Seismological Society of America*, **110**(4), 1736–1751.

Marafi, N. A., Berman, J. W., and Eberhard, M. O. 2016. Ductility-Dependent Intensity Measure That Accounts for Ground-Motion Spectral Shape and Duration. *Earthquake Engineering & Structural Dynamics*, **45**(4), 653–672.

Markhvida, M., Ceferino, L., and Baker, J. W. 2018. Modeling Spatially Correlated Spectral Accelerations at Multiple Periods Using Principal Component Analysis and Geostatistics. *Earthquake Engineering & Structural Dynamics*, **47**(5), 1107–1123.

Marone, C. 1998. The Effect of Loading Rate on Static Friction and the Rate of Fault Healing during the Earthquake Cycle. *Nature*, **391**, 69–72.

Marzocchi, W., and Jordan, T. H. 2014. Testing for Ontological Errors in Probabilistic Forecasting Models of Natural Systems. *Proceedings of the National Academy of Sciences*, **111**(33), 11973–11978.

Marzocchi, W., and Jordan, T. H. 2017. A Unified Probabilistic Framework for Seismic Hazard Analysis. *Bulletin of the Seismological Society of America*, **107**(6), 2738–2744.

Matasovic, N., and Hashash, Y. 2012. *Practices and Procedures for Site-Specific Evaluations of Earthquake Ground Motions: A Synthesis of Highway Practice*. National Academies of Sciences, Engineering, and Medicine. Washington, DC.

Matthews, M. V., Ellsworth, W. L., and Reasenberg, P. A. 2002. A Brownian Model for Recurrent Earthquakes. *Bulletin of the Seismological Society of America*, **92**(6), 2233–2250.

Maufroy, E., Chaljub, E., Hollender, F., Kristek, J., Moczo, P., Klin, P., Priolo, E., Iwaki, A., Iwata, T., Etienne, V., De Martin, F., Theodoulidis, N. P., Manakou, M., GuyonnetBenaize, C., Pitilakis, K., and Bard, P. 2015. Earthquake Ground Motion in the Mygdonian Basin, Greece: The E2VP Verification and Validation of 3D Numerical Simulation up to 4Hz. *Bulletin of the Seismological Society of America*, **105**(3), 1398–1418.

Mazzieri, I., Stupazzini, M., Guidotti, R., and Smerzini, C. 2013. SPEED: SPectral Elements in Elastodynamics with Discontinuous Galerkin: A Non-Conforming Approach for 3D Multi-scale Problems. *International Journal for Numerical Methods in Engineering*, **95**(12), 991–1010.

McCalpin, J. P. 2009. *Paleoseismology*, 2nd ed. International Geophysics Series, vol. 95. Academic Press.

McGinty, P. 2001. Preparation of the New Zealand Earthquake Catalogue for a Probabilistic Seismic Hazard Analysis. *Bulletin of the New Zealand Society for Earthquake Engineering*, **34**(1), 60–67.

McGuire, R., Cornell, C., and Toro, G. 2005a. The Case for Mean Seismic Hazard. *Earthquake Spectra*, **21**(3), 879–886.

McGuire, R. K. 1995. Probabilistic Seismic Hazard Analysis and Design Earthquakes: Closing the Loop. *Bulletin of the Seismological Society of America*, **85**(5), 1275–1284.

McGuire, R. K. 2004. *Seismic Hazard and Risk Analysis*. Berkeley, CA Earthquake Engineering Research Institute.

McGuire, R. K., Cornell, C. A., and Toro, G. R. 2005b. The Case for Using Mean Seismic Hazard. *Earthquake Spectra*, **21**(3), 879–886.

Melchers, R. E. 2007. Structural Reliability Theory in the Context of Structural Safety. *Civil Engineering and Environmental Systems*, **24**(1), 55–69.

Michael, A. J., and Werner, M. J. 2018. Preface to the Focus Section on the Collaboratory for the Study of Earthquake Predictability (CSEP): New Results and Future Directions. *Seismological Research Letters*, **89**(4), 1226–1228.

Miller, A. C., and Rice, T. R. 1983. Discrete Approximations of Probability Distributions. *Management Science*, **29**(3), 352–362.

Ming, D., Huang, C., Peters, G. W., and Galasso, C. 2019. An Advanced Estimation Algorithm for Ground-Motion Models with Spatial Correlation. *Bulletin of the Seismological Society of America*, **109**(2), 541–566.

Miyazawa, M., and Mori, J. 2009. Test of Seismic Hazard Map from 500 Years of Recorded Intensity Data in Japan. *Bulletin of the Seismological Society of America*, **99**(6), 3140–3149.

Moehle, J., and Deierlein, G. G. 2004. A Framework Methodology for Performance-Based Earthquake Engineering. In: *Proceedings, 13th World Conference on Earthquake Engineering*.

Moehle, J. P., Hamburger, R. O., Baker, J. W., Bray, J. D., Crouse, C. B., Deierlein, G. G., Hooper, J. D., Lew, M., Maffei, J. R., Mahin, S. A., Malley, J., Naeim, F., Stewart, J. P., and Wallace, J. W. 2017. *Guidelines for Performance-Based Seismic Design of Tall Buildings Version 2.0*. PEER Report 2017/06. Berkeley, CA.

Mohr, O. 1914. *Abhandlugen aus dem Gebiete der Technische Mechanik*, 2nd ed. Berlin: Ernst und Sohn.

Molkenthin, C., Scherbaum, F., Griewank, A., Kuehn, N., Stafford, P. J., and Leovey, H. 2015. Sensitivity of Probabilistic Seismic Hazard Obtained by Algorithmic Differentiation: A Feasability Study. *Bulletin of the Seismological Society of America*, **105**(3), 1810–1822.

Mosca, I., Console, R., and D'Addezio, G. 2012. Renewal Models of Seismic Recurrence Applied to Paleoseismological and Historical Observations. *Tectonophysics*, **564–565**, 54–67.

Moss, R. 2011. Reduced Sigma of Ground-Motion Prediction Equations through Uncertainty Propagation. *Bulletin of the Seismological Society of America*, **101**(1), 250–257.

Motazedian, D., and Atkinson, G. M. 2005. Stochastic Finite-Fault Modeling Based on a Dynamic Corner Frequency. *Bulletin of the Seismological Society of America*, **95**(3), 995–1010.

Musson, R. 2010. Ground Motion and Probabilistic Hazard. *Bulletin of Earthquake Engineering*, **7**(3), 575–589.

Musson, R. 2012b. On the Nature of Logic Trees in Probabilistic Seismic Hazard Assessment. *Earthquake Spectra*, **28**(3), 1291–1296.

Musson, R., and Winter, P. 2012. Objective Assessment of Source Models for Seismic Hazard Studies: With a Worked Example from UK Data. *Bulletin of Earthquake Engineering*, **10**(2), 367–378.

Musson, R. M. W. 2000. The Use of Monte Carlo Simulations for Seismic Hazard Assessment in the U.K. *Annals of Geophysics*, **43**(1), 1–9.

Musson, R. M. W. 2005. Against Fractiles. *Earthquake Spectra*, **21**(3), 887–891.

Musson, R. M. W. 2012a. PSHA Validated by Quasi Observational Means. *Seismological Research Letters*, **83**(1), 130–134.

Myung, I. J. 2003. Tutorial on Maximum Likelihood Estimation. *Journal of Mathematical Psychology*, **47**(1), 90–100.

National Institute of Standards and Technology. 2012. *Selecting and Scaling Earthquake Ground Motions for Performing Response-History Analyses*. NIST Report NIST GCR 11-917-15. National Institute of Standards and Technology.

National Research Council. 1996. *Understanding Risk: Informing Decisions in a Democratic Society*. National Academy Press.

NCEDC. 2014. *Northern California Earthquake Data Center*. UC Berkeley Seismological Laboratory.

NIST. 2011. *Selecting and Scaling Earthquake Ground Motions for Performing Response-History Analyses*. NIST GCR 11-917-15. Prepared by the NEHRP Consultants Joint Venture for the National Institute of Standards and Technology, Gaithersburg, MD.

NIST. 2012. *Soil–Structure Interaction for Building Structures*. Prepared by NEHRP Consultants Joint Venture (a Partnership of the Applied Technology Council and the Consortium of Universities for Research in Earthquake Engineering) GCR 12-917-21. National Institute of Standards and Technology, Gaithersburg, MD.

NRC. 1975. *Reactor Safety Study. An Assessment of Accident Risks in US Commercial Nuclear Power Plants. Executive Summary*. WASH-1400. United States Nuclear Regulatory Commission.

NRC. 2012. *Practical Implementation Guidelines for SSHAC Level 3 and 4 Hazard Studies*. US Nuclear Regulatory Commission NUREG-2117. Washington, DC.

NRC. 2018. *Updated Implementation Guidelines for SSHAC Hazard Studies*. US Nuclear Regulatory Commission NUREG-2213. Washington, DC.

NUREG. 1983. *PRA Procedures Guide—A Guide to the Performance of Probabilistic Risk Assessment for Nuclear Power Plants*. NUREG/CR-2300. Division of Engineering Technology Office of Nuclear Regulatory Research, US Nuclear Regulatory Commission, Washington, DC.

NZS. 2004. *Structural Design Actions–Part 5: Earthquake Design Actions*. NZS 1170.5:2004. Standards New Zealand.

Oberkampf, W. L., Trucano, T. G., and Hirsch, C. 2004. Verification, Validation, and Predictive Capability in Computational Engineering and Physics. *Applied Mechanics Reviews*, **57**(5), 345–384.

Ogata, Y. 1988. Statistical Models for Earthquake Occurrences and Residual Analysis for Point Processes. *Journal of the American Statistical Association*, **83**(401), 9–27.

Okada, Y. 1992. Internal Deformation Due to Shear and Tensile Faults in a Half-Space. *Bulletin of the Seismological Society of America*, **82**(2), 1018–1040.

Olami, Z., Feder, H. J. S., and Christensen, K. 1992. Self-Organized Criticality in a Continuous, Nonconservative Cellular Automaton Modeling Earthquakes. *Physical Review Letters*, **68**(8), 1244–1248.

Olsen, K. 2000. Site Amplification in the Los Angeles Basin from Three-Dimensional Modelling of Ground Motions. *Bulletin of the Seismological Society of America*, **90**(6B), S77–S94.

Olsen, K., and Takedatsu, R. 2015. The SDSU Broadband Ground-Motion Generation Module BBtoolbox Version 1.5. *Seismological Research Letters*, **86**(1), 81–88.

Olsen, K. B., and Archuleta, R. J. 1996. Three-Dimensional Simulation of Earthquakes on the Los Angeles Fault System. *Bulletin of the Seismological Society of America*, **86**(3), 575–596.

Olsen, K. B., Pechmann, J. C., and Schuster, G. T. 1995. Simulation of 3D Elastic Wave Propagation in the Salt Lake Basin. *Bulletin of the Seismological Society of America*, **85**(6), 1688–1710.

Omori, F. 1894. On After-Shocks. *Seismological Journal of Japan*, **19**, 71–80.

Ordaz, M., and Reyes, C. 1999. Earthquake Hazard in Mexico City: Observations versus Computations. *Bulletin of the Seismological Society of America*, **89**(5), 1379–1383.

Ou, G.-B., and Herrmann, R. B. 1990. A Statistical Model for Ground Motion Produced by Earthquakes at Local and Regional Distances. *Bulletin of the Seismological Society of America*, **80**(6A), 1397–1417.

Pagani, M., Monelli, D., Weatherill, G., Danciu, L., Crowley, H., Silva, V., Henshaw, P., Butler, L., Nastasi, M., Panzeri, L., et al. 2014. OpenQuake Engine: An Open Hazard (and Risk) Software for the Global Earthquake Model. *Seismological Research Letters*, **85**(3), 692–702.

Page, M., and Felzer, K. 2015. Southern San Andreas Fault Seismicity Is Consistent with the Gutenberg–Richter Magnitude–Frequency Distribution. *Bulletin of the Seismological Society of America*, **105**(4), 2070–2080.

Page, M. T., and van der Elst, N. J. 2018. TuringStyle Tests for UCERF3 Synthetic Catalogs. *Bulletin of the Seismological Society of America*, **108**(2), 729–741.

Pagni, C. A., and Lowes, L. N. 2006. Fragility Functions for Older Reinforced Concrete Beam-Column Joints. *Earthquake Spectra*, **22**(1), 215–238.

Papaspiliou, M., Kontoe, S., and Bommer, J. J. 2012. An Exploration of Incorporating Site Response into PSHA—Part I: Issues Related to Site Response Analysis Methods. *Soil Dynamics and Earthquake Engineering*, **42**(Nov.), 302–315.

Park, C. B., Miller, R. D., and Xia, J. 1999. Multichannel Analysis of Surface Waves. *Geophysics*, **64**(3), 800–808.

Park, J., Bazzurro, P., and Baker, J. W. 2007. Modeling Spatial Correlation of Ground Motion Intensity Measures for Regional Seismic Hazard and Portfolio Loss Estimation. In: *10th International Conference on Application of Statistics and Probability in Civil Engineering (ICASP10)*.

Parsons, T. 2008. Earthquake Recurrence on the South Hayward Fault Is Most Consistent with a Time Dependent, Renewal Process. *Geophysical Research Letters*, **35**(21), B08313.

Parsons, T., Console, R., Falcone, G., Murru, M., and Yamashina, K. 2012. Comparison of Characteristic and Gutenberg–Richter Models for Time-Dependent M 7.9 Earthquake Probability in the Nankai-Tokai Subduction Zone, Japan. *Geophysical Journal International*, **190**(3), 1673–1688.

Parsons, T., and Geist, E. L. 2009. Is There a Basis for Preferring Characteristic Earthquakes over a Gutenberg–Richter Distribution in Probabilistic Earthquake Forecasting? *Bulletin of the Seismological Society of America*, **99**(3), 2012–2019.

Parsons, T., Johnson, K. M., Bird, P., Bormann, J., Dawson, T. E., Field, E. H., Hammond, W. C., Herring, T. A., McCaffrey, R., Shen, Z.-K., Thatcher, W. R., Weldon, R. J. I., and Zeng, Y. 2013. *Appendix C: Deformation Models for UCERF3*. Open File Report 2013-1165. United States Geological Survey.

Pate-Cornell, M. E. 1994. Quantitative Safety Goals for Risk Management of Industrial Facilities. *Structural Safety*, **13**(3), 145–157.

Petersen, M. D., Frankel, A. D., Harmsen, S. C., Mueller, C. S., Haller, K. M., Wheeler, R. L., Wesson, R. L., Zeng, Y., Boyd, O. S., Perkins, D. M., Luco, N., Field, E. H., Wills, C. J., and Rukstales, K. S. 2008. *Documentation for the 2008 Update of the United States National Seismic Hazard Maps*. Open-File Report 2008–1128. US Geological Survey.

Petersen, M. D., Moschetti, M. P., Powers, P. M., Mueller, C. S., Haller, K. M., Frankel, A. D., Zeng, Y., Rezaeian, S., Harmsen, S. C., Boyd, O. S., Field, E. H., Chen, R., Rukstales, K. S., Luco, N., Wheeler, R. L., Williams, R. A., and Olsen, A. H. 2014. *Documentation for the 2014 Update of the United States National Seismic Hazard Maps*. Open-File Report 2014–1091. US Geological Survey.

Petersen, M. D., Mueller, C. S., Moschetti, M. P., Hoover, S. M., Shumway, A. M., McNamara, D. E., Williams, R. A., Llenos, A. L., Ellsworth, W. L., Michael, A. J., Rubinstein, J. L., McGarr, A. F., and Rukstales, K. S. 2017. 2017 One-Year Seismic-Hazard Forecast for the Central and Eastern United States from Induced and Natural Earthquakes. *Seismological Research Letters*, **88**(3), 772–783.

Peterson, J. R., and Hutt, C. R. 2014. *World-Wide Standardized Seismograph Network: A Data Users Guide*. Open File Report 2014-1218. United States Geological Survey.

Porter, K. 2020. *A Beginner's Guide to Fragility, Vulnerability, and Risk*. University of Colorado Boulder.

Porter, K., Jones, L., Cox, D., Goltz, J., Hudnut, K., Mileti, D., Perry, S., Ponti, D., Reichle, M., Rose, A. Z., Scawthorn, C. R., Seligson, H. A., Shoaf, K. I., Treiman, J., and Wein, A. 2011. The ShakeOut Scenario: A Hypothetical Mw7.8 Earthquake on the Southern San Andreas Fault. *Earthquake Spectra*, **27**(2), 239–261.

Porter, K., Kennedy, R., and Bachman, R. 2007. Creating Fragility Functions for Performance-Based Earthquake Engineering. *Earthquake Spectra*, **23**(2), 471–489.

Porter, K. A., Kiremidjian, A., and LeGrue, J. 2001. Assembly-Based Vulnerability of Buildings and Its Use in Performance Evaluation. *Earthquake Spectra*, **17**(2), 291–313.

Powers, P. M., and Field, E. H. 2013. *Appendix O: Gridded Seismicity Sources*. Open File Report 2013-1165. United States Geological Survey.

Prevost, J. 1978. Plasticity Theory for Soil Stress–Strain Behaviour. *Journal of Engineering Mechanics (ASCE)*, **104**(5), 1177–1194.

Proakis, J., and Manolakis, D. 1996. *Digital Signal Processing: Principles, Algorithms, and Applications*, 3rd ed. Prentice-Hall.

Qi, H., and Sun, D. 2006. A Quadratically Convergent Newton Method for Computing the Nearest Correlation Matrix. *SIAM Journal on Matrix Analysis and Applications*, **28**(2), 360–385.

Ramírez-Guzmán, L., Boyd, O. S., Hartzell, S., and Williams, R. A. 2012. Seismic Velocity Model of the Central United States (Version 1): Description and Simulation of the 18 April 2008 Mt. Carmel, Illinois, Earthquake. *Bulletin of the Seismological Society of America*, **102**(6), 2622–2645.

Raoof, M., Herrmann, R. B., and Malagnini, L. 1999. Attenuation and Excitation of Three-Component Ground Motion in Southern California. *Bulletin of the Seismological Society of America*, **89**(4), 888–902.

Reasenberg, P. 1985. Second-Order Moment of Central California Seismicity, 1969–1982. *Journal of Geophysical Research: Solid Earth*, **90**(B7), 5479–5495.

Reasenberg, P. A., and Jones, L. M. 1989. Earthquake Hazard after a Mainshock in California. *Science*, **243**(4895), 1173–1176.

Reasenberg, P. A., and Jones, L. M. 1990. California Aftershock Hazard Forecasts. *Science*, **247**(4940), 345–346.

Reasenberg, P. A., and Jones, L. M. 1994. Earthquake Aftershocks: Update. *Science*, **265**(5176), 1251–1252.

Reiter, L. 1990. *Earthquake Hazard Analysis: Issues and Insights*. New York: Columbia University Press.

Restrepo, D., Bielak, J., Serrano, R., Gómez, J., and Jaramillo, J. 2016. Effects of Realistic Topography on the Ground Motion of the Colombian Andes: A Case Study at the Aburrá Valley, Antioquia. *Geophysical Journal International*, **204**(3), 1801–1816.

Rezaeian, S., and Der Kiureghian, A. 2008. A Stochastic Ground Motion Model with Separable Temporal and Spectral Nonstationarities. *Earthquake Engineering & Structural Dynamics*, **37**(13), 1565–1584.

Rezaeian, S., and Der Kiureghian, A. 2011. Simulation of Orthogonal Horizontal Ground Motion Components for Specified Earthquake and Site Characteristics. *Earthquake Engineering & Structural Dynamics*, **41**(2), 335–353.

Rezaeian, S., Petersen, M. D., Moschetti, M. P., Powers, P., Harmsen, S. C., and Frankel, A. D. 2014. Implementation of NGA-West2 Ground Motion Models in the 2014 U.S. National Seismic Hazard Maps. *Earthquake Spectra*, **30**(3), 1319–1333.

Rhoades, D. 1997. Estimation of Attenuation Relations for Strong-Motion Data Allowing for Individual Earthquake Magnitude Uncertainties. *Bulletin of the Seismological Society of America*, **87**(6), 1674–1678.

Rhoades, D., Liukis, M., Christophersen, A., and Gerstenberger, M. 2016. Retrospective Tests of Hybrid Operational Earthquake Forecasting Models for Canterbury. *Geophysical Journal International*, **204**(1), 440–456.

Rhoades, D. A. 1996. Estimation of the Gutenberg–Richter Relation Allowing for Individual Earthquake Magnitude Uncertainties. *Tectonophysics*, **258**(1–4), 71–83.

Rhoades, D. A. 2007. Application of the EEPAS Model to Forecasting Earthquakes of Moderate Magnitude in Southern California. *Seismological Research Letters*, **78**(1), 110–115.

Rhoades, D. A., and Christophersen, A. 2019. Time-Varying Probabilities of Earthquake Occurrence in Central New Zealand Based on the EEPAS Model Compensated for Time-Lag. *Geophysical Journal International*, **219**(1), 417–429.

Rhoades, D. A., and Evison, F. F. 2004. Long-Range Earthquake Forecasting with Every Earthquake a Precursor According to Scale. *Pure and Applied Geophysics*, **161**(1), 47–72.

Rhoades, D. A., and Evison, F. F. 2006. The EEPAS Forecasting Model and the Probability of Moderate-to-Large Earthquakes in Central Japan. *Tectonophysics*, **417**(1–2), 119–130.

Rhoades, D. A., Gerstenberger, M. C., Christophersen, A., Zechar, J. D., Schorlemmer, D., Werner, M. J., and Jordan, T. H. 2014. Regional Earthquake Likelihood Models II: Information Gains of Multiplicative Hybrids. *Bulletin of the Seismological Society of America*, **104**(6), 3072–3083.

Robinson, R., Van Dissen, R., and Litchfield, N. 2011. Using Synthetic Seismicity to Evaluate Seismic Hazard in the Wellington Region, New Zealand. *Geophysical Journal International*, **187**(1), 510–528.

Rodgers, A. J., Anders Petersson, N., Pitarka, A., McCallen, D. B., Sjogreen, B., and Abrahamson, N. 2019. Broadband (0–5 Hz) Fully Deterministic 3D Ground-Motion Simulations of a Magnitude 7.0 Hayward Fault Earthquake: Comparison with Empirical Ground-Motion Models and 3D Path and Site Effects from Source Normalized Intensities. *Seismological Research Letters*, **90**(3), 1268–1284.

Rodriguez-Marek, A., Rathje, E. M., Bommer, J. J., Scherbaum, F., and Stafford, P. J. 2014. Application of Single-Station Sigma and Site-Response Characterization in a Probabilistic Seismic-Hazard Analysis for a New Nuclear Site. *Bulletin of the Seismological Society of America*, **104**(4), 1601–1619.

Rogers, G., and Dragert, H. 2003. Episodic Tremor and Slip on the Cascadia Subduction Zone: The Chatter of Silent Slip. *Science*, **300**(5627), 1942–1943.

Rojahn, C., and Sharpe, R. L. 1985. *Earthquake Damage Evaluation Data for California*. ATC-13. Applied Technology Council.

Roselli, P., Marzocchi, W., and Faenza, L. 2016. Toward a New Probabilistic Framework to Score and Merge Ground-Motion Prediction Equations: The Case of the Italian Region. *Bulletin of the Seismological Society of America*, **106**(2), 720–733.

Ross, S. M. 2014. *Introduction to Probability and Statistics for Engineers and Scientists*. Academic Press.

Rossetto, T., and Elnashai, A. 2005. A New Analytical Procedure for the Derivation of Displacement-Based Vulnerability Curves for Populations of RC Structures. *Engineering Structures*, **27**(3), 397–409.

Rossetto, T., and Ioannou, I. 2018. Empirical Fragility and Vulnerability Assessment: Not Just a Regression. Pages 79–103 of: *Risk Modeling for Hazards and Disasters*. Elsevier.

Roten, D., Cui, Y., Olsen, K., Day, S., Withers, K., Savran, W., Wang, P., and Mu, D. 2016 (Nov.). High-Frequency Nonlinear Earthquake Simulations on Petascale Heterogeneous Supercomputers. In: *Procs. Supercomputing Conference*.

Roten, D., Olsen, K., Day, S., Cui, Y., and Fah, D. 2014. Expected Seismic Shaking in Los Angeles Reduced by San Andreas Fault Zone Plasticity. *Geophysical Research Letters*, **41**, 2769–2777.

Roten, D., Olsen, K. B., and Pechmann, J. C. 2012. 3D Simulations of M 7 Earthquakes on the Wasatch Fault, Utah, Part II: Broadband (0–10 Hz) Ground Motions and Nonlinear Soil Behavior. *Bulletin of the Seismological Society of America*, **102**(5), 2008–2030.

Rougier, J., Hill, L. J., Sparks, S., and Sparks, R. S. J. 2013. *Risk and Uncertainty Assessment for Natural Hazards*. Cambridge University Press.

Ruff, L. J. 1999. Dynamic Stress Drop of Recent Earthquakes: Variations within Subduction Zones. *Pure and Applied Geophysics*, **154**, 409–431.

Ruina, A. 1983. Slip Instability and State Variable Friction Laws. *Journal of Geophysical Research*, **88**(B12), 10359–10370.

Rundle, P. B., Rundle, J. B., Tiampo, K. F., Donnellan, A., and Turcotte, D. L. 2006. Virtual California: Fault Model, Frictional Parameters, Applications. *Pure and Applied Geophysics*, **163**(9), 1819–1846.

Russo, J. E., and Schoemaker, P. J. 1992. Managing Overconfidence. *Sloan Management Review*, **33**(2), 7–17.

Sakaguchi, H., and Okamura, K. 2015. Aftershocks and Omor's Law in a Modified Carlson–Langer Model with Nonlinear Viscoelasticity. *Physical Review E*, **91**(052914), 1–6.

Saragoni, G., and Hart, G. C. 1973. Simulation of Artificial Earthquakes. *Earthquake Engineering & Structural Dynamics*, **2**(3), 249–267.

Scasserra, G., Stewart, J. P., Bazzurro, P., Lanzo, G., and Mollaioli, F. 2009. A Comparison of NGA Ground-Motion Prediction Equations to Italian Data. *Bulletin of the Seismological Society of America*, **99**(5), 2961–2978.

Scherbaum, F., Cotton, F., and Smit, P. 2004. On the Use of Response Spectral-Reference Data for the Selection and Ranking of Ground-Motion Models for Seismic-Hazard Analysis in Regions of Moderate Seismicity: The Case of Rock Motion. *Bulletin of the Seismological Society of America*, **94**(6), 2164–2185.

Scherbaum, F., and Kuehn, N. M. 2011. Logic Tree Branch Weights and Probabilities: Summing Up to One Is Not Enough. *Earthquake Spectra*, **27**(4), 1237–1251.

Schiappapietra, E., and Douglas, J. 2020. Modelling the Spatial Correlation of Earthquake Ground Motion: Insights from the Literature, Data from the 2016–2017 Central Italy Earthquake Sequence and Ground-Motion Simulations. *Earth-Science Reviews*, **203**(Feb.), 103139.

Schmedes, J., Archuleta, R. J., and Lavallée, D. 2013. A Kinematic Rupture Model Generator Incorporating Spatial Interdependency of Earthquake Source Parameters. *Geophysical Journal International*, **2013**(192), 1116–1131.

Schnabel, P. B., Lysmer, J., and Seed, H. B. 1972 (Dec.). *SHAKE: A Computer Program for Earthquake Response Analysis of Horizontally Layered Sites*. EERC 72-12. Earthquake Engineering Research Center/University of California, Berkeley.

Scholz, C. H. 2002. *The Mechanics of Earthquakes and Faulting*, 2nd ed. Cambridge University Press.

Schorlemmer, D., Werner, M. J., Marzocchi, W., Jordan, T. H., Ogata, Y., Jackson, D. D., Mak, S., Rhoades, D. A., Gerstenberger, M. C., Hirata, N., Liukis, M., Maechling, P. J., Strader, A., Taroni, M., Wiemer, S., Zechar, J. D., and Zhuang, J. 2018. The Collaboratory for the Study of Earthquake Predictability: Achievements and Priorities. *Seismological Research Letters*, **89**(4), 1305–1313.

Schotanus, M. I. J., Franchin, P., Lupoi, A., and Pinto, P. E. 2004. Seismic Fragility Analysis of 3D Structures. *Structural Safety*, **26**(4), 421–441.

Schwartz, D. P., and Coppersmith, K. J. 1984. Fault Behavior and Characteristic Earthquakes: Examples from the Wasatch and San Andreas Faults. *Journal of Geophysical Research*, **89**, 5681–5698.

Seifried, A. E., and Baker, J. W. 2016. Spectral Variability and Its Relationship to Structural Response Estimated from Scaled and Spectrum-Matched Ground Motions. *Earthquake Spectra*, **32**(4), 2191–2205.

Seyhan, E., Stewart, J. P., Ancheta, T. D., Darragh, R. B., and Graves, R. W. 2014. NGA-West2 Site Database. *Earthquake Spectra*, **30**(3), 1007–1024.

Sgobba, S., Stafford, P. J., Marano, G. C., and Guaragnella, C. 2011. An Evolutionary Stochastic Ground-Motion Model Defined by a Seismological Scenario and Local Site Conditions. *Soil Dynamics and Earthquake Engineering*, **31**(11), 1465–1479.

Shahi, S. K., and Baker, J. W. 2014. NGA-West2 Models for Ground-Motion Directionality. *Earthquake Spectra*, **30**(3), 1285–1300.

Shaw, B. E. 2009. Constant Stress Drop from Small to Great Earthquakes in Magnitude-Area Scaling. *Bulletin of the Seismological Society of America*, **99**(2A), 871–875.

Shaw, B. E. 2013. Earthquake Surface Slip-Length Data Is Fit by Constant Stress Drop and Is Useful for Seismic Hazard Analysis. *Bulletin of the Seismological Society of America*, **103**(2A), 876–893.

Shaw, B. E., and Scholz, C. H. 2001. Slip-Length Scaling in Large Earthquakes: Observations and Theory and Implications for Earthquake Physics. *Geophysical Research Letters*, **28**(15), 2995–2998.

Shcherbakov, R., Turcotte, D. L., and Rundle, J. B. 2004. A Generalized Omori's Law for Earthquake Aftershock Decay. *Geophysical Research Letters*, **31**(L11613), 1–5.

Shearer, P. M. 2009. *Introduction to Seismology*, 2nd ed. Cambridge University Press.

Shi, X., Wang, Y., Liu-Zeng, J., Weldon, R., Wei, S., Wang, T., and Sieh, K. E. 2017. How Complex Is the 2016 M_w 7.8 Kaikoura Earthquake, South Island, New Zealand? *Science Bulletin*, **62**(5), 309–311.

Shi, Y., and Stafford, P. J. 2018. Markov Chain Monte Carlo Ground-Motion Selection Algorithms for Conditional Intensity Measure Targets. *Earthquake Engineering & Structural Dynamics*, **47**(12), 2468–2489.

Shi, Z., and Day, S. M. 2013. Rupture Dynamics and Ground Motion from 3-D Rough-Fault Simulations. *Journal of Geophysical Research: Solid Earth*, **118**(3), 1122–1141.

Shinozuka, M., Feng, M. Q., Lee, J., and Naganuma, T. 2000. Statistical Analysis of Fragility Curves. *Journal of Engineering Mechanics*, **126**(12), 1224–1231.

Shiraki, N., Shinozuka, M., Moore, J. E., Chang, S. E., Kameda, H., and Tanaka, S. 2007. System Risk Curves: Probabilistic Performance Scenarios for Highway Networks Subject to Earthquake Damage. *Journal of Infrastructure Systems*, **13**(1), 43–54.

Shome, N., and Cornell, C. 1999. *Probabilistic Seismic Demand Analysis of Nonlinear Structures*. Report No. RMS-35, RMS Program. Stanford University, Stanford, CA.

Shome, N., and Cornell, C. A. 2000. Structural Seismic Demand Analysis: Consideration of Collapse. In: *8th ASCE Specialty Conference on Probabilistic Mechanics and Structural Reliability*.

Shome, N., Cornell, C. A., Bazzurro, P., and Carballo, J. E. 1998. Earthquakes, Records, and Nonlinear Responses. *Earthquake Spectra*, **14**(3), 469–500.

Shome, N., Jayaram, N., and Rahnama, M. 2012. Uncertainty and Spatial Correlation Models for Earthquake Losses. In: *15th World Conference on Earthquake Engineering (WCEE), Lisbon*.

Sibson, R. H. 1973. Interactions between Temperature and Pore Fluid Pressure during Earthquake Faulting and a Mechanism for Partial or Total Stress Relief. *Nature*, **243**, 66–68.

Sibson, R. H. 1985. A Note on Fault Reactivation. *Journal of Structural Geology*, **7**(6), 751–754.

Sibson, R. H., and Xie, G. 1998. Dip Range for Intracontinental Reverse Fault Rupture: Truth Not Stranger than Friction? *Bulletin of the Seismological Society of America*, **88**(4), 1014–1022.

Silva, V. 2019. Uncertainty and Correlation in Seismic Vulnerability Functions of Building Classes. *Earthquake Spectra*, **35**(4), 1515–1539.

Silva, V., Akkar, S., Baker, J., Bazzurro, P., Castro, J. M., Crowley, H., Dolsek, M., Galasso, C., Lagomarsino, S., Monteiro, R., Perrone, D., Pitilakis, K., and Vamvatsikos, D. 2019. Current Challenges and Future Trends in Analytical Fragility and Vulnerability Modelling. *Earthquake Spectra*, **35**(4), 1927–1952.

Silva, V., Crowley, H., and Bazzurro, P. 2016. Exploring Risk-Targeted Hazard Maps for Europe. *Earthquake Spectra*, **32**(2), 1165–1186.

Silva, V., Crowley, H., Pagani, M., Monelli, D., and Pinho, R. 2013. Development of the OpenQuake Engine, the Global Earthquake Model's Open-Source Software for Seismic Risk Assessment. *Natural Hazards*, **72**(3), 1409–1427.

Silva, V., Crowley, H., Varum, H., Pinho, R., and Sousa, R. 2014. Evaluation of Analytical Methodologies Used to Derive Vulnerability Functions. *Earthquake Engineering & Structural Dynamics*, **43**(2), 181–204.

Simons, M., Minson, S. E., Sladen, A., Ortega, F., Jiang, J., Owen, S. E., Meng, L., Ampuero, J.-P., Wei, S., Chu, R., Helmberger, D. V., Kanamori, H., Hetland, E., Moore, A. W., and Webb, F. H. 2011. The 2011 Magnitude 9.0 Tohoku-Oki Earthquake: Mosaicking the Megathrust from Seconds to Centuries. *Science*, **332**(6036), 1421.

Singhal, A., and Kiremidjian, A. 1996. Method for Probabilistic Evaluation of Seismic Structural Damage. *Journal of Structural Engineering*, **122**(12), 1459–1467.

So, E. K. M., and Pomonis, A. 2012. Derivation of Globally Applicable Casualty Rates for Use in Earthquake Loss Estimation Models. In: *Proceedings of 15th World Conference on Earthquake Engineering*.

Sokolov, V., and Wenzel, F. 2011. Influence of Spatial Correlation of Strong Ground Motion on Uncertainty in Earthquake Loss Estimation. *Earthquake Engineering & Structural Dynamics*, **40**(9), 993–1009.

Sokolov, V., and Wenzel, F. 2013a. Spatial Correlation of Ground Motions in Estimating Seismic Hazards to Civil Infrastructure. Pages 57–78 of: Tesfamariam, S., and Goda, K. (eds.), *Handbook of Seismic Risk Analysis and Management of Civil Infrastructure Systems*. Woodhead Publishing Series in Civil and Structural Engineering. Woodhead Publishing.

Sokolov, V., and Wenzel, F. 2013b. Further Analysis of the Influence of Site Conditions and Earthquake Magnitude on Ground-Motion Within-Earthquake Correlation: Analysis of PGA and PGV Data from the K-NET and the KiK-Net (Japan) Networks. *Bulletin of Earthquake Engineering*, **11**, 1909–1926.

Sokolov, V., and Wenzel, F. 2014. On the Modeling of Ground-Motion Field for Assessment of Multiple-Location Hazard, Damage, and Loss: Example of Estimation of Electric Network Performance during Scenario Earthquake. *Natural Hazards*, **74**, 1555–1575.

Sokolov, V., Wenzel, F., Wen, K.-L., and Jean, W.-Y. 2012. On the Influence of Site Conditions and Earthquake Magnitude on Ground-Motion Within-Earthquake Correlation: Analysis of PGA Data from TSMIP (Taiwan) Network. *Bulletin of Earthquake Engineering*, **10**(5), 1401–1429.

Somerville, P., Irikura, K., Graves, R., Sawada, S., Wald, D., Abrahamson, N., Iwasaki, Y., Kagawa, T., Smith, N., and Kowada, A. 1999. Characterizing Crustal Earthquake Slip Models for the Prediction of Strong Ground Motion. *Seismological Research Letters*, **70**(1), 59–80.

Somerville, P. G., McLaren, J. P., Saikia, C. K., and Helmberger, D. V. 1988. Site-Specific Estimation of Spatial Incoherence of Strong Ground Motion. Pages 188–202 of: *Earthquake Engineering and Soil Dynamics II: Recent Advances in Ground-Motion Evaluation*. Park City, UT: American Society of Civil Engineers.

Somerville, P. G., McLaren, J. P., Sen, M. K., and Helmberger, D. V. 1991. The Influence of Site Conditions on the Spatial Incoherence of Ground Motions. *Structural Safety*, **10**(1), 1–13.

Somerville, P. G., Smith, N. F., Graves, R. W., and Abrahamson, N. A. 1997. Modification of Empirical Strong Ground Motion Attenuation Relations to Include the Amplitude and Duration Effects of Rupture Directivity. *Seismological Research Letters*, **68**(1), 199–222.

Spassiani, I., and Marzocchi, W. 2018. How Likely Does an Aftershock Sequence Conform to a Single Omori Law Behavior? *Seismological Research Letters*, **89**(3), 1118–1128.

Spudich, P., and Chiou, B. S. J. 2008. Directivity in NGA Earthquake Ground Motions: Analysis Using Isochrone Theory. *Earthquake Spectra*, **24**(1), 279–298.

Spudich, P., Rowshandel, B., Shahi, S. K., and Baker, J. W. 2014. Overview and Comparison of the NGA-West2 Directivity Models. *Earthquake Spectra*, **30**(3), 1199–1221.

SSHAC. 1997. *Recommendations for Probabilistic Seismic Hazard Analysis: Guidance on Uncertainty and Use of Experts*. US Nuclear Regulatory Commission Report NUREG/CR-6372. Washington, DC.

Stafford, P. J. 2013. Uncertainties in Ground Motion Prediction in Probabilistic Seismic Hazard Analysis (PSHA) of Civil Infrastructure. Pages 29–56 of: *Handbook of Seismic Risk Analysis and Management of Civil Infrastructure Systems*. Woodhead Publishing.

Stafford, P. J. 2014a. Crossed and Nested Mixed-Effects Approaches for Enhanced Model Development and Removal of the Ergodic Assumption in Empirical Ground-Motion Models. *Bulletin of the Seismological Society of America*, **104**(2), 702–719.

Stafford, P. J. 2014b. Source-Scaling Relationships for the Simulation of Rupture Geometry within Probabilistic Seismic-Hazard Analysis. *Bulletin of the Seismological Society of America*, **104**(4), 1620–1635.

Stafford, P. J. 2015. Extension of the Random-Effects Regression Algorithm to Account for the Effects of Nonlinear Site Response. *Bulletin of the Seismological Society of America*, **105**(6), 3196–3202.

Stafford, P. J. 2017. Interfrequency Correlations among Fourier Spectral Ordinates and Implications for Stochastic Ground-Motion Simulation. *Bulletin of the Seismological Society of America*, **107**(6), 2774–2791.

Stafford, P. J. 2019. Continuous Integration of Data into Ground-Motion Models Using Bayesian Updating. *Journal of Seismology*, **23**(1), 39–57.

Stafford, P. J., Pettinga, J. R., and Berrill, J. B. 2008. Seismic Source Identification and Characterisation for Probabilistic Seismic Hazard Analyses Conducted in the Buller-NW Nelson Region, South Island, New Zealand. *Journal of Seismology*, **12**, 477–498.

Stafford, P. J., Rodriguez-Marek, A., Edwards, B., Kruiver, P. P., and Bommer, J. J. 2017. Scenario Dependence of Linear Site-Effect Factors for Short-Period Response Spectral Ordinates. *Bulletin of the Seismological Society of America*, **107**(6), 2859–2872.

Stafford, P. J., Sgobba, S., and Marano, G. C. 2009. An Energy-Based Envelope Function for the Stochastic Simulation of Earthquake Accelerograms. *Soil Dynamics and Earthquake Engineering*, **29**(7), 1123–1133.

Stafford, P. J., Zurek, B. D., Ntinalexis, M., and Bommer, J. J. 2019. Extensions to the Groningen Ground-Motion Model for Seismic Risk Calculations: Component-to-Component Variability and Spatial Correlation. *Bulletin of Earthquake Engineering*, **17**, 4417–4439.

Stein, M. L. 2012. *Interpolation of Spatial Data: Some Theory for Kriging*. Springer Science & Business Media.

Stein, S., Geller, R., and Liu, M. 2011. Bad Assumptions or Bad Luck: Why Earthquake Hazard Maps Need Objective Testing. *Seismological Research Letters*, **82**(5), 623–626.

Stewart, J. P., Abrahamson, N. A., Atkinson, G. M., Baker, J., Boore, D. M., Bozorgnia, Y., Campbell, K. W., Comartin, C. D., Idriss, I. M., Lew, M., Mehrain, M., Moehle, J. P., Naeim, F., and Sabol, T. A. 2011. Representation of Bi-directional Ground Motions for Design Spectra in Building Codes. *Earthquake Spectra*, **27**(3), 927–937.

Stewart, J. P., Afshari, K., and Goulet, C. A. 2017. Non-ergodic Site Response in Seismic Hazard Analysis. *Earthquake Spectra*, **33**(4), 1385–1414.

Stewart, J. P., Chiou, S.-J., Bray, J. D., Graves, R. W., Somerville, P. G., and Abrahamson, N. A. 2001. *Ground Motion Evaluation Procedures for Performance-Based Design*. PEER Report 2001/09. Pacific Earthquake Engineering Research Center.

Stewart, J. P., Luco, N., Hooper, J. D., and Crouse, C. B. 2020. Risk-Targeted Alternatives to Deterministic Ground Motion Caps in U.S. Seismic Provisions. *Earthquake Spectra*, **36**(2), 904–923.

Stirling, M., and Gerstenberger, M. C. 2018. Applicability of the Gutenberg–Richter Relation for Major Active Faults in New Zealand. *Bulletin of the Seismological Society of America*, **108**(2), 718–728.

Stirling, M., McVerry, G., Gerstenberger, M., Litchfield, N., Van Dissen, R., Berryman, K., Barnes, P., Wallace, L., Villamor, P., Langridge, R., Lamarche, G., Nodder, S., Reyners, M., Bradley, B., Rhoades, D., Smith, W., Nicol, A., Pettinga, J., Clark, K., and Jacobs, K. 2012. National Seismic Hazard Model for New Zealand: 2010 Update. *Bulletin of the Seismological Society of America*, **102**(4), 1514–1542.

Stirling, M. W., Wesnousky, S. G., and Shimazaki, K. 1996. Fault Trace Complexity, Cumulative Slip, and the Shape of the Magnitude–Frequency Distribution for Strike-Slip Faults: A Global Survey. *Geophysical Journal International*, **124**(3), 833–868.

Stock, C., and Smith, E. G. C. 2002. Comparison of Seismicity Models Generated by Different Kernel Estimations. *Bulletin of the Seismological Society of America*, **92**(3), 913–922.

Stoica, M., Medina, R. A., and McCuen, R. H. 2007. Improved Probabilistic Quantification of Drift Demands for Seismic Evaluation. *Structural Safety*, **29**(2), 132–145.

Strasser, F., Bommer, J. J., and Abrahamson, N. A. 2008. Truncation of the Distribution of Ground-Motion Residuals. *Journal of Seismology*, **12**(1), 79–105.

Strasser, F. O., Abrahamson, N. A., and Bommer, J. J. 2009. Sigma: Issues, Insights, and Challenges. *Seismological Research Letters*, **80**(1), 40–56.

Strasser, F. O., and Bommer, J. J. 2009. Review: Strong Ground Motions—Have We Seen the Worst? *Bulletin of the Seismological Society of America*, **99**(5), 2613–2637.

Sun, X., Clayton, B., Hartzell, S., and Rezaeian, S. 2018. Estimation of Ground-Motion Variability in the Central and Eastern United States Using Deterministic Physics-Based Synthetics. *Bulletin of the Seismological Society of America*, **108**(6), 3368–3383.

Sykes, L. R., and Menke, W. 2006. Repeat Times of Large Earthquakes: Implications for Earthquake Mechanics and Long-Term Prediction. *Bulletin of the Seismological Society of America*, **96**(5), 1569–1596.

Süss, M. P., and Shaw, J. H. 2003. P Wave Seismic Velocity Structure Derived from Sonic Logs and Industry Reflection Data in the Los Angeles Basin, California. *Journal of Geophysical Research: Solid Earth*, **108**(B3), 2170.

Taborda, R., and Bielak, J. 2013. Ground-Motion Simulation and Validation of the 2008 Chino Hills, California, Earthquake. *Bulletin of the Seismological Society of America*, **103**(1), 131–156.

Taborda, R., Bielak, J., and Restrepo, D. 2012. Earthquake GroundMotion Simulation including Nonlinear Soil Effects under Idealized Conditions with Application to Two Case Studies. *Seismological Research Letters*, **83**(6), 1047–1060.

Taborda, R., and Roten, D. 2015. Physics-Based Ground-Motion Simulation. In: *Encyclopedia of Earthquake Engineering*. Berlin: Springer-Verlag.

Tarbali, K., and Bradley, B. A. 2015. Ground Motion Selection for Scenario Ruptures Using the Generalised Conditional Intensity Measure (GCIM) Method. *Earthquake Engineering & Structural Dynamics*, **44**(10), 1601–1621.

Tarbali, K., and Bradley, B. A. 2016. The Effect of Causal Parameter Bounds in PSHA-Based Ground Motion Selection. *Earthquake Engineering & Structural Dynamics*, **45**(9), 1515–1535.

Tarbali, K., Bradley, B. A., and Baker, J. W. 2018. Consideration and Propagation of Ground Motion Selection Epistemic Uncertainties to Seismic Performance Metrics. *Earthquake Spectra*, **34**(2), 587–610.

Tarbali, K., Bradley, B., Huang, J., Lee, R., Lagrava, D., Bae, S., Polak, V., Motha, J., and Zhu, M. 2019. Cybershake NZ V18.5: New Zealand Simulation-Based Probabilistic Seismic Hazard Analysis. Pages 5224–5231 of: *Earthquake Geotechnical Engineering for Protection and Development of Environment and Constructions*.

Thomas, J. 1995. *Numerical Partial Differential Equations: Finite Difference Methods*. New York: Springer-Verlag.

Thomas, P., Wong, I., and Abrahamson, N. 2010. *Verification of Probabilistic Seismic Hazard Analysis Computer Programs*. PEER Report 2010/106.

Thompson, E. M., Baise, L. G., and Kayen, R. E. 2007. Spatial Correlation of Shear-Wave Velocity in the San Francisco Bay Area Sediments. *Soil Dynamics and Earthquake Engineering*, **27**, 144–152.

Tinti, E., Fukuyama, E., Piatanesi, A., and Cocco, M. 2005. A Kinematic Source-Time Function Compatible with Earthquake Dynamics. *Bulletin of the Seismological Society of America*, **95**(4), 1211–1223.

Tinti, S., and Mulargia, F. 1985. Effects of Magnitude Uncertainties on Estimating the Parameters in the Gutenberg–Richter Frequency–Magnitude Law. *Bulletin of the Seismological Society of America*, **75**(6), 1681–1697.

Toda, S., and Stein, R. S. 2002. Response of the San Andreas Fault to the 1983 Coalinga-Nuñez Earthquakes: An Application of Interaction-Based Probabilities for Parkfield. *Journal of Geophysical Research*, **107**(B6), 764.

Toda, S., Stein, R. S., Richards-Dinger, K., and Bozkurt, S. 2005. Forecasting the Evolution of Seismicity in Southern California: Animations Built on Earthquake Stress Transfer. *Journal of Geophysical Research*, **110**(B05S16).

Toda, S., Stein, R. S., Sevilgen, V., and Jian, L. 2011. *Coulomb 3.3 Graphic-Rich Deformation and Stress-Change Software for Earthquake, Tectonic, and Volcano Research and Teaching: User Guide*. Open File Report 2011-1060. United States Geological Survey.

Toro, G. R., Abrahamson, N. A., and Schneider, J. F. 1997. Model of Strong Ground Motions from Earthquakes in Central and Eastern North America: Best Estimates and Uncertainties. *Seismological Research Letters*, **68**(1), 41–57.

Toro, G. R., and Silva, W. J. 2001. *Scenario Earthquakes for Saint Louis, MO, Memphis, TN, and Seismic Hazard Maps for the Central United States Region Including the Effect of Site Conditions*. Boulder, CO: Risk Engineering.

Tothong, P., and Cornell, C. A. 2006. An Empirical Ground-Motion Attenuation Relation for Inelastic Spectral Displacement. *Bulletin of the Seismological Society of America*, **96**(6), 2146–2164.

Townend, J., and Zoback, M. D. 2000. How Faulting Keeps the Crust Strong. *Geology*, **28**(5), 399.

Trafimow, D., and Marks, M. 2015. Editorial. *Basic and Applied Social Psychology*, **37**(1), 1–2.

Trugman, D. T., and Dunham, E. M. 2014. A 2D Pseudodynamic Rupture Model Generator for Earthquakes on Geometrically Complex Faults. *Bulletin of the Seismological Society of America*, **104**(1), 95–112.

Tu, B. S., Holt, W. E., and Haines, A. J. 1998. Contemporary Kinematics of the Western United States Determined from Earthquake Moment Tensors, Very Long Baseline Interferometry, and GPS Observations. *Journal of Geophysical Research: Solid Earth*, **103**(B8), 18087–18117.

Tu, B. S., Holt, W. E., and Haines, A. J. 1999. Deformation Kinematics in the Western United States Determined from Quaternary Fault Slip Rates and Recent Geodetic Data. *Journal of Geophysical Research: Solid Earth*, **104**(B12), 28927–28955.

Tullis, T. E., Richards-Dinger, K., Barall, M., Dieterich, J. H., Field, E. H., Heien, E. M., Kellogg, L. H., Pollitz, F. F., Rundle, J. B., Sachs, M. K., Turcotte, D. L., Ward, S. N., and Burak Yikilmaz, M. 2012a. A Comparison among Observations and Earthquake Simulator Results for the Allcal2 California Fault Model. *Seismological Research Letters*, **83**(6), 994–1006.

Tullis, T. E., Richards-Dinger, K., Barall, M., Dieterich, J. H., Field, E. H., Heien, E. M., Kellogg, L. H., Pollitz, F. F., Rundle, J. B., Sachs, M. K., Turcotte, D. L., Ward, S. N., and Yikilmaz, M. B. 2012b. Generic Earthquake Simulator. *Seismological Research Letters*, **83**(6), 959–963.

USNRC. 2012. *Central and Eastern United States Seismic Source Characterization for Nuclear Facilities*. NUREG-2115. US Nuclear Regulatory Commission.

Utsu, T. 1961. A Statistical Study on the Occurrence of Aftershocks. *Geophysical Magazine*, **30**, 521–605.

Utsu, T. 1999. Representation and Analysis of the Earthquake Size Distribution: A Historical Review and Some New Approaches. *Pure and Applied Geophysics*, **155**, 509–535.

Utsu, T. 2002. Statistical Features of Seismicity. In: Lee, W. H. K., Kanamori, H., Jennings, P. C., and Kissinger, C. (eds.), *International Handbook of Earthquake Engineering Seismology*. Academic Press.

Vamvatsikos, D., and Cornell, C. A. 2002. Incremental Dynamic Analysis. *Earthquake Engineering & Structural Dynamics*, **31**(3), 491–514.

Veneziano, D., Casciati, F., and Faravelli, L. 1983. Method of Seismic Fragility for Complicated Systems. Pages 67–88 of: *Probabilistic Methods in Seismic Risk Assessment for Nuclear Power Plants*.

Veneziano, D., and Van Dyck, J. 1985. *Seismic Hazard Methodology for Nuclear Facilities in the Eastern U.S.* P101-29. EPRI/SOG.

Villani, M., and Abrahamson, N. A. 2015. Repeatable Site and Path Effects on the Ground-Motion Sigma Based on Empirical Data from Southern California and Simulated Waveforms from the CyberShake Platform. *Bulletin of the Seismological Society of America*, **105**(5), 2681–2695.

Virieux, J. 1984. SH-Wave Propagation in Heterogeneous Media: Velocity–Stress Finite-Difference Method. *Exploration Geophysics*, **15**(4), 265.

Vyas, J. C., Mai, P. M., and Galis, M. 2016. Distance and Azimuthal Dependence of Ground-Motion Variability for Unilateral Strike-Slip Ruptures. *Bulletin of the Seismological Society of America*, **106**(4), 1584–1599.

Wald, D. J., and Allen, T. I. 2007. Topographic Slope as a Proxy for Seismic Site Conditions and Amplification. *Bulletin of the Seismological Society of America*, **97**(5), 1379–1395.

Wald, D. J., and Heaton, T. H. 1994. Spatial and Temporal Distribution of Slip for the 1992 Landers, California, Earthquake. *Bulletin of the Seismological Society of America*, **84**(3), 668–691.

Wald, D. J., Helmberger, D. V., and Heaton, T. H. 1991. Rupture Model of the 1989 Loma Prieta Earthquake from the Inversion of Strong-Motion and Broadband Teleseismic Data. *Bulletin of the Seismological Society of America*, **81**(5), 1540–1572.

Wang, F., and Jordan, T. H. 2014. Comparison of Probabilistic Seismic-Hazard Models Using Averaging-Based Factorization. *Bulletin of the Seismological Society of America*, **104**(3), 1230–1257.

Wang, G. 2011. A Ground Motion Selection and Modification Method Capturing Response Spectrum Characteristics and Variability of Scenario Earthquakes. *Soil Dynamics and Earthquake Engineering*, **31**(4), 611–625.

Wang, M., and Takada, T. 2005. Macrospatial Correlation Model of Seismic Ground Motions. *Earthquake Spectra*, **21**(4), 1137–1156.

Wang, Q., Jackson, D. D., and Kagan, Y. Y. 2009. California Earthquakes, 1800–2007: A Unified Catalog with Moment Magnitudes, Uncertainties, and Focal Mechanisms. *Seismological Research Letters*, **80**(3), 446–457.

Wasserman, L. 2006. *All of Non-parametric Statistics*. New York: Springer.

Watanabe, K., Pisanó, F., and Jeremi, B. 2017. Discretization effects in the finite element simulation of seismic waves in elastic and elastic-plastic media. *Engineering with Computers*, **33**(3), 519–545.

Watson-Lamprey, J., and Abrahamson, N. A. 2006. Selection of Ground Motion Time Series and Limits on Scaling. *Soil Dynamics and Earthquake Engineering*, **26**(5), 477–482.

Weatherill, G. A., Silva, V., Crowley, H., and Bazzurro, P. 2015. Exploring the Impact of Spatial Correlations and Uncertainties for Portfolio Analysis in Probabilistic Seismic Loss Estimation. *Bulletin of Earthquake Engineering*, **13**(4), 957–981.

Wegener, A. 1912. Die Entstehung der Kontinente. *Geologische Rundschau*, **3**(4), 276–292.

Weichert, D. H. 1980. Estimation of the Earthquake Recurrence Parameters for Unequal Observation Periods for Different Magnitudes. *Bulletin of the Seismological Society of America*, **70**(4), 1337–1346.

Wells, D. L., and Coppersmith, K. J. 1994. New Empirical Relationships among Magnitude, Rupture Length, Rupture Width, Rupture Area, and Surface Displacement. *Bulletin of the Seismological Society of America*, **84**(4), 974–1002.

Wesnousky, S. G. 1994. The Gutenberg–Richter or Characteristic Earthquake Distribution, Which Is It? *Bulletin of the Seismological Society of America*, **84**(6), 1940–1959.

Wesson, R. L., Bakun, W. H., and Perkins, D. M. 2003. Association of Earthquakes and Faults in the San Francisco Bay Area Using Bayesian Inference. *Bulletin of the Seismological Society of America*, **93**(3), 1306–1332.

Wesson, R. L., and Perkins, D. M. 2001. Spatial Correlation of Probabilistic Earthquake Ground Motion and Loss. *Bulletin of the Seismological Society of America*, **91**(6), 1498–1515.

Wesson, R. L., Perkins, D. M., Luco, N., and Karaca, E. 2009. Direct Calculation of the Probability Distribution for Earthquake Losses to a Portfolio. *Earthquake Spectra*, **25**(3), 687–706.

Wills, C. J., Gutierrez, C. I., Perez, F. G., and Branum, D. M. 2015. A Next Generation VS30 Map for California Based on Geology and Topography. *Bulletin of the Seismological Society of America*, **105**(6), 3083–3091.

Withers, K. B., Olsen, K. B., and Day, S. M. 2015. Memory-Efficient Simulation of Frequency-Dependent Q. *Bulletin of the Seismological Society of America*, **105**(6), 3129–3142.

Withers, K. B., Olsen, K. B., Day, S. M., and Shi, Z. 2018. Ground Motion and Intraevent Variability from 3D Deterministic Broadband (0–7.5 Hz) Simulations along a Nonplanar Strike-Slip Fault. *Bulletin of the Seismological Society of America*, **109**(1), 229–250.

Working Group on California Earthquake Probabilities. 2003. *Earthquake Probabilities in the San Francisco Bay Region: 2002–2031*. Open File Report 03-214. United States Geological Survey.

Working Group on California Earthquake Probabilities. 2008. *The Uniform California Earthquake Rupture Forecast, Version 2 (UCERF2)*. Open File Report 2007-1437. United States Geological Survey.

Wu, R.-S., and Aki, K. 1988. Introduction: Seismic Wave Scattering in Three-Dimensionally Heterogeneous Earth. *Pure and Applied Geophysics*, **128**(1), 1–6.

Wyss, M. 2020. Return Times of Large Earthquakes Cannot Be Estimated Correctly from Seismicity Rates: 1906 San Francisco and 1717 Alpine Fault Ruptures. *Seismological Research Letters*, **91**(4), 2163–2169.

Yeats, R. S., Sieh, K. E., and Allen, C. R. 1997. *Geology of Earthquakes*. Oxford University Press.

Yeo, G. L., and Cornell, C. A. 2009. A Probabilistic Framework for Quantification of Aftershock Ground-Motion Hazard in California: Methodology and Parametric Study. *Earthquake Engineering & Structural Dynamics*, **38**(1), 45–60.

Yepes-Estrada, C., Silva, V., Rossetto, T., D'Ayala, D., Ioannou, I., Meslem, A., and Crowley, H. 2016. The Global Earthquake Model Physical Vulnerability Database. *Earthquake Spectra*, **32**(4), 2567–2585.

Youngs, R. R., and Coppersmith, K. J. 1985. Implications of Fault Slip Rates and Earthquake Recurrence Models to Probabilistic Seismic Hazard Estimates. *Bulletin of the Seismological Society of America*, **75**(4), 939–964.

Zeng, Y. 1993. Theory of Scattered P- and S-Wave Energy in a Random Isotropic Scattering Medium. *Bulletin of the Seismological Society of America*, **83**(4), 1264–1276.

Zerva, A. 2009. *Spatial Variation of Seismic Ground Motions: Modeling and Engineering Applications*. CRC Press.

Zerva, A., and Zhang, O. 1996. Estimation of Signal Characteristics in Seismic Ground Motions. *Probabilistic Engineering Mechanics*, **11**(4), 229–242.

Zhang, L., Werner, M. J., and Goda, K. 2018. Spatiotemporal Seismic Hazard and Risk Assessment of Aftershocks of M 9 Megathrust Earthquakes. *Bulletin of the Seismological Society of America*, **108**(6), 3313–3335.

Zhao, J., Irikura, K., Zhang, J., Fukushima, Y., Somerville, P., Asano, A., Ohno, Y., Oouchi, T., Takahashi, T., and Ogawa, H. 2006. An empirical site-classification method for strong-motion stations in Japan using H/V response spectral ratio. *Bulletin of the Seismological Society of America*, **96**(3), 914–925.

List of Symbols and Abbreviations

This list describes a number of symbols that are used throughout the book. Some symbols that are used in only a single location are not listed. Because this book covers topics in geophysics, engineering, probability, and risk, and because we have attempted to utilize standard notation from these fields, a single symbol is sometimes used for multiple concepts. In such cases, multiple definitions are provided, along with selected references to corresponding equations in the book.

β	Fragility function logarithmic standard deviation (dispersion)
Σ	Covariance matrix
θ	Vector of generic variables
M	Mean vector, Equation 11.10
	Moment magnitude, Equation 2.5
	Moment tensor, Equation 5.1
$\Delta\sigma$	Brune stress parameter, Equation 5.24
	Static stress drop, Equation 2.6
δB	Between-event residual
δW	Within-event residual
δ	Dip angle, Figure 2.4
	Partitioned epistemic residuals, Equation 8.13
$\delta()$	Dirac delta function
$\dot{\bar{u}}$	Average slip rate
$\dot{M}_0$	Seismic moment rate
$\dot{u}$	Slip velocity
γ	Cross-spectral density, Equation 11.40
	Semivariogram, Equation 11.17
κ_0	Near-surface crustal attenuation
λ	Rake angle, Figure 2.5
	Wavelength, Equation 5.5
	Lamé parameter, Equation 5.17
$\lambda()$	Occurrence rate of the specified event
μ	Crustal rigidity, Equation 2.3
	Mean, Equation A.33
	Shear modulus/Lamé parameter, Equation 5.17
ν	Poisson's ratio
ω	Angular/circular frequency
ϕ	Within-event standard deviation
$\Phi()$	Standard Gaussian CDF
$\phi()$	Standard Gaussian PDF
ψ	Partitioned aleatory residuals

ρ	Correlation coefficient, Equation A.53
	Density, Equation 5.9
σ	Normal stress, Equation 5.8
	Standard deviation, Equation A.35
σ^2	Variance
τ	Between-event standard deviation, Equation 4.29
	Shear stress, Figure 2.8
θ	Azimuth, Figure 2.4
	Fragility function median, Equation 9.1
	Incidence angle, Equation 5.8
ε	Standard normal variate
A	Rupture area, Equation 2.4
	Seismic wave amplitude, Equation 5.7
$a(t)$	Acceleration time series
AF	Site amplification factor
AI	Arias intensity
C	Consequence metric, Equation 9.7
	Covariance, Equation 11.25
CAV	Cumulative absolute velocity
D_B	Bracketed duration
D_U	Uniform duration
D_{S5-75}	5–75% significant duration
D_{S5-95}	5–95% significant duration
DS	Damage state
E	Young's modulus
$E[\ ]$	Expectation of the argument
EDP	Engineering demand parameter
F	Failure
f	Frequency
f_c	Corner frequency
$F_X(\ )$	CDF of X
$f_X(\ )$	PDF of X
G	Shear modulus
$G_X(\ )$	CCDF of X
$I[\ ]$	Indicator function
IM	Ground motion intensity measure
k	Wavenumber
L	Rupture length
M	Magnitude (implicitly moment magnitude unless otherwise defined)
M_0	Seismic moment
m_{max}	Maximum magnitude
m_{min}	Minimum magnitude
$P(\ )$	Probability of the specified event
PGA	Peak ground acceleration
PGD	Peak ground displacement

PGV	Peak ground velocity
Q	Anelastic quality factor
R	Source-to-site distance (generic description)
$R_{\theta\phi}$	Radiation pattern
R_{jb}	Source-to-site distance, measured to the surface projection of the rupture plane ("Joyner–Boore distance")
R_{rup}	Source-to-site distance, measured to the closest point on the rupture plane
R_x	Source-to-site distance, measured from the surface projection of the top edge of the rupture plane in the direction perpendicular to the fault strike
RI	Recurrence interval
$RotD100$	100th percentile (maximum) value of an IM that varies with orientation in the horizontal plane
$RotD50$	50th percentile value of an IM that varies with orientation in the horizontal plane
RP	Return period
rup	Earthquake rupture
SA	Pseudo-spectral acceleration
SD	Spectral displacement
$site$	Site of interest
T	Oscillator vibration period, Equation 4.4
	Time period, Equation 3.15
	Vibration period, Equation 5.5
t	Time
u	Slip displacement
V_L	Love-wave velocity
V_P	P-wave velocity
V_R	Rayleigh-wave velocity
$V_{S,30}$	Time-averaged shear wave velocity measured for $z = 0$–30 m below a site
V_S	S-wave velocity
$Var[\,]$	Variance of the argument
W	Rupture width
z	Depth measured relative to a datum (mean sea level or ground surface)
$Z_{1.0}$	Depth to shear-wave velocity of 1.0 km/s
$Z_{2.5}$	Depth to shear-wave velocity of 2.5 km/s
Z_{seis}	Depth to the base of the seismogenic layer
Z_{tor}	Depth to the top of rupture plane
AAL	Average annual loss
CCDF	Complementary cumulative distribution function
CDF	Cumulative distribution function
CMS	Conditional mean spectrum
CS	Conditional spectra
ERF	Earthquake rupture forecast
GCIM	Generalized conditional intensity measure
GM	Ground motion
GMM	Ground-motion model
MLE	Maximum likelihood estimation

MRD	Mean rate density
PDF	Probability density function
PSHA	Probabilistic seismic hazard analysis
PSRA	Probabilistic seismic risk analysis
SSM	Seismic-source model
UHS	Uniform hazard spectrum

Notation Conventions

The following conventions are used throughout the book. Many of these conventions are also noted when first introduced, but this section provides an additional brief high-level reference.

Scalars and Vectors

We utilize lightface italics variable names (e.g., x) to denote scalar values, and boldface italics variable names (e.g., $\boldsymbol{x}$) to denote vectors and matrices.

Moment magnitude is a notable exception, where geophysics convention is to use a bold $\boldsymbol{M}$ to denote this scalar quantity, and we sometimes adopt that convention when introducing moment magnitude in Chapter 2. In these situations it should be apparent from context that the variable is a scalar. Through the rest of book, when we are not comparing between magnitude scales, we use M to denote magnitude (even though it would implicitly be a moment magnitude) to avoid confusion between scalars and vectors.

Random Variables

Capital letters (e.g., X) represent random variables, and lowercase letters (e.g., x) represent a numerical value. Note, for the following section, that when writing x in lower case we are implicitly reflecting that $X = x$.

There several examples in the List of Symbols and Abbreviations where capital letters (e.g., Q) are the standard convention for a specific parameter, and we have typically used disciplinary conventions in those cases.

Probability Distributions

Standard conventions for marginal and conditional probability distributions are utilized. Marginal probability distributions are denoted as follows:

- Probability density functions: $f_X(x)$
- Cumulative distribution functions: $F_X(x)$
- Complementary cumulative distribution functions: $G_X(x)$.

Conditional probability distributions (for $Y = y$) are denoted as follows:

- Probability density functions: $f_{X|Y}(x \mid y)$
- Cumulative distribution functions: $F_{X|Y}(x \mid y)$
- Complementary cumulative distribution functions: $G_{X|Y}(x \mid y)$.

Conditioning

Conditional probabilities by default utilize an equality condition. With this default, the explicit equality conditioning is shortened. That is, $P(X > x \mid y)$ implies conditioning on $Y = y$. Whenever deviating from an equality conditioning, the notation is explicit. For example, $P(X > x \mid Y > y)$ is used to denote conditioning on an inequality.

A number of long equations in Chapter 10 include terms with conditioning on $\ln IM_1$. Because there is a 1:1 equivalence between $\ln IM_1$ and IM_1, the conditional result can be obtained using either $\ln IM_1$ or IM_1. Because of this, we have simplified expressions such as $\sigma_{\ln IM_2 \mid \ln IM_1}$ to $\sigma_{\ln IM_2 \mid IM_1}$. The first term is retained as $\ln IM_2$ to reflect the fact that we are computing the standard deviation of the logarithmic quantity, but "ln" on the conditioning terms is not needed to maintain mathematical consistency.

Dependence Rather than Conditioning

In some cases we wish to reflect the dependence upon a parameter (or set of parameters), but are not strictly conditioning in the sense of the parameter being a random variable taking a specific value. In such cases we write $P(IM > im; \theta)$ to represent dependence upon the parameter θ. In a case with probabilistic conditioning and parameter dependence, we write $P(IM > im \mid m; \theta)$ to represent a conditioning upon $M = m$, but a dependence upon the parameter θ.

Rate Notation

We define $\lambda(x)$ to read: *the rate of occurrence of the event x*. Therefore:

- $\lambda(rup)$ reads *the rate of occurrence of rup*.
- $\lambda(IM > im)$ reads *the rate of occurrence of IM > im*.
- $\lambda(IM > im \mid m)$ reads *the rate of occurrence of IM > im given M = m*.
- $\lambda(IM > im \mid M > m)$ reads *the rate of occurrence of IM > im given M > m*.

Index